W9-CDQ-805

SAXON MATH™
Course 1

Student Edition

Stephen Hake

A Harcourt Achieve Imprint

www.SaxonPublishers.com
1-800-284-7019

ACKNOWLEDGEMENTS

This book was made possible by the significant contributions of many individuals and the dedicated efforts of a talented team at Harcourt Achieve.

Special thanks to:

- Melody Simmons and Chris Braun for suggestions and explanations for problem solving in Courses 1–3,
- Elizabeth Rivas and Bryon Hake for their extensive contributions to lessons and practice in Course 3,
- Sue Ellen Fealko for suggested application problems in Course 3.

The long hours and technical assistance of John and James Hake on Courses 1–3, Robert Hake on Course 3, Tom Curtis on Course 3, and Roger Phan on Course 3 were invaluable in meeting publishing deadlines. The saintly patience and unwavering support of Mary is most appreciated.

– Stephen Hake

Staff Credits

Editorial: Jean Armstrong, Shelley Farrar-Coleman, Marc Connolly, Hirva Raj, Brooke Butner, Robin Adams, Roxanne Picou, Cecilia Colome, Michael Ota

Design: Alison Klassen, Joan Cunningham, Deborah Diver, Alan Klemp, Andy Hendrix, Rhonda Holcomb

Production: Mychael Ferris-Pacheco, Heather Jernt, Greg Gaspard, Donna Brawley, John-Paxton Gremillion

Manufacturing: Cathy Voltaggio

Marketing: Marilyn Trow, Kimberly Sadler

E-Learning: Layne Hedrick, Karen Stitt

ISBN 1-5914-1783-X

2 3 4 5 6 7 8 9 048 12 11 10 09 08 07 06

ABOUT THE AUTHOR

Stephen Hake has authored five books in the Saxon Math series. He writes from 17 years of classroom experience as a teacher in grades 5 through 12 and as a math specialist in El Monte, California. As a math coach, his students won honors and recognition in local, regional, and statewide competitions.

Stephen has been writing math curriculum since 1975 and for Saxon since 1985. He has also authored several math contests including Los Angeles County's first Math Field Day contest. Stephen contributed to the 1999 National Academy of Science publication on the Nature and Teaching of Algebra in the Middle Grades.

Stephen is a member of the National Council of Teachers of Mathematics and the California Mathematics Council. He earned his BA from United States International University and his MA from Chapman College.

EDUCATIONAL CONSULTANTS

Nicole Hamilton
Consultant Manager
Richardson, TX

Joquita McKibben
Consultant Manager
Pensacola, FL

John Anderson
Lowell, IN

Beckie Fulcher
Gulf Breeze, FL

Heidi Graviette
Stockton, CA

Brenda Halulka
Atlanta, GA

Marilyn Lance
East Greenbush, NY

Ann Norris
Wichita Falls, TX

Melody Simmons
Nogales, AZ

Benjamin Swagerty
Moore, OK

Kristyn Warren
Macedonia, OH

Mary Warrington
East Wenatchee, WA

CONTENTS OVERVIEW

TABLE OF CONTENTS

Integrated and Distributed Units of Instruction

Section 1 *Lessons 1–10, Investigation 1*

Math Focus:
Number & Operations • Algebra

Distributed Strands:
Number & Operations • Algebra • Geometry • Measurement • Data Analysis & Probability

Section 2 Lessons 11–20, Investigation 2

Math Focus:
Number & Operations • Problem Solving

Distributed Strands:
Number & Operations • Algebra • Measurement • Data Analysis & Probability • Problem Solving

Maintaining & Extending

Power Up

Enrichment

Section 3 *Lessons 21–30, Investigation 3*

Math Focus:
Number & Operations • Geometry

Distributed Strands:
Number & Operations • Geometry • Measurement • Problem Solving

Section 4 Lessons 31–40, Investigation 4

Math Focus:
Number & Operations • Measurement

Distributed Strands:
Number & Operations • Geometry • Measurement • Data Analysis & Probability • Problem Solving

Maintaining & Extending

Power Up

Enrichment

Section 5 *Lessons 41–50, Investigation 5*

Math Focus:
Number & Operations

Distributed Strands:
Number & Operations • Algebra • Geometry • Measurement • Data Analysis & Probability • Problem Solving

Maintaining & Extending

Section 6 Lessons 51–60, Investigation 6

Math Focus:
Number & Operations • Geometry

Distributed Strands:
Number & Operations • Geometry • Measurement • Data Analysis & Probability • Problem Solving

Section 7 Lessons 61–70, Investigation 7

Math Focus:
Number & Operations • Geometry

Distributed Strands:
Number & Operations • Algebra • Geometry • Measurement • Problem Solving

Section 8 Lessons 71–80, Investigation 8

Math Focus:
Number & Operations • Geometry

Distributed Strands:
Number & Operations • Geometry • Measurement • Problem Solving

Section 9 Lessons 81–90, Investigation 9

Math Focus:
Algebra • Measurement

Distributed Strands:
Number & Operations • Algebra • Geometry • Measurement • Data Analysis & Probability • Problem Solving

Section 10 — Lessons 91–100, Investigation 10

Math Focus:
Numbers & Operations • Geometry

Distributed Strands:
Number & Operations • Algebra • Geometry • Measurement • Data Analysis & Probability

Maintaining & Extending

Power Up

Enrichment

Section 11 — *Lessons 101–110, Investigation 11*

Math Focus:
Algebra • Geometry

Distributed Strands:
Number & Operations • Algebra • Geometry • Measurement • Problem Solving

Section 12 — Lessons 111–120, Investigation 12

Math Focus:
Measurement • Problem Solving

Distributed Strands:
Number & Operations • Algebra • Geometry • Measurement • Problem Solving

Dear Student,

We study mathematics because of its importance to our lives. Our school schedule, our trip to the store, the preparation of our meals, and many of the games we play involve mathematics. You will find that the word problems in this book are often drawn from everyday experiences.

As you grow into adulthood, mathematics will become even more important. In fact, your future in the adult world may depend on the mathematics you have learned. This book was written to help you learn mathematics and to learn it well. For this to happen, you must use the book properly. As you work through the pages, you will see that similar problems are presented over and over again.
Solving each problem day after day is the secret to success.

Your book is made up of daily lessons and investigations. Each lesson has three parts.

1. The first part is a Power Up that includes practice of basic facts and mental math. These exercises improve your speed, accuracy, and ability to do math "in your head." The Power Up also includes a problem-solving exercise to familiarize you with strategies for solving complicated problems.

2. The second part of the lesson is the New Concept. This section introduces a new mathematical concept and presents examples that use the concept. The Practice Set provides a chance to solve problems involving the new concept. The problems are lettered a, b, c, and so on.

3. The final part of the lesson is the Written Practice. This problem set reviews previously taught concepts and prepares you for concepts that will be taught in later lessons. Solving these problems helps you remember skills and concepts for a long time.

Investigations are variations of the daily lesson. The investigations in this book often involve activities that fill an entire class period. Investigations contain their own set of questions instead of a problem set.

Remember, solve every problem in every practice set, written practice set, and investigation. Do not skip problems. With honest effort, you will experience success and true learning that will stay with you and serve you well in the future.

Stephen D. Hake

Temple City, California

Saxon Math Course 1 is unlike any math book you have used! It doesn't have colorful photos to distract you from learning. The Saxon approach lets you see the beauty and structure within math itself. You will understand more mathematics, become more confident in doing math, and will be well prepared when you take high school math classes.

Power Yourself Up!

Start off each lesson by practicing your basic skills and concepts, mental math, and problem solving. Make your math brain stronger by exercising it every day. Soon you'll know these facts by memory!

Learn Something New!

Each day brings you a new concept, but you'll only have to learn a small part of it now. You'll be building on this concept throughout the year so that you understand and remember it by test time.

LESSON **27**

• Measures of a Circle

Power Up *Building Power*

facts Power Up E

mental math

a. **Number Sense:** 7×52

b. **Number Sense:** 6×33

c. **Number Sense:** $63 + 19$

d. **Number Sense:** $256 + 50$

e. **Money:** $10.00 − $7.25

f. **Fractional Parts:** $\frac{1}{2}$ of 86

g. **Geometry:** The perimeter of a square is 16 ft. What is the length of the sides of the square?

h. **Calculation:** 8×8, $- 1$, $\div 7$, $\times 2$, $+ 2$, $\div 2$

problem solving If ⬚ + ⬚ = 25, and ⬚ − ⬚ = 5, then ⬚ × ⬚ = ?

New Concept *Increasing Knowledge*

Thinking Skill
Verify
Why is the diameter of a circle twice the length of the radius?

There are several ways to measure a circle. We can measure the distance around the circle, the distance across the circle, and the distance from the center of the circle to the circle itself. The pictures below identify these measures.

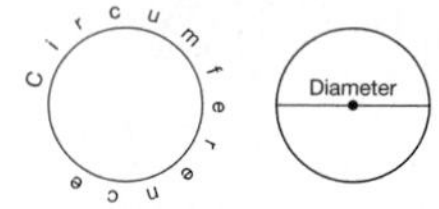

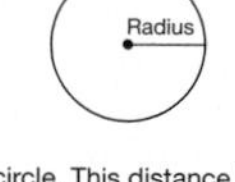

The **circumference** is the distance **around** the circle. This distance is the same as the perimeter of a circle. The **diameter** is the distance **across** a circle through its center. The **radius** is the distance from the center to the circle. The plural of *radius* is **radii.** For any circle, the diameter is twice the length of the radius.

142

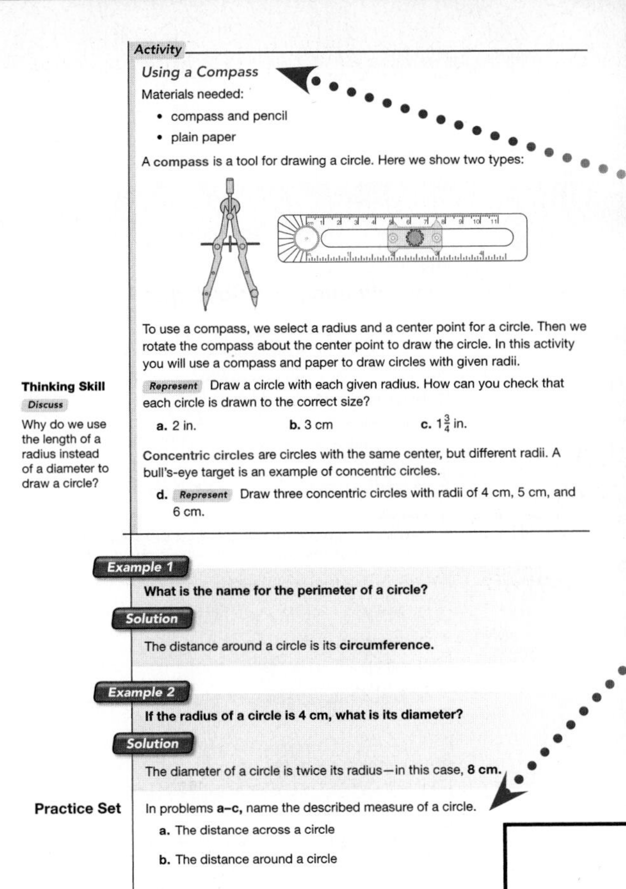

Activity

Using a Compass

Materials needed:

- compass and pencil
- plain paper

A compass is a tool for drawing a circle. Here we show two types:

To use a compass, we select a radius and a center point for a circle. Then we rotate the compass about the center point to draw the circle. In this activity you will use a compass and paper to draw circles with given radii.

Thinking Skill
Discuss
Why do we use the length of a radius instead of a diameter to draw a circle?

Represent Draw a circle with each given radius. How can you check that each circle is drawn to the correct size?

a. 2 in. **b.** 3 cm **c.** $1\frac{3}{4}$ in.

Concentric circles are circles with the same center, but different radii. A bull's-eye target is an example of concentric circles.

d. *Represent* Draw three concentric circles with radii of 4 cm, 5 cm, and 6 cm.

Example 1

What is the name for the perimeter of a circle?

Solution

The distance around a circle is its **circumference.**

Example 2

If the radius of a circle is 4 cm, what is its diameter?

Solution

The diameter of a circle is twice its radius—in this case, **8 cm.**

Practice Set In problems **a–c,** name the described measure of a circle.

a. The distance across a circle

b. The distance around a circle

Get Active!

Dig into math with a hands-on activity. Explore a math concept with your friends as you work together and use manipulatives to see new connections in mathematics.

Check It Out!

The Practice Set lets you check to see if you understand today's new concept.

Exercise Your Mind!

When you work the Written Practice exercises, you will review both today's new concept and also math you learned in earlier lessons. Each exercise will be on a different concept — you never know what you're going to get! It's like a mystery game — unpredictable and challenging.

As you review concepts from earlier in the book, you'll be asked to use higher-order thinking skills to show what you know and why the math works.

The mixed set of Written Practice is just like the mixed format of your state test. You'll be practicing for the "big" test every day!

c. The distance from the center to the circle.

d. *Explain* If the diameter of a circle is 10 in., what is its radius? Describe how you know.

Written Practice *Strengthening Concepts*

1. (12) *Analyze* What is the product of the sum of 55 and 45 and the difference of 55 and 45?

*** 2.** (22) *Model* Potatoes are three-fourths water. If a sack of potatoes weighs 20 pounds, how many pounds of water are in the potatoes? Draw a diagram to illustrate the problem.

3. (11) *Formulate* There were 306 students in the cafeteria. After some went outside, there were 249 students left in the cafeteria. How many students went outside? Write an equation and solve the problem.

*** 4.** (27) *Explain* **a.** If the diameter of a circle is 5 in., what is the radius of the circle?

b. What is the relationship of the diameter of a circle to its radius?

5. (21) *Classify* Which of these numbers is divisible by both 2 and 3?

A 122 **B** 123 **C** 132

6. (16) Round 1,234,567 to the nearest ten thousand.

7. (15) *Formulate* If ten pounds of apples costs $12.90, what is the price per pound? Write an equation and solve the problem.

8. (6) What is the denominator of $\frac{23}{24}$?

*** 9.** (22) *Model* What number is $\frac{3}{5}$ of 65? Draw a diagram to illustrate the problem.

*** 10.** (22) *Model* How much money is $\frac{2}{3}$ of $15? Draw a diagram to illustrate the problem.

Model Use your fraction manipulatives to help answer problems 11–18.

11. (Inv. 2) $\frac{1}{6} + \frac{2}{6} + \frac{3}{6}$ **12.** (Inv. 2) $\frac{7}{8} - \frac{3}{8}$

13. (Inv. 2) $\frac{6}{6} - \frac{5}{6}$ **14.** (Inv. 2) $\frac{2}{8} + \frac{5}{8}$

15. (Inv. 2) **a.** How many $\frac{1}{8}$s are in 1?

b. How many $\frac{1}{8}$s are in $\frac{1}{2}$?

*** 16.** (26) Reduce: $\frac{4}{6}$

17. (Inv. 2) What fraction is half of $\frac{1}{4}$?

18. (Inv. 2) What fraction of a circle is 50% of a circle?

144 *Saxon Math Course 1*

Become an Investigator!

Dive into math concepts and explore the depths of math connections in the Investigations.

Continue to develop your mathematical thinking through applications, activities, and extensions.

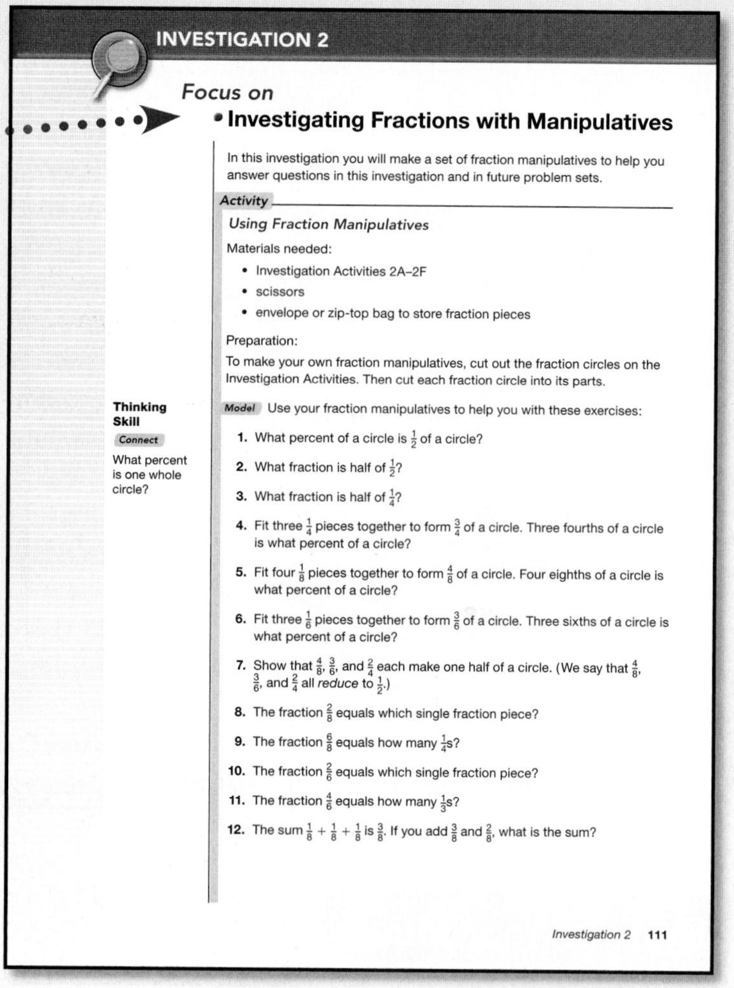

INVESTIGATION 2

Focus on

Investigating Fractions with Manipulatives

In this investigation you will make a set of fraction manipulatives to help you answer questions in this investigation and in future problem sets.

Activity

Using Fraction Manipulatives

Materials needed:

- Investigation Activities 2A–2F
- scissors
- envelope or zip-top bag to store fraction pieces

Preparation:

To make your own fraction manipulatives, cut out the fraction circles on the Investigation Activities. Then cut each fraction circle into its parts.

Thinking Skill
Connect
What percent is one whole circle?

Model Use your fraction manipulatives to help you with these exercises:

1. What percent of a circle is $\frac{1}{2}$ of a circle?
2. What fraction is half of $\frac{1}{2}$?
3. What fraction is half of $\frac{1}{4}$?
4. Fit three $\frac{1}{4}$ pieces together to form $\frac{3}{4}$ of a circle. Three fourths of a circle is what percent of a circle?
5. Fit four $\frac{1}{8}$ pieces together to form $\frac{4}{8}$ of a circle. Four eighths of a circle is what percent of a circle?
6. Fit three $\frac{1}{6}$ pieces together to form $\frac{3}{6}$ of a circle. Three sixths of a circle is what percent of a circle?
7. Show that $\frac{4}{8}$, $\frac{3}{6}$, and $\frac{2}{4}$ each make one half of a circle. (We say that $\frac{4}{8}$, $\frac{3}{6}$, and $\frac{2}{4}$ all *reduce* to $\frac{1}{2}$.)
8. The fraction $\frac{2}{8}$ equals which single fraction piece?
9. The fraction $\frac{6}{8}$ equals how many $\frac{1}{4}$s?
10. The fraction $\frac{2}{6}$ equals which single fraction piece?
11. The fraction $\frac{4}{6}$ equals how many $\frac{1}{3}$s?
12. The sum $\frac{1}{8} + \frac{1}{8} + \frac{1}{8}$ is $\frac{3}{8}$. If you add $\frac{3}{8}$ and $\frac{2}{8}$, what is the sum?

Investigation 2 111

Focus on
• Problem Solving

As we study mathematics we learn how to use tools that help us solve problems. We face mathematical problems in our daily lives, in our work, and in our efforts to advance our technological society. We can become powerful problem solvers by improving our ability to use the tools we store in our minds. In this book we will practice solving problems every day.

This lesson has three parts:

Problem-Solving Process The four steps we follow when solving problems.

Problem-Solving Strategies Some strategies that can help us solve problems.

Writing and Problem Solving Describing how we solved a problem or formulating a problem.

four-step problem-solving process

Solving a problem is like arriving at a destination, so the process of solving a problem is similar to the process of taking a trip. Suppose we are on the mainland and want to reach a nearby island.

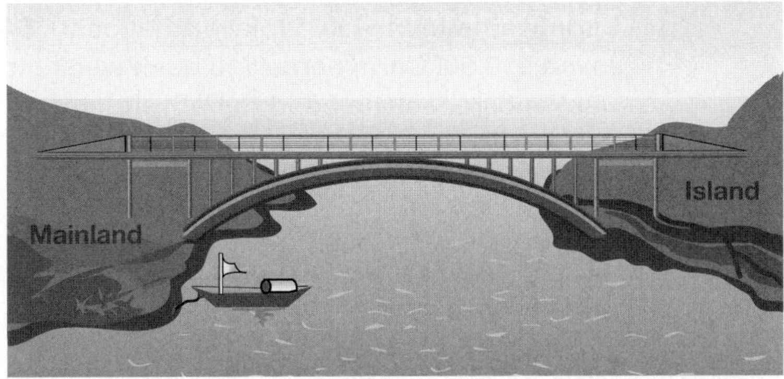

Problem-Solving Process	Taking a Trip
Step 1: *Understand* Know where you are and where you want to go.	We are on the mainland and want to go to the island.
Step 2: *Plan* Plan your route.	We might use the bridge, the boat, or swim.
Step 3: *Solve* Follow the plan.	Take the journey to the island.
Step 4: *Check* Check that you have reached the right place.	Verify that you have reached your desired destination.

When we solve a problem, it helps to ask ourselves some questions along the way.

Follow the Process	Ask Yourself Questions
Step 1: Understand	What information am I given? What am I asked to find or do?
Step 2: Plan	How can I use the given information to solve the problem? What strategy can I use to solve the problem?
Step 3: Solve	Am I following the plan? Is my math correct?
Step 4: Check ***(Look Back)***	Does my solution answer the question that was asked? Is my answer reasonable?

Below we show how we follow these steps to solve a word problem.

Example 1

Carla wants to buy a CD player that costs $48.70 including tax. She has saved $10.50. Carla earns $10 each weekend babysitting. How many weekends does she need to babysit to earn enough money to buy the CD player?

Solution

Step 1: Understand the problem. The problem gives the following information:

- The CD player costs $48.70.
- Carla has saved $10.50.
- Carla earns $10.00 every weekend.

We are asked to find out how many weekends Carla needs to babysit to have enough money to buy the CD player.

Step 2: Make a plan. We see that we cannot get to the answer in one step. We plan how to use the given information in a manner that will lead us toward the solution. One way to solve the problem is:

- Find out how much more money Carla needs.
- Then find out how many weekends it will take to earn the needed amount.

Step 3: Solve the problem. (Follow the plan.) First we subtract $10.50 from $48.70 to find out how much more money Carla needs.

$$\begin{array}{rl} \$48.70 & \text{cost} \\ -\ \$10.50 & \text{Carla has} \\ \hline \$38.20 & \text{Carla needs} \end{array}$$

Carla needs $38.20 more than she has.

Thinking Skill

Verify

What strategy did we use to find the number of weekends?

Now we find the number of weekends Carla needs to work. One way is to divide $38.20 by $10.00. Another way is to find the multiple of $10.00 that gives Carla enough money. We can make a table to do this.

Weekend 1	Weekend 2	Weekend 3	Weekend 4
$10	$20	$30	$40

After one weekend Carla earns $10, after two weekends $20, three weekends $30, and four weekends $40.

Carla needs an additional $38.20. She will need to work **four weekends** to have enough money to buy the CD player.

Step 4: Check your answer. (Look back.) We read the problem again to see if our solution answers the question. We decide if our answer is reasonable.

The problem asks how many weekends Carla will need to work to earn the rest of the money for the CD player. Our solution, 4 weekends, answers the question. Our solution is reasonable because $40 is just a little over the $38.20 that Carla needs.

After four weekends Carla has $10.50 + $40 = $50.50.

This is enough money to buy the CD player.

Example 2

Howard is planning to tile the top of an end table. The tabletop is 1 ft by 2 ft. He wants to show his initial "H" using the tiles. The 4-inch tiles come in black and white. How many tiles of each color does Howard need to cover the table with his initial?

Solution

Step 1: Understand the problem. The problem gives the following information:

- The table is 1 ft × 2 ft.
- The tiles are 4-inches on a side.
- The tiles are black and white.
- We need to model the letter "H".

We are asked to find how many tiles of each color we need to show the letter "H" on the tabletop.

Step 2: Make a plan. We see that we cannot get to the answer in one step. We plan how to use the given information in a manner that will lead us toward the solution.

- Change the table dimensions to inches and determine how many tiles will cover the table.
- Use tiles to model the tabletop and decide how many tiles are needed to show the letter "H".

Thinking Skill

Verify

What strategy did we use to design the tabletop?

Step 3: Solve the problem. (Follow the plan.) There are 12 inches in 1 foot, so the table is 12 in. by 24 in.

12 in. ÷ 4 in. = 3 24 in. ÷ 4 in. = 6 $3 \times 6 = 18$

Howard needs 18 tiles to cover the table.

How can we use 18 tiles to model the letter "H"?

Here are two possibilities.

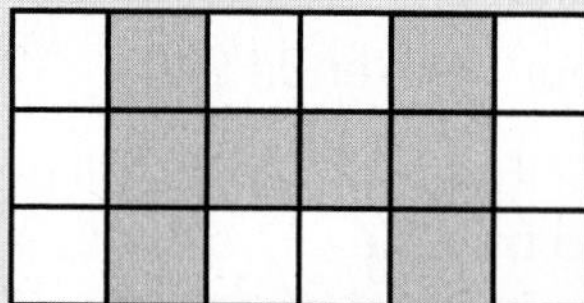
10 white tiles, 8 black tiles

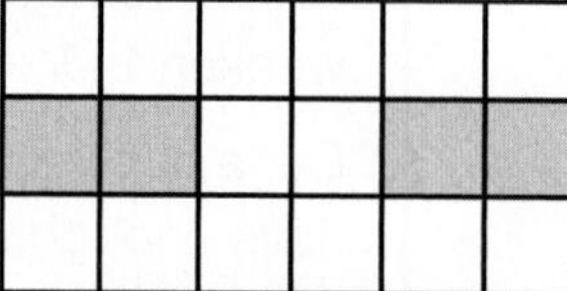
14 white tiles, 4 black tiles

Since the problem does not specify the number of each color, we can decide based on the design we choose. Howard needs 10 white tiles and 8 black tiles.

Step 4: Check your answer. (Look back.) We read the problem again to see if our solution answers the question. We decide if our answer is reasonable.

The problem asks for the number of tiles of each color Howard needs to show his initial on the tabletop. Our solution shows the letter "H" using 10 white tiles and 8 black tiles.

1. List in order the four steps in the problem-solving process.
2. What two questions do we answer to understand a problem?

Refer to the following problem to answer questions **3–8.**

Mrs. Rojas is planning to take her daughter Lena and her friend Natalie to see a movie. The movie starts at 4:30 p.m. She wants to arrive at the theater 20 minutes before the movie starts. It will take 15 minutes to drive to Natalie's house. It is 10 minutes from Natalie's house to the theater. At what time should Mrs. Rojas leave her house?

3. **Connect** What information are we given?
4. **Verify** What are you asked to find?
5. Which step of the four-step problem-solving process did you complete when you answered questions 3 and 4?
6. Describe your plan for solving the problem.
7. **Explain** Solve the problem by following your plan. Show your work. Write your solution to the problem in a way someone else will understand.
8. Check your work and your answer. Look back to the problem. Be sure you use the information correctly. Be sure you found what you were asked to find. Is your answer reasonable?

problem-solving strategies

As we consider how to solve a problem we choose one or more strategies that seem to be helpful. Referring to the picture at the beginning of this lesson, we might choose to swim, to take the boat, or to cross the bridge to travel from the mainland to the island. Other strategies might not be as effective for the illustrated problem. For example, choosing to walk or bike across the water are strategies that are not reasonable for this situation.

When solving mathematical problems we also select strategies that are appropriate for the problem. Problem-solving **strategies** are types of plans we can use to solve problems. Listed below are ten strategies we will practice in this book. You may refer to these descriptions as you solve problems throughout the year.

Act it out or make a model. Moving objects or people can help us visualize the problem and lead us to the solution.

Use logical reasoning. All problems require reasoning, but for some problems we use given information to eliminate choices so that we can close in on the solution. Usually a chart, diagram, or picture can be used to organize the given information and to make the solution more apparent.

Draw a picture or diagram. Sketching a picture or a diagram can help us understand and solve problems, especially problems about graphs or maps or shapes.

Write a number sentence or equation. We can solve many word problems by fitting the given numbers into equations or number sentences and then finding the unknown numbers.

Make it simpler. We can make some complicated problems easier by using smaller numbers or fewer items. Solving the simpler problem might help us see a pattern or method that can help us solve the complex problem.

Find a pattern. Identifying a pattern that helps you to predict what will come next as the pattern continues might lead to the solution.

Make an organized list. Making a list can help us organize our thinking about a problem.

Guess and check. Guessing the answer and trying the guess in the problem might start a process that leads to the answer. If the guess is not correct, use the information from the guess to make a better guess. Continue to improve your guesses until you find the answer.

Make or use a table, chart, or graph. Arranging information in a table, chart, or graph can help us organize and keep track of data. This might reveal patterns or relationships that can help us solve the problem.

Work backwards. Finding a route through a maze is often easier by beginning at the end and tracing a path back to the start. Likewise, some problems are easier to solve by working back from information that is given toward the end of the problem to information that is unknown near the beginning of the problem.

9. Name some strategies used in this lesson.

The chart below shows where each strategy is first introduced in this textbook.

Strategy	Lesson
Act It Out or Make a Model	Problem Solving Overview Example 2
Use Logical Reasoning	Lesson 3
Draw a Picture or Diagram	Lesson 17
Write a Number Sentence or Equation	Lesson 17
Make It Simpler	Lesson 4
Find a Pattern	Lesson 1
Make an Organized List	Lesson 8
Guess and Check	Lesson 5
Make or Use a Table, Chart, or Graph	Problem Solving Overview Example 1
Work Backwards	Lesson 84

writing and problem solving

Sometimes, a problem will ask us to explain our thinking. This helps us measure our understanding of math.

- Explain how you solved the problem.
- Explain how you know your answer is correct.
- Explain why your answer is reasonable.

For these situations, we can describe the way we followed our plan. This is a description of the way we solved example 1.

> *Subtract \$10.50 from \$48.70 to find out how much more money Carla needs. \$48.70 − \$10.50 = \$38.20. Make a table and count by 10s to determine that Carla needs to work 4 weekends so she can earn enough money for the CD player.*

10. Write a description of how we solved the problem in example 2.

Other times, we will be asked to write a problem for a given equation. Be sure to include the correct numbers and operations to represent the equation.

11. Write a word problem for the equation $32 + 32 = 64$.

LESSON 1

• Adding Whole Numbers and Money
• Subtracting Whole Numbers and Money
• Fact Families, Part 1

Power Up[1]

Building Power

facts Power Up A

mental math

a. **Number Sense:** 30 + 30

b. **Number Sense:** 300 + 300

c. **Number Sense:** 80 + 40

d. **Number Sense:** 800 + 400

e. **Number Sense:** 20 + 30 + 40

f. **Number Sense:** 200 + 300 + 400

g. **Measurement:** How many inches are in a foot?

h. **Measurement:** How many millimeters are in a centimeter?

problem solving

Sharon made three square patterns using 4 coins, 9 coins, and 16 coins. If she continues forming larger square patterns, how many coins will she need for each of the next three square patterns?

Understand We are given 4, 9, and 16 as the first three square numbers. We are asked to extend the sequence an additional three terms.

Plan We will *find the pattern* in the first three terms of the sequence, then use the pattern to extend the sequence an additional three terms.

Solve We see that the number of coins in each square can be found by multiplying the number of coins in each row by the number of rows: $2 \times 2 = 4$, $3 \times 3 = 9$, and $4 \times 4 = 16$. We use this rule to find the next three terms: $5 \times 5 = 25$, $6 \times 6 = 36$, and $7 \times 7 = 49$.

Check We found that Sharon needs 25, 36, and 49 coins to build each of the next three squares in the pattern. We can verify our answers by drawing pictures of each of the next three terms in the pattern and counting the coins.

[1] For instructions on how to use the Power Up, please consult the preface.

New Concepts *Increasing Knowledge*

adding whole numbers and money

To combine two or more numbers, we add. The numbers that are added together are called **addends.** The answer is called the **sum.** Changing the order of the addends does not change the sum. For example,

$$3 + 5 = 5 + 3$$

This property of addition is called the **Commutative Property of Addition.**

When adding numbers, we add digits that have the same place value.

Example 1

Add: 345 + 67

Solution

When we add whole numbers on paper, we write the numbers so that the place values are aligned. Then we add the digits by column.

$$\begin{array}{rl} \overset{1\,1}{345} & \text{addend} \\ +\ \ 67 & \text{addend} \\ \hline \mathbf{412} & \text{sum} \end{array}$$

Changing the order of the addends does not change the sum. One way to check an addition answer is to change the order of the addends and add again.

$$\begin{array}{rl} \overset{1\,1}{67} & \\ +\ 345 & \\ \hline 412 & \text{check} \end{array}$$

Example 2

Thinking Skill

Connect

$5 means five dollars and no cents. Why does writing $5 as $5.00 help when adding money amounts?

Add: $1.25 + $12.50 + $5

Solution

When we add money, we write the numbers so that the decimal points are aligned. We write $5 as $5.00 and add the digits in each column.

$$\begin{array}{r} \$1.25 \\ \$12.50 \\ +\ \ \$5.00 \\ \hline \mathbf{\$18.75} \end{array}$$

If one of two addends is zero, the sum of the addends is identical to the nonzero addend. This property of addition is called the **Identity Property of Addition.**

$$5 + 0 = 5$$

subtracting whole numbers and money

We subtract one number from another number to find the **difference** between the two numbers. In a subtraction problem, the **subtrahend** is taken from the **minuend.**

$$5 - 3 = 2$$

In the problem above, 5 is the minuend and 3 is the subtrahend. The difference between 5 and 3 is 2.

Verify Does the Commutative Property apply to subtraction? Give an example to support your answer.

Example 3

Subtract: 345 − 67

Solution

When we subtract whole numbers, we align the digits by place value. We subtract the bottom number from the top number and regroup when necessary.

```
  2 13 1
  3̸ 4̸ 5
−   6 7
  2 7 8  ← difference
```

Example 4

Jim spent $1.25 for a hamburger. He paid for it with a five-dollar bill. Find how much change he should get back by subtracting $1.25 from $5.

Solution

Thinking Skill

When is it necessary to line up decimals?

Order matters when we subtract. The starting amount is put on top. We write $5 as $5.00. We line up the decimal points to align the place values. Then we subtract. Jim should get back **$3.75.**

```
   4 9 1
  $5̸.0̸ 0
− $1.2 5
  $3.7 5
```

We can check the answer to a subtraction problem by adding. If we add the answer (difference) to the amount subtracted, the total should equal the starting amount. We do not need to rewrite the problem. We just add the two bottom numbers to see whether their sum equals the top number.

Subtract Down To find the difference ↓	$5.00 − $1.25 $3.75	↑ **Add Up** To check the answer

fact families, part 1

Addition and subtraction are called **inverse operations.** We can "undo" an addition by subtracting one addend from the sum. The three numbers that form an addition fact also form a subtraction fact. For example,

$$4 + 5 = 9 \qquad 9 - 5 = 4$$

The numbers 4, 5, and 9 are a **fact family.** They can be arranged to form the two addition facts and two subtraction facts shown below.

$$\begin{array}{r}4\\+\,5\\\hline 9\end{array}\qquad\begin{array}{r}5\\+\,4\\\hline 9\end{array}\qquad\begin{array}{r}9\\-\,5\\\hline 4\end{array}\qquad\begin{array}{r}9\\-\,4\\\hline 5\end{array}$$

Example 5

Rearrange the numbers in this addition fact to form another addition fact and two subtraction facts.

$$\mathbf{11 + 14 = 25}$$

Solution

We form another addition fact by reversing the addends.

$$14 + 11 = 25$$

We form two subtraction facts by making the sum, 25, the first number of each subtraction fact. Then each remaining number is subtracted from 25.

$$25 - 11 = 14$$

$$25 - 14 = 11$$

Example 6

Rearrange the numbers in this subtraction fact to form another subtraction fact and two addition facts.

$$\begin{array}{r} 11 \\ -\ 6 \\ \hline 5 \end{array}$$

Solution

The Commutative Property does not apply to subtraction, so we may not reverse the first two numbers of a subtraction problem. However, we may reverse the last two numbers.

$$\begin{array}{r} 11 \\ -\ 6 \\ \hline 5 \end{array} \quad \begin{array}{r} 11 \\ -\ 5 \\ \hline 6 \end{array}$$

For the two addition facts, 11 is the sum.

$$\begin{array}{r} 5 \\ +\ 6 \\ \hline 11 \end{array} \quad \begin{array}{r} 6 \\ +\ 5 \\ \hline 11 \end{array}$$

Practice Set

Simplify:

a. 3675 + 426 + 1357

b. \$6.25 + \$8.23 + \$12

c. 5374 − 168

d. \$5 − \$1.35

e. *Represent* Arrange the numbers 6, 8, and 14 to form two addition facts and two subtraction facts.

f. *Connect* Rearrange the numbers in this subtraction fact to form another subtraction fact and two addition facts.

$$25 - 10 = 15$$

Written Practice

Strengthening Concepts

1. What is the sum of 25 and 40?

2. At a planetarium show, Johnny counted 137 students and 89 adults. He also counted 9 preschoolers. How many people did Johnny count in all?

3. **Generalize** What is the difference when 93 is subtracted from 387?

4. Keisha paid $5 for a movie ticket that cost $3.75. Find how much change Keisha should get back by subtracting $3.75 from $5.

5. **Explain** Tatiana had $5.22 and earned $4.15 more by taking care of her neighbor's cat. How much money did she have then? Explain how you found the answer.

6. The soup cost $1.25, the fruit cost $0.70, and the drink cost $0.60. To find the total price of the lunch, add $1.25, $0.70, and $0.60.

7. $\begin{array}{r} 63 \\ 47 \\ +\ 50 \\ \hline \end{array}$

8. $\begin{array}{r} 632 \\ 57 \\ +\ 198 \\ \hline \end{array}$

9. $\begin{array}{r} 78 \\ 9 \\ +\ 987 \\ \hline \end{array}$

10. $\begin{array}{r} 432 \\ 579 \\ +\ 3604 \\ \hline \end{array}$

11. 345 − 67

12. 678 − 416

13. 3764 − 96

14. 875 + 1086 + 980

15. 10 + 156 + 8 + 27

16. $\begin{array}{r} \$3.47 \\ -\ \$0.92 \\ \hline \end{array}$

17. $\begin{array}{r} \$24.15 \\ -\ \$1.45 \\ \hline \end{array}$

18. $\begin{array}{r} \$0.75 \\ +\ \$0.75 \\ \hline \end{array}$

19. $\begin{array}{r} \$0.12 \\ \$0.46 \\ +\ \$0.50 \\ \hline \end{array}$

20. What is the name for the answer when we add?

21. What is the name for the answer when we subtract?

* 22. **Represent** The numbers 5, 6, and 11 are a fact family. Form two addition facts and two subtraction facts with these three numbers.

* 23. **Connect** Rearrange the numbers in this addition fact to form another addition fact and two subtraction facts.

$$27 + 16 = 43$$

* 24. **Connect** Rearrange the numbers in this subtraction fact to form another subtraction fact and two addition facts.

$$50 - 21 = 29$$

25. Describe a way to check the correctness of a subtraction answer.

---

* We encourage students to work first on the exercises on which they might want help, saving the easier exercises for last. Beginning in this lesson, we star the exercises that cover challenging or recently presented content. We suggest that these exercises be worked first.

LESSON 2

• Multiplying Whole Numbers and Money
• Dividing Whole Numbers and Money
• Fact Families, Part 2

Power Up — *Building Power*

facts Power Up A

mental math

a. **Number Sense:** 500 + 40

b. **Number Sense:** 60 + 200

c. **Number Sense:** 30 + 200 + 40

d. **Number Sense:** 70 + 300 + 400

e. **Number Sense:** 400 + 50 + 30

f. **Number Sense:** 60 + 20 + 400

g. **Measurement:** How many inches are in 2 feet?

h. **Measurement:** How many millimeters are in 2 centimeters?

problem solving

Sam thought of a number between ten and twenty. Then he gave a clue: You say the number when you count by twos and when you count by threes, but not when you count by fours. Of what number was Sam thinking?

New Concepts — *Increasing Knowledge*

multiplying whole numbers and money

Courtney wants to enclose a square garden to grow vegetables. How many feet of fencing does she need?

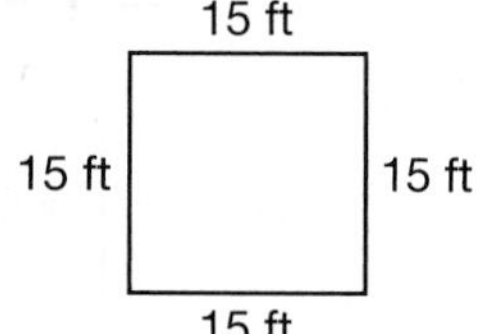

When we add the same number several times, we get a sum. We can get the same result by multiplying.

$$\underbrace{15 + 15 + 15 + 15}_{\text{Four 15s equal 60.}} = 60$$

$$4 \times 15 = 60$$

Numbers that are multiplied together are called **factors.** The answer is called the **product.**

To indicate multiplication, we can use a times sign, a dot, or write the factors side by side without a sign. Each of these expressions means that l and w are multiplied: $l \times w$ $\quad l \cdot w$ $\quad lw$

Notice that in the form $l \cdot w$ the multiplication dot is elevated and is not in the position of a decimal point. The form lw can be used to show the multiplication of two or more letters or of a number and letters, as we show below.

$$lwh \qquad 4s \qquad 4st$$

The form lw can also be used to show the multiplication of two or more numbers. To prevent confusion, however, we use parentheses to separate the numbers in the multiplication. Each of the following is a correct use of parentheses to indicate "3 times 5," although the first form is most commonly used. Without the parentheses, we would read each of these simply as the number 35.

$$3(5) \qquad (3)(5) \qquad (3)5$$

Thinking Skill

Discuss

Why do we multiply 28 by 4, by 10, and then add to find the product?

When we multiply by a two-digit number on paper, we multiply twice. To multiply 28 by 14, we first multiply 28 by 4. Then we multiply 28 by 10. For each multiplication we write a partial product. We add the partial products to find the final product.

28	factor
× 14	factor
112	partial product (28 × 4)
280	partial product (28 × 10)
392	product (14 × 28)

When multiplying dollars and cents by a whole number, the answer will have a dollar sign and a decimal point with two places after the decimal point.

$$\begin{array}{r} \$1.35 \\ \times \quad 6 \\ \hline \$8.10 \end{array}$$

Example 1

Find the cost of two dozen pencils at 35¢ each.

Solution

Two dozen is two 12s, which is 24. To find the cost of 24 pencils, we multiply 35¢ by 24.

$$\begin{array}{r} 35¢ \\ \times 24 \\ \hline 140 \\ 700 \\ \hline 840¢ \end{array}$$

The cost of two dozen pencils is 840¢, which is **$8.40.**

The **Commutative Property** applies to multiplication as well as addition, so changing the order of the factors does not change the product. For example,

$$4 \times 2 = 2 \times 4$$

One way to check multiplication is to reverse the order of factors and multiply.

$$\begin{array}{r} 23 \\ \times\ 14 \\ \hline 92 \\ 230 \\ \hline 322 \end{array} \qquad \begin{array}{r} 14 \\ \times\ 23 \\ \hline 42 \\ 280 \\ \hline 322 \end{array} \quad \text{check}$$

The **Identity Property of Multiplication** states that if one of two factors is 1, the product equals the other factor. The **Zero Property of Multiplication** states that if zero is a factor of a multiplication, the product is zero.

Represent Give an example for each property.

Example 2

Thinking Skill

Discuss

Why does writing trailing zeros not change the product?

Multiply: $\begin{array}{r} 400 \\ \times\ 874 \\ \hline \end{array}$

Solution

To simplify the multiplication, we reverse the order of the factors and write trailing zeros so that they "hang out" to the right.

$$\begin{array}{r} {\scriptstyle 21} \\ 874 \\ \times\quad 400 \\ \hline \mathbf{349{,}600} \end{array}$$

dividing whole numbers and money

When we separate a number into a certain number of equal parts, we divide. We can indicate division with a division symbol (÷), a division box ($\overline{)\quad}$), or a division bar (—). Each of the expressions below means "24 divided by 2":

$$24 \div 2 \qquad 2\overline{)24} \qquad \frac{24}{2}$$

The answer to a division problem is the **quotient.** The number that is divided is the **dividend.** The number by which the dividend is divided is the **divisor.**

dividend ÷ divisor = quotient
$\begin{array}{r} \text{quotient} \\ \text{divisor}\overline{)\text{dividend}} \end{array}$
$\dfrac{\text{dividend}}{\text{divisor}} = \text{quotient}$

When the dividend is zero, the quotient is zero. The divisor may not be zero. When the dividend and divisor are equal (and not zero), the quotient is 1.

Example 3

Divide: 3456 ÷ 7

Solution

On the next page, we show both the long-division and short-division methods.

Thinking Skill

Discuss

Why must the remainder always be less than the divisor?

Long Division	Short Division
$\begin{array}{r} 493 \text{ R } 5 \\ 7\overline{)3456} \\ \underline{28} \\ 65 \\ \underline{63} \\ 26 \\ \underline{21} \\ 5 \end{array}$	$\begin{array}{r} 4\,9\,3 \text{ R } 5 \\ 7\overline{)34^{6}5^{2}6} \end{array}$

Using the short-division method, we perform the multiplication and subtraction steps mentally, recording only the result of each subtraction.

To check our work, we multiply the quotient by the divisor. Then we add the remainder to this answer. The result should be the dividend. For this example we multiply 493 by 7. Then we add 5.

$$\begin{array}{r} {}^{6\,2} \\ 493 \\ \times \quad 7 \\ \hline 3451 \\ + \quad 5 \\ \hline 3456 \end{array}$$

When dividing dollars and cents, cents will be included in the answer. Notice that the decimal point in the quotient is directly above the decimal point in the division box, separating the dollars from the cents.

$$\begin{array}{r} \$.90 \\ 4\overline{)\$3.60} \\ \underline{3\,6} \\ 00 \\ \underline{0} \\ 0 \end{array}$$

fact families, part 2

Multiplication and division are inverse operations, so there are multiplication and division fact families just as there are addition and subtraction fact families. The numbers 5, 6, and 30 are a fact family. We can form two multiplication facts and two division facts with these numbers.

$5 \times 6 = 30 \qquad 30 \div 5 = 6$

$6 \times 5 = 30 \qquad 30 \div 6 = 5$

Example 4

Rearrange the numbers in this multiplication fact to form another multiplication fact and two division facts.

$$\mathbf{5 \times 12 = 60}$$

Solution

By reversing the factors, we form another multiplication fact.

$$\mathbf{12 \times 5 = 60}$$

By making 60 the dividend, we can form two division facts.

$$\mathbf{60 \div 5 = 12}$$

$$\mathbf{60 \div 12 = 5}$$

Practice Set

a. $20 \times 37¢$ **b.** $37 \cdot 0$ **c.** 407(37)

d. $5\overline{)\$8.40}$ **e.** $200 \div 12$ **f.** $\frac{234}{3}$

g. Which numbers are the divisors in problems **d, e,** and **f?**

h. **Represent** Use the numbers 8, 9, and 72 to form two multiplication facts and two division facts.

Written Practice[1] *Strengthening Concepts*

1. *(2)* If the factors are 7 and 11, what is the product?

2. *(1)* **Generalize** What is the difference between 97 and 79?

3. *(1)* If the addends are 170 and 130, what is the sum?

4. *(2)* If 36 is the dividend and 4 is the divisor, what is the quotient?

5. *(1)* Find the sum of 386, 98, and 1734.

6. *(1)* Fatima spent \$2.25 for a book. She paid for it with a five-dollar bill. Find how much change she should get back by subtracting \$2.25 from \$5.

7. *(1)* Luke wants to buy a \$70.00 radio for his car. He has \$47.50. Find how much more money he needs by subtracting \$47.50 from \$70.00.

8. *(2)* **Explain** Each energy bar costs 75¢. Find the cost of one dozen energy bars. Explain how you found your answer.

9. *(1)* $\begin{array}{r} 312 \\ -\ \ 86 \\ \hline \end{array}$

10. *(1)* $\begin{array}{r} 4106 \\ +\ 1398 \\ \hline \end{array}$

11. *(1)* $\begin{array}{r} 4000 \\ -\ 1357 \\ \hline \end{array}$

12. *(1)* $\begin{array}{r} \$10.00 \\ -\ \ \$2.83 \\ \hline \end{array}$

13. *(2)* 405(8)

14. *(2)* $25 \cdot 25$

15. *(2)* $\frac{288}{6}$

16. *(2)* $\frac{225}{15}$

17. *(2)* $\$1.25 \times 8$

18. *(2)* 400×50

19. *(2)* $1000 \div 8$

20. *(2)* $\$45.00 \div 20$

[1] The italicized numbers within parentheses underneath each problem number are called *lesson reference numbers.* These numbers refer to the lesson(s) in which the major concept of that particular problem is introduced. If additional assistance is needed, refer to the discussion, examples, or practice problems of that lesson.

* **21.** (2) **Represent** Use the numbers 6, 8, and 48 to form two multiplication facts and two division facts.

* **22.** (2) **Connect** Rearrange the numbers in this division fact to form another division fact and two multiplication facts.

$$4\overline{)36}\ \text{with quotient } 9$$

* **23.** (1) **Connect** Rearrange the numbers in this addition fact to form another addition fact and two subtraction facts.

$$12 + 24 = 36$$

24. (1) **a.** Find the sum of 9 and 6.

b. Find the difference between 9 and 6.

25. (2) The divisor, dividend, and quotient are in these positions when we use a division sign:

$$\text{dividend} \div \text{divisor} = \text{quotient}$$

On your paper, draw a division box and show the positions of the divisor, dividend, and quotient.

26. (2) Multiply to find the answer to this addition problem:

$$39¢ + 39¢ + 39¢ + 39¢ + 39¢ + 39¢$$

27. (2) 365×0 **28.** (2) $0 \div 50$ **29.** (2) $365 \div 365$

* **30.** (2) **Explain** How can you check the correctness of a division answer that has no remainder?

Early Finishers
Real-World Application

A customer at a bank deposits 2 one hundred-dollar bills, 8 twenty-dollar bills, 5 five-dollar bills, 20 one-dollar bills, 2 rolls of quarters, 25 dimes and 95 pennies. How much money will be deposited in all? Note: One roll of quarters = 40 quarters.

LESSON 3

• Unknown Numbers in Addition
• Unknown Numbers in Subtraction

Power Up — *Building Power*

facts Power Up B

mental math

a. Number Sense: 3000 + 4000

b. Number Sense: 600 + 2000

c. Number Sense: 20 + 3000

d. Number Sense: 600 + 300 + 20

e. Number Sense: 4000 + 300 + 200

f. Number Sense: 70 + 300 + 4000

g. Measurement: How many inches are in 3 feet?

h. Measurement: How many millimeters are in 3 centimeters?

problem solving

Tad picked up a number cube. His thumb and forefinger covered opposite faces. He counted the dots on the other four faces. How many dots did he count?

Understand We must first establish a base of knowledge about **standard number cubes.** The faces of a standard number cube are numbered with 1, 2, 3, 4, 5, or 6 dots. The number of dots on opposite faces of a number cube always total 7 (1 dot is opposite 6 dots, 2 dots are opposite 5 dots, and 3 dots are opposite 4 dots). Tad's thumb and forefinger covered opposite faces. We are asked to find how many dots were on the remaining four faces altogether.

Plan We will *use logical reasoning* about a number cube and *write an equation* to determine the number of dots Tad counted.

Solve Logical reasoning tells us that the four uncovered faces form two pairs of opposite faces. Each pair of opposite faces has 7 dots, so two pairs of opposite faces have 2 × 7, or 14 dots.

Check We determined that Tad counted 14 dots. We can check our answer by subtracting the number of dots Tad's fingers covered from the total number of dots on the number cube: 21 − 7 = 14 dots.

New Concepts — *Increasing Knowledge*

unknown numbers in addition

Below is an addition fact with three numbers. If one of the addends were missing, we could use the other addend and the sum to find the missing number.

$$\begin{array}{r l} 4 & \leftarrow \text{addend} \\ +\ 3 & \leftarrow \text{addend} \\ \hline 7 & \leftarrow \text{sum} \end{array}$$

Cover the 4 with your finger. How can you use the 7 and the 3 to find that the number under your finger is 4?

Now cover the 3 instead of the 4. How can you use the other two numbers to find that the number under your finger is 3?

Notice that we can find a missing addend by subtracting the known addend from the sum. We will use a letter to stand for a missing number.

Example 1

Find the value of *m*:

$$\begin{array}{r} 12 \\ + m \\ \hline 31 \end{array}$$

Solution

One of the addends is missing. The known addend is 12. The sum is 31. If we subtract 12 from 31, we find that the missing addend is **19.** We check our answer by using 19 in place of *m* in the original problem.

$$\begin{array}{r} \overset{2}{\cancel{3}}{}^{1}1 \\ -\ 12 \\ \hline 19 \end{array} \quad \xrightarrow{\text{Use 19 in place of } m.} \quad \begin{array}{r} \overset{1}{1}2 \\ +\ 19 \\ \hline 31 \end{array} \ \text{check}$$

Example 2

Find the value of *n*:

$$\mathbf{36 + 17 + 5 + n = 64}$$

Solution

First we add all the known addends.

$$\underbrace{36 + 17 + 5}_{58} + n = 64$$

$$58 \quad + n = 64$$

Then we find *n* by subtracting 58 from 64.

$64 - 58 = 6$ So *n* is **6.**

We check our work by using 6 in place of *n* in the original problem.

$36 + 17 + 5 + 6 = 64$ The answer checks.

unknown numbers in subtraction

Discuss Cover the 8 with your finger, and describe how to use the other two numbers to find that the number under your finger is 8.

$$\begin{array}{r} 8 \\ -\ 3 \\ \hline 5 \end{array} \begin{array}{l} \leftarrow \text{minuend} \\ \leftarrow \text{subtrahend} \\ \leftarrow \text{difference} \end{array}$$

Now cover the 3 instead of the 8. Describe how to use the other two numbers to find that the covered number is 3.

As we will show below, we can find a missing minuend by adding the other two numbers. We can find a missing subtrahend by subtracting the difference from the minuend.

Example 3

Find the value of *w*:

$$\begin{array}{r} w \\ -\ 16 \\ \hline 24 \end{array}$$

Solution

We can find the first number of a subtraction problem by adding the other two numbers. We add 16 and 24 to get **40.** We check our answer by using 40 in place of *w*.

$$\begin{array}{r} \overset{1}{1}6 \\ +\ 24 \\ \hline 40 \end{array} \quad \text{Use 40 in place of } w. \longrightarrow \quad \begin{array}{r} \overset{3}{\cancel{4}}\overset{1}{0} \\ -\ 16 \\ \hline 24 \end{array} \text{ check}$$

Example 4

Find the value of *y*:

$$236 - y = 152$$

Solution

One way to determine how to find a missing number is to think of a simpler problem that is similar. Here is a simpler subtraction fact:

$$5 - 3 = 2$$

In the problem, *y* is in the same position as the 3 in the simpler subtraction fact. Just as we can find 3 by subtracting 2 from 5, we can find *y* by subtracting 152 from 236.

$$\begin{array}{r} \overset{1}{\cancel{2}}\overset{1}{3}6 \\ -\ 152 \\ \hline 84 \end{array}$$

Thinking Skill

Discuss

What is another way we can check the subtraction?

We find that *y* is **84.** Now we check our answer by using 84 in place of *y* in the original problem.

$$\begin{array}{r} \overset{1}{\cancel{2}}\overset{1}{3}6 \\ -\ 84 \\ \hline 152 \end{array} \begin{array}{l} \\ \longleftarrow \text{Use 84 in place of } y. \\ \longleftarrow \text{The answer checks.} \end{array}$$

Statements such as $12 + m = 31$ are equations. An **equation** is a mathematical sentence that uses the symbol = to show that two quantities are equal. In algebra we refer to a missing number in an equation as an **unknown.** When asked to find the unknown in the exercises that follow, look for the number represented by the letter that makes the equation true.

Practice Set

Math Language

We can use a lowercase or an uppercase letter as an unknown:

$a + 3 = 5$

$A + 3 = 5$

The equations have the same meaning.

Analyze Find the unknown number in each problem. Check your work by using your answer in place of the letter in the original problem.

a. $\begin{array}{r} A \\ +\ 12 \\ \hline 45 \end{array}$ **b.** $\begin{array}{r} 32 \\ +\ B \\ \hline 60 \end{array}$ **c.** $\begin{array}{r} C \\ -\ 15 \\ \hline 24 \end{array}$ **d.** $\begin{array}{r} 38 \\ -\ D \\ \hline 29 \end{array}$

e. $e + 24 = 52$

f. $29 + f = 70$

g. $g - 67 = 43$

h. $80 - h = 36$

i. $36 + 14 + n + 8 = 75$

Written Practice

Strengthening Concepts

Math Language

Remember that **factors** are multiplied together to get a **product.**

1. (2) If the two factors are 25 and 12, what is the product?

2. (1) If the addends are 25 and 12, what is the sum?

3. (1) What is the difference of 25 and 12?

4. (2) Each of the 31 students brought 75 aluminum cans to class for a recycling drive. Find how many cans the class collected by multiplying 31 by 75.

5. (2) Find the total price of one dozen pizzas at \$7.85 each by multiplying \$7.85 by 12.

6. (1) *Explain* The basketball team scored 63 of its 102 points in the first half of the game. Find how many points the team scored in the second half. Explain how you found your answer.

7. (2) $\begin{array}{r} \$3.68 \\ \times\ \ \ \ 9 \\ \hline \end{array}$

8. (2) $\begin{array}{r} 407 \\ \times\ 80 \\ \hline \end{array}$

9. (2) $\begin{array}{r} 28¢ \\ \times\ 14 \\ \hline \end{array}$

10. (2) $\begin{array}{r} 370 \\ \times\ 140 \\ \hline \end{array}$

11. (2) $100 \cdot 100$

12. (2) $144 \div 12$

13. (2) $(12)(5)$

14. (1) $\begin{array}{r} 3627 \\ 598 \\ +\ 4881 \\ \hline \end{array}$

15. (1) $\begin{array}{r} 5010 \\ -\ 1376 \\ \hline \end{array}$

16. (1) $\begin{array}{r} \$10.00 \\ -\ \ \$0.26 \\ \hline \end{array}$

Find the unknown number in each problem.

17. (3) $\begin{array}{r} A \\ +\ 16 \\ \hline 48 \end{array}$

18. (3) $\begin{array}{r} 23 \\ +\ B \\ \hline 52 \end{array}$

19. (3)
$$\begin{array}{r} C \\ -\ 17 \\ \hline 31 \end{array}$$

20. (3)
$$\begin{array}{r} 42 \\ -\ D \\ \hline 25 \end{array}$$

21. (3) $x + 38 = 75$

22. (2) $x - 38 = 75$

23. (3) $75 - y = 38$

24. (3) $6 + 8 + w + 5 = 32$

* **25.** (1) **Connect** Rearrange the numbers in this addition fact to form another addition fact and two subtraction facts.

$$24 + 48 = 72$$

* **26.** (2) **Connect** Rearrange the numbers in this multiplication fact to form another multiplication fact and two division facts.

$$6 \times 15 = 90$$

Math Language
Remember the **divisor** is divided into the **dividend.** The resulting answer is the **quotient.**

27. (2) Find the quotient when the divisor is 20 and the dividend is 200.

28. (2) **Connect** Multiply to find the answer to this addition problem:

$$15 + 15 + 15 + 15 + 15 + 15 + 15 + 15$$

29. (2) $144 \div 144$

* **30.** (3) **Explain** How can you find a missing addend in an addition problem?

Early Finishers
Real-World Application

Petrov's family has a compact car that gets an average of 30 miles per gallon of gasoline. The family drives an average of 15,000 miles a year.

a. Approximately how many gallons of gas do they purchase every year?

b. If the average price of gas is $2.89 a gallon, how much should the family expect to spend on gas in a year?

LESSON 4

• Unknown Numbers in Multiplication
• Unknown Numbers in Division

Power Up

Building Power

facts Power Up A

mental math

a. **Number Sense:** 600 + 2000 + 300 + 20

b. **Number Sense:** 3000 + 20 + 400 + 5000

c. **Number Sense:** 7000 + 200 + 40 + 500

d. **Number Sense:** 700 + 2000 + 50 + 100

e. **Number Sense:** 60 + 400 + 30 + 1000

f. **Number Sense:** 900 + 8000 + 100 + 50

g. **Measurement:** How many feet are in a yard?

h. **Measurement:** How many centimeters are in a meter?

problem solving

The diagram below is called a *Jordan curve.* It is a simple *closed curve* (think of a clasped necklace that has been casually dropped on a table). Which letters are on the inside of the curve, and which letters are on the outside of the curve?

Understand We must determine if A, B, C, D, E and F are inside or outside the closed curve.

Plan We will make the problem simpler and use the simpler problems to find a pattern.

Solve We will draw less complicated closed curves, and place an X inside and a Y outside each of the closed curves.

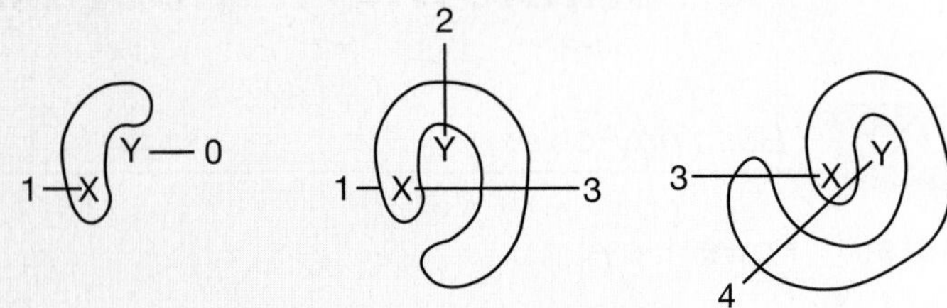

On our simpler curves, we notice that lines drawn from the outside of the curve to the X cross 1 or 3 lines. Lines drawn to the Y cross 0, 2, or 4 lines. We see this pattern: lines drawn to letters on the inside of the curve cross an odd number of lines, and lines drawn to letters on the outside of the curve cross an even number of lines.

We look at the Jordan curve again.

- A line drawn to A crosses 1 line, so it is inside the closed curve.
- A line drawn to B crosses 4 lines, so it is outside.
- A line drawn to C crosses 3 lines, so it is inside.
- A line drawn to D crosses 5 lines, so it is inside.
- A line drawn to E crosses 9 lines, so it is inside.
- A line drawn to F crosses 10 lines, so it is outside.

Check We determined that A, C, D, and E are inside the closed curve and that B and F are outside of the closed curve. We found a pattern that can help us quickly determine whether points are on the inside or outside of a closed curve.

New Concepts *Increasing Knowledge*

unknown numbers in multiplication

This multiplication fact has three numbers. If one of the **factors** were unknown, we could use the other factor and the product to figure out the unknown factor.

$$\begin{array}{r} 4 \\ \times\ 3 \\ \hline 12 \end{array}$$

Thinking Skill

Discuss

Why can we use division to find a missing factor?

Explain With your finger, cover the factors in this multiplication fact one at a time. Describe how you can use the two uncovered numbers to find the covered number. Notice that we can find an unknown factor by dividing the product by the known factor.

Example 1

Find the value of *A*:

$$\begin{array}{r} A \\ \times\ 6 \\ \hline 72 \end{array}$$

Solution

The unknown number is a factor. The product is 72. The factor that we know is 6. Dividing 72 by 6, we find that the unknown factor is **12.** We check our work by using 12 in the original problem.

$$6\overline{)7^{1}2} = 12 \longrightarrow \begin{array}{r} \overset{1}{1}2 \\ \times\ 6 \\ \hline 72 \end{array} \text{ check}$$

Example 2

Find the value of *w*: 6*w* = 84

Solution

Reading Math
In this problem 6*w* means "6 times *w*."

We divide 84 by 6 and find that the unknown factor is **14.** We check our work by multiplying.

$$6\overline{)8^{2}4} = 14 \longrightarrow \begin{array}{r} \overset{2}{1}4 \\ \times\ 6 \\ \hline 84 \end{array} \text{ check}$$

unknown numbers in division

This division fact has three numbers. If one of the numbers were unknown, we could figure out the third number.

$$6\overline{)24} = 4$$

Cover each of the numbers with your finger, and describe how to use the other two numbers to find the covered number. Notice that we can find the dividend (the number inside the division box) by multiplying the other two numbers. We can find either the divisor or quotient (the numbers outside of the box) by dividing.

Example 3

Find the value of *k*: $\frac{k}{6} = 15$

Solution

The letter *k* is in the position of the dividend. If we rewrite this problem with a division box, it looks like this:

$$6\overline{)k} = 15$$

We find an unknown dividend by multiplying the divisor and quotient. We multiply 15 by 6 and find that the unknown number is **90.** Then we check our work.

$$\begin{array}{r} \overset{3}{1}5 \\ \times\ 6 \\ \hline 90 \end{array} \longrightarrow 6\overline{)9^{3}0} = 15 \text{ check}$$

Example 4

Find the value of m: $126 \div m = 7$

Solution

The letter m is in the position of the divisor. If we were to rewrite the problem with a division box, it would look like this:

$$m\overline{)126}^{\,7}$$

We can find m by dividing 126 by 7.

$$7\overline{)12^{5}6}^{\,18}$$

We find that m is **18.** We can check our division by multiplying as follows:

$$\begin{array}{r} \scriptstyle 5 \\ 18 \\ \times\ 7 \\ \hline 126 \end{array}$$

In the original equation we can replace the letter with our answer and test the truth of the resulting equation.

$$126 \div 18 = 7$$
$$7 = 7$$

Practice Set

Analyze Find each unknown number. Check your work by using your answer in place of the letter in the original problem.

a. $\begin{array}{r} A \\ \times\ 7 \\ \hline 91 \end{array}$

b. $\begin{array}{r} 20 \\ \times\ B \\ \hline 440 \end{array}$

c. $7\overline{)\ C}^{\,15}$

d. $D\overline{)144}^{\,8}$

e. $7w = 84$

f. $112 = 8m$

g. $\frac{360}{x} = 30$

h. $\frac{n}{5} = 60$

i. **Formulate** Write a word problem using the equation in exercise **h.**

Written Practice

Strengthening Concepts

1. (2) Five dozen carrot sticks are to be divided evenly among 15 children. Find how many carrot sticks each child should receive by dividing 60 by 15.

2. (2) Matt separated 100 pennies into 4 equal piles. Find how many pennies were in each pile. Explain how you found your answer.

3. (2) Sandra put 100 pennies into stacks of 5 pennies each. Find how many stacks she formed by dividing 100 by 5.

4. (2) For the upcoming season, 294 players signed up for soccer. Find the number of 14-player soccer teams that can be formed by dividing 294 by 14.

5. (1) Angela is reading a 280-page book. She has just finished page 156. Find how many pages she still has to read by subtracting 156 from 280.

6. (2) Each month Bill earns $0.75 per customer for delivering newspapers. Find how much money he would earn in a month in which he had 42 customers by multiplying $0.75 by 42.

* **Analyze** Find each unknown number. Check your work.

7. (4) $\begin{array}{r} J \\ \times\ 5 \\ \hline 60 \end{array}$

8. (3) $\begin{array}{r} 27 \\ +\ K \\ \hline 72 \end{array}$

9. (3) $\begin{array}{r} L \\ +\ 36 \\ \hline 37 \end{array}$

10. (3) $\begin{array}{r} 64 \\ -\ M \\ \hline 46 \end{array}$

11. (3) $n - 48 = 84$

12. (4) $7p = 91$

13. (4) $q \div 7 = 0$

14. (4) $144 \div r = 6$

15. (2) $6\overline{)\$12.36}$

16. (2) $\frac{5760}{8}$

17. (2) $526 \div 18$

18. (1) $563 + 563 + 563 + 563$

19. (2) $\$3.75 \cdot 16$

20. (1) $\$3 + \$2.86 + \$0.98$

21. (1) $\$10 - \$6.43$

22. (4) If the divisor is 3 and the quotient is 12, what is the dividend?

23. (4) If the product is 100 and one factor is 5, what is the other factor?

* **24.** (1) **Connect** Rearrange the numbers in this subtraction fact to form another subtraction fact and two addition facts.

$$17 - 9 = 8$$

* **25.** (2) **Connect** Rearrange the numbers in this division fact to form another division fact and two multiplication facts.

$$72 \div 8 = 9$$

26. (3) $w + 6 + 8 + 10 = 40$

27. (2) Find the answer to this addition problem by multiplying:

$$23¢ + 23¢ + 23¢ + 23¢ + 23¢ + 23¢ + 23¢$$

28. (4) $25m = 25$

29. (4) $15n = 0$

* **30.** (4) **Explain** How can you find an unknown factor in a multiplication problem?

LESSON 5

Order of Operations, Part 1

Power Up

Building Power

facts Power Up B

mental math

a. **Number Sense:** 560 + 200

b. **Number Sense:** 840 + 30

c. **Number Sense:** 5200 + 2000

d. **Number Sense:** 650 + 140

e. **Number Sense:** 3800 + 2000

f. **Number Sense:** 440 + 200

g. **Measurement:** How many days are in a week?

h. **Measurement:** How many hours are in a day?

problem solving

Use the digits 5, 6, 7, and 8 to complete this addition problem. There are two possible arrangements.

$$\begin{array}{r} \_\_ \\ +\ \ 9 \\ \hline \_\_ \end{array}$$

Understand We are shown an addition problem with several digits missing. We are asked to complete the problem using the digits 5, 6, 7, and 8. Because the bottom addend is 9, we know that the ones digit of the sum will be one less than the ones digit of the top addend.

Plan We will intelligently guess and check for the ones place in the top addend by trying the numbers in an orderly way. We will then use logical reasoning to fill in the remaining digits of the problem.

Solve We quickly eliminate 5 as a possibility for the ones digit of the top addend because we do not have a 4 to place in the sum. We try 6 for the ones digit of the top addend. Six plus 9 is 15, so we write a 5 as the ones digit of the sum. If we write 7 as the tens digit of the top addend, we get 76 + 9. We add the two numbers and get 85. Placing an 8 in the sum, we see that we have used all the digits 5, 6, 7, and 8. We have found the first of two possible arrangements.

Next, we try 7 as the ones digit of the top addend. Seven plus 9 is 16, so we place a 6 in the sum. Now we must use the digits 5 and 8 in the tens column. We try 57 + 9 = 66. That does not work, because it does not use the 8. We try 87 + 9 = 96. That also does not work, because it omits the 5.

Finally, we try 8 in the top addend and 7 in the sum. This leaves 5 and 6 for the tens column. We try 58 + 9 = 67, and find the second solution to the problem.

Check The digits 5, 6, 7, and 8 can be used to form two solutions for our missing digit problem:

$$\begin{array}{r} 76 \\ +\ \ 9 \\ \hline 85 \end{array} \qquad \begin{array}{r} 58 \\ +\ \ 9 \\ \hline 67 \end{array}$$

New Concept *Increasing Knowledge*

Thinking Skill

Analyze

Why is it important to have rules for the order of operations?

When there is more than one addition or subtraction step within a problem, we take the steps in order from left to right. In this problem we first subtract 4 from 9. Then we add 3.

$$9 - 4 + 3 = 8$$

If a different order of steps is desired, parentheses are used to show which step is taken first. In the problem below, we first add 4 and 3 to get 7. Then we subtract 7 from 9.

$$9 - (4 + 3) = 2$$

These two rules are part of the rules for the **Order of Operations** in mathematics.

Example 1

a. 18 − 6 − 3 **b. 18 − (6 − 3)**

Solution

a. We subtract in order from left to right.

$\underbrace{18 - 6} - 3$	First subtract 6 from 18.
$12 - 3$	Then subtract 3 from 12.
9	The answer is 9.

b. We subtract within the parentheses first.

$18 - \underbrace{(6 - 3)}$	First subtract 3 from 6.
$18 - 3$	Then subtract 3 from 18.
15	The answer is 15.

When there is more than one multiplication or division step within a problem, we take the steps in order from left to right. In this problem we divide 24 by 6 and then multiply by 2.

$$24 \div 6 \times 2 = 8$$

If there are parentheses, then we first do the work within the parentheses. In the problem below, we first multiply 6 by 2 and get 12. Then we divide 24 by 12.

$$24 \div (6 \times 2) = 2$$

Example 2

a. 18 ÷ 6 ÷ 3 **b. 18 ÷ (6 ÷ 3)**

Solution

a. We take the steps in order from left to right.

$\underbrace{18 \div 6} \div 3$	First divide 18 by 6.
$3 \div 3$	Then divide 3 by 3.
1	The answer is 1.

b. We divide within the parentheses first.

$$18 \div \underbrace{(6 \div 3)} \quad \text{First divide 6 by 3.}$$
$$18 \div 2 \quad \text{Then divide 18 by 2.}$$
$$\mathbf{9} \quad \text{The answer is 9.}$$

Only two numbers are involved in each step of a calculation. If three numbers are added (or multiplied), changing the two numbers selected for the first addition (or first multiplication) does not change the final sum (or product).

$$(2 + 3) + 4 = 2 + (3 + 4) \qquad (2 \times 3) \times 4 = 2 \times (3 \times 4)$$

This property applies to addition and multiplication and is called the **Associative Property.** As shown by examples 1 and 2, the Associative Property does not apply to subtraction or to division.

Example 3

$$\frac{5 + 7}{1 + 2}$$

Solution

Before dividing we perform the operations above the bar and below the bar. Then we divide 12 by 3.

$$\frac{5 + 7}{1 + 2} = \frac{12}{3} = \mathbf{4}$$

Practice Set

a. $16 - 3 + 4$

b. $16 - (3 + 4)$

c. $24 \div (4 \times 3)$

d. $24 \div 4 \times 3$

e. $24 \div 6 \div 2$

f. $24 \div (6 \div 2)$

g. $\frac{6 + 9}{3}$

h. $\frac{12 + 8}{12 - 8}$

i. Connect Rewrite exercise **g** using parentheses instead of a bar.

Written Practice

Strengthening Concepts

1. (1) Jack paid \$5 for a sandwich that cost \$1.25 and milk that cost \$0.60. How much change should he get back?

2. (1) In one day the elephant ate 82 kilograms of hay, 8 kilograms of apples, and 12 kilograms of leaves and raw vegetables. How many kilograms of food did it eat in all?

3. (1) What is the difference of 110 and 25?

4. (2) What is the total price of one dozen apples that cost 25¢ each?

5. (3) What number must be added to 149 to total 516?

* **6.** (2) **Explain** Judy plans to read a 235-page book in 5 days. How can you find the average number of pages she needs to read each day.

7. (5) $5 + (3 \times 4)$

8. (5) $(5 + 3) \times 4$

9. (5) $800 - (450 - 125)$

10. (5) $600 \div (20 \div 5)$

11. (5) $800 - 450 - 125$

12. (5) $600 \div 20 \div 5$

13. (5) $144 \div (8 \times 6)$

14. (5) $144 \div 8 \times 6$

15. (5) $\$5 - (\$1.25 + \$0.60)$

* **16.** (2) **Represent** Use the numbers 63, 7, and 9 to form two multiplication facts and two division facts.

17. (4) If the quotient is 12 and the dividend is 288, what is the divisor?

Reading Math

Read expressions such as (4)(6) as "four times six." The parentheses indicate multiplication.

18. (2) $25\overline{)\$10.00}$

19. (2) (378)(64)

20. (2) $\begin{array}{r} 506 \\ \times\ 370 \\ \hline \end{array}$

21. (1) $\begin{array}{r} \$10.10 \\ -\ \ \$9.89 \\ \hline \end{array}$

* **Analyze** Find each unknown number. Check your work.

22. (3) $n - 63 = 36$

23. (3) $63 - p = 36$

24. (3) $56 + m = 432$

25. (4) $8w = 480$

26. (3) $5 + 12 + 27 + y = 50$

27. (4) $36 \div a = 4$

28. (4) $x \div 4 = 8$

* **29.** (1) **Represent** Use the numbers 7, 11, and 18 to form two addition facts and two subtraction facts.

30. (5) $3 \cdot 4 \cdot 5$

Early Finishers

Real-World Application

A painter is painting three exam rooms at a veterinarian's office. If each exam room requires 2 gallons of paint and the total cost of the paint is $270, how much does each gallon of paint cost?

LESSON 6

• Fractional Parts

Power Up

Building Power

facts

Power Up C

mental math

a. **Number Sense:** 2500 + 400

b. **Number Sense:** 6000 + 2400

c. **Number Sense:** 370 + 400

d. **Number Sense:** 9500 + 240

e. **Number Sense:** 360 + 1200

f. **Number Sense:** 480 + 2500

g. **Measurement:** How many seconds are in a minute?

h. **Measurement:** How many minutes are in an hour?

problem solving

Carrisa's school library received a gift of 500 new reference books. The books were arranged on a bookcase as shown in the diagram at right. How many books are on each shelf?

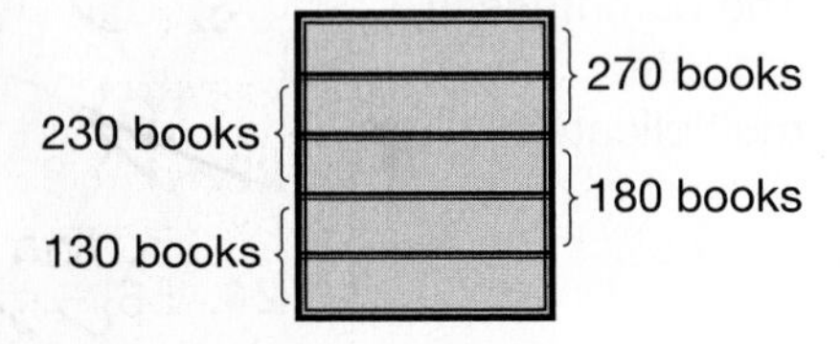

New Concept

Increasing Knowledge

As young children we learned to count objects using whole numbers. As we grew older, we discovered that there are parts of wholes—like sections of an orange—that cannot be named with whole numbers. We can name these parts with **fractions.** A common fraction is written with two numbers and a fraction bar. The "bottom" number is the **denominator.** The denominator shows the number of equal parts in the whole. The "top" number, the **numerator,** shows the number of the parts that are being represented.

Example 1

What fraction of this circle is shaded?

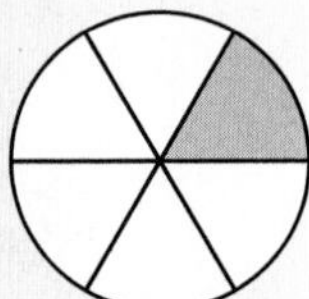

Solution

The circle has been divided into 6 equal parts. We use 6 for the bottom of the fraction. One of the parts is shaded, so we use 1 for the top of the fraction. The fraction of the circle that is shaded is one sixth, which we write as $\frac{1}{6}$.

We can also use a fraction to name a part of a group. There are 6 members in this group. We can divide this group in half by dividing it into two equal groups with 3 in each half. We write that $\frac{1}{2}$ of 6 is 3.

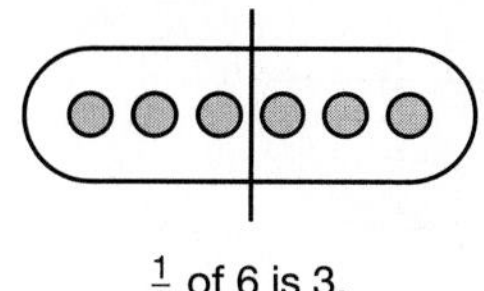

$\frac{1}{2}$ of 6 is 3.

Thinking Skill

Explain

How would we find $\frac{1}{6}$ of 6?

We can divide this group into thirds by dividing the 6 members into three equal groups. We write that $\frac{1}{3}$ of 6 is 2.

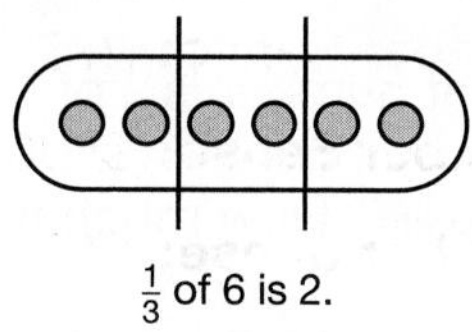

$\frac{1}{3}$ of 6 is 2.

Example 2

a. What number is $\frac{1}{2}$ of 450?

b. What number is $\frac{1}{3}$ of 450?

c. How much money is $\frac{1}{5}$ of \$4.50?

Solution

a. To find $\frac{1}{2}$ of 450, we divide 450 into two equal parts and find the amount in one of the parts. We find that $\frac{1}{2}$ of 450 is **225**.

$2\overline{)450}$ = 225 ⟶ $\frac{1}{2}$ of 450 is 225.

b. To find $\frac{1}{3}$ of 450, we divide 450 into three equal parts. Since each part is 150, we find that $\frac{1}{3}$ of 450 is **150**.

$3\overline{)450}$ = 150 ⟶ $\frac{1}{3}$ of 450 is 150.

c. To find $\frac{1}{5}$ of \$4.50, we divide \$4.50 by 5. We find that $\frac{1}{5}$ of \$4.50 is **\$0.90**.

$5\overline{)\$4.50}$ = \$0.90 ⟶ $\frac{1}{5}$ of \$4.50 is \$0.90.

Example 3

Copy the figure at right, and shade $\frac{1}{3}$ of it:

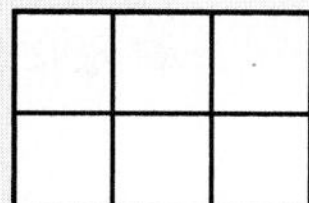

Solution

The rectangle has six parts of equal size. Since $\frac{1}{3}$ of 6 is 2, we shade any two of the parts.

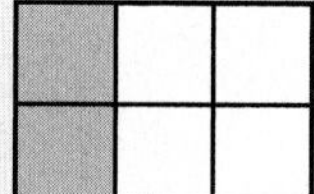

Practice Set Use both words and numbers to write the fraction that is shaded in problems **a–c.**

a. 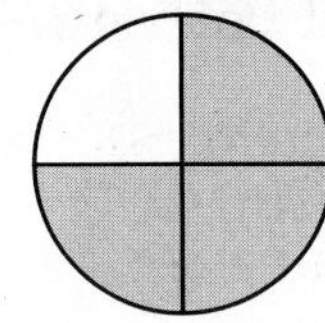

b. 

c. 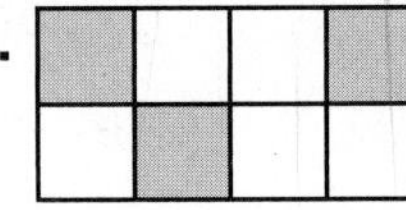

d. What number is $\frac{1}{2}$ of 72?

e. What number is $\frac{1}{2}$ of 1000?

f. What number is $\frac{1}{3}$ of 180?

g. **Explain** How much money is $\frac{1}{3}$ of \$3.60?

h. **Represent** Copy this figure and shade one half of it.

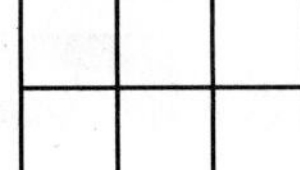

Written Practice

Strengthening Concepts

1. What number is $\frac{1}{2}$ of 540?
(6)

2. What number is $\frac{1}{3}$ of 540?
(6)

3. In four days of sight-seeing the Richmonds drove 346 miles, 417 miles, 289 miles, and 360 miles. How many miles did they drive in all?
(1)

4. Tanisha paid \$20 for a book that cost \$12.08. How much money should she get back?
(1)

5. How many days are in 52 weeks?
(2)

* **6.** **Analyze** How many \$20 bills would it take to make \$1000?
(2)

7. Use words and numbers to write the fraction of this circle that is shaded.
(6)

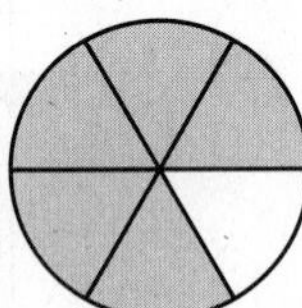

8. (1)
$$\begin{array}{r} 3604 \\ 5186 \\ +\ 7145 \\ \hline \end{array}$$

9. (1)
$$\begin{array}{r} \$30.01 \\ -\ \$15.76 \\ \hline \end{array}$$

10. (2)
$$\begin{array}{r} 376 \\ \times\ \ 87 \\ \hline \end{array}$$

11. (2)
$$\begin{array}{r} 470 \\ \times\ 203 \\ \hline \end{array}$$

12. $\$20 - \$11.98$ (1)

13. $596 - (400 - 129)$ (5)

14. $32 \div (8 \times 4)$ (5)

15. $8\overline{)4016}$ (2)

16. $15\overline{)6009}$ (2)

17. $36\overline{)9000}$ (2)

Find each unknown number. Check your work.

18. $8w = 480$ (4)

19. $x - 64 = 46$ (3)

20. $\frac{49}{N} = 7$ (4)

21. $\frac{M}{7} = 15$ (4)

22. $365 + P = 653$ (3)

23. $36¢ + 25¢ + m = 99¢$ (3)

* 24. (6) **Conclude** The square at right was divided in half. Then each half was divided in half. What fraction of the square is shaded?

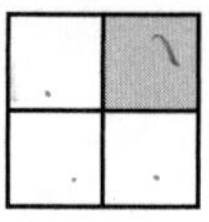

* 25. (6) **Represent** Copy this figure on your paper, and shade one fourth of it.

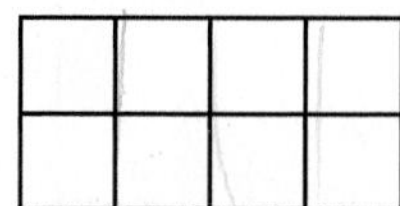

26. $\$6.35 \cdot 12$ (2)

27. (1) Use the numbers 2, 4, and 6 to form two addition facts and two subtraction facts.

28. (2) Write two multiplication facts and two division facts using the numbers 2, 4, and 8.

* 29. (2) **Connect** Write a multiplication equation to solve this addition problem.

$$38 + 38 + 38 + 38 + 38 + 38 + 38 + 38 + 38 + 38$$

* 30. (6) **Formulate** Make up a fractional-part question about money, as in Example 2 part c. Then find the answer.

LESSON 7

• Lines, Segments, and Rays
• Linear Measure

Power Up

Building Power

facts Power Up C

mental math

a. **Number Sense:** 800 − 300

b. **Number Sense:** 3000 − 2000

c. **Number Sense:** 450 − 100

d. **Number Sense:** 2500 − 300

e. **Number Sense:** 480 − 80

f. **Number Sense:** 750 − 250

g. **Measurement:** How many weeks are in a year?

h. **Measurement:** How many days are in a year?

problem solving

A pulley is in *equilibrium* when the total weight suspended from the left side is equal to the total weight suspended from the right side.

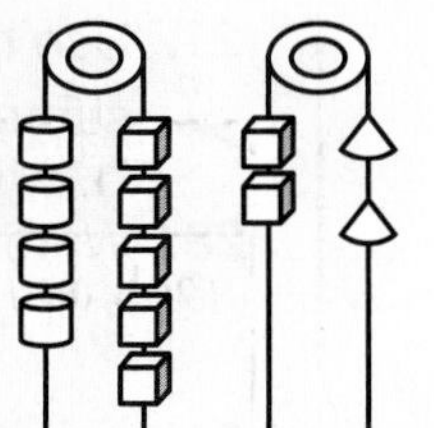

The two pulleys on the left are both in equilibrium. Is the pulley on the right in equilibrium, or is one side heavier than the other?

Understand We are shown three pulleys on which three kinds of weights are suspended. The first two pulleys are in equilibrium. We are asked to determine if the third pulley is in equilibrium or if one side is heavier than the other.

Plan We will *use logical reasoning* to determine whether the third pulley is in equilibrium.

Solve From the first pulley we see that four cylinders are equal in weight to five cubes. This means that cylinders are heavier than cubes. The second pulley shows that two cubes weigh the same as two cones. This means that cubes and cones weigh the same.

On the third pulley, the bottom cubes on either side have the same weight. We are left with two cones and two cubes on one side and four cylinders on the other. We know that cylinders are heavier than cubes and cones, so the pulley is not in equilibrium. The right side is heavier, so the pulley will pull to the right.

Check We can confirm our conclusion by looking at the third pulley as five cubes on the left (because the two cones are equal in weight to two cubes). From the first pulley, we know that five cubes are equal in weight to four cylinders. Another cube on the right side makes the right side heavier.

New Concepts *Increasing Knowledge*

lines, segments, and rays

Thinking Skill

Conclude

If two opposite-facing rays are joined at their endpoints, what is the result? What do those endpoints become?

In everyday language the following figure is often referred to as a line:

However, using mathematical terminology, we say that the figure represents a **segment,** or line segment. A segment is part of a line and has two **endpoints.** A mathematical **line** has no endpoints. To represent a line, we use arrowheads to indicate a line's unending quality.

A **ray** has one endpoint. We represent a ray with one arrowhead.

A ray is roughly represented by a beam of sunlight. The beam begins at the sun (which represents the endpoint of the ray) and continues across billions of light years of space.

linear measure

Line segments have length. In the United States we have two systems of units that we use to measure length. One system is the **U.S. Customary System.** Some of the units in this system are inches (in.), feet (ft), yards (yd), and miles (mi). The other system is the **metric system (International System).** Some of the units in the metric system are millimeters (mm), centimeters (cm), meters (m), and kilometers (km).

Some Units of Length and Benchmarks

U.S. Customary System		Metric System	
inch (in.)	width of thumb	millimeter (mm)	thickness of a dime
foot (ft)	length of ruler, 12 inches	centimeter (cm)	thickness of little finger tip, 10 millimeters
yard (yd)	a long step, 3 ft or 36 inches	meter (m)	a little over a yard, 100 centimeters
miles (mi)	distance walked in 20 minutes, 5280 feet	kilometer (km)	distance walked in 12 minutes, 1000 meters

In this lesson we will practice measuring line segments with an inch ruler and with a centimeter ruler, and we will select appropriate units for measuring lengths.

Activity

Inch Ruler

Materials needed:

- inch ruler
- narrow strip of tagboard about 6 inches long and 1 inch wide
- pencil

Model Use your pencil and ruler to draw inch marks on the strip of tagboard. Number the inch marks. When you are finished, the tagboard strip should look like this:

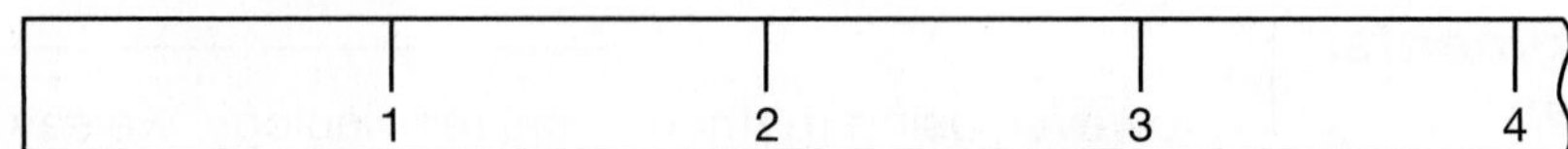

Estimate Now set aside your ruler. We will use estimation to make the rest of the marks on the tagboard strip. Estimate the halfway point between inch marks, and make the half-inch marks slightly shorter than the inch marks, as shown below.

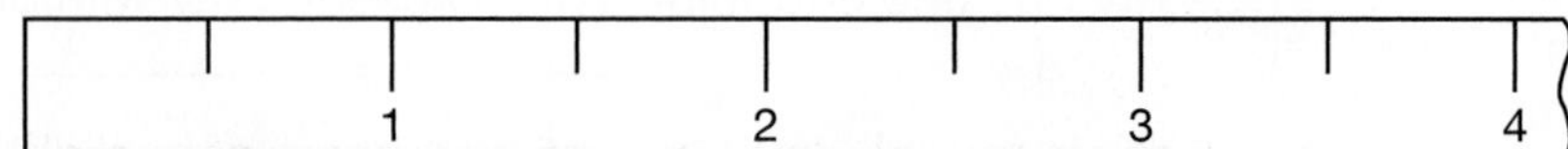

Now show every quarter inch on your tagboard ruler. To do this, estimate the halfway point between each mark on the ruler, and make the quarter-inch marks slightly shorter than the half-inch marks, as shown below.

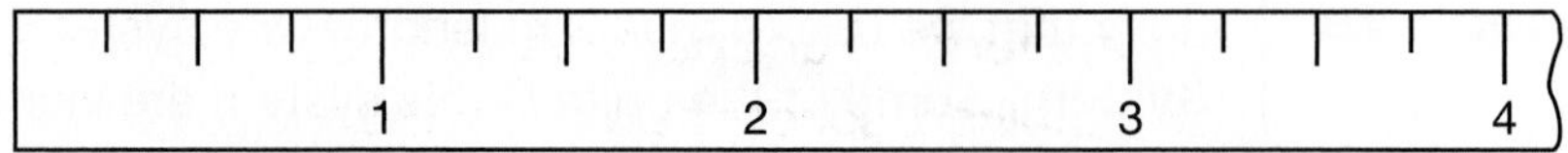

Save your tagboard ruler. We will be making more marks on it in a few days.

Connect A metric ruler is divided into centimeters. There are 100 centimeters in a meter. Each centimeter is divided into 10 millimeters. So 1 centimeter equals 10 millimeters, and 2 centimeters equals 20 millimeters.

By comparing an inch ruler with a centimeter ruler, we see that an inch is about $2\frac{1}{2}$ centimeters.

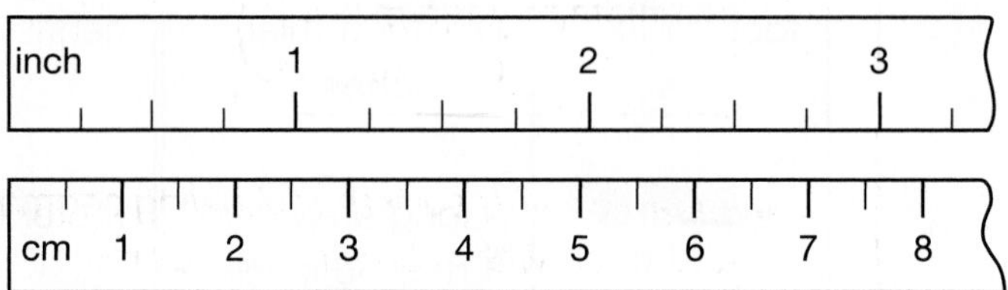

A cinnamon stick that is 3 inches long is about $7\frac{1}{2}$ cm long. A foot-long ruler is about 30 cm long.

Example 1

How long is the line segment?

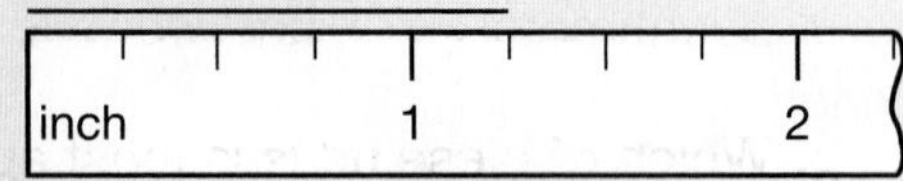

Solution

Reading Math

The abbreviation for inches (in.) ends with a period so it is not confused with the word *in*.

The line is one whole inch plus a fraction. The fraction is one fourth. So the length of the line is $\mathbf{1\frac{1}{4}}$ **in.**

Example 2

How long is the line segment?

cm 1 2 3

Solution

We simply read the scale to see that the line is **2 cm** long. The segment is also **20 mm** long.

Example 3

Select the appropriate unit for measuring the length of a soccer field.

A centimeters **B** meters **C** kilometers

Solution

An appropriate unit can give us a good sense of the measure of an object. Describing a soccer field as thousands of centimeters or a small fraction of a kilometer can be accurate without being appropriate. The best choice is **B meters** for measuring the length of a soccer field.

Practice Set

How long is each line segment?

a.

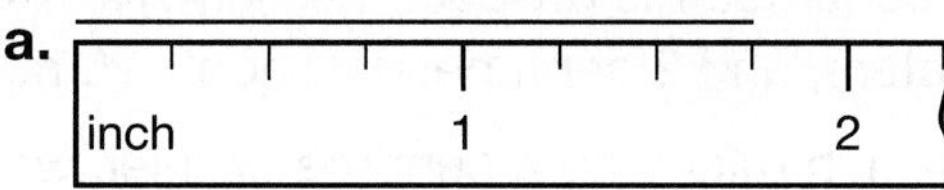

b.

mm 10 20 30 40

c. Connect Measure the following segment twice, once with an inch ruler and once with a centimeter ruler.

Use the words *line, segment,* or *ray* to describe each of these figures:

d.

e.

f.

g. Which of these units is most appropriate for measuring the length of a pencil?

A inches **B** yards **C** miles

h. Select the appropriate unit for measuring the distance between two towns.

A centimeters **B** meters **C** kilometers

Written Practice *Strengthening Concepts*

1. To earn money for gifts, Debbie sold decorated pinecones. If she sold 100 pinecones at \$0.25 each, how much money did she earn? (2)

2. There are 365 days in a common year. April 1 is the 91st day. How many days are left in the year after April 1? (1)

3. The Cardaso family is planning to complete a 1890-mile trip in 3 days. If they drive 596 miles the first day and 612 miles the second day, how far must they travel the third day? (*Hint:* This is a two-step problem. First find how far they traveled the first two days.) (5)

4. What number is $\frac{1}{2}$ of 234? (6)

5. How much money is $\frac{1}{3}$ of \$2.34? (6)

6. Use words and digits to write the fraction of this circle that is shaded. (6)

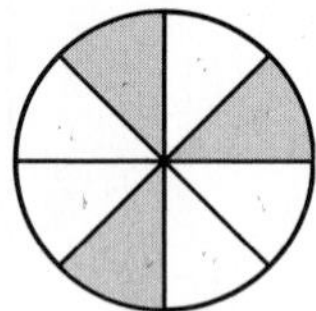

7. (1)
$$\begin{array}{r} 3654 \\ 2893 \\ +\ 5614 \\ \hline \end{array}$$

8. (1)
$$\begin{array}{r} \$41.01 \\ -\ \$15.76 \\ \hline \end{array}$$

9. (2)
$$\begin{array}{r} 28¢ \\ \times\ 74 \\ \hline \end{array}$$

10. (2)
$$\begin{array}{r} 906 \\ \times\ \ 47 \\ \hline \end{array}$$

11. $6\overline{)5000}$ (2)

12. $800 \div 16$ (2)

13. $60\overline{)3174}$ (2)

14. $3 + 6 + 5 + w + 4 = 30$ (3)

15. $300 - 30 + 3$ (5)

16. $300 - (30 + 3)$ (5)

17. $\$4.32 \cdot 20$ (2)

18. 24(48¢) (2)

19. $\$8.75 \div 25$ (2)

Find each unknown number. Check your work.

20. $W \div 6 = 7$ (4)

21. $6n = 96$ (4)

22. $58 + r = 213$ (4)

* 23. **Connect** Rearrange the numbers in this subtraction fact to form another subtraction fact and two addition facts. (1)

$$60 - 24 = 36$$

24. How long is the line segment below? (7)

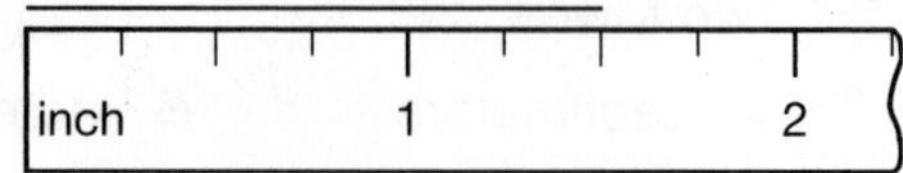

25. Find the length, in centimeters and in millimeters, of the line segment below.
(7)

cm 1 2 3 4

* **26.** **Connect** Use the numbers 9, 10, and 90 to form two multiplication facts and two division facts.
(2)

* **27.** **Explain** How can you find a missing dividend in a division problem?
(4)

28. $w - 12 = 8$
(3)

29. $12 - x = 8$
(3)

30. **a.** A meterstick is 100 centimeters long. One hundred centimeters is how many millimeters?
(7)

b. The length of which of the following would most likely be measured in meters?

A a pencil **B** a hallway **C** a highway

Early Finishers
Real-World Application

The district championship game will be played on an artificial surface. One-fifth of the team needs new shoes for the game. There are 40 players on the team, and each pair of shoes sells for $45.

a. How many players need new shoes?

b. How much money must the booster club raise to cover the entire cost of the shoes?

LESSON 8

• Perimeter

Power Up

Building Power

facts Power Up A

mental math

a. **Number Sense:** 400 + 2400

b. **Number Sense:** 750 + 36

c. **Number Sense:** 8400 + 520

d. **Number Sense:** 980 − 60

e. **Number Sense:** 4400 − 2000

f. **Number Sense:** 480 − 120

g. **Measurement:** How many feet are in 2 yards?

h. **Measurement:** How many centimeters are in 2 meters?

problem solving

The digits 2, 4, and 6 can be arranged to form six different three-digit numbers. Each ordering is called a **permutation** of the three digits. The smallest permutation of 2, 4, and 6 is 246. What are the other five permutations? List the six numbers in order from least to greatest.

Understand We have been asked to find five of the six permutations that exist for three digits, and then list the permutations from least to greatest.

Plan To make sure we find all permutations possible, we will make an organized list.

Solve We first write the permutations of 2, 4, and 6 that begin with 2, then those that begin with 4, then those that begin with 6: 246, 264, 426, 462, 624, 642.

Check We found all six permutations of the digits 2, 4, and 6. Writing them in an organized way helped us ensure we did not overlook any permutations. Because we wrote the numbers from least to greatest as we went along, we did not have to re-order our list to solve the problem.

New Concept

Increasing Knowledge

The distance around a shape is its **perimeter.** The perimeter of a square is the distance around it. The perimeter of a room is the distance around the room.

Activity

Perimeter

Model Walk the perimeter of your classroom. Start at a point along a wall of the classroom, and, staying close to the walls, walk around the room until you return to your starting point. Count your steps as you travel around the room. How many of your steps is the perimeter of the room?

Discuss

a. Did everyone count the same number of steps?

b. Does the perimeter depend upon who is measuring it?

c. Which of these is the best real-world example of perimeter?

1. The tile or carpet that covers the floor.

2. The molding along the base of the wall.

Here we show a rectangle that is 3 cm long and 2 cm wide.

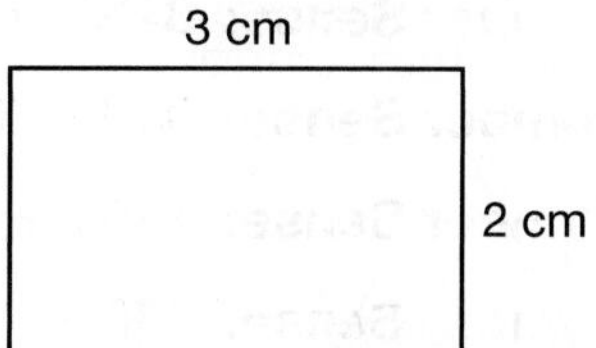

If we were to start at one corner and trace the perimeter of the rectangle, our pencil would travel 3 cm, then 2 cm, then 3 cm, and then 2 cm to get all the way around. We add these lengths to find the perimeter of the rectangle.

$$3 \text{ cm} + 2 \text{ cm} + 3 \text{ cm} + 2 \text{ cm} = \mathbf{10 \text{ cm}}$$

Example 1

What is the perimeter of this triangle?

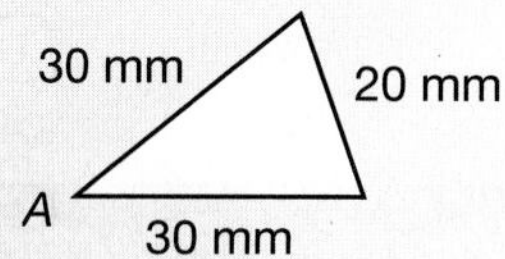

Solution

The perimeter of a shape is the distance around the shape. If we trace around the triangle from point *A,* the point of the pencil would travel 30 mm, then 20 mm, and then 30 mm. Adding these distances, we find that the perimeter is **80 mm.**

Example 2

The perimeter of a square is 20 cm. What is the length of each side?

Solution

The four sides of a square are equal in length. So we divide the perimeter by 4 to find the length of each side. We find that the length of each side is **5 cm.**

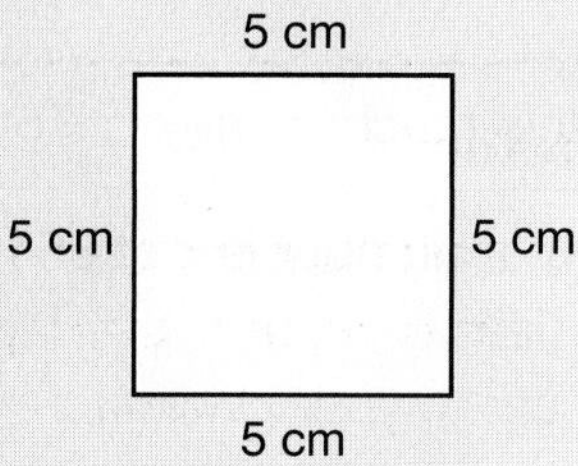

Practice Set

Thinking Skill

Verify

Why can we find the perimeter of a regular polygon when we know the length of one side?

What is the perimeter of each shape?

a. square

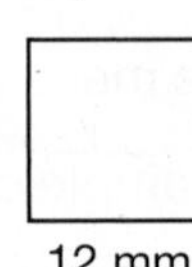

b. rectangle

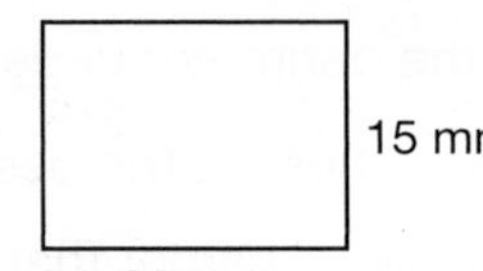

c. trapezoid

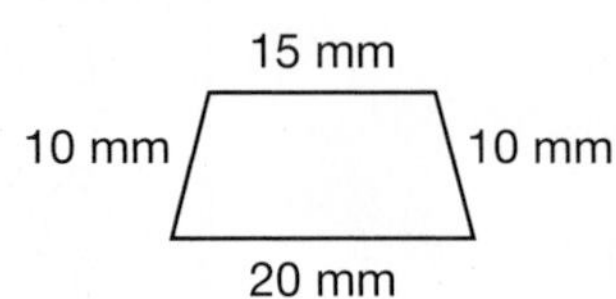

Figures **d** and **e** below are regular polygons because all of their sides are the same length and all of their angles are the same size. Find the perimeter of each shape.

d. equilateral triangle

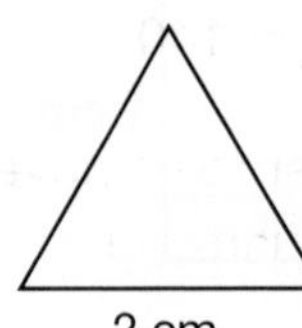

e. pentagon

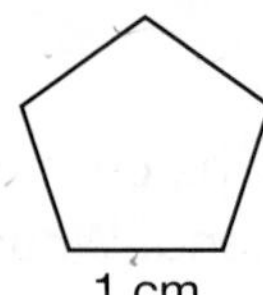

f. *Conclude* The perimeter of a square is 60 cm. How long is each side of the square?

g. *Represent* Draw two different figures that have perimeters that are the same length.

h. Select the appropriate unit for measuring the perimeter of a classroom.

A inches **B** feet **C** miles

Written Practice

Strengthening Concepts

1. (2) In an auditorium there are 25 rows of chairs with 18 chairs in each row. How many chairs are in the auditorium?

* **2.** (1) *Explain* The sixth-graders collected 765 cans of food for the food pantry last year. This year they collected 1750 cans. How many fewer cans did they collect last year than this year? Explain how you found the answer.

3. (2) A basketball team is made up of 5 players. Suppose there are 140 players signed up for a tournament. How many teams will there be of 5 players per team?

4. (8) What is the perimeter of this triangle?

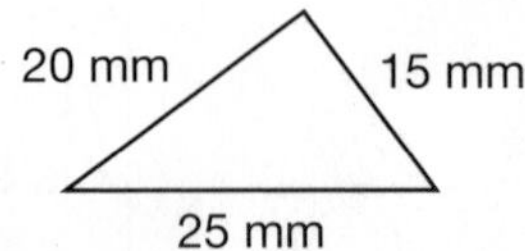

5. (6) How much money is $\frac{1}{2}$ of $6.54?

6. (6) What number is $\frac{1}{3}$ of 654?

* 7. (6) **Represent** What fraction of this rectangle is shaded?

8. (2) $4\overline{)\$9.00}$

9. (2) $10\overline{)373}$

10. (2) $12\overline{)1500}$

11. (2) $39\overline{)800}$

12. (5) $400 \div 20 \div 4$

13. (5) $400 \div (20 \div 4)$

* 14. (2) **Connect** Use the numbers 240, 20, and 12 to form two multiplication facts and two division facts.

15. (1) Rearrange the numbers in this addition fact to form another addition fact and two subtraction facts.

$$60 + 80 = 140$$

16. (8) The ceiling tiles used in many classrooms have sides that are 12 inches long. What is the perimeter of a square tile with sides 12 inches long?

17. (1, 2) **a.** Find the sum of 6 and 4.

b. Find the product of 6 and 4.

18. (3) $\$5 - M = \$1.48$

19. (5) $10 \times 20 \times 30$

20. (2) $825 \div 8$

Find each unknown number. Check your work.

21. (3) $w - 63 = 36$

22. (3) $150 + 165 + a = 397$

23. (4) $12w = 120$

24. (4) If the divisor is 8 and the quotient is 24, what is the dividend?

* 25. (7) **Estimate** **a.** Measure the length of the line segment below to the nearest centimeter.

b. Measure the length of the segment in millimeters.

* 26. (7) **Model** Use a ruler to draw a line segment that is $2\frac{3}{4}$ in. long.

27. (3) $w - 27 = 18$

28. (3) $27 - x = 18$

29. (2) Multiply to find the answer to this addition problem:

$$35 + 35 + 35 + 35$$

* 30. (8) **Explain** How can you calculate the perimeter of a rectangle?

LESSON 9

The Number Line: Ordering and Comparing

Power Up

Building Power

facts Power Up C

mental math

a. **Number Sense:** 48 + 120

b. **Number Sense:** 76 + 10 + 3

c. **Number Sense:** 7400 + 320

d. **Number Sense:** 860 − 50

e. **Number Sense:** 960 − 600

f. **Number Sense:** 365 − 200

g. **Geometry:** A square has a length of 5 inches. What is the perimeter of the square?

h. **Measurement:** How many days are in a leap year?

problem solving

As you sit at your desk facing forward, you can describe the locations of people and objects in your classroom compared to your position. Perhaps a friend is two seats in front and one row to the left. Perhaps the door is directly to your right about 6 feet. Describe the location of your teacher's desk, the pencil sharpener, and a person or object of your choice.

New Concept

Increasing Knowledge

Thinking Skill

Connect

Name some real-life situations in which we would use negative numbers.

A **number line** is a way to show numbers in order.

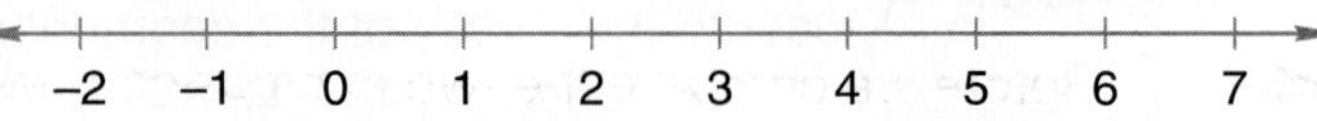

The arrowheads show that the line continues without end and that the numbers continue without end. The small marks crossing the horizontal line are called *tick marks.* Number lines may be labeled with various types of numbers. The numbers we say when we count (1, 2, 3, 4, and so on) are called **counting numbers.** All the counting numbers along with the number zero make up the **whole numbers.**

To the left of zero on this number line are **negative numbers,** which will be described in later lessons. As we move to the right on this number line, the numbers are greater in value. As we move to the left, the numbers are lesser in value.

Example 1

Arrange these numbers in order from least to greatest:

121 112 211

Solution

On a number line, these three numbers appear in order from least (on the left) to greatest (on the right).

112 121 211

For our answer, we write

112 121 211

Thinking Skill

Discuss

When using place value to compare two numbers, what do you need to do first?

When we **compare** two numbers, we decide whether the numbers are equal; if they are not equal, we determine which number is greater and which is lesser. We show a comparison with symbols. If the numbers are equal, the comparison symbol we use is the **equal sign** (=).

$$1 + 1 = 2$$

If the numbers are not equal, we use one of the **greater than/less than symbols** (> or <). When properly placed between two numbers, the small end of the symbol points to the lesser number.

Example 2

Compare: 5012 ◯ 5102

Solution

In place of the circle we should write =, >, or < to make the statement true. Since 5012 is less than 5102, we point the small end to the 5012.

$$\mathbf{5012 < 5102}$$

Example 3

Compare: 16 ÷ 8 ÷ 2 ◯ 16 ÷ (8 ÷ 2)

Solution

Reading Math

Remember to follow the order of operations. Do the work within the parentheses first. Then divide in order from left to right.

Before we compare the two expressions, we find the value of each expression.

$$\underbrace{16 \div 8 \div 2}_{1} \bigcirc \underbrace{16 \div (8 \div 2)}_{4}$$

Since 1 is less than 4, the comparison symbol points to the left.

$$\mathbf{16 \div 8 \div 2 < 16 \div (8 \div 2)}$$

Example 4

Use digits and symbols to write this comparison:

One fourth is less than one half.

Solution

We write the numbers in the order stated.

$$\frac{1}{4} < \frac{1}{2}$$

Practice Set

a. Arrange these amounts of money in order from least to greatest.

12¢ $12 $1.20

b. Compare: 16 − 8 − 2 ◯ 16 − (8 − 2)

c. Compare: 8 ÷ 4 × 2 ◯ 8 ÷ (4 × 2)

d. 2 × 3 ◯ 2 + 3

e. 1 × 1 × 1 ◯ 1 + 1 + 1

f. *Represent* Use digits and symbols to write this comparison:

One half is greater than one fourth.

g. Compare the lengths: 10 inches ◯ 1 foot

Written Practice

Strengthening Concepts

1. (2) Tamara arranged 144 books into 8 equal stacks. How many books were in each stack?

2. (1) *Generalize* Find how many years there were from 1492 to 1603 by subtracting 1492 from 1603.

*** 3.** (2) *Conclude* Martin is carrying groceries in from the car. If he can carry 2 bags at a time, how many trips will it take him to carry in 9 bags?

*** 4.** (7, 8) *Conclude* Use a centimeter ruler to measure the length and width of the rectangle below. Then calculate the perimeter of the rectangle.

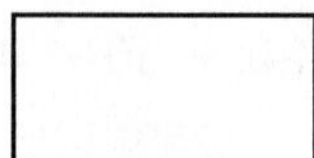

5. (6) How much money is $\frac{1}{2}$ of $5.80?

6. (6) How many cents is $\frac{1}{4}$ of a dollar?

*** 7.** (6) *Represent* Use words and digits to name the fraction of this triangle that is shaded.

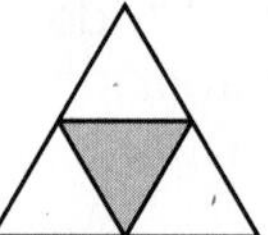

8. (7, 9) Compare:

a. 5012 ◯ 5120

b. 1 mm ◯ 1 cm

9. (9) Arrange these numbers in order from least to greatest:

1, 0, $\frac{1}{2}$

10. (9) Compare: 100 − 50 − 25 ◯ 100 − (50 − 25)

11. (1) $\begin{array}{r} 478 \\ 3692 \\ +\ \ 45 \\ \hline \end{array}$

12. (1) $\begin{array}{r} \$50.00 \\ -\ \$31.76 \\ \hline \end{array}$

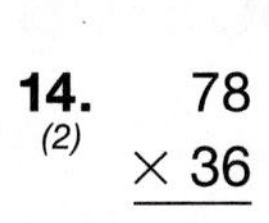

13. (2) $\begin{array}{r} \$4.20 \\ \times\ \ \ 60 \\ \hline \end{array}$

14. (2) $\begin{array}{r} 78 \\ \times\ 36 \\ \hline \end{array}$

15. (2) $9\overline{)7227}$

16. (2) $25\overline{)7600}$

17. (2) $20\overline{)8014}$

18. (2) $7136 \div 100$

19. (2) $736 \div 736$

Find each unknown number. Check your work.

20. (3) $165 + a = 300$

21. (3) $b - 68 = 86$

22. (4) $9c = 144$

23. (4) $\frac{d}{15} = 7$

* **24.** (7) **Estimate** Use an inch ruler to draw a line segment two inches long. Then use a centimeter ruler to find the length of the segment to the nearest centimeter.

25. (7) Which of the figures below represents a ray?

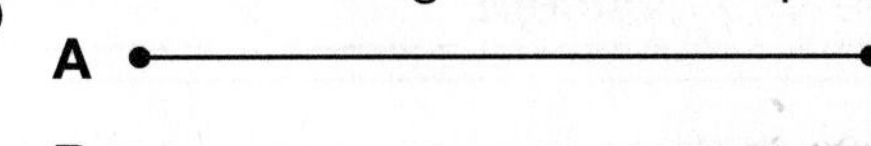

A

B

C

* **26.** (9) **Represent** Use digits and symbols to write this comparison:

One half is greater than one third.

* **27.** (2) **Connect** Arrange the numbers 9, 11, and 99 to form two multiplication facts and two division facts.

28. (9) Compare: $25 + 0 \bigcirc 25 \times 0$

29. (3) $100 = 20 + 30 + 40 + x$

* **30.** (9) **Explain** How did you choose the positions of the small and large ends of the greater than/less than symbols that you used in problem 8?

LESSON 10

• Sequences
• Scales

Power Up

Building Power

facts Power Up C

mental math

a. **Number Sense:** 43 + 20 + 5

b. **Number Sense:** 670 + 200

c. **Number Sense:** 254 + 20 + 5

d. **Number Sense:** 100 − 50

e. **Number Sense:** 300 − 50

f. **Number Sense:** 3600 − 400

g. **Measurement:** How many feet are in 3 yards?

h. **Measurement:** How many centimeters are in 3 meters?

problem solving

A ***parallax*** is an error that can result from an observer changing their position and not reading a measurement tool while directly in front of it.

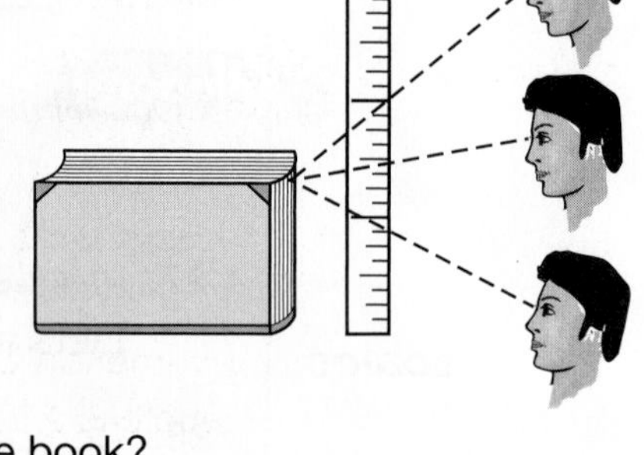

Hold a ruler near the width of your textbook on your desk. Then read the ruler from several angles as the diagram shows. How short can you make the width of the book appear? How tall can you make it appear? What is your best measure of the width of the book?

New Concepts

Increasing Knowledge

sequences

A **sequence** is an ordered list of numbers, called **terms,** that follows a certain rule. Here are two different sequences:

a. 5, 10, 15, 20, 25, ...

b. 5, 10, 20, 40, 80, ...

Sequence **a** is an **addition sequence** because the same number is added to each term of the sequence to get the next term. In this case, we add 5 to the value of a term to find the next term. Sequence **b** is a **multiplication sequence** because each term of the sequence is multiplied by the same number to get the next term. In **b** we find the value of a term by multiplying the preceding term by 2. When we are asked to find unknown numbers in a sequence, we inspect the numbers to discover the rule for the sequence. Then we use the rule to find other numbers in the sequence.

Example 1

Generalize **Describe the following sequence as an addition sequence or a multiplication sequence. State the rule of the sequence, and find the next term.**

1, 3, 9, 27, _____, ...

Solution

The sequence is a **multiplication sequence** because **each term in the sequence can be multiplied by 3 to find the next term.** Multiplying 27 by 3, we find that the term that follows 27 in the sequence is **81.**

Thinking Skill

Generalize

Is the sum of an even number and an odd number even or odd? Give some examples to support your answer.

The numbers ..., 0, 2, 4, 6, 8, ... form a special sequence called **even numbers.** We say even numbers when we "count by twos." Notice that zero is an even number. Any whole number with a ones digit of 0, 2, 4, 6, or 8 is an even number. Whole numbers that are not even numbers are **odd numbers.** The odd numbers form the sequence ..., 1, 3, 5, 7, 9, ... An even number of objects can be divided into two equal groups. An odd number of objects cannot be divided into two equal groups.

Example 2

Think of a whole number. Double that number. Is the answer even or odd?

Solution

The answer is **even.** Doubling any whole number—odd or even—results in an even number.

scales

Numerical information is often presented to us in the form of a **scale.** A scale is a display of numbers with an indicator to show the value of a certain measure. To read a scale, we first need to discover the value of the tick marks on the scale. Marks on a scale may show whole units or divisions of a unit, such as one fourth (as with the inch ruler from Lesson 7). We study the scale to find the value of the units before we try to read the indicated number.

Two commonly used scales on thermometers are the **Fahrenheit scale** and the **Celsius scale.** A cool room may be 68°F (20°C). The temperature at which water freezes under standard conditions is 32 degrees Fahrenheit (abbreviated 32°F) and zero degrees Celsius (0°C). The boiling temperature of water is 212°F, which is 100°C. Normal body temperature is 98.6°F (37°C). The thermometer on the right shows three temperatures that are helpful to know:

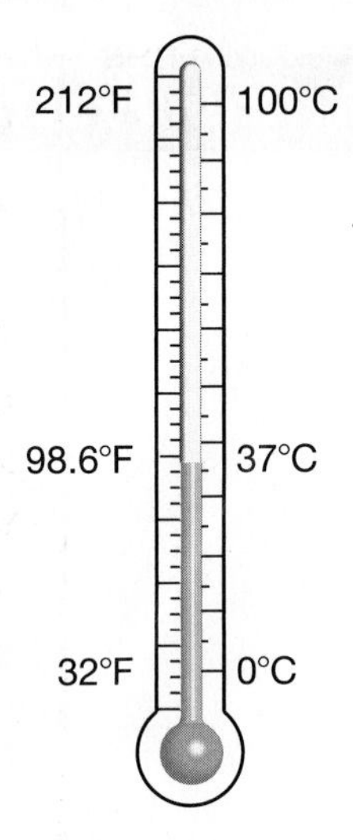

Boiling temperature of water	212°F	100°C
Normal body temperature	98.6°F	37°C
Freezing temperature of water	32°F	0°C

Example 3

What temperature is shown on this thermometer?

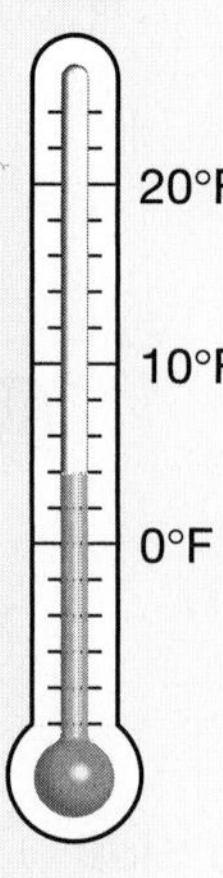

Solution

As we study the scale on this Fahrenheit thermometer, we see that the tick marks divide the distance from 0°F to 10°F into five equal sections. So the number of degrees from one tick mark to the next must be 2°F. Since the fluid in the thermometer is two marks above 0°F, the temperature shown is **4°F.**

Practice Set

Generalize For the sequence in problems **a** and **b,** determine whether the sequence is an addition sequence or a multiplication sequence, state the rule of the sequence, and find the next three terms.

a. ..., 18, 27, 36, 45, ______, ______, ______, ...

b. 1, 2, 4, 8, ______, ______, ______, ...

c. Think of a whole number. Double that number. Then add 1 to the answer. Is the final number even or odd?

Thinking Skill

Explain

Why is the thermometer shown with a broken scale?

d. Estimate This thermometer indicates a comfortable room temperature. Find the temperature indicated on this thermometer to the nearest degree Fahrenheit and to the nearest degree Celsius.

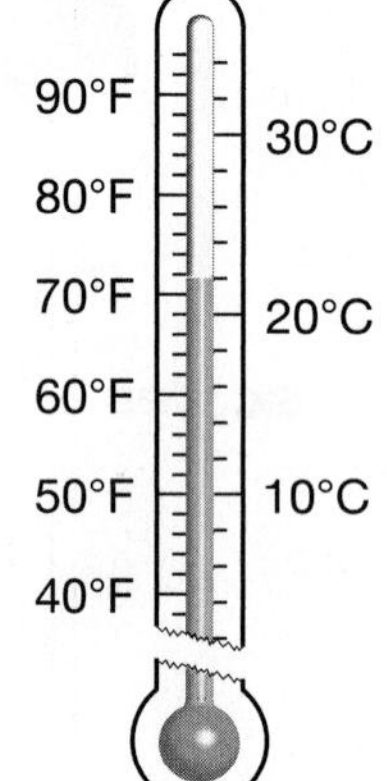

Written Practice

Strengthening Concepts

* **1.** (10) Generalize State the rule of the following sequence. Then find the next three terms.

16, 24, 32, ______, ______, ______, ...

2. (1) Find how many years there were from the year the Pilgrims landed in 1620 to the year the colonies declared their independence in 1776.

* **3.** (10) Explain Is the number 1492 even or odd? How can you tell?

4. What weight is indicated on this scale?
(10)

*** 5.** **Conclude** If the perimeter of a square is 40 mm, how long is each side of the square?
(8)

6. How much money is $\frac{1}{2}$ of \$6.50?
(6)

7. Compare: $4 \times 3 + 2 \bigcirc 4 \times (3 + 2)$
(9)

*** 8.** **Represent** Use words and digits to write the fraction of this circle that is **not** shaded.
(6)

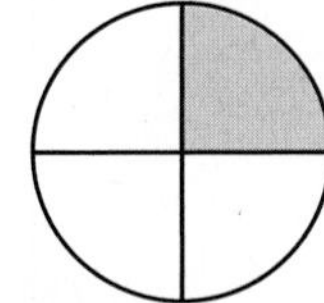

9. What is the
(1, 2)

a. product of 100 and 100?

b. sum of 100 and 100?

10. (2) $\begin{array}{r} 365 \\ \times\ 100 \\ \hline \end{array}$

11. (2) $\begin{array}{r} 146 \\ \times\ 240 \\ \hline \end{array}$

12. (2) $\begin{array}{r} 78¢ \\ \times\ 48 \\ \hline \end{array}$

13. (2) $\begin{array}{r} 907 \\ \times\ 36 \\ \hline \end{array}$

14. (2) $\frac{4260}{10}$

15. (2) $\frac{4260}{20}$

16. (2) $\frac{4260}{15}$

17. (1) 28,347 − 9,637

18. (3) \$8 + w = \$11.49

19. (1) \$10 − \$0.75

20. (2) \$0.56 × 60

21. (2) \$6.20 ÷ 4

Find each unknown number. Check your work.

22. (3) $56 + 28 + 37 + n = 200$

23. (3) $a - 67 = 49$

24. (3) $67 - b = 49$

25. (4) $8c = 120$

26. (4) $\frac{d}{8} = 24$

27. Here are three ways to write "12 divided by 4."
(2)

$4\overline{)12} \qquad 12 \div 4 \qquad \frac{12}{4}$

Show three ways to write "20 divided by 5."

28. What number is one third of 36?
(6)

*** 29.** **Connect** Arrange the numbers 346, 463, and 809 to form two addition equations and two subtraction equations.
(1)

30. At what temperature on the Fahrenheit scale does water freeze?
(10)

INVESTIGATION 1

Focus on

- **Frequency Tables**
- **Histograms**
- **Surveys**

frequency tables

Mr. Lawson made a frequency table to record student scores on a math test. He listed the intervals of scores (bins) he wanted to tally. Then, as he graded each test, he made a tally mark for each test in the corresponding row.

Frequency Table

Number Correct	Tally	Frequency
19–20	卌 ||||	9
17–18	卌 ||	7
15–16	||||	4
13–14	||	2

When Mr. Lawson finished grading the tests, he counted the number of tally marks in each row and then recorded the count in the frequency column. For example, the table shows that nine students scored either 19 or 20 on the test. A **frequency table** is a way of pairing selected data, in this case specified test scores, with the number of times the selected data occur.

1. *Justify* Can you tell from this frequency table how many students had 20 correct answers on the test? Why or why not?

2. *Conclude* Mr. Lawson tallied the number of scores in each interval. How wide is each interval? Suggest a reason why Mr. Lawson arranged the scores in such intervals.

3. *Represent* Show how to make a tally for 12.

4. *Represent* As a class activity, make a frequency table of the birth months of the students in the class. Make four bins by grouping the months: Jan.–Mar., Apr.–Jun., Jul.–Sep., Oct.–Dec.

histograms

Using the information in the frequency table, Mr. Lawson created a histogram to display the results of the test.

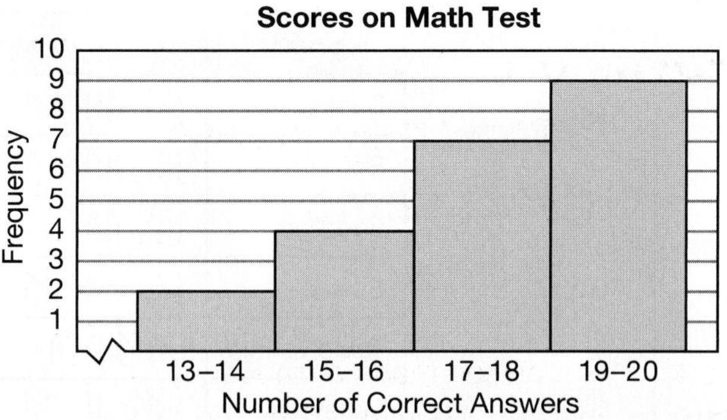

Bar graphs display numerical information with shaded rectangles (bars) of various lengths. Bar graphs are often used to show comparisons.

A **histogram** is a special type of bar graph. This histogram displays the data (test scores) in equal-size intervals (ranges of scores). There are no spaces between the bars. The break in the horizontal scale (-√^-) indicates that the portion of the scale between 0 and 13 has been omitted. The height of each bar indicates the number of test scores in each interval.

Refer to the histogram to answer problems **5–7.**

5. Which interval had the lowest frequency of scores?

6. Which interval had the highest frequency of scores?

7. Which interval had exactly twice as many scores as the 13–14 interval?

8. *Represent* Make a frequency table and a histogram for the following set of test scores. (Use 50–59, 60–69, 70–79, 80–89, and 90–99 for the intervals.)

63, 75, 58, 89, 92, 84, 95, 63, 78, 88,
96, 67, 59, 70, 83, 89, 76, 85, 94, 80

surveys

A **survey** is a way of collecting data about a population. Rather than collecting data from every member of a population, a survey might focus on only a small part of the population called a **sample.** From the sample, conclusions are formed about the entire population.

Mrs. Patterson's class conducted a survey of 100 students to determine what sport middle school students most enjoyed playing. Survey participants were given six different sports from which to choose. The surveyors displayed the results on the frequency table shown on the following page.

Frequency Table

Sport	Tally	Frequency
Basketball	𝍸 𝍸 𝍸 |	16
Bowling	𝍸 𝍸 ||	12
Football	𝍸 𝍸 𝍸	15
Softball	𝍸 𝍸 𝍸 𝍸 𝍸 |	26
Table Tennis	𝍸 𝍸 ||	12
Volleyball	𝍸 𝍸 𝍸 ||||	19

From the frequency table, Mrs. Patterson's students constructed a bar graph to display the results.

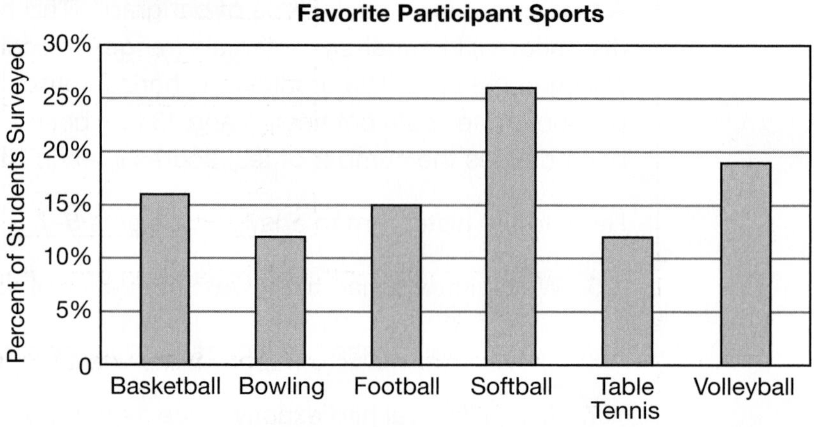

Since 16 out of 100 students selected basketball as their favorite sport to play, basketball was the choice of 16% (which means "16 out of 100") of the students surveyed. Refer to the frequency table and bar graph for this survey to answer problems **9–12.**

9. Which sport was the favorite sport of about $\frac{1}{4}$ of the students surveyed?

10. Which sport was the favorite sport of the girls who were surveyed?

A softball **B** volleyball

C basketball **D** cannot be determined from given information

11. **Evaluate** How might changing the sample group change the results of the survey?

12. How might changing the survey question—the choice of sports—change the results of the survey?

This survey was a **closed-option survey** because the responses were limited to the six choices offered. An **open-option survey** does not limit the choices. An example of an open-option survey question is "What is your favorite sport?"

extensions

a. *Represent* Make a histogram based on the frequency table created in problem **4.** What intervals did you use? What questions can be answered by referring to the histogram?

b. *Formulate* Conduct a survey of favorite foods of class members. If you choose to conduct a closed-option survey, determine which food choices will be offered. What will be the size of the sample? How will the data gathered by the survey be displayed?

c. *Analyze* The table below uses negative integers to express the estimated greatest depth of each of the Great Lakes.

U.S. Great Lakes
Est. Greatest Depth (in meters)

Lake	Depth
Erie	−65 m
Huron	−230 m
Michigan	−280 m
Ontario	−245 m
Superior	−400 m

- Write the depths of the lakes in order from the deepest to shallowest.
- Is a bar graph an appropriate way to represent the data in the table?
- Are the data in the table displayed correctly on the bar graph below? Explain why or why not.

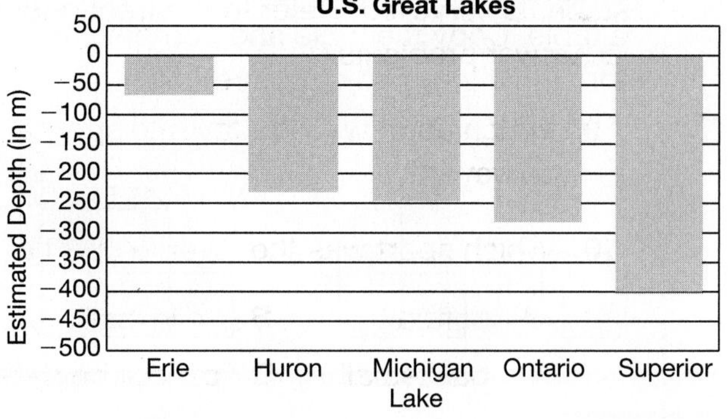

d. *Analyze* Choose between mental math, paper and pencil, or estimation to answer each of the following questions. Explain your choice. Use the table in extension **c** as a reference.

- Lake Superior is how much deeper than Lake Erie?
- Is the depth of Lake Michigan closer to the depth of Lake Ontario or Lake Huron?
- How much deeper is Lake Michigan than Lake Huron?

LESSON 11

• Problems About Combining
• Problems About Separating

Power Up

Building Power

facts Power Up D

mental math

a. **Number Sense:** 3 × 40

b. **Number Sense:** 3 × 400

c. **Number Sense:** \$4.50 + \$1.25

d. **Number Sense:** 451 + 240

e. **Number Sense:** 4500 − 400

f. **Number Sense:** \$5.00 − \$1.50

g. **Geometry:** A rectangle has a length of 3 cm and a width of 2 cm. What is the perimeter of the rectangle?

h. **Calculation:** Start with 10. Add 2; divide by 2; add 2; divide by 2; then subtract 2. What is the answer?

problem solving

Sitha began building stair-step structures with blocks. She used one block for a one-step structure, three blocks for a two-step structure, and six blocks for a three-step structure. She wrote the information in a table. Copy the table and complete it through a ten-step structure.

Number of Steps	Total Number of Blocks	Blocks Added to Previous Structure
1	1	N/A
2	3	2
3	6	3

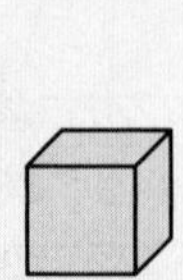
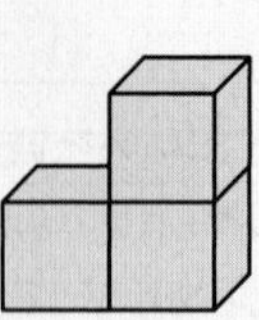

New Concepts

Increasing Knowledge

Like stories in your reading books, many of the stories we analyze in mathematics have plots. We can use the plot of a word problem to write an equation for the problem. Problems with the same plot are represented by the same equation. That is why we say there are **patterns** for certain plots.

problems about combining

Many word problems are about **combining.** Here is an example:

Before he went to work, Pham had \$24.50. He earned \$12.50 more putting up a fence. Then Pham had \$37.00. (Plot: Pham had some money and then he earned some more money.)

Problems about combining have an **addition pattern.**

$$\text{Some} + \text{some more} = \text{total}$$
$$s + m = t$$

There are three numbers in the pattern. In a word problem one of the numbers is unknown, as in the story below.

Katya had 734 stamps in her collection. Then her uncle gave her some more stamps. Now Katya has 813 stamps. How many stamps did Katya's uncle give her?

This problem has a plot similar to the previous one. (Plot: Katya had some stamps and was given some more stamps.) In this problem, however, one of the numbers is unknown. Katya's uncle gave her some stamps, but the problem does not say how many. We use the four-step process to solve the problem.

Step 1: *Understand* Since this problem is about combining, it has an addition pattern.

Step 2: *Plan* We use the pattern to set up an equation.

Pattern: Some + some more = total

Equation: 734 stamps + m = 813 stamps

Step 3: *Solve* The answer to the question is the unknown number in the equation. Since the unknown number is an addend, we subtract 734 from 813 to find the number.

Find answer:	Check answer:
813 stamps	813 stamps
− 734 stamps	− 79 stamps
79 stamps	734 stamps

Step 4: *Check* We review the question and write the answer. Katya's uncle gave her 79 stamps.

Thinking Skill

Connect

How is an odometer the same as a ruler? How is it different?

Example 1

Jenny rode her bike on a trip with her bicycling club. After the first day Jenny's trip odometer showed that she had traveled 86 miles. After the second day the trip odometer showed that she had traveled a total of 163 miles. How far did Jenny ride the second day?

Solution

Step 1: Jenny rode some miles and then rode some more miles. The distances from the two days combine to give a total. Since this is a problem about combining, it has an **addition pattern.**

Step 2: The trip odometer showed how far she traveled the first day and the total of the first two days. We record the information in the pattern.

Some	86 miles
+ Some more	+ m miles
Total	163 miles

Step 3: We solve the equation by finding the unknown number. From Lesson 3 we know that we can find the missing addend by subtracting 86 miles from 163 miles. We check the answer.

Find answer:		Check answer:
163 miles		86 miles
− 86 miles	→	+ 77 miles
77 miles		163 miles

Step 4: We review the question and write the answer. Jenny rode **77 miles** on the second day of the trip.

problems about separating

Another common plot in word problems is **separating.** There is a beginning amount, then some goes away, and some remains. Problems about separating have a **subtraction pattern.**

Beginning amount − some went away = what remains

$$b - a = r$$

Here is an example:

Waverly took $37.00 to the music store. She bought headphones for $26.17. Then Waverly had $10.83. (Plot: Waverly had some money, but some of her money went away when she spent it.)

This is a problem about separating. Thus it has a subtraction pattern.

Pattern:

Beginning amount − some went away = what remains

Equation:

$37.00 − $26.17 = $10.83

Example 2

On Saturday 47 people volunteered to clean up the park. Some people chose to remove trash from the lake. The remaining 29 people left to clean up the hiking trails. How many people chose to remove trash from the lake?

Solution

Step 1: There were 47 people. Then some went away. This problem has a **subtraction pattern.**

Step 2: We use the pattern to write an equation for the given information. We show the equation written vertically.

Beginning amount	47 people
− Some went away	− p people
What remains	29 people

Step 3: We find the unknown number by subtracting 29 from 47. We check the answer.

Find answer:	Check answer:
47 people	47 people
− 29 people	− 18 people
18 people	29 people

Step 4: We review the question and write the answer. There were **18 people** who chose to remove trash from the lake.

Practice Set

Formulate Follow the four-step method to solve each problem. Along with each answer, include the equation you use to solve the problem.

a. When Tim finished page 129 of a 314-page book, how many pages did he still have to read?

b. The football team scored 19 points in the first half of the game and 42 points by the end of the game. How many points did the team score in the second half of the game?

c. Formulate Write a word problem about combining. Solve the problem and write the answer.

Written Practice

Strengthening Concepts

* **1.** (11) Formulate Juan ran 8 laps and rested. Then he ran some more laps. If Juan ran 21 laps in all, how many laps did he run after he rested? Write an equation and solve the problem.

2. (1, 2) **a.** Find the product of 8 and 4.

b. Find the sum of 8 and 4.

3. (5) The expression below means "the product of 6 and 4 divided by the difference of 8 and 5." What is the quotient?

$$(6 \times 4) \div (8 - 5)$$

* **4.** (11) Marcia went to the store with $20.00 and returned home with $7.75. How much money did Marcia spend at the store? Write an equation and solve the problem.

* **5.** (11) When Franklin got his Labrador Retriever puppy, it weighed 8 pounds. A year later, it weighed 74 pounds. How much weight did Franklin's dog gain in that year? Write the equation and solve the problem.

6. (1) $0.65 + $0.40

Analyze Find each unknown number. Check your work.

7. (3) $87 + w = 155$

8. (3) $1000 - x = 386$

9. (3) $y - 1000 = 386$

10. (3) $42 + 596 + m = 700$

* **11.** (9) Compare: $1000 - (100 - 10)$ ◯ $1000 - 100 - 10$

12. $8\overline{)1000}$
(2)

13. $10\overline{)987}$
(2)

14. $12\overline{)W}$ (quotient 35)
(4)

15. 600×300
(2)

16. $365w = 365$
(4)

* 17. **Predict** What are the next three numbers in the following sequence?
(10)

2, 6, 10, ______, ______, ______, ...

18. $2 \times 3 \times 4 \times 5$
(5)

19. What number is $\frac{1}{2}$ of 360?
(6)

20. What number is $\frac{1}{4}$ of 360?
(6)

21. What is the product of eight and one hundred twenty-five?
(2)

* 22. **Connect** How long is the line segment below?
(7)

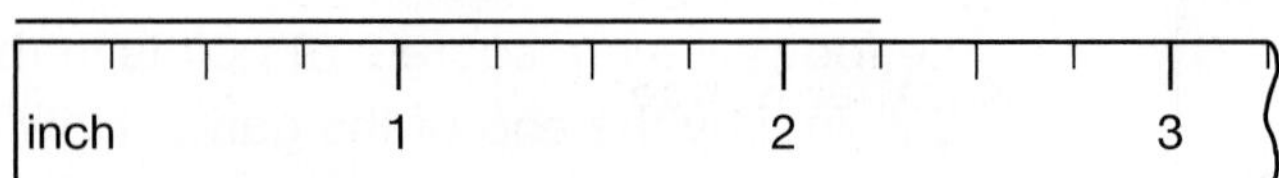

23. What fraction of the circle at right is not shaded?
(6)

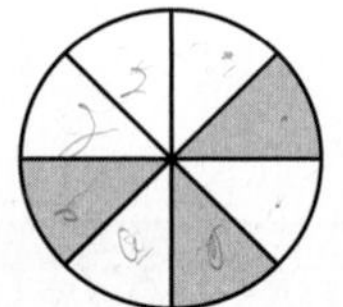

24. What is the perimeter of the square shown?
(6)

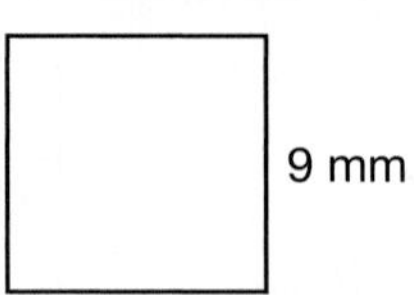

* 25. What is the sum of the first five odd numbers greater than zero?
(10)

26. Here are three ways to write "24 divided by 4":
(2)

$4\overline{)24}$ $\quad 24 \div 4 \quad \frac{24}{4}$

Show three ways to write "30 divided by 6."

27. Seventeen of the 30 students in the class are girls. So $\frac{17}{30}$ of the students in the class are girls. What fraction of the students in the class are boys?
(6)

* 28. At what temperature on the Celsius scale does water freeze?
(10)

29. **Represent** Use the numbers 24, 6, and 4 to write two multiplication facts and two division facts.
(2)

* 30. **Formulate** In the second paragraph of this lesson there is a problem with an addition pattern. Rewrite the problem by removing one of the numbers from the problem and asking a question instead.
(11)

LESSON 12

• Place Value Through Trillions
• Multistep Problems

Power Up

Building Power

facts Power Up D

mental math

a. **Number Sense:** 6×40

b. **Number Sense:** 6×400

c. **Number Sense:** \$12.50 + \$5.00

d. **Number Sense:** $451 + 24$

e. **Number Sense:** $7500 - 5000$

f. **Number Sense:** \$10.00 − \$2.50

g. **Measurement:** How many inches are in a yard?

h. **Calculation:** Start with 12. Divide by 2; subtract 2; divide by 2; then subtract 2. What is the answer?

problem solving

When he was a boy, German mathematician Karl Friedrich Gauss (1777–1855) developed a method for quickly adding a sequence of numbers. Like Gauss, we can sometimes solve difficult problems by *making the problem simpler.*

What is the sum of the first ten natural numbers?

Understand We are asked to find the sum of the first ten natural numbers.

Plan We will begin by making the problem simpler. If the assignment had been to add the first *four* natural numbers, we could simply add 1 + 2 + 3 + 4. However, adding columns of numbers can be time consuming. We will try to find a pattern that will help add the natural numbers 1–10 more quickly.

Solve We can find pairs of addends in the sequence that have the same sum and multiply by the number of pairs. We try this pairing technique on the sequence given in the problem:

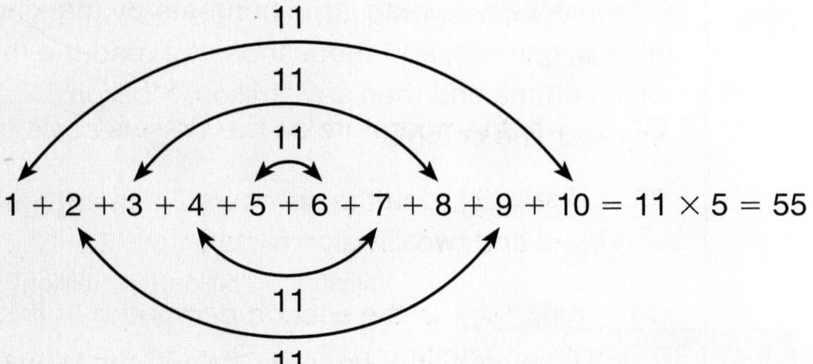

Check We found the sum of the first ten natural numbers by pairing the addends and multiplying. We can verify our solution by adding the numbers one-by-one with pencil and paper or a calculator.

New Concepts

Increasing Knowledge

place value through trillions

Math Language

Our number system is based on a pattern of tens. In a place value chart, each place has a value ten times greater than the place to its right.

In our number system the value of a digit depends upon its position. The value of each position is called its **place value.**

Whole-Number Place Values

hundred trillions	ten trillions	trillions		hundred billions	ten billions	billions		hundred millions	ten millions	millions		hundred thousands	ten thousands	thousands		hundreds	tens	ones
___	___	___	,	___	___	___	,	___	___	___	,	___	___	___	,	___	___	___

Example 1

In the number 123,456,789,000, which digit is in the ten-millions place?

Solution

Either by counting places from the right or looking at the chart, we find that the digit in the ten-millions place is **5.**

Example 2

In the number 5,764,283, what is the place value of the digit 4?

Solution

By counting places from the right or looking at the chart, we can see that the place value of 4 is **thousands.**

Thinking Skill

Connect

Make a list of real-life situations in which large number are used.

Large numbers are easy to read and write if we use commas to group the digits. To place commas, we begin at the right and move to the left, writing a comma after every three digits.

Putting commas in 1234567890, we get 1,234,567,890.

Commas help us read large numbers by marking the end of the trillions, billions, millions, and thousands. We read the three-digit number in front of each comma and then say "trillion," "billion," "million," or "thousand" when we reach the comma.

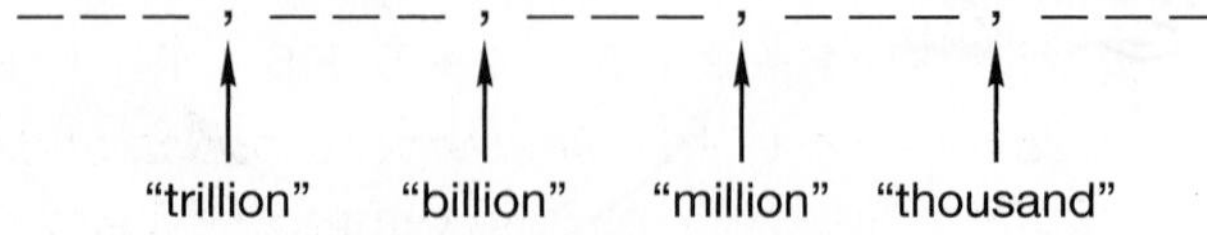

Example 3

Use words to write the number 1024305.

Solution

First we insert commas.

1,024,305

We write **one million, twenty-four thousand, three hundred five.**

Note: We write commas after the words *trillion, billion, million,* and *thousand.* We hyphenate compound numbers from 21 through 99. We do not say or write "and" when naming whole numbers.

Example 4

Use digits to write the number one trillion, two hundred fifty billion.

Solution

When writing large numbers, it may help to sketch the pattern before writing the digits.

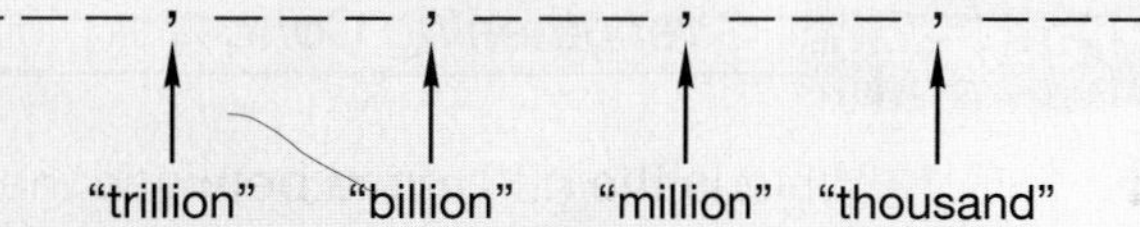

We write a 1 to the left of the trillions comma and 250 in the three places to the left of the billions comma. The remaining places are filled with zeros.

1,250,000,000,000

multistep problems

The **operations of arithmetic** are addition, subtraction, multiplication, and division. In this table we list the terms for the answers we get when we perform these operations:

Sum	the answer when we add
Difference	the answer when we subtract
Product	the answer when we multiply
Quotient	the answer when we divide

We will use these terms in problems that have several steps.

Example 5

What is the difference between the product of 6 and 4 and the sum of 6 and 4?

Solution

Math Language

Sometimes it is helpful to rewrite a question and underline the phrases that indicate operations of arithmetic.

We see the words *difference, product,* and *sum* in this question. We first look for phrases such as "the product of 6 and 4." We will rewrite the question, emphasizing these phrases.

What is the difference between the <u>product of 6 and 4</u> and the <u>sum of 6 and 4</u>?

For each phrase we find one number. "The product of 6 and 4" is 24, and "the sum of 6 and 4" is 10. So we can replace the two phrases with the numbers 24 and 10 to get this question:

What is the difference between 24 and 10?

We find this answer by subtracting 10 from 24. The difference between 24 and 10 is **14.**

Practice Set

a. Which digit is in the millions place in 123,456,789?

b. What is the place value of the 1 in 12,453,000,000?

c. Use words to write 21,350,608.

d. Use digits to write four billion, five hundred twenty million.

e. Analyze When the product of 6 and 4 is divided by the difference of 6 and 4, what is the quotient?

Written Practice *Strengthening Concepts*

* **1.** (12) What is the difference between the product of 1, 2, and 3 and the sum of 1, 2, and 3?

* **2.** (12) Earth is about ninety-three million miles from the Sun. Use digits to write that distance.

* **3.** (11) Formulate Gilbert and Kadeeja cooked 342 pancakes for the pancake breakfast. If Gilbert cooked 167 pancakes, how many pancakes did Kadeeja cook? Write an equation and solve the problem.

* **4.** (11) Formulate The two teams scored a total of 102 points in the basketball game. If the winning team scored 59 points, how many points did the losing team score? Write an equation and solve the problem.

* **5.** (8) What is the perimeter of the rectangle at right?

10 mm

18 mm

6. (4) $6m = 60$

7. (6) **a.** What number is $\frac{1}{2}$ of 100?

b. What number is $\frac{1}{4}$ of 100?

8. (9) Compare: $300 \times 1 \bigcirc 300 \div 1$

9. (5) $(3 \times 3) - (3 + 3)$

* **10.** (10) Predict What are the next three numbers in the following sequence?

1, 2, 4, 8, _____, _____, _____, …

11. (3) $1 + m + 456 = 480$

12. (3) $1010 - n = 101$

13. (2) $1234 \div 10$

14. (2) $1234 \div 12$

* **15.** (10) What is the sum of the first five even numbers greater than zero?

16. (7) **Connect** How many millimeters long is the line segment below?

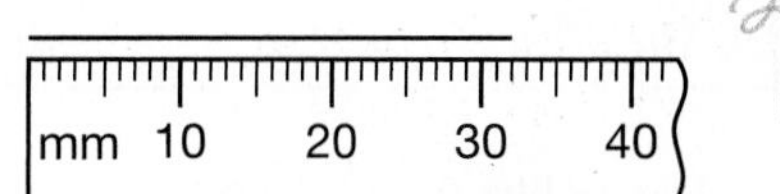

* **17.** (12) In the number 123,456,789,000, which digit is in the ten-billions place?

* **18.** (12) In the number 5,764,283,000, what is the place value of the digit 4?

* **19.** (12) Which digit is in the hundred-thousands place in the number 987,654,321?

20. (5) $1 \times 10 \times 100 \times 1000$

21. (2) $\$3.75 \times 3$

22. (4) $22y = 0$

23. (3) $100 + 200 + 300 + 400 + w = 2000$

24. (2) 24×26

25. (4) $m\overline{)625}$ with quotient 25

26. (4) If the divisor is 4 and the quotient is 8, what is the dividend?

27. (2) Show three ways to write "27 divided by 3."

28. (6) **Explain** Seven of the ten marbles in a bag are red. So $\frac{7}{10}$ of the marbles are red. What fraction of the marbles are not red? Explain why your answer is correct.

* **29.** (12) Use digits to write four trillion.

* **30.** (12) **Formulate** Using different numbers, make up a question similar to example 5 in this lesson. Then find the answer.

LESSON 13

• Problems About Comparing
• Elapsed-Time Problems

Power Up

Building Power

facts Power Up A

mental math

a. **Number Sense:** 5 × 300

b. **Number Sense:** 5 × 3000

c. **Number Sense:** \$7.50 + \$1.75

d. **Number Sense:** 3600 + 230

e. **Number Sense:** 4500 − 500

f. **Number Sense:** \$20.00 − \$5.00

g. **Measurement:** How many millimeters are in a meter?

h. **Measurement:** How many years are in a decade?

problem solving A pair of number cubes is tossed. The total number of dots on the two top faces is 6. What is the total of the dots on the bottom faces of the pair of number cubes?

New Concepts

Increasing Knowledge

We practiced using patterns to solve word problems in Lesson 11. For problems about combining, we used an addition pattern. For problems about separating, we used a subtraction pattern. In this lesson we will look at two other kinds of math word problems.

problems about comparing

Some word problems are about **comparing** the size of two groups. They usually ask questions such as "How many more are in the first group" and "How many fewer are in the second group?" Comparison problems such as these have a **subtraction pattern.** We write the numbers in the equation in this order:

Greater − lesser = difference

In place of the words, we can use letters. We use the first letter of each word.

$$g - l = d$$

Example 1

There were 324 girls and 289 boys in the school. How many fewer boys than girls were there in the school?

Solution

Again we use the four-step process to solve the problem.

Step 1: We are asked to compare the number of boys to the number of girls. The question asks "how many fewer?" This problem has a **subtraction pattern.**

Step 2: We use the pattern to write an equation. There are more girls than boys, so the number of girls replaces "greater" and the number of boys replaces "lesser."

Pattern: Greater − lesser = difference

Equation: $324 - 289 = d$

Step 3: We find the missing number by subtracting.

$$\begin{array}{r} 324 \text{ girls} \\ - \ 289 \text{ boys} \\ \hline 35 \text{ fewer boys} \end{array}$$

Step 4: We review the question and write the answer. There were **35 fewer boys** than girls in the school. We can also state that there were 35 more girls than boys in the school.

Explain How can we check the answer?

elapsed-time problems

Elapsed time is the length of time between two events. We illustrate this on the ray below.

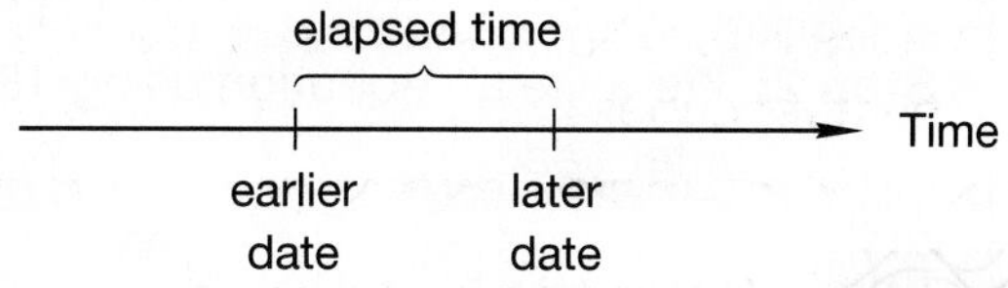

Thinking Skill

Verify

When you subtract your birth year from the current year, why don't you always get your exact age?

The time that has elapsed since the moment you were born until now is your age. Subtracting your birth date from today's date gives your age.

Today's date	(later)
− Your birth date	(earlier)
Your age	(difference)

Elapsed-time problems are like comparison problems. They have a **subtraction pattern.**

Later − earlier = difference

We use the first letter of each word to represent the word.

$$l - e = d$$

Example 2

How many years were there from the year Columbus landed in America in 1492 to the year the Pilgrims landed in 1620?

Solution

Step 1: This is an **elapsed-time** problem. It has a **subtraction pattern.** We use *l*, *e*, and *d* to stand for "later," "earlier," and "difference."

$$l - e = d$$

Step 2: The later year is 1620. The earlier year is 1492.

$$1620 - 1492 = d$$

Step 3: We find the missing number by subtracting. We can check the answer by adding.

$$\begin{array}{r} 1620 \\ -\ 1492 \\ \hline 128 \end{array} \quad \longrightarrow \quad \begin{array}{r} 1492 \\ +\ 128 \\ \hline 1620 \end{array}$$

Step 4: We review the question and write the answer. There were **128 years** from 1492 to 1620.

Example 3

Abraham Lincoln was born in 1809 and died in 1865. How many years did he live?

Solution

Step 1: This is an **elapsed-time** problem. It has a **subtraction pattern.** We use *l* for the later time, *e* for the earlier time, and *d* for the difference of the times.

$$l - e = d$$

Step 2: We write an equation using 1809 for the earlier year and 1865 for the later year.

$$1865 - 1809 = d$$

Step 3: We find the missing number by subtracting. We may add his age to the year of his birth to check the answer.

$$\begin{array}{r} 1865 \\ -\ 1809 \\ \hline 56 \end{array} \quad \longrightarrow \quad \begin{array}{r} 1809 \\ +\ 56 \\ \hline 1865 \end{array}$$

We also note that 56 is a reasonable age, so our computation makes sense.

Step 4: We review the question and write the answer. Abraham Lincoln lived **56 years.**

Practice Set

Formulate Follow the four-step method to solve each problem. Along with each answer, include the equation you use to solve the problem.

a. The population of Castor is 26,290. The population of Weston is 18,962. How many more people live in Castor than live in Weston?

b. Two important dates in British history are the Norman Conquest in 1066 and the signing of the Magna Carta in 1215, which limited the power of the king. How many years were there from 1066 to 1215?

Written Practice Strengthening Concepts

* **1.** When the sum of 8 and 5 is subtracted from the product of 8 and 5, what is the difference?
(12)

* **2.** The Moon is about two hundred fifty thousand miles from the Earth. Use digits to write that distance.
(12)

* **3.** Use words to write 521,000,000,000.
(12)

* **4.** Use digits to write five million, two hundred thousand.
(12)

5. **Explain** Robin entered a tennis tournament when she was three-score years old. How old was Robin when she entered the tournament? How do you know your answer is correct? (Recall that one score equals 20 years.)
(2)

* **6.** The auditorium at the Community Cultural Center has seats for 1000 people. For a symphony concert at the center, 487 tickets have already been sold. How many more tickets are still available? Write an equation and solve the problem.
(11)

* **7.** **Formulate** It is 405 miles from Minneapolis, Minnesota to Chicago, Illinois. It is 692 miles from Minneapolis to Cincinnati, Ohio. Cincinnati is how many miles farther from Minneapolis than Chicago? Write an equation and solve the problem.
(13)

Justify Use mental math to solve exercises **8** and **9.** Describe the mental math strategy you used for each exercise.

8. $99 + 100 + 101$
(1)

9. $9 \times 10 \times 11$
(5)

* **10.** Which digit is in the thousands place in 54,321?
(12)

* **11.** What is the place value of the 1 in 1,234,567,890?
(12)

12. The three sides of an equilateral triangle are equal in length. What is the perimeter of the equilateral triangle shown?
(8)

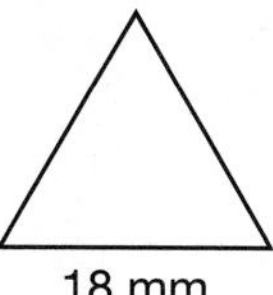

13. $5432 \div 100$
(2)

14. $\frac{60,000}{30}$
(2)

15. $1000 \div 7$
(2)

16. $\$4.56 \div 3$
(2)

17. Compare: $3 + 2 + 1 + 0 \bigcirc 3 \times 2 \times 1 \times 0$
(9)

* **18.** **Predict** The rule for the sequence below is different from the rules for addition sequences or multiplication sequences. What is the next number in the sequence?
(10)

1, 4, 3, 6, 5, 8, ______, ...

19. **Generalize** What is $\frac{1}{2}$ of 5280?
(6)

20. $365 \div w = 365$
(4)

21. $(5 + 6 + 7) \div 3$
(5)

22. Use a ruler to find the length in inches of the rectangle below.
(7)

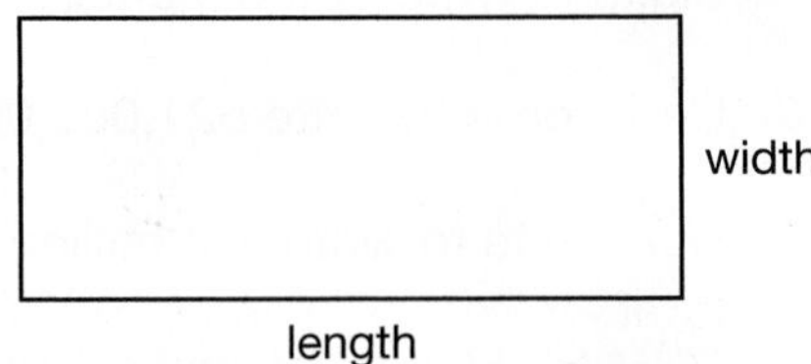

23. **Explain** Write two ways to find the perimeter of a square: one way by adding and the other way by multiplying.
(8)

24. Multiply to find the answer to this addition problem:
(2)

$$125 + 125 + 125 + 125 + 125 + 125$$

25. At what temperature on the Fahrenheit scale does water boil?
(10)

26. Show three ways to write "21 divided by 7."
(2)

Analyze Find each unknown number. Check your work.

27. $8a = 816$
(4)

28. $\frac{b}{4} = 12$
(4)

29. $\frac{12}{c} = 4$
(4)

30. $d - 16 = 61$
(3)

LESSON 14

The Number Line: Negative Numbers

Power Up

Building Power

facts

Power Up D

mental math

a. **Number Sense:** 8 × 400

b. **Number Sense:** 6 × 3000

c. **Number Sense:** $7.50 + $7.50

d. **Number Sense:** 360 + 230

e. **Number Sense:** 1250 − 1000

f. **Number Sense:** $10.00 − $7.50

g. **Measurement:** How many years are in a century?

h. **Calculation:** Start with 10. Add 2; divide by 3; multiply by 4; then subtract 5. What is the answer?

problem solving

It takes the local hardware store 8 seconds to cut through a piece of round galvanized steel pipe. How long will it take to cut a piece of pipe in half? Into quarters? Into six pieces? (Each cut must be perpendicular to the length of the pipe.)

New Concept

Increasing Knowledge

We have seen that a number line can be used to arrange numbers in order.

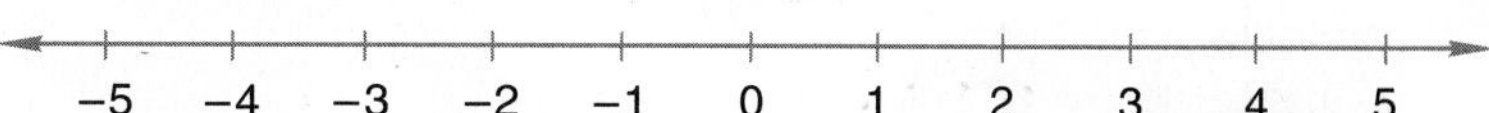

On the number line above, the points to the right of zero represent **positive numbers.** The points to the left of zero represent negative numbers. Zero is neither positive nor negative.

Reading Math

Negative numbers are represented by writing a minus sign before a number: −5.

Negative numbers are used in various ways. A temperature of five degrees below zero Fahrenheit may be written as −5°F. An elevation of 100 feet below sea level may be indicated as "elev. −100 ft." The change in a stock's price from $23.00 to $21.50 may be shown in a newspaper as −1.50.

Example 1

Arrange these numbers in order from least to greatest:

0, 1, −2

Solution

All negative numbers are less than zero. All positive numbers are greater than zero.

$$-2, 0, 1$$

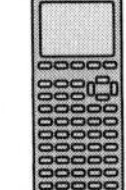

Example 2

Compare: −3 ○ −4

Solution

Visit www.SaxonPublishers.com/ActivitiesC1 *for a graphing calculator activity.*

Negative three is three less than zero, and negative four is four less than zero. So

$$-3 > -4$$

Math Language

Zero is neither positive nor negative. Zero has no opposite.

The number −5 is read "negative five." Notice that the points on the number line marked 5 and −5 are the same distance from zero but are on opposite sides of zero. We say that 5 and −5 are **opposites.** Other opposite pairs include −2 and 2, −3 and 3, and −4 and 4. The tick marks show the location of numbers called **integers.** Integers include all of the counting numbers and their opposites, as well as the number zero.

If you subtract a larger number from a smaller number (for example, 2 − 3), the answer will be a negative number. One way to find the answer to such questions is to use the number line. We start at 2 and count back (to the left) three integers. Maybe you can figure out a faster way to find the answer.

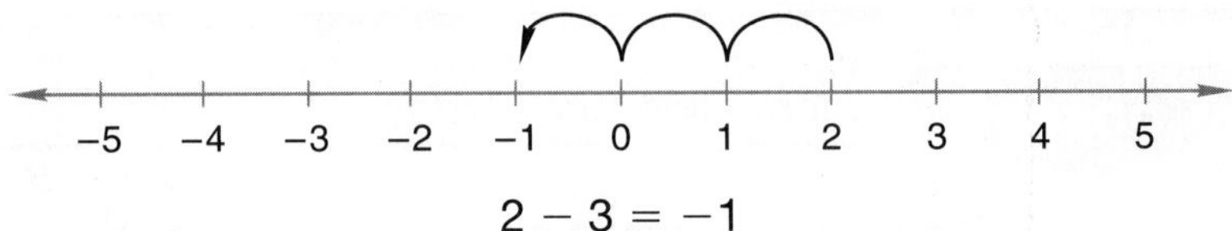

$$2 - 3 = -1$$

Example 3

Subtract 5 from 2.

Solution

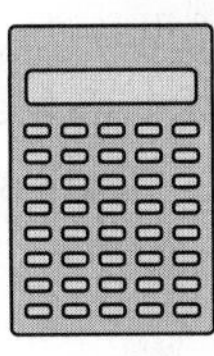

Order matters in subtraction. Start at 2 and count to the left 5 integers. You should end up at **−3.** Try this problem with a calculator by entering 2 − 5 =. What number is displayed after the = is pressed?

We see that the calculator displays −3 as the solution.

Example 4

Arrange these four numbers in order from least to greatest:

$$1, -2, 0, -1$$

Solution

A number line shows numbers in order. By arranging these numbers in the order they appear on a number line, we arrange them in order from least to greatest.

−2, −1, 0, 1

Example 5

What number is 7 less than 3?

Solution

The phrase "7 less than 3" means to start with 3 and subtract 7.

$$3 - 7$$

We count to the left 7 integers from 3. The answer is **−4.**

Practice Set

a. Compare: $-8 \bigcirc -6$

b. Use words to write this number: -8.

c. What number is the opposite of 3?

d. Arrange these numbers in order from least to greatest:

$$0, -1, 2, -3$$

e. What number is 5 less than 0?

f. What number is 10 less than 5?

g. $5 - 8$ **h.** $1 - 5$

i. *Verify* All five of the numbers below are integers. True or false?

$$-3, 0, 2, -10, 50$$

j. The temperature was twelve degrees below zero Fahrenheit. Use a negative number to write the temperature.

k. The desert floor was 186 feet below sea level. Use a negative number to indicate that elevation.

l. The stock's price dropped from \$18.50 to \$16.25. Use a negative number to express the change in the stock's value.

Written Practice *Strengthening Concepts*

* **1.** (12) *Connect* What is the quotient when the sum of 15 and 12 is divided by the difference of 15 and 12?

* **2.** (12) What is the place value of the 7 in 987,654,321,000?

B M Th

* **3.** (12) Light travels at a speed of about one hundred eighty-six thousand miles per second. Use digits to write that speed.

* **4.** (14) **Connect** What number is three integers to the left of 2 on the number line?

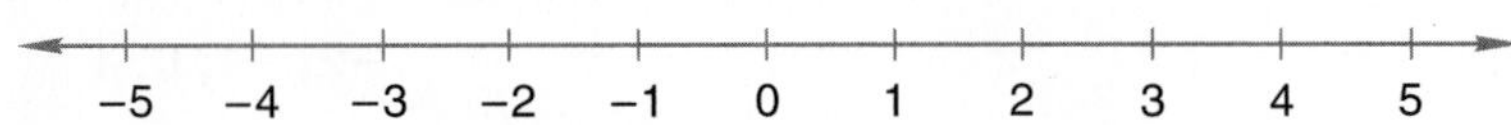

* **5.** (14) **Connect** Arrange these numbers in order from least to greatest:

$$5, -3, 1, 0, -2$$

* **6.** (14) What number is halfway between -4 and 0 on the number line?

* **7.** (11) **Formulate** There are 140 sixth-grade students in the school. Seventy-two play on school sports teams. How many are not on school sports teams? Write an equation and solve the problem.

8. (9) Compare: $1 + 2 + 3 + 4 \bigcirc 1 \times 2 \times 3 \times 4$

9. (8) What is the perimeter of this right triangle?

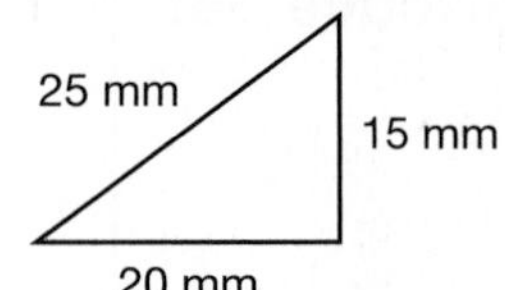

10. (10) **Predict** What are the next two numbers in the following sequence?

$$\ldots, 16, 8, 4, \_\_\_\_, \_\_\_\_, \ldots$$

* **11.** (13) **Formulate** There are 365 days in a common year. How much less than 500 is 365? Write an equation and solve the problem.

* **12.** (14) What number is 8 less than 6?

13. (2) $1020 \div 100$

14. (2) $\dfrac{36{,}180}{12}$

15. (2) $18\overline{)564}$

16. (1) $1234 + 567 + 89$

17. (3) $n - 310 = 186$

18. (5) $10 \cdot 11 \cdot 12$

19. (3) $\$3.05 - m = \$2.98$

20. (7) **Estimate** About how long is this nail in centimeters? Use a centimeter ruler to find its length to the nearest centimeter and to the nearest millimeter.

21. (5) (100)(100)(100)

22. (12) What digit in 123,456,789 is in the ten-thousands place?

23. (7) **Verify** If you know the length of an object in centimeters, how can you figure out the length of the object in millimeters without remeasuring?

24. (2) **Connect** Use the numbers 19, 21, and 399 to write two multiplication facts and two division facts.

25. (9) Compare: $12 \div 6 \times 2 \bigcirc 12 \div (6 \times 2)$

26. (2) Show three ways to write "60 divided by 6."

* **27.** (12) In January, 2005, the world's population was about six billion, four hundred million people. Use digits to write this number of people.

28. (6) One third of the 12 eggs in the carton were cracked. How many eggs were cracked?

* **29.** (14) What number is the opposite of 10?

* **30.** (14) Arrange these numbers in order from least to greatest:

$$1, 0, -1, \tfrac{1}{2}$$

Early Finishers
Real-World Application

At dawn the temperature was 42°F. By 5:00 p.m. the temperature had risen 33°F to its highest value for the day. Between 5:00 p.m. and dusk the temperature fell 12°F.

a. What was the temperature at dusk?

b. A cold front passes through during the night causing the temperature to drop 32°F just before dawn. What is the temperature at that time?

LESSON 15

• Problems About Equal Groups

Power Up

Building Power

facts Power Up C

mental math

a. **Number Sense:** 7×4000

b. **Number Sense:** 8×300

c. **Number Sense:** $12.50 + $12.50

d. **Number Sense:** $80 + 12$

e. **Number Sense:** $6250 - 150$

f. **Number Sense:** $20.00 − $2.50

g. **Measurement:** How many decades are in a century?

h. **Calculation:** Start with a dozen. Subtract 3; divide by 3; subtract 3; then multiply by 3. What is the answer?

problem solving

Copy this subtraction problem and fill in the missing digits:

```
  4_7
− _9_
   21
```

New Concept

Increasing Knowledge

We have studied several types of mathematical word problems. Problems about combining have an addition pattern. Problems about separating, comparing, and elapsed-time have subtraction patterns. Another type of mathematical problem is the **equal groups** problem. Here is an example:

> *In the auditorium there were 15 rows of chairs with 20 chairs in each row. Altogether, there were 300 chairs in the auditorium.*

The chairs were arranged in 15 groups (rows) with 20 chairs in each group. Here is how we write the pattern:

15 rows × 20 chairs in each row = 300 chairs

Number of groups × number in group = total

$$n \times g = t$$

In a problem about equal groups, any one of the numbers might be unknown. We multiply to find the **unknown** total. We divide to find an unknown **factor.**

Example

At Russell Middle School there were 232 seventh-grade students in 8 classrooms. If there were the same number of students in each classroom, how many students would be in each seventh-grade classroom at Russell Middle School?

Solution

Step 1: A number of students is divided into equal groups (classrooms). This is a problem about **equal groups.** The words *in each* often appear in "equal groups" problems.

Step 2: We draw the pattern and record the numbers, writing a letter in place of the unknown number.

Pattern	Equation
Number in each group	n in each classroom
× Number of groups	× 8 classrooms
Number in all groups	232 in all classrooms

Step 3: We find the unknown factor by dividing. Then we check our work.

$$8\overline{)232} = 29 \qquad \begin{array}{r} 29 \\ \times\ 8 \\ \hline 232 \end{array}$$

Step 4: We review the question and write the answer. If there were the same number of students in each classroom, there would be **29 students** in each seventh-grade classroom at Russell Middle School.

Practice Set

Reading Math
Sometimes it is helpful to write dollars and cents in cents-only form.

Formulate Follow the four-step method to solve each problem. Along with each answer, include the equation you use to solve the problem.

a. Marcie collected $4.50 selling lemonade at 25¢ for each cup. How many cups of lemonade did Marcie sell? (*Hint:* Record $4.50 as 450¢.)

b. In the store parking lot there were 18 parking spaces in each row, and there were 12 rows of parking spaces. Altogether, how many parking spaces were in the parking lot?

Written Practice

Strengthening Concepts

* **1.** (15) **Formulate** The second paragraph of this lesson contains an "equal groups" situation. Write a word problem by removing one of the numbers in the problem and writing an "equal groups" question.

* **2.** (13) **Explain** On the Fahrenheit scale, water freezes at 32°F and boils at 212°F. How many degrees difference is there between the freezing and boiling points of water? Write an equation and solve the problem. Explain why your answer is reasonable.

*** 3.** (15) **Formulate** There are about three hundred twenty little O's of cereal in an ounce. About how many little O's are there in a one-pound box? Write an equation and solve the problem (1 pound = 16 ounces).

*** 4.** (11) There are 31 days in August. How many days are left in August after August 3? Write an equation and solve the problem.

*** 5.** (14) Compare: $3 - 1 \bigcirc 1 - 3$

*** 6.** (14) **Represent** Subtract 5 from 2. Use words to write the answer.

*** 7.** (14) The stock's value dropped from $28.00 to $25.50. Use a negative number to show the change in the stock's value.

*** 8.** (10) **Predict** What are the next three numbers in the following sequence?

…, 6, 4, 2, 0, _____, _____, _____, …

*** 9.** (10, 14) What is the temperature reading on this thermometer? Write the answer twice, once with digits and an abbreviation and once with words.

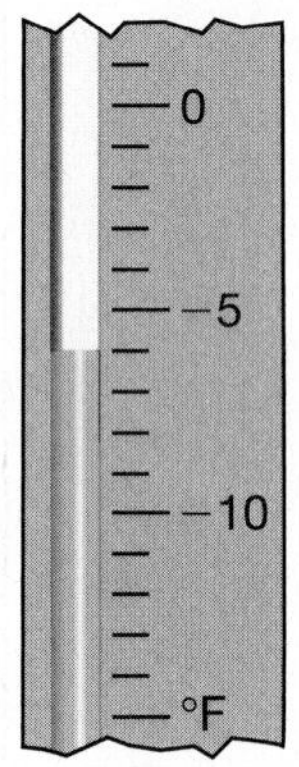

10. (1) $10 − 10¢

11. (6) How much money is $\frac{1}{2}$ of $3.50?

12. (9) **Connect** To which hundred is 587 closest?

13. (1) $9 + 87 + 654 + 3210$

14. (2) 574×76

15. (2) $\frac{4320}{9}$

16. (2) $36\overline{)493}$

Analyze Find each unknown number. Check your work.

17. (4) $1200 \div w = 300$

18. (4) $63w = 63$

19. (4) $\frac{76}{m} = 1$

20. (3) $w + \$65 = \$1000$

21. (3) $3 + n + 12 + 27 = 50$

22. *(7)* There are 10 millimeters in 1 centimeter. How many millimeters long is this paper clip?

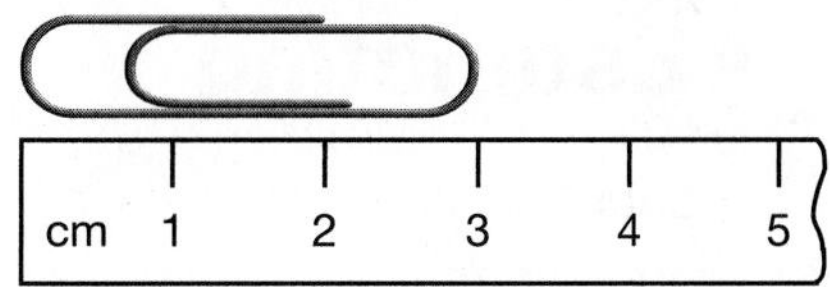

23. *(5)* $(8 + 9 + 16) \div 3$

24. *(12)* What is the place value of the 5 in 12,345,678?

25. *(12)* Which digit occupies the ten-billions place in 123,456,789,000?

26. *(1)* **Represent** Use the numbers 19, 21, and 40 to write two addition facts and two subtraction facts.

* **27.** *(14)* Arrange these numbers in order from least to greatest:

$$0, -1, 2, -3$$

28. *(6)* Of the seventeen students in Angela's class, eight play in the school band. What fraction of the total number of students in Angela's class are in the band?

* **29.** *(15)* **Analyze** Reggie sold buttons with a picture of his school's mascot for 75¢ each. If Reggie sold seven buttons, how much money did he receive?

* **30.** *(14)* What number is neither positive nor negative?

Early Finishers

Real-World Application

Franklin D. Roosevelt, the thirty-second president of the United States, was born on January 30, 1882 in Hyde Park, New York. His presidency began March 4, 1933 and ended April 12, 1945.

a. How old was Franklin D. Roosevelt when he took office?

b. How many years did he serve as president?

LESSON 16

• Rounding Whole Numbers
• Estimating

Power Up

Building Power

facts Power Up D

mental math

a. Number Sense: 3×30 plus 3×2

b. Number Sense: 4×20 plus 4×3

c. Number Sense: $150 + 20$

d. Number Sense: $75 + 9$

e. Number Sense: $800 - 50$

f. Number Sense: $8000 - 500$

g. Measurement: How many yards are in 6 feet?

h. Calculation: Start with 1. Add 2; multiply by 3; subtract 4; then divide by 5. What is the answer?

problem solving Find the next four numbers in this sequence: 2, 3, 5, 8, 9, 11, 14, 15, ...

New Concepts

Increasing Knowledge

rounding whole numbers

When we **round** a whole number, we are finding another whole number, usually ending in zero, that is close to the number we are rounding. The number line can help us visualize rounding.

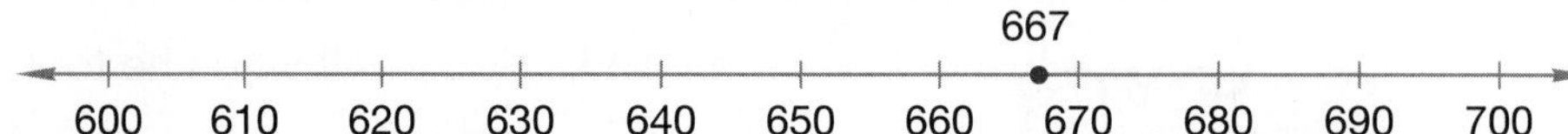

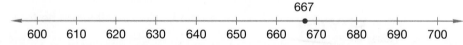

In order to round 667 to the nearest ten, we recognize that 667 is closer to 670 than it is to 660. In order to round 667 to the nearest hundred, we recognize that 667 is closer to 700 than to 600.

Example 1

Round 6789 to the nearest thousand.

Thinking Skill

Verify

Draw a number line to verify the answer is correct.

Solution

The number we are rounding is between 6000 and 7000. It is closer to **7000.**

Example 2

Round 550 to the nearest hundred.

Solution

The number we are to round is halfway between 500 and 600. When the number we are rounding is halfway between two round numbers, we round **up.** So 550 rounds to **600.**

estimating

Rounding can help us **estimate** the answer to a problem. Estimating is a quick way to "get close" to the answer. It can also help us decide whether an answer is reasonable. In some situations an estimate is sufficient to solve a problem because an exact answer is not needed. To estimate, we round the numbers before we add, subtract, multiply, or divide.

Example 3

Estimate the sum of 467 and 312.

Solution

Estimating is a skill we can learn to do in our head. First we round each number. Since both numbers are in the hundreds, we will round each number to the nearest hundred.

467 rounds to 500

312 rounds to 300

To estimate the sum, we add the rounded numbers.

$$\begin{array}{r} 500 \\ +\ 300 \\ \hline 800 \end{array}$$

We estimate the sum of 467 and 312 to be **800.**

Example 4

Stephanie stopped at the store to pick up a few items she needs. She has a $10 bill and a couple of quarters. She needs to buy milk for $2.29, her favorite cereal for $4.78, and orange juice for $2.42. Does Stephanie have enough money to buy what she needs?

Solution

An estimate is probably good enough to solve the problem. Milk and juice are less than $2.50 each, so they total less than $5. Cereal is less than $5.00, so all three items are less than $10. If tax is not charged on food, she has enough money. Even if tax is charged she probably has enough.

Math Language

Words such as *about* and *approximately* indicate that an estimate, not an exact answer, is needed.

Example 5

According to this graph, about how many more people lived in Ashton in 2000 than in 1980?

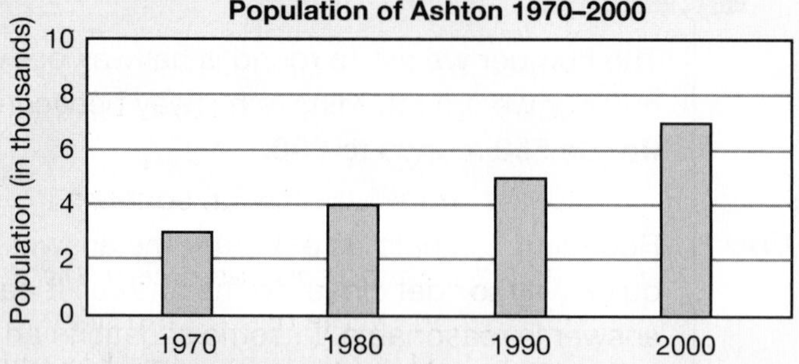

Solution

We often need to use estimation skills when reading graphs. The numbers along the left side of the graph (the vertical axis) indicate the population in thousands. The bar for the year 2000 is about halfway between the 6000 and 8000 levels, so the population was about 7000. In 1980 the population was about 4000. This problem has a subtraction pattern. We subtract and find that about **3000 more people** lived in Ashton in 2000 than in 1980.

Practice Set

Round each of these numbers to the nearest ten:

a. 57 **b.** 63 **c.** 45

Round each of these numbers to the nearest hundred:

d. 282 **e.** 350 **f.** 426

Round each of these numbers to the nearest thousand:

g. 4387 **h.** 7500 **i.** 6750

Estimate Use rounded numbers to estimate each answer.

j. $397 + 206$ **k.** $703 - 598$

l. 29×31 **m.** $29\overline{)591}$

Use the graph in example 5 to answer problems **n** and **o.**

n. **Estimate** About how many fewer people lived in Ashton in 1980 than in 1990?

o. **Predict** The graph shows an upward trend in the population of Ashton. If the population grows the same amount from 2000 to 2010 as it did from 1990 to 2000, what would be a reasonable projection for the population in 2010?

Written Practice *Strengthening Concepts*

* **1.** (12) What is the difference between the product of 20 and 5 and the sum of 20 and 5?

* **2.** (13) Walter Raleigh began exploring the coastline of North America in 1584. Lewis and Clark began exploring the interior of North America in 1803. How many years after Raleigh did Lewis and Clark begin exploring North America? Write an equation and solve the problem.

* **3.** (15) **Explain** Jacob separated his 140 trading cards into 5 equal groups. He placed four of the groups into binders. The remaining cards he placed in a box. How many cards did he put in the box? Write an equation and solve the problem. Explain why your answer is reasonable.

4. (12) Which digit in 159,342,876 is in the hundred-thousands place?

5. (12) In the 2004 U.S. presidential election, 121,068,715 votes were tallied for president. Use words to write that number of votes.

6. (9) What number is halfway between 5 and 11 on the number line?

* **7.** (16) Round 56,789 to the nearest thousand.

* **8.** (16) Round 550 to the nearest hundred.

* **9.** (16) Estimate the product of 295 and 406 by rounding each number to the nearest hundred before multiplying.

10. (1) $45 + 5643 + 287$

11. (1) $40{,}312 - 14{,}908$

12. (2) $\frac{7308}{12}$

13. (2) $100\overline{)5367}$

14. (5) $(5 + 11) \div 2$

15. (6) How much money is $\frac{1}{2}$ of \$5?

16. (6) How much money is $\frac{1}{4}$ of \$5?

17. (2) $\$0.25 \times 10$

18. (5) $325(324 - 323)$

19. (9) Compare: $1 + (2 + 3) \bigcirc (1 + 2) + 3$

* **20.** (14) **Explain** *Wind chill* describes the effect of temperature and wind combining to make it feel colder outside. At 3 p.m. in Minneapolis, Minnesota, the wind chill was $-10°$ Fahrenheit. At 11 p.m. the wind chill was $-3°$ Fahrenheit. At which time did it feel colder outside, 3 p.m. or 11 p.m.? Explain how you arrived at your answer.

* **21.** (15) **Formulate** Your heart beats about 72 times per minute. At that rate, how many times will it beat in one hour? (Write an equation and solve the problem.)

22. (8) **Estimate** The distance between bases on a major league baseball diamond is 90 feet. A player who runs around the diamond runs about how many feet?

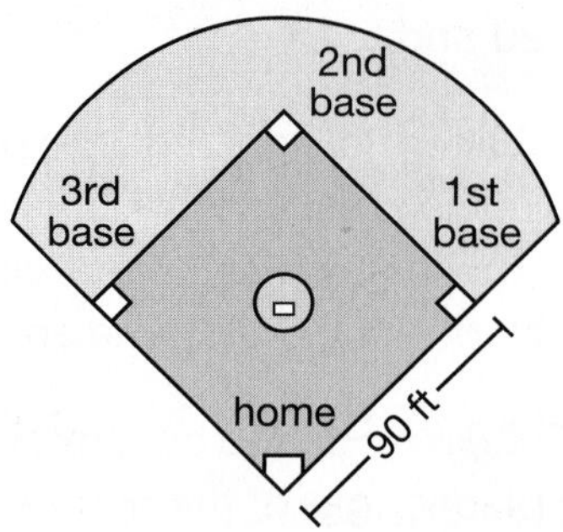

Refer to the bar graph shown below to answer problems **23–26.**

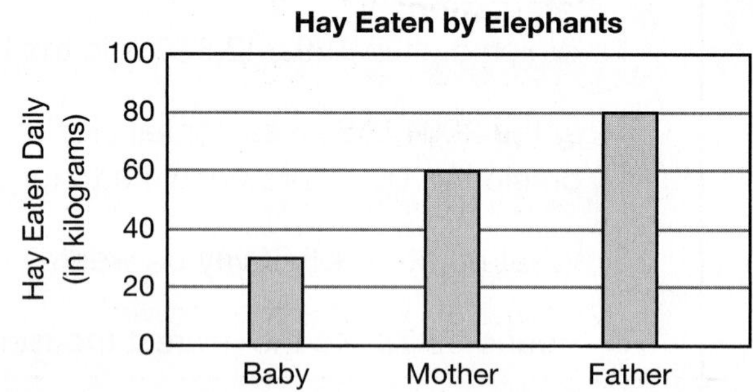

* **23.** (16) How many more kilograms of hay does the father elephant eat each day than the baby elephant?

* **24.** (16) Altogether, how many kilograms of hay do the three elephants eat each day?

* **25.** (16) How many kilograms of hay would the mother elephant eat in one week?

* **26.** (16) **Formulate** Using the information in this graph, write a comparison word problem.

Analyze Find each unknown number. Check your work.

27. (4) $6w = 66$

28. (3) $m - 60 = 37$

29. (3) $60 - n = 37$

* **30.** (Inv. 1) **Represent** Each day Chico, Fuji, and Rolo drink 6, 8, and 9 glasses of water respectively. Draw a bar graph to illustrate this information.

LESSON 17

The Number Line: Fractions and Mixed Numbers

Power Up

Building Power

facts Power Up E

mental math

a. **Number Sense:** 5×30 plus 5×4

b. **Number Sense:** 4×60 plus 4×4

c. **Number Sense:** $180 + 12$

d. **Calculation:** $64 + 9$

e. **Number Sense:** $3000 - 1000 - 100$

f. **Calculation:** \$10.00 − \$7.50

g. **Measurement:** How many millimeters are in 2 meters?

h. **Calculation:** Start with 5. Multiply by 4; add 1; divide by 3; then subtract 2. What is the answer?

problem solving

If an 8 in.-by-8 in. pan of lasagna serves four people, how many 12 in.-by-12 in. pans of lasagna should be purchased to serve 56 people? (*Hint:* You may have "leftovers.")

Understand An eight-inch-square pan of lasagna will serve four people. We are asked to find how many 12-inch-square pans of lasagna are needed to feed 56 people.

Plan We will *draw a diagram* to help us visualize the problem. Then we will *write an equation* to find the number of 12-inch-square pans of lasagna needed.

Solve First, we find the size of each serving by "cutting" the 8-inch-square pan into four pieces. Then we see how many pieces of the same size can be made from the 12-inch-square pan:

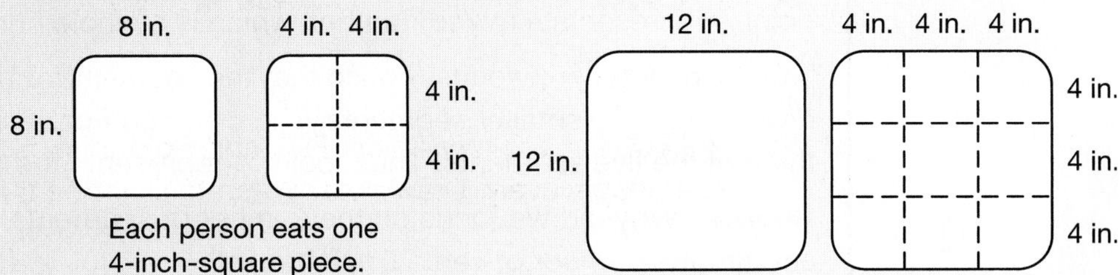

Each person eats one 4-inch-square piece.

One 12-inch-square pan of lasagna can be cut into nine 4-inch-square pieces.

One 12-inch-square pan of lasagna can serve 9 people. We use this information to write an equation: $N \times 9 = 56$. We divide to find that $N = 6$ R2. Six pans of lasagna would only provide 54 pieces, so we must buy seven pans of lasagna in order to serve 56 people.

Check We found that we need to buy seven 12-inch-square pans of lasagna to serve 56 people. It would take 14 8-inch-square pans of lasagna to feed 56 people, and one 12-inch-square lasagna feeds about twice as many people as one 8-inch-square pan, so our answer is reasonable.

New Concept

Increasing Knowledge

Math Language

Recall that **integers** are the set of counting numbers, their opposites, and zero.

On this number line the tick marks show the location of the integers:

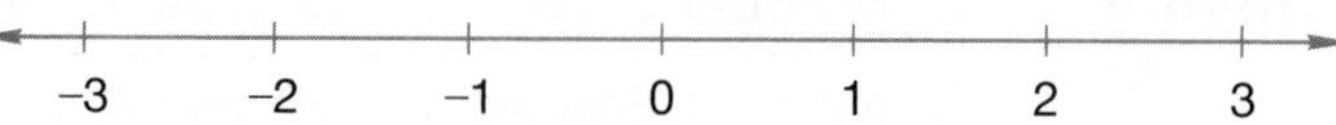

There are points on the number line between the integers that can be named with fractions or **mixed numbers.** A mixed number is a whole number plus a fraction. Halfway between 0 and 1 is $\frac{1}{2}$. Halfway between 1 and 2 is $1\frac{1}{2}$. Halfway between −1 and −2 is $-1\frac{1}{2}$.

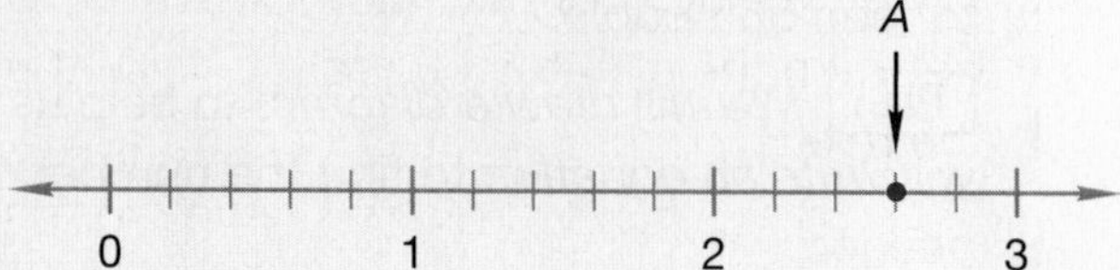

We count from zero.

The distance between consecutive integers on a number line may be divided into halves, thirds, fourths, fifths, or any other number of equal divisions. To determine which fraction or mixed number is represented by a point on the number line, we follow the steps described in the next example.

Example 1

Point *A* represents what mixed number on this number line?

A

0 1 2 3

Solution

We see that point *A* represents a number greater than 2 but less than 3. So point *A* represents a mixed number, which is a whole number plus a fraction.

To find the fraction, we first notice that the segment from 2 to 3 has been divided into five smaller segments. The distance from 2 to point *A* crosses three of the five segments. Thus, point *A* represents the mixed number $\mathbf{2\frac{3}{5}}$.

Analyze Why did we focus on the number of segments on the number line and not the number of vertical tick marks?

Activity

Inch Ruler to Sixteenths

Materials needed:

- inch ruler made in Lesson 7

In Lesson 7 we made an inch ruler divided into fourths. In this activity we will divide the ruler into eighths and sixteenths. First we will review what we did in Lesson 7.

We used a ruler to make one-inch divisions on a strip of tagboard.

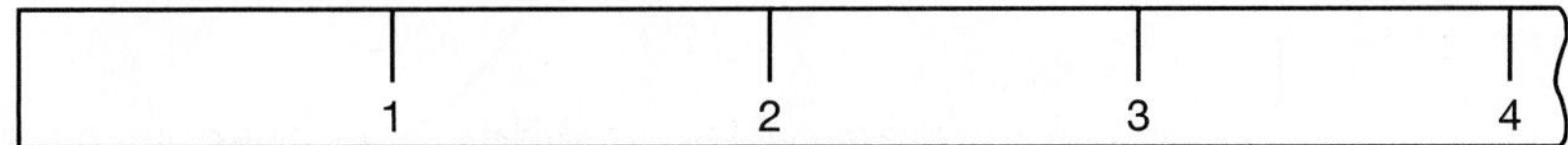

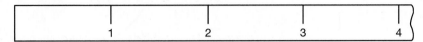

Then we estimated the halfway point between inch marks and drew new marks. The new marks were half-inch divisions. Then we estimated the halfway point between the half-inch marks and made quarter-inch divisions.

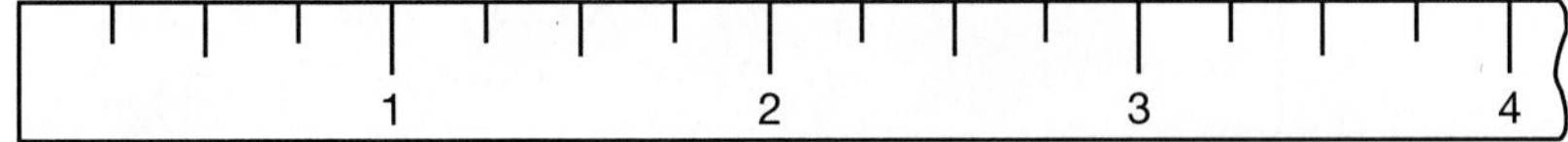

We made the half-inch marks a little shorter than the inch marks and the quarter-inch marks a little shorter than the half-inch marks.

Now divide your ruler into eighths of an inch by estimating the halfway point between the quarter-inch marks. Make these eighth-inch marks a little shorter than the quarter-inch marks.

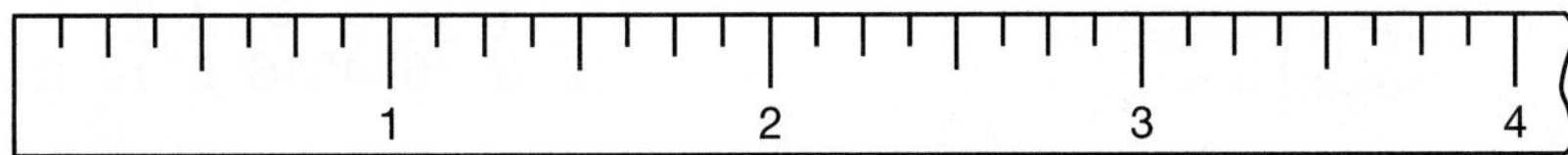

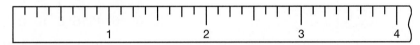

Finally, divide your ruler into sixteenths by estimating the halfway point between the eighth-inch marks. Make these marks the shortest marks on the ruler.

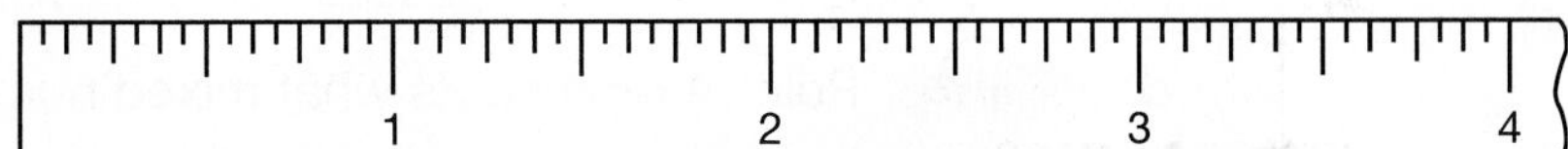

Example 2

Estimate the length of this line segment in inches. Then use your ruler to find its length to the nearest sixteenth of an inch.

Solution

The line segment is about 3 inches long. The ruler has been divided into sixteenths. We align the zero mark (or end of the ruler) with one end of the line segment. Then we find the mark on the ruler closest to the other end of the line segment and read this mark. As shown on the next page, we will enlarge a portion of a ruler to show how each mark is read.

Thinking Skill

Connect

How is a number line like a ruler?

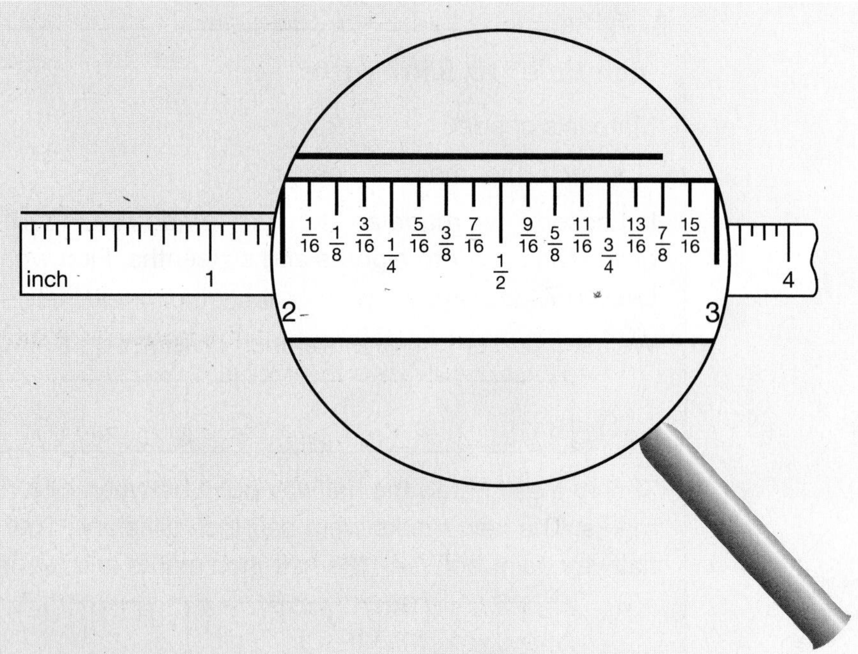

Estimate We find that the line segment is about $2\frac{7}{8}$ **inches long.** This is the nearest sixteenth because the end of the segment aligns more closely to the $\frac{7}{8}$ mark (which equals $\frac{14}{16}$) than it does to the $\frac{13}{16}$ mark or to the $\frac{15}{16}$ mark.

Practice Set

a. **Generalize** Continue this sequence to $1\frac{1}{2}$:

$$\frac{1}{16}, \frac{1}{8}, \frac{3}{16}, \frac{1}{4}, \frac{5}{16}, \frac{3}{8}, \frac{7}{16}, \frac{1}{2}, \ldots$$

b. What number is halfway between −2 and −3?

c. **Represent** What number is halfway between 2 and 5? Draw a number line to show the number that is halfway between 2 and 5.

d. **Connect** Point A represents what mixed number on this number line?

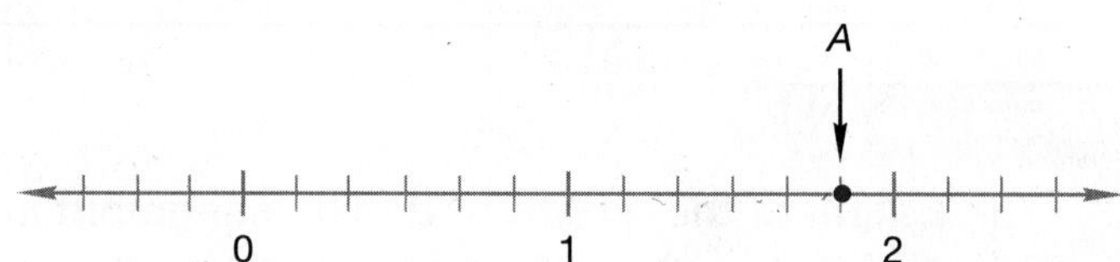

Use your ruler to find the length of each of these line segments to the nearest sixteenth of an inch:

e. ________

f. ________________________

g. __________________________________

Written Practice

Strengthening Concepts

1. (12) What is the sum of twelve thousand, five hundred and ten thousand, six hundred ten?

2. (13) In 1903 the Wright brothers made the first powered airplane flight. In 1969 Americans first landed on the moon. How many years was it from the first powered airplane flight to the first moon landing? (Write an equation and solve the problem.)

*** 3.** (15) Linda can run about 6 yards in one second. About how far can she run in 12 seconds? Write an equation and solve the problem.

*** 4.** (15) A coin collector has a collection of two dozen rare coins. If the value of each coin is $1000, what is the value of the entire collection? Write an equation and solve the problem.

*** 5.** (16) **Estimate** Find the sum of 5280 and 1760 by rounding each number to the nearest thousand before adding.

6. (2) $\frac{480}{3}$

7. (5) $\frac{6 - 6}{3}$

8. (1) **Represent** The letters *a, b,* and *c* represent three different numbers. The sum of *a* and *b* is *c*.

$$a + b = c$$

Rearrange the letters to form another addition equation and two subtraction equations. Hint: To be sure you arranged the letters in the correct order, choose numbers for *a, b,* and *c* that make $a + b = c$ true. Then try those numbers in place of the letters in your three equations.

9. (2) Rewrite $2 \div 3$ with a division bar, but do not divide.

10. (8) **Explain** A square has sides 10 cm long. Describe how to find its perimeter.

*** 11.** (17) **Connect** Use a ruler to find the length of the line segment below to the nearest sixteenth of an inch.

Find each unknown number. Check your work.

12. (3) $\$3 - y = \$1.75$

13. (3) $m - 20 = 30$

14. (4) $12n = 0$

15. (3) $16 + 14 = 14 + w$

16. (9) Compare: $19 \times 21 \bigcirc 20 \times 20$

17. (5) $100 - (50 - 25)$

18. (2) $\frac{5280}{44}$

19. (1) $365 + 4576 + 50{,}287$

20. (10) What number is missing in the following sequence?

5, 10, ______, 20, 25, …

21. (12) Which digit in 987,654,321 is in the hundred-millions place?

22. (2) $250{,}000 \div 100$

23. (2) $\$3.75 \times 10$

* **24.** (15) **Estimate** An 8-ounce serving of 2% milk contains 26 grams of protein, fat, and carbohydrates. About half the grams are carbohydrates. About how many grams are not from carbohydrates?

25. (10) **Analyze** What is the sum of the first six positive odd numbers?

26. (6) **Explain** How can you find $\frac{1}{4}$ of 52?

27. (6) A quarter is $\frac{1}{4}$ of a dollar.

a. How many quarters are in one dollar?

b. How many quarters are in three dollars?

* **28.** (17) On an inch ruler, which mark is halfway between the $\frac{1}{4}$-inch mark and the $\frac{1}{2}$-inch mark?

* **29.** (17) Point *A* represents what mixed number on the number line below?

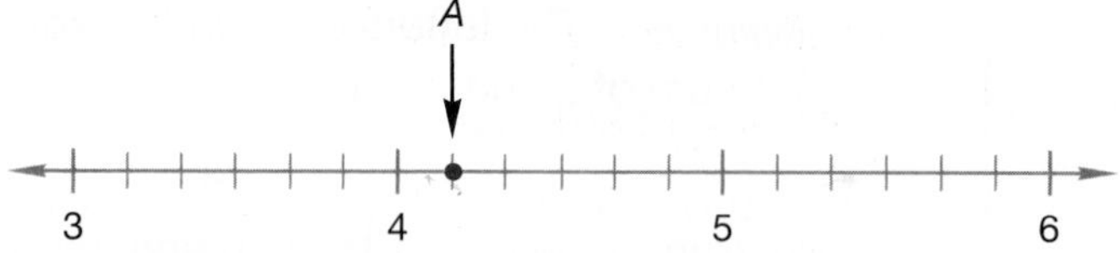

* **30.** (17) A segment that is $\frac{1}{2}$ of an inch long is how many sixteenths of an inch long?

Early Finishers
Real-World Application

Read the problem below. Then decide whether you can use an estimate to answer the question or if you need to compute an exact answer. Explain how you find your answer.

Manny has one gallon of paint. The label states that it covers about 350–450 square feet. Manny wants to paint the living room, dining room and family room of his home. The lateral surface area of all of the rooms totals about 800 square feet. How much more paint should Manny buy?

LESSON 18

• Average
• Line Graphs

Power Up — *Building Power*

facts Power Up B

mental math

a. **Number Sense:** 4 × 23 equals 4 × 20 plus 4 × 3. Find 4 × 23.

b. **Number Sense:** 4 × 32

c. **Number Sense:** 3 × 42

d. **Number Sense:** 3 × 24

e. **Geometry:** A hexagon has sides that measure 15 ft. What is the perimeter of the hexagon?

f. **Measurement:** How many days are in 2 weeks?

g. **Measurement:** How many hours are in 2 days?

h. **Calculation:** Start with a half dozen. Add 2; multiply by 3; divide by 4; then subtract 5.What is the answer?

problem solving

In his pocket Alex had seven coins totaling exactly one dollar. Name a possible combination of coins in his pocket. How many different combinations of coins are possible?

New Concepts — *Increasing Knowledge*

average

Here we show three stacks of books; the stacks contain 8 books, 7 books, and 3 books respectively. Altogether there are 18 books, but the number of books in each stack is not equal.

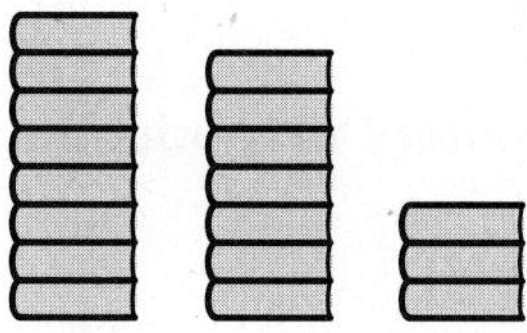

If we move some of the books from the taller stacks to the shortest stack, we can make the three stacks the same height. Then there will be 6 books in each stack.

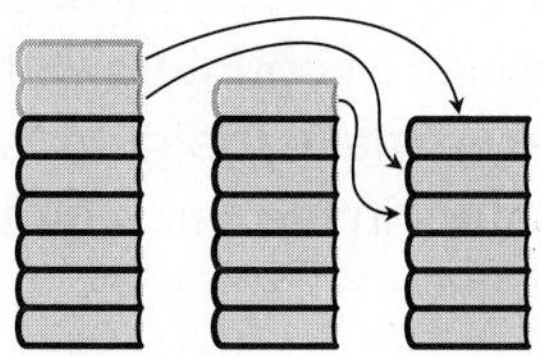

By making the stacks equal, we have found the **average** number of books in the three stacks. Notice that the average number of books in each stack is greater than the number in the smallest stack and less than the number in the largest stack. One way we can find an average is by making equal groups.

Example 1

In four classrooms there were 28 students, 27 students, 26 students, and 31 students respectively. What was the average number of students per classroom?

Solution

The average number of students per classroom is how many students there would be in each room if we made the numbers equal. So we will take the total number of students and make four equal groups. To find the total number of students, we add the numbers in each classroom.

$$\begin{array}{r} 28 \text{ students} \\ 27 \text{ students} \\ 26 \text{ students} \\ +\ 31 \text{ students} \\ \hline 112 \text{ students in all} \end{array}$$

We make four equal groups by dividing the total number of students by four.

$$4\overline{)112 \text{ students}} = 28 \text{ students}$$

If the groups were equal, there would be 28 students in each classroom. The average number of students per classroom would be **28.**

Notice that an average problem is a "combining" problem and an "equal groups" problem. First we found the total number of students in all the classrooms ("combining" problem). Then we found the number of students that would be in each group if the groups were equal ("equal groups" problem).

Example 2

Use counters to model the average of 3, 7, and 8.

Solution

Thinking Skill

Connect

What are some other real-life situations where we might need to find the average of a set of numbers?

This question does not tell us whether the numbers 3, 7, and 8 refer to books or students or coins or quiz scores. Still, we can find the average of these numbers by combining and then making equal groups. We can model the problem with counters. Since there are three numbers, there will be three groups of counters with 3, 7, and 8 counters in the groups. To find the average we make the groups equal. One way to do this is to first combine all the counters in the three groups. That gives us a total of 18 counters.

$$3 + 7 + 8 = 18$$

Then we divide the total into three equal groups.

$$18 \div 3 = 6$$

That gives us three groups of six counters. We find that the average of 3, 7, and 8 is **6.**

Example 3

What number is halfway between 27 and 81?

Solution

The number halfway between two numbers is also the average of the two numbers. For example, the average of 7 and 9 is 8, and 8 is halfway between 7 and 9. So the average of 27 and 81 will be the number halfway between 27 and 81. We add 27 and 81 and divide by 2.

$$\text{Average of 27 and 81} = \frac{27 + 81}{2}$$

$$= \frac{108}{2}$$

$$= 54$$

The number halfway between 27 and 81 is **54.**

The average we have talked about in this lesson is also called the **mean.** We will learn more about average and mean in later lessons.

line graphs

Line graphs display numerical information as points connected by line segments. Whereas bar graphs often display comparisons, line graphs often show how a measurement changes over time.

Example 4

Reading Math

On this line graph, the *horizontal axis* is divided into equal segments that represent years. The *vertical axis* is divided into equal segments that represent inches. The labels on the axes tell us how many years and how many inches.

This line graph shows Margie's height in inches from her eighth birthday to her fourteenth birthday. During which year did Margie grow the most?

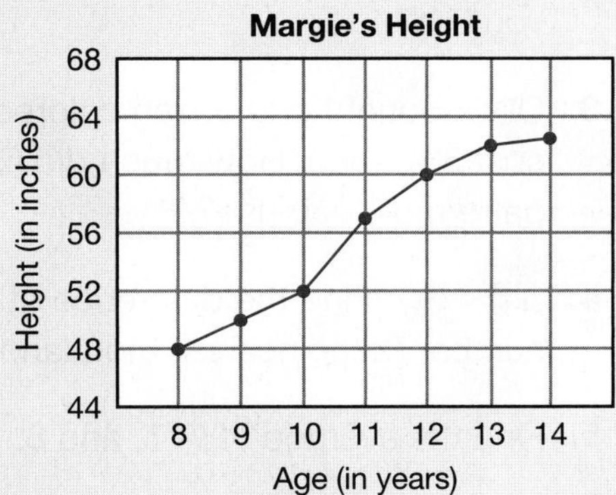

Solution

From Margie's eighth birthday to her ninth birthday, she grew about two inches. She also grew about two inches from her ninth to her tenth birthday. From her tenth to her eleventh birthday, Margie grew about five inches. Notice that this is the steepest part of the growth line. So the year Margie grew the most was **the year she was ten.**

Your teacher might ask you to keep a line graph of your math test scores. The **Lesson Activity 2** Test Scores Line Graph can be used for this purpose.

Practice Set

a. There were 26 books on the first shelf, 36 books on the second shelf, and 43 books on the third shelf. Velma rearranged the books so that there were the same number of books on each shelf. After Velma rearranged the books, how many were on the first shelf ?

b. What is the average of 96, 44, 68, and 100?

c. What number is halfway between 28 and 82?

d. What number is halfway between 86 and 102?

e. Find the average of 3, 6, 9, 12, and 15.

Use the information in the graph in example 4 to answer these questions:

f. How many inches did Margie grow from her eighth to her twelfth birthday?

g. During which year did Margie grow the least?

h. *Predict* Based on the information in the graph, would you predict that Margie will grow to be 68 inches tall?

Written Practice *Strengthening Concepts*

1. (11) Jumbo ate two thousand, sixty-eight peanuts in the morning and three thousand, nine hundred forty in the afternoon. How many peanuts did Jumbo eat in all? What kind of pattern did you use?

2. (11) Jimmy counted his permanent teeth. He had 11 on the top and 12 on the bottom. An adult has 32 permanent teeth. How many more of Jimmy's teeth need to grow in? What kind of pattern did you use?

*** 3.** (15) Olivia bought one dozen colored pencils for an art project. Each pencil cost 53¢ each. How much did Olivia spend on pencils? What kind of pattern did you use?

*** 4.** (16) *Estimate* Find the difference of 5035 and 1987 by rounding each number to the nearest thousand before subtracting.

*** 5.** (18) Find the average of 9, 7, and 8.

*** 6.** (18) What number is halfway between 59 and 81?

*** 7.** (14) What number is 6 less than 2?

8. (2) $0.35 × 100 **9.** (2) 10,010 ÷ 10 **10.** (2) 34,180 ÷ 17

11. (1) \$3.64 + \$94.28 + 87¢

12. (1) $41{,}375 - 13{,}576$

13. (2) 125×16

14. (5) $4 \cdot 3 \cdot 2 \cdot 1 \cdot 0$

Analyze Find each unknown number. Check your work.

15. (3) $w - 84 = 48$

16. (4) $\frac{234}{n} = 6$

17. (3, 5) $(1 + 2) \times 3 = (1 \times 2) + m$

18. (8) **Model** Draw a rectangle 5 cm long and 3 cm wide. What is its perimeter?

19. (10) What is the sum of the first six positive even numbers?

20. (10) **Generalize** Describe the rule of the following sequence. Then find the missing term.

1, 2, 4, ______, 16, 32, 64, ...

21. (9) Compare: $500 \times 1 \bigcirc 500 \div 1$

22. (6) **Generalize** What number is $\frac{1}{2}$ of 1110?

23. (12) What is the place value of the 7 in 987,654,321?

Refer to the line graph shown below to answer problems **24–26.**

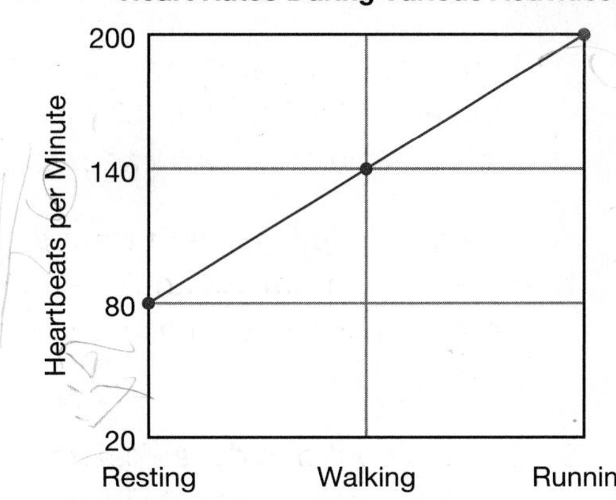

* **24.** (18) **Estimate** Running increases a resting person's heart rate by about how many heartbeats per minute?

* **25.** (18) **Estimate** About how many times would a person's heart beat during a 10-minute run?

* **26.** (18) **Formulate** Using the information in the line graph, write a word problem about comparing. Then answer the problem.

* **27.** (18) **Explain** In three classrooms there are 24, 27, and 33 students respectively. How many students will be in each classroom if some students are moved from one classroom to the other classrooms so that the number of students in each classroom is equal? How do you know that your answer is reasonable?

28. A dime is $\frac{1}{10}$ of a dollar.
(6)

a. How many dimes are in a dollar?

b. How many dimes are in three dollars?

*** 29.** **Model** Use a ruler to draw a rectangle that is $2\frac{1}{4}$ inches long and $1\frac{3}{4}$ inches wide.
(17)

*** 30.** **Formulate** Word problems about finding an average include which two types of problems? (Select two)
(18)

combining separating comparing equal groups

Early Finishers
Real-World Application

The cost of one cubic foot of natural gas in my town is $0.67. My meter reading on July 31 was 1518. On August 31 it was 1603. The meter measures natural gas in cubic feet.

a. How many cubic feet of gas did I use in the month of August (between the two meter readings)?

b. What was my total gas cost for the month?

LESSON 19

• Factors
• Prime Numbers

Power Up — *Building Power*

facts Power Up D

mental math

a. **Number Sense:** 3 × 64

b. **Number Sense:** 3 × 46

c. **Number Sense:** 120 + 18

d. **Calculation:** 34 + 40 + 9

e. **Calculation:** 34 + 50 − 1

f. **Calculation:** \$20.00 − \$12.50

g. **Measurement:** How many decades are in 2 centuries?

h. **Calculation:** Start with 100. Divide by 2; subtract 1; divide by 7; then add 3. What is the answer?

problem solving

Doubles tennis is played on a rectangular-shaped court that is 78 feet long and 36 feet wide. In the drawing of the doubles tennis court, how many rectangles can you find?

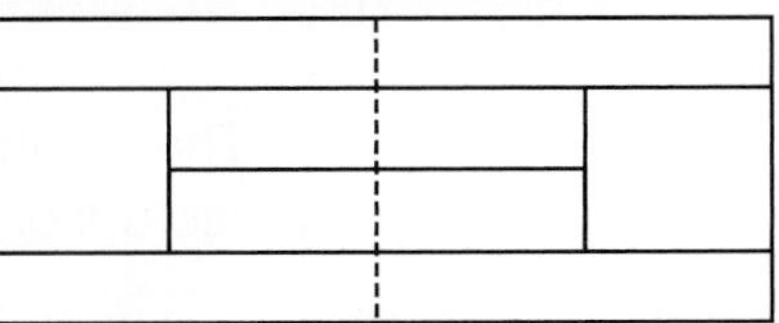

New Concepts — *Increasing Knowledge*

factors

Recall from Lesson 2 that a factor is one of the numbers multiplied to form a product.

$2 \times 3 = 6$ Both 2 and 3 are factors.

$1 \times 6 = 6$ Both 1 and 6 are factors.

Thinking Skill
How are the two definitions of **factor** connected?

We see that each of the numbers 1, 2, 3, and 6 are factors of 6. Notice that when we divide 6 by 1, 2, 3, or 6, the resulting quotient has no remainder (that is, it has a remainder of zero). We say that 6 is "divisible by" 1, 2, 3, and 6.

$$\begin{array}{r}6\\1\overline{)6}\\\underline{6}\\0\end{array}\qquad\begin{array}{r}3\\2\overline{)6}\\\underline{6}\\0\end{array}\qquad\begin{array}{r}2\\3\overline{)6}\\\underline{6}\\0\end{array}\qquad\begin{array}{r}1\\6\overline{)6}\\\underline{6}\\0\end{array}$$

This leads us to another definition of *factor*.

> The **factors** of a given number are the whole numbers that divide the given number without a remainder.

We can illustrate the factors of 6 by arranging 6 tiles to form rectangles. With 6 tiles we can make a 1-by-6 rectangle. We can also make a 2-by-3 rectangle.

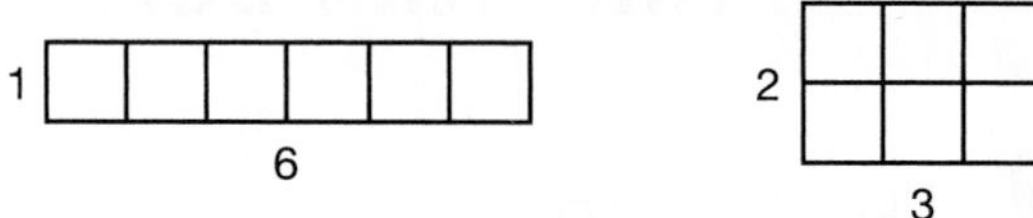

The number of tiles along the sides of these two rectangles (1, 6, 2, 3) are the four factors of 6.

Example 1

What are the factors of 10?

Solution

The factors of 10 are all the numbers that divide 10 evenly (with no remainder). They are **1, 2, 5,** and **10.**

$$\begin{array}{r}10\\ \mathbf{1}\overline{)10}\end{array} \qquad \begin{array}{r}5\\ \mathbf{2}\overline{)10}\end{array} \qquad \begin{array}{r}2\\ \mathbf{5}\overline{)10}\end{array} \qquad \begin{array}{r}1\\ \mathbf{10}\overline{)10}\end{array}$$

We can illustrate the factors of 10 with two rectangular arrays of tiles.

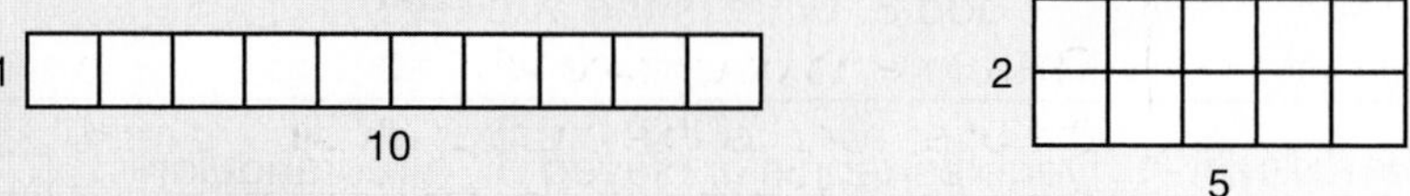

The number of tiles along the sides of the two rectangles (1, 10, 2, 5) are the factors of 10.

Example 2

How many different whole numbers are factors of 12?

Solution

Twelve can be divided evenly by 1, 2, 3, 4, 6, and 12. The question asked "How many?" Counting factors, we find that 12 has **6** different whole-number factors.

Twelve tiles can be arranged to form three different shapes of rectangles. The lengths of the sides illustrate that the six factors of 12 are 1, 12, 2, 6, 3, and 4.

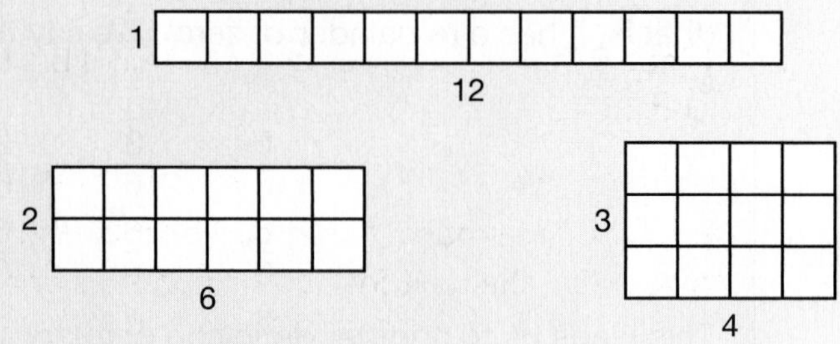

Model Draw tiles to illustrate the factors of 18. How many different shapes of rectangles can you make?

prime numbers

Here we list the first ten counting numbers and their factors. Which of the numbers have exactly two factors?

Number	Factors
1	1
2	1, 2
3	1, 3
4	1, 2, 4
5	1, 5
6	1, 2, 3, 6
7	1, 7
8	1, 2, 4, 8
9	1, 3, 9
10	1, 2, 5, 10

Counting numbers that have exactly two factors are **prime numbers.** The first four prime numbers are 2, 3, 5, and 7. The only factors of a prime number are the number itself and 1. The number 1 is not a prime number, because it has only one factor, itself.

Therefore, to determine whether a number is prime, we may ask ourselves the question, "Is this number divisible by any number other than the number itself and 1?" If the number is divisible by any other number, the number is not prime.

Example 3

The first four prime numbers are 2, 3, 5, and 7. What are the next four prime numbers?

Solution

We will consider the next several numbers and eliminate those that are not prime.

8, 9, 10, 11, 12, 13, 14, 15, 16, 17, 18, 19, 20

All even numbers have 2 as a factor. So no even numbers greater than two are prime numbers. We can eliminate the even numbers from the list.

~~8~~, 9, ~~10~~, 11, ~~12~~, 13, ~~14~~, 15, ~~16~~, 17, ~~18~~, 19, ~~20~~

Since 9 is divisible by 3, and 15 is divisible by 3 and by 5, we can eliminate 9 and 15 from the list.

~~8~~, ~~9~~, ~~10~~, 11, ~~12~~, 13, ~~14~~, ~~15~~, ~~16~~, 17, ~~18~~, 19, ~~20~~

Each of the remaining four numbers on the list is divisible only by itself and by 1. Thus the next four prime numbers after 7 are **11, 13, 17,** and **19.**

Activity

Prime Numbers

List the counting numbers from 1 to 100 (or use **Lesson Activity 3** Hundred Number Chart). Then follow these directions:

Step 1: Draw a line through the number 1. The number 1 is not a prime number.

Step 2: Circle the prime number 2. Draw a line through all the other multiples of 2 (4, 6, 8, etc.).

Step 3: Circle the prime number 3. Draw a line through all the other multiples of 3 (6, 9, 12, etc.).

Step 4: Circle the prime number 5. Draw a line through all the other multiples of 5 (10, 15, 20, etc.).

Step 5: Circle the prime number 7. Draw a line through all the other multiples of 7 (14, 21, 28, etc.).

Step 6: Circle all remaining numbers on your list (the numbers that do not have a line drawn through them).

When you have finished, all the prime numbers from 1 to 100 will be circled on your list.

Thinking Skill

Conclude

Why don't the directions include "Circle the number 4" and "Circle the number 6"?

Practice Set

List the factors of the following numbers:

a. 14 **b.** 15

c. 16 **d.** 17

Justify Which number in each group is a prime number? Explain how you found your answer.

e. 21, 23, 25

f. 31, 32, 33

g. 43, 44, 45

Classify Which number in each group is not a prime number?

h. 41, 42, 43

i. 31, 41, 51

j. 23, 33, 43

Prime numbers can be multiplied to make whole numbers that are not prime. For example, $2 \cdot 2 \cdot 3$ equals 12 and $3 \cdot 5$ equals 15. (Neither 12 nor 15 are prime.) Show which prime numbers we multiply to make these products:

k. 16 **l.** 18

Written Practice *Strengthening Concepts*

Math Language
The **dividend** is the number that is to be divided.

1. (4) If two hundred fifty-two is the dividend and six is the quotient, what is the divisor?

*** 2.** (11, 15) In 1863, President Abraham Lincoln gave his Gettysburg Address, which began "Fourscore and seven years ago… ." A *score* equals twenty. What year was Lincoln referring to in his speech? Explain how you found your answer.

*** 3.** (14) The temperature in Barrow, Alaska was –46°F on January 22, 2002. It was 69°F on July 15, 2002. How many degrees warmer was it on July 15 than on January 22?

*** 4.** (15) If 203 turnips are to be shared equally among seven rabbits, how many should each receive? Write an equation and solve the problem.

*** 5.** (18) What is the average of 1, 2, 4, and 9?

6. (10) **Predict** What is the next number in the following sequence?

1, 4, 9, 16, 25, ______, …

7. (8) A regular hexagon has six sides of equal length. If each side of a hexagon is 25 mm, what is the perimeter?

8. (7) One centimeter equals ten millimeters. How many millimeters long is the line segment below?

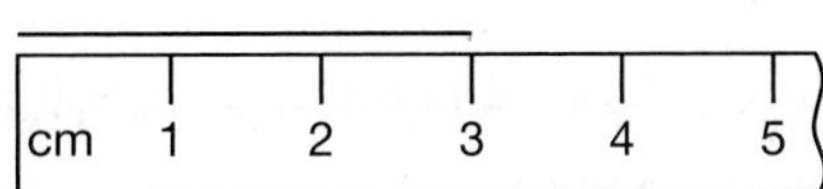

*** 9.** (19) What are the whole-number factors of 20?

*** 10.** (19) How many different whole numbers are factors of 15?

*** 11.** (19) **Classify** Which of the numbers below is a prime number?

A 25 **B** 27 **C** 29

12. (12) 250,000 ÷ 100

13. (2) 1234 ÷ 60

*** 14.** (15) $\frac{6 + 18 + 9}{3}$

15. (2) \$3.45 × 10

Find each unknown number. Check your work.

16. (3) \$10.00 − w = \$1.93

17. (4) $\frac{w}{3} = 4$

18. (2) **Represent** The letters a, b, and c represent three different numbers. The product of a and b is c.

$$ab = c$$

Rearrange the letters to form another multiplication equation and two division equations.

* **19.** (17) Arrange these numbers in order from least to greatest:

$$3, -2, 1, \tfrac{1}{2}, 0$$

20. (9) Compare: $123 \div 1 \bigcirc 123 - 1$

21. (12) Which digit in 135,792,468,000 is in the ten-millions place?

* **22.** (16) *Connect* Round 123,456,789 to the nearest million.

23. (6) How much money is $\frac{1}{2}$ of \$11.00?

24. (8) If a square has a perimeter of 48 inches, how long is each side of the square?

25. (5) $(51 + 49) \times (51 - 49)$

* **26.** (19) *Classify* Which of the numbers below is a prime number?

A 2 **B** 22 **C** 222

* **27.** (19) Prime numbers can be multiplied to make whole numbers that are not prime. To make 18, we perform the multiplication $2 \cdot 3 \cdot 3$. Show which prime numbers we multiply to make 20.

* **28.** (18) The dictionaries are placed in three stacks. There are 6 dictionaries in one stack and 12 dictionaries in each of the other two stacks. How many dictionaries will be in each stack if some dictionaries are moved from the taller stacks to the shortest stack so that there are the same number of dictionaries in each stack?

* **29.** (17) *Model* Draw a square with sides that are $1\frac{3}{8}$ inches long.

* **30.** (10, 19) *Explain* How can you use the concepts of "even" and "odd" numbers to determine whether a number is divisible by 2.

Early Finishers
Real-World Application

The school newspaper reported that middle school students own an average of 47 CDs. Yolanda asked seven of her friends how many CDs they owned and got these results.

32, 49, 21, 59, 37, 44, 52

a. What is the average number of CDs owned by Yolanda's seven friends?

b. How does this average compare to the average reported in the school newspaper?

LESSON 20

• Greatest Common Factor (GCF)

Power Up

Building Power

facts Power Up C

mental math

a. **Number Sense:** 6 × 23

b. **Number Sense:** 6 × 32

c. **Number Sense:** 640 + 1200

d. **Calculation:** 63 + 20 + 9

e. **Calculation:** 63 + 30 − 1

f. **Number Sense:** $100.00 − $75.00

g. **Measurement:** How many minutes are in 2 hours?

h. **Calculation:** Start with 10. Multiply by 10; subtract 1; divide by 9; then add 1. What is the answer?

problem solving

A card with a triangle on it in the position shown is rotated 90° clockwise three times. Sketch the pattern and draw the triangles in the correct positions.

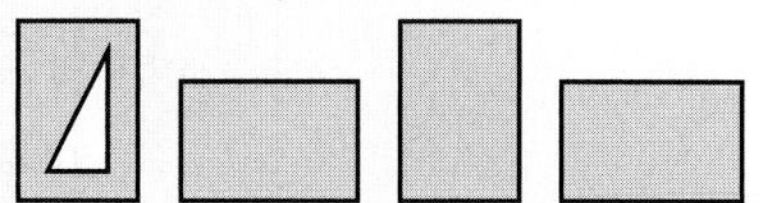

New Concept

Increasing Knowledge

The factors of 8 are

1, 2, 4, and 8

The factors of 12 are

1, 2, 3, 4, 6, and 12

Math Language
Recall that a **factor** is a whole number that divides another whole number without a remainder.

We see that 8 and 12 have some of the same factors. They have three factors in common. These common factors are 1, 2, and 4. Their **greatest common factor**—the largest factor that they both have—is 4. *Greatest common factor* is often abbreviated **GCF.**

Example 1

Find the greatest common factor of 12 and 18.

Solution

The factors of 12 are: 1, 2, 3, 4, 6, and 12.

The factors of 18 are: 1, 2, 3, 6, 9, and 18.

We see that 12 and 18 share four common factors: 1, 2, 3, and 6. The greatest of these is **6.**

Example 2

Find the GCF of 6, 9, and 15.

Solution

The factors of 6 are: 1, 2, 3, and 6.

The factors of 9 are: 1, 3, and 9.

The factors of 15 are: 1, 3, 5, and 15.

The GCF of 6, 9, and 15 is **3.**

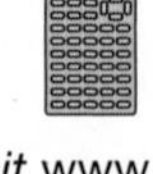

Visit www.SaxonPublishers.com/ActivitiesC1 *for a graphing calculator activity.*

Note: The search for the greatest common factor of two or more numbers is a search for the **largest** number that evenly divides each of them. In this problem we can quickly determine that 3 is the largest number that evenly divides 6, 9, and 15. A complete listing of the factors might be helpful but is not required.

Generalize What is the GCF of any two prime numbers? Explain your answer.

Practice Set

Find the greatest common factor (GCF) of the following:

a. 10 and 15

b. 18 and 27

c. 18 and 24

d. 12, 18, and 24

e. 15 and 25

f. 20, 30, and 40

g. 12 and 15

h. 20, 40, and 60

i. **Analyze** Write a list of three numbers whose GCF is 7.

Written Practice

Strengthening Concepts

1. (12) What is the difference between the product of 12 and 8 and the sum of 12 and 8?

2. (12) Saturn's average distance from the Sun is one billion, four hundred twenty-nine million kilometers. Use digits to write that distance.

3. (12) Which digit in 497,325,186 is in the ten-millions place?

* **4.** (16) **Estimate** Jill has exactly 427 beads, but when Dwayne asked her how many beads she has, Jill rounded the amount to the nearest hundred. How many beads did Jill say she had?

* **5.** (14) The morning temperature was −3°C. By afternoon it had warmed to 8°C. How many degrees had the temperature risen?

*** 6.** (18) In three basketball games Allen scored 31, 52, and 40 points. What was the average number of points Allen scored per game?

*** 7.** (20) Find the greatest common factor of 12 and 20.

*** 8.** (20) Find the GCF of 9, 15, and 21.

9. (6) **Connect** How much money is $\frac{1}{4}$ of $3.24?

10. (2) $5432 \div 10$

11. (5) $\frac{28 + 42}{14}$

12. (1) $56{,}042 + 49{,}985$

13. (2) $37{,}080 \div 12$

14. (2) $\$6.47 \times 10$

15. (5) $5 \times 4 \times 3 \times 2 \times 1$

Analyze Find each unknown number. Check your work.

16. (3) $w - 76 = 528$

17. (3) $14{,}009 - w = 9670$

18. (4) $6w = 90$

19. (3) $q - 365 = 365$

20. (3) $365 - p = 365$

21. (10) **Generalize** Find the missing number in the following sequence:

_______, 10, 16, 22, 28, ...

22. (9) Compare: $50 - 1 \bigcirc 49 + 1$

23. (10) **Predict** The first positive odd number is 1. What is the tenth positive odd number?

24. (8) **Explain** The perimeter of a square is 100 cm. Describe how to find the length of each side.

*** 25.** (17) **Estimate** Estimate the length of this key to the nearest inch. Then use a ruler to find the length of the key to the nearest sixteenth of an inch.

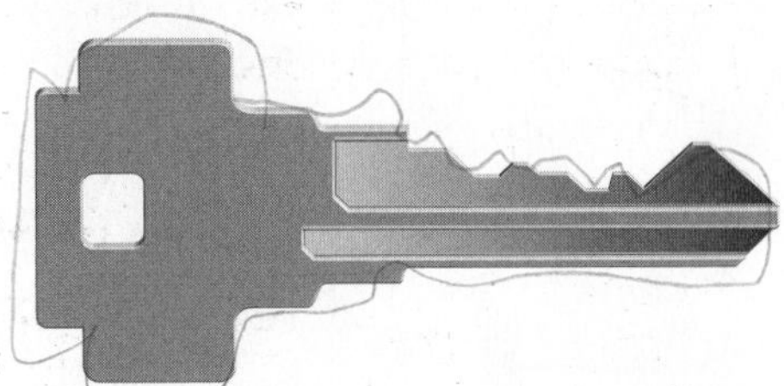

26. (6) A "bit" is $\frac{1}{8}$ of a dollar.

a. How many bits are in a dollar?

b. How many bits are in three dollars?

*** 27.** (18) In four boxes there are 12, 24, 36, and 48 golf balls respectively. If the golf balls are rearranged so that there are the same number of golf balls in each of the four boxes, how many golf balls will be in each box?

* **28.** (19) *Classify* Which of the numbers below is a prime number?

A 5 **B** 15 **C** 25

* **29.** (19) *Explain* List the whole-number factors of 24. How did you find your answer?

* **30.** (12) Ten billion is how much less than one trillion?

Early Finishers
Real-World Application

Tino's parents like mathematics, especially prime numbers. So they made a plan for his allowance. They will number the weeks of the year from 1 to 52. On week 1, they will pay Tino $1. On weeks that are prime numbers, they will pay him $3. On weeks that are composite numbers, they will pay him $5.

a. How much money will Tino receive for one year? Show your work.

b. Would Tino receive more or less if his parents paid him $5 on "prime weeks" and $3 on "composite weeks"? Show your work.

INVESTIGATION 2

Focus on

• Investigating Fractions with Manipulatives

In this investigation you will make a set of fraction manipulatives to help you answer questions in this investigation and in future problem sets.

Activity

Using Fraction Manipulatives

Materials needed:

- Investigation Activities 4–8
- scissors
- envelope or zip-top bag to store fraction pieces

Preparation:

To make your own fraction manipulatives, cut out the fraction circles on the Investigation Activities. Then cut each fraction circle into its parts.

Thinking Skill

Connect

What percent is one whole circle?

Model Use your fraction manipulatives to help you with these exercises:

1. What percent of a circle is $\frac{1}{2}$ of a circle?
2. What fraction is half of $\frac{1}{2}$?
3. What fraction is half of $\frac{1}{4}$?
4. Fit three $\frac{1}{4}$ pieces together to form $\frac{3}{4}$ of a circle. Three fourths of a circle is what percent of a circle?
5. Fit four $\frac{1}{8}$ pieces together to form $\frac{4}{8}$ of a circle. Four eighths of a circle is what percent of a circle?
6. Fit three $\frac{1}{6}$ pieces together to form $\frac{3}{6}$ of a circle. Three sixths of a circle is what percent of a circle?
7. Show that $\frac{4}{8}$, $\frac{3}{6}$, and $\frac{2}{4}$ each make one half of a circle. (We say that $\frac{4}{8}$, $\frac{3}{6}$, and $\frac{2}{4}$ all *reduce* to $\frac{1}{2}$.)
8. The fraction $\frac{2}{8}$ equals which single fraction piece?
9. The fraction $\frac{6}{8}$ equals how many $\frac{1}{4}$s?
10. The fraction $\frac{2}{6}$ equals which single fraction piece?
11. The fraction $\frac{4}{6}$ equals how many $\frac{1}{3}$s?
12. The sum $\frac{1}{8} + \frac{1}{8} + \frac{1}{8}$ is $\frac{3}{8}$. If you add $\frac{3}{8}$ and $\frac{2}{8}$, what is the sum?

13. **Connect** Form a whole circle using six of the $\frac{1}{6}$ pieces. Then remove (subtract) $\frac{1}{6}$. What fraction of the circle is left? What equation represents your model?

14. Demonstrate subtracting $\frac{1}{3}$ from 1 by forming a circle of $\frac{3}{3}$ and then removing $\frac{1}{3}$. What fraction is left?

15. Use four $\frac{1}{4}$s to demonstrate the subtraction $1 - \frac{1}{4}$. Then write the answer.

16. Eight $\frac{1}{8}$s form one circle. If $\frac{3}{8}$ of a circle is removed from one circle $(1 - \frac{3}{8})$, then what fraction of the circle remains?

17. What percent of a circle is $\frac{1}{3}$ of a circle?

18. What percent of a circle is $\frac{1}{6}$ of a circle?

Fraction manipulatives can help us compare fractions. Since $\frac{1}{2}$ of a circle is larger than $\frac{1}{3}$ of a circle, we can see that

$$\frac{1}{2} > \frac{1}{3}$$

Model For problems **19** and **20,** use your fraction manipulatives to construct models of the fractions. Use the models to help you write the correct comparison for each problem.

Reading Math

< means is *less than;*
= means *is equal to;*
> means *is greater than.*

19. Compare: $\frac{2}{3} \bigcirc \frac{3}{4}$

20. Compare: $\frac{2}{3} \bigcirc \frac{3}{8}$

Represent We can also draw pictures to help us compare fractions.

21. Draw two rectangles of the same size. Shade $\frac{1}{3}$ of one rectangle and $\frac{1}{5}$ of the other rectangle. What fraction represents the rectangle that has the larger amount shaded?

22. Draw and shade rectangles to illustrate this comparison:

$$\frac{3}{5} > \frac{3}{10}$$

Problems **23–29** involve **improper fractions.** Improper fractions are fractions that are equal to or greater than 1. In a fraction equal to 1 the numerator equals the denominator (as in $\frac{3}{3}$). In a fraction greater than 1 the numerator is greater than the denominator (as in $\frac{4}{3}$).

Work in groups of two or three students for the remaining problems.

23. Show that the improper fraction $\frac{5}{4}$ equals the mixed number $1\frac{1}{4}$ by combining four of the $\frac{1}{4}$ pieces to make a whole circle.

24. The improper fraction $\frac{7}{4}$ equals what mixed number?

25. The improper fraction $\frac{3}{2}$ equals what mixed number?

26. Form $1\frac{1}{2}$ circles using only $\frac{1}{4}$s. How many $\frac{1}{4}$ pieces are needed to make $1\frac{1}{2}$?

27. **Explain** How many $\frac{1}{3}$ pieces are needed to make two whole circles? How do you know your answer is correct?

28. The improper fraction $\frac{4}{3}$ equals what mixed number?

29. Convert $\frac{11}{6}$ to a mixed number.

30. **Evaluate** An analog clock can serve as a visual reference for twelfths. At 1 o'clock the hands mark off $\frac{1}{12}$ of a circle; at 2 o'clock the hands mark off $\frac{2}{12}$ of a circle, and so on. How many twelfths are in each of these fractions of a circle?

$$\frac{1}{2}, \frac{1}{4}, \frac{1}{3}, \frac{1}{6}$$

Hint: Try holding each fraction piece at arm's length in the direction of the clock (as an artist might extend a thumb toward a subject.)

After you have completed the exercises, gather and store your fraction manipulatives for later use.

---

extensions

a. **Justify** Use examples and **nonexamples** to support or disprove that the Commutative Property can be applied to the addition and subtraction of two fractions.

b. **Analyze** Choose between mental math or estimation and use that method to answer each of the following questions. Explain your choice.

1. One-fourth of a pizza was eaten. How much of the pizza was not eaten?

2. Each of 3 students ate $\frac{1}{8}$ of a new box of cereal. What amount of cereal in the box was eaten?

3. More than half of the students in the class are girls. What fraction of the students in the class are boys?

c. **Estimate** Copy this number line.

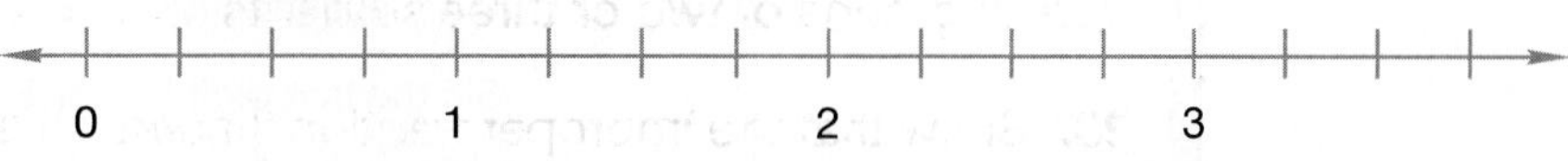

Then estimate the placement of the following fractions on the number line. Use your fraction pieces to help you if you need to.

$$\frac{3}{4} \quad \frac{3}{2} \quad \frac{4}{8} \quad 3\frac{3}{4} \quad \frac{9}{5} \quad \frac{10}{3} \quad 2\frac{1}{8} \quad \frac{7}{7}$$

LESSON

21

• Divisibility

Power Up

Building Power

facts

Power Up D

mental math

a. **Number Sense:** 4×42

b. **Number Sense:** 3×76

c. **Number Sense:** $64 + 19$

d. **Number Sense:** $450 + 37$

e. **Calculation:** \$10.00 − \$6.50

f. **Fractional Parts:** $\frac{1}{2}$ of 24

g. **Measurement:** How many months are in a year?

h. **Calculation:** Start with 25, × 2, − 1, ÷ 7, + 1, ÷ 2

problem solving

Here is part of a randomly-ordered multiplication table. What is the missing product?

48	30	42
32	?	28
56	35	49

New Concept

Increasing Knowledge

Thinking Skill

Discuss

Without dividing by 2, how can you tell that a number is even?

There are ways of discovering whether some numbers are factors of other numbers without actually dividing. For instance, even numbers can be divided by 2. Therefore, 2 is a factor of every even counting number. Since even numbers are "able" to be divided by 2, we say that even numbers are "divisible" by 2.

Tests for **divisibility** can help us find the factors of a number. Here we list divisibility tests for the numbers 2, 3, 5, 9, and 10.

Last-Digit Tests

Inspect the last digit of the number. A number is divisible by …

2 if the last digit is even.

5 if the last digit is 0 or 5.

10 if the last digit is 0.

Sum-of-Digits Tests

Add the digits of the number and inspect the total. A number is divisible by …
3 if the sum of the digits is divisible by 3.
9 if the sum of the digits is divisible by 9.

Example 1

Which of these numbers is divisible by 2?

365 1179 1556

Solution

To determine whether a number is divisible by 2, we inspect the last digit of the number. If the last digit is an even number, then the number is divisible by 2. The last digits of these three numbers are 5, 9, and 6. Since 5 and 9 are not even numbers, neither 365 nor 1179 is divisible by 2. Since 6 is an even number, 1556 is divisible by 2. It is not necessary to perform the division to answer the question. By inspecting the last digit of each number, we see that the number that is divisible by 2 is **1556.**

Example 2

Which of these numbers is divisible by 3?

365 1179 1556

Solution

To determine whether a number is divisible by 3, we add the digits of the number and then inspect the sum. If the sum of the digits is divisible by 3, then the number is also divisible by 3.

The digits of 365 are 3, 6, and 5. The sum of these is 14.

$$3 + 6 + 5 = 14$$

We try to divide 14 by 3 and find that there is a remainder of 2. Since 14 is not divisible by 3, we know that 365 is not divisible by 3 either.

The digits of 1179 are 1, 1, 7, and 9. The sum of these digits is 18.

$$1 + 1 + 7 + 9 = 18$$

We divide 18 by 3 and get no remainder. We see that 18 is divisible by 3, so 1179 is also divisible by 3.

The sum of the digits of 1556 is 17.

$$1 + 5 + 5 + 6 = 17$$

Since 17 is not divisible by 3, the number 1556 is not divisible by 3.

By using the divisibility test for 3, we find that the number that is divisible by 3 is **1179.**

Discuss Is 9536 divisible by 2? by 3? How do you know?

Example 3

Which of the numbers 2, 3, 5, 9, and 10 are factors of 135?

Solution

First we will use the last-digit tests. The last digit of 135 is 5, so 135 is divisible by 5 but not by 2 or by 10. Next we use the sum-of-digits tests. The sum of the digits in 135 is 9 ($1 + 3 + 5 = 9$). Since 9 is divisible by both 3 and 9, we know that 135 is also divisible by 3 and 9. So **3, 5,** and **9** are factors of 135.

Predict Will the product of any two prime factors of 135 given above also be a factor of 135?

Practice Set

a. Which of these numbers is divisible by 2?

123   234   345

b. Which of these numbers is divisible by 3?

1234   2345   3456

Use the divisibility tests to decide which of the numbers 2, 3, 5, 9, and 10 are factors of the following numbers:

c. 120   **d.** 102

Written Practice

Strengthening Concepts

Math Language
The word **product** is related to multiplication, the word **sum** is related to addition, and the word **difference** is related to subtraction.

1. *(12)* What is the product of the sum of 8 and 5 and the difference of 8 and 5?

2. *(13)* **Formulate** In 1787 Delaware became the first state. In 1959 Hawaii became the fiftieth state admitted to the Union. How many years were there between these two events? Write an equation and solve the problem.

*** 3.** *(15)* **Formulate** Maria figured that the bowling balls on the rack weighed a total of 240 pounds. How many 16-pound bowling balls weigh a total of 240 pounds? Write an equation and solve the problem.

4. *(6)* An apple pie was cut into four equal slices. One slice was quickly eaten. What fraction of the pie was left?

5. *(6)* There are 17 girls in a class of 30 students. What fraction of the class is made up of girls?

6. *(6)* Use digits to write the fraction three hundredths.

7. *(6)* How much money is $\frac{1}{2}$ of \$2.34?

8. *(6)* What is the place value of the 7 in 987,654,321?

9. (10) **Generalize** Describe the rule of the following sequence. Then find the next term.

1, 4, 16, 64, ______, ...

10. (9) Compare: $64 \times 1 \bigcirc 64 + 1$

* 11. (21) Which of these numbers is divisible by 9?

A 365 B 1179 C 1556

* 12. (16) **Estimate** Find the sum of 396, 197, and 203 by rounding each number to the nearest hundred before adding.

* 13. (20) What is the greatest common factor (GCF) of 12 and 16?

14. (2) $100\overline{)4030}$

15. (2) $48{,}840 \div 24$

16. (2) $\frac{678}{6}$

17. (2) $\$4.75 \times 10$

Find each unknown number. Check your work.

18. (3) $\$10 - w = 87¢$

19. (3) $463 + 27 + m = 500$

* 20. (17) Arrange these numbers in order from least to greatest:

$1, \frac{1}{2}, 0, -2, \frac{1}{4}$

* 21. (18) What is the average of 12, 16, and 23?

* 22. (19) List the whole numbers that are factors of 28.

* 23. (19) What whole numbers are factors of both 20 and 30?

24. (7) Use an inch ruler to draw a line segment four inches long. Then use a centimeter ruler to find the length to the nearest centimeter.

25. (5) $(12 \times 12) - (11 \times 13)$

* 26. (Inv. 2) **Represent** To divide a circle into thirds, John first imagined the face of a clock. From the center of the "clock," he drew one segment up to the 12. Then, starting from the center, John drew two other segments. To which two numbers on the "clock" did John draw the two segments when he divided the circle into thirds?

* 27. (Inv. 2) **Model** Draw and shade rectangles to illustrate this comparison:

$$\frac{2}{3} < \frac{3}{4}$$

*** 28.** (6) A "bit" is $\frac{1}{8}$ of a dollar.

a. How many bits are in a dollar?

b. How many bits are in a half-dollar?

29. (8) A regular octagon has eight sides of equal length. What is the perimeter of a regular octagon with sides 18 cm long?

*** 30.** (Inv. 2) **Represent** Describe a method for dividing a circle into eight equal parts that involves drawing a plus sign and a times sign. Illustrate the explanation.

Early Finishers
Real-World Application

The 22 sixth grade students at the book fair want to buy a mystery or science fiction novel. Ten of the 15 students who want a mystery book changed their minds and decided to look for a humorous book. How many sixth grade students are NOT looking for mystery books?

Write one equation and use it to solve the problem.

LESSON 22

• "Equal Groups" Problems with Fractions

Power Up

Building Power

facts Power Up C

mental math

a. **Number Sense:** 4 × 54

b. **Number Sense:** 3 × 56

c. **Number Sense:** 36 + 29

d. **Calculation:** 359 − 42

e. **Calculation:** \$10.00 − \$3.50

f. **Fractional Parts:** $\frac{1}{2}$ of 48

g. **Measurement:** How many yards are in 9 feet?

h. **Calculation:** Start with 100, − 1, ÷ 9, + 1, ÷ 2, − 1, × 5

problem solving

Truston has 16 tickets, Sergio has 8 tickets, and Melina has 6 tickets. How many tickets should Truston give to Sergio and to Melina so that they all have the same number of tickets?

New Concept

Increasing Knowledge

Here we show a collection of six objects. The collection is divided into three equal groups. We see that there are two objects in $\frac{1}{3}$ of the collection. We also see that there are four objects in $\frac{2}{3}$ of the collection.

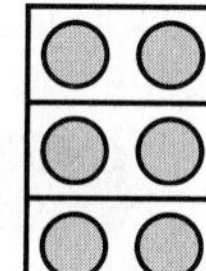

This collection of twelve objects is divided into four equal groups. There are three objects in $\frac{1}{4}$ of the collection, so there are nine objects in $\frac{3}{4}$ of the collection.

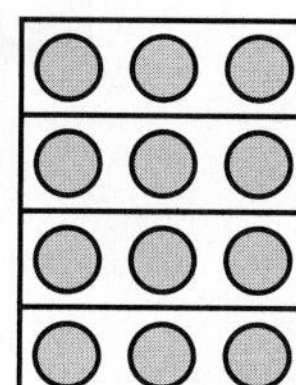

Example 1

Thinking Skill

Infer

Why do you divide the musicians by 3?

Two thirds of the 12 musicians played guitars. How many of the musicians played guitars?

Solution

This is a two-step problem. First we divide the 12 musicians into three equal groups (thirds). Each group contains 4 musicians. Then we count the number of musicians in two of the three groups.

	12 musicians
$\frac{1}{3}$ did not play guitars.	4 musicians
$\frac{2}{3}$ played guitars.	4 musicians
	4 musicians

Since there are 4 musicians in each third, the number of musicians in two thirds is 8. We find that **8 musicians** played guitars.

Example 2

Cory has finished $\frac{3}{4}$ of the 28 problems on the assignment. How many problems has Cory finished?

Solution

First we divide the 28 problems into four equal groups (fourths). Then we find the number of problems in three of the four groups. Since 28 ÷ 4 is 7, there are 7 problems in each group (in each fourth).

	28 problems
$\frac{1}{4}$ are not finished.	7 problems
$\frac{3}{4}$ are finished.	7 problems
	7 problems
	7 problems

In each group there are 7 problems. So in two groups there are 14 problems, and in three groups there are 21 problems. We see that Cory has finished **21 problems.**

Example 3

How much money is $\frac{3}{5}$ of \$3.00?

Solution

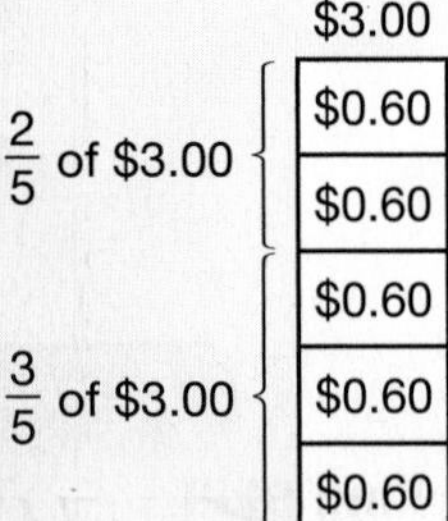

First we divide \$3.00 into five equal groups. Then we find the amount of money in three of the five groups. We divide \$3.00 by 5 to find the amount of money in each group.

$$5\overline{)\$3.00} = \$0.60 \text{ in each group}$$

Now we multiply \$0.60 by 3 to find the amount of money in three groups.

$$\begin{array}{r} \$0.60 \\ \times \quad 3 \\ \hline \$1.80 \end{array}$$

We find that $\frac{3}{5}$ of \$3.00 is **\$1.80.**

Example 4

What number is $\frac{3}{4}$ of 100?

Solution

We divide 100 into four equal groups. Since $100 \div 4$ is 25, there are 25 in each group. We will find the total of three of the parts.

$$3 \times 25 = \mathbf{75}$$

100
$\frac{1}{4}$ of 100: 25
$\frac{3}{4}$ of 100: 25, 25, 25

Example 5

a. What percent of a whole circle is $\frac{1}{5}$ of a circle?

b. What percent of a whole circle is $\frac{3}{5}$ of a circle?

Solution

A whole circle is 100%. We divide 100% into five equal groups.

a. One of the five parts $\left(\frac{1}{5}\right)$ is **20%.**

b. Three of the five parts $\left(\frac{3}{5}\right)$ is $3 \times 20\%$, which equals **60%.**

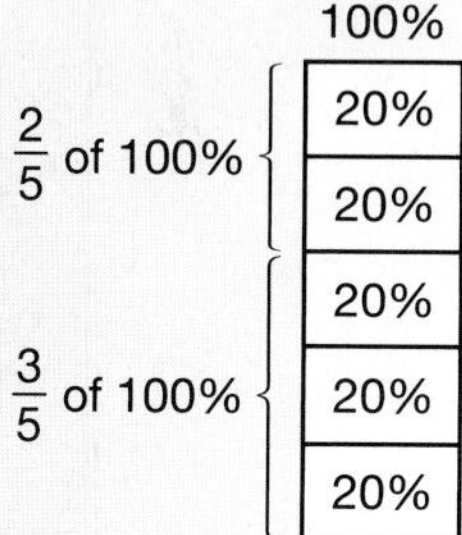

Practice Set

Model Draw a diagram to illustrate each problem.

a. Three fourths of the 12 musicians could play the piano. How many of the musicians could play the piano?

b. How much money is $\frac{2}{3}$ of $4.50?

c. What number is $\frac{4}{5}$ of 60?

d. What number is $\frac{3}{10}$ of 80?

e. Five sixths of 24 is what number?

f. Giovanni answered $\frac{9}{10}$ of the questions correctly. What percent of the questions did Giovanni answer correctly?

Written Practice

Strengthening Concepts

1. (12) When the sum of 15 and 12 is subtracted from the product of 15 and 12, what is the difference?

2. (6) There were 13 original states. There are now 50 states. What fraction of the states are the original states?

*** 3.** (11, 15) A marathon race is 26 miles plus 385 yards long. A mile is 1760 yards. Altogether, how many yards long is a marathon? (First use a multiplication pattern to find the number of yards in 26 miles. Then use an addition pattern to include the 385 yards.)

*** 4.** (22) **Model** If $\frac{2}{3}$ of the 12 apples were eaten, how many were eaten? Draw a diagram to illustrate the problem.

*** 5.** (22) **Model** What number is $\frac{3}{4}$ of 16? Draw a diagram to illustrate the problem.

*** 6.** (22) **Model** How much money is $\frac{3}{10}$ of $3.50? Draw a diagram to illustrate the problem.

*** 7.** (14) As Shannon rode her bike out of the low desert, the elevation changed from −100 ft to 600 ft. What was the total elevation change for her ride?

Find each unknown number. Check your work.

8. (3) $w - 15 = 8$

9. (4) $\frac{w}{15} = 345$

10. (1) 36¢ + $4.78 + $34.09

11. (2) $12.45 ÷ 3

12. (2) $35\overline{)1000}$

13. (5) $\frac{7 + 9 + 14}{3}$

*** 14.** (16) **Estimate** Shannon bought three dozen party favors for $1.24 each. To estimate the total cost she thought of 36 as 9 × 4, and she thought of $1.24 as $1.25. Then she multiplied the three numbers.

a. What was Shannon's estimate of the cost?

b. To find the cost quickly, which two numbers should she multiply first?

15. (12) Which digit in 375,426,198,000 is in the ten-millions place?

*** 16.** (20) Find the greatest common factor of 12 and 15.

*** 17.** (19) List the whole numbers that are factors of 30.

*** 18.** (21) The number 100 is divisible by which of these numbers: 2, 3, 5, 9, 10?

*** 19.** (22) **Model** Jeb answered $\frac{4}{5}$ of the questions correctly. What percent of the questions did Jeb answer correctly? Draw a diagram to illustrate the problem.

20. (9) Compare: $\frac{1}{3} \bigcirc \frac{1}{2}$

*** 21.** (19) **Classify** Which of these numbers is not a prime number?

A 19 **B** 29 **C** 39

22. (5, 14) $(3 + 3) - (3 \times 3)$

23. (18) **Generalize** Find the number halfway between 27 and 43.

Math Language

Perimeter is the distance around a closed, flat shape.

24. (8) What is the perimeter of the rectangle below?

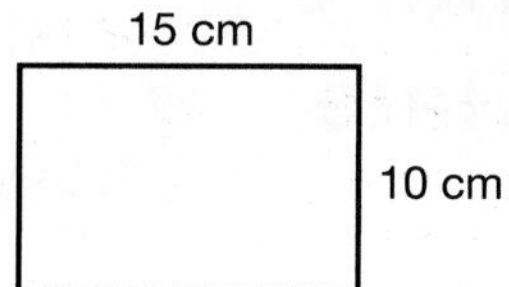

25. (7) Use an inch ruler to find the length of the line segment below.

* **26.** (Inv. 2) *Analyze* Corn bread and wheat bread were baked in pans of equal size. The corn bread was cut into six equal slices. The wheat bread was cut into five equal slices. Which was larger, a slice of corn bread or a slice of wheat bread?

* **27.** (Inv. 2) Compare these fractions. Draw and shade rectangles to illustrate the comparison.

$$\frac{2}{4} \bigcirc \frac{3}{5}$$

* **28.** (22) *Model* A quarter of a year is $\frac{1}{4}$ of a year. There are 12 months in a year. How many months are in a quarter of a year? Draw a diagram to illustrate the problem.

29. (6) A "bit" is one eighth of a dollar.

a. How many bits are in a dollar?

b. How many bits are in a quarter of a dollar?

* **30.** (1) *Represent* The letters *c, p,* and *t* represent three different numbers. When *p* is subtracted from *c,* the answer is *t.*

$$c - p = t$$

Use these letters to write another subtraction equation and two addition equations. *Hint:* To be sure you arranged the letters in the correct order, choose numbers for *c, p,* and *t* that make $c - p = t$ true. Then try those numbers with these letters for your three equations.

LESSON

23

• Ratio
• Rate

Power Up

Building Power

facts Power Up D

mental math

a. Number Sense: 5×62

b. Number Sense: 5×36

c. Number Sense: $87 + 9$

d. Number Sense: $1200 + 350$

e. Calculation: \$20.00 − \$15.50

f. Fractional Parts: $\frac{1}{2}$ of 84

g. Measurement: How many millimeters are in 3 meters?

h. Calculation: 10×3, $+ 2$, $\div 4$, $+ 1$, $\div 3$, $\times 4$, $\div 6$

problem solving How many different bracelets can be made from 7 white beads and 2 gray ones?

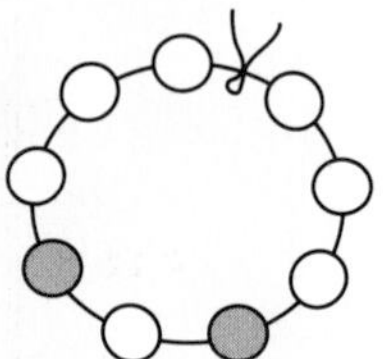

New Concepts

Increasing Knowledge

ratio

A **ratio** is a way to describe a relationship between numbers. If there are 13 boys and 15 girls in a classroom, then the ratio of boys to girls is 13 to 15. Ratios can be written in several forms. Each of these forms is a way to write the boy-girl ratio:

Math Language
13 and 15 are the terms of the ratio.

$$13 \text{ to } 15 \qquad 13{:}15 \qquad \frac{13}{15}$$

Each of these forms is read the same: "Thirteen to fifteen."

In this lesson we will focus on the fraction form of a ratio. When writing a ratio in fraction form, we keep the following points in mind:

1. We write the terms of the ratio in the order we are asked to give them.

2. We reduce ratios in the same manner as we reduce fractions.

3. We leave ratios in fraction form. We do not write ratios as mixed numbers.

Example 1

A team lost 3 games and won 7 games. What was the team's win-loss ratio?

Solution

The question asks for the ratio in the order of wins, then losses. The team's win-loss ratio was 7 to 3, which we write as the fraction $\frac{7}{3}$.

$$\frac{\text{number of games won}}{\text{number of games lost}} = \mathbf{\frac{7}{3}}$$

We leave the ratio in fraction form.

Discuss Is the win-loss ratio the same as the loss-win ratio?

Example 2

In a class of 28 students, there are 13 boys. What is the ratio of boys to girls in the class?

Solution

To write the ratio, we need to know the number of girls. If 13 of the 28 students are boys, then 15 of the students are girls. We are asked to write the ratio in "boys to girls" order.

$$\frac{\text{number of boys}}{\text{number of girls}} = \mathbf{\frac{13}{15}}$$

Model Use tiles of two colors to model this ratio.

rate

A **rate** is a ratio of measures. Below are some commonly used rates. Notice that per means "for each" and substitutes for the division sign.

Common Rates

Name	Rate of Measures	Example	Alternate Form
Speed	$\frac{\text{distance}}{\text{time}}$	$\frac{\text{55 miles}}{\text{1 hour}}$	55 miles per hour
Mileage	$\frac{\text{distance}}{\text{fuel used}}$	$\frac{\text{28 miles}}{\text{1 gallon}}$	28 miles per gallon
Unit price	$\frac{\text{price}}{\text{quantity}}$	$\frac{\$2.89}{\text{1 pound}}$	$2.89 per pound

Rate problems are a type of equal groups problem. The rate is the number in each group. The problems often involve three numbers. One number is the rate, and the other two numbers are about the measures that form the rate. The three numbers are related by multiplication or division as we show below.

Pattern: $\text{distance} = \frac{\text{distance}}{\text{time}} \times \text{time}$

Example: 165 miles = 55 miles per hour × 3 hours

In a rate problem one of the numbers is unknown. We find an unknown product by multiplying, and we find an unknown factor by dividing the product by the known factor.

Example 3

On a bike trip Jeremy rode 60 miles in 4 hours. What was his average speed in miles per hour?

Solution

We are given the distance and time. We are asked for the speed, which is distance divided by time.

$$\frac{\text{distance}}{\text{time}} \quad \frac{60 \text{ miles}}{4 \text{ hour}} = \textbf{15 miles per hour}$$

Example 4

Mr. Moscal's car averages 32 miles per gallon on the highway. Predict about how far he can expect to travel on a road trip using 10 gallons of fuel.

Solution

This is a problem about gas mileage. Notice the similar pattern.

$$\text{distance} = \frac{\text{distance}}{\text{fuel used}} \times \text{fuel used}$$

We are given the rate and the fuel used. We are asked for the distance, which is the product.

$$\begin{aligned} \text{distance} &= 32 \text{ miles per gallon} \times 10 \text{ gallons} \\ &= 320 \text{ miles} \end{aligned}$$

Mr. Moscal can expect to travel about **320 miles** on 10 gallons of fuel.

Making a table can help us solve some rate problems. Here is the beginning of a table for example 4.

Distance Traveled at 32 Miles per Gallon

Fuel Used (gallon)	1	2	3	4	5
Distance (miles)	32	64	96	128	160

Practice Set

a. What is the ratio of dogs to cats in a neighborhood that has 19 cats and 12 dogs?

b. Analyze What is the girl-boy ratio in a class of 30 students with 17 boys?

c. If the ratio of cars to trucks in the parking lot is 7 to 2, what is the ratio of trucks to cars in the parking lot?

d. How long will it take a trucker to drive 400 miles at 50 miles per hour?

e. If a four-quart container of milk costs $2.48, what is the cost per quart?

Written Practice *Strengthening Concepts*

1. (7) How many millimeters long is a ruler that is 30 cm long?

*** 2.** (22) **Model** Dan has finished $\frac{2}{3}$ of the 30 problems on an assignment during class. How many problems did Dan finish during class? Draw a diagram to illustrate the problem.

3. (7) Diego walked the length of a football field in 100 large paces. About how long was the football field?

*** 4.** (23) On the open highway the car traveled 245 miles on 7 gallons of gas. What was the car's gas mileage for the trip in miles per gallon?

*** 5.** (22) **Model** What number is $\frac{3}{5}$ of 25? Draw a diagram to illustrate the problem.

*** 6.** (22) **Model** How much money is $\frac{7}{10}$ of $36.00? Draw a diagram to illustrate the problem.

Use your fraction manipulatives to help answer problems **7–9.**

*** 7.** (Inv. 2) What is the sum of $\frac{3}{8}$ and $\frac{4}{8}$?

*** 8.** (Inv. 2) The improper fraction $\frac{9}{8}$ equals what mixed number?

*** 9.** (Inv. 2) Two eighths of a circle is what percent of a circle?

10. (2) $3.75 · 16

11. (2) $\frac{\$3.75}{25}$

12. (12) What is the place value of the 6 in 36,174,591?

*** 13.** (22) **Explain** How can you find $\frac{2}{3}$ of a number?

Find each unknown number. Check your work.

14. (4) $\$0.35n = \$35.00$

15. (3) $\$10.20 - m = \$3.46$

16. (17) Compare: $\frac{3}{4}$ ◯ 1

*** 17.** (8) **Analyze** The length of a rectangle is 20 inches. The width of the rectangle is half its length. What is the perimeter of the rectangle?

18. (10) **Generalize** Describe the rule for the following sequence. Then, write the sixth number in the sequence.

2, 4, 8, 16, ...

19. (10) Yesterday it snowed. The meteorologist on the radio said that it was 14° outside. What scale was the meteorologist reading? How do you know?

20. (9) Compare: $12 \div 6 - 2$ ◯ $12 \div (6 - 2)$

*** 21.** (20) What is the greatest common factor (GCF) of 24 and 32?

22. (10) What is the sum of the first seven positive odd numbers?

Model Use your fraction manipulatives to help answer problems **23–25.**

* **23.** (Inv. 2) **a.** How many $\frac{1}{4}$s are in 1?

b. How many $\frac{1}{4}$s are in $\frac{1}{2}$?

* **24.** (Inv. 2) One eighth of a circle is what percent of a circle?

* **25.** (Inv. 2) Write a fraction with a denominator of 8 that is equal to $\frac{1}{2}$.

* **26.** (23) There were 16 members of the 2004–2005 men's national swim team. Five of them competed in the freestyle. What is the ratio of those who competed in the freestyle to those who did not?

* **27.** (19) **Classify** Which prime numbers are greater than 20 but less than 30?

28. (7) Which of the figures below represents a line?

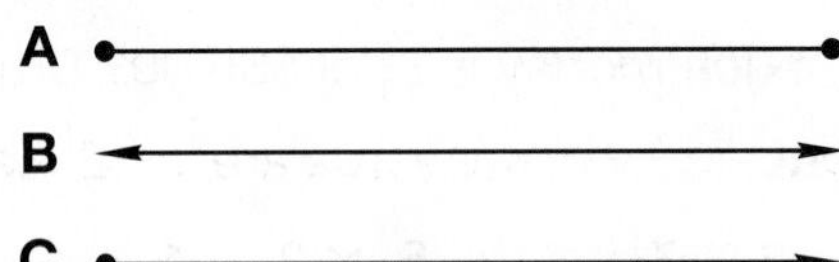

* **29.** (21) **Classify** Which of these numbers is divisible by both 2 and 5?

A 252 **B** 525 **C** 250

* **30.** (23) If a team lost 9 games and won 5 games, then what is the team's win-loss ratio?

Early Finishers
Real-World Application

Mrs. Akiba bought 3 large bags of veggie sticks for her students. Each bag contains 125 veggie sticks. One sixth of Mrs. Akiba's students did not eat any veggie sticks. The remaining students split the veggie sticks evenly and ate them. How many veggie sticks did each of the remaining students eat?

LESSON 24

Adding and Subtracting Fractions That Have Common Denominators

Power Up

Building Power

facts

Power Up C

mental math

a. **Number Sense:** 6×24

b. **Number Sense:** 4×75

c. **Number Sense:** $47 + 39$

d. **Number Sense:** $1500 - 250$

e. **Calculation:** \$20.00 − \$14.50

f. **Fractional Parts:** $\frac{1}{2}$ of 68

g. **Measurement:** How many yards are in 12 feet?

h. **Calculation:** 6×7, $- 2$, $\div 5$, $\times 2$, $- 1$, $\div 3$

problem solving

Tom followed the directions on the treasure map. Starting at the big tree, he walked five paces north, turned right, and walked seven more paces. He turned right again and walked nine paces, turned left, and walked three more paces. Finally, he turned left, and took four paces. In which direction was Tom facing, and how many paces was he from the big tree?

New Concept

Increasing Knowledge

Using our fraction manipulatives, we see that when we add $\frac{2}{8}$ to $\frac{3}{8}$ the sum is $\frac{5}{8}$.

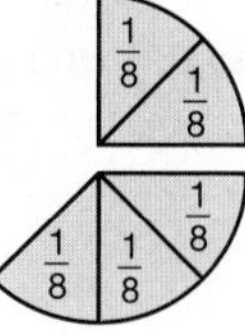

$$\frac{3}{8} + \frac{2}{8} = \frac{5}{8}$$

Three eighths plus two eighths equals five eighths.

Math Language
The **denominator** tells you into how many parts the whole is divided.

Likewise, if we subtract $\frac{2}{8}$ from $\frac{5}{8}$, then $\frac{3}{8}$ are left.

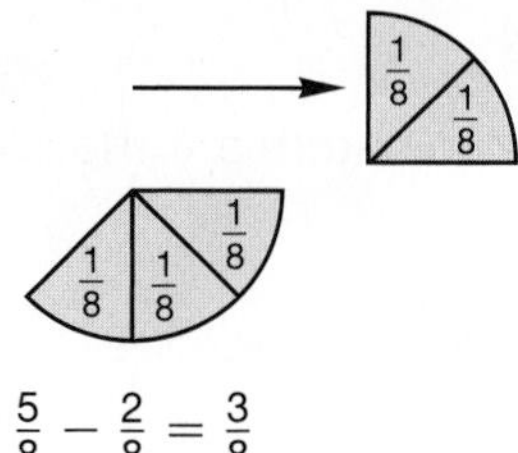

$$\frac{5}{8} - \frac{2}{8} = \frac{3}{8}$$

Five eighths minus two eighths equals three eighths.

Notice that we add the numerators when we add fractions that have the same denominator, and we subtract the numerators when we subtract fractions that have the same denominator. The denominators of the fractions do not change when we add or subtract fractions that have the same denominator.

Thinking Skill

Justify

Why does the denominator stay the same when we add fractions with the same denominator?

Example 1

Add: $\mathbf{\frac{1}{4} + \frac{1}{4} + \frac{1}{4}}$

Solution

The denominators are the same. We add the numerators.

$$\frac{1}{4} + \frac{1}{4} + \frac{1}{4} = \mathbf{\frac{3}{4}}$$

Example 2

Add: $\mathbf{\frac{1}{2} + \frac{1}{2}}$

Solution

One half plus one half is two halves, which is one whole.

$$\frac{1}{2} + \frac{1}{2} = \frac{2}{2} = \mathbf{1}$$

Example 3

Add: $\mathbf{\frac{3}{4} + \frac{3}{4} + \frac{3}{4} + \frac{3}{4}}$

Solution

Thinking Skill

Discuss

Why can we write $\frac{12}{4}$ as 3?

The denominators are the same. We add the numerators.

$$\frac{3}{4} + \frac{3}{4} + \frac{3}{4} + \frac{3}{4} = \frac{12}{4} = \mathbf{3}$$

Example 4

Subtract: $\mathbf{\frac{7}{8} - \frac{2}{8}}$

Solution

The denominators are the same. We subtract the numerators.

$$\frac{7}{8} - \frac{2}{8} = \mathbf{\frac{5}{8}}$$

Example 5

Subtract: $\frac{1}{2} - \frac{1}{2}$

Solution

If we start with $\frac{1}{2}$ and subtract $\frac{1}{2}$, then what is left is zero.

$$\frac{1}{2} - \frac{1}{2} = \frac{0}{2} = \mathbf{0}$$

Practice Set

Find each sum or difference:

a. $\frac{3}{8} + \frac{4}{8}$

b. $\frac{3}{4} + \frac{1}{4}$

c. $\frac{1}{8} + \frac{1}{8} + \frac{1}{8}$

d. $\frac{4}{8} - \frac{1}{8}$

e. $\frac{3}{4} - \frac{2}{4}$

f. $\frac{1}{4} - \frac{1}{4}$

g. Connect Use words to write the subtraction problem in exercise **d.**

Written Practice *Strengthening Concepts*

1. (11, 15) Analyze Martin worked in the yard for five hours and was paid $6.00 per hour. Then he was paid $5.00 for washing the car. Altogether, how much money did Martin earn? What pattern did you use to find Martin's yard-work earnings? What pattern did you use to find his total earnings?

*** 2.** (22) Model Juan used $\frac{3}{4}$ of a dozen eggs to make omelets for his family. How many eggs did Juan use? Draw a diagram to illustrate the problem.

*** 3.** (22) Explain One mile is one thousand, seven hundred sixty yards. How many yards is $\frac{1}{8}$ of a mile? Explain how you found your answer.

Model Use your fraction manipulatives to help with exercises **4–8.** Then choose one of the exercises to write a word problem that is solved by the exercise.

*** 4.** (24) $\frac{1}{4} + \frac{2}{4}$

*** 5.** (24) $\frac{7}{8} - \frac{4}{8}$

*** 6.** (24) $\frac{1}{2} + \frac{1}{2}$

*** 7.** (24) $\frac{1}{2} - \frac{1}{2}$

*** 8.** (Inv. 2) What percent of a circle is $\frac{1}{2}$ of a circle plus $\frac{1}{4}$ of a circle?

*** 9.** (23) In the classroom library there were 23 nonfiction books and 41 fiction books. What was the ratio of fiction to nonfiction books in the library?

10. (18) Explain How can you find the number halfway between 123 and 321?

11. (8) Mr. Chen wanted to fence in a square corral for his horse. Each side needed to be 25 feet long. How many feet of fence did Mr. Chen need for the corral?

*** 12.** (19) *Classify* Which of these numbers is not a prime number?

A 21 **B** 31 **C** 41

13. (2) $9\overline{)1000}$

14. (2) 22,422 ÷ 32

15. (2) $350.00 ÷ 100

16. (17) Compare: $\frac{1}{2} \bigcirc \frac{1}{4}$

*** 17.** (16) *Conclude* Mr. Johnson rented a moving van and will drive from Seattle, Washington, to San Francisco, California. On the way to San Francisco he will go through Portland, Oregon. The distance from Seattle to Portland is 172 miles, and the distance from Portland to San Francisco is 636 miles. The van rental company charges extra if a van is driven more than 900 miles. If Mr. Johnson stays with his planned route, will he be charged extra for the van? To solve this problem, do you need an exact answer or an estimate? Explain your thinking.

*** 18.** (23) If Mr. Johnson drives 172 miles from Seattle to Portland in 4 hours, then the average speed of the rental van for that portion of the trip is how many miles per hour?

19. (10) *Connect* What temperature is shown on the thermometer at right?

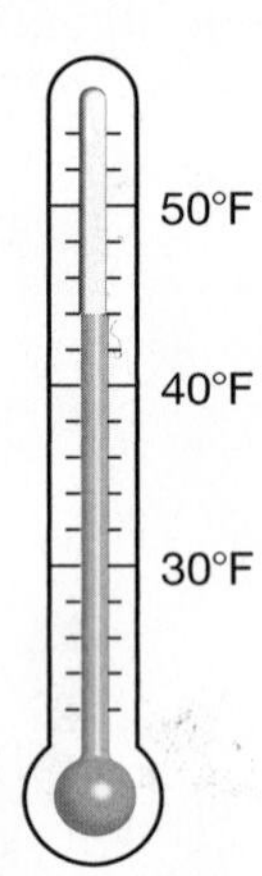

20. (16) Round 32,987,145 to the nearest million.

Math Language
The abbreviation GCF stands for *greatest common factor.*

*** 21.** (20) What is the GCF of 21 and 28?

*** 22.** (21) *Classify* Which of these numbers is divisible by 9?

A 123 **B** 234 **C** 345

23. (Inv. 2) Write a fraction equal to 1 that has 4 as the denominator.

Find each unknown number. Check your work.

24. (4) $\frac{w}{8} = 20$

25. (4) $7x = 84$

26. (3) $376 + w = 481$

27. (3) $m - 286 = 592$

Refer to the bar graph shown below to answer problems **28–30.**

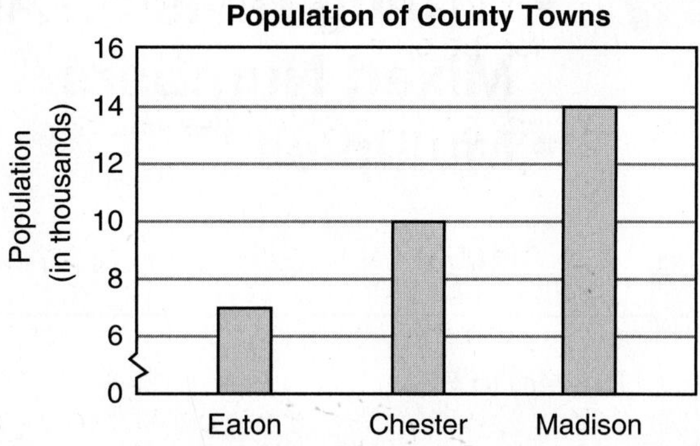

28. (16) *Estimate* Which town has about twice the population of Eaton?

29. (16) *Estimate* About how many more people live in Madison than in Chester?

30. (Inv. 1) *Represent* Copy this graph on your paper, and add a fourth town to your graph: Wilson, population 11,000.

Early Finishers
Real-World Application

Gina's dance team is performing at a local charity event on Saturday. Some members of the team will ride in vans to the event, while others will ride in cars. Seven-eighths of the 112 members will travel to the event in vans.

a. How many members will be traveling in vans?

b. If each van can carry 11 passengers, how many vans will they need?

LESSON 25

• Writing Division Answers as Mixed Numbers
• Multiples

Power Up — *Building Power*

facts Power Up F

mental math

a. **Number Sense:** 6×43

b. **Number Sense:** 3×75

c. **Number Sense:** $57 + 29$

d. **Calculation:** $2650 - 150$

e. **Calculation:** \$10.00 − \$6.25

f. **Fractional Parts:** $\frac{1}{2}$ of 30

g. **Measurement:** Which is greater, 5 millimeters or one centimeter?

h. **Calculation:** 10×2, $+ 1$, $\div 3$, $+ 2$, $\div 3$, $\times 4$, $\div 3$

problem solving The digits 1 through 9 are used in this subtraction problem. Copy the problem and fill in the missing digits.

```
  ___
− 452
 3__
```

New Concepts — *Increasing Knowledge*

writing division answers as mixed numbers

We have been writing division answers with remainders. However, not all questions involving division can be appropriately answered using remainders. Some word problems have answers that are mixed numbers, as we will see in the following example.

Example 1

A 15-inch length of ribbon was cut into four equal lengths. How long was each piece of ribbon?

Solution

We divide 15 by 4 and write the answer as a mixed number.

```
   3¾  ←┐
4)15    │
  12    │
   3  ──┘
```

Notice that the remainder is the numerator of the fraction, and the divisor is the denominator of the fraction. We find that the length of each piece of ribbon is $3\frac{3}{4}$ **inches.**

Example 2

A whole circle is 100% of a circle. One third of a circle is what percent of a circle?

Solution

If we divide 100% by 3, we will find the percent equivalent of $\frac{1}{3}$.

$$\begin{array}{r} 33\frac{1}{3}\% \\ 3\overline{)100\%} \\ \underline{9} \\ 10 \\ \underline{9} \\ 1 \end{array}$$

Connect One third of a circle is $\mathbf{33\frac{1}{3}\%}$ of a circle. Notice that our answer matches our fraction manipulative piece for $\frac{1}{3}$.

Example 3

Reading Math

$\frac{25}{6}$ is read "25 divided by 6".

Write $\frac{25}{6}$ as a mixed number.

Solution

The fraction bar in $\frac{25}{6}$ serves as a division symbol. We divide 25 by 6 and write the remainder as the numerator of the fraction.

$$\begin{array}{r} 4\frac{1}{6} \\ 6\overline{)25} \\ \underline{24} \\ 1 \end{array}$$

We find that the improper fraction $\frac{25}{6}$ equals the mixed number $\mathbf{4\frac{1}{6}}$.

multiples

We find **multiples** of a number by multiplying the number by 1, 2, 3, 4, 5, 6, and so on.

The first six multiples of 2 are 2, 4, 6, 8, 10, and 12.

The first six multiples of 3 are 3, 6, 9, 12, 15, and 18.

The first six multiples of 4 are 4, 8, 12, 16, 20, and 24.

The first six multiples of 5 are 5, 10, 15, 20, 25, and 30.

Example 4

What are the first four multiples of 8?

Solution

Multiplying 8 by 1, 2, 3, and 4 gives the first four multiples: **8, 16, 24,** and **32.**

Example 5

What number is the eighth multiple of 7?

Solution

The eighth multiple of 7 is 8×7, which is **56.**

Practice Set

a. A 28-inch long ribbon was cut into eight equal lengths. How long was each piece of ribbon?

b. A whole circle is 100% of a circle. What percent of a circle is $\frac{1}{7}$ of a circle?

c. Divide 467 by 10 and write the quotient as a mixed number.

d. What are the first four multiples of 12?

e. What are the first six multiples of 8?

f. Classify What number is both the third multiple of 8 and the second multiple of 12?

Write each of these improper fractions as a mixed number:

g. $\frac{35}{6}$ **h.** $\frac{49}{10}$ **i.** $\frac{65}{12}$

Written Practice

Strengthening Concepts

* **1.** (Inv. 2) What is the difference between the sum of $\frac{1}{2}$ and $\frac{1}{2}$ and the sum of $\frac{1}{3}$ and $\frac{1}{3}$?

2. (18) In three tries Carlos punted the football 35 yards, 30 yards, and 37 yards. How can Carlos find the average distance of his punts?

3. (12) Earth's average distance from the Sun is one hundred forty-nine million, six hundred thousand kilometers. Use digits to write that distance.

* **4.** (8, 24) Connect What is the perimeter of the rectangle?

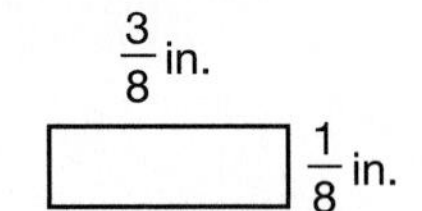

* **5.** (25) Formulate A 30-inch length of ribbon was cut into 4 equal lengths. How long was each piece of ribbon? Write an equation and solve the problem.

6. (Inv. 2) Two thirds of the class finished the test on time. What fraction of the class did not finish the test on time?

* **7.** (22) Compare: $\frac{1}{2}$ of 12 ◯ $\frac{1}{3}$ of 12

* **8.** (Inv. 2) Evaluate What fraction is half of the fraction that is half of $\frac{1}{2}$?

* **9.** (25) A whole circle is 100% of a circle. What percent of a circle is $\frac{1}{9}$ of a circle?

Math Language

What operation do the words *"are in"* tell us to perform?

* **10.** (Inv. 2) **a.** How many $\frac{1}{6}$s are in 1?

b. How many $\frac{1}{6}$s are in $\frac{1}{2}$?

* **11.** (25) What fraction of a circle is $33\frac{1}{3}\%$ of a circle?

* **12.** (25) Divide 365 by 7 and write the answer as a mixed number.

* **13.** (24) $\frac{2}{3} + \frac{2}{3} + \frac{2}{3}$

* **14.** (24) $\frac{6}{6} - \frac{5}{6}$

15. (5) $30 \times 40 \div 60$

* **16.** (24) $\frac{5}{12} - \frac{5}{12}$

* **17.** (23) A team won seven of the twenty games played and lost the rest. What was the team's win-loss ratio?

18. (15) **Formulate** Cheryl bought 10 pens for 25¢ each. How much did she pay for all 10 pens? Write an equation and solve the problem.

19. (20) What is the greatest common factor (GCF) of 24 and 30?

20. (22) What number is $\frac{1}{100}$ of 100?

Find each unknown number. Check your work.

21. (Inv. 2) $\frac{5}{8} + m = 1$

22. (4) $\frac{144}{n} = 12$

23. (16) **Estimate** What is the sum of 3142, 6328, and 4743 to the nearest thousand?

* **24.** (22) **Model** Two thirds of the 60 students liked peaches. How many of the students liked peaches? Draw a diagram that illustrates the problem.

25. (17) **Estimate** Estimate the length in inches of the line segment below. Then use an inch ruler to find the length of the line segment to the nearest sixteenth of an inch.

26. (Inv. 2) **Represent** To divide a circle into thirds, Jan imagined the circle was the face of a clock. Describe how Jan could draw segments to divide the circle into thirds.

* **27.** (25) Write $\frac{15}{4}$ as a mixed number.

28. (Inv. 2) **Model** Draw and shade rectangles to illustrate and complete this comparison:

$$\frac{3}{4} \bigcirc \frac{4}{5}$$

* **29.** (25) What are the first four multiples of 25?

30. (21) **Classify** Which of these numbers is divisible by both 9 and 10? How do you know?

A 910 **B** 8910 **C** 78,910

LESSON 26

• Using Manipulatives to Reduce Fractions
• Adding and Subtracting Mixed Numbers

Power Up — *Building Power*

facts Power Up C

mental math

a. **Number Sense:** 7×34

b. **Number Sense:** 4×56

c. **Number Sense:** $74 + 19$

d. **Calculation:** $475 + 125$

e. **Money:** $5.00 − $1.75

f. **Fractional Parts:** $\frac{1}{2}$ of 32

g. **Statistics:** Find the average of the following: 20, 25, 30

h. **Calculation:** 7×5, $+ 1$, $\div 6$, $\times 3$, $\div 2$, $+ 1$, $\div 5$

problem solving James was thinking of a prime number between 75 and 100 that did **not** have 9 as one of its digits. Of what number was he thinking?

New Concepts — *Increasing Knowledge*

using manipulatives to reduce fractions

You can use fraction manipulatives to model these fractions:

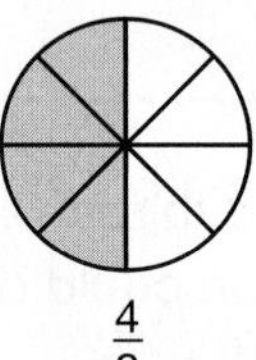

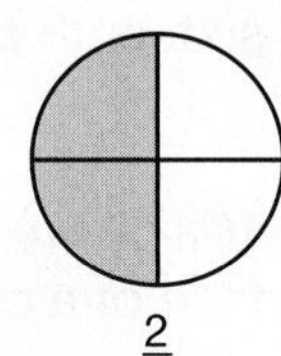

$\frac{4}{8}$ $\frac{3}{6}$ $\frac{2}{4}$ $\frac{1}{2}$

We see that each picture illustrates half of a circle. The model that uses the fewest pieces is $\frac{1}{2}$. We say that each of the other fractions **reduces** to $\frac{1}{2}$.

Thinking Skill
Verify
Why doesn't $\frac{2}{3}$ reduce to $\frac{1}{2}$?

Model We can use our fraction manipulatives to reduce a given fraction by making an equivalent model that uses fewer pieces.

Example 1

Use your fraction manipulatives to reduce $\frac{2}{6}$.

Solution

First we use our manipulatives to form $\frac{2}{6}$.

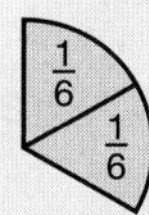

Then we search for a fraction piece equivalent to the $\frac{2}{6}$ model. We find $\frac{1}{3}$.

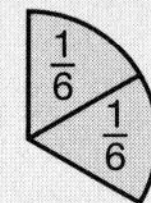

The models illustrate that $\frac{2}{6}$ reduces to $\frac{1}{3}$.

adding and subtracting mixed numbers

When adding mixed numbers, we first add the fraction parts, and then we add the whole-number parts. Likewise, when subtracting mixed numbers, we first subtract the fraction parts, and then we subtract the whole-number parts.

Example 2

Thinking Skill

Justify

Explain how to change $\frac{1}{3}$ to $33\frac{1}{3}\%$.

Two thirds of a circle is what percent of a circle?

Solution

Model Use your fraction manipulatives to represent $\frac{1}{3} + \frac{1}{3}$.

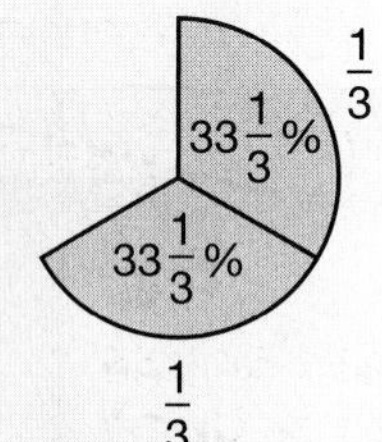

One third equals $33\frac{1}{3}\%$. So two thirds can be found by adding $33\frac{1}{3}\%$ and $33\frac{1}{3}\%$.

$$\begin{array}{r} 33\frac{1}{3}\% \\ +\ 33\frac{1}{3}\% \\ \hline \mathbf{66\frac{2}{3}\%} \end{array}$$

Example 3

Two sixths of a circle is what percent of a circle?

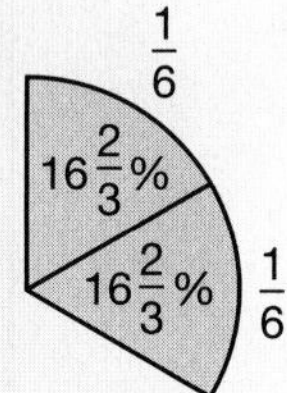

Solution

We add $16\frac{2}{3}\%$ and $16\frac{2}{3}\%$.

$$\begin{array}{r} 16\frac{2}{3}\% \\ +\ 16\frac{2}{3}\% \\ \hline 32\frac{4}{3}\% \end{array}$$

Math Language
Recall that an **improper fraction** is a fraction with a numerator equal to or greater than the denominator.

We notice that the fraction part of the answer, $\frac{4}{3}$, is an improper fraction that equals $1\frac{1}{3}$.

So $32\frac{4}{3}\%$ equals $32\% + 1\frac{1}{3}\%$, which is $\mathbf{33\frac{1}{3}\%}$. This makes sense because $\frac{2}{6}$ reduces to $\frac{1}{3}$, which is the same as $33\frac{1}{3}\%$.

Example 4

Rory lives $2\frac{3}{4}$ miles from school. He rode his bike from home to school and back to home. How far did Rory ride?

Solution

This problem has an addition pattern.

$$\begin{array}{r} 2\frac{3}{4}\text{ mi} \\ +\ 2\frac{3}{4}\text{ mi} \\ \hline 4\frac{6}{4}\text{ mi} \end{array}$$

The fraction part of the answer reduces to $1\frac{1}{2}$ $\left(\frac{6}{4} = 1\frac{2}{4} = 1\frac{1}{2}\right)$. So we add $1\frac{1}{2}$ to the whole-number part of the answer and find that Rory rode his bike $\mathbf{5\frac{1}{2}}$ **miles.**

Example 5

Subtract: $\mathbf{5\frac{3}{8} - 1\frac{1}{8}}$

Solution

We subtract $\frac{1}{8}$ from $\frac{3}{8}$, and we subtract 1 from 5. The resulting difference is $4\frac{2}{8}$.

$$5\frac{3}{8} - 1\frac{1}{8} = 4\frac{2}{8}$$

We reduce the fraction $\frac{2}{8}$ to $\frac{1}{4}$ and write the answer as $\mathbf{4\frac{1}{4}}$.

Practice Set

Model Use your fraction manipulatives to reduce these fractions:

a. $\frac{2}{8}$ **b.** $\frac{6}{8}$

Add. Reduce the answer when possible.

c. $12\frac{1}{2}\% + 12\frac{1}{2}\%$ **d.** $16\frac{2}{3}\% + 66\frac{2}{3}\%$

e. $3\frac{3}{4} + 2\frac{3}{4}$ **f.** $1\frac{1}{8} + 2\frac{7}{8}$

g. $3 + 2\frac{2}{3}$

h. $\frac{3}{4} + 4$

i. Use words to write the addition problem in exercise **f.**

Written Practice *Strengthening Concepts*

* **1.** (26) Maya rode her bike to the park and back. If the trip was $3\frac{3}{4}$ miles each way, how far did she ride in all?

2. (15) The young elephant was 36 months old. How many years old was the elephant?

3. (6, 15) **Justify** Mrs. Ling bought $2\frac{1}{2}$ dozen balloons for the party. Is this enough balloons for 30 children to each get one balloon? Explain your thinking.

4. (15) There are 100 centimeters in a meter. There are 1000 meters in a kilometer. How many centimeters are in a kilometer?

* **5.** (8, 24) What is the perimeter of the equilateral triangle shown?

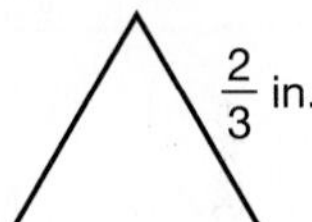

* **6.** (Inv. 2) Compare: $\frac{1}{2}$ plus $\frac{1}{2}$ ◯ $\frac{1}{2}$ of $\frac{1}{2}$

* **7.** (26) $5\frac{7}{8} + 7\frac{5}{8}$

* **8.** (26) One eighth of a circle is $12\frac{1}{2}$% of a circle. What percent of a circle is $\frac{3}{8}$ of a circle?

9. (Inv. 2) Write a fraction equal to 1 that has a denominator of 12.

10. (20) What is the greatest common factor of 15 and 25?

11. (10) **Generalize** Describe the rule of the following sequence. Then find the *seventh* term.

8, 16, 24, 32, 40, ...

* **12.** (25) Write $\frac{14}{5}$ as a mixed number.

* **13.** (26) Add and simplify: $\frac{2}{5} + \frac{4}{5}$

Find the unknown number. Remember to check your work.

14. (Inv. 2) $\frac{2}{3} + n = 1$

15. (20) **Classify** What is the greatest factor of both 12 and 18?

16. (Inv. 2) $1 - \frac{3}{4}$

* **17.** (26) $3\frac{3}{4} + 3$

* **18.** (26) $2\frac{1}{2} - 2\frac{1}{2}$

19. (21) **Classify** Which of the numbers below is divisible by both 2 and 3?

A 4671 **B** 3858 **C** 6494

20. List the prime numbers between 30 and 40.
(19)

21. **Estimate** Find the difference of 5063 and 3987 to the nearest thousand.
(16)

* **22.** **Model** Use your fraction manipulatives to reduce $\frac{6}{8}$.
(26)

23. At $2.39 per pound, what is the cost of four pounds of grapes?
(23)

* **24.** **Model** How much money is $\frac{3}{5}$ of $30? Draw a diagram to illustrate the problem.
(22)

25. **Connect** **a.** How many millimeters long is the line segment below?
(7, 17)

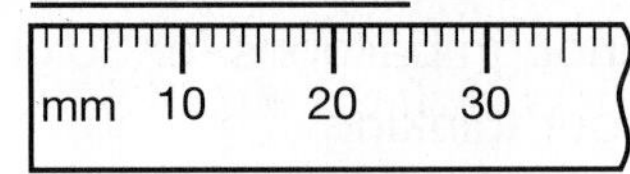

b. Use an inch ruler to find the length of the segment to the nearest sixteenth of an inch.

26. Arrange these numbers in order from least to greatest:
(17)

$$\frac{1}{2}, 0, -1, 1$$

Adriana began measuring rainfall when she moved to her new home. The bar graph below shows the annual rainfall near Adriana's home during her first three years there. Refer to this graph to answer problems **27–30.**

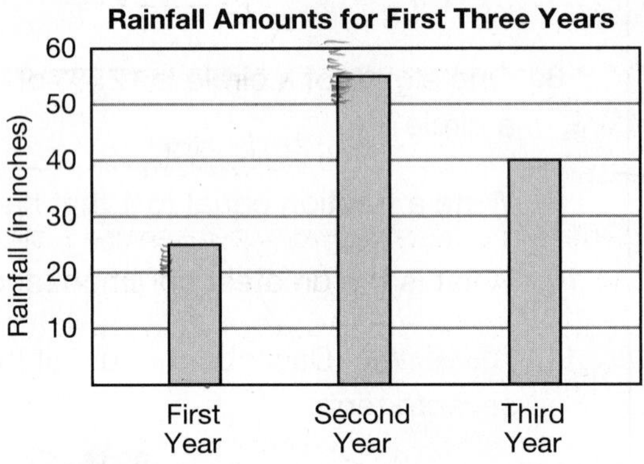

27. About how many more inches of rain fell during the second year than during the first year?
(16)

28. What was the approximate average annual rainfall during the first three years?
(18)

29. The first year's rainfall was about how many inches below the average annual rainfall of the first three years?
(18)

30. **Formulate** Write a problem with an addition pattern that relates to the graph. Then answer the problem.
(11)

LESSON 27

• Measures of a Circle

Power Up — *Building Power*

facts Power Up E

mental math

a. **Number Sense:** 7 × 52

b. **Number Sense:** 6 × 33

c. **Number Sense:** 63 + 19

d. **Number Sense:** 256 + 50

e. **Money:** \$10.00 − \$7.25

f. **Fractional Parts:** $\frac{1}{2}$ of 86

g. **Geometry:** The perimeter of a square is 16 ft. What is the length of the sides of the square?

h. **Calculation:** 8 × 8, − 1, ÷ 7, × 2, + 2, ÷ 2

problem solving

If + = 25, and − = 5,

then × = ?

New Concept — *Increasing Knowledge*

Thinking Skill

Verify

Why is the diameter of a circle twice the length of the radius?

There are several ways to measure a circle. We can measure the distance around the circle, the distance across the circle, and the distance from the center of the circle to the circle itself. The pictures below identify these measures.

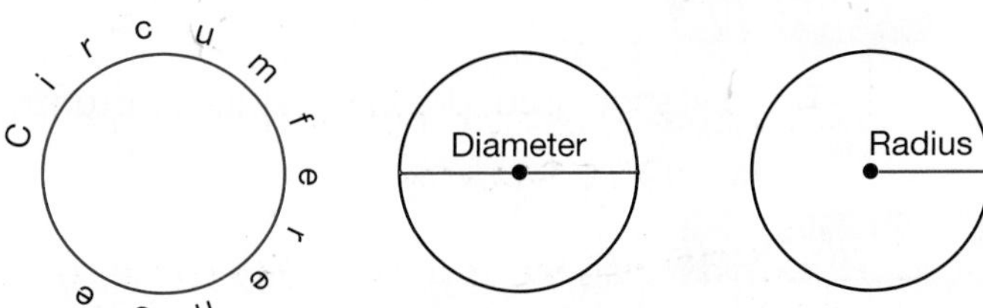

The **circumference** is the distance **around** the circle. This distance is the same as the perimeter of a circle. The **diameter** is the distance **across** a circle through its center. The **radius** is the distance from the center to the circle. The plural of *radius* is **radii.** For any circle, the diameter is twice the length of the radius.

Activity

Using a Compass

Materials needed:

- compass and pencil
- plain paper

A **compass** is a tool for drawing a circle. Here we show two types:

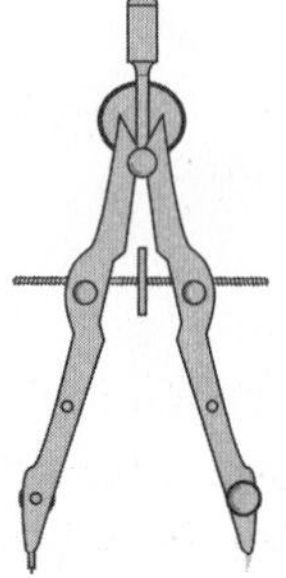

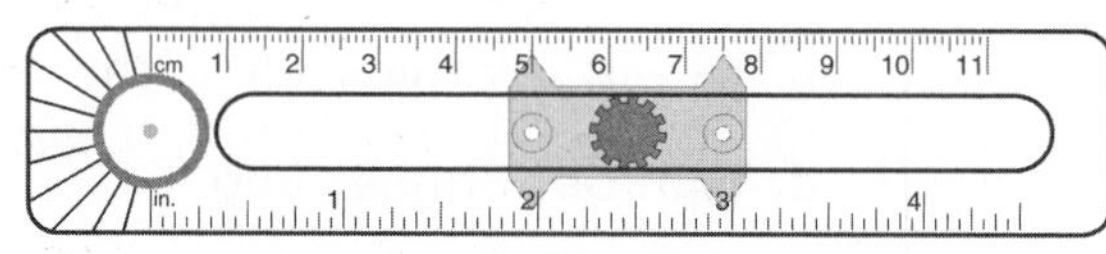

To use a compass, we select a radius and a center point for a circle. Then we rotate the compass about the center point to draw the circle. In this activity you will use a compass and paper to draw circles with given radii.

Thinking Skill

Discuss

Why do we use the length of a radius instead of a diameter to draw a circle?

Represent Draw a circle with each given radius. How can you check that each circle is drawn to the correct size?

a. 2 in. **b.** 3 cm **c.** $1\frac{3}{4}$ in.

Concentric circles are circles with the same center. A bull's-eye target is an example of concentric circles.

d. *Represent* Draw three concentric circles with radii of 4 cm, 5 cm, and 6 cm.

Example 1

What is the name for the perimeter of a circle?

Solution

The distance around a circle is its **circumference.**

Example 2

If the radius of a circle is 4 cm, what is its diameter?

Solution

The diameter of a circle is twice its radius—in this case, **8 cm.**

Practice Set

In problems **a–c,** name the described measure of a circle.

a. The distance across a circle

b. The distance around a circle

c. The distance from the center to the circle

d. **Explain** If the diameter of a circle is 10 in., what is its radius? Describe how you know.

Written Practice

Strengthening Concepts

1. (12) **Analyze** What is the product of the sum of 55 and 45 and the difference of 55 and 45?

*** 2.** (22) Potatoes are three-fourths water. If a sack of potatoes weighs 20 pounds, how many pounds of water are in the potatoes? Draw a diagram to illustrate the problem.

3. (11) **Formulate** There were 306 students in the cafeteria. After some went outside, there were 249 students left in the cafeteria. How many students went outside? Write an equation and solve the problem.

*** 4.** (27) **a.** If the diameter of a circle is 5 in., what is the radius of the circle?

b. What is the relationship of the diameter of a circle to its radius?

5. (21) **Classify** Which of these numbers is divisible by both 2 and 3?

A 122 **B** 123 **C** 132

6. (16) Round 1,234,567 to the nearest ten thousand.

7. (15) **Formulate** If ten pounds of apples costs $12.90, what is the price per pound? Write an equation and solve the problem.

8. (6) What is the denominator of $\frac{23}{24}$?

*** 9.** (22) **Model** What number is $\frac{3}{5}$ of 65? Draw a diagram to illustrate the problem.

*** 10.** (22) **Model** How much money is $\frac{2}{3}$ of $15? Draw a diagram to illustrate the problem.

Model Use your fraction manipulatives to help answer problems **11–18.**

11. (Inv. 2) $\frac{1}{6} + \frac{2}{6} + \frac{3}{6}$

12. (Inv. 2) $\frac{7}{8} - \frac{3}{8}$

13. (Inv. 2) $\frac{6}{6} - \frac{5}{6}$

14. (Inv. 2) $\frac{2}{8} + \frac{5}{8}$

15. (Inv. 2) **a.** How many $\frac{1}{8}$s are in 1?

b. How many $\frac{1}{8}$s are in $\frac{1}{2}$?

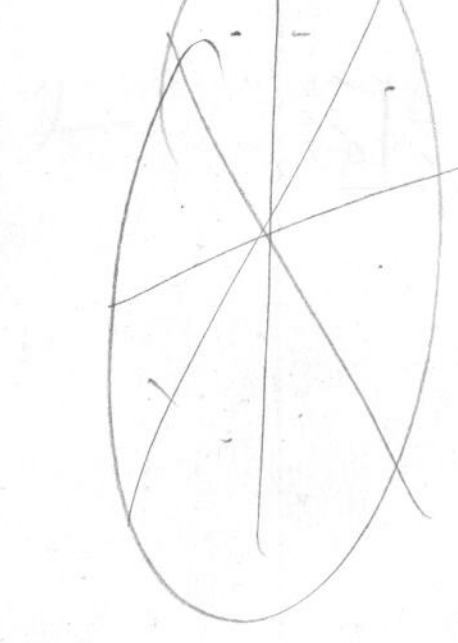

*** 16.** (26) Reduce: $\frac{4}{6}$

17. (Inv. 2) What fraction is half of $\frac{1}{4}$?

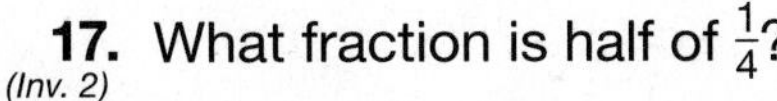

18. (Inv. 2) What fraction of a circle is 50% of a circle?

19. (2) Divide 2100 by 52 and write the answer with a remainder.

20. (8) If a 36-inch-long string is made into the shape of a square, how long will each side be?

* **21.** (25) Convert $\frac{7}{6}$ to a mixed number.

22. (2) $\frac{432}{18}$

23. (5) $(55 + 45) \div (55 - 45)$

24. (21) **Classify** Which of these numbers is divisible by both 2 and 5?

A 502 **B** 205 **C** 250

25. (21) **Justify** Describe a method for determining whether a number is divisible by 9.

26. (19) Which prime number is not an odd number?

* **27.** (27) What is the name for the perimeter of a circle?

28. (23) **Explain** What is the ratio of even numbers to odd numbers in the square below? Explain your thinking.

1	2	3
4	5	6
7	8	9

* **29.** (26) $37\frac{1}{2}\% - 12\frac{1}{2}\%$

* **30.** (26) $33\frac{1}{3}\% + 16\frac{2}{3}\%$

Early Finishers
Real-World Application

While the neighbors are on vacation, Jason is taking care of their dogs Max, Fifi, and Tinker. Max needs $2\frac{1}{4}$ cups of food. Fifi needs $\frac{3}{4}$ cup of food, and Tinker needs $1\frac{3}{4}$ cups. How much food will Jason need to feed all three dogs? Show your work.

LESSON 28

• Angles

Power Up

Building Power

facts Power Up F

mental math

a. **Number Sense:** 8×42

b. **Number Sense:** 3×85

c. **Number Sense:** $36 + 49$

d. **Number Sense:** $1750 - 500$

e. **Money:** $10.00 – $8.25

f. **Fractional Parts:** $\frac{1}{2}$ of 36

g. **Measurement:** Which is greater, 9 inches or one foot?

h. **Calculation:** 8×4, $+ 1$, $\div 3$, $+ 1$, $\times 2$, $+ 1$, $\div 5$

problem solving

Zuna has 2¢ stamps, 3¢ stamps, 10¢ stamps, and 37¢ stamps. She wants to mail a package that requires $1.29 postage. In order to pay exactly the expected postage, what is the smallest number of stamps Zuna can use? What is the largest number of stamps she can use? If Zuna only has two 37¢ stamps, what is the fewest number of stamps she can use?

New Concept

Increasing Knowledge

In mathematics, a **plane** is a flat surface, such as a tabletop or a sheet of paper. When two lines are drawn in the same plane, they will either cross at one point or they will not cross at all. When lines do not cross but stay the same distance apart, we say that the lines are **parallel.** When lines cross, we say that they **intersect.** When they intersect and make square angles, we call the lines **perpendicular.** If lines intersect at a point but are not perpendicular, then the lines are **oblique.**

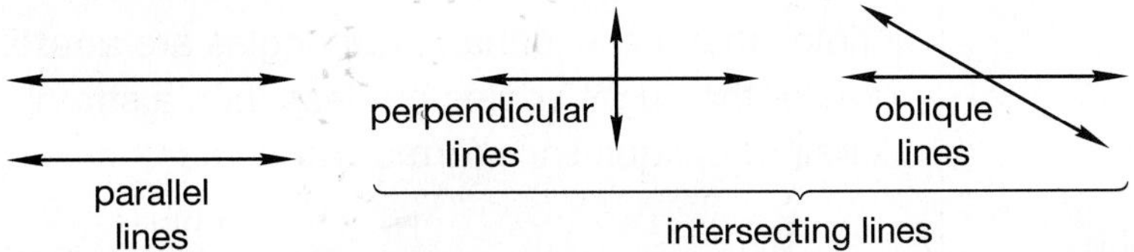

Where lines intersect, **angles** are formed. We show several angles below.

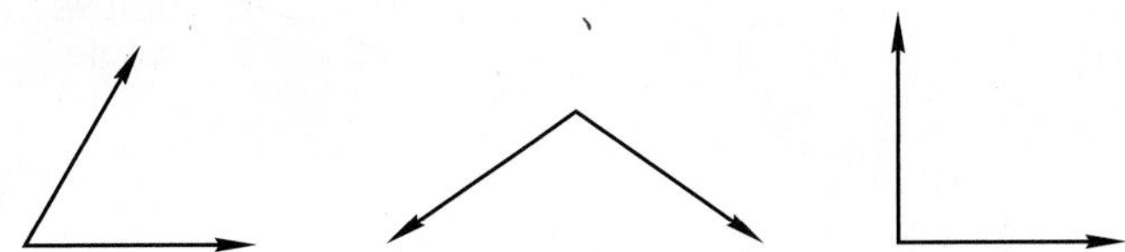

Math Language

A **ray** has one endpoint and continues in one direction without end.

Rays make up the sides of the angles. The rays of an angle originate at a point called the **vertex** of the angle.

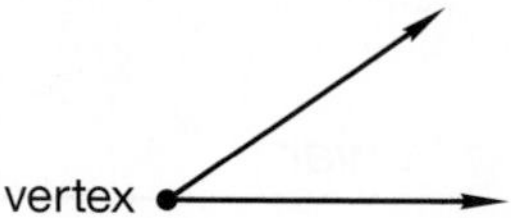

Angles are named in a variety of ways. When there is no chance of confusion, an angle may be named with only one letter: the letter of its vertex. Here is angle *B* (abbreviated $\angle B$):

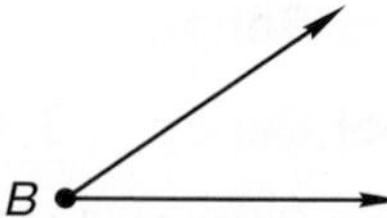

An angle may also be named with three letters, using a point from one side, the vertex, and a point from the other side. Here is angle *ABC* ($\angle ABC$):

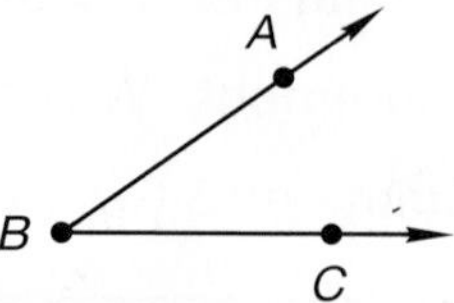

This angle may also be named angle *CBA* ($\angle CBA$). However, it may not be named $\angle BAC$, $\angle BCA$, $\angle CAB$, or $\angle ACB$. The vertex must be in the middle. Angles may also be named with a number or letter in the interior of the angle. In the figure below we see $\angle 1$ and $\angle 2$.

Thinking Skill

Analyze

What figure is created when we add the measures of $\angle 1$ and $\angle 2$?

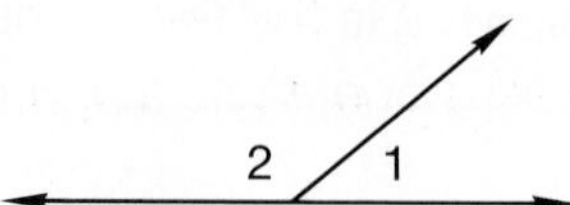

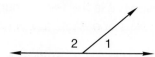

The square angles formed by perpendicular lines, rays, or segments are called **right angles.** We may mark a right angle with a small square.

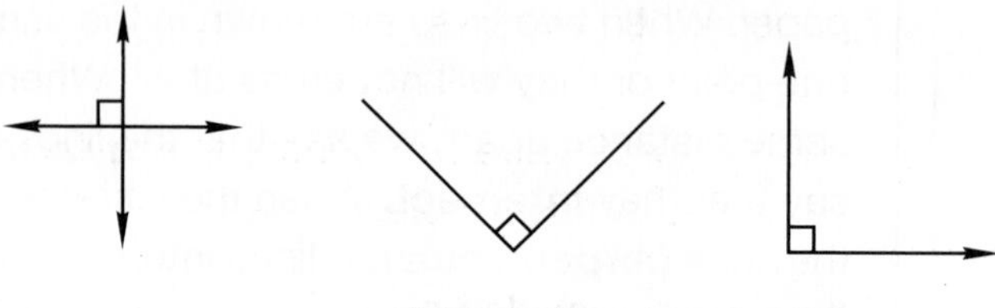

right angles

Angles that are less than right angles are **acute angles.** Angles that are greater than right angles but less than a straight line are **obtuse angles.** A pair of oblique lines forms two acute angles and two obtuse angles.

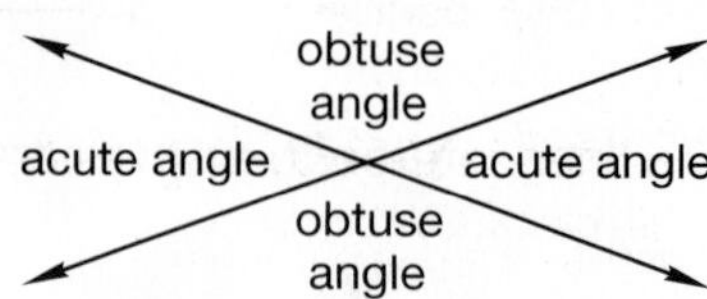

Example 1

a. Name the acute angle in this figure.

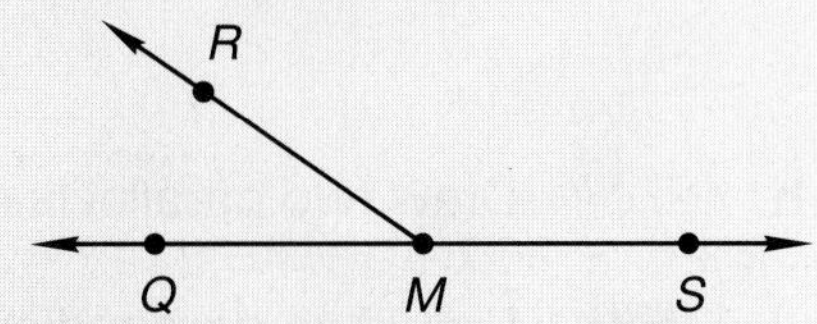

b. Name the obtuse angle in this figure.

Solution

Reading Math

When we use three letters to name an angle, the middle letter is the vertex of the angle.

To avoid confusion, we use three letters to name the angles.

a. The acute angle is $\angle QMR$ (or $\angle RMQ$).

b. The obtuse angle is $\angle RMS$ (or $\angle SMR$).

Example 2

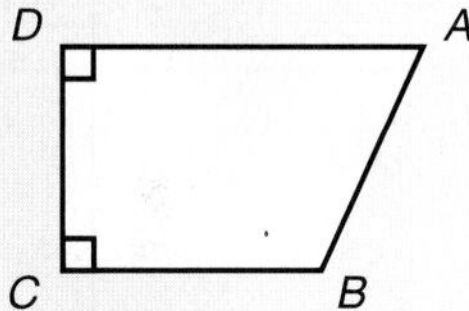

In this figure, angle *D* is a right angle.

a. Which other angle is a right angle?

b. Which angle is acute?

c. Which angle is obtuse?

Solution

Since there is one angle at each vertex, we may use a single letter to name each angle.

a. Angle *C* is a right angle.

b. Angle *A* is acute.

c. Angle *B* is obtuse.

Practice Set

a. Represent Use two pencils to approximate an acute angle, a right angle, and an obtuse angle.

Describe each angle below as acute, right, or obtuse.

b. **c.**

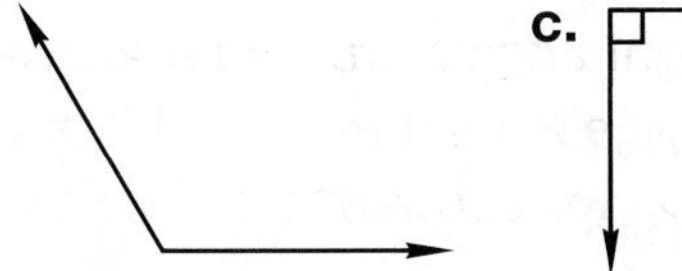

d.

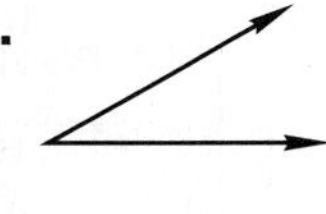

e. Connect What type of angle is formed by the hands of a clock at 4 o'clock?

f. Connect What type of angle is formed at the corner of a door in your classroom?

g. Which two angles formed by these oblique lines are acute angles?

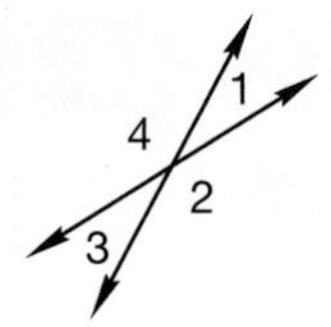

h. **Model** Draw two parallel line segments.

i. **Model** Draw two perpendicular lines.

Refer to the triangle to answer problems **j** and **k**.

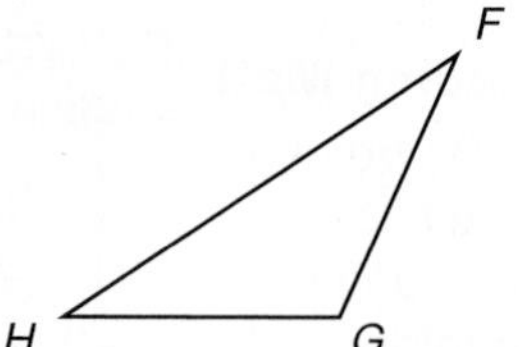

j. Angle H is an acute angle. Name another acute angle.

k. Name an obtuse angle.

Written Practice — *Strengthening Concepts*

* **1.** (24) What is the sum of $\frac{1}{3}$ and $\frac{2}{3}$ and $\frac{3}{3}$?

* **2.** (22) **Explain** According to the 2000 census, about $\frac{2}{5}$ of the 20 million people who lived in Texas at the time were under the age of 24. About how many Texans were under 24 years old in 2000? Explain how you found your answer.

3. (15) **Formulate** Seven hundred sixty-eight peanuts are to be shared equally by the thirty-two children at the party. How many peanuts should each child receive? Write an equation and solve the problem.

4. (13) **Formulate** The Declaration of Independence was signed in 1776. How many years ago was that? Write an equation and solve the problem.

* **5.** (25) Convert $\frac{23}{3}$ to a mixed number.

* **6.** (26) $1\frac{2}{3} + 1\frac{2}{3}$

* **7.** (26) $3 + 4\frac{2}{3}$

* **8.** (26) $3\frac{5}{6} - 1\frac{4}{6}$

* **9.** (26) **Model** Use your fraction manipulatives to reduce $\frac{4}{8}$.

10. (22) **Model** How much money is $\frac{2}{3}$ of $24.00? Draw a diagram to illustrate the problem.

11. (22) **a.** What is $\frac{1}{10}$ of 100%?

b. What is $\frac{3}{10}$ of 100%?

12. (Inv. 2) Twenty-five percent of a circle is what fraction of a circle?

13. (23) At 26 miles per gallon, how far can Ms. Olsen expect to drive on 11 gallons of gas?

Find each missing number. Remember to check your work.

14. (Inv. 2) $\frac{1}{4} + m = 1$

15. (3) $423 - w = 297$

* **16.** (28) **Represent** Refer to the figure below to answer **a** and **b**.

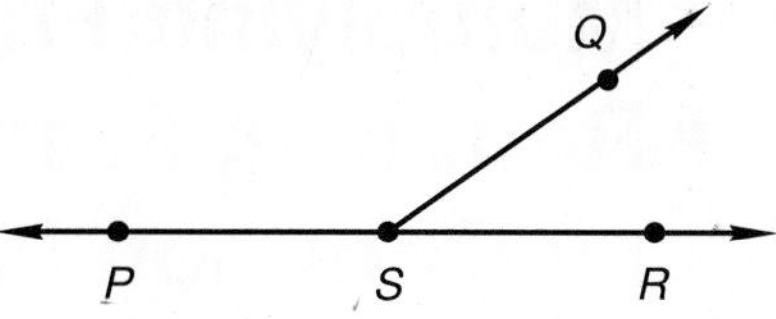

a. Name an obtuse angle.

b. Name an acute angle.

17. (18) On the last four tests the number of questions Christie answered correctly was 22, 20, 23, and 23 respectively. She averaged how many correct answers on each test?

18. (8) The three sides of an equilateral triangle are of equal length. If a 36-inch-long string is formed into the shape of an equilateral triangle, how long will each side of the triangle be?

19. (20) What is the greatest common factor (GCF) of 24, 36, and 60?

20. (1) 10,010 − 9909

21. (5) (100 × 100) − (100 × 99)

22. (22) **Model** If $\frac{1}{10}$ of the class was absent, what percent of the class was absent? Draw a diagram to illustrate the problem.

* **23.** (25) Divide 5097 by 10 and write the answer as a mixed number.

24. (22) **Model** Three fourths of two dozen eggs is how many eggs? Draw a diagram to illustrate the problem.

* **25.** (7, 17) **a.** Use a ruler to find the length of the line segment below to the nearest sixteenth of an inch.

---

b. Use a centimeter ruler to find the length of the line segment to the nearest centimeter.

* **26.** (25) **Analyze** List the first five multiples of 6 and the first five multiples of 8. Circle any numbers that are multiples of both 6 and 8.

27. (Inv. 2) **Model** Which fraction manipulative covers $\frac{1}{2}$ of $\frac{1}{3}$?

28. (23) There are thirteen stripes on the United States flag. Seven of the stripes are red, and the rest of the stripes are white. What is the ratio of red stripes to white stripes on the United States flag?

29. (19) Here we show 24 written as a product of prime numbers:

$$2 \cdot 2 \cdot 2 \cdot 3$$

Show how prime numbers can be multiplied to equal 27.

30. (21) **Classify** Which of the numbers below is not divisible by 9?

A 234 **B** 345 **C** 567

LESSON

29

• Multiplying Fractions
• Reducing Fractions by Dividing by Common Factors

Power Up

Building Power

facts Power Up B

mental math

a. **Number Sense:** 7×43

b. **Number Sense:** 4×64

c. **Number Sense:** $53 + 39$

d. **Number Sense:** $325 + 50$

e. **Money:** \$20.00 − \$17.25

f. **Fractional Parts:** $\frac{1}{2}$ of 70

g. **Measurement:** Which is greater, 10 millimeters or one centimeter?

h. **Calculation:** 4×5, $-\ 6$, $\div\ 7$, $\times\ 8$, $+\ 9$, $\times\ 2$

problem solving Vivian bought a pizza and ate one fourth of it. Then her sister ate one-third of what was left. Then their little brother ate half of what his sisters had left. What fraction of the whole pizza did Vivian's little brother eat?

New Concepts

Increasing Knowledge

multiplying fractions

Below we have shaded $\frac{1}{2}$ of $\frac{1}{2}$ of a circle.

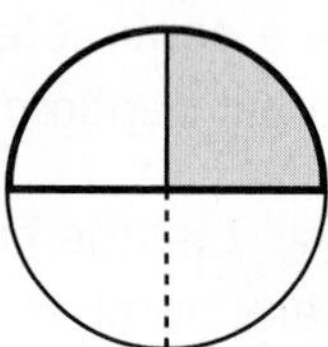

We see that $\frac{1}{2}$ of $\frac{1}{2}$ is $\frac{1}{4}$.

When we find $\frac{1}{2}$ of $\frac{1}{2}$, we are actually multiplying.

$$\frac{1}{2} \times \frac{1}{2} = \frac{1}{4}$$

When we multiply fractions, we multiply the numerators to find the numerator of the product, and we multiply the denominators to find the denominator of the product.

Reading Math

The "of" in "$\frac{1}{2}$ of $\frac{1}{2}$" means to multiply.

Example 1

What fraction is $\frac{1}{2}$ of $\frac{3}{4}$?

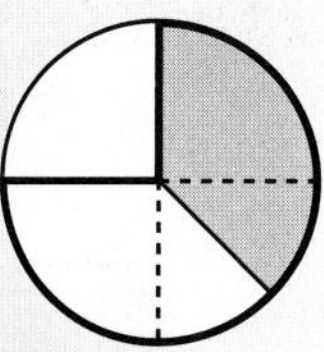

Solution

The word *of* in the question means to multiply. We multiply $\frac{1}{2}$ and $\frac{3}{4}$ to find $\frac{1}{2}$ of $\frac{3}{4}$.

$$\frac{1}{2} \times \frac{3}{4} = \frac{3}{8} \begin{array}{l} \longleftarrow (1 \times 3 = 3) \\ \longleftarrow (2 \times 4 = 8) \end{array}$$

Model We find that $\frac{1}{2}$ of $\frac{3}{4}$ is $\mathbf{\frac{3}{8}}$. You can illustrate this with your fraction manipulatives by using three $\frac{1}{4}$s to make $\frac{3}{4}$ of a circle, then covering half of that area with three $\frac{1}{8}$s.

Example 2

Multiply: $\frac{3}{4} \times \frac{2}{3}$

Solution

By performing this multiplication, we will find $\frac{3}{4}$ of $\frac{2}{3}$. We multiply the numerators to find the numerator of the product, and we multiply the denominators to find the denominator of the product.

$$\frac{3}{4} \times \frac{2}{3} = \frac{6}{12}$$

The fraction $\frac{6}{12}$ can be reduced to $\mathbf{\frac{1}{2}}$, as we can see in this figure:

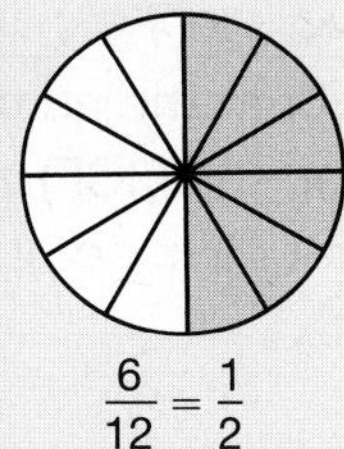

$\frac{6}{12} = \frac{1}{2}$

A whole number can be written as a fraction by writing the whole number as the numerator of the fraction and 1 as the denominator of the fraction. Thus, the whole number 2 can be written as the fraction $\frac{2}{1}$. Writing whole numbers as fractions is helpful when multiplying whole numbers by fractions.

Example 3

Thinking Skill

Discuss

How do you change $\frac{8}{3}$ to $2\frac{2}{3}$? Explain your thinking.

Multiply: $4 \times \frac{2}{3}$

Solution

We write 4 as $\frac{4}{1}$ and multiply.

$$\frac{4}{1} \times \frac{2}{3} = \frac{8}{3}$$

Then we convert the improper fraction $\frac{8}{3}$ to a mixed number.

$$\frac{8}{3} = \mathbf{2\frac{2}{3}}$$

Example 4

Three pennies are placed side by side as shown below. The diameter of one penny is $\frac{3}{4}$ inch. How long is the row of pennies?

$\vdash \frac{3}{4}$ in. $\dashv$

Solution

We can find the answer by adding or by multiplying. We will show both ways.

$$\text{Adding: } \frac{3}{4}\text{ in.} + \frac{3}{4}\text{ in.} + \frac{3}{4}\text{ in.} = \frac{9}{4}\text{ in.} = 2\frac{1}{4}\text{ in.}$$

$$\text{Multiplying: } \frac{3}{1} \times \frac{3}{4}\text{ in.} = \frac{9}{4}\text{ in.} = 2\frac{1}{4}\text{ in.}$$

We find that the row of pennies is $\mathbf{2\frac{1}{4}}$ **inches** long.

reducing fractions by dividing by common factors

We can reduce fractions by dividing the numerator and the denominator by a factor of both numbers. To reduce $\frac{6}{12}$, we will divide both the numerator and the denominator by 6.

$$\frac{6 \div 6}{12 \div 6} = \frac{1}{2}$$

Math Language
$\frac{6}{12}$ and $\frac{1}{2}$ are called *equivalent fractions* or equal fractions.

We divided both the numerator and the denominator by 6 because 6 is the largest factor (the GCF) of 6 and 12. If we had divided by 2 instead of by 6, we would not have completely reduced the fraction.

$$\frac{6 \div 2}{12 \div 2} = \frac{3}{6}$$

The fraction $\frac{3}{6}$ can be reduced by dividing the numerator and the denominator by 3.

$$\frac{3 \div 3}{6 \div 3} = \frac{1}{2}$$

It takes two or more steps to reduce fractions if we do not divide by the greatest common factor in the first step.

Example 5

Reduce: $\frac{8}{12}$

Solution

We will show two methods.

Method 1: Divide both numerator and denominator by 2.

$$\frac{8 \div 2}{12 \div 2} = \frac{4}{6}$$

Again divide both numerator and denominator by 2.

$$\frac{4 \div 2}{6 \div 2} = \frac{2}{3}$$

Method 2: Divide both numerator and denominator by 4.

$$\frac{8 \div 4}{12 \div 4} = \frac{2}{3}$$

Either way, we find that $\frac{8}{12}$ reduces to $\mathbf{\frac{2}{3}}$. Since the greatest common factor of 8 and 12 is 4, we reduced $\frac{8}{12}$ in one step in Method 2 by dividing the numerator and denominator by 4.

Example 6

Multiply: $2 \times \frac{5}{12}$

Solution

We write 2 as $\frac{2}{1}$ and multiply.

$$\frac{2}{1} \times \frac{5}{12} = \frac{10}{12}$$

We can reduce $\frac{10}{12}$ because both 10 and 12 are divisible by 2.

$$\frac{10 \div 2}{12 \div 2} = \mathbf{\frac{5}{6}}$$

Example 7

There were 8 boys and 12 girls in the class. What was the ratio of boys to girls in the class?

Solution

We reduce ratios the same way we reduce fractions. The ratio 8 to 12 reduces to $\frac{2}{3}$.

$$\frac{\text{number of boys}}{\text{number of girls}} = \frac{8}{12} = \mathbf{\frac{2}{3}}$$

Practice Set

Multiply; then reduce if possible.

a. $\frac{1}{2}$ of $\frac{4}{5}$ **b.** $\frac{1}{4}$ of $\frac{2}{3}$ **c.** $\frac{2}{3} \times \frac{3}{4}$

Multiply; then convert each answer from an improper fraction to a whole number or to a mixed number.

d. $\frac{5}{6} \times \frac{6}{5}$ **e.** $5 \times \frac{2}{3}$ **f.** $2 \times \frac{4}{3}$

Reduce each fraction:

g. $\frac{9}{12}$ **h.** $\frac{6}{10}$ **i.** $\frac{18}{24}$

j. In a class of 30, there were 20 girls. What was the ratio of boys to girls?

Written Practice *Strengthening Concepts*

1. (15) The African elephant can weigh eight tons. A ton is two thousand pounds. How many pounds can an African elephant weigh?

2. (15) If sixteen dried beans weigh one ounce, then how many dried beans weigh one pound (1 pound = 16 ounces)?

*** 3.** (Inv. 2, 29) **Analyze** If the product of $\frac{1}{2}$ and $\frac{1}{2}$ is subtracted from the sum of $\frac{1}{2}$ and $\frac{1}{2}$, what is the difference?

*** 4.** (29) A team won 6 games and lost 8 games. What was the team's win-loss ratio?

*** 5.** (29) Reduce: $\frac{16}{24}$

*** 6.** (24, 29) $\frac{1}{8} + \frac{3}{8}$

*** 7.** (29) $\frac{1}{2} \times \frac{2}{3}$

*** 8.** (24, 29) $\frac{7}{12} - \frac{3}{12}$

9. (22) The Nobel Prize is a famous international award that recognizes important work in physics, chemistry, medicine, economics, peacemaking, and literature. Ninety-six Nobel Prizes in Literature were awarded from 1901 to 1999. One eighth of the prizes were given to Americans. How many Nobel Prizes in Literature were awarded to Americans from 1901 to 1999?

10. (10) **Predict** Find the next three numbers in the sequence below:

1, 4, 7, 10, ______, ______, ______, . . .

11. (Inv. 2) **Connect** When five months have passed, what fraction of the year remains?

12. (2) $3.60 × 100

13. (2) 50,000 ÷ 100

*** 14.** (25) Convert $\frac{18}{4}$ to a mixed number. Remember to reduce the fraction part of the mixed number.

15. (14) The temperature rose from –8°F to 15°F. This was an increase of how many degrees?

Find each unknown number. Remember to check your work.

16. (3) $m + 496 + 2684 = 3217$

17. (3) $1000 - n = 857$

18. (4) $24x = 480$

19. (5) $7 \cdot 11 \cdot 13$

20. (16) **Estimate** Explain how to estimate the quotient of 4963 ÷ 39.

* **21.** (29) Compare: $\frac{2}{3} \times \frac{3}{2} \bigcirc 1$

22. (8) **Analyze** The perimeter of the rectangle shown is 60 mm. The width of the rectangle is 10 mm. What is its length?

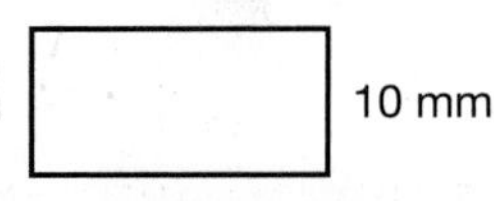

23. (14) 12 − 40

* **24.** (29) $\left(\frac{1}{2} \times \frac{1}{2}\right) - \frac{1}{4}$

* **25.** (28) **a.** Which angles in the figure at right are acute angles?

b. Which angles are obtuse angles?

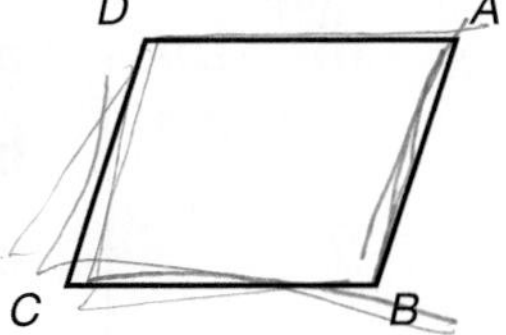

* **26.** (29) What fraction is $\frac{2}{3}$ of $\frac{3}{5}$?

* **27.** (29) What is the product of $\frac{3}{4}$ and $\frac{4}{3}$?

* **28.** (27, 29) **Connect** If the diameter of a bicycle wheel is 24 inches, what is the ratio of the radius of the wheel to the diameter of the wheel?

* **29.** (28) **Represent** What type of an angle is formed by the hands of a clock at 2 o'clock?

30. (22) What percent of a circle is $\frac{2}{5}$ of a circle? Explain why your answer is correct.

Early Finishers
Real-World Application

The high school basketball team has 14 players: $\frac{2}{7}$ are guards, $\frac{1}{2}$ are forwards, and the rest are centers. Find the number of players in each position on the team. Show your work.

LESSON

30

• Least Common Multiple (LCM)
• Reciprocals

Power Up — *Building Power*

facts Power Up E

mental math

a. **Number Sense:** 9×32

b. **Number Sense:** 5×42

c. **Number Sense:** $45 + 49$

d. **Number Sense:** $436 + 99$

e. **Money:** $20.00 − $12.75

f. **Fractional Parts:** $\frac{1}{2}$ of 72

g. **Statistics:** Find the average: 120, 99, 75

h. **Calculation:** 7×7, $- 1$, $\div 6$, $\times 3$, $+ 1$, $\times 2$, $- 1$

problem solving

A *pentomino* is a geometric shape made of five equal squares joined by their edges. There are twelve different pentominos, and they are named after letters of the alphabet: F, I, L, N, P, T, U, V, W, X, Y, and Z. (The rotation or reflection of a pentomino does not count as a different pentomino.) Can you create the remaining ten pentominos?

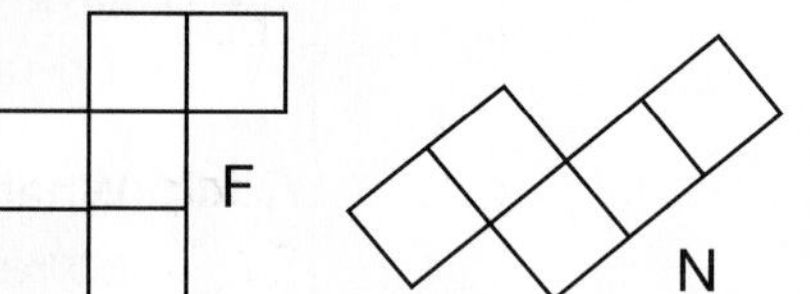

New Concepts — *Increasing Knowledge*

least common multiple (LCM)

A number that is a multiple of two or more numbers is called a *common multiple* of those numbers. Here we show some multiples of 2 and 3. We have circled the common multiples.

Multiples of 2: 2, 4, (6), 8, 10, (12), 14, 16, (18), 20, ...

Multiples of 3: 3, (6), 9, (12), 15, (18), 21, ...

We see that 6, 12, and 18 are common multiples of 2 and 3. Since the number 6 is the least of these common multiples, it is called the **least common multiple.** The term *least common multiple* is abbreviated LCM.

Math Language
Recall that a **multiple** is the product of a counting number and another number.

Example 1

What is the least common multiple of 3 and 4?

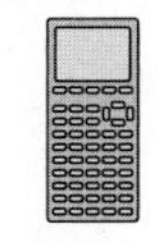

Visit www. SaxonPublishers. com/ActivitiesC1 *for a graphing calculator activity.*

Solution

We will list some multiples of each number and emphasize the common multiples.

Multiples of 3: 3, 6, 9, (12), 15, 18, 21, (24), ...

Multiples of 4: 4, 8, (12), 16, 20, (24), 28, ...

We see that the number 12 and 24 are in both lists. Both 12 and 24 are common multiples of 3 and 4. The least common multiple is **12.** When we list the multiples in order, the first number that is a common multiple is always the least common multiple.

Example 2

What is the LCM of 2 and 4?

Solution

We will list some multiples of 2 and 4.

Multiples of 2: 2, (4), 6, (8), 10, (12), 14, (16), ...

Multiples of 4: (4), (8), (12), (16), ...

The first number that is a common multiple of both 2 and 4 is **4.**

reciprocals

Reciprocals are two numbers whose product is 1. For example, the numbers 2 and $\frac{1}{2}$ are reciprocals because $2 \times \frac{1}{2} = 1$.

$$2 \times \frac{1}{2} = 1$$

reciprocals

We say that 2 is the reciprocal of $\frac{1}{2}$ and that $\frac{1}{2}$ is the reciprocal of 2. Sometimes we want to find the reciprocal of a certain number. One way we will practice finding the reciprocal of a number is by solving equations like this:

$$3 \times \square = 1$$

The number that goes in the box is $\frac{1}{3}$ because 3 times $\frac{1}{3}$ is 1. One third is the reciprocal of 3.

Reciprocals also answer questions like this:

How many $\frac{1}{4}$s are in 1?

The answer is the reciprocal of $\frac{1}{4}$, which is 4.

Fractions have two **terms,** the numerator and the denominator. To form the reciprocal of a fraction, we reverse the terms of the fraction.

$$\frac{3}{4} \rightarrow \frac{4}{3}$$

The new fraction, $\frac{4}{3}$, is the reciprocal of $\frac{3}{4}$.

If we multiply $\frac{3}{4}$ by $\frac{4}{3}$, we see that the product, $\frac{12}{12}$, equals 1.

$$\frac{3}{4} \times \frac{4}{3} = \frac{12}{12} = 1$$

Example 3

How many $\frac{2}{3}$s are in 1?

Solution

To find the number of $\frac{2}{3}$s in 1, we need to find the reciprocal of $\frac{2}{3}$. The easiest way to find the reciprocal of $\frac{2}{3}$ is to reverse the positions of the 2 and the 3. The reciprocal of $\frac{2}{3}$ is $\mathbf{\frac{3}{2}}$. (We may convert $\frac{3}{2}$ to $1\frac{1}{2}$, but we usually write reciprocals as fractions rather than as mixed numbers.)

Example 4

What number goes into the box to make the equation true?

$$\frac{5}{6} \times \square = 1$$

Solution

When $\frac{5}{6}$ is multiplied by its reciprocal, the product is 1. So the answer is the reciprocal of $\frac{5}{6}$, which is $\mathbf{\frac{6}{5}}$. When we multiply $\frac{5}{6}$ by $\frac{6}{5}$, we get $\frac{30}{30}$.

$$\frac{5}{6} \times \frac{6}{5} = \frac{30}{30}$$

The fraction $\frac{30}{30}$ equals 1.

Example 5

What is the reciprocal of 5?

Solution

Recall that a whole number can be written as a fraction that has a denominator of 1. So 5 can be written as $\frac{5}{1}$. (This means "five wholes.") Reversing the positions of the 5 and the 1 gives us the reciprocal of 5, which is $\mathbf{\frac{1}{5}}$. This makes sense because five $\frac{1}{5}$s make 1, and $\frac{1}{5}$ of 5 is 1.

Practice Set

Find the least common multiple of each pair of numbers:

a. 6 and 8 **b.** 3 and 5 **c.** 5 and 10

Write the reciprocal of each number:

d. 6 **e.** $\frac{2}{3}$ **f.** $\frac{8}{5}$ **g.** $\frac{1}{3}$

Analyze For problems **h–k,** find the number that goes into the box to make the equation true.

h. $\frac{3}{8} \times \square = 1$ **i.** $4 \times \square = 1$

j. $\square \times \frac{1}{6} = 1$ **k.** $\square \times \frac{7}{8} = 1$

l. How many $\frac{2}{5}$s are in 1?

m. How many $\frac{5}{12}$s are in 1?

Written Practice *Strengthening Concepts*

* **1.** (12, 25) **Analyze** If the fourth multiple of 3 is subtracted from the third multiple of 4, what is the difference?

2. (22) **Model** About $\frac{2}{3}$ of a person's body weight is water. Albert weighs 117 pounds. About how many pounds of Albert's weight is water? Draw a diagram to illustrate the problem.

3. (15, 23) **Formulate** Cynthia ate 42 pieces of popcorn during the first 15 minutes of a movie. If she kept eating at the same rate, how many pieces of popcorn did she eat during the 2-hour movie? Write an equation and solve the problem.

4. (25) What are the first four multiples of 12?

* **5.** (30) What is the least common multiple (LCM) of 4 and 6?

6. (23) **Connect** There were 12 minutes of commercials during the one-hour program. What was the ratio of commercial to noncommercial time during the one-hour program? Explain how you found your answer.

7. (24) $\frac{2}{5} + \frac{2}{5} + \frac{2}{5}$

8. (Inv. 2) $1 - \frac{1}{10}$

* **9.** (24, 29) $\frac{11}{12} - \frac{1}{12}$

* **10.** (29) $\frac{3}{4} \times \frac{4}{3}$

* **11.** (29) $5 \times \frac{3}{4}$

* **12.** (29) $\frac{5}{2} \times \frac{5}{3}$

13. (19) The number 24 has how many different whole-number factors?

14. (1) \$3 + \$24 + \$6.50

15. (1) \$5 − \$1.50

16. (16) Estimate the product: 596 × 405

* **17.** (28) Which angle of the triangle at right is an obtuse angle?

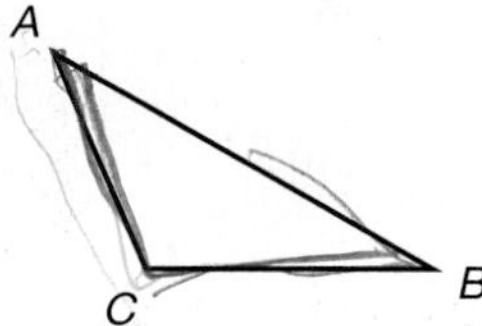

* **18.** (29) Compare: $\frac{2}{3} \times \frac{2}{3} \bigcirc \frac{2}{3} \times 1$

* **19.** (29) 500,000 ÷ 100

20. (2) $35\overline{)8540}$

* **21.** (25) $\frac{100\%}{7}$

* **22.** (29) Reduce: $\frac{4}{12}$

23. (18) What is the average of 375, 632, and 571?

24. (8) A regular hexagon has six sides of equal length. If a regular hexagon is made from a 36-inch-long string, what is the length of each side?

* **25.** (30) What is the product of a number and its reciprocal?

* **26.** (30) How many $\frac{2}{5}$s are in 1?

* **27.** (30) **Analyze** What number goes into the box to make the equation true?

$$\frac{3}{8} \times \square = 1$$

* **28.** (19, 30) **Connect** What is the reciprocal of the only even prime number?

29. (25) Convert $\frac{45}{10}$ to a mixed number. Remember to reduce the fraction part of the mixed number.

* **30.** (29) Four pennies are placed side by side as shown below. The diameter of one penny is $\frac{3}{4}$ inch. What is the length of the row of pennies?

$\vdash \frac{3}{4}$ in. $\dashv$

Early Finishers

Real-World Application

Fernando's class is going to make cheese sandwiches for their school picnic. They want to have at least 80 sandwiches. Each package of bread contains enough slices for 10 sandwiches. Each package of cheese contains enough slices for 18 sandwiches. The class wants to buy the fewest packages of cheese and bread with no slices left over.

a. How many sandwiches should the class make? Explain how you found your answer.

b. How many packages of bread and cheese should they buy? Show your work.

INVESTIGATION 3

Focus on

• Measuring and Drawing Angles with a Protractor

Reading Math

We abbreviate 360 degrees as 360°.

One way to measure angles is with units called **degrees.** A full circle measures 360 degrees. A tool to help us measure angles is a **protractor.** To measure an angle, we place the center point of the protractor on the vertex of the angle, and we place one of the zero marks on one ray of the angle. Where the other ray of the angle passes through the scale, we can read the degree measure of the angle.

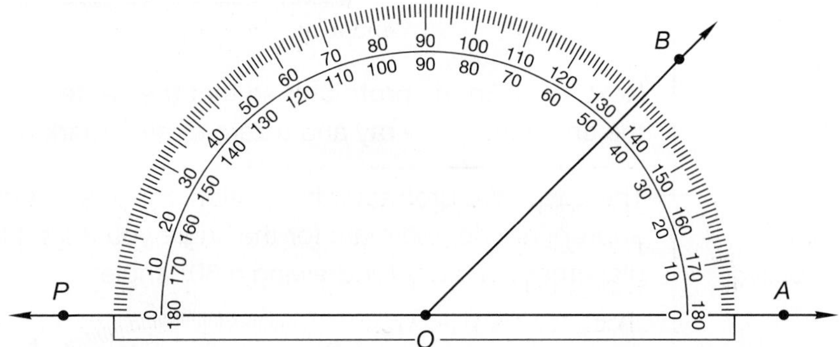

The scale on a protractor has two sets of numbers. One set is for measuring angles starting from the right side, and the other set is for measuring angles starting from the left side. The easiest way to ensure that we are reading from the correct scale is to decide whether the angle we are measuring is acute or obtuse. Looking at $\angle AOB$, we read the numbers 45° and 135°. Since the angle is less than 90° (acute), it must measure 45°, not 135°. We say that "the measure of angle *AOB* is 45°," which we may write as follows:

$$m\angle AOB = 45°$$

Classify Practice reading a protractor by finding the measures of these angles. Then tell whether each is obtuse, acute, or right.

Thinking Skill

Extend

In exercises **1** and **3,** we found $m\angle AOC$ and $m\angle AOF$. Tell how to find $m\angle COF$ without using the protractor.

1. $\angle AOC$
2. $\angle AOE$
3. $\angle AOF$
4. $\angle AOH$
5. $\angle IOH$
6. $\angle IOE$

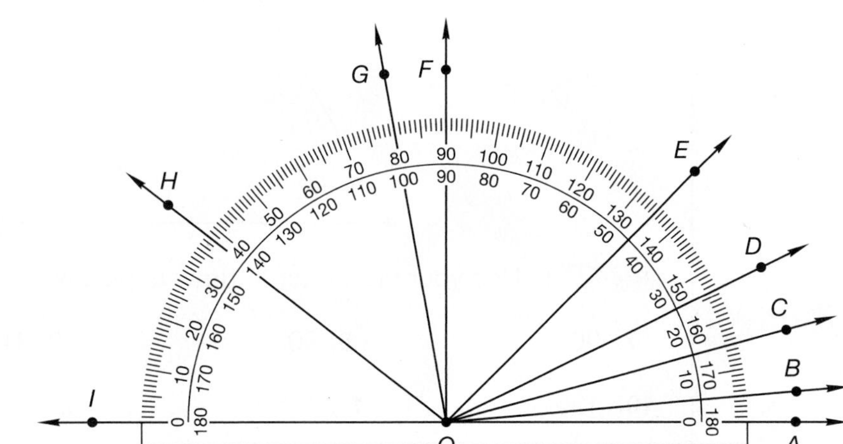

Activity

Measuring Angles

Materials needed:

- Investigation Activity 10
- protractor

Use a protractor to find the measures of the angles.

To draw angles with a protractor, follow these steps. Begin by drawing a horizontal ray. The sketch of the ray should be longer than half the diameter of the protractor.

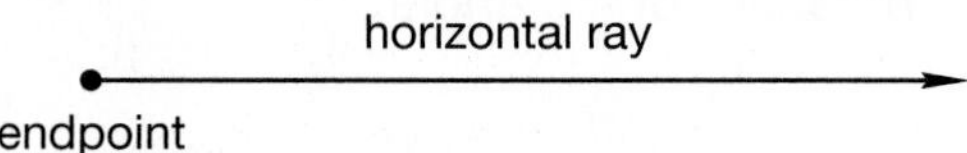

Next, position the protractor so that the center point of the protractor is on the endpoint of the ray and a zero degree mark of the protractor is on the ray.

Then, with the protractor in position, make a dot on the paper at the appropriate degree mark for the angle you intend to draw. Here we show the placement of a dot for drawing a 60° angle:

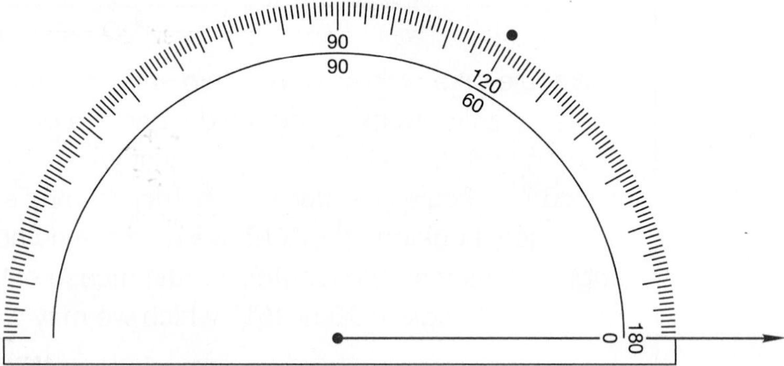

Finally, remove the protractor and draw a ray from the endpoint of the first ray through the dot you made.

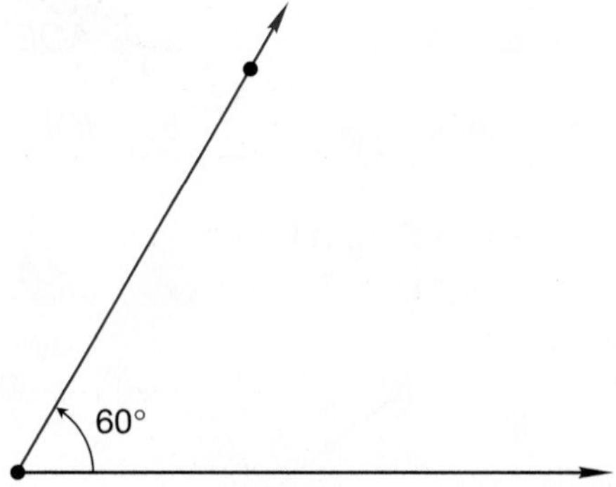

Represent Use your protractor to draw angles with these measures:

7. 30° **8.** 80° **9.** 110°

10. 135° **11.** 45° **12.** 15°

13. **Represent** Draw triangle *ABC* by first drawing segment *BC* six inches long. Then draw a 60° angle at vertex *B* and a 60° angle at vertex *C*. Extend the segments so that they intersect at point *A*.

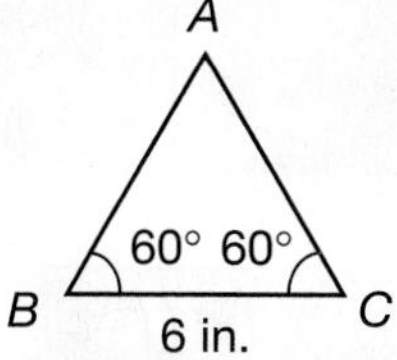

Refer to the triangle you drew in problem **13** to answer problems **14** and **15.**

14. Use a ruler to find the lengths of segment *AB* and segment *AC* in triangle *ABC*.

15. Use a protractor to find the measure of angle *A* in triangle *ABC*.

16. **Represent** Draw triangle *STU* by first drawing angle *S* so that angle *S* is 90° and segments *ST* and *SU* are each 10 cm long. Complete the triangle by drawing segment *TU*.

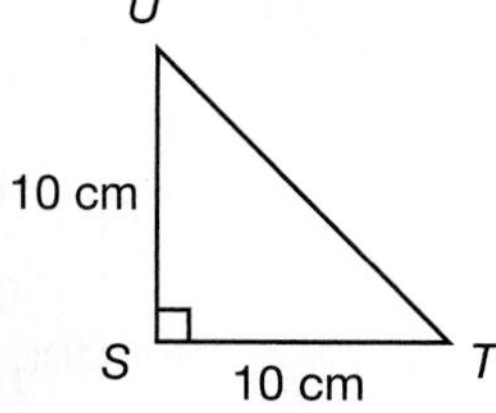

Refer to the triangle you drew in problem **16** to answer problems **17** and **18.**

17. Use a protractor to find the measures of angle *T* and angle *U*.

18. **Estimate** Use a centimeter ruler to measure segment *TU* to the nearest centimeter.

extensions

a. **Analyze** The building code for this staircase requires that the inclination be between 30° and 35°. Does this staircase meet the building code? Explain your thinking.

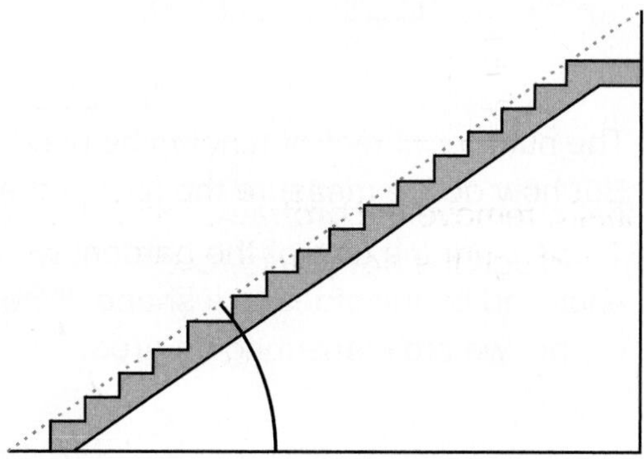

b. **Conclude** Look at the two sets of polygons. Set 1 contains something not found in Set 2. Name another figure that would fit in Set 1. Support your choice.

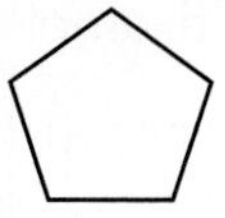 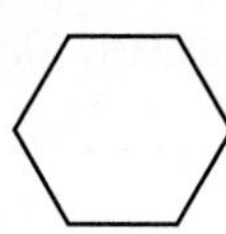 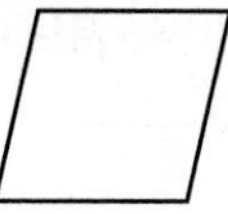

Set 1

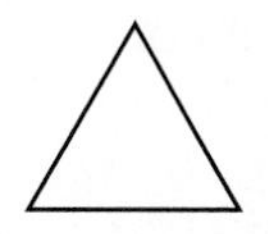  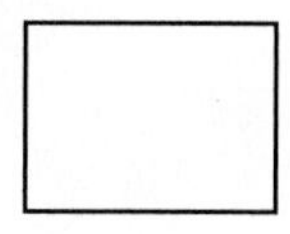

Set 2

LESSON 31

• Areas of Rectangles

Power Up

Building Power

facts Power Up F

mental math

a. **Number Sense:** 4×25

b. **Calculation:** 6×37

c. **Number Sense:** $28 + 29$

d. **Money:** $6.25 + $2.50

e. **Fractional Parts:** $\frac{1}{3}$ of 63

f. **Number Sense:** $\frac{600}{10}$

g. **Measurement:** A minute is how many seconds?

h. **Calculation:** 10×10, $-\ 20$, $+\ 1$, $\div\ 9$, $\times\ 2$, $\div\ 3$, $\times\ 5$, $+\ 2$, $\div\ 4$

problem solving Franki has 7 coins in her hand totaling 50¢. What are the coins?

New Concept

Increasing Knowledge

Mr. McGregor fenced in an area for a garden.

The perimeter of a shape is the distance around it.

The number of feet of fencing he used was the perimeter of the rectangle. But how do we measure the size of the garden?

To measure the size of the garden, we measure how much surface is enclosed by the sides of a shape. When we measure the "inside" of a flat shape, we are measuring its **area.**

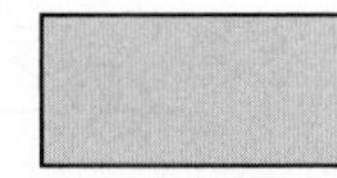
The area of a shape is the amount of surface enclosed by its sides.

We use a different kind of unit to measure area than we use to measure perimeter. To measure perimeter, we use units of length such as centimeters. Units of area are called **square units.** One example is a square centimeter.

This is 1 centimeter.

This is 1 square centimeter.

Math Language
We use "sq." to abbreviate the word *square.*

Other common units of area are square inches, square feet, square yards, and square meters. Very large areas may be measured in square miles. We can think of units of area as floor tiles. The area of a shape is the number of "floor tiles" of a certain size that completely cover the shape.

Example 1

How many floor tiles, 1 foot on each side, are needed to cover the floor of a room that is 8 feet wide and 12 feet long?

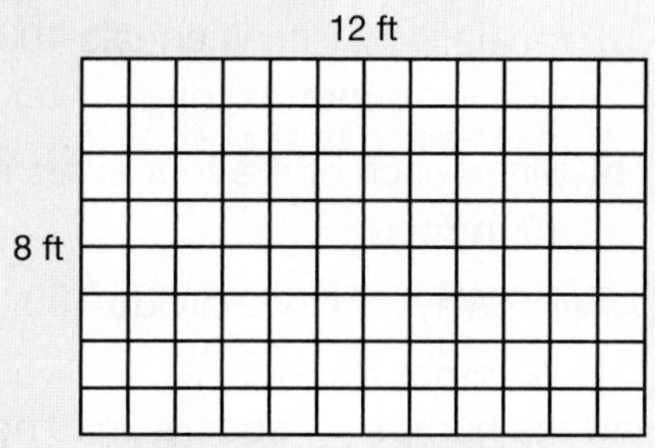

Solution

The surface of the floor is covered with tiles. By answering this question, we are finding the area of the room in square feet. We could count the tiles, but a faster way to find the number of tiles is to multiply. There are 8 rows of tiles with 12 tiles in each row.

$$\begin{array}{rl} & 12 \text{ tiles in each row} \\ \times & 8 \text{ rows} \\ \hline & 96 \text{ tiles} \end{array}$$

To cover the floor, **96 tiles** are needed. The area of the room is 96 sq. ft.

Discuss Why do we use square feet as the unit of measure?

Example 2

What is the area of this rectangle?

8 cm

4 cm

Solution

The diagram shows the length and width of the rectangle in centimeters. Therefore, we will use square centimeters to measure the area of the rectangle. We calculate the number of square-centimeter tiles needed to cover the rectangle by multiplying the length by the width.

Length × width = area

8 cm × 4 cm = **32 sq. cm**

Example 3

The area of a square is 100 square inches.

a. How long is each side of the square?

b. What is the perimeter of the square?

Solution

a. The length and width of a square are equal. So we think, "What number multiplied by itself equals 100?" Since $10 \times 10 = 100$, we find that each side is **10 inches** long.

b. Since each of the four sides is 10 inches, the perimeter of the square is **40 inches.**

Justify Why can't we divide 100 by 4 to find the length of each side?

Practice Set Find the number of square units needed to cover the area of these rectangles. For reference, square units have been drawn along the length and width of each rectangle.

a.

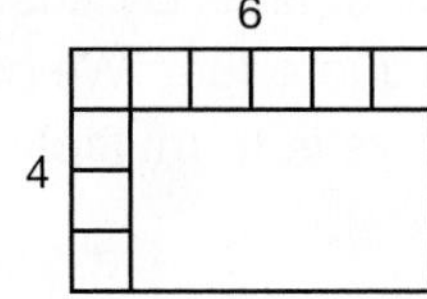

b.

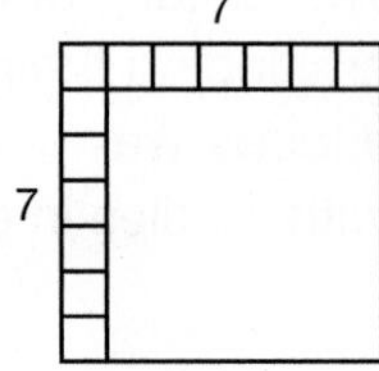

Find the area of these rectangles:

c.

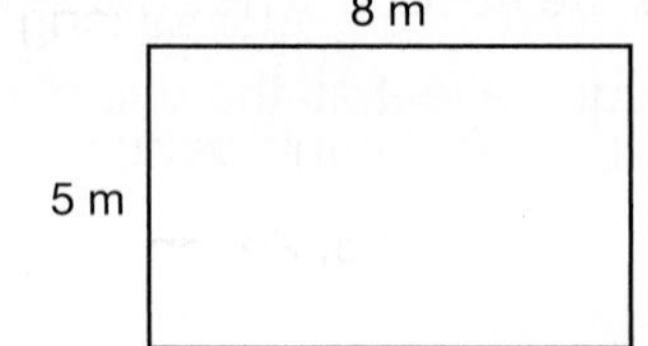

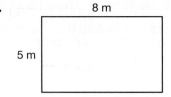

d.

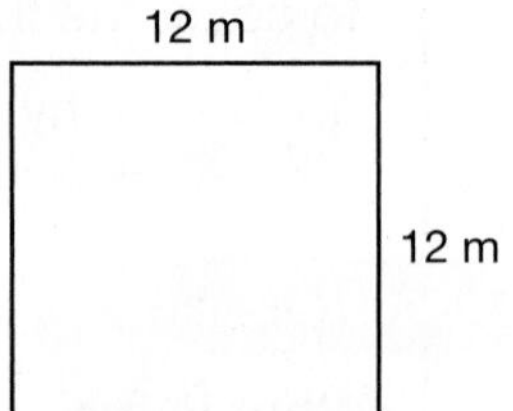

Analyze The area of a square is 25 square inches.

e. How long is each side of the square?

f. What is the perimeter of the square?

g. *Analyze* Find the area of Mr. McGregor's garden described at the beginning of this lesson.

h. *Analyze* Choose the appropriate unit for the area of a room in a home.

A square inches **B** square feet **C** square miles

Written Practice *Strengthening Concepts*

1. (12, 25) **Analyze** When the third multiple of 4 is divided by the fourth multiple of 3, what is the quotient?

2. (12) The distance the Earth travels around the Sun each year is about five hundred eighty million miles. Use digits to write that distance.

3. (25) Convert $\frac{10}{3}$ to a mixed number.

*** 4.** (31) **Generalize** How many square stickers with sides 1 centimeter long would be needed to cover the rectangle below?

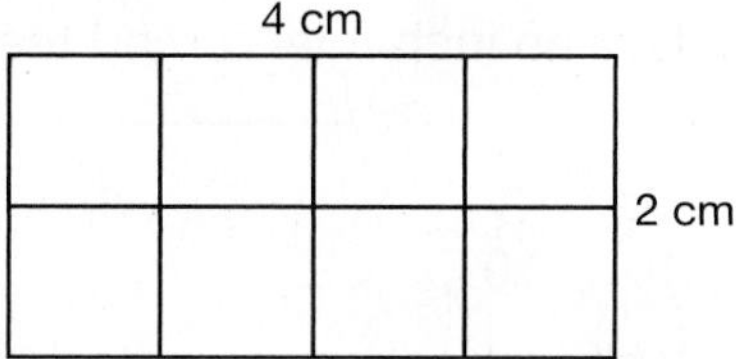

*** 5.** (31) How many floor tiles with sides 1 foot long would be needed to cover the square below?

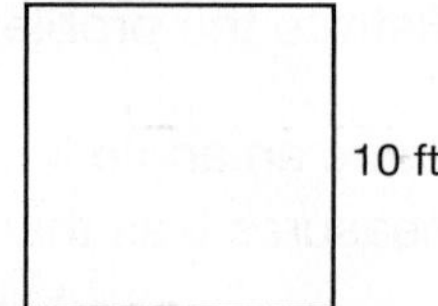

*** 6.** (31) What is the area of a rectangle 12 inches long and 8 inches wide?

7. (10) **Generalize** Describe the rule for this sequence. What is the next term?

$$1, 4, 9, 16, 25, 36, \ldots$$

8. (22) **Model** What number is $\frac{2}{3}$ of 24? Draw a diagram to illustrate the problem.

9. (3) Find the unknown number. Remember to check your work.

$$24 + f = 42$$

Write each answer in simplest form:

10. (24) $\frac{1}{8} + \frac{1}{8}$

11. (24) $\frac{5}{6} - \frac{1}{6}$

*** 12.** (29) $\frac{2}{3} \cdot \frac{1}{2}$

*** 13.** (29) $\frac{2}{3} \times 5$

14. (16) Estimate the product of 387 and 514.

15. (2) \$20.00 ÷ 10

16. (2) (63)47¢

17. (2) 4623 ÷ 22

*** 18.** (19, 30) What is the reciprocal of the smallest odd prime number?

*** 19.** (26) Two thirds of a circle is what percent of a circle?

20. (9) **Estimate** Which of these numbers is closest to 100?

A 90 **B** 89 **C** 111 **D** 109

21. (12) For most of its orbit, Pluto is the farthest planet from the Sun in our solar system. Pluto's average distance from the Sun is about three billion, six hundred seventy million miles. Use digits to write the average distance between Pluto and the Sun.

*** 22.** (23, 27) The diameter of the pizza was 14 inches. What was the ratio of the radius to the diameter of the pizza?

*** 23.** (29) Three of the nine softball players play outfield. What fraction of the players play outfield? Write the answer as a reduced fraction.

24. (17) Use an inch ruler to find the length of the line segment below.

____

*** 25.** (29) $\frac{3}{10} \times \frac{3}{10}$

*** 26.** (30) How many $\frac{3}{4}$s are in 1?

*** 27.** (29) Write a fraction equal to 1 with a denominator of 8.

28. (22) **Model** Five sixths of the 24 students in the class scored 80% or higher on the test. How many students scored 80% or higher? Draw a diagram to illustrate the problem.

*** 29.** (28) **a.** Name an angle in the figure at right that measures less than 90°.

b. Name an obtuse angle in the figure at right.

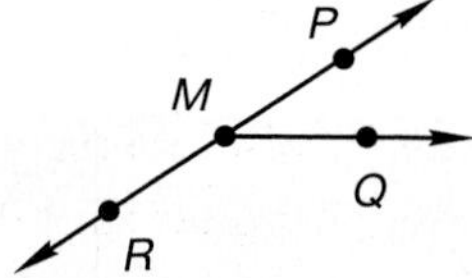

*** 30.** (31) **Evaluate** Using a ruler, how could you calculate the floor area of your classroom?

LESSON
32

• Expanded Notation
• More on Elapsed Time

Power Up

Building Power

facts Power Up G

mental math

a. Number Sense: 4×75

b. Number Sense: $380 + 1200$

c. Number Sense: $54 + 19$

d. Money: $8.00 − $1.50

e. Fractional Parts: $\frac{1}{2}$ of 240

f. Number Sense: $\frac{600}{100}$

g. Geometry: A square has a length of 4 ft. What is the area of the square?

h. Calculation: 12×3, $- 1$, $\div 5$, $\times 2$, $+ 1$, $\div 3$, $\times 2$

problem solving The product of $10 \times 10 \times 10$ is 1000. Find three prime numbers whose product is 1001.

New Concepts

Increasing Knowledge

expanded notation

The price of a new car is $27,000. The price of a house is $270,000. Which price is more, or are the prices the same? How do you know?

Recall that in our number system the location of a digit in a number has a value called its *place value.* Consider the value of the 2 in these two numbers:

27,000   270,000

In 27,000 the value of the 2 is $2 \times 10{,}000$. In 270,000 the value of the 2 is $2 \times 100{,}000$. Therefore, $270,000 is greater than $27,000.

To find a digit's value within a number, we multiply the digit by the value of the place occupied by the digit. To write a number in **expanded notation,** we write each nonzero digit times its place value.

Example 1

Write 27,000 in expanded notation.

Solution

The 2 is in the ten-thousands place, and the 7 is in the thousands place. In expanded notation we write

$$(2 \times 10{,}000) + (7 \times 1000)$$

Since zero times any number equals zero, it is not necessary to include zeros when writing numbers in expanded notation.

Example 2

Write (5 × 1000) + (2 × 100) + (8 × 10) in standard notation.

Solution

Standard notation is our usual way of writing numbers. One way to think about this number is 5000 + 200 + 80. Another way to think about this number is 5 in the thousands place, 2 in the hundreds place, and 8 in the tens place. We may assume a 0 in the ones place. Either way we think about the number, the standard form is **5280.**

Verify How would the expanded notation change if we added 20 to 5280?

more on elapsed time

The hours of the day are divided into two parts: **a.m.** and **p.m.** The 12 "a.m." hours extend from midnight (12:00 a.m.) to the moment just before noon (12:00 p.m.). The 12 "p.m." hours extend from noon to the moment just before midnight. Recall from Lesson 13 that when we calculate the amount of time between two events, we are calculating elapsed time (the amount of time that has passed). We can use the later-earlier-difference pattern to solve elapsed-time problems about hours and minutes.

Example 3

Jason started the marathon at 7:15 a.m. He finished the race at 11:10 a.m. How long did it take Jason to run the marathon?

Solution

This problem has a subtraction pattern. We find Jason's race time (elapsed time) by subtracting the earlier time from the later time.

Later	11:10 a.m.
− Earlier	− 7:15 a.m.
Difference	

Since we cannot subtract 15 minutes from 10 minutes, we rename one hour as 60 minutes. Those 60 minutes plus 10 minutes equal 70 minutes. (This means 70 minutes after 9:00, which is the same as 10:10.)

$$\begin{array}{r} \scriptstyle 10:70 \\ \not{1}\not{1}:\not{1}\not{0} \\ -\ 7:15 \\ \hline 3:55 \end{array}$$

We find that it took Jason **3 hours 55 minutes** to run the marathon.

Example 4

What time is two and a half hours after 10:43 a.m.?

Solution

This is an elapsed-time problem, and it has a subtraction pattern. The elapsed time, $2\frac{1}{2}$ hours, is the difference. We write the elapsed time as 2:30. The earlier time is 10:43 a.m.

Later	Later
− Earlier	− 10:43 a.m.
Difference	2:30

We need to find the later time, so we add $2\frac{1}{2}$ hours to 10:43 a.m. We will describe two methods to do this: a mental calculation and a pencil-and-paper calculation. For the mental calculation, we could first count two hours after 10:43 a.m. One hour later is 11:43 a.m. Another hour later is 12:43 p.m. (Note the switch from a.m. to p.m.) From 12:43 p.m., we count 30 minutes (one half hour). To do this, we can count 10 minutes at a time from 12:43 p.m.: 12:53 p.m., 1:03 p.m., 1:13 p.m. We find that $2\frac{1}{2}$ hours after 10:43 a.m. is **1:13 p.m.**

Thinking Skill

Discuss

What is another way to solve this problem using mental math?

To perform a pencil-and-paper calculation, we add 2 hours 30 minutes to 10:43 a.m.

```
   10:43 a.m.
+   2:30
   12:73 p.m.
```

Notice that the time switches from a.m. to p.m. and that the sum, 12:73 p.m., is improper. Seventy-three minutes is more than an hour. We think of 73 minutes as "one hour plus 13 minutes." We add 1 to the number of hours and write 13 as the number of minutes. So $2\frac{1}{2}$ hours after 10:43 a.m. is **1:13 p.m.**

Practice Set

Write each of these numbers in expanded notation:

a. 270,000

b. 1760

c. 8050

Write each of these numbers in standard form:

d. $(6 \times 1000) + (4 \times 100)$ **e.** $(7 \times 100) + (5 \times 1)$

f. *Explain* George started the marathon at 7:15 a.m. He finished the race at 11:05 a.m. How long did it take George to run the marathon? How do you know your answer is correct?

g. What time is $3\frac{1}{2}$ hours after 11:50 p.m.?

h. *Analyze* Dakota got home from soccer practice $4\frac{1}{2}$ hours before she went to sleep. If she went to sleep at 10:00 p.m., at what time did she get home?

Written Practice *Strengthening Concepts*

1. (12) **Analyze** When the sum of 24 and 7 is multiplied by the difference of 18 and 6, what is the product?

2. (13) **Formulate** Davy Crockett was born in Tennessee in 1786 and died at the Alamo in 1836. How many years did he live? Write an equation and solve the problem.

3. (15, 23) A 16-ounce box of a certain cereal costs $2.24. What is the cost per ounce of the cereal?

*** 4.** (32) What time is 3 hours 30 minutes after 6:50 a.m.?

*** 5.** (29) Forty percent equals $\frac{40}{100}$. Reduce $\frac{40}{100}$.

*** 6.** (31) A baseball diamond is the square section formed by the four bases on a baseball field. On a major league field the distance between home plate and 1st base is 90 feet. What is the area of a baseball diamond?

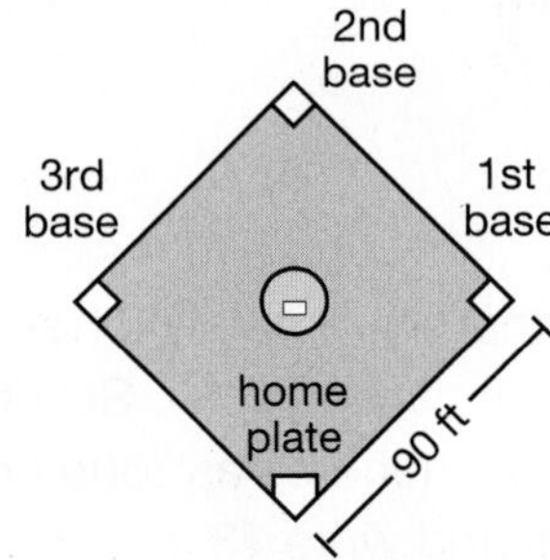

7. (8) What is the perimeter of a baseball diamond, as described in problem 6?

8. (10) **Generalize** Describe the sequence below. Then find the **eighth** term.

1, 3, 5, 7, . . .

*** 9.** (32) Write 7500 in expanded notation.

10. (9) **Estimate** Which of these numbers is closest to 1000?

A 990 **B** 909 **C** 1009 **D** 1090

11. (18) In three separate bank accounts Sumi has $623, $494, and $380. What is the average amount of money she has per account?

12. (2) $0.05 × 100

*** 13.** (30) How many $\frac{2}{5}$s are in 1?

14. (22) **Model** How much money is $\frac{3}{4}$ of $24? Draw a diagram to illustrate the problem.

Write each answer in simplest form:

15. (24) $\frac{3}{5} + \frac{3}{5}$

16. (24) $\frac{3}{4} - \frac{1}{4}$

*** 17.** (29) $\frac{3}{4} \times \frac{1}{3}$

* **18.** $\frac{3}{10} \times \frac{7}{10}$
(29)

* **19.** $1\frac{2}{3} - 1\frac{1}{3}$
(26)

20. **Connect** Three fourths of a circle is what percent of a circle?
(Inv. 2)

Find each unknown number. Remember to check your work.

21. $w - 53 = 12$
(3)

22. $8q = 240$
(4)

23. Fifteen of the three dozen students in the science club were boys. What was the ratio of boys to girls in the club?
(23)

* **24.** What is the least common multiple of 4 and 6?
(30)

* **25.** **Represent** Draw triangle *ABC* so that $\angle C$ measures 90°, side *AC* measures 3 in., and side *BC* measures 4 in. Then draw and measure the length of side *AB.*
(Inv. 3)

* **26.** If 24 of the 30 students finished the assignment in class, what fraction of the students finished in class?
(29)

* **27.** Ajani and Sharon began the hike at 6:45 a.m. and finished at 11:15 a.m. For how long did they hike?
(32)

* **28.** Compare: $(3 \times 100) + (5 \times 1) \bigcirc 350$
(32)

29. **Connect** What fraction is represented by point *A* on the number line below?
(17)

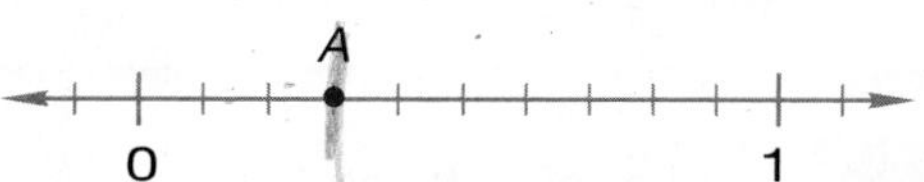

30. **Explain** Some grocery stores post the price per ounce of different cereals to help customers compare costs. How can we find the cost per ounce of a box of cereal?
(15, 23)

LESSON 33

• Writing Percents as Fractions, Part 1

Power Up

Building Power

facts Power Up G

mental math

a. **Order of Operations:** $(4 \times 100) + (4 \times 25)$

b. **Number Sense:** 7×29

c. **Calculation:** $56 + 28$

d. **Money:** $5.50 + $1.75

e. **Number Sense:** Double 120.

f. **Number Sense:** $\frac{120}{10}$

g. **Geometry:** A rectangle has a length of 6 in. and a width of 3 in. What is the area of the rectangle?

h. **Calculation:** 2×3, $+ 1$, $\times 8$, $+ 4$, $\div 6$, $\times 2$, $+ 1$, $\div 3$

problem solving

Monica picked up a number cube and held it so that she could see the dots on three adjoining faces. Monica said that she could see a total of 7 dots. How many dots were on each of the faces she could see? What was the total number of dots on the three faces she could not see?

New Concept

Increasing Knowledge

Our fraction manipulatives describe parts of circles as fractions and as percents. The manipulatives show that 50% is equivalent to $\frac{1}{2}$ and that 25% is equivalent to $\frac{1}{4}$. A **percent** is actually a fraction with a denominator of 100. The word *percent* and its symbol, %, mean "per hundred."

We can use a grid with 100 squares to model percent.

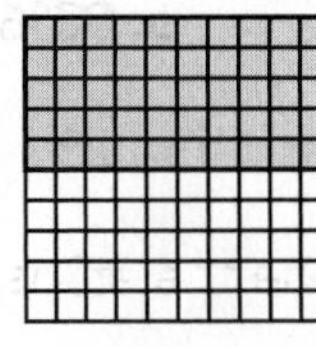

$50\% = \frac{50}{100} = \frac{1}{2}$ $25\% = \frac{25}{100} = \frac{1}{4}$

To write a percent as a fraction, we remove the percent sign and write the number as the numerator and 100 as the denominator. Then we reduce if possible.

Example 1

In Benjamin's class, 60% of the students walk to school. Write 60% as a fraction.

Solution

We remove the percent sign and write 60 over 100.

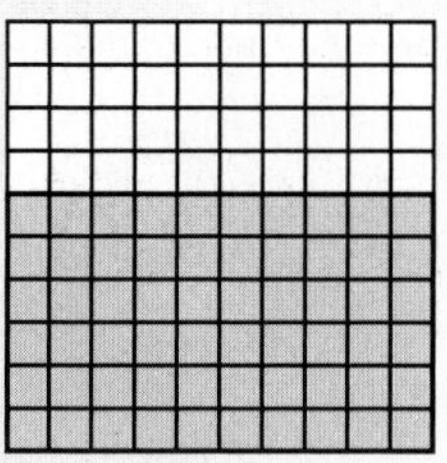

$$60\% = \frac{60}{100}$$

Reading Math

GCF stands for *greatest common factor.* The greatest common factor is the greatest number that is a factor of each of two or more numbers.

We can reduce $\frac{60}{100}$ in one step by dividing 60 and 100 by their GCF, which is 20. If we begin by dividing by a number smaller than 20, it will take more than one step to reduce the fraction.

$$\frac{60 \div 20}{100 \div 20} = \frac{3}{5}$$

We find that 60% is equivalent to the fraction $\frac{3}{5}$.

Example 2

Find the reduced fraction that equals 4%.

Solution

Thinking Skill

Discuss

How do you find the greatest common factor of two numbers?

We remove the percent sign and write 4 over 100.

$$4\% = \frac{4}{100}$$

We reduce the fraction by dividing both the numerator and denominator by 4, which is the GCF of 4 and 100.

$$\frac{4 \div 4}{100 \div 4} = \frac{1}{25}$$

We find that 4% is equivalent to the fraction $\frac{1}{25}$.

Practice Set

Write each percent as a fraction. Reduce when possible.

a. 80% **b.** 5% **c.** 25%

d. 24% **e.** 23% **f.** 10%

g. 20% **h.** 2% **i.** 75%

j. *Justify* Describe the steps you would take to write 40% as a reduced fraction.

Written Practice

Strengthening Concepts

1. (12) *Analyze* When the product of 10 and 15 is divided by the sum of 10 and 15, what is the quotient?

2. (13) The Nile River is 6690 kilometers long. The Mississippi River is 3792 kilometers long. How much longer is the Nile River than the Mississippi? Write an equation and solve the problem.

3. Some astronomers think that the universe is about fourteen billion years
(12) old. Use digits to write that number of years.

*** 4.** Write 3040 in expanded notation.
(32)

*** 5.** Write $(6 \times 100) + (2 \times 1)$ in standard notation.
(32)

*** 6.** **Connect** Write two fractions equal to 1, one with a denominator of 10
(29) and the other with a denominator of 100.

*** 7.** By what number should $\frac{5}{3}$ be multiplied for the product to be 1?
(30)

8. What is the perimeter of this rectangle?
(8)

*** 9.** How many square tiles with sides 1 inch
(31) long would be needed to cover this rectangle?

12 in.

8 in.

10. **Classify** Which of these numbers is divisible by both 2 and 3?
(21)

A 56 **B** 75 **C** 83 **D** 48

11. Estimate the difference of 4968 and 2099.
(16)

12. $4.30 × 100
(2)

13. $402.00 ÷ 25
(2)

14. What is $\frac{3}{5}$ of 20?
(29)

Write each answer in simplest form:

15. $\frac{4}{5} + \frac{4}{5}$
(24)

16. $\frac{5}{8} - \frac{1}{8}$
(24)

*** 17.** $\frac{5}{2} \times \frac{3}{2}$
(29)

*** 18.** $\frac{3}{10} \times \frac{3}{100}$
(29)

19. **Generalize** Describe the sequence below. Then find the **tenth** term.
(10)

2, 4, 6, 8, ...

Find each unknown number. Remember to check your work.

20. $q - 24 = 23$
(3)

*** 21.** $\frac{1}{2}w = 1$
(30)

22. Here we show 16 written as a product of prime numbers:
(19)

$$2 \cdot 2 \cdot 2 \cdot 2$$

Write 15 as a product of prime numbers.

23. **Estimate** A meter is a little longer than a yard. About how many meters
(7) tall is a door?

*** 24.** Five of the 30 students in the class were absent. What fraction of the
(29) class was absent? Write the answer as a reduced fraction. How do you know your answer is correct?

* **25.** (17) **Connect** To what mixed number is the arrow pointing on the number line below?

* **26.** (33) Write each percent as a reduced fraction:

a. 70% **b.** 30%

27. (22) **Model** Four fifths of Gina's 20 answers were correct. How many of Gina's answers were correct? Draw a diagram to illustrate the problem.

28. (Inv. 2) **Explain** By looking at the numerator and denominator of a fraction, how can you tell whether the fraction is greater than or less than $\frac{1}{2}$?

* **29.** (32) **Analyze** What time is $6\frac{1}{2}$ hours after 8:45 p.m.?

30. (17) Arrange these fractions in order from least to greatest:

$$\frac{1}{8}, \frac{1}{4}, \frac{1}{16}, \frac{1}{2}$$

Early Finishers
Real-World Application

The Parthenon was built in Ancient Greece over 2500 years ago. The Parthenon's base is a rectangle measuring approximately 31 meters by 70 meters.

a. What is the approximate area of the Parthenon's base?

b. What is the approximate perimeter of the Parthenon's base?

LESSON 34

• Decimal Place Value

Power Up — *Building Power*

facts Power Up D

mental math

a. **Order of Operations:** $(4 \times 200) + (4 \times 25)$

b. **Number Sense:** $1480 - 350$

c. **Number Sense:** $45 + 18$

d. **Money:** $12.00 – $2.50

e. **Number Sense:** Double 250.

f. **Number Sense:** $\frac{1500}{100}$

g. **Measurement:** Which is greater, 3 feet or 1 yard?

h. **Calculation:** 3×3, $\times 9$, $- 1$, $\div 2$, $+ 2$, $\div 7$, $\times 2$

problem solving Jeanna folded a square piece of paper in half so that the left edge aligned with the right edge. Then she folded the paper again so that the top edge aligned with the bottom edge. With scissors, Jeanna cut off the top, left corner of the square. Which diagram will the paper look like when it is unfolded?

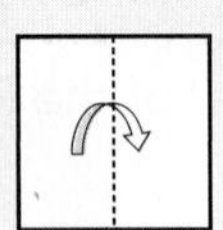
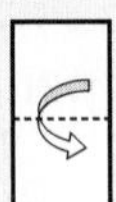
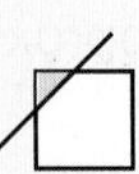
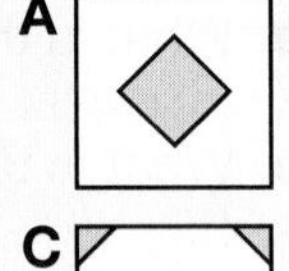

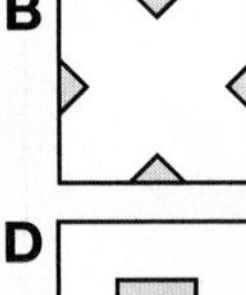

New Concept — *Increasing Knowledge*

Since Lesson 12 we have studied place value from the ones place leftward to the hundred trillions place. As we move to the left, each place is ten times as large as the preceding place. If we move to the right, each place is one tenth as large as the preceding place.

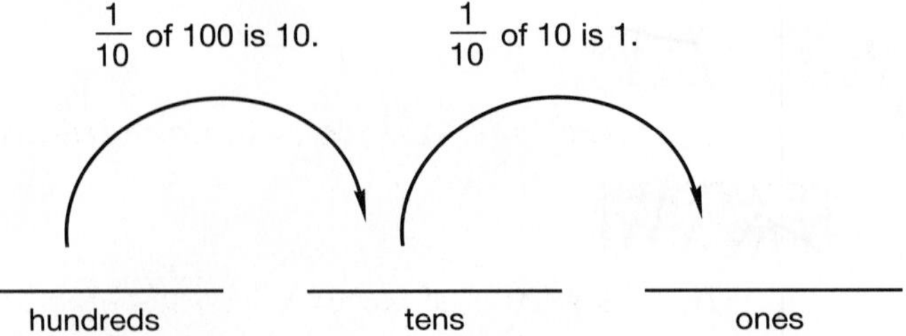

Each place to the right of the ones place also has a value one tenth the value of the place to its left. Each of these places has a value less than one (but more than zero). We use a **decimal point** to mark the separation between the ones place and places with values less than one. Places to the right of a decimal point are often called **decimal places.** Here we show three decimal places:

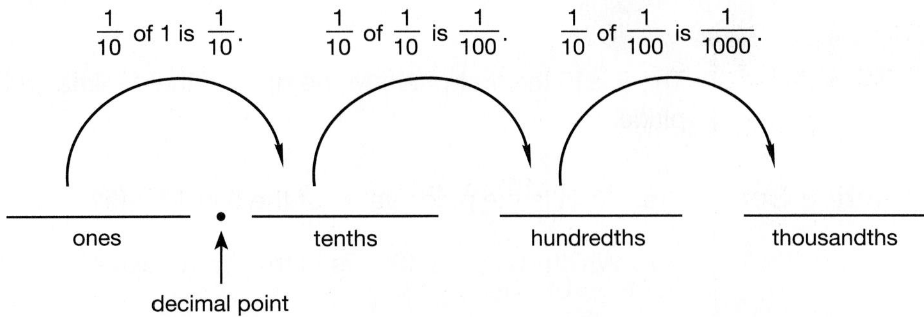

Thinking Skill

Connect

Why is thinking about money a helpful way to remember decimal place values?

Thinking about money is a helpful way to remember decimal place values.

A mill is $\frac{1}{1000}$ of a dollar and $\frac{1}{10}$ of a cent. We do not have a coin for a mill. However, purchasers of gasoline are charged mills at the gas pump. A price of $\$2.29\frac{9}{10}$ per gallon is one mill less than \$2.30 and nine mills more than \$2.29.

Of course, decimal place values extend beyond the thousandths place. The chart below shows decimal place values from the millions place through the millionths place. Moving to the right, the place values get smaller and smaller; each place has one tenth the value of the place to its left.

Discuss What pattern do you see?

Decimal Place Values

millions	1,000,000
,	
hundred thousands	100,000
ten thousands	10,000
thousands	1000
,	
hundreds	100
tens	10
ones	1
decimal point	.
tenths	$\frac{1}{10}$
hundredths	$\frac{1}{100}$
thousandths	$\frac{1}{1000}$
ten-thousandths	$\frac{1}{10,000}$
hundred-thousandths	$\frac{1}{100,000}$
millionths	$\frac{1}{1,000,000}$

Example 1

Which digit in 123.45 is in the hundredths place?

Solution

The *-ths* ending of *hundredths* indicates that the hundredths place is to the right of the decimal point. The first place to the right of the decimal point is the tenths place. The second place is the hundredths place. The digit in the hundredths place is **5.**

Example 2

What is the place value of the 8 in 67.89?

Solution

The 8 is in the first place to the right of the decimal point, which is the **tenths** place.

Practice Set

a. What is the place value of the 5 in 12.345?

b. Which digit in 5.4321 is in the tenths place?

c. In 0.0123, what is the digit in the thousandths place?

d. Connect What is the value of the place held by zero in 50.375?

e. What is the name for one hundredth of a dollar?

f. Conclude What is the name for one thousandth of a dollar?

Written Practice

Strengthening Concepts

1. (22) Model Three eighths of the 24 choir members were tenors. How many tenors were in the choir? Draw a diagram to illustrate the problem.

2. (15) Mom wants to triple a recipe for fruit salad. If the recipe calls for 8 ounces of pineapple juice, how many ounces of pineapple juice should she use?

*** 3.** (32) The mayfly has the shortest known adult life span of any animal on the planet. The mayfly grows underwater in a lake or stream for two or three years, but it lives for as little as an hour after it sprouts wings and becomes an adult. If a mayfly sprouts wings at 8:47 a.m. and lives for one hour and fifteen minutes, at what time does it die?

*** 4.** (33) Connect Write each percent as a reduced fraction:

a. 60% **b.** 40%

*** 5.** (29) Compare: $\frac{100}{100} \bigcirc \frac{10}{10}$

*** 6.** (32) Analyze Write $(6 \times 100) + (5 \times 1)$ in standard notation.

*** 7.** (34) Which digit is in the ones place in $42,876.39?

*** 8.** (31) If the perimeter of a square is 24 inches,

a. how long is each side of the square?

b. what is the area of the square?

*** 9.** (Inv. 3) Represent Draw triangle *ABC* so that $\angle C$ is a right angle, side *AC* is $1\frac{1}{2}$ inches, and side *BC* is 2 inches. Then measure the length of side *AB*.

*** 10.** (30) What is the least common multiple of 6 and 8?

11. $5.60 ÷ 10
(2)

12. $\frac{9}{10} \cdot \frac{9}{10}$
(29)

13. Estimate the quotient when 898 is divided by 29.
(16)

14. Round 36,847 to the nearest hundred.
(16)

Find each unknown number. Remember to check your work.

15. $6d = 144$
(4)

16. $\frac{d}{6} = 144$
(4)

* **17.** Compare: $\frac{5}{2} + \frac{5}{2} \bigcirc 2 \times \frac{5}{2}$
(29)

18. $\frac{3}{8} + \frac{3}{8}$
(24)

19. $\frac{11}{12} - \frac{1}{12}$
(24)

* **20.** $\frac{5}{4} \times \frac{3}{2}$
(29)

21. **Predict** What is the ratio of the first term to the fifth term of the sequence below?
(10, 23)

6, 12, 18, 24, ...

22. **Formulate** The movie theater sold 88 tickets to the afternoon show for $7.50 per ticket. What was the total of the ticket sales for the show? Write an equation and solve the problem. Explain why your answer is reasonable.
(15)

23. **Connect** To what number is the arrow pointing on the number line below?
(14)

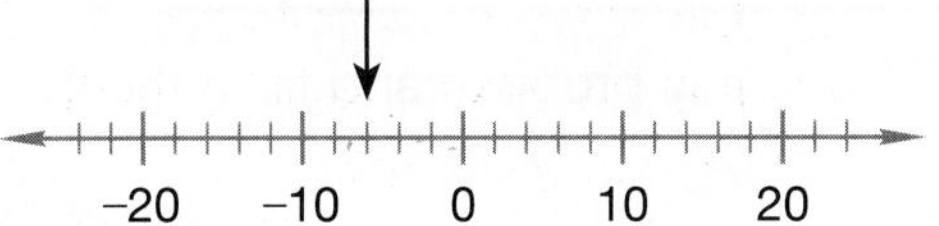

24. $(80 \div 40) - (8 \div 4)$
(5)

25. Which digit in 2,345.678 is in the thousandths place?
(24)

26. **Model** Draw a circle and shade $\frac{2}{3}$ of it.
(Inv. 2)

27. Divide 5225 by 12 and write the quotient as a mixed number.
(25)

28. **Evaluate** The first glass contained 12 ounces of water. The second glass contained 11 ounces of water. The third glass contained 7 ounces of water. If water was poured from the first and second glasses into the third glass until each glass contained the same amount, then how many ounces of water would be in each glass?
(18)

29. **Represent** The letters *r*, *t*, and *d* represent three different numbers. The product of *r* and *t* is *d*.
(2)

$$rt = d$$

Arrange the letters to form another multiplication equation and two division equations.

30. **Connect** Instead of dividing 75 by 5, Sandy mentally doubled both numbers and divided 150 by 10. Find the quotient of 75 ÷ 5 and the quotient of 150 ÷ 10.
(2)

LESSON 35

• Writing Decimal Numbers as Fractions, Part 1
• Reading and Writing Decimal Numbers

Power Up

Building Power

facts Power Up G

mental math

a. **Order of Operations:** $(4 \times 300) + (4 \times 25)$

b. **Number Sense:** 8×43

c. **Number Sense:** $37 + 39$

d. **Money:** $7.50 + $7.50

e. **Fractional Parts:** $\frac{1}{3}$ of 360

f. **Number Sense:** $\frac{3600}{10}$

g. **Measurement:** Which is greater, 10 centimeters or 10 millimeters?

h. **Calculation:** $5 \times 5, - 1, \div 3, \times 4, + 1, \div 3, + 1, \div 3$

problem solving

Copy this problem and fill in the missing digits:

$$\begin{array}{r} \_\_\_ \\ \times \quad 9 \\ \hline \_\_2 \end{array}$$

New Concepts

Increasing Knowledge

writing decimal numbers as fractions, part 1

Decimal numbers are actually fractions. Their denominators come from the sequence 10, 100, 1000, ... The denominator of a decimal fraction is not written. Instead, it is indicated by the number of decimal places.

One decimal place indicates that the denominator is 10.

$$0.3 = \frac{3}{10}$$

$\frac{3}{10}$

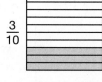

Two decimal places indicate that the denominator is 100.

$$0.03 = \frac{3}{100}$$

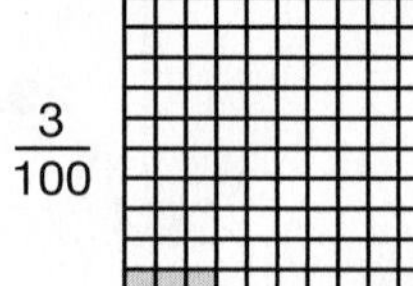

Three decimal places indicate that the denominator is 1000.

$$0.003 = \frac{3}{1000}$$

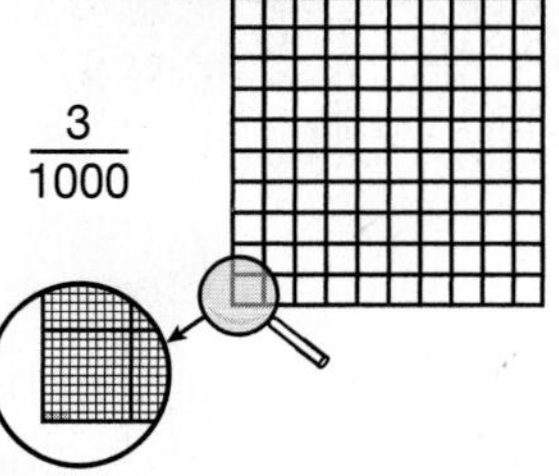

Notice that the number of zeros in the denominator equals the number of decimal places in the decimal number.

Example 1

A quart is 0.25 gallons. Write 0.25 as a fraction.

Solution

The decimal number 0.25 has two decimal places, so the denominator is 100. The numerator is 25.

$$0.25 = \mathbf{\frac{25}{100}}$$

The fraction $\frac{25}{100}$ reduces to $\mathbf{\frac{1}{4}}$.

Example 2

A kilometer is about $\frac{6}{10}$ of a mile. Write $\frac{6}{10}$ as a decimal number.

Solution

The denominator is 10, so the decimal number has one decimal place. We write the digit 6 in this place.

$$\frac{6}{10} \longrightarrow 0.\_ \longrightarrow \mathbf{0.6}$$

reading and writing decimal numbers

We read numbers to the right of a decimal point the same way we read whole numbers, and then we say the place value of the last digit. We read 0.23 as "twenty-three hundredths" because the last digit is in the hundredths place. To read a mixed decimal number like 20.04, we read the whole number part, say "and," and then read the decimal part.

Example 3

The length of a football field is about 0.057 miles. Write 0.057 with words.

Solution

We see 57 and three decimal places. We write **fifty-seven thousandths.**

Example 4

An inch is 2.54 centimeters. Use words to write 2.54.

Solution

The decimal point separates the whole number part of the number from the decimal part of the number. We name the whole number part, write "and," and then name the decimal part.

two and fifty-four hundredths

Example 5

Write twenty-one hundredths

a. as a fraction.

b. as a decimal number.

Solution

The same words name both a fraction form and a decimal form of the number.

a. The word *hundredths* indicates that the denominator is 100.

$$\mathbf{\frac{21}{100}}$$

b. The word *hundredths* indicates that the decimal number has two decimal places.

0.21

Example 6

Write fifteen and two tenths as a decimal number.

Solution

The whole number part is fifteen. The fractional part is two tenths, which we write in decimal form.

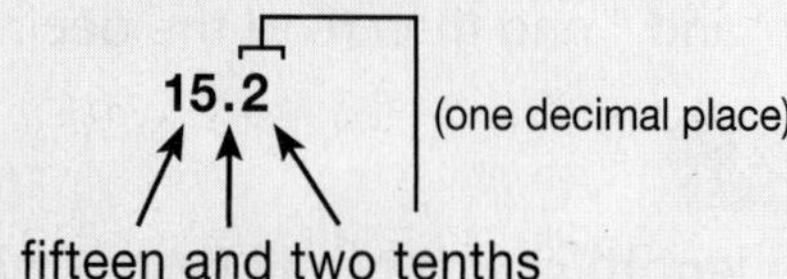

Verify How would you write 15.2 as a mixed number?

Practice Set Write each decimal number as a fraction:

a. 0.1 **b.** 0.31 **c.** 0.321

Write each fraction as a decimal number:

d. $\frac{3}{10}$ **e.** $\frac{17}{100}$ **f.** $\frac{123}{1000}$

Use words to write each number:

g. 0.05 **h.** 0.015 **i.** 1.2

Connect Write each number first as a fraction, then as a decimal number:

j. seven tenths

k. thirty-one hundredths

l. seven hundred thirty-one thousandths

Write each number as a decimal number:

m. five and six tenths

n. eleven and twelve hundredths

o. one hundred twenty-five thousandths

Written Practice *Strengthening Concepts*

1. (29) What is the product of three fourths and three fifths?

2. (22) **Model** Thomas planted 360 carrot seeds in his garden. Three fourths of them sprouted. How many carrot seeds sprouted? Draw a diagram to illustrate the problem.

* **3.** (32) Sakari's casserole must bake for 2 hours 15 minutes. If she put it into the oven at 11:45 a.m., at what time will it be done?

* **4.** (35) **Represent** Write twenty-three hundredths

a. as a fraction.

b. as a decimal number.

* **5.** (35) **Represent** Write 10.01 with words.

* **6.** (35) Write ten and five tenths as a decimal number.

* **7.** (33) **Connect** Write each percent as a reduced fraction:

a. 25% **b.** 75%

* **8.** (32) Write $(5 \times 1000) + (6 \times 100) + (4 \times 10)$ in standard notation.

* **9.** (34) **Connect** Which digit in 1.23 has the same place value as the 5 in 0.456?

10. (31) What is the area of the rectangle below?

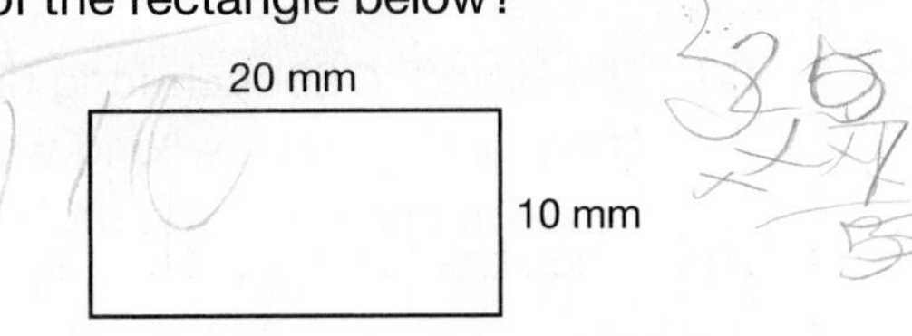

11. (8) In problem **10,** what is the perimeter of the rectangle?

12. (7) There are 100 centimeters in a meter. How many centimeters are in 10 meters?

* **13.** (34) Arrange these numbers in order from least to greatest:

$$0.001, 0.1, 1.0, 0.01, 0, -1$$

14. (7) **Estimate** A meter is about one big step. About how many meters wide is a door?

15. (24) $\frac{3}{5} + \frac{2}{5}$

16. (24) $\frac{5}{8} - \frac{5}{8}$

17. (29) $\frac{2}{3} \times \frac{3}{4}$

* **18.** (30) **a.** How many $\frac{2}{5}$s are in 1?

b. **Connect** Use the answer to part **a** to find the number of $\frac{2}{5}$s in 2.

19. (25) Convert $\frac{20}{6}$ to a mixed number. Remember to reduce the fraction part of the number.

20. (25) $\frac{100\%}{6}$

21. (26) $3\frac{4}{4} - 1\frac{1}{4}$

22. (29) Compare: $5 \bigcirc 4\frac{4}{4}$

23. (Inv. 2) **Represent** One sixth of a circle is what percent of a circle?

24. (9) Compare: $3 \times 18 \div 6 \bigcirc 3 \times (18 \div 6)$

25. (14) **Connect** To what number is the arrow pointing on the number line below?

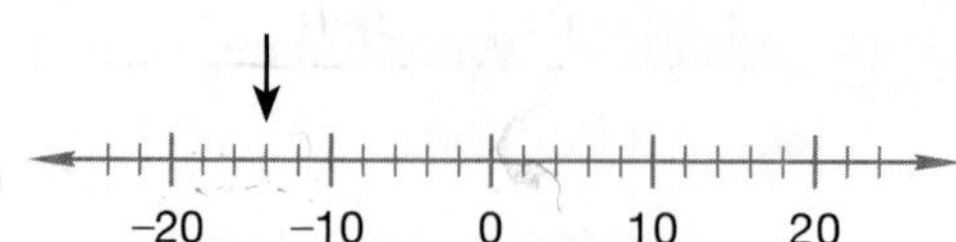

26. (2) **Conclude** Which of these division problems has the greatest quotient?

A $\frac{6}{2}$ 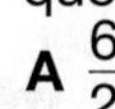 **B** $\frac{60}{20}$ **C** $\frac{12}{4}$ **D** $\frac{25}{8}$

* **27.** (35) **Conclude** Write 0.3 and 0.7 as fractions. Then multiply the fractions. What is the product?

* **28.** (33, 35) Write 21% as a fraction. Then write the fraction as a decimal number.

29. (2) Instead of solving the division problem $400 \div 50$, Minh doubled both numbers to form the division $800 \div 100$. Find both quotients.

30. (23, 25) **Formulate** A 50-inch-long ribbon was cut into four shorter ribbons of equal length. How long was each of the shorter ribbons? Write an equation and solve the problem. Explain why your answer is reasonable.

LESSON 36

Subtracting Fractions and Mixed Numbers from Whole Numbers

Power Up

Building Power

facts Power Up F

mental math

a. **Order of Operations:** $(4 \times 400) + (4 \times 25)$

b. **Number Sense:** $2500 + 375$

c. **Number Sense:** $86 - 39$

d. **Money:** \$15.00 − \$2.50

e. **Fractional Parts:** $\frac{1}{2}$ of 320

f. **Number Sense:** $\frac{4800}{100}$

g. **Measurement:** Which is greater, 6 months or 52 weeks?

h. **Calculation:** 2×4, $\times 5$, $+ 10$, $\times 2$, $- 1$, $\div 9$, $\times 3$, $- 1$, $\div 4$

problem solving The playground is filled with bicycles and wagons. If there are 24 vehicles and 80 wheels altogether, how many bicycles are on the playground? How many wagons?

New Concept

Increasing Knowledge

Read this "separating" word problem about pies.

There were four pies on the shelf. The server sliced one of the pies into sixths and took $2\frac{1}{6}$ pies from the shelf. How many pies were left on the shelf?

We can illustrate this problem with circles. There were four pies on the shelf.

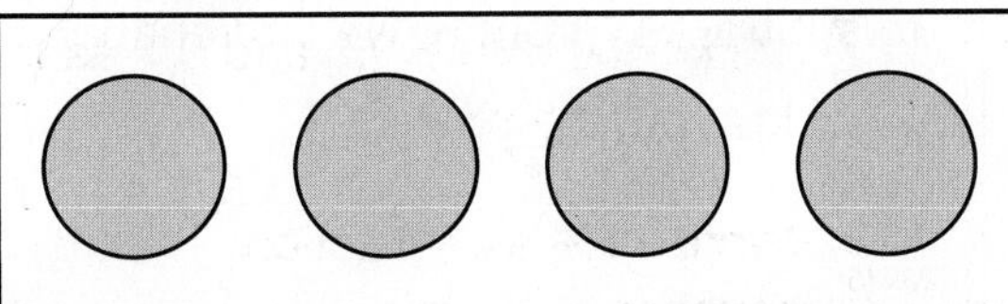

The server sliced one of the pies into sixths. (Then there were $3\frac{6}{6}$ pies, which is another name for 4 pies.)

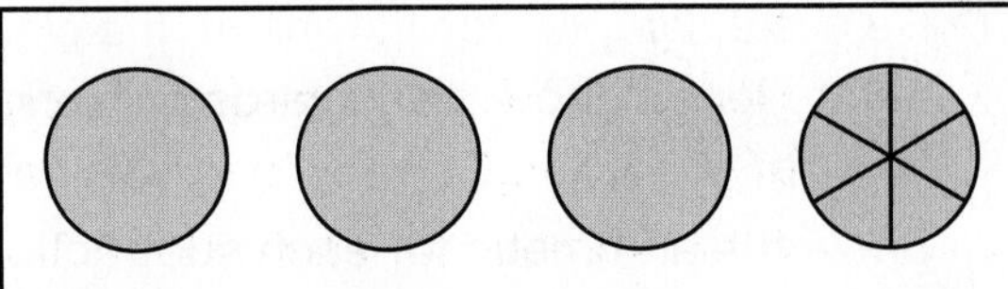

The server took $2\frac{1}{6}$ pies from the shelf.

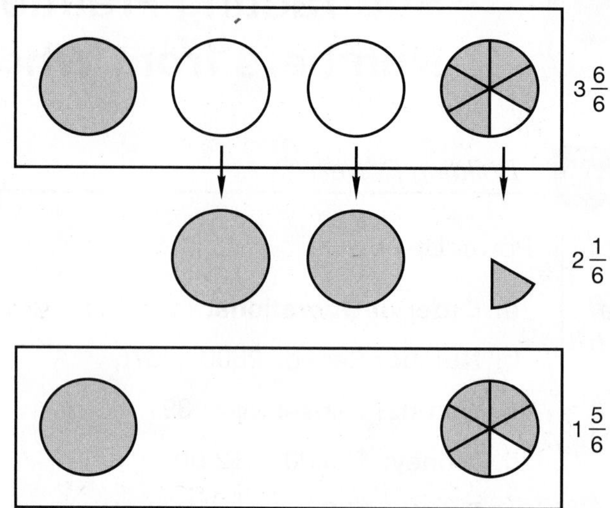

We see $1\frac{5}{6}$ pies left on the shelf.

Now we show the arithmetic for subtracting $2\frac{1}{6}$ from 4.

$$\begin{array}{r} 4 \text{ pies} \\ -\ 2\frac{1}{6} \text{ pies} \\ \hline \end{array}$$

The server sliced one of the pies into sixths, so we change 4 wholes into 3 wholes plus 6 sixths. Then we subtract.

$$\begin{array}{r} 4 \\ -\ 2\frac{1}{6} \\ \hline \end{array} \xrightarrow{3 + \frac{6}{6}} \begin{array}{r} 3\frac{6}{6} \\ -\ 2\frac{1}{6} \\ \hline \mathbf{1\frac{5}{6}} \end{array}$$

Example

Subtract: $5 - 1\frac{2}{3}$. Then write a word problem that is solved by the subtraction.

Solution

To subtract $1\frac{2}{3}$ from 5, we first change 5 to 4 plus $\frac{3}{3}$. Then we subtract.

Thinking Skill

Discuss

Why do we need to change 5 to $4\frac{3}{3}$ before we subtract $1\frac{2}{3}$ from 5?

$$\begin{array}{r} 5 \\ -\ 1\frac{2}{3} \\ \hline \end{array} \xrightarrow{4 + \frac{3}{3}} \begin{array}{r} 4\frac{3}{3} \\ -\ 1\frac{2}{3} \\ \hline \mathbf{3\frac{1}{3}} \end{array}$$

Common subtraction word problems are about comparing or separating. Here is a sample word problem. If a clerk cuts $1\frac{2}{3}$ yards of fabric from a 5-yard length, how many yards of fabric remain?

Practice Set

Show the arithmetic for each subtraction:

a. $3 - 2\frac{1}{2}$

b. $2 - \frac{1}{4}$

c. $4 - 2\frac{1}{4}$

d. $3 - \frac{5}{12}$

e. $10 - 2\frac{1}{2}$ **f.** $6 - 1\frac{3}{10}$

g. **Model** Select one of the exercises to model with a drawing.

h. **Formulate** Select another exercise and write a word problem that is solved by the subtraction.

i. There were four whole pies on the shelf. The server took $1\frac{5}{6}$ pies. How many pies were left on the shelf?

j. **Formulate** Write a word problem similar to problem **i,** and then find the answer.

Written Practice *Strengthening Concepts*

*** 1.** (33) **Analyze** Twenty-five percent of the students played musical instruments. What fraction of the students played musical instruments?

*** 2.** (36) About $\frac{3}{4}$ of the Earth's surface is covered with water. What fraction of Earth's surface is not covered with water?

3. (15) A mile is 5280 feet. There are 3 feet in a yard. How many yards are in a mile?

*** 4.** (34) Which digit in 23.47 has the same place value as the 6 in 516.9?

*** 5.** (35) Write 1.3 with words.

*** 6.** (35) Write the decimal number five hundredths.

*** 7.** (35) Write thirty-one hundredths

a. as a fraction.

b. as a decimal number.

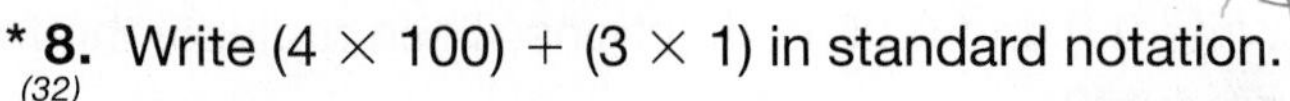

*** 8.** (32) Write $(4 \times 100) + (3 \times 1)$ in standard notation.

*** 9.** (34) Which digit in 4.375 is in the tenths place?

*** 10.** (8, 31) **Analyze** If the area of a square is 9 square inches,

a. how long is each side of the square?

b. what is the perimeter of the square?

11. (28) Name two obtuse angles in the figure below.

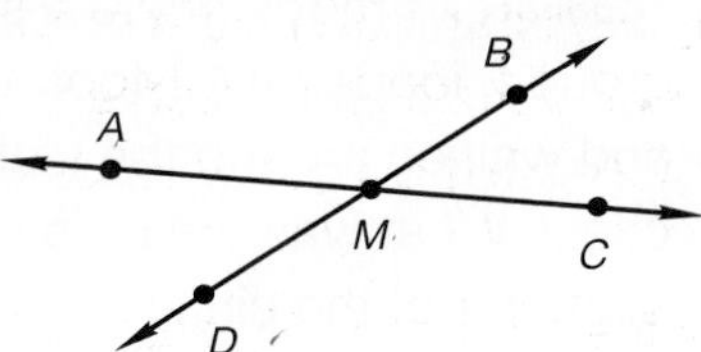

12. (26) $3\frac{1}{4} + 2\frac{1}{4}$ *** 13.** (36) $3 - 1\frac{1}{4}$ **14.** (26) $3\frac{1}{3} + 2\frac{2}{3}$

15. (29) What is $\frac{3}{4}$ of 28?

16. (29) $\frac{3}{4} \times \frac{4}{6}$

17. (22) **Model** Chen went to the mall with $24. He spent $\frac{5}{6}$ of his money in the music store. How much money did he spend in the music store? Draw a diagram to illustrate the problem.

18. (18) What is the average of 42, 57, and 63?

19. (19) The factors of 6 are 1, 2, 3, and 6. List the factors of 20.

20. (20, 30) **a.** What is the least common multiple of 9 and 6?

b. What is the greatest common factor of 9 and 6?

Find each unknown number. Remember to check your work.

21. (4) $\frac{m}{12} = 6$

22. (4) $\frac{12}{n} = 6$

23. (16) Round 58,742,177 to the nearest million.

24. (16) Estimate the product of 823 and 680.

25. (7) How many millimeters long is the line segment shown below (1 cm = 10 mm)?

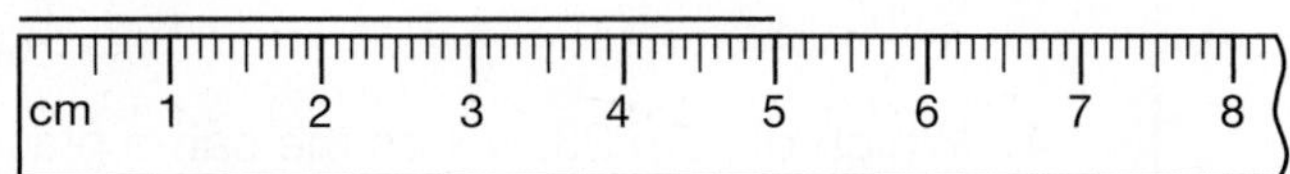

26. (Inv. 2) **Model** Using your fraction manipulatives, you can find that the sum of $\frac{1}{3}$ and $\frac{1}{6}$ is $\frac{1}{2}$.

$$\frac{1}{3} + \frac{1}{6} = \frac{1}{2}$$

Represent Arrange these fractions to form another addition equation and two subtraction equations.

* **27.** (35) Write 0.9 and 0.09 as fractions. Then multiply the fractions. What is the product?

* **28.** (35) **Represent** Write $\frac{81}{1000}$ as a decimal number. (*Hint:* Write a zero in the tenths place.)

* **29.** (30) **Analyze** **a.** How many $\frac{3}{4}$s are in 1?

b. Use the answer to part **a** to find the number of $\frac{3}{4}$s in 3.

Math Language

The **radius** is the distance from the center of a circle to a point on the circle. The **diameter** is the distance across a circle through its center.

30. (23, 27) **Connect** Freddy drove a stake into the ground, looped a 12-foot-long rope over it, and walked around the stake to mark off a circle. What was the ratio of the radius to the diameter of the circle?

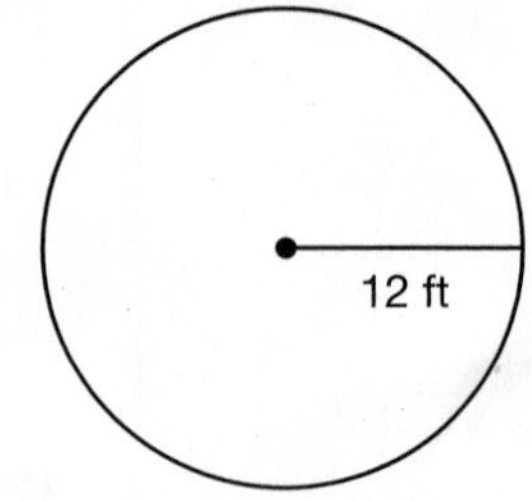

LESSON 37

• Adding and Subtracting Decimal Numbers

Power Up — *Building Power*

facts Power Up G

mental math

a. Order of Operations: $(4 \times 500) + (4 \times 25)$

b. Number Sense: 9×43

c. Number Sense: $76 - 29$

d. Money: \$17.50 + \$2.50

e. Fractional Parts: $\frac{1}{2}$ of 520

f. Number Sense: $\frac{2500}{10}$

g. Geometry: A rectangle has a length of 10 in. and the width is $\frac{1}{2}$ as long as the length. What is the perimeter of the rectangle?

h. Calculation: 6×8, $+ 1$, $\div 7$, $\times 3$, $- 1$, $\div 5$, $+ 1$, $\div 5$

problem solving If a piglet weighs the same as two ducks, three piglets weigh the same as a young goat, and a young goat weighs the same as two terriers, then how many of each animal weighs the same as a terrier?

New Concept — *Increasing Knowledge*

When we add or subtract numbers using pencil and paper, it is important to align digits that have the same place value. For whole numbers this means lining up the ending digits. When we line up the ending digits (which are in the ones place) we automatically align other digits that have the same place value.

```
    23
   241
 + 317
 -----
```

Lining up the ones place automatically aligns all other digits by their place value.

Thinking Skill

Analyze

In the number 2.41, in what place is the digit 4? the digit 1?

However, lining up the ending digits of decimal numbers might not properly align all the digits. We use another method for decimal numbers. **We line up decimal numbers for addition or subtraction by lining up the decimal points.** The decimal point in the answer is aligned with the other decimal points. Empty places are treated as zeros.

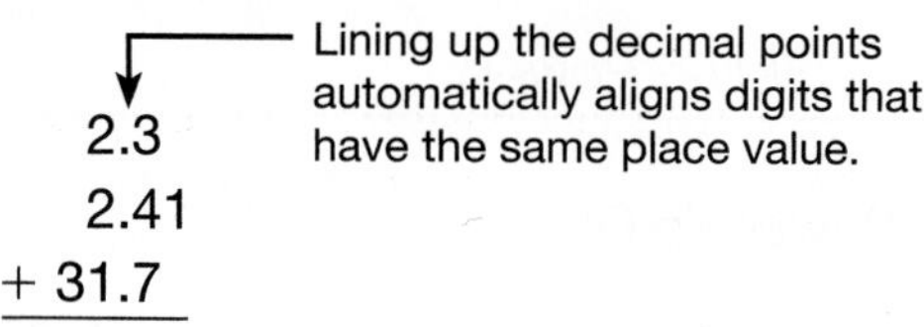
Lining up the decimal points automatically aligns digits that have the same place value.
2.3
2.41
+ 31.7
.

Example 1

The rainfall over a three-day period was 3.4 inches, 0.26 inches, and 0.3 inches. Altogether, how many inches of rain fell during the three days?

Solution

We line up the decimal points and add. The decimal point in the sum is placed in line with the other decimal points. In three days **3.96 inches** of rain fell.

$$\begin{array}{r} 3.4 \\ 0.26 \\ +\ 0.3 \\ \hline 3.96 \end{array}$$

Example 2

A gallon is about 3.78 liters. If Margaret pours 2.3 liters of milk from a one-gallon bottle, how much milk remains in the bottle?

Solution

We subtract 2.3 from 3.78. We line up the decimal points to subtract and find that **1.48 liters** of milk remains in the bottle.

$$\begin{array}{r} 3.78 \\ -\ 2.3 \\ \hline 1.48 \end{array}$$

Practice Set

Find each sum or difference. Remember to line up the decimal points.

a. 3.46 + 0.2 **b.** 8.28 − 6.1

c. 0.735 + 0.21 **d.** 0.543 − 0.21

e. 0.43 + 0.1 + 0.413 **f.** 0.30 − 0.27

g. 0.6 + 0.7 **h.** 1.00 − 0.24

i. 0.9 + 0.12 **j.** 1.23 − 0.4

Written Practice

Strengthening Concepts

* **1.** (33) *Represent* Sixty percent of the students in the class were girls. What fraction of the students in the class were girls?

2. (22) *Analyze* Penny broke 8 pencils during the math test. She broke half as many during the spelling test. How many pencils did she break in all?

3. **Analyze** What number must be added to three hundred seventy-five to get the number one thousand?
(3)

* 4. $3.4 + 0.62 + 0.3$
(37)

* 5. $4.56 - 3.2$
(37)

6. $\$0.37 + \$0.23 + \$0.48$
(1)

7. $\$5 - m = 5¢$
(3)

8. **Predict** What is the next number in this sequence?
(10)

1, 10, 100, 1000, ...

* 9. **Connect** Harriet used 100 square floor tiles with sides 1 foot long to cover the floor of a square room.
(8, 31)

a. What was the length of each side of the room?

b. What was the perimeter of the room?

10. Which digit is in the ten-millions place in 1,234,567,890?
(12)

* 11. **Classify** Three of the numbers shown below are equal. Which number is not equal to the others? How do you know?
(35)

A $\frac{1}{10}$ B 0.1 C $\frac{10}{100}$ D 0.01

12. Estimate the product of 29, 42, and 39.
(16)

13. $3210 \div 3$
(2)

14. $32{,}100 \div 30$
(2)

15. $\$10{,}000 - \$345$
(1)

16. $\frac{3}{4} + \frac{3}{4}$
(24)

* 17. **Analyze** $3 - 1\frac{3}{5}$
(36)

* 18. $\frac{3}{3} - \frac{2}{2}$
(29)

19. $1\frac{1}{3} + 2\frac{1}{3} + 3\frac{1}{3}$
(26)

* 20. **Analyze** Compare: $\frac{1}{4} + \frac{3}{4} \bigcirc \frac{1}{4} \times \frac{3}{4}$
(29)

21. Convert the improper fraction $\frac{100}{7}$ to a mixed number.
(25)

22. What is the average of 90 lb, 84 lb, and 102 lb?
(18)

23. What is the least common multiple of 4 and 5?
(30)

24. The stock's value dropped from \$38.50 to \$34.00. What negative number shows the change in value?
(14)

25. **Connect** To what mixed number is the arrow pointing on the number line below?
(17)

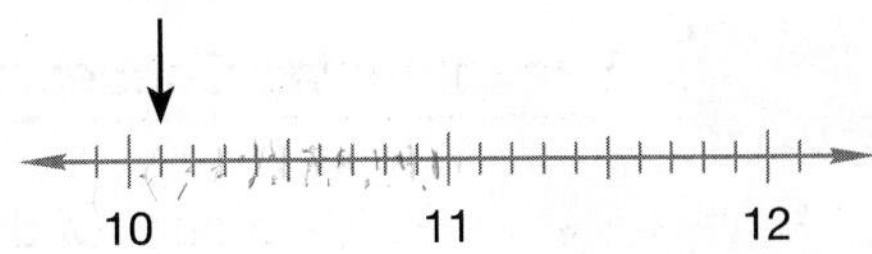

* **26.** (35) **Represent** Write 0.3 and 0.9 as fractions. Then multiply the fractions. Change the product to a decimal number.

27. (Inv. 2) **Explain** Write three different fractions equal to 1. How can you tell whether a fraction is equal to 1?

28. (2) **Justify** Instead of dividing 6 by $\frac{1}{2}$, Feodor doubled both numbers and divided 12 by 1. Do you think both quotients are the same? Write a one- or two-sentence reason for your answer.

* **29.** (32) The movie started at 2:50 p.m. and ended at 4:23 p.m. How long was the movie?

30. (22) **Model** Three fifths of the 25 students in the class were boys. How many boys were in the class? Draw a diagram to illustrate the problem.

Early Finishers
Real-World Application

Chamile is making a shirt and she needs to determine how many yards of fabric to buy. After calculating her measurements, Chamile determines she needs 90 inches of fabric (1 yard = 36 inches).

a. How many yards of fabric does it take to make the shirt?

b. If the fabric store only sells fabric in full yards, how much fabric will Chamile have leftover?

LESSON

38

• Adding and Subtracting Decimal Numbers and Whole Numbers
• Squares and Square Roots

Power Up

Building Power

facts Power Up A

mental math

a. **Order of Operations:** $(4 \times 600) + (4 \times 25)$

b. **Number Sense:** 875 − 125

c. **Number Sense:** 56 − 19

d. **Money:** \$10.00 − \$6.25

e. **Fractional Parts:** $\frac{1}{2}$ of 150

f. **Number Sense:** $\frac{\$40.00}{10}$

g. **Geometry:** A regular octagon has sides that measure 7 cm. What is the perimeter of the octagon?

h. **Calculation:** 10 + 10, − 2, ÷ 3, × 4, + 1, × 4, ÷ 2, + 6, ÷ 7

problem solving

Andre, Robert, and Carolina stood side-by-side for a picture. Then they changed their order for another picture. Then they changed their order again. List all possible side-by-side arrangements of the three friends.

New Concepts

Increasing Knowledge

adding and subtracting decimal numbers and whole numbers

Margie saw this sale sign in the clothing department. What is another way to write three dollars?

T-shirt close-out
\$3 each

Here we show two ways to write three dollars:

\$3 \$3.00

We see that we may write dollar amounts with or without a decimal point. We may also write whole numbers with or without a decimal point. The decimal point follows the ones place. Here are several ways to write the whole number three:

3 3. 3.0 3.00

As we will see in the following examples, it may be helpful to write a whole number with a decimal point when adding and subtracting with decimal numbers.

Example 1

Paper used in school is often 11 inches long and 8.5 inches wide. Find the perimeter of a sheet of paper by adding the lengths of the four sides.

Solution

When adding decimal numbers, we align decimal points so that we add digits with the same place values. The whole number 11 may be written with a decimal point to the right. We line up the decimal points and add. The perimeter is **39.0 inches.**

$$\begin{array}{r} 11. \\ 8.5 \\ 11. \\ +\ 8.5 \\ \hline \mathbf{39.0} \end{array}$$

Example 2

Subtract: 12.75 − 5

Solution

We write the whole number 5 with a decimal point to its right. Then we line up the decimal points and subtract.

$$\begin{array}{r} 12.75 \\ -\ 5. \quad \\ \hline \mathbf{7.75} \end{array}$$

squares and square roots

Recall that we find the area of a square by multiplying the length of a side of the square by itself. For example, the area of a square with sides 5 cm long is 5 cm × 5 cm, which equals 25 sq. cm.

Area is
25 sq. cm

Side is 5 cm

From the model of the square comes the expression "squaring a number." We square a number by multiplying the number by itself.

"Five squared" is 5 × 5, which is 25.

To indicate squaring, we use the **exponent** 2.

$$5^2 = 25$$

"Five squared equals 25."

An exponent shows how many times the other number, the **base,** is to be used as a factor. In this case, 5 is to be used as a factor twice.

Reading Math

An exponent is elevated and written to the right of a number. Read 5^2 as "five squared."

Example 3

a. What is twelve squared?

b. Simplify: $3^2 + 4^2$

Solution

a. "Twelve squared" is 12 × 12, which is **144.**

b. We apply exponents before adding, subtracting, multiplying, or dividing.

$$3^2 + 4^2 = 9 + 16 = \mathbf{25}$$

Analyze What step is not shown in the solution for **b?**

Example 4

Reading Math

We can use an exponent of 2 with a unit of length to indicate square units for measuring area. 1 cm^2 = 1 square centimeter

What is the area of a square with sides 5 meters long?

Solution

We multiply 5 meters by 5 meters. Both the units and the numbers are multiplied.

$$5 \text{ m} \cdot 5 \text{ m} = (5 \cdot 5)(\text{m} \cdot \text{m}) = \mathbf{25 \text{ m}^2}$$

We read 25 m^2 as "twenty-five square meters."

If we know the area of a square, we can find the length of each side. We do this by determining the length whose square equals the area. For example, a square whose area is 49 cm^2 has side lengths of 7 cm because 7 cm · 7 cm equals 49 cm^2.

Reading Math

The square root symbol looks like this: $\sqrt{}$.

Determining the length of a side of a square from the area of the square is a model for finding the principal **square root** of a number. Finding the square root of a number is the inverse of squaring a number.

6 squared is 36.

The principal square root of 36 is 6.

We read $\sqrt{100}$ as "the square root of 100." This expression means, "What positive number, when multiplied by itself, has a product of 100?" Since 10 × 10 equals 100, the principal square root of 100 is 10.

$$\sqrt{100} = 10$$

A number is a **perfect square** if it has a square root that is a whole number. Starting with 1, the first four perfect squares are 1, 4, 9, and 16, as illustrated below.

1 × 1 = 1

$\sqrt{1} = 1$

2 × 2 = 4

$\sqrt{4} = 2$

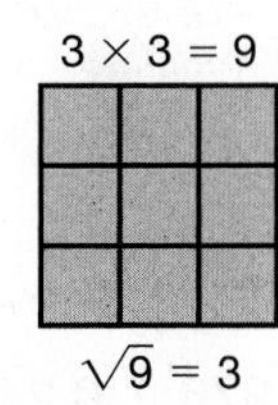

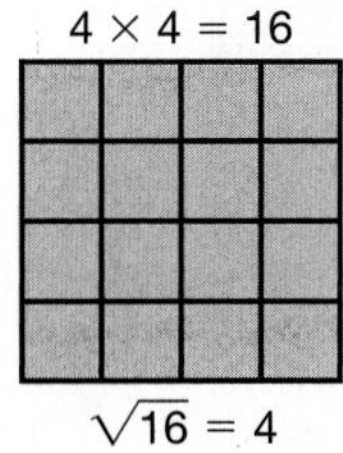

Generalize What is the fifth perfect square? Draw it.

Example 5

Simplify: $\sqrt{64}$

Solution

The square root of 64 can be thought of in two ways:

1. What is the side length of a square that has an area of 64 square units?
2. What positive number multiplied by itself equals 64?

With either approach, we find that $\sqrt{64}$ equals **8.**

Practice Set

Simplify:

a. $4 + 2.1$

b. $4.3 - 2$

c. $3 + 0.4$

d. $43.2 - 5$

e. $0.23 + 4 + 3.7$

f. $6.3 - 6$

g. $12.5 + 10$

h. $75.25 - 25$

i. 9^2

j. $\sqrt{81}$

k. $6^2 + 8^2$

l. $\sqrt{100} - \sqrt{49}$

m. 15^2

n. $\sqrt{144}$

o. 6 ft · 6 ft

p. $\sqrt{64 \text{ m}^2}$

q. *Predict* Starting with 1, the first four perfect squares are as follows:

1, 4, 9, 16

What are the next four perfect squares? Explain how you know.

Written Practice

Strengthening Concepts

1. (20) What is the greatest factor of both 54 and 45?

2. (15) *Formulate* Roberto began saving $3 each week for a bicycle, which costs $126. How many weeks will it take him to save that amount of money? Write an equation and solve the problem.

3. (13) *Formulate* Gandhi was born in 1869. About how old was he when he was assassinated in 1948? Write an equation and solve the problem.

* **4.** (38) $\sqrt{9} + 1.2$

* **5.** (38) $3.6 + \sqrt{16}$

* **6.** (37) $5.63 - 1.2$

* **7.** (37) $5.376 + 0.24$

* **8.** (37) $4.75 - 0.6$

* **9.** (38) $\sqrt{16} - \sqrt{9}$

* **10.** (35) Write forty-seven hundredths

a. as a fraction.

b. as a decimal number.

11. Write $(9 \times 1000) + (4 \times 10) + (3 \times 1)$ in standard notation.
(32)

* **12.** Which digit is in the hundredths place in $123.45?
(34)

* **13.** The area of a square is 81 square inches.
(8, 38)

a. What is the length of each side?

b. What is the perimeter of the square?

14. What is the least common multiple of 2, 3, and 4?
(30)

15. $1\frac{2}{3} + 2\frac{2}{3}$
(26)

* **16.** $3^2 - 1\frac{1}{4}$
(36, 38)

17. What is $\frac{3}{4}$ of $\frac{4}{5}$?
(29)

18. $\frac{7}{10} \times \frac{11}{10}$
(29)

19. **a.** How many $\frac{2}{3}$s are in 1?
(30)

b. Use the answer to part **a** to find the number of $\frac{2}{3}$s in 2.

20. Six of the nine players got on base. What fraction of the players got on base?
(29)

21. List the factors of 30.
(19)

* **22.** Write each percent as a reduced fraction:
(33)

a. 35% **b.** 65%

23. Round 186,282 to the nearest thousand.
(16)

24. $\frac{1}{3}m = 1$
(30)

25. $\frac{22 + 23 + 24}{3}$
(5)

26. Compare: $24 \div 8 \bigcirc 240 \div 80$
(9)

* **27.** Write 0.7 and 0.21 as fractions. Then multiply the fractions. Change the product to a decimal number. Explain why your answer is reasonable. (*Hint:* Round the original problem.)
(35)

28. Peter bought ten carrots for $0.80. What was the cost for each carrot?
(15)

* **29.** **Estimate** Which of these fractions is closest to 1?
(38)

A $\frac{1}{5}$ **B** $\frac{2}{5}$ **C** $\frac{3}{5}$ **D** $\frac{4}{5}$

* **30.** **Justify** If you know the perimeter of a square, you can find the area of the square in two steps. Describe the two steps.
(36)

LESSON 39

• Multiplying Decimal Numbers

Power Up

Building Power

facts Power Up G

mental math

a. **Order of Operations:** (4 × 700) + (4 × 25)

b. **Number Sense:** 6 × 45

c. **Number Sense:** 67 − 29

d. **Money:** \$8.75 + \$0.75

e. **Fractional Parts:** $\frac{1}{2}$ of 350

f. **Number Sense:** $\frac{2500}{100}$

g. **Statistics:** Find the average of 68, 124, 98, and 42

h. **Calculation:** 8 × 5, ÷ 2, + 1, ÷ 7, × 3, + 1, ÷ 10, ÷ 2

problem solving

Sarah used eight sugar cubes to make a larger cube. The cube she made was two cubes deep, two cubes wide, and two cubes high. How many cubes will she need to make a new cube that has three cubes along each edge?

New Concept

Increasing Knowledge

Doris hung a stained-glass picture in front of her window to let the light shine through the design. The picture frame is 0.75 meters long and 0.5 meters wide. How much of the area of the window is covered by the stained-glass picture?

To find the area of a rectangle that is 0.75 meters long and 0.5 meters wide, we multiply 0.75 m by 0.5 m.

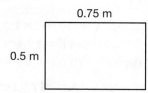

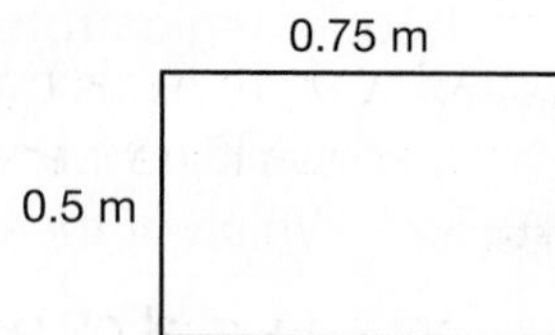

One way to multiply these numbers is to write each decimal number as a proper fraction and then multiply the fractions.

$$0.75 \times 0.5$$

$$\downarrow \quad \downarrow$$

$$\frac{75}{100} \times \frac{5}{10} = \frac{375}{1000}$$

The product $\frac{375}{1000}$ can be written as the decimal number 0.375. We find that the picture covers 0.375 square meters of the window area.

Thinking Skill

Discuss

Why can we count the decimal places to multiply 0.75 and 0.5?

Notice that the product 0.375 has three decimal places and that three is the **total** number of decimal places in the factors, 0.75 (two) and 0.5 (one). When we multiply decimal numbers, the product has the same number of decimal places as there are in all of the factors combined. This fact allows us to multiply decimal numbers as if they were whole numbers.

After multiplying, we count the total number of decimal places in the factors. Then we place a decimal point in the product to give it the same number of decimal places as there are in the factors.

$$\begin{array}{r} 0.75 \\ \times\ \ 0.5 \\ \hline 0.375 \end{array}$$

Three decimal places in the factors (0.75 and 0.5). We do not align decimal points. We multiply and then count decimal places.

Three decimal places in the product

Example 1

Multiply: 0.25 × 0.7

Solution

We set up the problem as though we were multiplying whole numbers, initially ignoring the decimal points. Then we multiply.

$$\begin{array}{r} 0.25 \\ \times\ \ 0.7 \\ \hline 0.175 \end{array}$$

3 places

Next we count the digits to the right of the decimal points in the two factors. There are three, so we place a decimal point in the product three places from the right-hand end. We write .175 as **0.175.**

Predict How could we have predicted that the answer would be less than one?

Example 2

Simplify: $(2.5)^2$

Solution

We square 2.5 by multiplying 2.5 by 2.5. We set up the problem as if we were multiplying whole numbers.

$$\begin{array}{r} 2.5 \\ \times\ 2.5 \\ \hline 125 \\ 50\ \ \\ \hline 6.25 \end{array}$$

2 places

Next we count decimal places in the factors. There are two, so we place a decimal point in the product two places from the right-hand end. We see that $(2.5)^2$ equals **6.25.**

Verify Why is 6.25 a reasonable answer?

Example 3

A mile is about 1.6 kilometers. Maricruz ran a 3-mile cross-country race. About how many kilometers did she run?

Solution

We multiply as though we were multiplying whole numbers. Then we count decimal places in the factors. There is only one, so we place a decimal point in the product one place from the right-hand end. The answer is **4.8 kilometers.**

$$\begin{array}{r} 1.6 \\ \times\ 3 \\ \hline 4.8 \end{array} \Big\} \text{ 1 place}$$

Practice Set

Simplify:

a. 15×0.3

b. 1.5×3

c. 1.5×0.3

d. 0.15×3

e. 1.5×1.5

f. 0.15×10

g. 0.25×0.5

h. 0.025×100

i. $(0.8)^2$

j. $(1.2)^2$

k. Conclude How are exercises **a–d** similar? How are they different?

Written Practice

Strengthening Concepts

1. (12) Mount Everest, the world's tallest mountain, rises to an elevation of twenty-nine thousand, twenty-nine feet above sea level. Use digits to write that elevation.

2. (7, 25) There are three feet in a yard. About how many yards above sea level is Mount Everest's peak?

*** 3.** (6) Connect If you had lived in the 1800's, you may have seen a sign like this in a barber shop:

Shave and a Haircut **six bits**

A bit is $\frac{1}{8}$ of a dollar. How many cents is 6 bits?

*** 4.** (39) 0.25×0.5

5. (2) $\$1.80 \times 10$

*** 6.** (39) 63×0.7

*** 7.** (38) $1.23 + \sqrt{16} + 0.5$

*** 8.** (37) $12.34 - 5.6$

*** 9.** (39) $(1.1)^2$

* **10.** (35) Write ten and three tenths

a. as a decimal number.

b. as a mixed number.

11. (17) **Evaluate** Think of two different fractions that are greater than zero but less than one. Multiply the two fractions to form a third fraction. For your answer, write the three fractions in order from least to greatest.

* **12.** (35) Write the decimal number one hundred twenty-three thousandths.

13. (32) Write $(6 \times 100) + (4 \times 10)$ in standard form.

* **14.** (38) **Connect** The perimeter of a square is 40 inches. How many square tiles with sides 1 inch long are needed to cover its area?

15. (30) What is the least common multiple (LCM) of 2, 3, and 6?

16. (25) Convert $\frac{20}{8}$ to a mixed number. Simplify your answer. Remember to reduce the fraction part of the mixed number.

17. (24) $\left(\frac{1}{3} + \frac{2}{3}\right) - 1$ **18.** (29) $\frac{3}{5} \times \frac{2}{3}$ **19.** (29) $\frac{8}{9} \times \frac{9}{8}$

* **20.** (36) **Represent** A pie was cut into six equal slices. Two slices were eaten. What fraction of the pie was left? Write the answer as a reduced fraction.

21. (32) What time is $2\frac{1}{2}$ hours before 1 a.m.?

22. (18) On Hiroshi's last four assignments he had 26, 29, 28, and 25 correct answers. He averaged how many correct answers on these papers? Explain how you found your answer.

23. (16) Estimate the quotient of 7987 divided by 39.

24. (14) Compare: $365 - 364 \bigcirc 364 - 365$

* **25.** (34) Which digit in 3.675 has the same place value as the 4 in 14.28?

26. (17) **Estimate** Use an inch ruler to find the length of the segment below to the nearest sixteenth of an inch.

27. (30) **a.** How many $\frac{3}{5}$s are in 1?

b. Use the answer to part **a** to find the number of $\frac{3}{5}$s in 2.

28. (2) **Connect** Instead of solving the division problem $390 \div 15$, Roosevelt divided both numbers by 3 to form the division $130 \div 5$. Then he multiplied both of those numbers by 2 to get $260 \div 10$. Find all three quotients.

* **29.** (39) **Analyze** Find the area of the rectangle below.

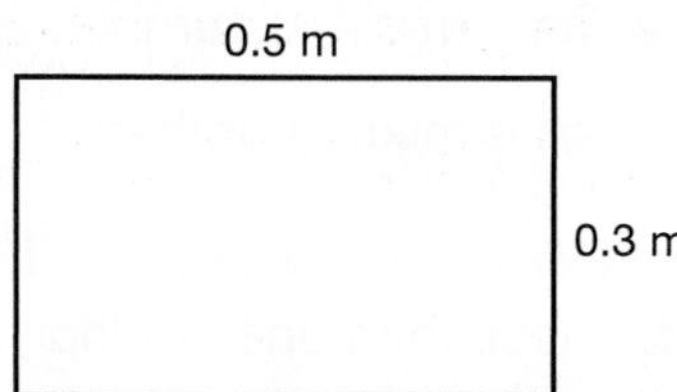

30. (23) **Conclude** In the figure below, what is the ratio of the measure of $\angle ABC$ to the measure of $\angle CBD$?

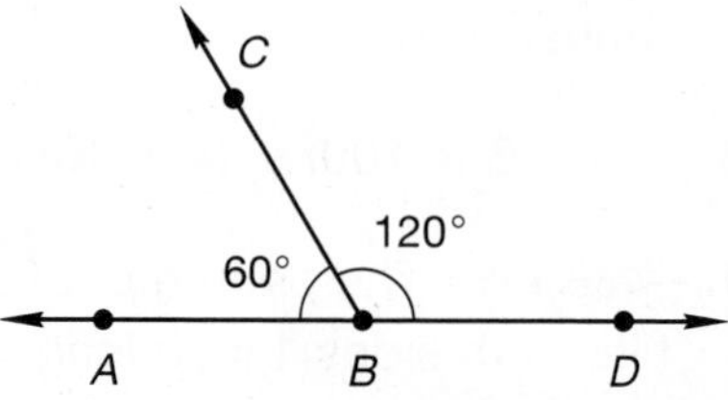

Early Finishers
Real-World Application

On Samuel's birthday, his mother wants to cook his favorite meat, turkey. She purchased a frozen 9-pound turkey from the grocery store. The turkey takes 2 days to thaw in the refrigerator, 30 minutes to prepare and $2\frac{1}{2}$ hours to cook in the oven. If the turkey starts thawing at 9 a.m. on November 3, when will the turkey be ready to eat? Show your work.

LESSON 40

• Using Zero as a Placeholder
• Circle Graphs

Power Up — *Building Power*

facts Power Up D

mental math

a. **Order of Operations:** $(4 \times 800) + (4 \times 25)$

b. **Number Sense:** $1500 + 750$

c. **Number Sense:** $74 - 39$

d. **Money:** $8.25 − $1.50

e. **Number Sense:** Double 240.

f. **Number Sense:** $\frac{480}{10}$

g. **Measurement:** Convert 180 minutes into hours.

h. **Calculation:** 4×4, $-\ 1$, $\div\ 5$, $\times\ 6$, $+\ 2$, $\times\ 2$, $+\ 2$, $\div\ 6$

problem solving Using five 2s and any symbols or operations, write an expression that is equal to 5.

New Concepts — *Increasing Knowledge*

using zero as a placeholder

When subtracting, multiplying, and dividing decimal numbers, we often encounter empty decimal places.

$$\begin{array}{r} 0.5\_ \\ -\ 0.32 \\ \hline \end{array} \qquad \begin{array}{r} 0.2 \\ \times\ 0.3 \\ \hline \_.\_6 \end{array} \qquad \begin{array}{r} \$0.\_4 \\ 3\overline{)\$0.12} \end{array}$$

When this occurs, we will fill each empty decimal place with a zero.

In order to subtract, it is sometimes necessary to attach zeros to the top number.

Example 1

Subtract: 0.5 − 0.32

Solution

We write the problem, making sure to line up the decimal points. We fill the empty place with zero and subtract.

$$\begin{array}{r} 0.5\_ \\ -\ 0.32 \\ \hline \end{array} \longrightarrow \begin{array}{r} \overset{4}{0.\cancel{5}}\overset{1}{0} \\ -\ 0.32 \\ \hline \mathbf{0.18} \end{array}$$

Discuss How can you check subtraction?

Example 2

Subtract: 3 − 0.4

Solution

We place a decimal point on the back of the whole number and line up the decimal points. We fill the empty place with zero and subtract.

$$\begin{array}{r} 3.\_ \\ -\ 0.4 \\ \hline \end{array} \longrightarrow \begin{array}{r} \overset{2}{\cancel{3}}.\overset{1}{0} \\ -\ 0.4 \\ \hline \mathbf{2.6} \end{array}$$

When multiplying decimal numbers, we may need to insert one or more zeros between the multiplication answer and the decimal point to hold the other digits in their proper places.

Example 3

Multiply: 0.2 × 0.3

Solution

We multiply and count two places from the right. We fill the empty place with zero, and we write a zero in the ones place.

$$\begin{array}{r} 0.2 \\ \times\ 0.3 \\ \hline \_.\_6 \end{array} \longrightarrow \begin{array}{r} 0.2 \\ \times\ 0.3 \\ \hline \mathbf{0.06} \end{array}$$

Verify Why must the product of 0.2 and 0.3 have two decimal places?

Example 4

Use digits to write the decimal number twelve thousandths.

Solution

The word *thousandths* tells us that there are three places to the right of the decimal point.

.___

We fit the two digits of twelve into the last two places.

._12

Then we fill the empty place with zero.

0.012

circle graphs

Circle graphs, which are sometimes called **pie graphs** or **pie charts,** display quantitative information in fractions of a circle. The next example uses a circle graph to display information about students' pets.

Example 5

Brett collected information from his classmates about their pets. He displayed the information about the number of pets in a circle graph.

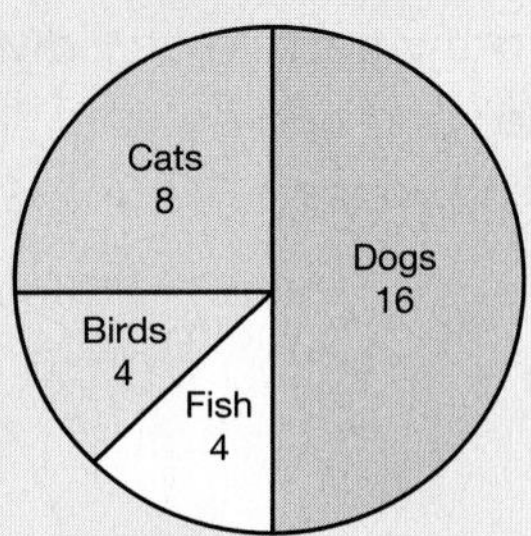

Use the graph to answer the following questions:

a. How many pets are represented in the graph?

b. What fraction of the pets are birds?

c. What percent of the pets are dogs?

Thinking Skill

Discuss

What information would we need to find the number of each kind of pet if we were given the percentages?

Solution

a. We add the number of dogs, cats, birds, and fish. The total is **32.**

b. Birds are 4 of the 32 pets. The fraction $\frac{4}{32}$ reduces to $\frac{1}{8}$. (The bird portion of the circle is $\frac{1}{8}$ of the whole circle.)

c. Dogs are 16 of the 32 pets, which means that $\frac{1}{2}$ of the pets are dogs. From our fraction manipulatives we know that $\frac{1}{2}$ equals **50%.** Circle graphs often express portions in percent form. Instead of showing the number of each kind of animal, the graph could have labeled each portion with a percent.

Example 6

A newspaper polled likely voters to survey support for a local bond measure on the November ballot. Display the data shown below with a circle graph. Then compare the two graphs.

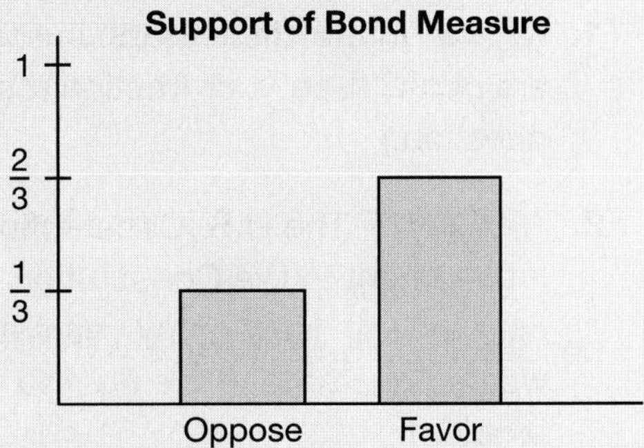

Solution

The bar graph shows us that $\frac{1}{3}$ of the voters oppose the bond measure and $\frac{2}{3}$ are in favor of it. Using the fraction circle manipulatives from Investigation 2, we can draw a circle and divide it into 3 sectors. We label one sector "Oppose" and two sectors "Favor." We know from our fraction manipulatives that $\frac{1}{3}$ is $33\frac{1}{3}\%$ and $\frac{2}{3}$ is $66\frac{2}{3}\%$, so we add these percentages to our labels.

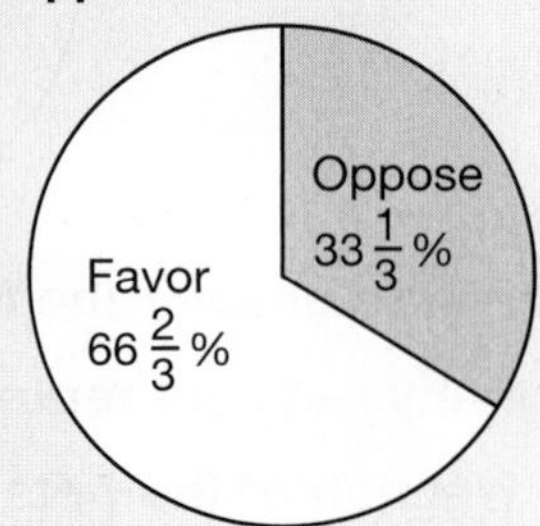

The bar graph helps us visualize the quantities relative to each other, and the circle graph helps us see their relationship to the whole.

Practice Set Simplify:

a. 0.2×0.3

b. $4.6 - 0.46$

c. 0.1×0.01

d. $0.4 - 0.32$

e. 0.12×0.4

f. $1 - 0.98$

g. $(0.3)^2$

h. $(0.12)^2$

i. Write the decimal number ten and eleven thousandths.

j. **Connect** In the circle graph in example 5, what percent of the pets are cats?

Written Practice

Strengthening Concepts

1. (40) **Model** In the circle graph in example 5, what percent of the pets are birds? (Use your fraction manipulatives to help you answer the question.)

2. (13) **Formulate** The U.S. Constitution was ratified in 1788. In 1920 the 19th amendment to the Constitution was ratified, guaranteeing women the right to vote. How many years after the Constitution was ratified were women guaranteed the right to vote? Write an equation and solve the problem.

3. (32) **Analyze** White Rabbit is three-and-a-half-hours late for a very important date. If the time is 2:00 p.m., what was the time of his date?

Predict Look at problems **4–9.** Predict which of the answers to those problems will be greater than 1. Then simplify each expression and check your predictions.

* **4.** (38, 40) $\sqrt{9} - 0.3$ * **5.** (40) $1.2 - 0.12$ * **6.** (40) $1 - 0.1$

* **7.** (40) 0.12×0.2 * **8.** (40) $(0.1)^2$ * **9.** (39) 4.8×0.23

* **10.** (40) Write one and two hundredths as a decimal number.

11. (32) Write $(6 \times 10{,}000) + (8 \times 100)$ in standard form.

12. (8, 31) **Connect** A square room has a perimeter of 32 feet. How many square floor tiles with sides 1 foot long are needed to cover the floor of the room?

13. (30) What is the least common multiple (LCM) of 2, 4, and 8?

14. (26) $6\frac{2}{3} + 4\frac{2}{3}$ **15.** (36) $5 - 3\frac{3}{8}$

16. (29) $\frac{5}{8} \times \frac{2}{3}$ **17.** (26) $2\frac{5}{6} + 5\frac{2}{6}$

18. (29) Compare: $\frac{1}{2} \times \frac{2}{2} \bigcirc \frac{1}{2} \times \frac{3}{3}$

19. (3) $1000 - w = 567$

20. (19) **Classify** Nine whole numbers are factors of 100. Two of the factors are 1 and 100. List the other seven factors.

* **21.** (38) $9^2 + \sqrt{9}$

22. (16) Round $4167 to the nearest hundred dollars.

* **23.** (40) The circle graph below displays the favorite sports of a number of students in a recent survey. Use the graph to answer **a–c.**

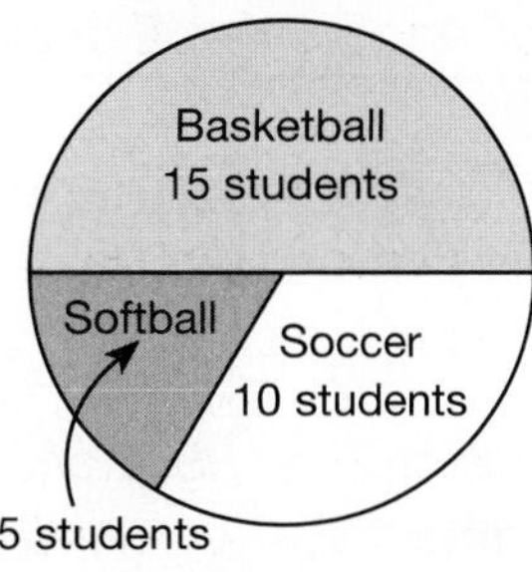

a. **Analyze** How many students responded to the survey?

b. What fraction of the students named softball as their favorite sport?

c. What percent of the students named basketball as their favorite sport?

24. (35) Jamal earned $5.00 walking his neighbor's dog for one week. He was given $\frac{1}{5}$ of the $5.00 at the beginning of the week and $\frac{1}{5}$ in the middle of the week. How much of the $5.00 was Jamal given at the end of the week? Express your answer as a fraction and as a dollar amount.

*** 25.** (23, 40) **Formulate** Write a ratio problem that relates to the circle graph in problem **23.** Then answer the problem.

*** 26.** (2) **Represent** Arrange the numbers in this multiplication fact to form another multiplication fact and two division facts.

$$0.2 \times 0.3 = 0.06$$

27. (2) **Connect** To solve the division problem $240 \div 15$, Elianna divided both numbers by 3 to form the division $80 \div 5$. Then she doubled 80 and 5 to get $160 \div 10$. Find all three quotients.

*** 28.** (23, 33) **Analyze** Forty percent of the 25 students in the class are boys. Write 40% as a reduced fraction. Then find the ratio of girls to boys in the class.

29. (17) **Connect** What mixed number is represented by point *A* on the number line below?

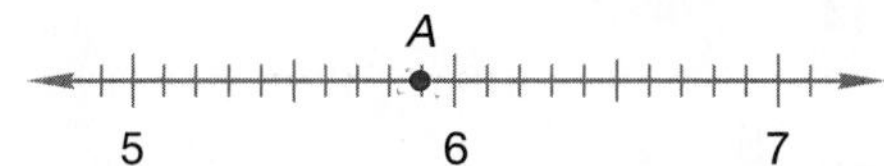

30. (40) Make a circle graph that shows the portion of a full day spent in various ways. Include activities such as sleeping, attending school, and eating. Label each sector with the activity name and its percent of the whole. Draw a circle and divide it into 24 equal sections. Use the circle below as a model.

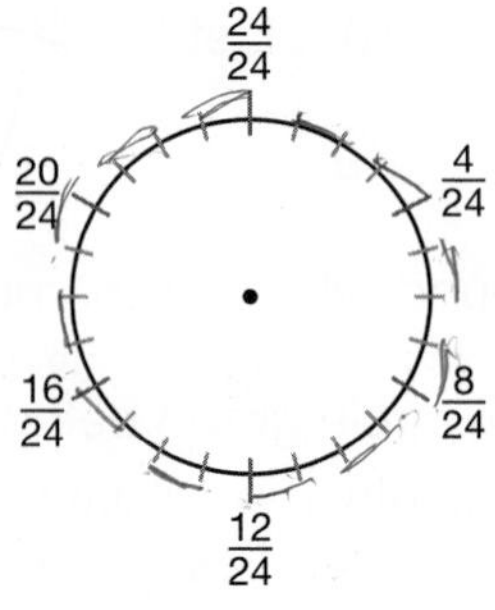

INVESTIGATION 4

Focus on

• Collecting, Organizing, Displaying, and Interpreting Data

Math Language
The term *data* is plural because it refers to a collection of individual figures and facts. Its singular form is *datum.*

Statistics is the science of gathering and organizing **data** (a plural word meaning information) in such a way that we can draw conclusions from the data.

For example, Patricia wondered which of three activities—team sports, dance, or walking/jogging—was most popular among her classmates. She gathered data by asking each classmate to select his or her favorite. Then she displayed the data with a **bar graph.**

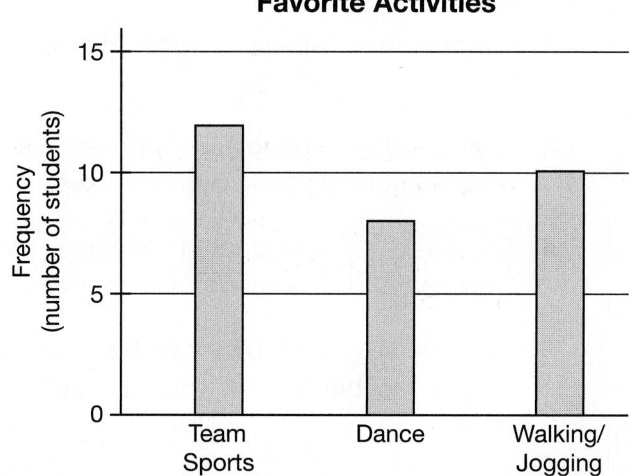

Analyze Which type of exercise is most popular among Patricia's classmates?

Roger wondered how frequently the residents on his street visit the city park. He went to every third house on his street and asked, "How many times per week do you visit the city park?" Roger displayed the data he collected with a **line plot,** which shows individual data points. For each of the 16 responses, he placed an "x" above the corresponding number.

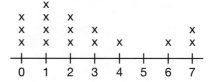

Number of Visits to the City Park per Week

```
   x
x  x  x
x  x  x  x           x
x  x  x  x  x     x  x
+--+--+--+--+--+--+--+
0  1  2  3  4  5  6  7
```

Analyze For most of the residents surveyed, the number of visits to the city park are between what numbers?

In this Investigation we will focus on ways in which statistical data can be collected. We also will practice collecting, organizing, displaying, and interpreting data. Data can be either quantitative or qualitative in nature. **Quantitative data** come in numbers: the population of a city, the number of pairs of shoes someone owns, or the number of hours per week someone watches television. **Qualitative data** come in categories: the month in which someone is born or a person's favorite flavor of ice cream.

Roger collected quantitative data when he asked about the number of visits to the park each week. Patricia collected qualitative data when she asked about the student's favorite sport.

Classify In problems **1–5** below, determine what information is collected. Then decide whether the data are qualitative or quantitative.

1. Jagdish collects 50 bags of clothing for a clothing drive and counts the number of items in each bag.
2. For one hour Carlos notes the color of each car that drives past his house.
3. Sharon rides a school bus home after school. For two weeks she measures the time the bus trip takes.
4. Brigit asks each student in her class, "Which is your favorite holiday—New Year's, Thanksgiving, or Independence Day?"
5. Marcello asks each player on his little league team, "Which major league baseball team is your favorite? Which team do you like the least?"
6. Write a survey about television viewing with two questions: one that collects quantitative data and one that collects qualitative data. Conduct the survey in your class.

Represent For question **6** above, organize your data. Then use a line plot to display the quantitative data and a bar graph to display the qualitative data. Interpret the results.

This question will give you quantitative data: "How many hours a week do you spend watching television?" We record the number of hours a person watches TV each week. Then we organize the quantitative data from least to greatest.

0 1 1 2 2 2 2 3 3 3 3 3 4 4 4 4 4 4
5 5 6 6 7 7 7 7 8 8 8 10 12 15

We display the data with a line plot. (We may use an "x" or dot for this.)

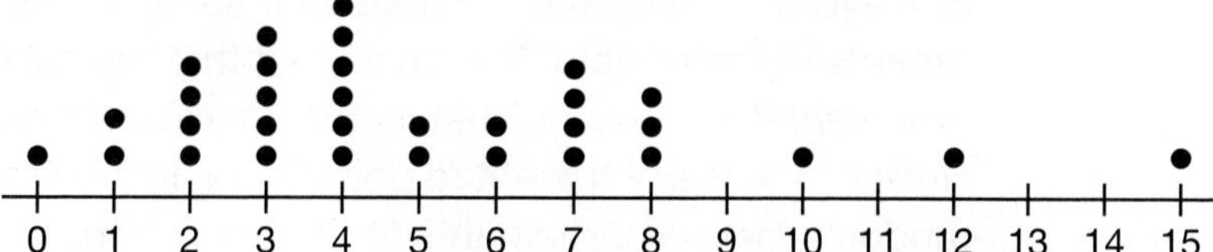

The line plot illustrates that the 32 students surveyed watch between 0 and 15 hours of TV each week. Many students watch either 2–4 or 7–8 hours each week.

This question will give you qualitative data: "Which type of TV shows do you prefer to watch–sports, news, animation, sit-coms, or movies?" We organize the qualitative data by category and tally the frequency of each category.

Category	Tally	Frequency
Sports	𝍸 𝍸 I	11
News	III	3
Animation	𝍸 I	6
Sit-coms	𝍸 IIII	9
Movies	III	3

We can display the data with a bar graph.

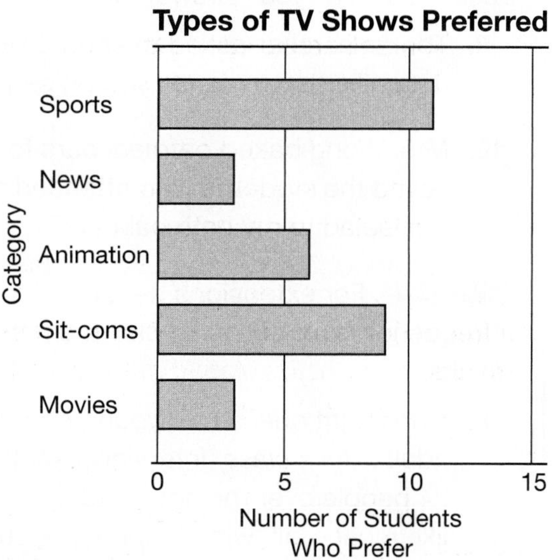

The most popular type of TV show among the students surveyed is sports.

Surveys can be designed to gather data about a certain group of people. This "target group" is called a **population.** For example, if a record company wants to know about teenagers' music preferences in the United States, it would not include senior citizens in the population it studies.

Often it is not realistic to poll an entire population. In these cases, a small part of the population is surveyed. We call this small part a sample. Surveyors must carefully select their samples, because different samples will provide different data. It is important that the sample for a population be a *representative sample*. That is, the characteristics of the sample should be similar to those of the entire population. In order to do this, researchers often randomly select participants for a survey from the entire population.

Validate In problems **7** and **8** below, explain why each sample is not representative. How would you expect the sample's responses to differ from those of the general population?

7. To determine public opinion in the city of Miami about a proposed leash law for dogs, Sally interviews shoppers in several Miami pet stores.

8. Tamika wants to know the movie preferences of students in her middle school. Since she is in the school orchestra, she chooses to survey orchestra members.

Often the results of a survey depend on the way its questions are worded or who is asking the questions. These factors can introduce **bias** into a survey. When a survey is biased, the people surveyed might be influenced to give certain answers over other possible answers.

Verify For problems **9** and **10,** identify the bias in the survey that is described. Is a "yes" answer *more likely* or *less likely* because of the bias?

9. The researcher asked the group of adults, "If you were lost in an unfamiliar town, would you be sensible and ask for directions?"

10. Mrs. Wong baked oatmeal bars for her daughter's fundraising sale. She asked the students who attended the sale, "Would you have preferred fruit salad to my oatmeal bars?"

extensions

Represent For extensions **a–c,** conduct a survey and organize the data with a frequency table. Then display the results with a circle graph. Interpret the results.

Thinking Skill

Verify

Why would a good survey for **a** and **b** include data from the same number of young people and adults?

a. One might guess that young people prefer different seasons than adults. *As a class,* interview exactly 24 people under the age of 15 and 24 people over the age of 20. Ask, "Which season of the year do you like most—fall, winter, spring, or summer?"

b. One might guess that young people drink different beverages for breakfast than adults. *As a class,* interview exactly 24 people under 15 and 24 people over 20. Ask, "Which of the following beverages do you most often drink at breakfast—juice, coffee, milk, or something else?" If the choice is "something else," record the person's preferred breakfast drink.

c. For ten consecutive days, count the number of students in your class who wear something green (or some other color of your choosing). Count the same color all ten days.

d. **Formulate** With a friend, construct a six-question, true-false quiz on a topic that interests you both (for example, music, animals, or geography). Have your classmates take the quiz; then record the number of questions that each student answers correctly. To encourage participation, ask students not to write their names on the quiz.

e. **Analyze** Use the menu to answer the questions below.

Seafood Cafe	
Appetizers Shrimp Cocktail..................$7.00 Zucchini Fingers................ $5.00 **Soup** Seafood Gumbo................ $4.50 Lobster Bisque.................. $4.50	**Main Course** Halibut.............................. $15.75 Swordfish .... $18.00 Flounder .... $13.75 Crab Cakes .... $12.50 **Dessert** Sorbet.................................$3.25

- What is the most expensive item on the menu? The least expensive?
- What is the average price of the main course dinners?
- What is the range of prices on the menu?

f. **Evaluate** A number of bicyclists participated in a 25-mile bicycle race. The winner completed the race in 45 minutes and 27 seconds. The table below shows the times of the next four riders expressed in the number of minutes and seconds that they placed behind the winner.

Rider Number	Time Behind Winner (minutes and seconds in hundredths)
021	−1:09.02
114	−0.04.64
008	−1:29.77
065	−0:13.45

The least time in the table represents the rider who finished second. The greatest time represents the rider who finished fifth. Write the four rider numbers in the order of their finish.

g. **Analyze** The average rate of speed for rider 008 was 33 miles per hour. At that rate, did this rider finish more than 1 mile or less than 1 mile behind the winner? Show your work. (Hint: Change the average rate in miles per hour to miles per minute, and round the rider number 008's time to the nearest second.)

LESSON

41

• Finding a Percent of a Number

Power Up

Building Power

facts Power Up G

mental math

a. Number Sense: 4×250

b. Number Sense: $625 + 50$

c. Calculation: $47 + 8$

d. Money: $\$3.50 + \$1.75$

e. Fractional Parts: $\frac{1}{2}$ of 700

f. Number Sense: $\frac{600}{10}$

g. Measurement: How many feet are in 60 inches?

h. Calculation: $5 \times 3, + 1, \div 2, + 1, \div 3, \times 8, \div 2$

problem solving

Tennis is played on a rectangular surface that usually has two different courts drawn on it. Two players compete on a singles court, and four players (two per team) compete on a doubles court. The singles court is 78 feet long and 27 feet wide. The doubles court has the same length, but is 9 feet wider. What is the perimeter and the area of a singles tennis court? How wide is a doubles court? What is the perimeter and the area of the doubles court?

New Concept

Increasing Knowledge

Math Language
Recall that a **percent** is really a fraction whose denominator is 100.

To describe part of a group, we often use a fraction or a percent. Here are a couple of examples:

Three fourths of the students voted for Imelda.
Tim answered 80% of the questions correctly.

We also use percents to describe financial situations.

Music CDs were on sale for 30% off the regular price.

The sales-tax rate is 7%.

The bank pays 3% interest on savings accounts.

When we are asked to find a certain percent of a number, we usually change the percent to either a fraction or a decimal before performing the calculation.

$$25\% \text{ means } \frac{25}{100}, \text{ which reduces to } \frac{1}{4}.$$
$$5\% \text{ means } \frac{5}{100}, \text{ which reduces to } \frac{1}{20}.$$

A percent is also easily changed to a decimal number. Study the following changes from percent to fraction to decimal:

$$25\% \rightarrow \frac{25}{100} \rightarrow 0.25 \qquad 5\% \rightarrow \frac{5}{100} \rightarrow 0.05$$

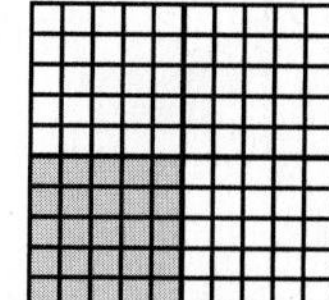

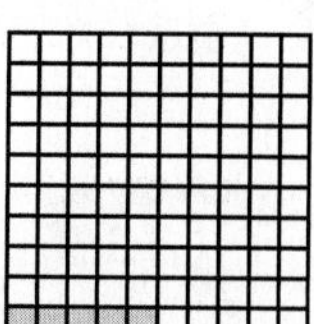

We see that the same nonzero digits are in both the decimal and percent forms of a number. In the decimal form, however, the decimal point is shifted two places to the left.

Example 1

Write 15% in decimal form.

Solution

Fifteen percent means $\frac{15}{100}$, which can be written **0.15.**

Model Use a grid to show 15%.

Example 2

Write 75% as a reduced fraction.

Solution

We write 75% as $\frac{75}{100}$ and reduce.

$$\frac{75}{100} = \mathbf{\frac{3}{4}}$$

Example 3

What number is 75% of 20?

Solution

We can translate this problem into an equation, changing the percent into either a fraction or a decimal. We use a letter for "what number," an equal sign for "is," and a multiplication sign for "of."

Percent To Fraction

What number is 75% of 20?

$$n = \frac{3}{4} \times 20$$

Percent To Decimal

What number is 75% of 20?

$$n = 0.75 \times 20$$

We show both the fraction form and the decimal form. Often, one form is easier to calculate than the other form.

Thinking Skill

Conclude

Since 75% of 20 is 15, what is 25% of 20?

$$\frac{3}{4} \times 20 = 15 \qquad 0.75 \times 20 = 15.00$$

We find that 75% of 20 is **15.**

Discuss Which is easier to compute, $\frac{3}{4} \times 20$ or 0.75×20? Why?

Example 4

Jamaal correctly answered 80% of the 25 questions. How many questions did he answer correctly?

Solution

Thinking Skill

Verify

How do you change 80% to a fraction? To a decimal?

We want to find 80% of 25. We can change 80% to a fraction ($\frac{80}{100} = \frac{4}{5}$) or to a decimal number (80% = 0.80).

Percent To a Fraction	Percent To a Decimal
80% of 25	80% of 25
$\frac{4}{5} \times 25$	0.80×25

Then we calculate.

$$\frac{4}{5} \times 25 = 20 \qquad 0.80 \times 25 = 20.00$$

We find that Jamaal correctly answered **20 questions.**

Example 5

The sales-tax rate was 6%. Find the tax on a $12.00 purchase. Then find the total price including tax.

Solution

Math Language

Sales tax is a tax charged on the sale of an item. It is some percent of the item's purchase price.

We can change 6% to a fraction ($\frac{6}{100} = \frac{3}{50}$) or to a decimal number (6% = 0.06). It seems easier for us to multiply $12.00 by 0.06 than by $\frac{3}{50}$, so we will use the decimal form.

6% of $12.00

$0.06 \times \$12.00 = \$0.72$

So the tax on the $12.00 purchase was **$0.72.** To find the total price including tax, we add $0.72 to $12.00.

$$\begin{array}{r} \$12.00 \\ +\ \ \ 0.72 \\ \hline \mathbf{\$12.72} \end{array}$$

Practice Set

Write each percent in problems **a–f** as a reduced fraction:

a. 50% **b.** 10% **c.** 25%

d. 75% **e.** 20% **f.** 1%

Write each percent in problems **g–l** as a decimal number:

g. 65% **h.** 7% **i.** 30%

j. 8% **k.** 60% **l.** 1%

m. Explain Mentally find 10% of 350. Describe how to perform the mental calculation.

n. Explain Mentally find 25% of 48. Describe how to perform the mental calculation.

o. How much money is 8% of $15.00?

p. Estimate The sales-tax rate is $9\frac{1}{2}$%. Estimate the tax on a $9.98 purchase. How did you arrive at your answer?

q. Erika sold 80% of her 30 baseball cards. How many baseball cards did she sell?

Written Practice *Strengthening Concepts*

* **1.** *(41)* A student correctly answered 80% of the 20 questions on the test. How many questions did the student answer correctly?

* **2.** *(41)* Ramon ordered items from the menu totaling $8.50. If the sales-tax rate is 8%, how much should be added to the bill for sales tax? Explain why your answer is reasonable.

3. *(8)* Analyze The ten-acre farm is on a square piece of land 220 yards on each side. A fence surrounds the land. How many yards of fencing surrounds the farm?

* **4.** *(41)* Explain Describe how to find 20% of 30.

5. *(11)* The dinner cost $9.18. Jeb paid with a $20 bill. How much money should he get back? Write an equation and solve the problem.

6. *(15)* Formulate Two hundred eighty-eight chairs were arranged in 16 equal rows. How many chairs were in each row? Write an equation and solve the problem.

7. *(18)* Yuki's bowling scores for three games were 126, 102, and 141. What was her average score for the three games?

* **8.** *(31, 39)* What is the area of this rectangle?

2.5 m

2 m

9. *(14, 17)* Arrange these numbers in order from least to greatest:

$$\frac{3}{2}, 0, -1, \frac{2}{3}$$

10. List the first eight prime numbers.
(19)

11. **Classify** By which of these numbers is 600 not divisible?
(21)

A 2 **B** 3 **C** 5 **D** 9

12. The fans were depressed, for their team had won only 15 of the first 60 games. What was the team's win-loss ratio after 60 games?
(27)

13. To loosen her shoulder, Mary swung her arm around in a big circle. If her arm was 28 inches long, what was the diameter of the circle?
(27)

14. **Connect** The map shows three streets in town.
(28)

a. Name a street parallel to Vine.

b. Name a street perpendicular to Vine.

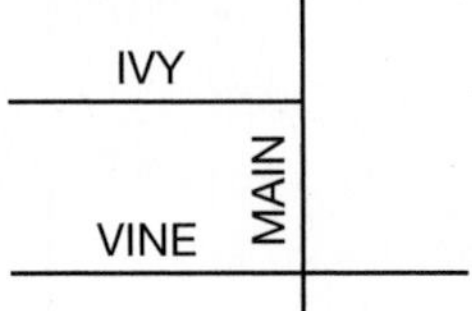

15. Rob remembers that an acute angle is "a cute little angle." Which of the following could be the measure of an acute angle?
(28, Inv. 3)

A 0° **B** 45° **C** 90° **D** 135°

* **16.** $(2.5)^2$
(38, 39)

* **17.** $\sqrt{81}$
(38)

* **18.** Write 40% as a reduced fraction.
(41)

* **19.** Write 9% as a decimal number. Then find 9% of $10.
(41)

20. What is the reciprocal of $\frac{2}{3}$?
(30)

Find each unknown number. Remember to check your work.

21. $7m = 3500$
(4)

22. $\$6.25 + w = \$10.00$
(3)

* **23.** $\frac{2}{3}n = 1$
(30)

24. $x - 37 = 76$
(3)

25. $6.25 + (4 - 2.5)$
(37)

26. $3\frac{3}{4} + 2\frac{3}{4}$
(26)

27. $\frac{4}{4} - \frac{3}{3}$
(Inv. 2)

28. $\frac{5}{6} \cdot \frac{3}{5}$
(29)

29. What is $\frac{3}{4}$ of 48?
(29)

* **30.** **Justify** Fran estimated 9% of $21.90 by first rounding 9% to 10% and rounding $21.90 to $20. She then mentally calculated 10% of $20 and got the answer $2. Use Fran's method to estimate 9% of $32.17. Describe the steps.
(41)

LESSON 42

Renaming Fractions by Multiplying by 1

Power Up — *Building Power*

facts Power Up G

mental math

a. **Number Sense:** 4×125

b. **Number Sense:** $825 + 50$

c. **Calculation:** $67 + 8$

d. **Money:** $6.75 + $2.50

e. **Fractional Parts:** $\frac{1}{2}$ of 1000

f. **Number Sense:** $\frac{580}{10}$

g. **Measurement:** How many millimeters are in 10 centimeters?

h. **Calculation:** 3×4, $-\ 2$, $\times\ 5$, $-\ 2$, $\div\ 6$, $+\ 1$, $\div\ 3$

problem solving

Victor dropped a rubber ball and found that each bounce was half as high as the previous bounce. He dropped the ball from 8 feet, measured the height of each bounce, and recorded the results in a table. Copy this table and complete it through the fifth bounce.

Heights of Bounces

First	4 ft
Second	
Third	
Fourth	
Fifth	

New Concept — *Increasing Knowledge*

With our fraction manipulatives we have seen that the same fraction can be named many different ways. Here we show six ways to name the fraction $\frac{1}{2}$:

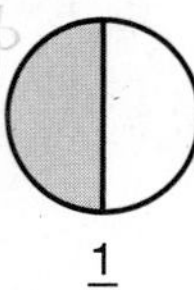
$\frac{1}{2}$

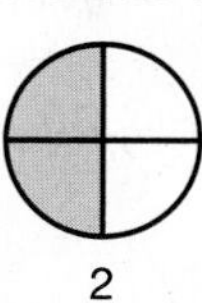
$\frac{2}{4}$

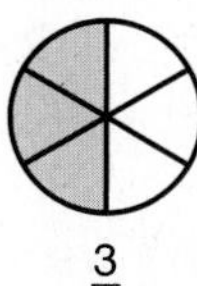
$\frac{3}{6}$

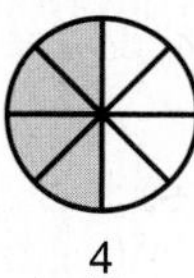
$\frac{4}{8}$

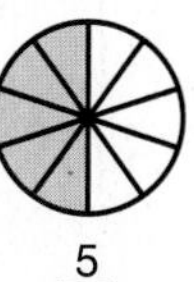
$\frac{5}{10}$

$\frac{6}{12}$

In this lesson we will practice renaming fractions by multiplying them by a fraction equal to 1. Here we show six ways to name 1 as a fraction:

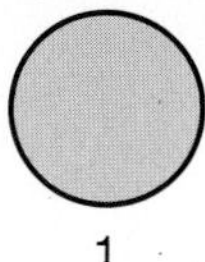
$\frac{1}{1}$

$\frac{2}{2}$

$\frac{3}{3}$

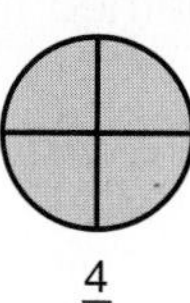
$\frac{4}{4}$

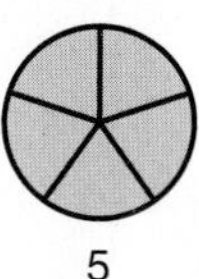
$\frac{5}{5}$

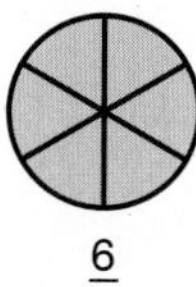
$\frac{6}{6}$

Math Language

The **Identity Property of Multiplication** states that if one of two factors is 1, the product equals the other factor.

We know that when we multiply a number by 1, the product equals the number multiplied. So if we multiply $\frac{1}{2}$ by 1, the answer is $\frac{1}{2}$.

$$\frac{1}{2} \times 1 = \frac{1}{2}$$

However, if we multiply $\frac{1}{2}$ by fractions equal to 1, we find different names for $\frac{1}{2}$. Here we show $\frac{1}{2}$ multiplied by $\frac{2}{2}$, $\frac{3}{3}$, and $\frac{4}{4}$:

$$\frac{1}{2} \times \frac{2}{2} = \frac{2}{4} \qquad \frac{1}{2} \times \frac{3}{3} = \frac{3}{6} \qquad \frac{1}{2} \times \frac{4}{4} = \frac{4}{8}$$

Fractions with the same value but different names are called **equivalent fractions.** The fractions $\frac{2}{4}$, $\frac{3}{6}$, and $\frac{4}{8}$ are all equivalent to $\frac{1}{2}$.

Example 1

Write a fraction equal to $\frac{1}{2}$ that has a denominator of 20.

$$\mathbf{\frac{1}{2} = \frac{?}{20}}$$

Solution

To rename a fraction, we multiply the fraction by a fraction equal to 1. The denominator of $\frac{1}{2}$ is 2. We want to make an equivalent fraction with a denominator of 20.

$$\frac{1}{2} = \frac{?}{20}$$

Since we need to multiply the denominator by 10, we multiply $\frac{1}{2}$ by $\frac{10}{10}$.

$$\frac{1}{2} \times \frac{10}{10} = \mathbf{\frac{10}{20}}$$

Example 2

Write $\frac{1}{2}$ and $\frac{1}{3}$ as fractions with denominators of 6. Then add the renamed fractions.

Solution

Thinking Skill

Represent

Draw a diagram to show that $\frac{1}{2} = \frac{3}{6}$.

We multiply each fraction by a fraction equal to 1 to form fractions that have denominators of 6.

$$\frac{1}{2} = \frac{?}{6} \qquad \frac{1}{3} = \frac{?}{6}$$

We multiply $\frac{1}{2}$ by $\frac{3}{3}$, and we multiply $\frac{1}{3}$ by $\frac{2}{2}$.

$$\frac{1}{2} \times \frac{3}{3} = \mathbf{\frac{3}{6}} \qquad \frac{1}{3} \times \frac{2}{2} = \mathbf{\frac{2}{6}}$$

The renamed fractions are $\frac{3}{6}$ and $\frac{2}{6}$. We are told to add these fractions.

$$\frac{3}{6} + \frac{2}{6} = \mathbf{\frac{5}{6}}$$

Practice Set

In problems **a–d,** multiply by a fraction equal to 1 to complete each equivalent fraction.

a. $\frac{1}{3} = \frac{?}{12}$

b. $\frac{2}{3} = \frac{?}{6}$

c. $\frac{3}{4} = \frac{?}{8}$

d. $\frac{3}{4} = \frac{?}{12}$

e. *Model* On your paper draw two rectangles that look like this one. Shade $\frac{2}{3}$ of the squares of one rectangle and shade $\frac{1}{4}$ of the squares of the other rectangle. Then name each shaded rectangle as a fraction with a denominator of 12.

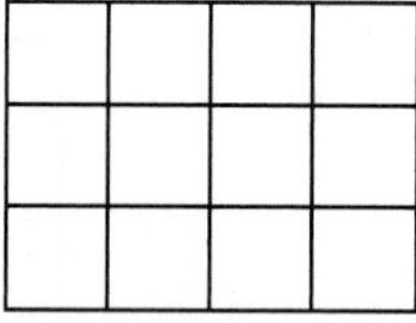

f. Write $\frac{2}{3}$ and $\frac{1}{4}$ as fractions with denominators of 12. Then add the renamed fractions.

g. Describe how the rectangles you drew for exercise **e** can help you add the fractions in **f.**

h. Write $\frac{1}{6}$ as a fraction with 12 as the denominator. Subtract the renamed fraction from $\frac{5}{12}$. Reduce the subtraction answer.

Written Practice

Strengthening Concepts

* **1.** (42) *Analyze* Write $\frac{1}{2}$ and $\frac{2}{3}$ as fractions with denominators of 6. Then add the renamed fractions. Write the answer as a mixed number.

2. (13) According to some estimates, our own galaxy, the Milky Way, contains about two hundred billion stars. Use digits to write that number of stars.

3. (31) *Explain* The rectangular school yard is 120 yards long and 40 yards wide. How many square yards is its area? Explain why your answer is reasonable.

* **4.** (41) *Analyze* What number is 40% of 30?

In problems **5** and **6,** multiply $\frac{1}{2}$ by a fraction equal to 1 to complete each equivalent fraction.

* **5.** (42) $\frac{1}{2} = \frac{?}{8}$

* **6.** (42) $\frac{1}{2} = \frac{?}{10}$

* **7.** (37, 38) $4.32 + 0.6 + \sqrt{81}$

* **8.** (37) $6.3 - 0.54$

* **9.** (38, 40) $(0.15)^2$

10. (30) What is the reciprocal of $\frac{6}{7}$?

11. (34) Which digit in 12,345 has the same place value as the 6 in 67.89?

12. (30) What is the least common multiple of 3, 4, and 6?

13. $5\frac{3}{5} + 4\frac{4}{5}$
(26)

* **14.** $\sqrt{36} - 4\frac{2}{3}$
(36, 38)

15. $\frac{8}{3} \times \frac{1}{2}$
(29)

16. $\frac{6}{5} \times 3$
(29)

* **17.** $1 - \frac{1}{4}$
(36)

18. $\frac{10}{10} - \frac{5}{5}$
(Inv. 2)

* **19.** Form three different fractions that are equal to $\frac{1}{3}$. (*Hint:* Multiply $\frac{1}{3}$ by three different fraction names for 1).
(42)

20. **Analyze** The prime numbers that multiply to form 35 are 5 and 7. Which prime numbers can be multiplied to form 34?
(19)

21. **Estimate** In three games Alma's scores were 12,143; 9870; and 14,261. Describe how to estimate her average score per game.
(18)

22. **Explain** Estimate the quotient of $\frac{8176}{41}$. Describe how you performed the estimate.
(16)

23. **Model** How many eggs are in $\frac{2}{3}$ of a dozen? Draw a diagram to illustrate the problem.
(22)

* **24.** Write $\frac{3}{4}$ with a denominator of 8. Subtract the renamed fraction from $\frac{7}{8}$.
(42)

25. What is the perimeter of this rectangle?
(8)

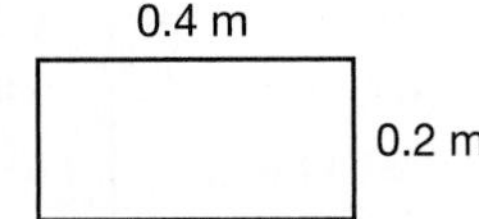

26. What is the area of this rectangle?
(31)

* **27.** **Represent** The regular price r minus the discount d equals the sale price s.
(1)

$$r - d = s$$

Arrange these letters to form another subtraction equation and two addition equations.

28. Below we show the same division problem written three different ways. Identify which number is the divisor, which is the dividend, and which is the quotient.
(2)

$$\frac{20}{4} = 5 \qquad 4\overline{)20}^{\,5} \qquad 20 \div 4 = 5$$

29. What time is $2\frac{1}{2}$ hours after 11:45 a.m.?
(32)

* **30.** **a.** How many $\frac{5}{6}$s are in 1?
(30)

b. Use the answer to part **a** to find the number of $\frac{5}{6}$s in 3.

LESSON 43

• Equivalent Division Problems
• Finding Unknowns in Fraction and Decimal Problems

Power Up — *Building Power*

facts Power Up C

mental math

a. **Calculation:** 4×225

b. **Number Sense:** $720 - 200$

c. **Number Sense:** $37 + 28$

d. **Money:** \$200 − \$175

e. **Fractional Parts:** $\frac{1}{2}$ of 1200

f. **Number Sense:** $\frac{\$70.00}{10}$

g. **Probability:** How many different three digit numbers can be made with the digits 3, 5, and 9?

h. **Calculation:** 8×4, -2, $\times 2$, $+3$, $\div 7$, $\times 2$, $\div 3$

problem solving Teresa wanted to paint each face of a cube so that the adjacent faces (the faces next to each other) were different colors. She wanted to use fewer than six different colors. What is the fewest number of colors she could use? Describe how the cube could be painted.

New Concepts — *Increasing Knowledge*

equivalent division problems

Math Language

Recall that equivalent numbers or expressions have the same value.

The following two division problems have the same quotient. We call them **equivalent division problems.** Which problem seems easier to perform mentally?

a. $700 \div 14$

b. $350 \div 7$

We can change problem **a** to problem **b** by dividing both 700 and 14 by 2.

$700 \div 14$

↓

Divide both 700 and 14 by 2.

↓

$350 \div 7$

By dividing both the dividend and divisor by the same number (in this case, 2), we formed an equivalent division problem that was easier to divide mentally. This process simply reduces the terms of the division as we would reduce a fraction.

$$\frac{700}{14} = \frac{350}{7} \quad \begin{matrix}(700 \div 2 = 350) \\ (14 \div 2 = 7)\end{matrix}$$

We may also form equivalent division problems by multiplying the dividend and divisor by the same number. Consider the following equivalent problems:

c. $7\frac{1}{2} \div \frac{1}{2}$

d. $15 \div 1$

We changed problem **c** to problem **d** by doubling both $7\frac{1}{2}$ and $\frac{1}{2}$; that is, by multiplying both numbers by 2.

$$7\frac{1}{2} \div \frac{1}{2}$$

↓

Multiply both $7\frac{1}{2}$ and $\frac{1}{2}$ by 2.

↓

$$15 \div 1$$

This process forms an equivalent division problem in the same way we would form an equivalent fraction.

$$\frac{7\frac{1}{2}}{\frac{1}{2}} \cdot \frac{2}{2} = \frac{15}{1}$$

Example 1

Form an equivalent division problem for the division below. Then calculate the quotient.

$$\mathbf{1200 \div 16}$$

Solution

Instead of dividing 1200 by the two-digit number 16, we can divide both the dividend and the divisor by 2 to form the equivalent division of 600 ÷ 8. We then calculate.

$$16\overline{)1200} \xrightarrow{\text{Divide both numbers by 2.}} \begin{array}{r} \mathbf{75} \\ 8\overline{)600} \\ \underline{56} \\ 40 \\ \underline{40} \\ 0 \end{array}$$

Both quotients are 75, but dividing by 8 is easier than dividing by 16.

Notice that several equivalent division problems can be formed from the original problem 1200 ÷ 16:

$$1200 \div 16 \longrightarrow 600 \div 8 \longrightarrow 300 \div 4 \longrightarrow 150 \div 2$$

All of these problems have the same quotient.

Example 2

Form an equivalent division problem for the division problem below. Then calculate the quotient.

$$7\frac{1}{2} \div 2\frac{1}{2}$$

Solution

Instead of performing the division with these mixed numbers, we will double both numbers to form a whole-number division problem.

$$7\frac{1}{2} \div 2\frac{1}{2}$$

↓

Multiply both $7\frac{1}{2}$ and $2\frac{1}{2}$ by 2.

↓

$$\mathbf{15 \div 5 = 3}$$

finding unknowns in fraction and decimal problems

Since Lessons 3 and 4 we have practiced finding **unknowns** in whole-number arithmetic problems. Beginning with this lesson we will find unknowns in fraction and decimal problems. If you are unsure how to find the solution to a problem, try making up a similar, easier problem to help you determine how to find the answer.

Example 3

Solve: $d - 5 = 3.2$

Solution

This problem is similar to the subtraction problem $d - 5 = 3$. We remember that we find the first number of a subtraction problem by adding the other two numbers. So we have the following:

$$\begin{array}{r} 5 \\ +\ 3.2 \\ \hline \mathbf{8.2} \end{array}$$

We check our work by replacing the letter with the solution and testing the result.

$$d - 5 = 3.2$$

$$8.2 - 5 = 3.2$$

$$3.2 = 3.2$$

Example 4

Solve: $f + \frac{1}{5} = \frac{4}{5}$

Solution

This problem is similar to $f + 1 = 4$. We can find an unknown addend by subtracting the known addend from the sum.

$$\frac{4}{5} - \frac{1}{5} = \frac{3}{5}$$

$$f = \mathbf{\frac{3}{5}}$$

We check the solution by substituting it into the original equation.

$$f + \frac{1}{5} = \frac{4}{5}$$

$$\frac{3}{5} + \frac{1}{5} = \frac{4}{5}$$

$$\frac{4}{5} = \frac{4}{5}$$

Example 5

Solve: $\frac{3}{5}n = 1$

Solution

In this problem two numbers are multiplied, and the product is 1. This can only happen when the two factors are reciprocals. So we want to find the reciprocal of the known factor, $\frac{3}{5}$. Switching the terms of $\frac{3}{5}$ gives us the fraction $\frac{5}{3}$. We check our answer by substituting $\frac{5}{3}$ into the original equation.

$$\frac{3}{5} \cdot \frac{5}{3} = \frac{15}{15} = 1 \quad \text{check}$$

$$n = \mathbf{\frac{5}{3}}$$

Practice Set

a. Connect Form an equivalent division problem for $5 \div \frac{1}{3}$ by multiplying both the dividend and divisor by 3. Then find the quotient.

b. Connect Form an equivalent division problem for $266 \div 14$ that has a one-digit divisor. Then find the quotient.

Solve:

c. $5 - d = 3.2$

d. $f - \frac{1}{5} = \frac{4}{5}$

e. $m + 1\frac{1}{5} = 4$

f. $\frac{3}{8}w = 1$

Written Practice *Strengthening Concepts*

* **1.** (41) **Analyze** The bike cost \$120. The sales-tax rate was 8%. What was the total cost of the bike including sales tax?

2. (11) **Formulate** If one hundred fifty knights could sit at the Round Table and only one hundred twenty-eight knights were seated, how many empty places were at the table? Write an equation and solve the problem.

3. (32, 37) During the 1996 Summer Olympics in Atlanta, Georgia, the American athlete Michael Johnson set an Olympic and world record in the men's 200-meter run. He finished the race in 19.32 seconds, breaking the previous Olympic record of 19.73 seconds. By how much did Michael Johnson break the previous Olympic record?

In problems **4** and **5,** multiply by a fraction equal to 1 to complete each equivalent fraction.

* **4.** (42) $\frac{2}{3} = \frac{?}{6}$

* **5.** (42) $\frac{1}{2} = \frac{?}{6}$

Find each unknown number. Remember to check your work.

* **6.** (43) $\frac{2}{3}n = 1$

* **7.** (43) $6 - w = 1\frac{4}{5}$

* **8.** (43) $m - 4\frac{1}{4} = 6\frac{3}{4}$

* **9.** (43) $c - 2.45 = 3$

* **10.** (43) $12 - d = 1.43$

11. (29) $\frac{5}{8} \times \frac{1}{5}$

12. (29) $\frac{3}{4} \times 5$

13. (26) $3\frac{7}{8} - 1\frac{3}{8}$

14. (19) **Classify** Which of these numbers is not a prime number?

A 23 **B** 33 **C** 43

15. (29) Compare: $\frac{2}{2} \bigcirc \frac{2}{2} \times \frac{2}{2}$

16. (14) In football a loss of yardage is often expressed as a negative number. If a quarterback is sacked for a 5-yard loss, the yardage change on the play can be shown as −5. How would a 12-yard loss be shown using a negative number?

17. (35) Write the decimal number for nine and twelve hundredths.

18. (16) Round 67,492,384 to the nearest million.

* **19.** (38, 39) **Analyze** 0.37×10^2

* **20.** (40) **Analyze** $0.6 \times 0.4 \times 0.2$

21. (38) The perimeter of a square room is 80 feet. The area of the room is how many square feet?

22. (25) Divide 100 by 16 and write the answer as a mixed number. Reduce the fraction part of the mixed number.

* **23.** (43) **Connect** **a.** Instead of dividing 100 by 16, Sandy divided the dividend and divisor by 4. What new division problem did Sandy make? What is the quotient?

b. Form an equivalent division problem for $4\frac{1}{2} \div \frac{1}{2}$ by doubling both the dividend and divisor. Then find the quotient.

24. (30) What is the least common multiple (LCM) of 4, 6, and 8?

25. (17) **Predict** What are the next three numbers in this sequence?

$$\frac{1}{16}, \frac{1}{8}, \frac{3}{16}, \frac{1}{4}, \frac{5}{16}, \frac{3}{8}, \frac{7}{16}, \_\_\_\_, \_\_\_\_, \_\_\_\_, \ldots$$

26. (17) Find the length of the segment below to the nearest eighth of an inch.

____

27. (17) What mixed number is indicated on the number line below?

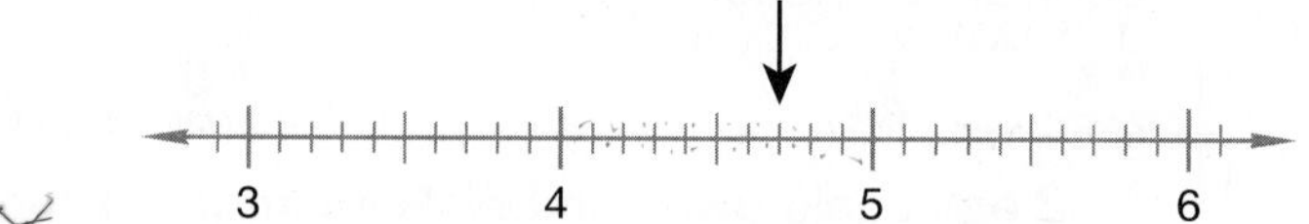

* **28.** (42) Write $\frac{1}{2}$ and $\frac{1}{5}$ as fractions with denominators of 10. Then add the renamed fractions.

29. (22, 33) **Model** Forty percent of the 20 seats on the bus were occupied. Write 40% as a reduced fraction. Then find the number of seats that were occupied. Draw a diagram to illustrate the problem.

30. (Inv. 3) Describe each angle in the figure as acute, right, or obtuse.

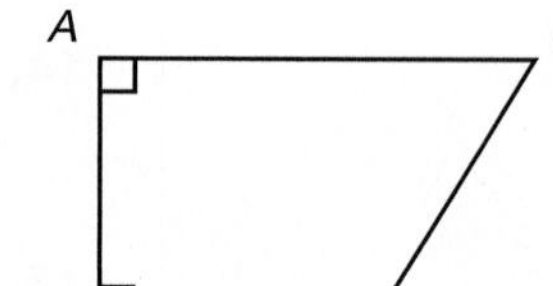

a. angle *A*

b. angle *B*

c. angle *C*

d. angle *D*

LESSON 44

• Simplifying Decimal Numbers
• Comparing Decimal Numbers

Power Up — *Building Power*

facts Power Up G

mental math

a. **Calculation:** 4 × 325

b. **Calculation:** 426 + 35

c. **Calculation:** 28 + 57

d. **Money:** \$8.50 + \$2.75

e. **Fractional Parts:** $\frac{1}{2}$ of 1400

f. **Number Sense:** $\frac{\$15.00}{100}$

g. **Probability:** How many different one-topping pizzas can be made with 2 types of crust and 4 types of toppings?

h. **Calculation:** 6 × 8, − 3, ÷ 5, + 1, × 6, + 3, ÷ 9

problem solving

Jeanna folded a square piece of paper in half from top to bottom. Then she folded the paper in half from left to right so that the four corners were together at the lower right. Then she cut off the lower right corner as shown. Which diagram will the paper look like when it is unfolded?

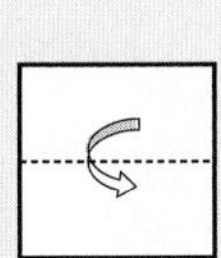

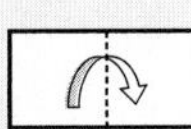

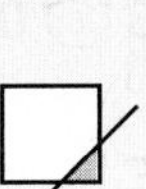

A 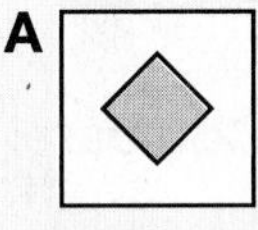

B 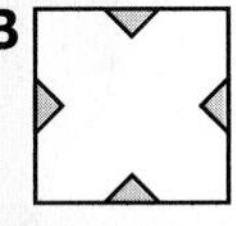

C 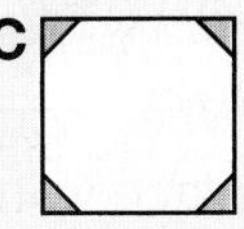

D 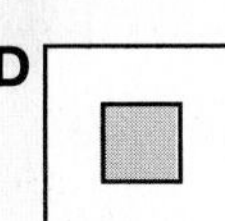

New Concepts — *Increasing Knowledge*

simplifying decimal numbers

Perform these two subtractions with a calculator. Which calculator answer differs from the printed answers below?

$$\begin{array}{r} 425 \\ -\ 125 \\ \hline 300 \end{array} \qquad \begin{array}{r} 4.25 \\ -\ 1.25 \\ \hline 3.00 \end{array}$$

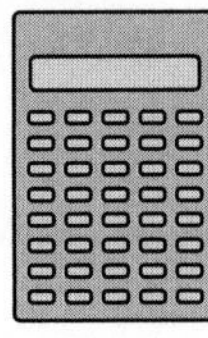

Calculators automatically simplify decimal numbers with zeros at the end by removing the extra zeros. Many calculators show a decimal point at the end of a whole number, although we usually remove the decimal point when we write the whole number. So "3.00" simplifies to "3." on a calculator. We remove the decimal point and write "3" only.

Example 1

Multiply 0.25 by 0.04 and simplify the product.

Solution

Thinking Skill

Generalize

How many decimal places are in the product when each of two factors has two decimal places?

We multiply.

$$\begin{array}{r} 0.25 \\ \times\ 0.04 \\ \hline 0.0100 \end{array}$$

If we perform this multiplication on a calculator, the answer 0.01 is displayed. The calculator simplifies the answer by removing zeros at the end of the decimal number.

0.0100 simplifies to **0.01**

In this book decimal answers are printed in simplified form unless otherwise stated.

comparing decimal numbers

Zeros at the end of a decimal number do not affect the value of the decimal number. Each of these decimal numbers has the same value because the 3 is in the tenths place:

0.3 0.30 0.300

Although 0.3 is the simplified form, sometimes it is useful to attach extra zeros to a decimal number. For instance, comparing decimal numbers can be easier if the numbers being compared have the same number of decimal places.

Example 2

Compare: 0.3 ◯ 0.303

Solution

When comparing decimal numbers, it is important to pay close attention to place values. Writing both numbers with the same number of decimal places can make comparing easier. We will attach two zeros to 0.3 so that it has the same number of decimal places as 0.303.

0.3 ◯ 0.303

↓

0.300 ◯ 0.303

We see that 300 thousandths is less than 303 thousandths. We write our answer like this:

0.3 < 0.303

Example 3

Arrange these numbers in order from least to greatest:

0.3 0.042 0.24 0.235

Solution

We write each number with three decimal places.

0.300   0.042   0.240   0.235

Then we arrange the numbers in order, omitting ending zeros.

0.042   0.235   0.24   0.3

Practice Set

Write these numbers in simplified form:

a. 0.0500

b. 50.00

c. 1.250

d. 4.000

Compare:

e. 0.2 ◯ 0.15

f. 12.5 ◯ 1.25

g. 0.012 ◯ 0.12

h. 0.31 ◯ 0.039

i. 0.4 ◯ 0.40

j. Write these numbers in order from least to greatest:

0.12   0.125   0.015   0.2

Written Practice

Strengthening Concepts

1. (12, 25) **Analyze** What is the sum of the third multiple of four and the third multiple of five?

2. (15) One mile is 5280 feet. How many feet is five miles?

The summit of Mt. Everest is 29,035 feet above sea level. The summit of Mt. Whitney is 14,495 feet above sea level. Use this information to answer problems **3** and **4.**

3. (13) Mt. Everest is how many feet taller than Mt. Whitney?

*** 4.** (13) **Analyze** The summit of Mt. Everest is how many feet higher than 5 miles above sea level? (Refer to problem 2.)

Find each unknown number. Remember to check your work.

*** 5.** (43) $5\frac{1}{3} - w = 4$

*** 6.** (43) $m - 6\frac{4}{5} = 1\frac{3}{5}$

*** 7.** (43) $6.74 + 0.285 + f = 11.025$

*** 8.** (43) $0.4 - d = 0.33$

9. (26) Wearing shoes, Fiona stands $67\frac{3}{4}$ inches tall. If the heels of her shoes are $1\frac{1}{4}$ inches thick, then how tall does Fiona stand without shoes?

*** 10.** (41) Form an equivalent division problem for $8\frac{1}{2} \div \frac{1}{2}$ by doubling both the dividend and divisor. Then find the quotient.

11. (35) Write thirty-two thousandths as a decimal number.

12. (29) What number is $\frac{1}{6}$ of 24,042?

* **13.** (44) Compare:

a. 0.25 ◯ 0.125 **b.** 25% ◯ 12.5%

14. (32) Write the standard numeral for $(6 \times 100) + (4 \times 1)$.

* **15.** (43) **Justify** A $36 dress is on sale for 10% off the regular price. Mentally calculate 10% of $36. Describe the method you used to arrive at your answer.

16. (30) **a.** How many $\frac{5}{8}$s are in 1?

b. Use the answer to part **a** to find the number of $\frac{5}{8}$s in 3.

17. (30) What is the least common multiple of 2, 3, 4, and 6?

18. (39) $(1.3)^2$

19. (42) $\frac{3}{4} = \frac{?}{12}$

20. (42) $\frac{2}{3} = \frac{?}{12}$

21. (18) Find the average of 26, 37, 42, and 43.

22. (16) Round 364,857 to the nearest thousand.

23. (23, 29) Twelve of the 30 students in the classroom were girls. What was the ratio of boys to girls in the classroom?

24. (19) **a.** List the factors of 100.

b. **Classify** Which of the factors of 100 are prime numbers?

25. (33, 35) Write 9% as a fraction. Then write the fraction as a decimal number.

* **26.** (42) Write $\frac{3}{4}$ and $\frac{2}{3}$ as fractions with denominators of 12. Then add the renamed fractions.

27. (Inv. 2) **Estimate** Which percent best describes the shaded portion of this rectangle? Explain why.

A 80% **C** 60%

B 40% **D** 20%

28. (32) Shelby started working at 10:30 a.m. and finished working at 2:15 p.m. How long did Shelby work?

* **29.** (44) **Estimate** Which of these numbers is closest to 1?

A 0.1 **B** 0.8 **C** 1.1 **D** 1.2

30. (17) **Connect** What mixed number corresponds to point X on the number line below?

X

9 10 11

LESSON 45

Dividing a Decimal Number by a Whole Number

Power Up

Building Power

facts Power Up F

mental math

a. **Calculation:** 4×425

b. **Number Sense:** $375 + 500$

c. **Calculation:** $77 + 18$

d. **Money:** \$12.00 − \$1.25

e. **Fractional Parts:** $\frac{1}{2}$ of 1500

f. **Money:** $\frac{\$40.00}{10}$

g. **Geometry:** A square has a length of 9 cm. What is the area of the square?

h. **Calculation:** 4 × 8, − 2, ÷ 3, + 2, ÷ 3, × 5, + 1, ÷ 3

problem solving Copy this problem and fill in the missing digits:

$$\begin{array}{r} \_\,\_\,\_ \\ +\ \ \ \ \ \_ \\ \hline \_\,\_\,\_\,8 \end{array}$$

New Concept

Increasing Knowledge

Dividing a decimal number by a whole number is similar to dividing dollars and cents by a whole number.

$$5\overline{)\$2.25}\ \text{quotient } \$0.45 \qquad 5\overline{)2.25}\ \text{quotient } 0.45$$

Notice that the decimal point in the quotient is directly above the decimal point in the **dividend.**

Example 1

Divide: $3\overline{)4.2}$

Solution

The decimal point in the quotient is directly above the decimal point in the dividend.

$$\begin{array}{r} \mathbf{1.4} \\ 3\overline{)4.2} \\ \underline{3} \\ 1\,2 \\ \underline{1\,2} \\ 0 \end{array}$$

Example 2

Divide: $3\overline{)0.24}$

Solution

The decimal point in the quotient is directly above the decimal point in the dividend. We fill the empty place with zero.

$$\begin{array}{r} \mathbf{0.08} \\ 3\overline{)0.24} \\ \underline{24} \\ 0 \end{array}$$

Decimal division answers are not written with remainders. Instead, we attach zeros to the end of the dividend and continue dividing.

Example 3

Divide: $5\overline{)0.6}$

Solution

Thinking Skill

Verify

Why are 0.6 and 0.60 equivalent?

The decimal point in the quotient is directly above the decimal point in the dividend. To complete the division, we attach a zero to 0.6, making the equivalent decimal number 0.60. Then we continue dividing.

$$\begin{array}{r} \mathbf{0.12} \\ 5\overline{)0.60} \\ \underline{5} \\ 10 \\ \underline{10} \\ 0 \end{array}$$

Practice Set

a. The distance from Margaret's house to school and back is 3.6 miles. How far does Margaret live from school?

b. The perimeter of a square is 6.4 meters. How long is each side of the square? How can you check that your answer is reasonable?

Divide:

c. $\frac{4.5}{3}$ **d.** $0.6 \div 4$ **e.** $2\overline{)0.14}$

f. $0.4 \div 5$ **g.** $4\overline{)0.3}$ **h.** $\frac{0.012}{6}$

i. $10\overline{)1.4}$ **j.** $\frac{0.7}{5}$ **k.** $0.1 \div 4$

Written Practice

Strengthening Concepts

1. (30) By what fraction must $\frac{5}{3}$ be multiplied to get a product of 1?

2. (15) How many $20 bills equal one thousand dollars?

3. (29) Cindy made $\frac{2}{3}$ of her 24 shots at the basket. Each basket was worth 2 points. How many points did she score?

* **4.** (45) $3\overline{)4.5}$ * **5.** (45) $8\overline{)0.24}$ * **6.** (45) $5\overline{)0.8}$

7. (30) What is the least common multiple (LCM) of 2, 4, 6, and 8?

Find each unknown number. Remember to check your work.

* **8.** (43) $\sqrt{36} - m = 2\frac{3}{10}$

* **9.** (43) $g - 2\frac{2}{5} = 5\frac{4}{5}$

* **10.** (43) $m - 1.56 = 1.44$

* **11.** (38, 43) $3^2 - n = 5.39$

12. (26) $4\frac{3}{8} - 2\frac{1}{8}$

13. (29) $\frac{8}{3} \cdot \frac{5}{2}$

14. (16) Estimate the product of 694 and 412.

15. (39) $0.7 \times 0.6 \times 0.5$

* **16.** (40) 0.46×0.17

* **17.** (15, 45) **Formulate** Mrs. Lopez's car traveled 177.6 miles on 8 gallons of gas. Her car traveled an average of how many miles per gallon? Use a multiplication pattern. Write an equation and solve the problem.

18. (29) What number is $\frac{3}{8}$ of 6? What operation did you use to find your answer?

* **19.** (41) **Justify** A shirt regularly priced at \$40 is on sale for 25% off. Mentally calculate 25% of \$40. Explain the method you used to arrive at your answer.

* **20.** (42) Write a fraction equal to $\frac{5}{6}$ that has 12 as the denominator. Then subtract $\frac{7}{12}$ from the fraction. Reduce the answer.

21. (38) **Analyze** The area of a square is 36 ft^2.

a. How long is each side of the square?

b. What is the perimeter of the square?

22. (33, 35) Write 27% as a fraction. Then write the fraction as a decimal number.

23. (17) Use a ruler to find the length of this rectangle to the nearest eighth of an inch.

24. (22, 33) **Model** Seventy-five percent of the 20 answers were correct. Write 75% as a reduced fraction. Then find the number of answers that were correct. Draw a diagram to illustrate this fractional-parts problem.

25. (29) The product of $\frac{1}{2}$ and $\frac{2}{3}$ is $\frac{1}{3}$.

$$\frac{1}{2} \times \frac{2}{3} = \frac{1}{3}$$

Arrange these fractions to form another multiplication fact and two division facts.

26. (Inv. 2) **Estimate** Which percent best describes the shaded portion of this circle? Explain why.

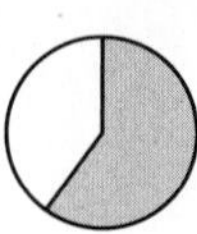

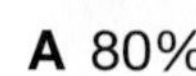

A 80% **C** 40%

B 60% **D** 20%

27. Write nine hundredths
(35)

a. as a fraction.

b. as a decimal number.

* **28.** Form an equivalent division problem for $5 \div \frac{1}{3}$ by multiplying both the dividend and divisor by 3. Then find the quotient.
(43)

29. The average number of students in three classrooms was 24. Altogether, how many students were in the three classrooms?
(18)

* **30.** *Analyze* Coach O'Rourke has a measuring wheel that records the distance the wheel is rolled along the ground. The circumference of the wheel is one yard. If the wheel is pushed half a mile, how many times will the wheel go around (1 mi = 5280 ft)?
(27)

Early Finishers
Real-World Application

Decide whether you can use an estimate to answer the question or if you need to compute an exact amount. Explain how you found your answer.

Emily and Jacob are equally sharing the $13.65 cost of a lunch. Tax is 5% and they want to leave a 15% tip. What is each person's share of the cost?

LESSON 46

• Writing Decimal Numbers in Expanded Notation
• Mentally Multiplying Decimal Numbers by 10 and by 100

Power Up

Building Power

facts Power Up G

mental math

a. **Calculation:** 4×525

b. **Number Sense:** $567 - 120$

c. **Number Sense:** $38 + 17$

d. **Money:** $\$5.75 + \$2.50$

e. **Fractional Parts:** $\frac{1}{2}$ of 950

f. **Number Sense:** $\frac{2000}{100}$

g. **Geometry:** The perimeter of a regular hexagon is 36 mm. What is the length of the sides of the hexagon?

h. **Calculation:** 9×7, $+ 1$, $\div 8$, $\times 3$, $+ 1$, $\times 2$, $- 1$, $\div 7$

problem solving

Copy this problem and fill in the missing digits:

$$\begin{array}{r} 9\_ \\ 9\overline{)\_9\_} \\ \underline{\_\_} \\ \_\_ \\ \underline{\_\_} \\ 0 \end{array}$$

New Concepts

Increasing Knowledge

writing decimal numbers in expanded notation

We may use **expanded notation** to write decimal numbers just as we have used expanded notation to write whole numbers. The values of some decimal places are shown in this table:

Decimal Place Values

1		$\frac{1}{10}$	$\frac{1}{100}$	$\frac{1}{1000}$
___	.	___	___	___
ones		tenths	hundredths	thousandths

We write 4.025 in expanded notation this way:

$$(4 \times 1) + \left(2 \times \frac{1}{100}\right) + \left(5 \times \frac{1}{1000}\right)$$

The zero that serves as a placeholder is usually not included in expanded notation.

Example 1

Reading Math

We say the word *and* when we see a decimal point. Read 5.06 as "five and six hundredths."

Write 5.06 in expanded notation.

Solution

The 5 is in the ones place, and the 6 is in the hundredths place.

$$(5 \times 1) + \left(6 \times \frac{1}{100}\right)$$

Example 2

Write $(4 \times \frac{1}{10}) + (5 \times \frac{1}{1000})$ as a decimal number.

Solution

We write the decimal number with a 4 in the tenths place and a 5 in the thousandths place. No digits in the ones place or the hundredths place are indicated, so we write zeros in those places.

0.405

mentally multiplying decimal numbers by 10 and by 100

Thinking Skill

Predict

How many zeros are in the product of 600 × 400?

When we multiply whole numbers by 10 or by 100, we can find the product mentally by attaching zeros to the whole number we are multiplying.

$$24 \times 10 = 240$$

$$24 \times 100 = 2400$$

It may seem that we are just attaching zeros, but we are actually shifting the digits to the left. When we multiply 24 by 10, the digits shift one place to the left. When we multiply 24 by 100, the digits shift two places to the left. In each product zeros hold the 2 and the 4 in their proper places.

1000s	100s	10s	1s	
		2	4	24
	2	4	0	24 × 10 (one-place shift)
2	4	0	0	24 × 100 (two-place shift)

When we multiply a decimal number by 10, the digits shift one place to the left. When we multiply a decimal number by 100, the digits shift two places to the left. Here we show the products when 0.24 is multiplied by 10 and by 100.

10s	1s	.	$\frac{1}{10}$s	$\frac{1}{100}$s	
	0	.	2	4	0.24
	2	.	4		0.24 × 10 (one-place shift)
2	4	.			0.24 × 100 (two-place shift)

Although it is the digits that are shifting one or two places to the left, we get the same effect by shifting the decimal point one or two places to the right.

$0.24 \times 10 = 2.4$ (one-place shift)

$0.24 \times 100 = 24. = 24$ (two-place shift)

Example 3

Multiply: 3.75 × 10

Solution

Since we are multiplying by 10, the product will have the same digits as 3.75, but the digits will be shifted one place. The product will be ten times as large, so we mentally shift the decimal point one place to the right.

$3.75 \times 10 = \mathbf{37.5}$ (one-place shift)

We do not need to attach any zeros, because the decimal point serves to hold the digits in their proper places.

Example 4

Multiply: 3.75 × 100

Solution

When multiplying by 100, we mentally shift the decimal point two places to the right.

$3.75 \times 100 = 375. = \mathbf{375}$ (two-place shift)

We do not need to attach zeros. Since there are no decimal places, we may leave off the decimal point.

Example 5

Multiply: $\mathbf{\frac{1.2}{0.4} \times \frac{10}{10}}$

Solution

Multiplying both 1.2 and 0.4 by 10 shifts each decimal point one place.

$$\frac{1.2}{0.4} \times \frac{10}{10} = \frac{12}{4}$$

The expression $\frac{12}{4}$ means "12 divided by 4."

$$\frac{12}{4} = \mathbf{3}$$

Practice Set

Write these numbers in expanded notation:

a. 2.05

b. 20.5

c. 0.205

Write these numbers in decimal form:

d. $(7 \times 10) + \left(8 \times \frac{1}{10}\right)$

e. $\left(6 \times \frac{1}{10}\right) + \left(4 \times \frac{1}{100}\right)$

Mentally calculate each product:

f. 0.35×10

g. 0.35×100

h. 2.5×10

i. 2.5×100

j. 0.125×10

k. 0.125×100

Conclude For the following statements, answer "true" or "false":

l. If 0.04 is multiplied by 10, the product is a whole number.

m. If 0.04 is multiplied by 100, the product is a whole number.

Multiply as shown. Then complete the division.

n. $\frac{1.5}{0.5} \times \frac{10}{10}$

o. $\frac{2.5}{0.05} \times \frac{100}{100}$

Written Practice

Strengthening Concepts

1. (25) **Analyze** When a fraction with a numerator of 30 and a denominator of 8 is converted to a mixed number and reduced, what is the result?

2. (13) **Formulate** Normal body temperature is 98.6° on the Fahrenheit scale. A person with a temperature of 100.2°F would have a temperature how many degrees above normal? Write an equation and solve the problem.

3. (15) Four and twenty is how many dozen?

4. (46) Write $(5 \times 10) + (6 \times \frac{1}{10}) + (7 \times \frac{1}{1000})$ in decimal form.

5. (33, 35) Twenty-one percent of the earth's atmosphere is oxygen. Write 21% as a fraction. Then write the fraction as a decimal number.

6. (33) Twenty-one percent is slightly more than 20%. Twenty percent is equivalent to what reduced fraction?

*** 7.** (45) $5\overline{)6.35}$

*** 8.** (45) $4\overline{)0.5}$

*** 9.** (45) $8\overline{)1.0}$

Find each unknown number:

*** 10.** (43) $x + 3\frac{5}{8} = 9$

*** 11.** (43) $y - 16\frac{1}{4} = 4\frac{3}{4}$

*** 12.** (43) $1 - q = 0.235$

*** 13.** (43) $26.9 + 12 + w = 49.25$

14. (31, 41) **Verify** Fifty percent of the area of this rectangle is shaded. What is the area of the shaded region? Explain your thinking.

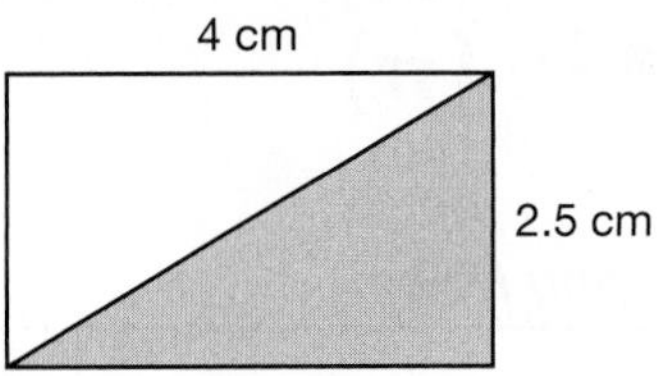

15. (23) **Connect** What is the ratio of the value of a dime to the value of a quarter?

16. (39) 3.7×0.25

* **17.** (42) $\frac{3}{4} = \frac{?}{12}$

18. (30) What is the least common multiple of 3, 4, and 8?

* **19.** (40, 44) Compare:

a. $\frac{1}{10} \bigcirc 0.1$

b. $0.1 \bigcirc (0.1)^2$

20. (34) Which digit is in the thousandths place in 1,234.5678?

21. (16) Estimate the quotient when 3967 is divided by 48.

22. (38) The area of a square is 100 cm^2. What is its perimeter?

23. (14) John carried the football twice in the game. One play gained 6 yards. The other play lost 8 yards. Use a negative number to show John's total yardage for the game.

24. (29) $\frac{1}{2} \cdot \frac{4}{5}$

25. (29) $\left(\frac{3}{4}\right)\left(\frac{5}{3}\right)$

26. (27) **Connect** Laquesha bought a 24-inch-diameter wheel for her bicycle and measured it carefully. Arrange these measures in order from least to greatest:

circumference, radius, diameter

* **27.** (41) **Estimate** The chef's salad cost $6.95. The sales-tax rate was 8%. What was the total cost including tax? Explain how to use estimation to check whether your answer is reasonable.

28. (17) Use a ruler to find the width of this rectangle to the nearest eighth of an inch.

29. (30) **a.** How many $\frac{3}{8}$s are in 1?

b. Use the answer to part **a** to find the number of $\frac{3}{8}$s in 3.

* **30.** (42) Rename $\frac{1}{2}$ and $\frac{1}{3}$ so that the denominators of the renamed fractions are 6. Then add the renamed fractions.

LESSON 47

• Circumference
• Pi (π)

Power Up — Building Power

facts Power Up E

mental math

a. Calculation: 4×925

b. Calculation: 3×87

c. Number Sense: $56 - 19$

d. Money: $9.00 − $1.25

e. Fractional Parts: $\frac{1}{2}$ of $12.50

f. Money: $\frac{\$25.00}{10}$

g. Probability: How many different outfits can be made with 3 shirts and 5 pairs of pants?

h. Calculation: 6×8, $+ 2$, $\times 2$, $- 10$, $\div 9$, $+ 5$, $\div 3$, $+ 1$, $\div 6$

problem solving Radley held a number cube so that he could see three adjoining faces. Radley said that he could see a total of 8 dots. Could he be correct? Explain your answer.

New Concepts — Increasing Knowledge

circumference

Laquesha measured the diameter of her bicycle wheel with a yardstick and found that the diameter was 2 feet. She wondered whether she could find the circumference of the tire with only this information. In other words, she wondered how many diameters equal the circumference. In the following activity we will estimate and measure to find the number of diameters in a circumference.

Activity

Circumference

Materials needed:

- 2–4 different circular objects (e.g., paper plates, pie pans, flying disks, bicycle tires, plastic lids, trash cans)
- Lesson Activity 11
- string or masking tape
- scissors
- cloth tape measure(s)
- calculator(s)

Model This activity has two parts. In the first part you and your group will cut a length of string as long as the diameter of each object you will measure. (A length of masking tape may be used in place of string.) Then you will wrap the string around the object and estimate the number of diameters needed to reach all the way around. To do this, first mark a starting point on the object. Wrap the string around the object, and mark the point where the string ends. Repeat this process until you reach the starting point, counting the whole lengths of string and estimating any fractional part. Do this for each object you selected.

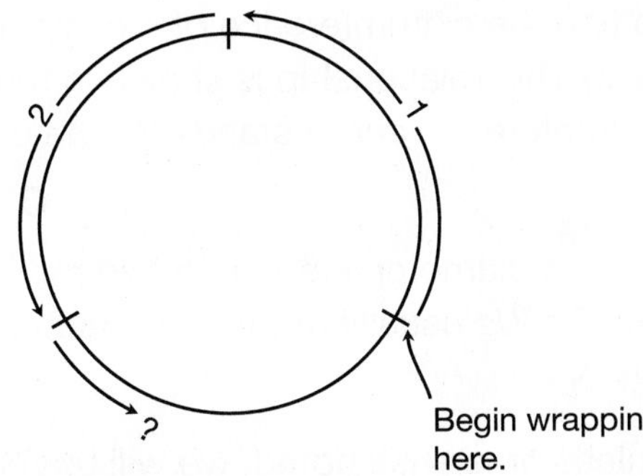

Thinking Skills

Connect

Remember that:

$\frac{1}{4} = 0.25$

$\frac{3}{4} = 0.75$

$\frac{1}{2} = 0.5$

Estimate In the second part of the activity, you will measure the circumference and diameter of the circular objects and record the measurements on a recording sheet. If you have a metric tape measure, record the measurements to the nearest centimeter. If you have a customary tape measure, record your answers to the nearest quarter inch in decimal form. Using a calculator, divide the circumference of each circle by its diameter to determine the number of diameters in the circumference. Round each quotient to the nearest hundredth.

Record your results on **Lesson Activity 11,** as shown below.

Part 1: Estimates

Object	Approximate number of diameters in the circumference
plate	$3\frac{1}{5}$
trash can	$3\frac{1}{4}$

Part 2: Measures

Object	Circumference	Diameter	$\frac{\text{Circumference}}{\text{Diameter}}$
plate	78 cm	25 cm	3.12
trash can	122 cm	38 cm	3.21

pi (π)

If we know the radius or diameter of a circle, we can calculate the approximate circumference of the circle. In the previous activity we found that for any given circle there are a little more than three diameters in the circumference. Some people use 3 as a very rough approximation of the number of diameters in a circumference. The actual number of diameters in a circumference is closer to $3\frac{1}{7}$, which is approximately 3.14. The exact number of diameters in a circumference cannot be expressed as a fraction or as a decimal number, so mathematicians use the Greek letter **π (pi)** to stand for this number.

Thinking Skill

Verify

Why can you use the formula $\pi 2r$ or $2\pi r$ and get the same answer?

To find the circumference of a circle, we multiply the diameter of the circle by π. This relationship is shown in the formula below, where *C* stands for the circumference and *d* stands for the diameter.

$$C = \pi d$$

Since a diameter is equal to two radii (2*r*), we may replace *d* in the formula with 2*r*. We usually arrange the factors this way:

$$C = 2\pi r$$

Reading Math

Symbols

We read ≈ as "is approximately equal to."

Unless otherwise noted, we will use 3.14 as an approximation for π. We may use a "wavy" equal sign to indicate that two numbers are approximately equal, as shown below.

$$\pi \approx 3.14$$

To use a formula such as $C = \pi d$, we **substitute** the measures or numbers we are given in place of the variables in the formula.

Example

Sidney drew a circle with a 2-inch radius. What is the circumference of the circle?

Solution

The radius of the circle is 2 inches, so the diameter is 4 inches. We multiply 4 inches by π (3.14) to find the circumference.

$$C = \pi d$$

$$C \approx (3.14)(4 \text{ in.})$$

$$C \approx 12.56 \text{ in.}$$

The circumference of the circle is about **12.56 inches.**

Justify Why is the answer about 12.56 inches reasonable for the circumference of the circle?

Practice Set

a. *Explain* In this lesson two formulas for the circumference of a circle are shown, $C = \pi d$ and $C = 2\pi r$. Why are these two formulas equivalent?

Find the circumference of each of these circles. (Use 3.14 for π.)

b.

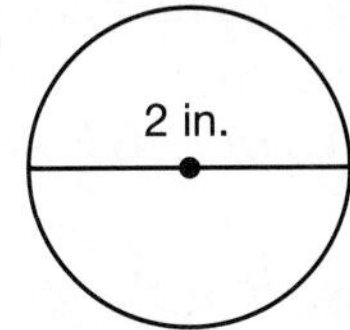

c.

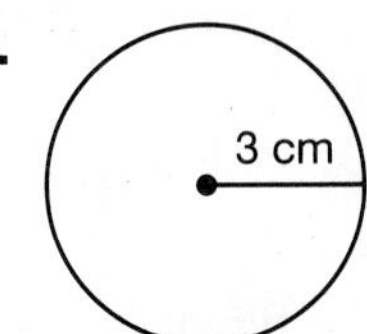

d. **Estimate** The diameter of a penny is about $\frac{3}{4}$ of an inch (0.75 inch). Find the circumference of a penny. Round your answer to two decimal places. Explain why your answer is reasonable.

e. **Model** Roll a penny through one rotation on a piece of paper. Mark the start and the end of the roll. How far did the penny roll in one rotation? Measure the distance to the nearest eighth of an inch.

f. **Justify** The radius of the great wheel was $14\frac{7}{8}$ ft. Which of these numbers is the best rough estimate of the wheel's circumference? Explain how you decided on your answer.

A 15 ft **B** 60 ft **C** 90 ft **D** 120 ft

g. Use the formula $C = 2\pi r$ to find the circumference of a circle with a radius of 5 inches. (Use 3.14 for π.)

Written Practice *Strengthening Concepts*

1. *(10)* The first positive odd number is 1. The second is 3. What is the tenth positive odd number?

2. *(15)* **Formulate** A passenger jet can travel 600 miles per hour. How long would it take a jet traveling at that speed to cross 3000 miles of ocean? Write an equation and solve the problem.

3. *(15)* José bought Carmen one dozen red roses, two for each month he had known her. How long had he known her?

* **4.** *(47)* **Conclude** If $A = bh$, what is A when $b = 8$ and $h = 4$?

5. *(2)* The Commutative Property of Multiplication allows us to rearrange factors without changing the product. So $3 \cdot 5 \cdot 2$ may be arranged $2 \cdot 3 \cdot 5$. Use the commutative property of multiplication to rearrange these prime factors in order from least to greatest:

$$3 \cdot 7 \cdot 2 \cdot 5 \cdot 2 \cdot 3 \cdot 3 \cdot 5$$

6. *(29)* If $s = \frac{1}{2}$, what number does $4s$ equal?

* **7.** *(46)* **Generalize** Write 6.25 in expanded notation.

8. *(33, 35)* Write 99% as a fraction. Then write the fraction as a decimal number.

* **9.** *(45)* $12\overline{)0.18}$

* **10.** *(45)* $10\overline{)12.30}$

Find each missing number:

11. $w \div \sqrt{36} = 6^2$ (4, 38)

* **12.** $5y = 1.25$ (43)

* **13.** $n + 5\frac{11}{12} = 10$ (43)

* **14.** $m - 6\frac{2}{5} = 3\frac{3}{5}$ (43)

15. $8\frac{3}{4} + 5\frac{3}{4}$ (26)

16. $\frac{5}{3} \times \frac{5}{4}$ (29)

* **17.** $\frac{3}{4} = \frac{?}{20}$ (42)

* **18.** $\frac{3}{5} = \frac{?}{20}$ (42)

19. (18) Bob's scores on his first five tests were 18, 20, 18, 20, and 20. His average score is closest to which of these numbers?

A 17 **B** 18 **C** 19 **D** 20

* **20.** (47) Robert's bicycle tires are 20 inches in diameter. What is the circumference of a 20-inch circle? (Use 3.14 for π.) Explain why your answer is reasonable.

21. (19) Which factors of 20 are also factors of 30?

* **22.** (46) **Explain** Mentally calculate the product of 6.25 and 10. Describe how you performed the mental calculation.

* **23.** (46) Multiply as shown. Then complete the division.

$$\frac{1.25}{0.5} \cdot \frac{10}{10}$$

* **24.** (41) Shelly answered 90% of the 40 questions correctly. What number is 90% of 40?

Refer to the chart shown below to answer problems **25** and **26.**

Planet	Number of Earth Days to Orbit Sun
Mercury	88
Venus	225
Earth	365
Mars	687

25. (13) Mars takes how many more days than Earth to orbit the Sun?

26. (15) **Estimate** In the time it takes Mars to orbit the Sun once, Venus orbits the Sun about how many times?

27. (17) Use an inch ruler to find the length and width of this rectangle.

28. Calculate the perimeter of the rectangle in problem 27.
(8)

29. Rename $\frac{2}{5}$ so that the denominator of the renamed fraction is 10. Then subtract the renamed fraction from $\frac{9}{10}$. Reduce the answer.
(42)

* **30.** **Verify** When we mentally multiply 15 by 10, we can simply attach a zero to 15 to make the product 150. When we multiply 1.5 by 10, why can't we attach a zero to make the product 1.50?
(46)

Early Finishers
Real-World Application

Decide whether you can use an estimate to answer the question or if you need to compute an exact amount. Explain how you found your answer.

A music store is having a sale. Hassan wants to buy 3 CDs. One CD costs $11.95. The other two CDs cost $14.99 each. Hassan has $50 but needs to use $10 of it to repay a loan from his brother. Does Hassan have enough money to buy the CDs and repay his brother?

LESSON 48

Subtracting Mixed Numbers with Regrouping, Part 1

Power Up

Building Power

facts Power Up G

mental math

a. **Calculation:** 8×25

b. **Number Sense:** $630 - 50$

c. **Number Sense:** $62 + 19$

d. **Money:** \$4.50 + 75¢

e. **Fractional Parts:** $\frac{1}{2}$ of \$15.00

f. **Money:** $\frac{\$25.00}{100}$

g. **Geometry:** What is the relationship between the radius and the diameter of a circle?

h. **Calculation:** 4×7, $- 1$, $\div 3$, $\times 4$, $\div 6$, $\times 3$, $\div 2$

problem solving

Josepha is making sack lunches. She has two kinds of sandwiches, three kinds of fruit, and two kinds of juice. If each sack lunch contains one kind of sandwich, one kind of fruit, and one kind of juice, how many different sack-lunch combinations can Josepha make?

New Concept

Increasing Knowledge

Here is another "separating" problem about pies:

There were $4\frac{1}{6}$ pies on the restaurant shelf. The server sliced one of the whole pies into sixths. Then the server removed $1\frac{2}{6}$ pies. How many pies were left on the shelf?

Thinking Skill

Discuss

What words or phrases in the problem give clues about which operation we should use to answer the question?

We may illustrate this problem with circles. There were $4\frac{1}{6}$ pies on the shelf.

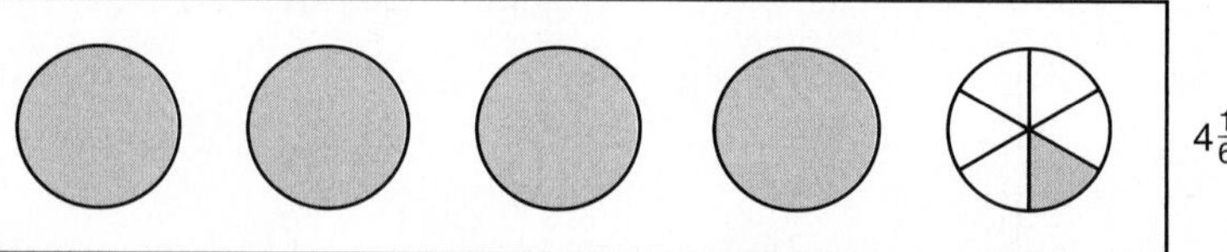

The server sliced one of the whole pies into sixths. This makes $3\frac{7}{6}$ pies, which equals $4\frac{1}{6}$ pies.

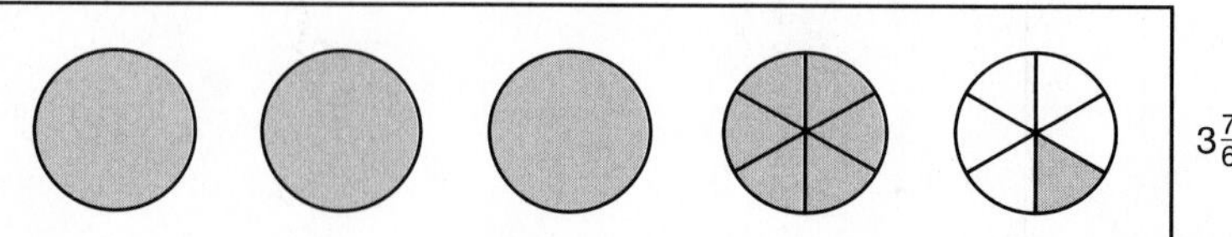

Then the server removed $1\frac{2}{6}$ pies. So $2\frac{5}{6}$ pies were left on the shelf.

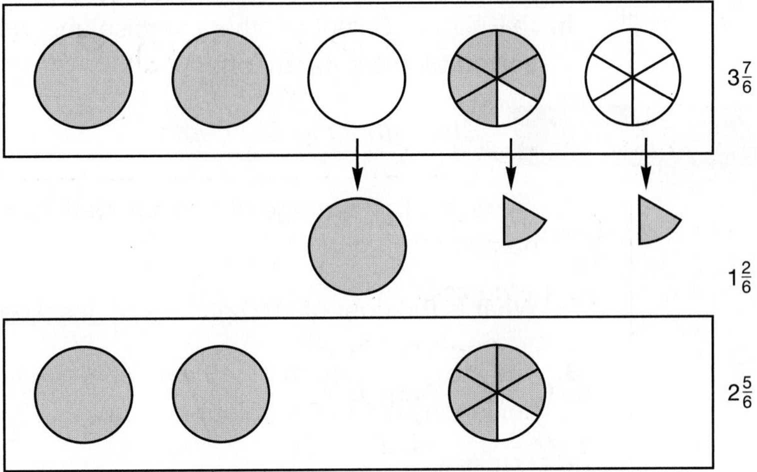

Now we show the arithmetic for subtracting $1\frac{2}{6}$ from $4\frac{1}{6}$.

$$\begin{array}{r} 4\frac{1}{6} \text{ pies} \\ -\ 1\frac{2}{6} \text{ pies} \\ \hline \end{array}$$

Thinking Skill

Verify

How can you check the answer to the subtraction problem?

We cannot subtract $\frac{2}{6}$ from $\frac{1}{6}$, so we rename $4\frac{1}{6}$. Just as the server sliced one of the pies into sixths, so we change one of the four wholes into $\frac{6}{6}$. This makes three whole pies plus $\frac{6}{6}$ plus $\frac{1}{6}$, which is $3\frac{7}{6}$. Now we can subtract.

$$\begin{array}{r} 4\frac{1}{6} \\ -\ 1\frac{2}{6} \\ \hline \end{array} \xrightarrow{3 + \frac{6}{6} + \frac{1}{6}} \begin{array}{r} 3\frac{7}{6} \text{ pies} \\ -\ 1\frac{2}{6} \text{ pies} \\ \hline 2\frac{5}{6} \text{ pies} \end{array}$$

Example

Subtract: $5\frac{1}{3} - 2\frac{2}{3}$

Solution

We cannot subtract $\frac{2}{3}$ from $\frac{1}{3}$, so we rename $5\frac{1}{3}$. We change one of the five wholes into $\frac{3}{3}$. Then we combine 4 and $\frac{3}{3} + \frac{1}{3}$ to get $4\frac{4}{3}$. Now we can subtract.

$$\begin{array}{r} 5\frac{1}{3} \\ -\ 2\frac{2}{3} \\ \hline \end{array} \xrightarrow{4 + \frac{3}{3} + \frac{1}{3}} \begin{array}{r} 4\frac{4}{3} \\ -\ 2\frac{2}{3} \\ \hline \mathbf{2\frac{2}{3}} \end{array}$$

Practice Set

Subtract.

a. $\begin{array}{r} 4\frac{1}{3} \\ -\ 1\frac{2}{3} \\ \hline \end{array}$ **b.** $\begin{array}{r} 3\frac{2}{5} \\ -\ 2\frac{3}{5} \\ \hline \end{array}$ **c.** $\begin{array}{r} 5\frac{2}{4} \\ -\ 1\frac{3}{4} \\ \hline \end{array}$

d. $\begin{array}{r} 5\frac{1}{8} \\ -\ 2\frac{4}{8} \\ \hline \end{array}$ **e.** $\begin{array}{r} 7\frac{3}{12} \\ -\ 4\frac{10}{12} \\ \hline \end{array}$ **f.** $\begin{array}{r} 6\frac{1}{4} \\ -\ 2\frac{3}{4} \\ \hline \end{array}$

g. Model Select one of the exercises to model with a drawing.

h. Formulate Select another exercise and write a word problem that is solved by the subtraction.

Written Practice *Strengthening Concepts*

1. (18) Analyze The average of two numbers is 10. What is the sum of the two numbers?

*** 2.** (46) What is the cost of 10.0 gallons of gasoline priced at $2.279 per gallon?

3. (32) The movie started at 11:45 a.m. and ended at 1:20 p.m. The movie was how many hours and minutes long?

*** 4.** (44) Classify Three of the numbers shown below are equal. Which number is not equal to the others?

A $\frac{1}{2}$ **B** 0.2 **C** 0.5 **D** $\frac{10}{20}$

*** 5.** (44) Analyze Arrange these numbers in order from least to greatest:

$$1.02, 0.102, 0.12, 1.20, 0, -1$$

6. (37) $0.1 + 0.2 + 0.3 + 0.4$

7. (39) $(8)(0.125)$

8. (11) Formulate Juan was hiking to a waterfall 3 miles away. After hiking 2.1 miles, how many more miles did he have to hike to reach the waterfall? Write an equation and solve the problem.

9. (16) Estimate the sum of 4967, 8142, and 6890.

*** 10.** (45) $8\overline{)0.144}$

*** 11.** (45) $6\overline{)0.9}$

*** 12.** (45) $4\overline{)0.9}$

*** 13.** (46) What is the cost of 100 pens priced at 39¢ each?

*** 14.** (46) Write $(5 \times 10) + (6 \times \frac{1}{10}) + (4 \times \frac{1}{100})$ in standard form.

15. (30) What is the least common multiple of 6 and 8?

Find each unknown number:

*** 16.** (43) $w - 7\frac{7}{12} = 5\frac{5}{12}$

*** 17.** (43) $12 - m = 5\frac{2}{3}$

*** 18.** (43) $n + 2\frac{3}{4} = 5\frac{1}{4}$

*** 19.** (43) $x + 3.21 = 4$

20. (29) What fraction is $\frac{2}{3}$ of $\frac{3}{4}$?

21. (14) Sam carried the football three times during the game. He had gains of 3 yards and 5 yards and a loss of 12 yards. Use a negative number to show Sam's overall yardage for the game.

*** 22.** (47) Justify If a spool for thread is 2 cm in diameter, then one wind of thread is about how many centimeters long? (Use 3.14 for π.) Explain how to mentally check whether your answer is reasonable.

23. (23, 29) If a rectangle is 12 inches long and 8 inches wide, what is the ratio of its length to its width?

24. (8) The perimeter of this square is 4 feet. What is its perimeter in inches?

25. (31) The area of this square is one square foot. What is its area in square inches?

12 in.

1 ft | 12 in.

1 ft

* **26.** (47) **Evaluate** If $d = rt$, and if $r = 60$ and $t = 4$, what does d equal?

27. (22, 33) Seventy-five percent of the 32 chairs in the room were occupied. Write 75% as a reduced fraction. Then find the number of chairs that were occupied.

* **28.** (42) Rename $\frac{1}{3}$ and $\frac{1}{4}$ as fractions with denominators of 12. Then add the renamed fractions.

* **29.** (46) **Analyze** Multiply as shown. Then simplify the answer.

$$\frac{3.5}{0.7} \cdot \frac{10}{10}$$

* **30.** (48) **Explain** There were $3\frac{1}{6}$ pies on the shelf. How can the server take $1\frac{5}{6}$ pies from the shelf?

Early Finishers
Real-World Application

Leonardo purchased 4 pounds of grapes. Leonardo gave his neighbor half a pound of grapes so he could make a fruit salad. Next, he gave his friend 0.09 lb of grapes for a snack. Lastly, Leonardo gave his brother 1.25 lb of grapes so his brother could make grape juice. What is the weight of the grapes Leonardo has left? Show your work.

LESSON 49

• Dividing by a Decimal Number

Power Up — *Building Power*

facts Power Up H

mental math

a. Order of Operations: $(8 \times 100) + (8 \times 25)$

b. Number Sense: $290 + 50$

c. Number Sense: $58 - 19$

d. Money: \$5.00 − \$3.25

e. Fractional Parts: $\frac{1}{2}$ of \$30.00

f. Number Sense: $\frac{4000}{100}$

g. Geometry: A circle has a radius of 3 cm. Using 3.14 for π, what is the circumference of the circle?

h. Calculation: 5×10, $\div 2$, $+ 5$, $\div 2$, $+ 5$, $\div 2$, $\div 2$

problem solving

Ned walked from his home to school following the path from H to I to J to K to L to M to S. After school he walked home from S to C to H. Compare the distance of Ned's walk to school to the distance of his walk home.

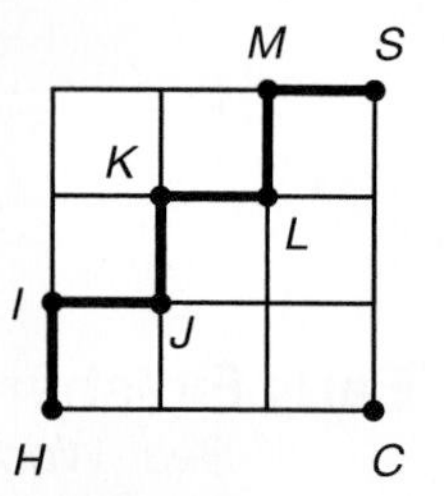

New Concept — *Increasing Knowledge*

When the divisor of a division problem is a decimal number, we change the problem so that the divisor is a whole number.

$$\frac{1.24}{0.4} \qquad 0.4\overline{)1.24}$$

The divisor is a decimal number. We change the problem before we divide.

One way to change a division problem is to multiply the divisor and the dividend by 10. Notice in the whole-number division problem below that multiplying both numbers by 10 does not change the quotient.

$$4\overline{)8}\text{ (quotient 2)} \longrightarrow 40\overline{)80}\text{ (quotient 2)}$$

Multiplying 4 and 8 by 10 does not change the quotient.

The quotient is not changed, because we formed an equivalent division problem just as we form equivalent fractions—by multiplying by a form of 1.

$$\frac{8}{4} \times \frac{10}{10} = \frac{80}{40}$$

Thinking Skill

Summarize

How can we mentally multiply a decimal number by 10 or 100?

We can also use this method to change a division by a decimal number to a division by a whole number.

If we multiply the divisor and dividend in $\frac{1.24}{0.4}$ by 10, the new problem has a whole-number divisor.

$$\text{decimal divisor} \longrightarrow \frac{1.24}{0.4} \times \frac{10}{10} = \frac{12.4}{4} \longleftarrow \text{whole-number divisor}$$

We divide 12.4 by 4 to find the quotient.

$$4\overline{)12.4}\quad 3.1$$

Example 1

Divide: $\frac{\mathbf{1.24}}{\mathbf{0.04}}$

Solution

The divisor, 0.04, is a decimal number with two decimal places. To make the divisor a whole number, we will multiply $\frac{1.24}{0.4}$ by $\frac{100}{100}$, which shifts each decimal point two places to the right.

$$\frac{1.24}{0.04} \times \frac{100}{100} = \frac{124}{4}$$

This forms an equivalent division problem in which the divisor is a whole number. Now we perform the division.

$$\begin{array}{r} \mathbf{31} \\ 4\overline{)124} \\ 12 \\ \hline 04 \\ 4 \\ \hline 0 \end{array}$$

Example 2

Divide: $\mathbf{0.6\overline{)1.44}}$

Solution

The divisor, 0.6, has one decimal place. If we multiply the divisor and dividend by 10, we will shift the decimal point one place to the right in both numbers.

$$06.\overline{)14.4}$$

This makes a new problem with a whole-number divisor, which we solve below.

$$\begin{array}{r} \mathbf{2.4} \\ 6\overline{)14.4} \\ 12 \\ \hline 2\,4 \\ 2\,4 \\ \hline 0 \end{array}$$

Some people use the phrase "over, over, and up" to keep track of the decimal points when dividing by decimal numbers.

$$06.\overline{)14.4}$$

up; over over

Practice Set

Thinking Skill

Discuss

What do you notice about the answers in exercises **a** and **b**?

a. We would multiply the divisor and dividend of $\frac{1.44}{1.2}$ by what number to make the divisor a whole number?

b. We would multiply the divisor and dividend of $0.12\overline{)0.144}$ by what number to make the divisor a whole number?

Change each problem so that the divisor is a whole number. Then divide.

c. $\frac{0.24}{0.4}$ **d.** $\frac{9}{0.3}$

e. $0.05\overline{)2.5}$ **f.** $0.3\overline{)12}$

g. $0.24 \div 0.8$ **h.** $0.3 \div 0.03$

i. $0.05\overline{)0.4}$ **j.** $0.2 \div 0.4$

k. Find how many nickels are in \$3.25 by dividing 3.25 by 0.05.

Written Practice

Strengthening Concepts

* **1.** (12, 39) When the product of 0.2 and 0.3 is subtracted from the sum of 0.2 and 0.3, what is the difference?

2. (22) *Model* Four fifths of a dollar is how many cents? Draw a diagram to illustrate the problem.

3. (31, 39) The rectangular, 99-piece "Nano" jigsaw puzzle is only 2.6 inches long and 2.2 inches wide. What is the area of the puzzle in square inches?

4. (39) Find the perimeter of the puzzle described in problem 3.

* **5.** (44) Compare:

a. 0.31 ◯ 0.301 **b.** 31% ◯ 30.1%

6. (38) $0.67 + 2 + 1.33$ **7.** (39) $12(0.25)$

* **8.** (49) $0.07\overline{)3.5}$ * **9.** (49) $0.5\overline{)12}$

* **10.** (45) $8\overline{)0.14}$ * **11.** (39) $(0.012)(1.5)$

Find each unknown number:

* **12.** (43) $n - 6\frac{1}{8} = 4\frac{3}{8}$ * **13.** (42) $\frac{4}{5} = \frac{x}{100}$

* **14.** (43) $5 - m = 1.37$ * **15.** (43) $m + 7\frac{1}{4} = 15$

16. Write the decimal number one and twelve thousandths.
(35)

17. $5\frac{7}{10} + 4\frac{9}{10}$
(26)

18. $\frac{5}{2} \cdot \frac{5}{3}$
(29)

19. How much money is 40% of $25.00?
(41)

20. There are 24 hours in a day. James sleeps 8 hours each night.
(29)

a. Eight hours is what fraction of a day?

b. What fraction of a day does James sleep?

c. What fraction of a day does James not sleep?

21. **List** What factors do 12 and 18 have in common (that is, the numbers that are factors of both 12 and 18).
(19)

* **22.** **Analyze** What is the average of 1.2, 1.3, and 1.7?
(18, 45)

23. **Estimate** Jan estimated that 49% of $19.58 is $10. She rounded 49% to 50% and rounded $19.58 to $20. Then she mentally calculated 50% of $20. Use Jan's method to estimate 51% of $49.78. Explain how to perform the estimate.
(41)

24. **a.** How many $\frac{3}{4}$s are in 1?
(30)

b. Use the answer to part **a** to find the number of $\frac{3}{4}$s in 4.

25. **Connect** Refer to the number line shown below to answer parts **a–c.**
(17)

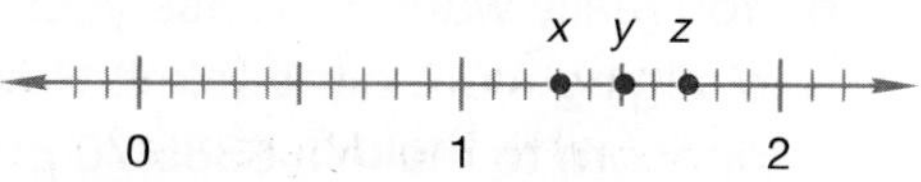

a. Which point is halfway between 1 and 2?

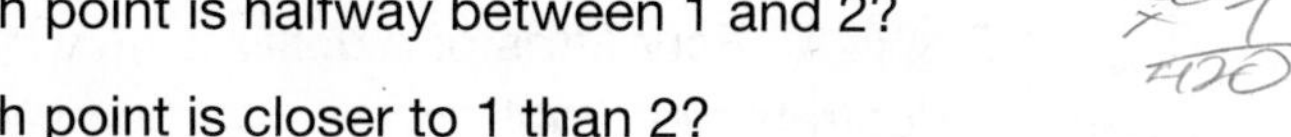

b. Which point is closer to 1 than 2?

c. Which point is closer to 2 than 1?

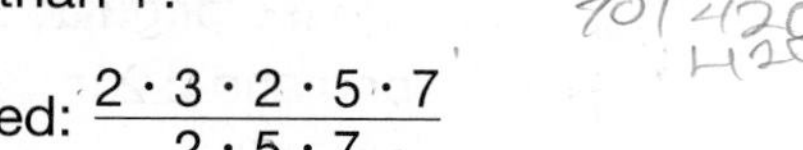

26. Multiply and divide as indicated: $\frac{2 \cdot 3 \cdot 2 \cdot 5 \cdot 7}{2 \cdot 5 \cdot 7}$
(5)

* **27.** **Represent** We can find the number of quarters in three dollars by dividing $3.00 by $0.25. Show this division using the pencil-and-paper method taught in this lesson.
(49)

28. **Connect** Use a ruler to find the length of each side of this square to the nearest eighth of an inch. Then calculate the perimeter of the square.
(8)

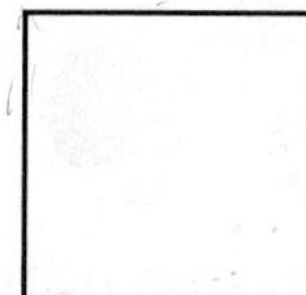

* **29.** A paper-towel tube is about 4 cm in diameter. The circumference of a paper-towel tube is about how many centimeters?
(47)
(Use 3.14 for π.)

* **30.** **Explain** Sam was given the following division problem:
(49)

$$\frac{2.5}{0.5}$$

Instead of multiplying the numerator and denominator by 10, he accidentally multiplied by 100, as shown below.

$$\frac{2.5}{0.5} \times \frac{100}{100} = \frac{250}{50}$$

Then he divided 250 by 50 and found that the quotient was 5. Did Sam find the correct answer to 2.5 ÷ 0.5? Why or why not?

Early Finishers
Real-World Application

After collecting tickets for three years, you are finally promoted to night manager of the local movie theater. You want to look good in your new job, so you try to increase profits.

The cost for a bucket of popcorn is as follows: the popcorn kernels and butter used in each bucket cost 5¢ and 2¢, and the bucket itself costs a quarter. Each bucket of popcorn sells for $3.00.

a. In one night you sold 115 buckets. What was your profit?

b. You really want to please your boss, so you decide to increase profits by charging more per bucket. How much must you charge for a bucket of popcorn to make a $365.70 profit from selling 115 buckets?

LESSON 50

• Decimal Number Line (Tenths)
• Dividing by a Fraction

Power Up — *Building Power*

facts Power Up G

mental math

a. **Order of Operations:** $(8 \times 200) + (8 \times 25)$

b. **Number Sense:** $565 - 250$

c. **Calculation:** $58 + 27$

d. **Money:** $1.45 + 99¢

e. **Fractional Parts:** $\frac{1}{2}$ of $25.00

f. **Number Sense:** $\frac{5000}{10}$

g. **Statistics:** Find the average of 134, 120, 96, and 98.

h. **Calculation:** 8×9, $+ 3$, $\div 3$, $- 1$, $\div 3$, $+ 1$, $\div 3$, $\div 3$

problem solving Half of a gallon is a half gallon. Half of a half gallon is a quart. Half of a quart is a pint. Half of a pint is a cup. Into an empty gallon container is poured a half gallon of milk, plus a quart of milk, plus a pint of milk, plus a cup of milk. How much more milk is needed to fill the gallon container?

New Concepts — *Increasing Knowledge*

decimal number line (tenths)

We can locate different kinds of numbers on the **number line.** We have learned to locate whole numbers, negative numbers, and fractions on the number line. We can also locate decimal numbers on the number line.

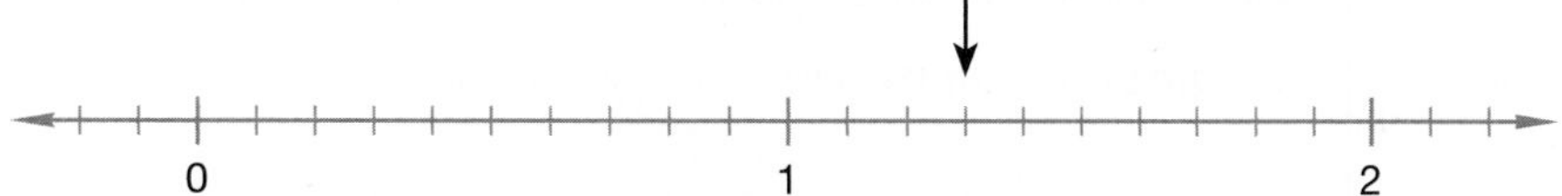

On the number line above, the distance between consecutive whole numbers has been divided into ten equal lengths. Each length is $\frac{1}{10}$ of the distance between consecutive whole numbers.

The arrow is pointing to a mark three spaces beyond the 1, so it is pointing to $1\frac{3}{10}$. We can rename $\frac{3}{10}$ as the decimal 0.3, so we can say that the arrow is pointing to the mark representing 1.3. When a unit has been divided into ten spaces, we normally use the decimal form instead of the fractional form to name the number represented by the mark.

Example 1

What decimal number is represented by point y on this number line?

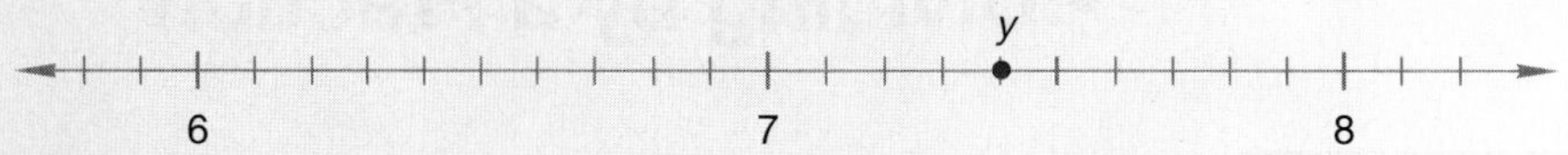

Solution

The distance from 7 to 8 has been divided into ten smaller segments. Point y is four segments to the right of the whole number 7. So point y represents $7\frac{4}{10}$. We write $7\frac{4}{10}$ as the decimal number **7.4.**

dividing by a fraction

The following question can be answered by dividing by a decimal number or by dividing by a fraction:

How many quarters are in three dollars?

If we think of a quarter as $\frac{1}{4}$ of a dollar, we have this division problem:

$$3 \div \frac{1}{4}$$

Math Language

Two numbers whose product is 1 are **reciprocals.** $\frac{1}{4} \times \frac{4}{1} = 1$, so $\frac{1}{4}$ and $\frac{4}{1}$ are reciprocals.

We solve this problem in two steps. First we answer the question, "How many quarters are in one dollar?" The answer is the reciprocal of $\frac{1}{4}$, which is $\frac{4}{1}$, which equals 4.

$$1 \div \frac{1}{4} = \frac{4}{1} = 4$$

For the second step, we use the answer to the question above to find the number of quarters in three dollars. There are four quarters in one dollar, and there are three times as many quarters in three dollars. We multiply 3 by 4 and find that there are 12 quarters in three dollars.

number of dollars → 3; number of quarters in one dollar → 4

$$3 \times 4 = 12$$

12 ← number of quarters in three dollars

We will review the steps we took to solve the problem.

Thinking Skill

Model

Draw a diagram to represent the number of quarters in $3.

Original problem: How many quarters are in $3?	$3 \div \frac{1}{4}$
Step 1: Find the number of quarters in $1.	$1 \div \frac{1}{4} = 4$
Step 2: Use the number of quarters in $1 to find the number in $3.	$3 \times 4 = 12$

Example 2

This row of pennies is $2\frac{1}{4}$ inches long.

The diameter of a penny is $\frac{3}{4}$ of an inch. How many pennies are needed to make a row of pennies 6 inches long?

Solution

In effect, this problem asks, "How many $\frac{3}{4}$-inch segments are in 6 inches?" We can write the question this way:

$$6 \div \frac{3}{4}$$

We will take two steps. First we will find the number of pennies (the number of $\frac{3}{4}$-inch segments) in 1 inch. The number of $\frac{3}{4}$s in 1 is the reciprocal of $\frac{3}{4}$, which is $\frac{4}{3}$.

$$1 \div \frac{3}{4} = \frac{4}{3}$$

We will not convert $\frac{4}{3}$ to the mixed number $1\frac{1}{3}$. Instead, we will use $\frac{4}{3}$ in the second step of the solution. Since there are $\frac{4}{3}$ pennies in 1 inch, there are six times as many in 6 inches. So we multiply 6 by $\frac{4}{3}$.

$$6 \times \frac{4}{3} = \frac{24}{3} = 8$$

We find there are **8 pennies** in a 6-inch row. We will review the steps of the solution.

Thinking Skill

Connect

How many pennies are in a row that is 1 foot long?

Original problem: How many $\frac{3}{4}$s are in 6? $\quad 6 \div \frac{3}{4}$

Step 1: Find the number of $\frac{3}{4}$s in 1. $\quad 1 \div \frac{3}{4} = \frac{4}{3}$

Step 2: Use the number of $\frac{3}{4}$s in 1 to find the number of $\frac{3}{4}$s in 6. Then simplify the answer. $\quad 6 \times \frac{4}{3} = \frac{24}{3} = 8$

Practice Set

Connect To which decimal number is each arrow pointing?

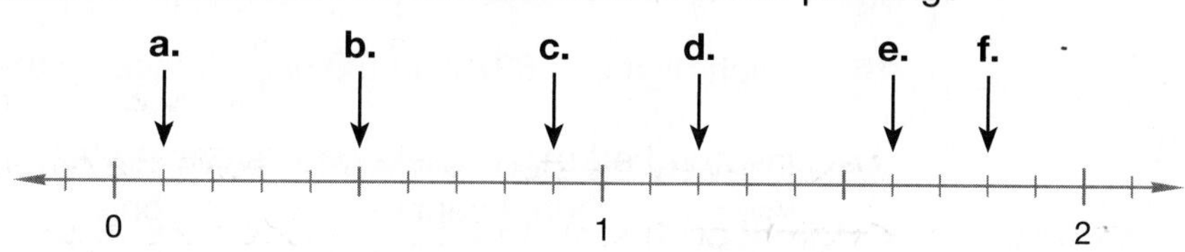

g. Formulate Write and solve a division problem to find the number of quarters in four dollars. Use $\frac{1}{4}$ instead of 0.25 for a quarter. Follow this pattern:

Original Problem

Step 1

Step 2

h. Formulate Write and solve a fraction division problem for this question: *Pads of writing paper were stacked 12 inches high on a shelf. The thickness of each pad was $\frac{3}{8}$ of an inch. How many pads were in a 12-inch stack?*

Written Practice *Strengthening Concepts*

* **1.** (10) **Predict** The first three positive odd numbers are 1, 3, and 5. Their sum is 9. The first five positive odd numbers are 1, 3, 5, 7, and 9. Their sum is 25. What is the sum of the first ten positive odd numbers? What strategy did you use to solve this problem?

* **2.** (50) **Connect** Jack keeps his music CDs stacked in plastic boxes $\frac{3}{8}$ inch thick. Use the method taught in this lesson to find the number of boxes in a stack 6 inches tall.

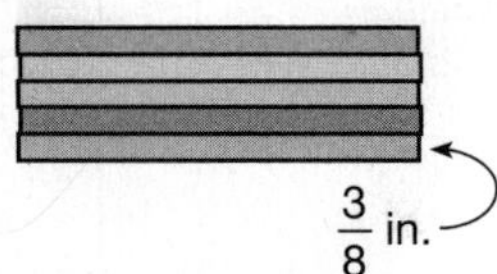

3. (15) The game has 12 three-minute rounds. If the players stop after two minutes of the twelfth round, for how many minutes did they play?

4. (44) Compare:

a. 3.4 ◯ 3.389 **b.** 0.60 ◯ 0.600

Find each unknown number:

Math Language

Symbols

$\sqrt{}$ means *square root.* The **square root** of a number is one of two equal factors of the number.

5. (15) $7.25 + 2 + w = \sqrt{100}$

6. (43) $6w = 0.144$

7. (43) $w + \frac{5}{12} = 1^2$

* **8.** (43) $6\frac{1}{8} - x = 1\frac{7}{8}$

9. (41) The book cost $20.00. The sales-tax rate was 7%. What was the total cost of the book including sales tax?

10. (38) $1 - 0.97$

* **11.** (49) $0.12\overline{)7.2}$

* **12.** (49) $0.4\overline{)7}$

13. (45) $6\overline{)0.138}$

14. (39) $(3.75)(2.4)$

15. (42) $\frac{3}{4} = \frac{?}{24}$

16. (34) Which digit in 4.637 is in the same place as the 2 in 85.21?

17. (31) One hundred centimeters equals one meter. How many square centimeters equal one square meter?

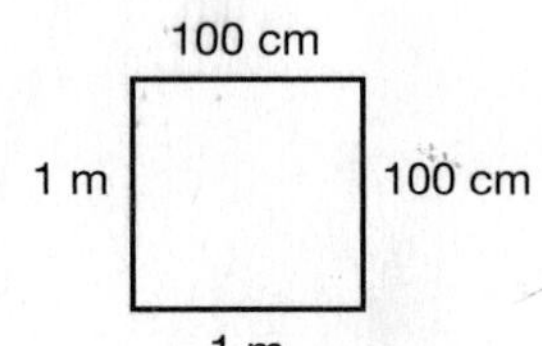

18. (30) What is the least common multiple of 6 and 9?

19. (26) $6\frac{5}{8} + 4\frac{5}{8}$

20. (29) $\frac{8}{3} \cdot \frac{3}{1}$

21. (29) $\frac{2}{3} \cdot \frac{3}{4}$

* **22.** (47) **Justify** The diameter of a soup can is about 7 cm. The label wraps around the can. About how many centimeters long must the label be to go all the way around the soup can? (Use $\frac{22}{7}$ for π.) Explain how to mentally check whether your answer is reasonable.

* **23.** (18) Find the average of 2.4, 6.3, and 5.7.

* **24.** (49) Find the number of quarters in $8.75 by dividing 8.75 by 0.25.

* **25.** (50) **Connect** What decimal number corresponds to point A on this number line?

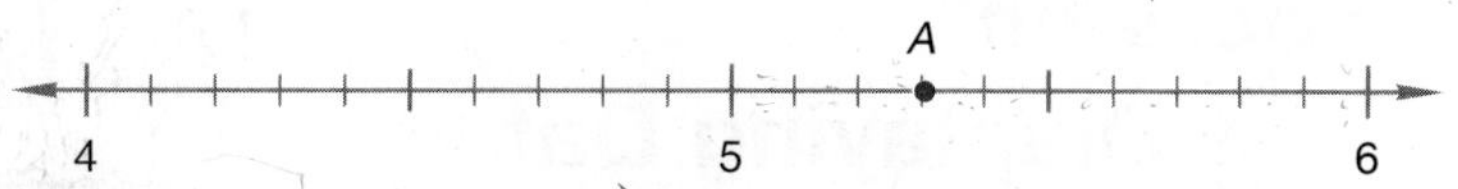

26. $\frac{2 \cdot 3 \cdot 5 \cdot 7}{2 \cdot 5}$

27. (46) 0.375×100

* **28.** (42) Rename $\frac{1}{3}$ as a fraction with 6 as the denominator. Then subtract the renamed fraction from $\frac{5}{6}$. Reduce the answer.

* **29.** (50) **Estimate** Points x, y, and z are three points on this number line. Refer to the number line to answer the questions that follow.

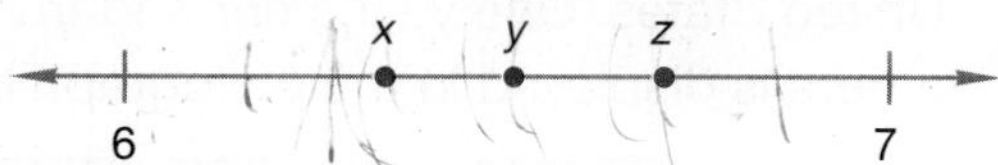

a. Which point is halfway between 6 and 7?

b. Which point corresponds to $6\frac{7}{10}$?

c. Of the points x, y, and z, which point corresponds to the number that is closest to 6?

30. (21) **Clarify** Which of these numbers is divisible by both 2 and 5?

A 552 **B** 255 **C** 250 **D** 525

Early Finishers
Real-World Application

Alex is planning a bicycle trip with his family. They want to ride a total of 195 miles through Texas seeing the sights. Alex and his family plan to ride five hours each day. They have been averaging 13 miles per hour on their training rides.

a. How many miles should they expect to ride in one day?

b. Should Alex's family be able to complete the ride in three days?

INVESTIGATION 5

Focus on

• Displaying Data

In this investigation we will compare various ways to display data.

part 1: displaying qualitative data

We have already displayed data with bar graphs and circle graphs. Now we will further investigate circle graphs and consider another graph called a **pictograph.**

There are four states that produce most of the cars and trucks made in the United States. One year's car and truck production from these states and others is displayed in the pictograph below.

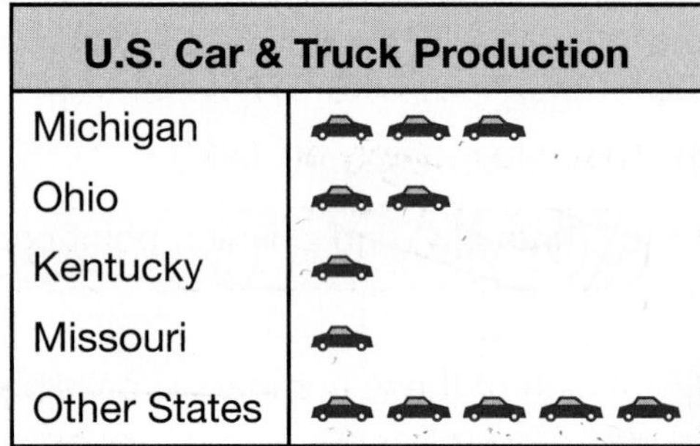

In a pictograph, pictured objects represent the data being counted. Each object represents a certain number of units of data, as indicated in the key. The two cars by Ohio, for example, indicate that 2,000,000 cars and trucks were produced in Ohio that year.

Conclude How many cars and trucks are produced in Michigan? In the four named states? In the nation?

1. **Conclude** Display the car and truck production data with a horizontal bar graph.

2. **Conclude** What fraction of U.S. car and truck production took place in Michigan?

Another way to display qualitative data is in a **circle graph.** In a circle graph each category corresponds to a **sector** of the circle. Think of a circle as a pie; a sector is simply a slice of the pie. We use circle graphs when we are interested in the fraction of the group represented by each category and not so interested in the particular number of units in each category. Another name for a circle graph is a **pie chart.**

We have used a template with equal sectors to help us sketch a circle graph. We can also sketch circle graphs with the help of a compass and protractor. We can calculate the angle of each sector (section) of the circle if we know the fraction of the whole each part represents.

The sectors of a circle graph form central angles. A **central angle** has its vertex on the center of the circle and its rays are radii of the circle. Since the central angles of a full circle total 360°, the number of degrees in a fraction of a circle is the fraction times 360°. For example, if each sector is $\frac{1}{4}$ of the circle, then each central angle measures 90°.

$$\frac{1}{4} \times 360° = 90°$$

To construct a circle graph, we need to determine how many degrees to assign each category. Our bar graph shows that a total of 12 million cars and trucks were produced. First we find the fraction of cars and trucks produced in each state. Then we multiply by 360°.

Category	Count (millions)	Fraction
Michigan	3	$\frac{3}{12}$
Ohio	2	$\frac{2}{12}$
Kentucky	1	$\frac{1}{12}$
Missouri	1	$\frac{1}{12}$
Other States	5	$\frac{5}{12}$
Total	12	$\frac{12}{12}$

The sector of the circle graph representing Michigan will cover 90°.

$$\frac{3}{12} \times 360° = 90°$$

3. Determine the central angle measures of sectors for each category.

4. Sketch the circle graph by following these instructions: Use a compass to draw a circle, then mark the center. Position the center of the protractor over the center of the circle and draw a 90° sector for Michigan. Continue around the circle drawing the appropriate angle measure for each category.

5. Compare the pictograph, bar graph, and circle graph. What are the benefits of each type of display?

part 2: displaying quantitative data

We now turn to quantitative data. Quantitative data consists of individual measurements or numbers called **data points.** When there are many possible values for data points, we can group them in intervals and display the data in a histogram as we did in Investigation 1.

When we group data in intervals, however, the individual data points disappear. In order to display the individual data points, we can use a **line plot** as we did in Investigation 4.

Suppose 18 students take a test that has 20 possible points. Their scores, listed in increasing order, are

5, 8, 8, 10, 10, 11, 12, 12, 12, 12, 13, 13, 14, 16, 17, 17, 18, 19

We represent these data in the line plot below.

Reading Math

Each x in a line plot represents one individual data point.

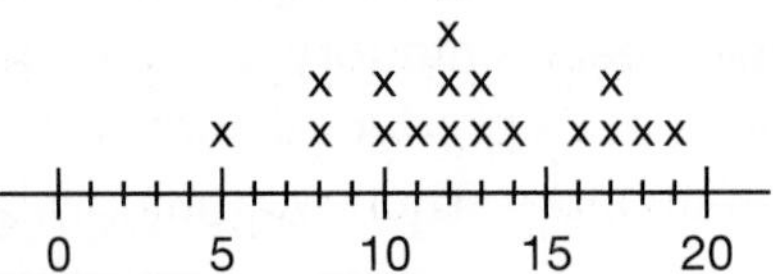

When describing numerical data, we often use terms such as **mean, median, mode,** and **range** which are defined below.

Mean: The average of the numbers.

Median: The middle number when the data are arranged in numerical order.

Mode: The most frequently occurring number.

Range: The difference between the greatest and least of the numbers.

To find the mean of the test scores above, we add the 18 scores and divide the sum by 18. The mean is about 12.6.

To find the median, we look for the middle score. If the number of scores were an odd number, we would simply select the middle score. But the number of scores is an even number, 18. Therefore, we use the average of the ninth and tenth scores for the median. We find that the ninth and tenth scores are both 12, so the median score is 12.

From the line plot we can easily see that the most common score is 12. So the mode of the test scores is 12.

We also see that the scores range from 5 to 19, so the range is 19–5, which is 14. That is, 14 points separate the lowest and highest score.

6. **Explain** The daily high temperatures in degrees Fahrenheit for 20 days in a row are listed below. Organize the data by writing the temperatures in increasing order and display them in a line plot. Explain how you chose the values for the scale on your line plot.

   60, 52, 49, 51, 47, 53, 62, 60, 57, 56,
   58, 56, 63, 58, 53, 50, 48, 60, 62, 53

7. What is the median of the temperatures in problem **6?**

8. The distribution of the temperatures in problem **6** is **bimodal** because there are two modes. What are the two modes?

9. What is the range of the temperatures? (In this case, the range is the difference between the lowest temperature and the highest temperature.)

10. Quantitative data can be displayed in **stem-and-leaf plots.** The beginning of a stem-and-leaf plot for the data in problem **6** is shown below. The "stems" are the tens digits of the data points. The "leaves" for each stem are all the ones digits in the data points that begin with that stem.

    We have plotted the data points for these heights: 47, 48, 49, 50, 51, 52, 53. Copy this plot. Then insert the rest of the temperatures from problem **6.**

Stem	Leaf
4	7 8 9
5	0 1 2 3

11. Compare the stem-and-leaf plot from problem **10** to the line plot of the same data. Discuss the benefits of each type of display.

extension

Consider a problem to solve by gathering data. State the problem. Then conduct a study. Organize the collected data. Select two ways to display the data and explain your choices. Compare the two displays. Which would you use to present your study to your class? Interpret the data. Does the gathered data help to solve a problem or is it inconclusive?

LESSON

51 • Rounding Decimal Numbers

Power Up

Building Power

facts Power Up H

mental math

a. **Calculation:** 8×125

b. **Calculation:** 4×68

c. **Number Sense:** $64 - 29$

d. **Money:** $4.64 + 99¢

e. **Fractional Parts:** $\frac{1}{2}$ of $150.00

f. **Money:** $\frac{\$100.00}{100}$

g. **Measurement:** Convert 36 hours to days.

h. **Calculation:** 8×8, $- 4$, $\div 2$, $+ 2$, $\div 4$, $+ 2$, $\div 5$, $\times 10$

problem solving

The smallest official set of dominos uses only the numbers 0 through 6. Each domino has two numbers on its face, and once a combination of numbers is used, it is not repeated. How many dominos are in the smallest official set of dominos? (*Note:* Combinations in which the two numbers are equal, called "doubles," are allowed. For example, the combination 3-3.)

New Concept

Increasing Knowledge

It is often necessary or helpful to round decimal numbers. For instance, money amounts are usually rounded to two places after the decimal point because we do not have a coin smaller than one hundredth of a dollar.

Example 1

Dan wanted to buy a book for $6.89. The sales-tax rate was 8%. Dan calculated the sales tax. He knew that 8% equaled the fraction $\frac{8}{100}$ and the decimal 0.08. To figure the amount of tax, he multiplied the price ($6.89) by the sales-tax rate (0.08).

$$\begin{array}{r} \$6.89 \\ \times \quad 0.08 \\ \hline \$0.5512 \end{array}$$

How much tax would Dan pay if he purchased the book?

Solution

Sales tax is rounded to the nearest cent, which is two places to the right of the decimal point. We mark the places that will be included in the answer.

$0.55|12

Next we consider the possible answers. We see that $0.5512 is a little more than $0.55 but less than $0.56. We decide whether $0.5512 is closer to $0.55 or $0.56 by looking at the next digit (in this case, the digit in the third decimal place). If the next digit is 5 or more, we round up to $0.56. If it is less than 5, we round down to $0.55. Since the next digit is 1, we round $0.5512 down. If Dan buys the book, he will need to pay **$0.55** in sales tax.

Example 2

Sheila pulled into the gas station and filled the car's tank with 10.381 gallons of gasoline. Round the amount of gasoline she purchased to the nearest tenth of a gallon.

Solution

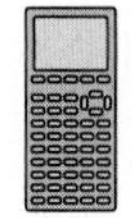

Visit www.SaxonPublishers.com/ActivitiesC1 *for a graphing calculator activity.*

The tenths place is one place to the right of the decimal point. We mark the places that will be included in the answer.

$$\underline{10.3}81$$

Next we consider the possible answers. The number we are rounding is more than 10.3 but less than 10.4. We decide that 10.381 is closer to 10.4 because the digit in the next place is 8, and we round up when the next digit is 5 or more. Sheila bought about **10.4 gallons** of gasoline.

Example 3

Estimate the product of 6.85 and 4.2 by rounding the numbers to the nearest whole number before multiplying.

Solution

We mark the whole-number places.

$$\underline{6}.85 \qquad \underline{4}.2$$

We see that 6.85 is more than 6 but less than 7. The next digit is 8, so we round 6.85 up to 7. The number 4.2 is more than 4 but less than 5. The next digit is 2, so we round 4.2 down to 4. We multiply the rounded numbers.

$$7 \cdot 4 = 28$$

We estimate that the product of 6.85 and 4.2 is about **28.**

Thinking Skill

Generalize

If you are rounding a whole number with two decimal places to the nearest whole number, what place do you look at to round?

Summarize Explain in your own words how to round a decimal number to the nearest whole number.

Practice Set Round to the nearest cent:

a. $6.6666 **b.** $0.4625 **c.** $0.08333

Round to the nearest tenth:

d. 0.12 **e.** 12.345 **f.** 2.375

Round to the nearest whole number or whole dollar:

g. 16.75 **h.** 4.875 **i.** $73.29

j. Estimate If the sales-tax rate is 6%, then how much sales tax is there on a $3.79 purchase? (Round the answer to the nearest cent.)

k. Estimate Describe how to estimate a 7.75% sales tax on a $7.89 item.

Written Practice

Strengthening Concepts

1. (12, 25) Analyze When the third multiple of 8 is subtracted from the fourth multiple of 6, what is the difference?

2. (37) From Mona's home to school is 3.5 miles. How far does Mona travel riding from home to school and back home?

3. (13) Formulate Napoleon I was born in 1769. How old was he when he was crowned emperor of France in 1804? Write an equation and solve the problem.

*** 4.** (41, 51) Shelly purchased a music CD for $12.89. The sales-tax rate was 8%.

a. What was the tax on the purchase?

b. What was the total price including tax?

*** 5.** (47, 51) Malcom used a compass to draw a circle with a radius of 3 inches.

a. Find the diameter of the circle.

b. Estimate Find the circumference of the circle. Round the answer to the nearest inch. (Use 3.14 for π.)

*** 6.** (51) Explain How can you round 12.75 to the nearest whole number?

7. (37) $0.125 + 0.25 + 0.375$

8. (43) $0.399 + w = 0.4$

*** 9.** (49) $\frac{4}{0.25}$

10. (45) $4\overline{)0.5}$

11. (45) $3.25 \div \sqrt{100}$

*** 12.** (48) $3\frac{5}{12} - 1\frac{7}{12}$

13. (42) $\frac{5}{8} = \frac{?}{24}$

*** 14.** (48) $5^2 - 17\frac{3}{4}$

15. (39) (0.19)(0.21)

16. (35) Write 0.01 as a fraction.

17. (46) Write $(6 \times 10) + (7 \times \frac{1}{100})$ as a decimal number.

18. (38) Analyze The area of a square is 64 cm^2. What is the perimeter of the square?

19. (30) What is the least common multiple of 2, 3, and 4?

20. (26) $5\frac{3}{10} + 6\frac{9}{10}$

21. (29) $\frac{10}{3} \times \frac{1}{2}$

*** 22.** (50) **Connect** A collection of paperback books was stacked 12 inches high. Each book in the stack was $\frac{3}{4}$ inch thick. Use the method described in Lesson 50 to find the number of books in the stack.

23. (16) Estimate the quotient when 4876 is divided by 98.

24. (19) What factors do 16 and 24 have in common?

*** 25.** (51) **Estimate** Find the product of 11.8 and 3.89 by rounding the factors to the nearest whole number before multiplying. Explain how you arrived at your answer.

*** 26.** (50) **Analyze** Find the average of the decimal numbers that correspond to points *x* and *y* on this number line.

27. (5) $\frac{2 \cdot 2 \cdot 3 \cdot 3 \cdot 5}{2 \cdot 2 \cdot 3 \cdot 5}$

28. (46) **Justify** Mentally calculate the total price of ten pounds of bananas at $0.79 per pound. Explain how you performed the mental calculation.

29. (42) Rename $\frac{2}{3}$ and $\frac{3}{4}$ as fractions with 12 as the denominator. Then add the renamed fractions. Write the sum as a mixed number.

*** 30.** (Inv. 5) **a.** **Represent** Jason's first nine test scores are shown below. Find the median and mode of the scores.

85, 80, 90, 75, 85, 100, 90, 80, 90

b. Sketch a graph of Jason's scores. The heights of the bars should indicate the scores. Title the graph and label the two axes.

LESSON

52 Mentally Dividing Decimal Numbers by 10 and by 100

Power Up

Building Power

facts Power Up G

mental math

a. **Number Sense:** 4×250

b. **Number Sense:** $368 - 150$

c. **Number Sense:** $250 + 99$

d. **Money:** \$15.00 + \$7.50

e. **Fractional Parts:** $\frac{1}{2}$ of 5

f. **Number Sense:** 20×40

g. **Geometry:** A rectangle has a width of 4 in. and a perimeter of 18 in. What is the length of the rectangle?

h. **Calculation:** 5×10, $+ 4$, $\div 6$, $\times 8$, $+ 3$, $\div 3$

problem solving

The monetary systems in Australia and New Zealand have six coins: 5¢, 10¢, 20¢, 50¢, \$1, and \$2. The price of any item is rounded to the nearest 5 cents. At the end of Ellen's vacation in New Zealand she had two \$2 coins. She wants to bring back at least one of each of the six coins. How many ways can she exchange one of the \$2 coins for the remaining five coins?

New Concept

Increasing Knowledge

Thinking Skill

Verify

How can we check if the quotients are correct?

When we divide a decimal number by 10 or by 100, the quotient has the same digits as the dividend. However, the position of the digits is shifted. Here we show 12.5 divided by 10 and by 100:

$$10\overline{)12.50} = 1.25 \qquad 100\overline{)12.500} = .125$$

When we divide by 10, the digits shift one place to the right. When we divide by 100, the digits shift two places to the right. Although it is the digits that are shifting places, we produce the shift by moving the decimal point. When we divide by 10, the decimal point moves one place to the left. When we divide by 100, the decimal point moves two places to the left.

Example 1

Divide: 37.5 ÷ 10

Solution

Since we are dividing by 10, the answer will be less than 37.5. We mentally shift the decimal point one place to the left.

$$37.5 \div 10 = \mathbf{3.75}$$

Example 2

Divide: 3.75 ÷ 100

Solution

Since we are dividing by 100, we mentally shift the decimal point two places to the left. This creates an empty place between the decimal point and the 3, which we fill with a zero. We also write a zero in the ones place.

$$3.75 \div 100 = \mathbf{0.0375}$$

Practice Set

Mentally calculate each quotient. Write each answer as a decimal number.

a. $2.5 \div 10$ **b.** $2.5 \div 100$

c. $87.5 \div 10$ **d.** $87.5 \div 100$

e. $0.5 \div 10$ **f.** $0.5 \div 100$

g. $25 \div 10$ **h.** $25 \div 100$

i. A stack of 10 pennies is 1.5 cm high. How thick is one penny in centimeters? In millimeters?

Written Practice

Strengthening Concepts

1. (29) What is the product of one half and two thirds?

2. (13) A piano has 88 keys. Fifty-two of the keys are white. How many more white keys are there than black keys?

3. (12) In the Puerto Rico Trench, the Atlantic Ocean reaches its greatest depth of twenty-eight thousand, two hundred thirty-two feet. Use digits to write that number of feet.

*** 4.** (46, 52) *Justify* Mentally calculate each answer. Explain how you performed each mental calculation.

a. 3.75×10 **b.** $3.75 \div 10$

5. (23) At Carver School there are 320 students and 16 teachers. What is the student-teacher ratio at Carver?

Simplify:

6. (5) $2 \cdot 2 \cdot 2 \cdot 2 \cdot 2$

7. (39) $(4)(0.125)$

8. (29) $\dfrac{150}{12}$

9. (49) $\dfrac{(1 + 0.2)}{(1 - 0.2)}$

10. $\frac{5}{2} \times \frac{4}{1}$
(29)

Find each unknown number:

11. $5\frac{1}{3} - m = 1\frac{2}{3}$
(43)

12. $m - 5\frac{1}{3} = 1\frac{2}{3}$
(43)

13. $\$10 - w = \$0.10$
(43)

* **14.** **Estimate** At a 6% sales-tax rate, what is the tax on an $8.59 purchase? Round the answer to the nearest cent.
(41, 51)

* **15.** **Estimate** The diameter of a tire on the car was 24 inches. Find the circumference of the tire to the nearest inch. (Use 3.14 for π.)
(47, 51)

* **16.** **Analyze** Arrange these numbers in order from least to greatest:
(44)

1.02, 1.2, 0.21, 0.201

17. What is the missing number in this sequence?
(10)

1, 2, 4, 7, 11, ______, 22, ...

18. The perimeter of a square room is 80 feet. How many floor tiles 1 foot square would be needed to cover the area of the room?
(38)

19. **Model** One foot is 12 inches. What fraction of a foot is 3 inches? Draw a diagram to illustrate the problem.
(22)

20. **Model** How many cents is $\frac{2}{5}$ of a dollar? Draw a diagram to illustrate the problem.
(22)

* **21.** **Connect** The diameter of a penny is $\frac{3}{4}$ inch. How many pennies are needed to form a row 12 inches long? Explain how you found your answer.
(50)

22. What is the least common multiple of 2, 4, and 6?
(30)

23. **a.** $\frac{4}{4} - \frac{2}{2}$ **b.** $\sqrt{4} - 2^2$
(29, 38)

24. **Estimate** About how many meters above the floor is the top of the chalkboard?
(7)

25. **Connect** To what decimal number is the arrow pointing on the number line below?
(50)

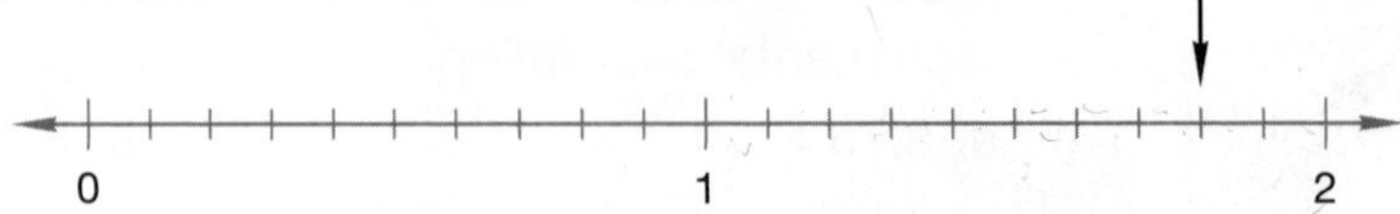

26. Rename $\frac{1}{2}$ and $\frac{2}{3}$ as fractions with denominators of 6. Then add the renamed fractions. Write the sum as a mixed number.
(42)

27. **Model** Draw a square with a perimeter of 4 inches. Then shade 50% of the square.
(8)

Use the data in the table to answer the problems **28–30.**

Gases in Earth's Atmosphere

Gas	Percent Composition
Nitrogen	78.08%
Oxygen	20.95%
Argon	0.93%
Other	0.04%

28. (33, 35) **a.** *Analyze* Nitrogen makes up what percent of Earth's atmosphere? Round to the nearest whole-number percent.

b. Write the answer to **a** as an unreduced fraction and as a decimal number.

29. (16, 35) **a.** About what percent of Earth's atmosphere consists of oxygen? Round to the nearest ten percent.

b. Write the answer to **a** as a reduced fraction.

30. (40, Inv. 5) *Model* Sketch a graph to display the data rounding to the nearest whole percent. Label the graph and explain why you chose that type of graph.

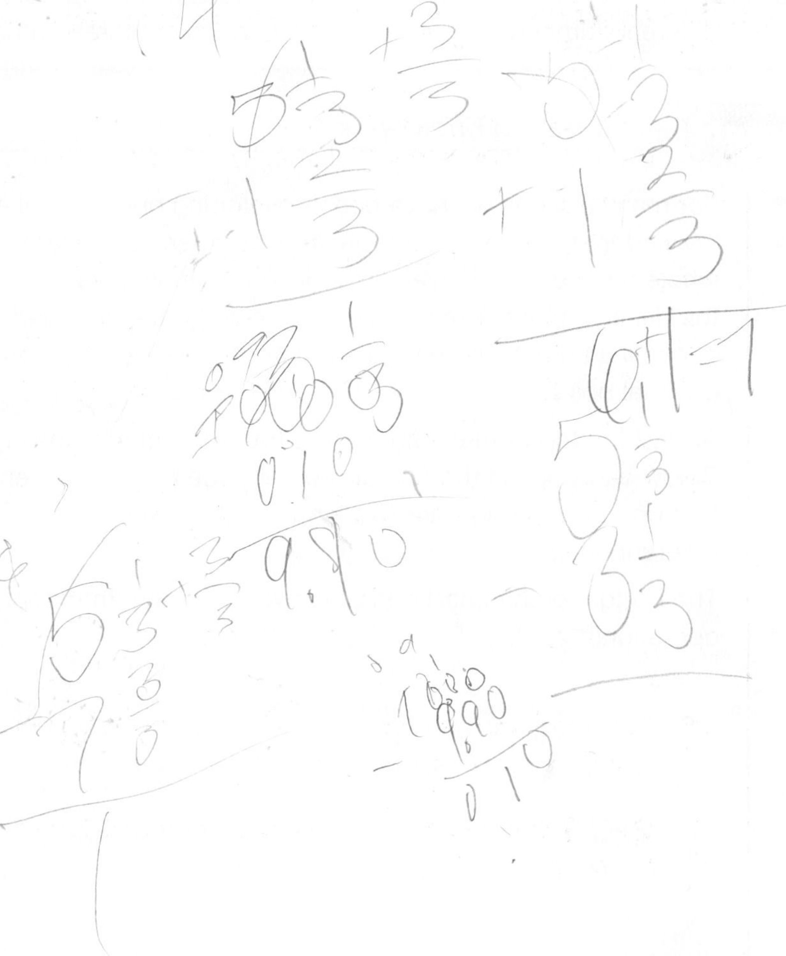

LESSON
53

• Decimals Chart
• Simplifying Fractions

Power Up

Building Power

facts Power Up D

mental math

a. Calculation: 8×225

b. Calculation: $256 + 34$

c. Number Sense: $250 - 99$

d. Money: \$25.00 − \$12.50

e. Number Sense: Double $2\frac{1}{2}$.

f. Number Sense: $\frac{800}{20}$

g. Geometry: A circle has a radius of 5 ft. What is the circumference of the circle?

h. Calculation: 10×10, $-\ 20$, $+\ 1$, $\div\ 9$, $\times\ 5$, $-\ 1$, $\div\ 4$

problem solving

You can roll six different numbers with one toss of one number cube. You can roll eleven different numbers with one toss of two number cubes. How many different numbers can you roll with one toss of three number cubes?

New Concepts

Increasing Knowledge

decimals chart

For many lessons we have been developing our decimal arithmetic skills. We find that arithmetic with decimal numbers is similar to arithmetic with whole numbers. However, in decimal arithmetic, we need to keep track of the decimal point. The chart on the facing page summarizes the rules for decimal arithmetic by providing memory cues to help you keep track of the decimal point.

Across the top of the chart are the four operation signs (+, −, ×, ÷). Below each sign is the rule or memory cue to follow when performing that operation. (There are two kinds of division problems, each with a different cue.)

The bottom of the chart contains two reminders that apply to all of the operations.

Decimal Arithmetic Reminders

Operation	+ or −	×	÷ by whole (*W*)	÷ by decimal (*D*)
Memory cue	line up	×; then count	up	over, over, up
You may need to … • Place a decimal point to the right of a whole number. • Fill empty places with zeros.				

simplifying fractions

We simplify fractions in two ways. We **reduce** fractions to lowest terms, and we convert improper fractions to mixed numbers. Sometimes a fraction can be reduced and converted to a mixed number.

Example

Simplify: $\frac{4}{6} + \frac{5}{6}$

Solution

Math Language
A fraction is in **lowest terms** if the only common factor of the numerator and denominator is 1.

By adding the fractions $\frac{4}{6}$ and $\frac{5}{6}$, we get the improper fraction $\frac{9}{6}$. We can simplify this fraction. We may reduce first and then convert the fraction to a mixed number, or we may convert first and then reduce. We show both methods below.

$$\begin{array}{r} \frac{4}{6} \\ + \frac{5}{6} \\ \hline \frac{9}{6} \end{array}$$

Reduce First

1. Reduce: $\frac{9}{6} = \frac{3}{2}$
2. Convert: $\frac{3}{2} = \mathbf{1\frac{1}{2}}$

Convert First

1. Convert: $\frac{9}{6} = 1\frac{3}{6}$
2. Reduce: $1\frac{3}{6} = \mathbf{1\frac{1}{2}}$

Practice Set

a. *Connect* Discuss how the rules in the decimals chart apply to each of these problems:

$5 - 4.2 \qquad 0.4 \times 0.2 \qquad 0.12 \div 3 \qquad 5 \div 0.4$

b. Draw the decimals chart on your paper.

Simplify:

c. $\frac{10}{12} + \frac{5}{12}$ **d.** $\frac{9}{10} + \frac{6}{10}$ **e.** $\frac{8}{12} + \frac{7}{12}$

Written Practice Strengthening Concepts

* **1.** (53) Explain The decimals chart in this lesson shows that we line up the decimal points when we add or subtract decimal numbers. Why do we do that?

2. (32) A turkey must cook for 4 hours 45 minutes. At what time must it be put into the oven in order to be done by 3:00 p.m.?

* **3.** (50) Billy won the contest by eating $\frac{1}{4}$ of a berry pie in 7 seconds. At this rate, how long would it take Billy to eat a whole berry pie?

* **4.** (Inv. 5) In four games the basketball team scored 47, 52, 63, and 66 points. What is the mean of these scores? What is the range of these scores?

Find each unknown number:

5. (43) $0.375x = 37.5$

* **6.** (52) $\frac{m}{10} = 1.25$

7. (33, 35) Write 1% as a fraction. Then write the fraction as a decimal number.

8. (38) $3.6 + 4 + 0.39$

* **9.** (49) $\frac{36}{0.12}$

10. (45) $\frac{0.15}{4}$

* **11.** (48) $6\frac{1}{4} - 3\frac{3}{4}$

12. (28) $\frac{2}{3} \times \frac{3}{5}$

13. (26) $5\frac{5}{8} + 7\frac{7}{8}$

14. (34) Which digit in 3456 has the same place value as the 2 in 28.7?

* **15.** (41, 51) The items Kameko ordered for lunch totaled \$5.20. The sales-tax rate was 8%.

a. Estimate Find the sales tax to the nearest cent.

b. Find the total price for the lunch including sales tax.

* **16.** (50) Which number is closest to 1?

A 1.2 **B** 0.9 **C** 0.1 **D** $\frac{1}{2}$

17. (47) Estimate The entire class held hands and formed a big circle. If the circle was 40 feet across the center, then it was how many feet around? Round the answer to the nearest foot. (Use 3.14 for π.)

18. (8) What is the perimeter of this square?

$\frac{3}{8}$ in.

19. (29) A yard is 36 inches. What fraction of a yard is 3 inches?

20. (19) **a.** List the factors of 11.

b. What is the name for a whole number that has exactly two factors?

21. Four squared is how much greater than the square root of 4?
(13, 38)

22. What is the smallest number that is a multiple of both 6 and 9?
(30)

23. The product of $\frac{2}{3}$ and $\frac{3}{2}$ is 1.
(30)

$$\frac{2}{3} \cdot \frac{3}{2} = 1$$

Use these numbers to form another multiplication fact and two division facts.

24. $\frac{2 \cdot 3 \cdot 2 \cdot 5 \cdot 2 \cdot 5}{2 \cdot 5 \cdot 2 \cdot 5}$
(5)

25. $\frac{5}{6} = \frac{?}{24}$
(42)

26. **Represent** Copy this rectangle on your paper, and shade two thirds of it.
(29)

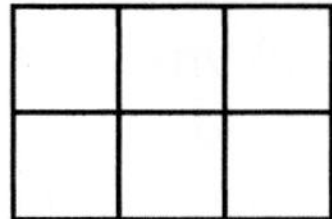

27. **Model** Thirty percent of the 350 students ride the bus to Thompson School. Find the number of students who ride the bus. Draw a diagram to illustrate the problem.
(22, 33)

28. Rename $\frac{1}{4}$ and $\frac{1}{6}$ as fractions with denominators of 12. Then add the renamed fractions.
(42)

* **29.** **Connect** The number that corresponds to point *A* is how much less than the number that corresponds to point *B*?
(50)

* **30.** **Connect** The classroom encyclopedia set fills a shelf that is 24 inches long. Each book is $\frac{3}{4}$ inch thick. How many books are in the classroom set? (To answer this question, write and solve a fraction division problem using the method shown in Lesson 50.)
(50)

Early Finishers
Real-World Application

At an online auction site, a model of a 1911 touring car was listed for sale. The twelve highest bids are shown below.

\$10 \$15 \$11 \$10 \$12 \$10 \$13 \$13 \$11 \$13 \$11 \$10

Which display—a stem-and-leaf plot or a line plot—is the most appropriate way to display this data? Draw the display and justify your choice.

LESSON 54

• Reducing by Grouping Factors Equal to 1
• Dividing Fractions

Power Up — *Building Power*

facts Power Up G

mental math

a. **Number Sense:** 6×250

b. **Number Sense:** $736 - 400$

c. **Number Sense:** $375 + 99$

d. **Money:** \$8.75 + \$5.00

e. **Fractional Parts:** $\frac{1}{2}$ of 9

f. **Number Sense:** 30×30

g. **Geometry:** Can you think of a time when a figure could have the same value for both its perimeter and its area?

h. **Calculation:** 8×8, $- 1$, $\div 9$, $\times 7$, $+ 1$, $\div 5$, $\times 10$

problem solving

The PE class ran counterclockwise around the school block, starting and finishing at point A. Instead of running all the way around the block, Nimah took what she called her "shortcut," shown by the dotted line. How many meters did Nimah save with her "shortcut?"

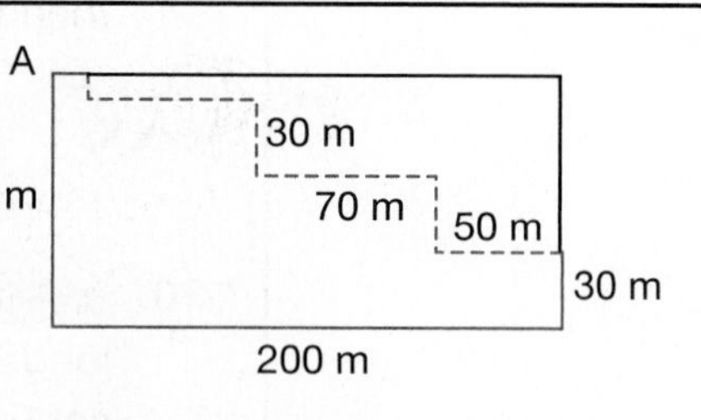

New Concepts — *Increasing Knowledge*

reducing by grouping factors equal to 1

The **factors** in the problem below are arranged in order from least to greatest. Notice that some factors appear in both the dividend and the divisor.

$$\frac{2 \cdot 2 \cdot 3 \cdot 5}{2 \cdot 2 \cdot 3}$$

Since $2 \div 2$ equals 1 and $3 \div 3$ equals 1, we will mark the combinations of factors equal to 1 in this problem.

$$\frac{\overset{1}{2} \cdot \overset{1}{2} \cdot \overset{1}{3} \cdot 5}{2 \cdot 2 \cdot 3}$$

Looking at the factors this way, the problem becomes $1 \cdot 1 \cdot 1 \cdot 5$, which is 5.

Verify Which property helps us know that $1 \cdot 5 = 5$?

Example 1

Reduce this fraction: $\mathbf{\dfrac{2 \cdot 2 \cdot 2 \cdot 5}{2 \cdot 2 \cdot 3 \cdot 5}}$

Solution

We will mark combinations of factors equal to 1.

$$\frac{2 \cdot 2 \cdot 2 \cdot 5}{2 \cdot 2 \cdot 3 \cdot 5}$$

By grouping factors equal to 1, the problem becomes $1 \cdot 1 \cdot 1 \cdot \frac{2}{3}$, which is $\mathbf{\frac{2}{3}}$.

dividing fractions

When we divide 10 by 5, we are answering the question "How many 5s are in 10?" When we divide $\frac{3}{4}$ by $\frac{1}{2}$ we are answering the same type of question. In this case the question is "How many $\frac{1}{2}$s are in $\frac{3}{4}$?" While it is easy to see how many 5s are in 10, it is not as easy to see how many $\frac{1}{2}$s are in $\frac{3}{4}$. We remember from Lesson 50 that when the divisor is a fraction, we take two steps to find the answer. We first find how many of the divisors are in 1. This is the reciprocal of the divisor. Then we use the reciprocal to answer the original division problem by multiplying.

Example 2

How many $\frac{1}{2}$s are in $\frac{3}{4}$? $\left(\frac{3}{4} \div \frac{1}{2}\right)$

Solution

Before we show the two-step process, we will solve the problem with our fraction manipulatives. The question can be stated this way:

How many $\frac{1}{2}$s are needed to make $\frac{3}{4}$?

We see that the answer is more than one but less than two. If we take one $\frac{1}{2}$ and cut another $\frac{1}{2}$ into two equal parts, then we can fit the first $\frac{1}{2}$ and one of the smaller parts together to make three fourths.

We see that we need $1\frac{1}{2}$ of the $\frac{1}{2}$ pieces to make $\frac{3}{4}$.

Now we will use arithmetic to show that $\frac{3}{4} \div \frac{1}{2}$ equals $1\frac{1}{2}$. The original problem asks, "How many $\frac{1}{2}$s are in $\frac{3}{4}$?"

$$\frac{3}{4} \div \frac{1}{2}$$

The first step is to find the number of $\frac{1}{2}$s in 1.

$$1 \div \frac{1}{2} = 2$$

The number of $\frac{1}{2}$s in 1 is 2, which is the reciprocal of $\frac{1}{2}$. So the number of $\frac{1}{2}$s in $\frac{3}{4}$ should be $\frac{3}{4}$ of 2. We find $\frac{3}{4}$ of 2 by multiplying.

$$\frac{3}{4} \times 2 = \frac{6}{4}, \text{ which equals } \mathbf{1\frac{1}{2}}$$

We simplified $\frac{6}{4}$ by reducing $\frac{6}{4}$ to $\frac{3}{2}$ and then converting $\frac{3}{2}$ to $1\frac{1}{2}$. We will review the steps we took to solve the problem.

Original problem:

How many $\frac{1}{2}$s are in $\frac{3}{4}$? $\qquad \frac{3}{4} \div \frac{1}{2}$

Step 1: Find the number of $\frac{1}{2}$s in 1. $\qquad 1 \div \frac{1}{2} = 2$

Step 2: Use the number of $\frac{1}{2}$s in 1 to find the number of $\frac{1}{2}$s in $\frac{3}{4}$. Then simplify the answer. $\qquad \frac{3}{4} \times 2 = \frac{6}{4} = 1\frac{1}{2}$

Example 3

How many $\frac{3}{4}$s are in $\frac{1}{2}$? $\left(\frac{1}{2} \div \frac{3}{4}\right)$

Solution

Using our fraction manipulatives, the question can be stated this way:

What fraction of $\frac{3}{4}$ is needed to make $\frac{1}{2}$?

The answer is less than 1. We need to cut off part of $\frac{3}{4}$ to make $\frac{1}{2}$. If we cut off one of the three parts of three fourths ($\frac{3}{4}$), we see that two of the three parts equal $\frac{1}{2}$. So $\frac{2}{3}$ of $\frac{3}{4}$ is needed to make $\frac{1}{2}$.

Now we will show the arithmetic. The original problem asks, "How many $\frac{3}{4}$s are in $\frac{1}{2}$?"

$$\frac{1}{2} \div \frac{3}{4}$$

First we find the number of $\frac{3}{4}$s in 1. The number is the reciprocal of $\frac{3}{4}$.

$$1 \div \frac{3}{4} = \frac{4}{3}$$

The number of $\frac{3}{4}$s in 1 is $\frac{4}{3}$. So the number of $\frac{3}{4}$s in $\frac{1}{2}$ is $\frac{1}{2}$ of $\frac{4}{3}$. We find $\frac{1}{2}$ of $\frac{4}{3}$ by multiplying.

$$\frac{1}{2} \times \frac{4}{3} = \frac{4}{6}, \text{ which equals } \mathbf{\frac{2}{3}}$$

Again, we will review the steps we took to solve the problem.

Original problem:

How many $\frac{3}{4}$s are in $\frac{1}{2}$? $\qquad \frac{1}{2} \div \frac{3}{4}$

Step 1: Find the number of $\frac{3}{4}$s in 1. $\qquad 1 \div \frac{3}{4} = \frac{4}{3}$

Step 2: Use the number of $\frac{3}{4}$s in 1 to find the number of $\frac{3}{4}$s in $\frac{1}{2}$. Then simplify the answer. $\qquad \frac{1}{2} \times \frac{4}{3} = \frac{4}{6} = \frac{2}{3}$

Practice Set

Reduce:

a. $\dfrac{2 \cdot 2 \cdot 3 \cdot 5}{2 \cdot 2 \cdot 5}$

b. $\dfrac{2 \cdot 2 \cdot 3 \cdot 3 \cdot 5}{2 \cdot 2 \cdot 3 \cdot 5 \cdot 5}$

c. How many $\frac{3}{8}$s are in $\frac{1}{2}$? $\left(\frac{1}{2} \div \frac{3}{8}\right)$

d. How many $\frac{1}{2}$s are in $\frac{3}{8}$? $\left(\frac{3}{8} \div \frac{1}{2}\right)$

Written Practice *Strengthening Concepts*

* **1.** (53) **Summarize** Draw the decimals chart from Lesson 53.

2. (45) If 0.4 is the dividend and 4 is the divisor, what is the quotient?

3. (16) **Estimate** In 1900 the U.S. population was 76,212,168. In 1950 the population was 151,325,798. Estimate the increase in population between 1900 and 1950 to the nearest million.

4. (48) **Formulate** Marjani was $59\frac{3}{4}$ inches tall when she turned 11 and $61\frac{1}{4}$ inches tall when she turned 12. How many inches did Marjani grow during the year? Write an equation and solve the problem.

5. (5) $1000 - (100 - 1)$

6. (25) $\frac{1000}{24}$

7. (18) What number is halfway between 37 and 143?

Find each unknown number:

8. (3) $\$3 - n = 24¢$

9. (43) $m + 3\frac{4}{5} = 6\frac{2}{5}$

* **10.** (45) $4.2 \div 10^2$

* **11.** (53) $(1.2 \div 0.12)(1.2)$

12. (29, 38) $\left(\frac{4}{3}\right)^2$

13. (38) $\sqrt{9} + \sqrt{16}$

14. (12) Which digit is in the hundred-thousands place in 123,456,789?

15. (41, 45) The television cost \$289.90. The sales-tax rate was 8%. How much was the sales tax on the television?

* **16.** (9) **Connect** Use rulers to compare:

two centimeters ◯ one inch

* **17.** (10) **Predict** Isadora found the sixth term of this sequence by doubling six and then subtracting one. She found the seventh term by doubling seven and subtracting one. What is the twelfth term of this sequence?

1, 3, 5, 7, 9, ...

18. (31) How many square feet of tile would be needed to cover the area of a room 14 feet long and 12 feet wide?

19. (23) Nine of the 30 students played basketball on the school team.

a. What fraction of the students played on the team?

b. What is the ratio of students who played basketball to students who did not play basketball?

20. (42) a. $\frac{5}{6} = \frac{?}{24}$ b. $\frac{5}{8} = \frac{?}{24}$

21. (30) What is the least common multiple of 3, 4, and 6?

* **22.** (54) **Represent** How many $\frac{1}{2}$s are in $\frac{2}{3}$? $\left(\frac{2}{3} \div \frac{1}{2}\right)$

23. (22, 23) **Model** Eighty percent of the 30 answers were correct. Write 80% as a reduced fraction. Then find the number of answers that were correct. Draw a diagram to illustrate the problem.

24. (46) **Connect** One inch equals 2.54 centimeters. A ribbon 100 inches long is how many centimeters long?

Reduce:

* **25.** (54) $\dfrac{2 \cdot 3 \cdot 5 \cdot 3 \cdot 2}{2 \cdot 3 \cdot 2 \cdot 5}$

* **26.** (54) $\dfrac{2 \cdot 3 \cdot 3 \cdot 5}{2 \cdot 2 \cdot 2 \cdot 3 \cdot 5}$

27. (42) Rename $\frac{2}{3}$ and $\frac{1}{2}$ as fractions with denominators of 6. Then add the renamed fractions, and convert the answer to a mixed number.

28. (47) **Justify** The diameter of a wheel is 15 inches. How far would your hand move in one full turn of the wheel if you do not let go of the wheel? Round the answer to the nearest inch. (Use 3.14 for π.) Describe how to tell whether your answer is reasonable.

29. (8, 31) Draw a rectangle that is $1\frac{1}{2}$ inches long and 1 inch wide.

a. What is the perimeter of the rectangle?

b. What is the area of the rectangle?

30. (43) **Analyze** Instead of dividing $2\frac{1}{2}$ by $\frac{1}{2}$, Sandra formed an equivalent division problem with whole numbers by doubling the dividend and the divisor. What equivalent problem did she form, and what is the quotient?

Early Finishers
Choose A Strategy

Brooke is planning to plant tulips around a circular flowerbed. The diameter of the flowerbed is about 4 feet. The tulips should be planted about 6 inches apart. She has 12 tulip bulbs. Her friend Jenna said she should buy 50% more. Is Jenna correct? Explain why or why not. You may use tiles to help you visualize the problem.

LESSON 55

• Common Denominators, Part 1

Power Up — *Building Power*

facts Power Up I

mental math

a. **Calculation:** 8×325

b. **Number Sense:** $329 + 50$

c. **Number Sense:** $375 - 99$

d. **Money:** \$12.50 − \$5.00

e. **Number Sense:** Double $3\frac{1}{2}$.

f. **Number Sense:** $\frac{600}{20}$

g. **Measurement:** How many yards are in 60 inches?

h. **Calculation:** 8×5, $+ 2$, $\div 6$, $\times 7$, $+ 7$, $\div 8$, $\times 4$, $\div 7$

problem solving Copy this problem and fill in the missing digits:

```
   _4_
6)__6
  _
  __
  __
   __
   3_
    0
```

New Concept — *Increasing Knowledge*

When the denominators of two or more fractions are equal, we say that the fractions have **common denominators.** The fractions $\frac{3}{5}$ and $\frac{2}{5}$ have common denominators.

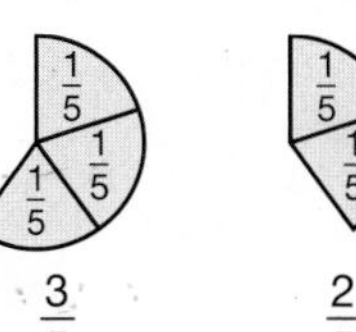

$\frac{3}{5}$ $\frac{2}{5}$

The common denominator is 5.

The fractions $\frac{3}{4}$ and $\frac{1}{2}$ do not have common denominators because the denominators 4 and 2 are not equal.

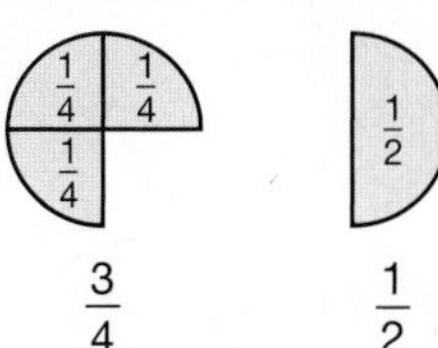

$\frac{3}{4}$ $\frac{1}{2}$

These fractions do not have common denominators.

Fractions that do not have common denominators can be renamed to form fractions that do have common denominators. Since $\frac{2}{4}$ equals $\frac{1}{2}$, we can rename $\frac{1}{2}$ as $\frac{2}{4}$. The fractions $\frac{3}{4}$ and $\frac{2}{4}$ have common denominators.

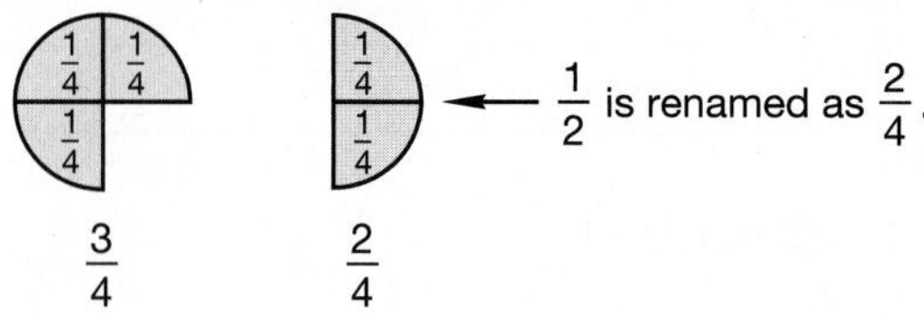

$\frac{3}{4}$ $\frac{2}{4}$

The common denominator is 4.

Fractions that have common denominators can be added by counting the number of parts, that is, by adding the numerators.

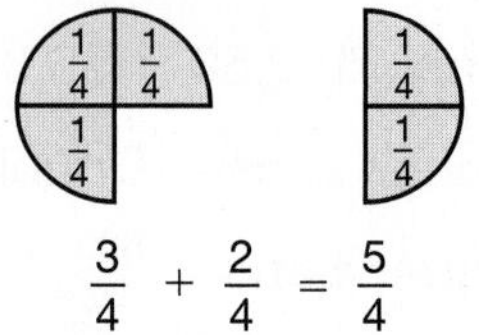

$$\frac{3}{4} + \frac{2}{4} = \frac{5}{4}$$

To add or subtract fractions that do not have common denominators, we rename one or more of them to form fractions that do have common denominators. Then we add or subtract. Recall that to rename a fraction, we multiply it by a fraction equal to 1. Here we rename $\frac{1}{2}$ by multiplying it by $\frac{2}{2}$. This forms the equivalent fraction $\frac{2}{4}$, which can be added to $\frac{3}{4}$.

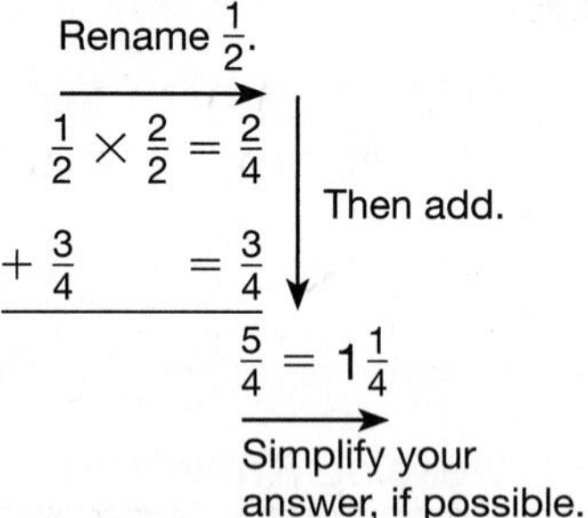

Math Language

Review

The **least common multiple** is the smallest whole number that is a multiple of two given numbers.

To find the common denominator of two fractions, we find a common multiple of the denominators. The least common multiple of the denominators is the **least common denominator** of the fractions.

Example

Subtract: $\frac{1}{2} - \frac{1}{6}$

Solution

The denominators are 2 and 6. The least common multiple of 2 and 6 is 6. So 6 is the least common denominator of the two fractions. We do not need to rename $\frac{1}{6}$. We change halves to sixths by multiplying by $\frac{3}{3}$ and then we subtract.

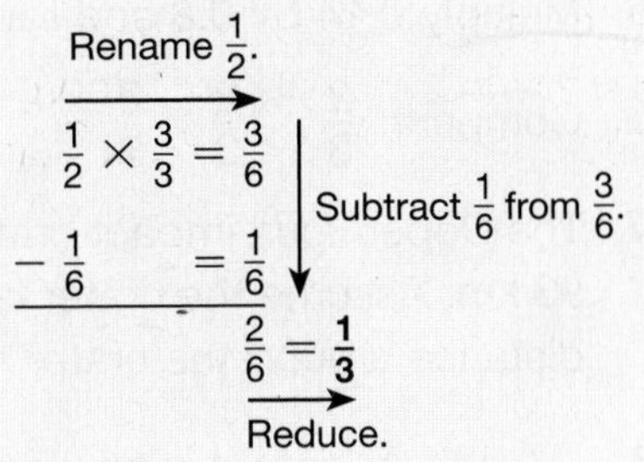

Rename $\frac{1}{2}$.
$\frac{1}{2} \times \frac{3}{3} = \frac{3}{6}$
$-\frac{1}{6} \quad = \frac{1}{6}$
Subtract $\frac{1}{6}$ from $\frac{3}{6}$.
$\frac{2}{6} = \frac{1}{3}$
Reduce.

Practice Set Find each sum or difference:

a. $\frac{1}{2} + \frac{3}{8}$

b. $\frac{3}{8} + \frac{1}{4}$

c. $\frac{3}{4} + \frac{1}{8}$

d. $\frac{1}{2} - \frac{1}{4}$

e. $\frac{5}{8} - \frac{1}{4}$

f. $\frac{3}{4} - \frac{3}{8}$

g. *Evaluate* How can we use mental math to check our answers to **a–f?**

Written Practice

Strengthening Concepts

*** 1.** (53) *Explain* In the decimals chart, the memory cue for dividing by a whole number is "up." What does that mean?

*** 2.** (50) *Connect* How many $\frac{3}{8}$-inch-thick CD cases will fit on a 12-inch-long shelf? (To answer the question, write and solve a fraction division problem using the method shown in Lesson 50.)

3. (15) The average pumpkin weighs 6 pounds. The prize-winning pumpkin weighs 324 pounds. The prize-winning pumpkin weighs as much as how many average pumpkins?

*** 4.** (55) $\frac{1}{8} + \frac{1}{2}$

*** 5.** (55) $\frac{1}{2} - \frac{1}{8}$

*** 6.** (55) $\frac{2}{3} - \frac{1}{6}$

7. (38) $6.28 + 4 + 0.13$

8. (49) $81 \div 0.9$

9. (52) $0.2 \div 10$

10. (46) $(0.17)(100)$

11. (26) $\frac{3}{4} + 3\frac{1}{4}$

12. (29) $\frac{5}{6} \cdot \frac{2}{3}$

13. (42) $\frac{5}{8} = \frac{?}{24}$

14. (32) Mt. McKinley is the tallest mountain in North America. Its height in feet is shown in expanded notation below. Write the height of Mt. McKinley in standard form.

$$(2 \times 10{,}000) + (3 \times 100) + (2 \times 10)$$

15. (38, 51) Multiply 0.14 by 0.8 and round the product to the nearest hundredth.

16. (29) Compare: $\frac{2}{3} \bigcirc \frac{2}{3} \times \frac{2}{2}$

* **17.** (47) The Copernicus impact crater on the moon has a diameter of about 93 km. Assume the crater is round, and use 3.14 for π. What is the distance around the crater rounded to the nearest ten kilometers?

* **18.** (38) **Model** A 20-foot rope was used to make a square. How many square feet of area were enclosed by the rope? Draw a diagram to help you solve the problem.

19. (29) What fraction of a dollar is six dimes?

20. (30) What is the least common multiple (LCM) of 3 and 4?

21. (19) **a.** List the factors of 23.

b. What is the name for a whole number that has exactly two factors?

22. (30) By what fraction should $\frac{2}{5}$ be multiplied to form the product 1?

23. (9, 38) Compare: $3^2 + 4^2 \bigcirc 5^2$

* **24.** (54) How many $\frac{2}{5}$s are in $\frac{1}{2}$? $(\frac{1}{2} \div \frac{2}{5})$

25. (15) How many 12s are in 1212?

26. (23) The window was 48 inches wide and 36 inches tall. What was the ratio of the height to the width of the window?

27. (29) What fraction of this group of circles is shaded?

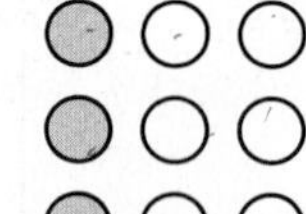

* **28.** (54) **Analyze** Reduce: $\frac{2 \cdot 3 \cdot 2 \cdot 5 \cdot 3 \cdot 7}{2 \cdot 2 \cdot 3 \cdot 5 \cdot 5 \cdot 5}$

29. (32) The performance began at 7:45 p.m. and concluded at 10:25 p.m. How long was the performance in hours and minutes?

30. (28) **Conclude** Triangle *ABC* has three acute angles.

a. Which triangle has one right angle?

b. Which triangle has one obtuse angle?

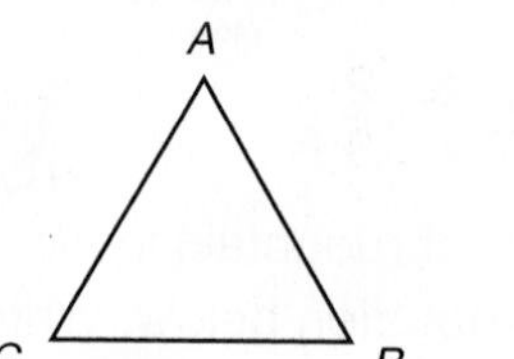

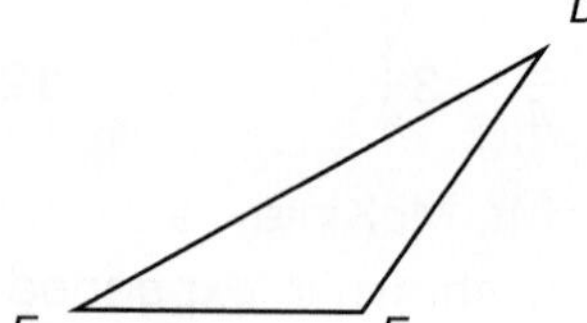

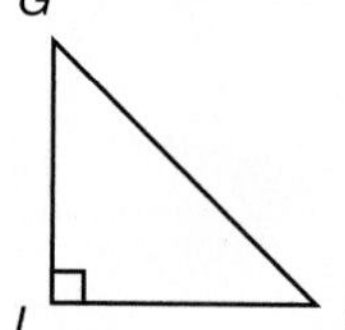

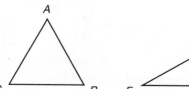

LESSON 56

Common Denominators, Part 2

Power Up

Building Power

facts Power Up G

mental math

a. **Number Sense:** 8×250

b. **Number Sense:** $462 - 350$

c. **Number Sense:** $150 + 49$

d. **Money:** $3.75 + $4.50

e. **Fractional Parts:** $\frac{1}{2}$ of 15

f. **Number Sense:** 30×40

g. **Measurement:** How many centimeters are in 10 meters?

h. **Calculation:** 10×8, $+ 1$, $\div 9$, $\times 3$, $+ 1$, $\div 4$, $\times 6$, $- 2$, $\div 5$

problem solving

Terry folded a square piece of paper in half diagonally to form a triangle. Then he folded the triangle in half as shown, making a smaller triangle. With scissors Terry cut off the upper corner of the triangle. What will the paper look like when it is unfolded?

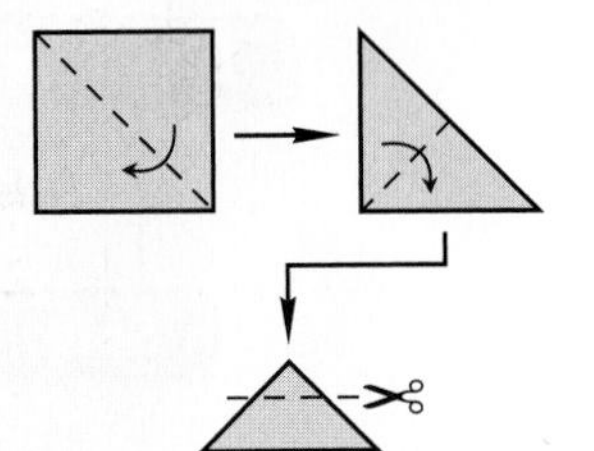

New Concept

Increasing Knowledge

In Lesson 55 we added and subtracted fractions that did not have common denominators. We renamed one of the fractions in order to add or subtract. In this lesson we will rename both fractions before we add or subtract. To see why this is sometimes necessary, consider the problem below.

> *Tony and Catherine ordered a pineapple pizza. Tony ate half and Catherine ate a third. What fraction of the pizza did Tony and Catherine eat?*

We cannot add the fractions $\frac{1}{2}$ and $\frac{1}{3}$ by simply counting the number of parts, because the parts are not the same size (that is, the denominators are different).

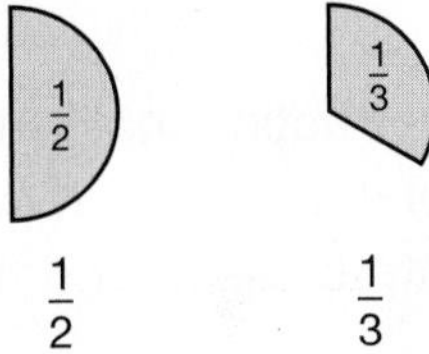

$\frac{1}{2}$ $\frac{1}{3}$

These fractions do not have common denominators.

Renaming $\frac{1}{2}$ as $\frac{2}{4}$ does not help us either. In the fractions $\frac{2}{4}$ and $\frac{1}{3}$, the parts are still of different sizes.

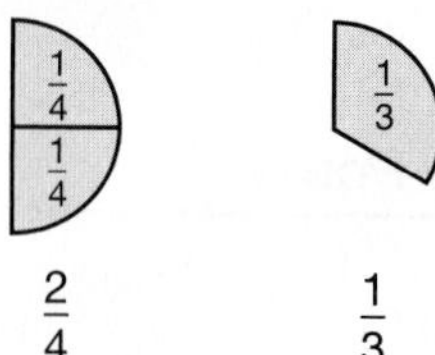

These fractions do not have common denominators.

We must rename both fractions in order to get parts that are the same size. The least common multiple of the denominators can be used as the common denominator of the renamed fractions. The least common multiple of 2 and 3 (the denominators of $\frac{1}{2}$ and $\frac{1}{3}$) is 6.

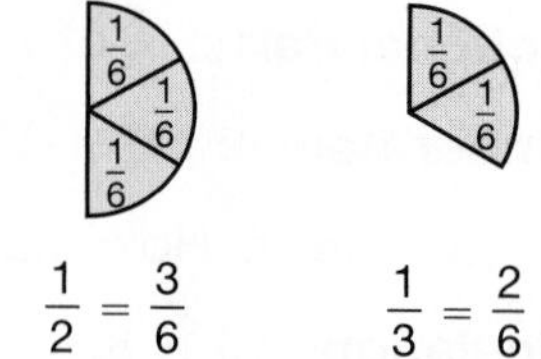

$\frac{1}{2} = \frac{3}{6}$ $\quad$ $\frac{1}{3} = \frac{2}{6}$

The common denominator is 6.

Example 1

Add: $\frac{1}{2} + \frac{1}{3}$

Solution

The denominators are 2 and 3. The least common multiple of 2 and 3 is 6. We rename each fraction so that 6 is the common denominator. Then we add.

Thinking Skill

Explain

How do you rename a fraction?

Rename $\frac{1}{2}$ and $\frac{1}{3}$.

$$\begin{array}{r} \frac{1}{2} \times \frac{3}{3} = \frac{3}{6} \\ + \frac{1}{3} \times \frac{2}{2} = \frac{2}{6} \\ \hline \mathbf{\frac{5}{6}} \end{array}$$

Then add.

Example 2

Risa saw that she had $\frac{3}{4}$ of a carton of 12 eggs. She used $\frac{2}{3}$ of a whole carton to make a batch of French toast for the family. What fraction of a carton of eggs was not used?

Solution

To find the fraction that remains, we subtract $\frac{2}{3}$ from $\frac{3}{4}$. The least common multiple of 4 and 3 is 12. We rename both fractions so that their denominators are 12 and then subtract.

Rename $\frac{3}{4}$ and $\frac{2}{3}$.

Then subtract.

$$\begin{array}{r} \frac{3}{4} \times \frac{3}{3} = \frac{9}{12} \\ -\ \frac{2}{3} \times \frac{4}{4} = \frac{8}{12} \\ \hline \mathbf{\frac{1}{12}} \end{array}$$

We can check for reasonableness by thinking about the number of eggs. If the carton holds 12 eggs, then there were nine eggs in the carton because $\frac{3}{4}$ of 12 is 9. Risa used $\frac{2}{3}$ of 12 eggs, which is 8 eggs. Subtracting 8 from 9 leaves 1 egg in the carton, which is $\frac{1}{12}$ of a carton.

Renaming one or more fractions can also help us compare fractions. To compare fractions that have common denominators, we simply compare the numerators.

$$\frac{4}{6} < \frac{5}{6}$$

To compare fractions that do not have common denominators, we can rename one or both fractions so that they do have common denominators.

Example 3

Compare: $\frac{3}{8} \bigcirc \frac{1}{2}$

Solution

We rename $\frac{1}{2}$ so that the denominator is 8.

$$\frac{1}{2} \cdot \frac{4}{4} = \frac{4}{8}$$

We see that $\frac{3}{8}$ is less than $\frac{4}{8}$.

$$\frac{3}{8} < \frac{4}{8}$$

Therefore, $\frac{3}{8}$ is less than $\frac{1}{2}$.

$$\mathbf{\frac{3}{8} < \frac{1}{2}}$$

The answer is reasonable because the numerator of $\frac{3}{8}$, which is 3, is less than half of the denominator 8, indicating that the fraction is less than $\frac{1}{2}$.

Example 4

Compare: $\frac{2}{3} \bigcirc \frac{3}{4}$

Solution

The denominators are 3 and 4. We rename both fractions with a common denominator of 12.

$$\frac{2}{3} \cdot \frac{4}{4} = \frac{8}{12} \qquad \frac{3}{4} \cdot \frac{3}{3} = \frac{9}{12}$$

We see that $\frac{8}{12}$ is less than $\frac{9}{12}$.

$$\frac{8}{12} < \frac{9}{12}$$

Therefore, $\frac{2}{3}$ is less than $\frac{3}{4}$.

$$\mathbf{\frac{2}{3} < \frac{3}{4}}$$

Generalize Use the answer to Example 4 to arrange these fractions in order from least to greatest. $\frac{3}{4}, \frac{1}{2}, \frac{2}{3}$

Practice Set

Find each sum or difference:

a. $\frac{2}{3} - \frac{1}{2}$

b. $\frac{1}{4} + \frac{2}{5}$

c. $\frac{3}{4} - \frac{1}{3}$

d. $\frac{2}{3} + \frac{1}{4}$

e. $\frac{1}{3} - \frac{1}{4}$

f. $\frac{1}{2} - \frac{1}{10}$

Before comparing, write each pair of fractions with common denominators:

g. $\frac{2}{3} \bigcirc \frac{1}{2}$

h. $\frac{4}{6} \bigcirc \frac{3}{4}$

i. $\frac{2}{3} \bigcirc \frac{3}{5}$

Written Practice

Strengthening Concepts

* **1.** (56) **Analyze** Add $\frac{1}{4}$ and $\frac{1}{3}$. Use 12 as the common denominator.

* **2.** (56) Subtract $\frac{1}{3}$ from $\frac{1}{2}$. Use 6 as the common denominator.

3. (29) Of the 88 keys on a piano, 52 are white.

a. What fraction of a piano's keys are white?

b. What is the ratio of black keys to white keys?

4. (29) If $7\frac{1}{2}$ apples are needed to make an apple pie, how many apples would be needed to make two apple pies?

* **5.** (56) Subtract $\frac{1}{4}$ from $\frac{2}{3}$. Use 12 as the common denominator.

6. (55) Add $\frac{1}{3}$ and $\frac{1}{6}$. Reduce your answer.

* **7.** (55) **Analyze** Subtract $\frac{1}{2}$ from $\frac{5}{6}$. Reduce your answer.

8. (1) $\$3 + \$1.75 + 65¢$

9. (39) $(0.625)(0.4)$

10. (49) $6^2 \div 0.08$

11. (48) $3\frac{1}{8} - 1\frac{7}{8}$

12. (29) $\frac{5}{8} \cdot \frac{2}{3}$

13. (22, 33) Forty percent of the 100 students in the World Language Club speak more than two languages.

a. Write 40% as a reduced fraction.

b. How many students in the World Language Club speak more than two languages?

14. Write the following as a decimal number:
(46)

$$(8 \times 10) + \left(6 \times \frac{1}{10}\right) + \left(5 \times \frac{1}{100}\right)$$

15. **Estimate** What is the sum of 3627 and 4187 to the nearest hundred?
(16)

16. **Justify** Molly measured the diameter of her bike tire and found that it was 2 feet across. She estimated that for every turn of the tire, the bike traveled about 6 feet. Was Molly's estimate reasonable? Why or why not?
(47)

* **17.** **Analyze** What is the mean of 1.2, 1.3, 1.4, and 1.5?
(Inv. 5)

18. The perimeter of a square is 36 inches. What is its area?
(38)

19. Here we show 24 written as a product of prime numbers:
(19)

$$2 \cdot 2 \cdot 2 \cdot 3$$

Show how 30 can be written as a product of prime numbers.

Find each unknown number:

20. $\frac{2}{3}w = 0$ (43)

21. $\frac{2}{3}m = 1$ (30)

22. $\frac{2}{3} - n = 0$ (43)

Refer to this bar graph to answer problems **23–26.** Before you answer the questions, be sure you understand what the scale on the graph represents.

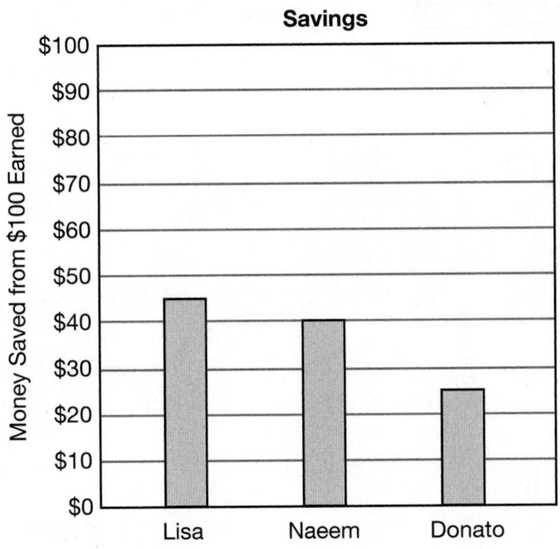

23. How much did Naeem spend?
(Inv. 5)

24. How much more did Lisa save than Donato?
(Inv. 4)

25. What fraction of his earnings did Donato save?
(29)

* **26.** **Formulate** Write a percent question that relates to the bar graph, and then answer the question.
(40)

* **27.** **Analyze** Reduce: $\frac{2 \cdot 3 \cdot 5}{2 \cdot 3 \cdot 5 \cdot 7}$
(54)

* **28.** How many $\frac{2}{3}$s are in $\frac{1}{2}$? ($\frac{1}{2} \div \frac{2}{3}$)
(54)

29. **Model** Draw three rectangles that are 2 cm long and 1 cm wide. On each rectangle show a different way to divide the rectangle in half. Then shade half of each rectangle.
(7)

* **30.** Compare: $\frac{2}{3} \bigcirc \frac{5}{6}$
(56)

Early Finishers
Real-World Application

Dalia and her three sisters make a pizza and decide to split it into four equal pieces (one for each person). Dalia finds out that her best friend is coming over, so she cuts her piece in half to share with her friend. What fraction of the whole pizza does Dalia have now? Show your work.

LESSON

57

Adding and Subtracting Fractions: Three Steps

Power Up

Building Power

facts Power Up H

mental math

a. **Calculation:** 8×425

b. **Number Sense:** $465 + 250$

c. **Number Sense:** $150 - 49$

d. **Money:** $9.75 − $3.50

e. **Number Sense:** Double $4\frac{1}{2}$.

f. **Number Sense:** $\frac{600}{30}$

g. **Measurement:** Which is greater, a century or 5 decades?

h. **Calculation:** $2 \times 2, \times 2, \times 2, \times 2, \times 2, \div 8, \div 8$

problem solving It takes ten hens one week to lay fifty eggs. How many eggs will sixty hens lay in four weeks?

New Concept

Increasing Knowledge

We follow three steps to solve fraction problems:

Step 1: Put the problem into the correct **shape** or form if it is not already. (When adding or subtracting fractions, the correct form is with common denominators.)

Step 2: Perform the **operation** indicated. (Add, subtract, multiply, or divide.)

Step 3: **Simplify** the answer if possible. (Reduce the fraction or write an improper fraction as a mixed number.)

Example 1

Add: $\frac{1}{2} + \frac{2}{3}$

Solution

We follow the steps described above.

Step 1: Shape: write the fractions with common denominators.

Step 2: Operate: add the renamed fractions.

Step 3: Simplify: convert the improper fraction to a mixed number.

1. Shape

$$\begin{array}{r} \frac{1}{2} \times \frac{3}{3} = \frac{3}{6} \\ + \frac{2}{3} \times \frac{2}{2} = \frac{4}{6} \\ \hline \frac{7}{6} = \mathbf{1\frac{1}{6}} \end{array}$$

2. Operate

3. Simplify

Example 2

Subtract: $\frac{5}{6} - \frac{1}{3}$

Solution

Step 1: Shape: write the fractions with common denominators.

Step 2: Operate: subtract the renamed fractions.

Step 3: Simplify: reduce the fraction.

1. Shape

$$\begin{array}{r} \frac{5}{6} \qquad = \frac{5}{6} \\ - \frac{1}{3} \times \frac{2}{2} = \frac{2}{6} \\ \hline \frac{3}{6} = \mathbf{\frac{1}{2}} \end{array}$$

2. Operate

3. Simplify

Practice Set

Find each sum or difference:

a. $\frac{1}{2} + \frac{1}{6}$ **b.** $\frac{2}{3} + \frac{3}{4}$ **c.** $\frac{1}{5} + \frac{3}{10}$

d. $\frac{5}{6} - \frac{1}{2}$ **e.** $\frac{7}{10} - \frac{1}{2}$ **f.** $\frac{5}{12} - \frac{1}{6}$

Written Practice

Strengthening Concepts

1. (12, 29) **Analyze** What is the difference between the sum of $\frac{1}{2}$ and $\frac{1}{2}$ and the product of $\frac{1}{2}$ and $\frac{1}{2}$?

2. (13) **Formulate** Thomas Jefferson was born in 1743. How old was he when he was elected president of the United States in 1800? Write an equation and solve the problem.

*** 3.** (56) Subtract $\frac{3}{4}$ from $\frac{5}{6}$. Use 12 as the common denominator.

*** 4.** (57) $\frac{1}{2} + \frac{2}{3}$ *** 5.** (57) $\frac{1}{2} + \frac{1}{6}$ *** 6.** (57) $\frac{5}{6} + \frac{2}{3}$

*** 7.** (54) **Represent** How many $\frac{3}{5}$s are in $\frac{3}{4}$? $\left(\frac{3}{4} \div \frac{3}{5}\right)$

8. (52) \$32.50 ÷ 10 **9.** (5, 38) $\sqrt{4} - (1^2 - 0.2)$

10. (49) 6 ÷ 0.12 **11.** (48) $5\frac{3}{8} - 2\frac{5}{8}$ **12.** (29) $\frac{3}{4} \cdot \frac{5}{3}$

13. (13, 33) Fifty percent of this rectangle is shaded. Write 50% as a reduced fraction. What is the area of the shaded part of the rectangle?

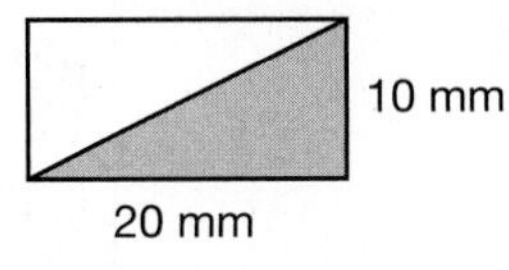

14. (34) What is the place value of the 7 in 3.567?

15. (45, 51) **Estimate** Divide 0.5 by 4 and round the quotient to the nearest tenth.

16. (44) Arrange these numbers in order from least to greatest:

0.3, 3.0, 0.03

17. (10) **Predict** In this sequence the first term is 2, the second term is 4, and the third term is 6. What is the twentieth term of the sequence?

2, 4, 6, 8, . . .

18. (23) In a deck of 52 cards there are four aces. What is the ratio of aces to all cards in a deck of cards?

* **19.** (31, 57) **a.** **Analyze** Calculate the perimeter of the rectangle shown.

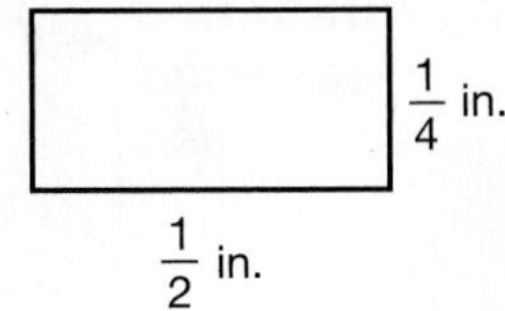

b. Multiply the length of the rectangle by its width to find the area of the rectangle in square inches.

20. (29) What number is $\frac{5}{8}$ of 80?

21. (19) **Analyze** List the factors of 29.

22. (30) What is the least common multiple of 12 and 18?

* **23.** (56) Compare: $\frac{5}{8} \bigcirc \frac{7}{10}$

24. (10) **Connect** What temperature is shown on this thermometer?

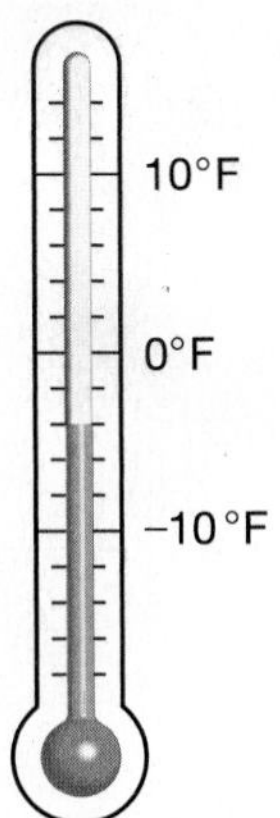

25. (14) If the temperature shown on this thermometer rose to 12°F, then how many degrees would the temperature have risen?

* **26.** (54) **Analyze** Reduce: $\frac{2 \cdot 2 \cdot 3 \cdot 3 \cdot 5 \cdot 7}{2 \cdot 2 \cdot 5 \cdot 5 \cdot 7 \cdot 7}$

27. (29) What fraction of the group of circles is shaded?

* **28.** (50) **Justify** Ling has a 9-inch stack of CDs on the shelf. Each CD is in a $\frac{3}{8}$-inch-thick plastic case. How many CDs are in the 9-inch stack? Write and solve a fraction division problem to answer the question. Explain why your answer is reasonable.

* **29.** (55) Subtract $\frac{1}{2}$ from $\frac{4}{5}$. Use 10 as the common denominator.

30. (47) The diameter of a regulation basketball hoop is 18 inches. What is the circumference of a regulation basketball hoop? (Use 3.14 for π.)

Early Finishers
Real-World Application

Misty is going to bake a fruit cake for her sister. These are the ingredients:

$\frac{3}{4}$ cup butter

1 cup sugar

2 cups flour

3 cups golden raisins

$\frac{1}{2}$ cup cherries

1 cup crushed pineapple (with juice)

How many total cups of ingredients are needed for this recipe?

LESSON **58**

• Probability and Chance

Power Up

Building Power

facts Power Up I

mental math

a. **Calculation:** 2×75

b. **Number Sense:** $315 - 150$

c. **Number Sense:** $250 + 199$

d. **Money:** \$7.50 + \$12.50

e. **Fractional Parts:** $\frac{1}{2}$ of 25

f. **Number Sense:** 20×50

g. **Patterns:** Find the next number in the pattern: 4, 7, 10, ______

h. **Calculation:** 10×10, $- 1$, $\div 11$, $\times 8$, $+ 3$, $\div 3$, $\div 5$

problem solving When Erinn was 12 years old, she walked six dogs in her neighborhood every day to earn spending money. She preferred to walk two dogs at a time. How many different combinations of dogs could Erinn take for a walk?

New Concept

Increasing Knowledge

We live in a world full of uncertainties.

What is the chance of rain on Saturday?
What are the odds of winning the big game?
What is the probability that I will roll the number needed to land on the winning space?

The study of **probability** helps us assign numbers to uncertain events and compare the likelihood that various events will occur.

Events that are certain to occur have a probability of one. Events that are certain not to occur have a probability of zero.

If I roll a number cube whose faces are numbered 1 through 6, the probability of rolling a 6 or less is one. The probability of rolling a number greater than 6 is zero.

Events that are uncertain have probabilities that fall anywhere between zero and one. The closer to zero the probability is, the less likely the event is to occur; the closer to one the probability is, the more likely the event is to occur. We typically express probabilities as fractions or as decimals.

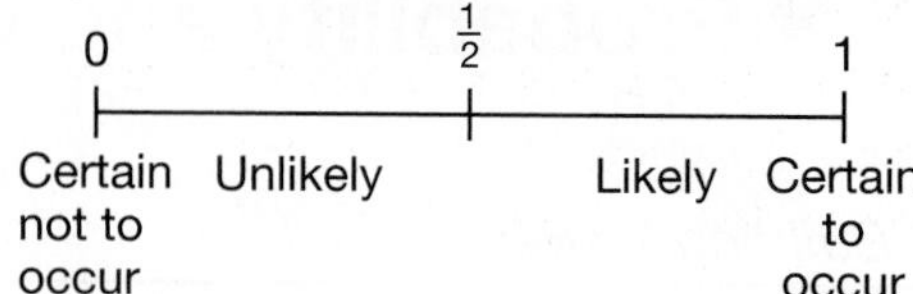

In this lesson we will practice assigning probabilities to specific events.

The set of possible outcomes for an event is called the **sample space.** If we flip a coin once, the possible outcomes are heads and tails, and the outcomes are equally likely.

Sample space = {heads, tails}

Using abbreviations, we might identify the sample space as {H, T}. If we flip a coin twice and list the results of each flip in order, then there are four equally likely possible outcomes.

Sample space = {HH, HT, TH, TT}

Imagine you spin the spinner below once. The spinner could land in sector A, in sector B, or in sector C. Since sector B and sector C have the same area, landing in either one is equally likely. Since sector A has the largest area, we can expect the spinner to land in sector A more often than in either sector B or sector C.

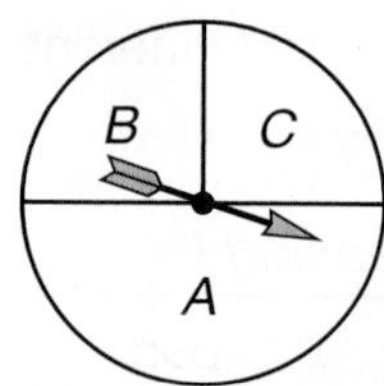

Discuss What is the sample space for the experiment?

Conclude Are the three outcomes equally likely? Why or why not?

The probability of a particular outcome is the fraction of spins we expect to result in that outcome, if we spin the spinner many times. Since sector A takes up $\frac{1}{2}$ of the area, the probability that the spinner will land in A is $\frac{1}{2}$, or 0.5. In terms of area, sector B is half the size of sector A ($\frac{1}{2}$ of $\frac{1}{2}$). Sector B takes up $\frac{1}{4}$ of the area of the spinner. Therefore, the probability that the spinner will land in sector B is $\frac{1}{4}$, or 0.25.

This is also the probability that the spinner will land in sector C, since sectors B and C are equal in size. We know that the spinner is certain to land in one of the sectors, so the probabilities of the three outcomes must add up to 1.

$$\frac{1}{2} + \frac{1}{4} + \frac{1}{4} = 1 \qquad \text{or} \qquad 0.5 + 0.25 + 0.25 = 1$$

If we spin the spinner a large number of times, we would expect about $\frac{1}{2}$ of the spins to land in sector A, about $\frac{1}{4}$ of the spins to land in sector B, and about $\frac{1}{4}$ of the spins to land in sector C.

Example 1

Meredith spins the spinner shown on the previous page 28 times. About how many times can she expect the spinner to land in sector A? In sector B? In sector C?

Solution

The spinner should land in sector A about $\frac{1}{2}$ of 28 times. Instead of multiplying by the fraction $\frac{1}{2}$, we can simply divide by 2.

$$28 \text{ times} \div 2 = 14 \text{ times}$$

The spinner should land in sector B about $\frac{1}{4}$ of 28 times. We divide the total number of spins by 4.

$$28 \text{ times} \div 4 = 7 \text{ times}$$

As we noted before, the probability that the spinner will land in sector B is equal to the probability that it will land in sector C. So Meredith can expect the spinner to land in sector A about **14 times,** in sector B about **7 times,** and in sector C about **7 times.**

It would be very unlikely for the spinner *never* to land in sector A in 28 spins. In 28 spins it also would be very unlikely for the spinner to *always* land in sector A. It would not be unusual, however, if the spinner were to land 12 times in sector A, 10 times in sector B, and 6 times in sector C. It is important to remember that probability indicates expectation; actual results may vary.

In the language of percent, we expect the spinner to land in sector A roughly 50% of the time, and we expect it to land in each of the other sectors roughly 25% of the time. When we express a probability as a percent, it is called a **chance.**

Example 2

A weather forecaster says that there is a 60% chance of rain tomorrow. Find the probability that it will rain tomorrow.

Solution

To find the probability that it will rain, we convert the chance, 60%, to a fraction and a decimal.

$$60\% = \frac{60}{100} = \frac{6}{10} = \mathbf{\frac{3}{5}}$$

$$60\% = \frac{60}{100} = 0.60 = \mathbf{0.6}$$

The probability of rain can be expressed as either $\frac{3}{5}$ or 0.6.

The **complement of an event** is the opposite of the event. The complement of event A is "not A." Consider the probability of rain for example. We are certain that it will either rain or not rain. The probabilities of these two possible outcomes must total 1. Subtracting the probability of rain from 1 gives us the probability that it will not rain.

$$1 - \frac{6}{10} = \frac{4}{10} = \mathbf{\frac{2}{5}}$$

$$1 - 0.6 = \mathbf{0.4}$$

So the probability that it will not rain can be expressed as either $\frac{2}{5}$ or 0.4.

The probability of an event and the probability of its complement total 1.

In some experiments or games, all the outcomes have the same probability. This is true if we flip a coin. The probability of the coin landing heads up is $\frac{1}{2}$, and the probability of the coin landing tails up is also $\frac{1}{2}$. Similarly, if we roll a number cube, each number has a probability of $\frac{1}{6}$ of appearing on the cube's upturned side.

To find the probability of an event, we simply add the probabilities of the outcomes that make up the event. If all outcomes of an experiment or game have the *same* probability, then the probability of an event is:

$$\frac{\text{number of outcomes in the event}}{\text{number of possible outcomes}}$$

Example 3

Math Language

A *number cube* is a six-sided cube. Its sides are marked 1–6 with numbers or with dots.

A number cube is rolled. Find the probability that the upturned number is greater than 4. Then find the probability that the number is not greater than 4.

Solution

There are six possible, equally likely outcomes.

$$\text{Sample space} = \{1, 2, 3, 4, 5, 6\}$$

Two of the outcomes are greater than 4.

$$\frac{\text{number of outcomes in the event}}{\text{number of possible outcomes}} = \frac{2}{6} = \mathbf{\frac{1}{3}}$$

Thus the probability of greater than 4 is $\frac{1}{3}$. The complement of greater than 4 is not "less than 4." Rather, the complement of greater than 4 is "not greater than 4." Four of the six outcomes are not greater than 4.

$$\text{Probability of not greater than 4} = \frac{4}{6} = \mathbf{\frac{2}{3}}$$

The calculation is reasonable because the probabilities of an event and its complement total 1, and $\frac{1}{3} + \frac{2}{3} = 1$.

Example 4

A bag contains 5 red marbles, 4 yellow marbles, 2 green marbles, and 1 orange marble. Without looking, Brendan draws a marble from the bag and notes its color.

What is the probability of drawing red?

What is the complement of this event?

What is the probability of the complement?

Solution

There are 12 possible outcomes in Brendan's experiment (each marble represents one outcome). The event we are considering is drawing a red marble. Since 5 of the possible outcomes are red, we see that the probability that the drawn marble is red is

$$\frac{\text{number of outcomes in the event}}{\text{number of possible outcomes}} = \mathbf{\frac{5}{12}}$$

The complement of drawing red is drawing **"not red."** Its probability can be found by subtracting $\frac{5}{12}$ from 1.

$$1 - \frac{5}{12} = \frac{7}{12}$$

So the probability of not red is $\mathbf{\frac{7}{12}}$.

Example 5

In the experiment in example 4, what is the probability that the marble Brendan draws is a primary color?

Solution

Red and yellow are primary colors. Green and orange are not. Since 5 possible outcomes are red and 4 are yellow, the probability that the drawn marble is a primary color is:

$$\frac{5 + 4}{12} = \frac{9}{12} = \mathbf{\frac{3}{4}}$$

Practice Set

Juan is waiting for the roller coaster at an amusement park. He has been told there is a 40% chance that he will have to wait more than 15 minutes.

a. Find the probability that Juan's wait will be more than 15 minutes. Write the probability both as a decimal and as a reduced fraction.

b. Find the probability that Juan's wait will not be more than 15 minutes. Express your answer as a decimal and as a fraction.

c. What word names the relationship between the events in **a** and **b?**

A number cube is rolled. The faces of the cube are numbered 1 through 6.

d. What is the sample space?

e. What is the probability that the number rolled will be odd? Explain.

f. What is the probability that the number rolled will be less than 6?

g. State the complement to the event in **f** and find its probability.

Refer to the spinner at right for problems **h–k.**

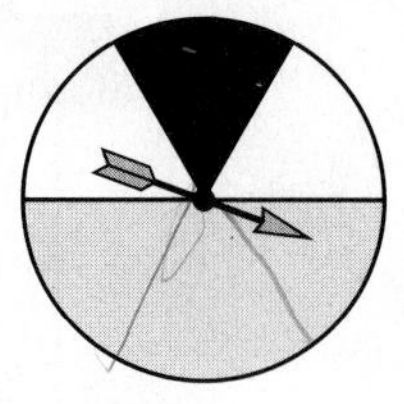

h. What is the probability that the spinner will land in the blue sector? In the black sector? In either of the white sectors? (Note that $\frac{1}{3}$ of $\frac{1}{2}$ is $\frac{1}{6}$.)

i. **Predict** What is the probability that the spinner will not land on white?

j. State the complement of the event in **i** and find its probability.

k. If you spin this spinner 30 times, roughly how many times would you expect it to land in each sector?

l. **Represent** Roll a number cube 24 times and make a frequency table for the 6 possible outcomes. Which outcomes occurred more than you expected?

Written Practice *Strengthening Concepts*

1. (12, 55) **Analyze** What is the difference between the sum of $\frac{1}{2}$ and $\frac{1}{3}$ and the product of $\frac{1}{2}$ and $\frac{1}{3}$?

2. (29) The flat of eggs held $2\frac{1}{2}$ dozen eggs. How many eggs are in $2\frac{1}{2}$ dozen?

3. (18) In three nights Rumpelstiltskin spun \$44,400 worth of gold thread. What was the average value of the thread he spun per night?

4. (56) Compare: $\frac{5}{8} \bigcirc \frac{1}{2}$

5. (38) **Analyze** Compare: $6^2 + 8^2 \bigcirc 10^2$

Find each unknown number:

***6.** (57) $m + \frac{3}{8} = \frac{1}{2}$	***7.** (57) $n - \frac{2}{3} = \frac{3}{4}$	***8.** (43) $3 - f = \frac{5}{6}$
9. (46) \$32.50 × 10	**10.** (39) (6.2)(0.48)	**11.** (49) 1.0 ÷ 0.8
12. (49) 120 ÷ 0.5	**13.** (30) $\frac{7}{8} \cdot \frac{8}{7}$	**14.** (29) $\frac{5}{6} \cdot \frac{3}{4}$

15. (43) **Connect** Instead of dividing $7\frac{1}{2}$ by $1\frac{1}{2}$, Julie doubled both numbers and then divided mentally. What is the division problem Julie performed mentally, and what is the quotient?

***16.** (57) **Analyze** Find the total price including 7% tax on a \$9.79 purchase.

17. (10) **Predict** What number is next in this sequence?

..., 0.6, 0.7, 0.8, 0.9, ...

18. (38) **Analyze** The perimeter of this square is 4 cm. What is the area of the square?

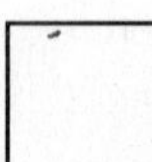

* **19.** How many $\frac{3}{5}$s are in $\frac{3}{4}$? $(\frac{3}{4} \div \frac{3}{5})$
(54)

Find each unknown number:

20. $0.32w = 32$
(43)

21. $x + 3.4 = 5$
(43)

* **22.** On one roll of a 1–6 number cube, what is the probability that the upturned face will show an even number of dots?
(58)

23. Arrange these measurements in order from shortest to longest:
(7)

1 in., 3 cm, 20 mm

24. Larry correctly answered 45% of the questions.
(29, 33)

a. *Explain* Did Larry correctly answer more than or less than half the questions? How do you know?

b. Write 45% as a reduced fraction.

25. *Justify* Describe how to mentally calculate $\frac{1}{10}$ of \$12.50.
(52)

26. Reduce: $\frac{2 \cdot 5 \cdot 2 \cdot 3 \cdot 3 \cdot 7}{2 \cdot 2 \cdot 2 \cdot 5 \cdot 5 \cdot 7}$
(54)

* **27.** *Analyze* What is the sum of the decimal numbers represented by points x and y on this number line?
(50)

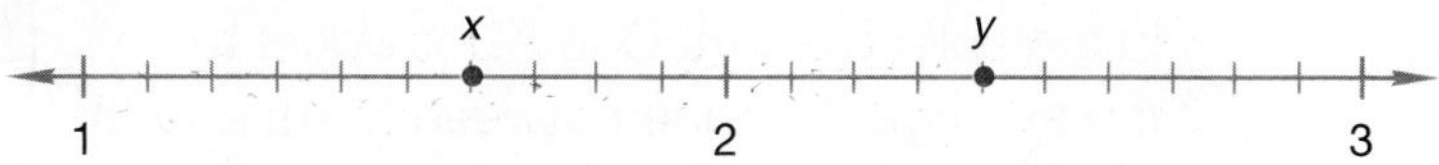

28. *Model* Draw a rectangle that is $1\frac{1}{2}$ inches long and $\frac{3}{4}$ inch wide. Then draw a segment that divides the rectangle into two triangles.
(7)

29. What is the perimeter of the rectangle drawn in problem 28?
(8)

* **30.** If $A = lw$, and if $l = 1.5$ and $w = 0.75$, what does A equal?
(47)

Early Finishers
Real-World Application

Millions of shares of stock are bought and sold each business day, and records are kept for stock prices for every trading day. Here are the closing prices, rounded to the nearest eighth, for one share of a corporation's stock during a week in 1978.

Mon.	Tu.	Wed.	Th.	Fri.
$\$14\frac{7}{8}$	$\$15\frac{1}{8}$	$\$15\frac{1}{4}$	$\$14\frac{3}{4}$	$\$14\frac{1}{2}$

Which display—a line graph or a bar graph—is the most appropriate way to display this data if you want to emphasize the changes in the daily closing prices? Draw the display and justify your choice.

LESSON

59 • Adding Mixed Numbers

Power Up

Building Power

facts Power Up G

mental math

a. Calculation: 4×75

b. Number Sense: $279 + 350$

c. Number Sense: $250 - 199$

d. Money: $15.00 − $7.75

e. Money: Double $1.50.

f. Number Sense: $\frac{800}{40}$

g. Patterns: Find the next number in the pattern: 12, 24, 36, ______

h. Calculation: 4×12, $\div 6$, $\times 8$, $- 4$, $\div 6$, $\times 3$, $\div 2$

problem solving

Astronomers use the astronomical unit (AU) to measure distances in the solar system. One AU is equal to the average distance between Earth and the Sun, about 93,000,000 miles. On average, how many miles farther from the Sun is Saturn than Mars?

Planet	AU from the Planet to the Sun
Mercury	0.39
Venus	0.72
Earth	1.00
Mars	1.52
Jupiter	5.20
Saturn	9.52

New Concept

Increasing Knowledge

We have been practicing adding mixed numbers since Lesson 26. In this lesson we will rename the fraction parts of the mixed numbers so that the fractions have common denominators. Then we will add.

Example 1

Add: $2\frac{1}{2} + 1\frac{1}{6}$

Solution

Step 1: Shape: write the fractions with common denominators.

Step 2: Operate: add the renamed fractions and add the whole numbers.

Thinking Skill

Discuss

When should you reduce a fraction?

Step 3: Simplify: reduce the fraction.

$$\begin{array}{r} 2\frac{1}{2} \times \frac{3}{3} = 2\frac{3}{6} \\ + 1\frac{1}{6} \quad\quad = 1\frac{1}{6} \\ \hline 3\frac{4}{6} = \mathbf{3\frac{2}{3}} \end{array}$$

1. Shape 2. Operate 3. Simplify

Example 2

Add: $1\frac{1}{2} + 2\frac{2}{3}$

Solution

Step 1: Shape: write the fractions with common denominators.

Thinking Skill

Verify

Why did we rename both fractions?

Step 2: Operate: add the renamed fractions and add the whole numbers.

Step 3: Simplify: convert the improper fraction to a mixed number, and combine the mixed number with the whole number.

$$\begin{array}{r} 1\frac{1}{2} \times \frac{3}{3} = 1\frac{3}{6} \\ + 2\frac{2}{3} \times \frac{2}{2} = 2\frac{4}{6} \\ \hline 3\frac{7}{6} = 3 + 1\frac{1}{6} = \mathbf{4\frac{1}{6}} \end{array}$$

1. Shape 2. Operate 3. Simplify

Practice Set

Add:

a. $1\frac{1}{2} + 1\frac{1}{3}$ **b.** $1\frac{1}{2} + 1\frac{2}{3}$ **c.** $5\frac{1}{3} + 2\frac{1}{6}$

d. $3\frac{3}{4} + 1\frac{1}{3}$ **e.** $5\frac{1}{2} + 3\frac{1}{6}$ **f.** $7\frac{1}{2} + 4\frac{5}{8}$

Written Practice *Strengthening Concepts*

1. (40) Connect What is the product of the decimal numbers four tenths and four hundredths?

2. (32) Larry looked at the clock. It was 9:45 p.m. The bus for his class trip leaves at 8:30 a.m. How many hours and minutes are there until the bus leaves?

3. (12) Pluto orbits the sun at an average distance of about five billion, nine hundred million kilometers. Use digits to write that distance.

* **4.** (59) $2\frac{1}{2} + 1\frac{1}{6}$ * **5.** (59) $1\frac{1}{2} + 2\frac{2}{3}$

6. (56) Compare: $\frac{1}{2} \bigcirc \frac{3}{5}$ **7.** (56) Compare: $\frac{2}{3} \bigcirc \frac{6}{9}$

8. (48) $8\frac{1}{5} - 3\frac{4}{5}$

9. (29) $\frac{3}{4} \cdot \frac{5}{2}$

* **10.** (54) How many $\frac{1}{2}$s are in $\frac{2}{5}$? $(\frac{2}{5} \div \frac{1}{2})$

11. (39) (0.875)(40)

12. (45) $0.07 \div 4$

13. (49) $30 \div d = 0.6$

14. (50) **Analyze** What number is halfway between 0.1 and 0.24?

15. (16) **Estimate** Round 36,428,591 to the nearest million.

16. (14) What temperature is 23° less than 8°F?

17. (47) **Estimate** Miguela wound a garden hose around a circular reel. If the diameter of the reel was 10 inches, how many inches of hose was wound on the first full turn of the reel? Round the answer to the nearest whole inch. (Use 3.14 for π.)

* **18.** (38) How many square inches are needed to cover a square foot?

19. (52) One centimeter is what fraction of one meter?

20. (46, 52) **Justify** Mentally calculate each answer. Describe how you performed each mental calculation.

a. 6.25×10

b. $6.25 \div 10$

* **21.** (58) **a.** With one toss of a single number cube, what is the probability of rolling a number less than three?

b. Write the complement of the event in **a** and find its probability.

22. (39, 44) Compare: $(0.8)^2 \bigcirc 0.8$

Refer to the line graph below to answer problems **23–25.**

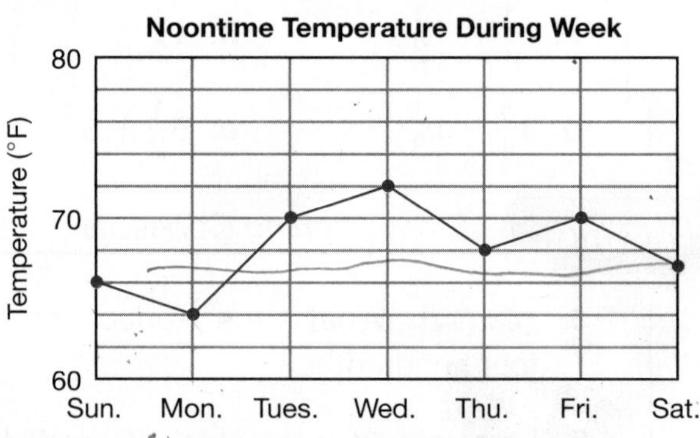

Math Language
Remember that **range** is the difference between the greatest and least numbers in a data set.

23. (18, Inv. 5) What was the range of the noontime temperatures during the week?

24. (18) What was the Saturday noontime temperature?

* **25.** (18) Write a question that relates to this line graph and answer the question.

* **26.** (50) **Connect** Nana can pack a bag of groceries in six tenths of a minute. At that rate, how many bags of groceries can she pack in 15 minutes? Write and solve a fraction division problem to answer the question.

27. (Inv. 2) One eighth is equivalent to $12\frac{1}{2}\%$. To what percent is three eighths equivalent?

* **28.** (46) **Justify** Mentally calculate the total cost of 10 gallons of gas priced at $2.299 per gallon. Describe the process you used.

* **29.** (56) **Analyze** Arrange these three numbers in order from least to greatest:

$$\frac{3}{4}, \text{ the reciprocal of } \frac{3}{4}, 1$$

30. (47) **Evaluate** If $P = 2l + 2w$, and if $l = 4$ and $w = 3$, what does P equal?

Early Finishers
Real-World Application

The Estevez family has three children born in different years.

a. List all the possible birth orders by gender of the children. For example, boy, boy, boy is one possible order and boy, girl, boy is another.

b. Use the answer to **a** to find the probability that the Estevez family has two boys and one girl in any order.

LESSON 60

• Polygons

Power Up

Building Power

facts Power Up I

mental math

a. **Number Sense:** 2×750

b. **Number Sense:** $429 - 250$

c. **Number Sense:** $750 + 199$

d. **Money:** \$9.50 + \$1.75

e. **Money:** $\frac{1}{2}$ of \$5

f. **Number Sense:** 40×50

g. **Primes and Composites:** Name the prime numbers between 1 and 10.

h. **Calculation:** 12×3, $+ 4$, $\times 2$, $+ 20$, $\div 10$, $\times 5$, $\div 2$

problem solving

Melina glued 27 small blocks together to make this cube. Then she painted each of the 6 faces of the cube a different color. Later the cube broke apart into the 27 small blocks it was made up of. How many of the smaller blocks had 3 painted faces? 2 painted faces? 1 painted face?

New Concept

Increasing Knowledge

Polygons are closed, flat shapes with straight sides.

Example 1

Which of the following is a polygon?

A

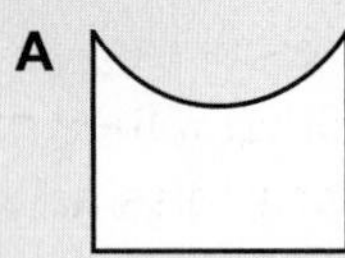

B

C

D

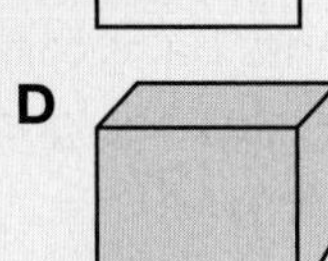

Solution

Only choice **C** is a polygon because it is the only closed, flat shape with straight sides.

Math Language
The term *poly-* means "many," and *-gon* means "angles." The prefix of a polygon's name tells how many angles and sides the polygon has: *tri-* means 3, *quad-* means 4, and so on.

Polygons are named by the number of sides they have. The chart below names some common polygons.

Polygons

Shape	Number of Sides	Name of Polygon
△	3	triangle
▭	4	quadrilateral
⬠	5	pentagon
⬡	6	hexagon
⯃	8	octagon

Two sides of a polygon meet, or **intersect,** at a **vertex** (plural: **vertices**). A polygon has the same number of vertices as it has sides.

Example 2

What is the name of a polygon that has four sides?

Solution

The answer is not "square" or "rectangle." Squares and rectangles do have four sides, but not all four-sided polygons are squares or rectangles. The correct answer is **quadrilateral.** A rectangle is one kind of quadrilateral. A square is a rectangle with sides of equal length.

If all the sides of a polygon have the same length and if all the angles have the same measure, then the polygon is called a **regular polygon.** A square is a regular quadrilateral, but a rectangle that is longer than it is wide is not a regular quadrilateral.

We often use the word **congruent** when describing geometric figures. We say that polygons are congruent to each other if they have the same shape and size. We may also refer to segments or angles as congruent. Congruent segments have the same length; congruent angles have the same measure. In a regular polygon all the sides are congruent and all the angles are congruent.

Example 3

A regular octagon has a perimeter of 96 inches. How long is each side?

Solution

An octagon has eight sides. The sides of a regular octagon are the same length. Dividing the perimeter of 96 inches by 8, we find that each side is **12 inches.** Many of the red stop signs on our roads are regular octagons with sides 12 inches long.

Practice Set

a. What is the name of this six-sided shape?

b. How many sides does a pentagon have?

c. Can a polygon have 19 sides?

d. What is the name for a corner of a polygon?

e. What is the name for a polygon with four vertices?

Written Practice *Strengthening Concepts*

1. (15) **Formulate** What is the cost per ounce of a 42-ounce box of oatmeal priced at $1.26? Write an equation and solve the problem.

* 2. (39, 51) **Estimate** Ling needs to purchase some gas so that she can mow her lawn. At the station she fills a container with 1.1 gallons of gas. The station charges $\$2.47\frac{9}{10}$ per gallon.

   a. How much will Ling spend on gas? Round your answer to the nearest cent.

   b. Explain how to use estimation to check whether your answer is reasonable.

3. (12) The smallest three-digit whole number is 100. What is the largest three-digit whole number?

* 4. (57) $\frac{3}{4} + \frac{5}{8}$

* 5. (59) $1\frac{1}{2} + 3\frac{1}{6}$

6. (29) $\frac{5}{5} \times \left(\frac{4}{4} - \frac{3}{3}\right)$

7. (29) $\frac{3}{5} \cdot \frac{1}{3}$

* 8. (54) How many $\frac{1}{3}$s are in $\frac{3}{5}$? $(\frac{3}{5} \div \frac{1}{3})$

9. (22) **Model** How much money is $\frac{5}{8}$ of $24? Draw a diagram to illustrate the problem.

10. (40) (0.65)(0.14)

11. (49) $65 \div 0.05$

12. (60) A quadrilateral has how many sides?

* 13. (51) **Estimate** Round the product of 0.24 and 0.26 to the nearest hundredth.

14. (18) What is the average of 1.3, 2, and 0.81?

15. (10) **Predict** What is the sum of the first seven numbers in this sequence?

1, 3, 5, 7, ...

16. (38) How many square feet are needed to cover a square yard?

17. (52) Ten centimeters is what fraction of one meter?

Find each unknown number:

18. (43) $3x = 1.2 + 1.2 + 1.2$

19. (30) $\frac{4}{3}y = 1$

20. (49) $m + 1\frac{3}{5} = 5$

* **21.** (58) $6\frac{1}{8} - w = 3\frac{5}{8}$

22. (19) **List** What are the prime numbers between 40 and 50?

* **23.** (47) **Analyze** Pedro cut a lime into thin slices. The largest slice was about 4 cm in diameter. Then he removed the outer peel from the slice. About how long was the outer peel? Round the answer to the nearest centimeter. (Use 3.14 for π.)

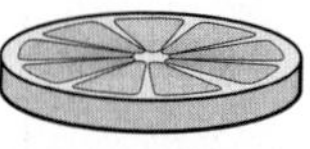

24. (50, 51) **Connect** **a.** To what decimal number is the arrow pointing?

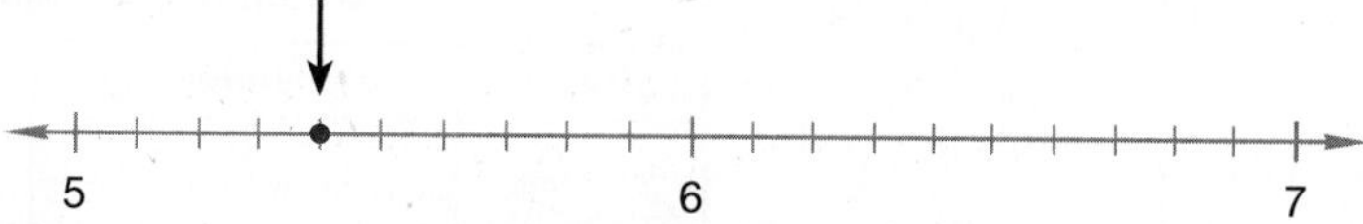

b. This decimal number rounds to what whole number?

* **25.** (58) **Explain** The face of this spinner is divided into 8 congruent sectors. What is the sample space? The spinner is spun once, what ratio expresses the probability that it will stop on a 3?

* **26.** (50) **Estimate** Mary found that the elm tree added about $\frac{3}{8}$ inch to the diameter of its trunk every year. If the diameter of the tree is about 12 inches, then the tree is about how many years old? Write and solve a fraction division problem to answer the question.

27. (29) Duncan's favorite TV show starts at 8 p.m. and ends at 9 p.m. Duncan timed the commercials and found that 12 minutes of commercials aired between 8 p.m. and 9 p.m. Commercials were shown for what fraction of the hour?

* **28.** (43) **Connect** Instead of dividing 400 by 16, Fede thought of an equivalent division problem that was easier to perform. Write an equivalent division problem that has a one-digit divisor and find the quotient.

29. (41) What is the total price of a \$6.89 item plus 6% sales tax?

* **30.** (56) Compare:

a. $3\frac{1}{2} \bigcirc \frac{6}{2} + \frac{1}{2}$

b. $\frac{5}{8} \bigcirc \frac{3}{4}$

INVESTIGATION 6

Focus on

• Attributes of Geometric Solids

Polygons are two-dimensional shapes. They have length and width, but they do not have height (depth). The objects we encounter in the world around us are three-dimensional. These objects, called **geometric solids,** have length, width, and height; in other words, they take up space. The table below illustrates some three-dimensional shapes. You should learn to recognize, name, and draw each of these shapes.

Notice that if every face of a solid is a polygon, then the solid is called a **polyhedron.** Polyhedrons do not have any curved surfaces. So rectangular prisms and pyramids are polyhedrons, but spheres and cylinders are not.

Thinking Skill

Verify

Why is a cube also a rectangular prism?

Geometric Solids

Shape	Name	Description
	Triangular Prism	Polyhedron
	Rectangular Prism	Polyhedron
	Cube	Polyhedron
	Pyramid	Polyhedron
	Cylinder	Not Polyhedron
	Cone	Not Polyhedron
	Sphere	Not Polyhedron

Name each shape using terms from the table above. Then name an object from the real world that has the same shape.

1. 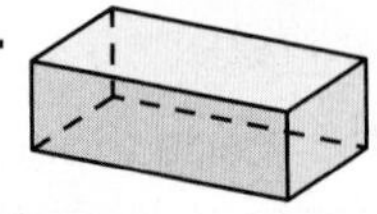

2. 

3. 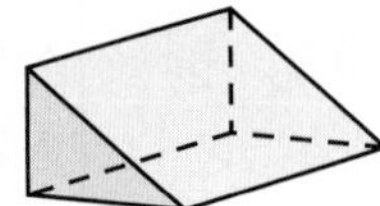

4. 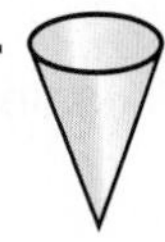

5. 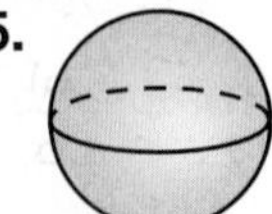

6. 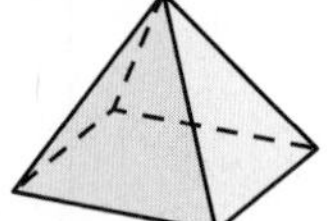

Solids can have **faces, edges,** and **vertices.** The illustration below points out a face, an edge, and a vertex of a cube.

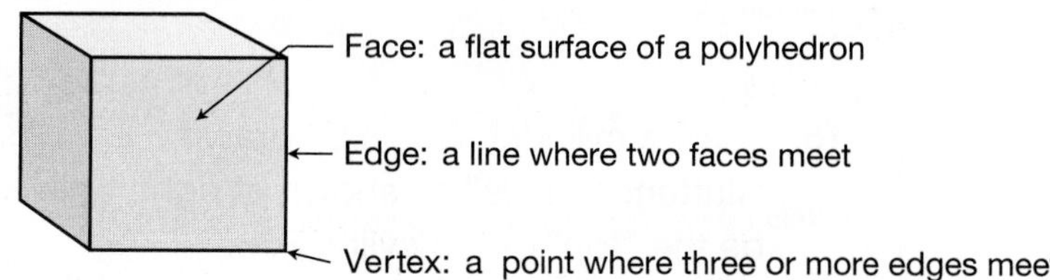
Face: a flat surface of a polyhedron
Edge: a line where two faces meet
Vertex: a point where three or more edges meet

7. A cube has how many faces?

8. A cube has how many edges?

9. A cube has how many vertices?

Activity

Comparing Geometric Solids

Materials: Relational GeoSolids

Using the solids, try identifying each shape by touch rather than by sight. Discuss the following questions.

- How can you tell if a solid is a polyhedron or not?
- How are a cone and a pyramid similar and different?
- How are a cone and a cylinder similar and different?
- How are cylinders and right prisms similar and different?

A pyramid with a square base is shown at right. One face is a square; the others are triangles.

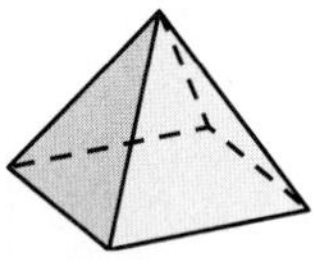

10. How many faces does this pyramid have?

11. How many edges does this pyramid have?

12. How many vertices does this pyramid have?

Represent When solids are drawn, the edges that are hidden from view can be indicated with dashes. To draw a cube, for example, we first draw two squares that overlap as shown.

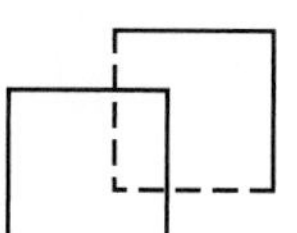

Then we connect the corresponding vertices of the two squares. In both steps we use dashes to represent the edges that are hidden from view. Practice drawing a cube.

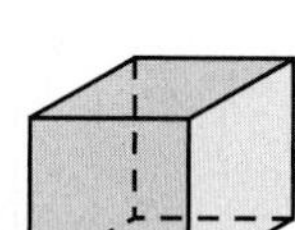

13. Draw a rectangular prism. Begin by drawing two rectangles as shown at right.

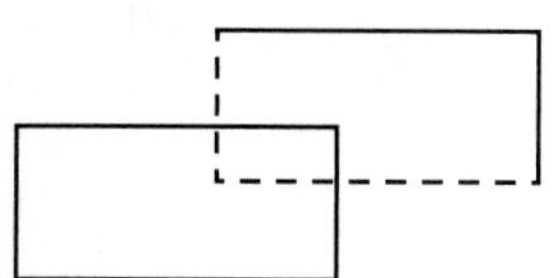

14. Draw a triangular prism. Begin by drawing two triangles as shown at right.

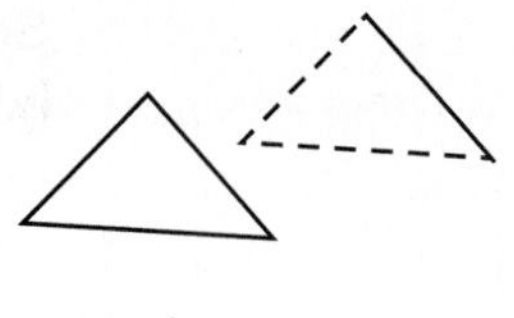

15. Draw a cylinder. Begin by drawing a "flattened circle" as shown at right. This will be the "top" of the cylinder.

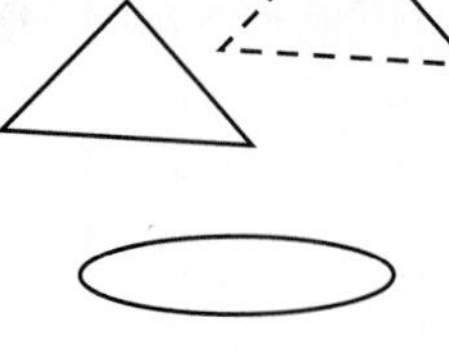

One way to measure a solid is to find the area of its surfaces. We can find how much surface a polyhedron has by adding the area of its faces. The sum of these areas is called the **surface area** of the solid.

Each edge of the cube at right is 5 inches long. So each face of the cube is a square with sides 5 inches long. Use this information to answer problems **16** and **17.**

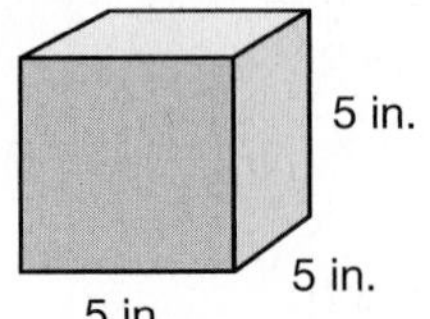

16. What is the area of each face of the cube?

17. Analyze What is the total surface area of the cube?

A cereal box has six faces, but not all the faces have the same area. The front and back faces have the same area; the top and bottom faces have the same area; and the left and right faces have the same area. Here we show a cereal box that is 10 inches tall, 7 inches wide, and 2 inches deep.

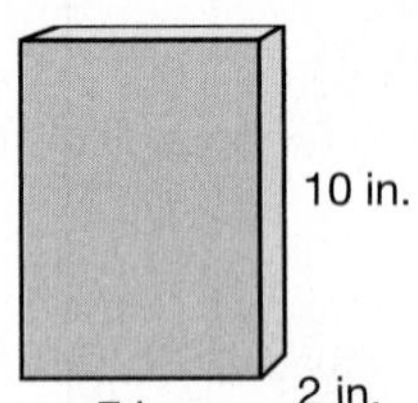

18. What is the area of the front of the box?

19. What is the area of the top of the box?

20. What is the area of the right panel of the box?

21. Combine the areas of all six faces to find the total surface area of the box.

A container such as a cereal box is constructed out of a flat sheet of cardboard that is printed, cut, folded, and glued to create a colorful three-dimensional container. By cutting apart a cereal box, you can see the six faces of the box at one time. Here we show one way to cut apart a box, but many arrangements are possible.

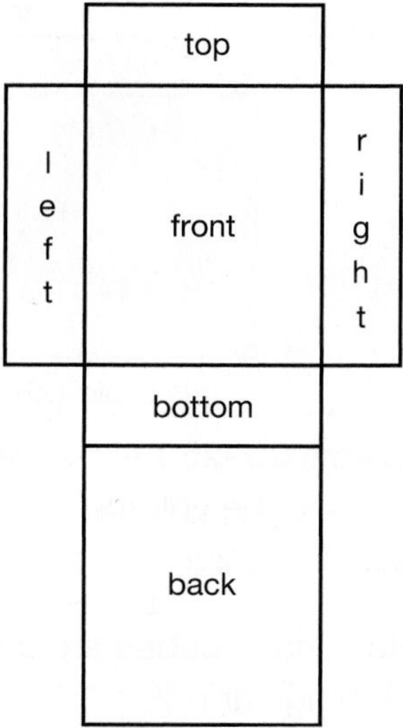

22. *Conclude* Here we show three ways to cut apart a box shaped like a cube. We have also shown an arrangement of six squares that does not fold into a cube. Which pattern below does not form a cube?

A

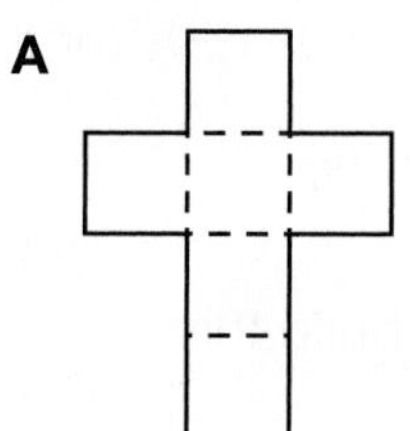

B

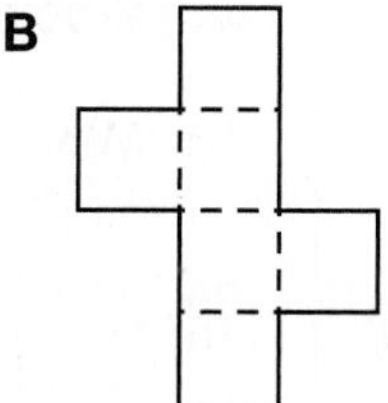

C

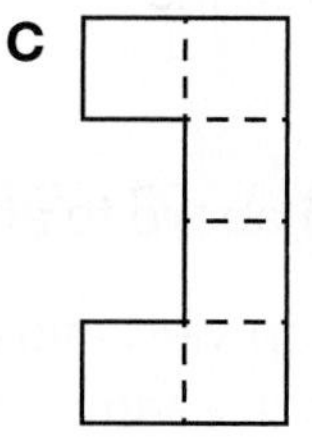

D

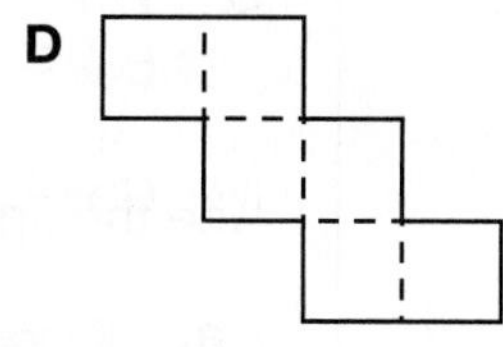

In addition to measuring the surface area of a solid, we can also measure its **volume.** The volume of a solid is the amount of space it occupies. To measure volume, we use units that occupy space, such as cubic centimeters, cubic inches, or cubic feet. Here we show two-dimensional images of a cubic inch and a cubic centimeter:

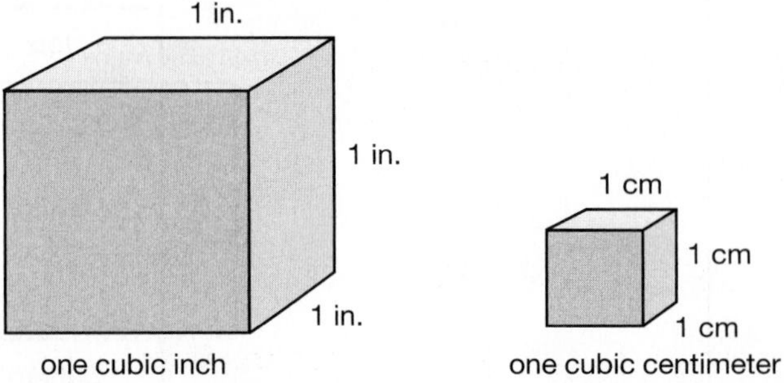

In problems **23–25** below, we will practice counting the number of cubes to determine the volume of a solid. In a later lesson we will expand our discussion of volume.

23. How many cubes are used to form this rectangular prism?

24. How many small cubes are used to form the larger cube at right?

25. How many cubes are used to build this solid?

Use the figures in problems **24** and **25** above to answer extension **a.**

extensions

a. *Represent* Draw the front view, top view, side view, and bottom view of each figure. Explain how they are the same and how they are different.

b. *Estimate* Bring an empty cereal container from home. Open the glue joints and unfold the box until it is flat. On the unprinted side of the box, label the front, back, side, top, and bottom faces. Identify the glue tabs or any overlapping areas.

Estimate the area of each of the six faces of the unfolded box. Do not include any glue tabs or overlapping areas in your estimate. Then estimate the amount of cardboard that was used to make the box.

Did you find the volume or the surface area of the cereal box? Explain your thinking.

Math Language

A two-dimensional pattern that folds to make a three-dimensional solid is called a **net.** Sometimes it is called a *map.*

c. *Model* In problem **22,** we show three nets that will form a cube. **Investigation Activities 12 and 13** show nets for a triangular prism and a square pyramid. Cut out and fold these nets to form solids. Use tape to hold the shapes together. Describe how the solids are alike and different.

d. *Represent* How many blocks are in the tenth term of this pattern? Explain how you will represent the pattern to find the answer.

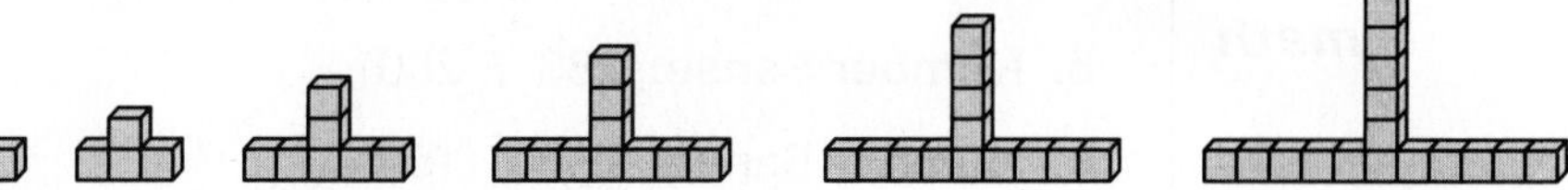

e. *Represent* Sketch the front, top, and bottom of each 3-dimensional figure.

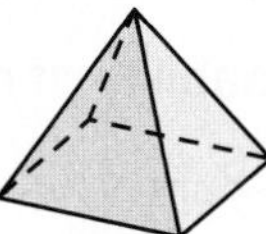

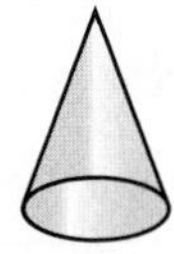

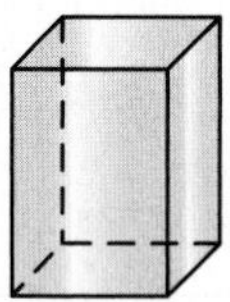

f. *Classify* Figures *A*, *B*, and *C* were sorted into a group based on one common attribute.

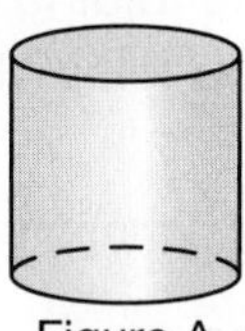

Figure A

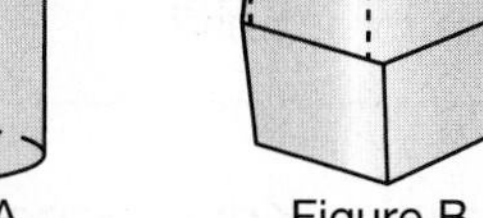

Figure B

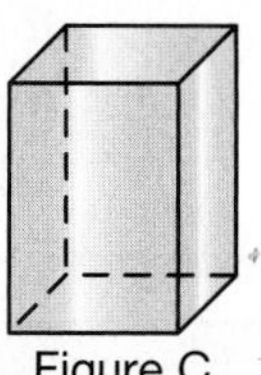

Figure C

Figures *D* and *E* do not belong in the group above.

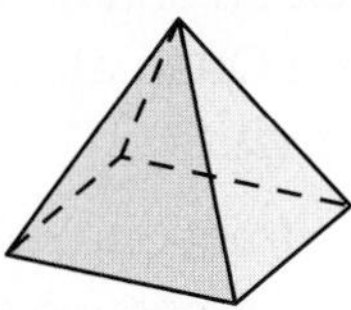

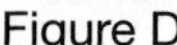

Figure D

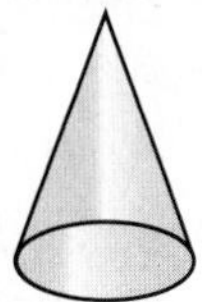

Figure E

What attribute is common to figures *A*, *B*, and *C* but not figures *D* and *E*?

Sketch a figure that would belong in the group with figures *A*, *B*, and *C*.

LESSON 61

Adding Three or More Fractions

Power Up

Building Power

facts Power Up H

mental math

a. **Number Sense:** 4×750

b. **Number Sense:** $283 + 250$

c. **Number Sense:** $750 - 199$

d. **Calculation:** $8.25 − $2.50

e. **Number Sense:** Double $12\frac{1}{2}$.

f. **Number Sense:** $\frac{900}{30}$

g. **Probability:** What is the probability of rolling a 3 on a number cube?

h. **Calculation:** 6×10, $\div 3$, $\times 2$, $\div 4$, $\times 5$, $\div 2$, $\times 4$

problem solving

A large piece of cardstock is 1 mm thick. If we fold it in half, and then fold it in half again, we have a stack of 4 layers of cardstock that is 4 mm high. If it were possible to continue folding the cardstock in half, how thick would the stack of layers be after 10 folds? Is that closest in height to a book, a table, a man, or a bus?

New Concept

Increasing Knowledge

Pedro, Leticia, and Quan shared a pizza. Pedro ate half the pizza, Leticia ate $\frac{1}{4}$ of the pizza, and Quan ate $\frac{1}{8}$ of the pizza. Together the three friends ate what fraction of the pizza?

To add three or more fractions, we find a common denominator for all the fractions being added. Once we determine a common denominator, we can rename the fractions and add. We usually use the **least common denominator,** which is the least common multiple of all the denominators.

Example 1

Add $\frac{1}{2} + \frac{1}{4} + \frac{1}{8}$ and draw a diagram illustrating the addition.

Solution

To add, we first we find a common denominator. The LCM of 2, 4, and 8 is 8, so we rename all fractions as eighths. Then we add and simplify if possible.

We illustrate the addition with fractions of a circle.

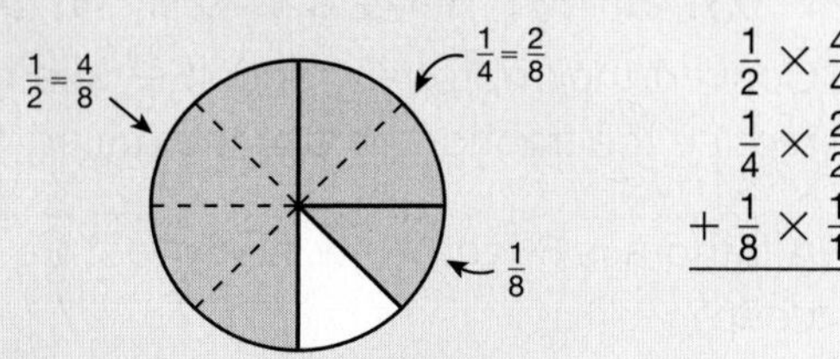

$$\begin{aligned} \tfrac{1}{2} \times \tfrac{4}{4} &= \tfrac{4}{8} \\ \tfrac{1}{4} \times \tfrac{2}{2} &= \tfrac{2}{8} \\ + \tfrac{1}{8} \times \tfrac{1}{1} &= \tfrac{1}{8} \\ \hline & \mathbf{\tfrac{7}{8}} \end{aligned}$$

Verify Why don't we need to simplify $\frac{7}{8}$?

Example 2

Add: $1\frac{1}{2} + 2\frac{1}{3} + 3\frac{1}{6}$

Solution

Thinking Skill

Predict

If we rename $\frac{1}{2}$, $\frac{1}{3}$, and $\frac{1}{6}$ as fractions with a common denominator that is not the LCM, will the sum be the same?

A common denominator is 6. We rename all fractions. Then we add the whole numbers, and we add the fractions. We simplify the result if possible.

$$\begin{aligned} 1\tfrac{1}{2} \times \tfrac{3}{3} &= 1\tfrac{3}{6} \\ 2\tfrac{1}{3} \times \tfrac{2}{2} &= 2\tfrac{2}{6} \\ + 3\tfrac{1}{6} \times \tfrac{1}{1} &= 3\tfrac{1}{6} \\ \hline 6\tfrac{6}{6} &= \mathbf{7} \end{aligned}$$

Justify What steps do you use to simplify $6\frac{6}{6}$?

Practice Set

Add:

a. $\frac{1}{2} + \frac{3}{4} + \frac{1}{8}$ **b.** $\frac{1}{2} + \frac{1}{3} + \frac{1}{6}$ **c.** $1\frac{1}{2} + 1\frac{1}{3} + 1\frac{1}{4}$

d. $\frac{1}{2} + \frac{2}{3} + \frac{5}{6}$ **e.** $\frac{1}{2} + \frac{3}{4} + \frac{7}{8}$ **f.** $1\frac{1}{4} + 1\frac{1}{8} + 1\frac{1}{2}$

g. Find the perimeter of the triangle.

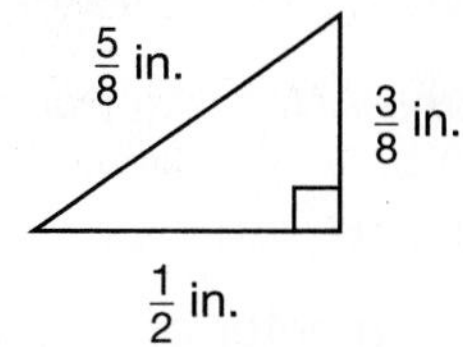

h. Select one of the exercises **a–f** and write a word problem that involves adding the fractions.

Written Practice *Strengthening Concepts*

1. *(25)* Convert the improper fraction $\frac{20}{6}$ to a mixed number. Remember to reduce the fraction part of the mixed number.

2. *(29)* A fathom is 6 feet. How many feet deep is water that is $2\frac{1}{2}$ fathoms deep?

3. *(18)* In 3 hours 2769 cars passed through a tollbooth. What is the average number of cars that pass through the tollbooth per hour?

* **4.** *(57)* $5\frac{1}{2} - 1\frac{2}{3}$

* **5.** *(57)* $5\frac{1}{3} - 2\frac{1}{2}$

* **6.** *(61)* $1\frac{1}{2} + 2\frac{1}{3} + 3\frac{1}{4}$

* **7.** *(59)* $3\frac{3}{4} + 3\frac{1}{3}$

8. *(38, 56)* Compare:

a. $\frac{2}{3} \bigcirc \frac{3}{5}$

b. $4^2 \bigcirc \sqrt{144}$

9. *(29, 38)* $\frac{5}{6} \times 6^2$

10. *(29)* $\frac{3}{8} \cdot \frac{2}{3}$

11. *(54)* How many $\frac{2}{3}$s are in $\frac{3}{8}$? ($\frac{3}{8} \div \frac{2}{3}$)

12. *(53)* $(4 - 0.4) \div 4$

13. *(53)* $4 - (0.4 \div 4)$

14. *(34)* Which digit in 49.63 has the same place value as the 7 in 8.7?

15. *(16)* **Estimate** Find the sum of \$642.23 and \$861.17 to the nearest hundred dollars. Explain how you arrived at your answer.

16. *(47)* Elizabeth used a compass to draw a circle with a radius of 4 cm.

a. What was the diameter of the circle? Describe how the radius and diameter are related.

b. What was the circumference of the circle? (Use 3.14 for π.)

* **17.** *(10)* **Predict** What is the next number in this sequence?

..., 100, 10, 1, ...

18. *(38)* The perimeter of a square is 1 foot. How many square inches cover its area?

19. *(23)* **Connect** What is the ratio of the value of a dime to the value of a quarter?

Find each unknown number:

20. *(4, 38)* $15m = 3 \cdot 10^2$

21. *(42)* $\frac{1}{10} = \frac{n}{100}$

22. *(42)* By what fraction name for 1 must $\frac{2}{3}$ be multiplied to form a fraction with a denominator of 15?

23. *(32)* What time is 5 hours 15 minutes after 9:50 a.m.?

24. (38) **Analyze** The area of a square is 16 square inches. What is its perimeter?

* **25.** (60) This figure shows the shape of home plate on a baseball field. What kind of a polygon is shown?

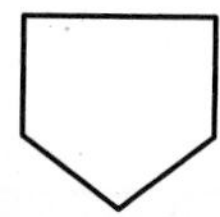

26. (41) **Explain** The sales-tax rate was 7%. Dexter bought two items, one for $4.95 and the other for $2.79. What was the total cost of the two items including sales tax? Describe how to use estimation to check whether your answer is reasonable.

27. (52) Ramla bought a sheet of 100 stamps from the post office for $39. What was the price of each stamp?

* **28.** (Inv. 6) **Represent** Draw a rectangular prism. A rectangular prism has how many

a. faces? **b.** edges? **c.** vertices?

Refer to the cube shown below to answer problems **29** and **30.**

* **29.** (Inv. 6) Each face of a cube is a square. What is the area of each face of this cube?

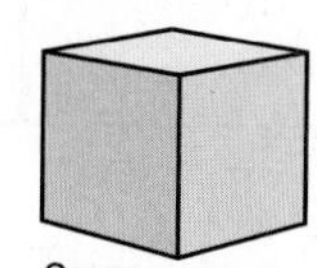

* **30.** (Inv. 6) Find the total surface area of the cube by adding up the area of all of the faces of the cube.

Early Finishers
Math and Architecture

The Pentagon in Washington, D.C. is the world's largest office building. Each of the five sides of the Pentagon is 921 feet long. What is the perimeter of the Pentagon in yards?

LESSON 62

Writing Mixed Numbers as Improper Fractions

Power Up

Building Power

facts Power Up G

mental math

a. **Number Sense:** 5×40

b. **Number Sense:** $475 + 1200$

c. **Calculation:** 3×84

d. **Calculation:** \$8.50 + \$2.50

e. **Fractional Parts:** $\frac{1}{3}$ of \$36.00

f. **Number Sense:** $\frac{\$25}{10}$

g. **Measurement:** Convert 240 seconds into minutes.

h. **Calculation:** 6×8, $- 4$, $\div 4$, $\times 2$, $+ 2$, $\div 6$, $\div 2$

problem solving There are approximately 520 nine-inch long noodles in a 1-pound package of spaghetti. Placed end-to-end, how many feet of noodles are in a pound of uncooked spaghetti?

New Concept

Increasing Knowledge

Thinking Skill

Model

Use your fraction manipulatives to represent $3\frac{5}{6}$.

Here is another word problem about pies. In this problem we will change a mixed number to an improper fraction.

There were $3\frac{5}{6}$ pies on the shelf. The restaurant manager asked the server to cut the whole pies into sixths. Altogether, how many slices of pie were there after the server cut the pies?

We illustrate this problem with circles. There were $3\frac{5}{6}$ pies on the shelf.

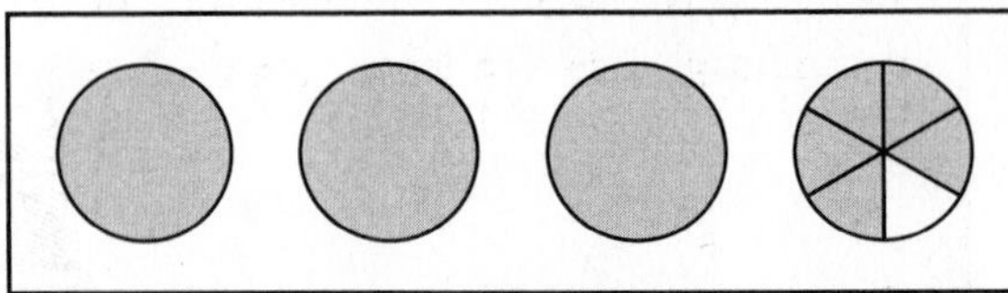

The server cut the whole pies into sixths. Each whole pie then had six slices.

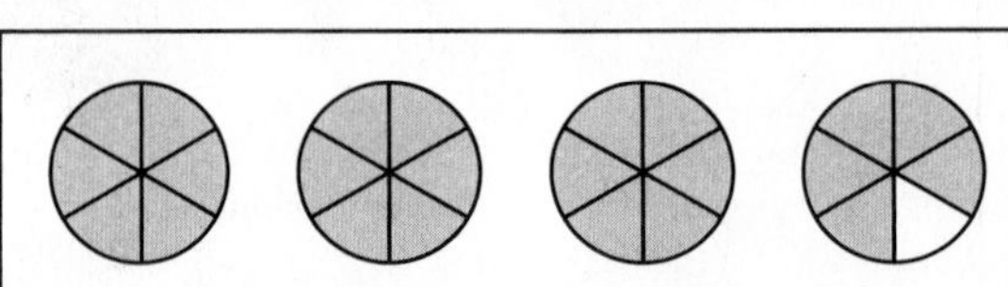

The three whole pies contain 18 slices ($3 \times 6 = 18$). The 5 additional slices from the $\frac{5}{6}$ of a pie bring the total to 23 slices (23 sixths). This problem illustrates that $3\frac{5}{6}$ is equivalent to $\frac{23}{6}$.

Now we describe the arithmetic for changing a mixed number such as $3\frac{5}{6}$ to an improper fraction. Recall that a mixed number has a whole-number part and a fraction part.

whole number → 3, fraction → $\frac{5}{6}$

$$3\frac{5}{6}$$

The denominator of the mixed number will also be the denominator of the improper fraction.

$$3\frac{5}{6} = \frac{}{6}$$

same denominator

The denominator indicates the size of the fraction "pieces." In this case the fraction pieces are sixths, so we change the whole number 3 into sixths. We know that one whole is $\frac{6}{6}$, so three wholes is $3 \times \frac{6}{6}$, which is $\frac{18}{6}$. Therefore, we add $\frac{18}{6}$ and $\frac{5}{6}$ to get $\frac{23}{6}$.

$$3\frac{5}{6}$$

$$\overbrace{\frac{6}{6} + \frac{6}{6} + \frac{6}{6}} + \frac{5}{6} = \frac{23}{6}$$

$$\frac{18}{6} + \frac{5}{6} = \frac{23}{6}$$

Example 1

Write $2\frac{3}{4}$ as an improper fraction.

Solution

The denominator of the fraction part of the mixed number is fourths, so the denominator of the improper fraction will also be fourths.

$$2\frac{3}{4} = \frac{}{4}$$

We change the whole number 2 into fourths. Since 1 equals $\frac{4}{4}$, the whole number 2 equals $2 \times \frac{4}{4}$, which is $\frac{8}{4}$. We add $\frac{8}{4}$ and $\frac{3}{4}$ to get $\frac{11}{4}$.

$$2\frac{3}{4}$$

$$\frac{8}{4} + \frac{3}{4} = \mathbf{\frac{11}{4}}$$

Represent Draw a model to show that $2\frac{3}{4} = \frac{11}{4}$.

Example 2

Write $5\frac{2}{3}$ as an improper fraction.

Solution

Thinking Skill

Evaluate

Which method of changing a mixed number to an improper fraction do you prefer, the method in example 1 or example 2? Why?

We see that the denominator of the improper fraction will be thirds.

$$5\frac{2}{3} = \frac{}{3}$$

Some people use a quick, mechanical method to find the numerator of the improper fraction. Looking at the mixed number, they multiply the denominator by the whole number and then add the numerator. The result is the numerator of the improper fraction.

$$5\frac{2}{3} = \mathbf{\frac{17}{3}}$$

Example 3

Write $1\frac{2}{3}$ and $2\frac{2}{5}$ as improper fractions. Then multiply the improper fractions. What is the product?

Solution

First we write $1\frac{2}{3}$ and $2\frac{2}{5}$ as improper fractions.

$$1\frac{2}{3}: \quad \frac{3}{3} + \frac{2}{3} = \mathbf{\frac{5}{3}} \qquad\qquad 2\frac{2}{5}: \quad \frac{10}{5} + \frac{2}{5} = \mathbf{\frac{12}{5}}$$

Next we multiply $\frac{5}{3}$ by $\frac{12}{5}$.

$$\frac{5}{3} \cdot \frac{12}{5} = \frac{60}{15}$$

The result is an improper fraction, which we simplify.

$$\frac{60}{15} = 4$$

So $1\frac{2}{3} \times 2\frac{2}{5}$ equals **4**.

Practice Set

Write each mixed number as an improper fraction:

a. $2\frac{4}{5}$ **b.** $3\frac{1}{2}$ **c.** $1\frac{3}{4}$

d. $6\frac{1}{4}$ **e.** $1\frac{5}{6}$ **f.** $3\frac{3}{10}$

g. $2\frac{1}{3}$ **h.** $12\frac{1}{2}$ **i.** $3\frac{1}{6}$

j. Write $1\frac{1}{2}$ and $3\frac{1}{3}$ as improper fractions. Then multiply the improper fractions. What is the product?

Written Practice — *Strengthening Concepts*

1. (54) In music there are whole notes, half notes, quarter notes, and eighth notes.

a. How many quarter notes equal a whole note?

b. How many eighth notes equal a quarter note?

*** 2.** (62) Don is 5 feet $2\frac{1}{2}$ inches tall. How many inches tall is that?

3. (19) **Classify** Which of these numbers is not a prime number?

A 11 **B** 21 **C** 31 **D** 41

*** 4.** (62) **Analyze** Write $1\frac{1}{3}$ and $1\frac{1}{2}$ as improper fractions, and multiply the improper fractions. What is the product?

5. (58) If the chance of rain is 20%, what is the chance that it will not rain?

6. (18) The prices for three pairs of skates were \$36.25, \$41.50, and \$43.75. What was the average price for a pair of skates? Estimate to show that your answer is reasonable.

7. (43) **Evaluate** Instead of dividing 15 by $2\frac{1}{2}$, Solomon doubled both numbers and then divided mentally. What was Solomon's mental division problem and its quotient?

Find each unknown number:

*** 8.** (43, 59) $m - 4\frac{3}{8} = 3\frac{1}{4}$

9. (43, 56) $n + \frac{3}{10} = \frac{3}{5}$

10. (43, 45) $6d = 0.456$

11. (43, 49) $0.04w = 1.5$

*** 12.** (61) $\frac{1}{2} + \frac{3}{4} + \frac{5}{8}$

*** 13.** (57) $\frac{5}{6} - \frac{1}{2}$

14. (29) $\frac{1}{2} \cdot \frac{4}{5}$

15. (54) $\frac{2}{3} \div \frac{1}{2}$

16. (40) $1 - (0.2 - 0.03)$

17. (39) $(0.14)(0.16)$

18. (49) One centimeter equals 10 millimeters. How many millimeters does 2.5 centimeters equal?

19. (19) List all of the common factors of 18 and 24. Then circle the greatest common factor.

*** 20.** (58) **Analyze** Ten marbles are in a bag. Four of the marbles are red.

a. If one marble is drawn from the bag, what is the probability that it will be red?

b. Write the complement of the event in **a** and state its probability.

c. Describe the relationship between the event and its complement.

21. (38) **Analyze** If the perimeter of a square is 40 mm, what is the area of the square?

22. (14) At 6 a.m. the temperature was $-6°F$. At noon the temperature was 14°F. From 6 a.m. to noon the temperature rose how many degrees?

23. (47) Lisa used a compass to draw a circle with a radius of $1\frac{1}{2}$ inches.

a. What was the diameter of the circle?

b. What was the circumference of the circle? (Use 3.14 for π.)

The circle graph below shows the favorite sport of 100 people. Refer to the graph to answer problems **24–27.**

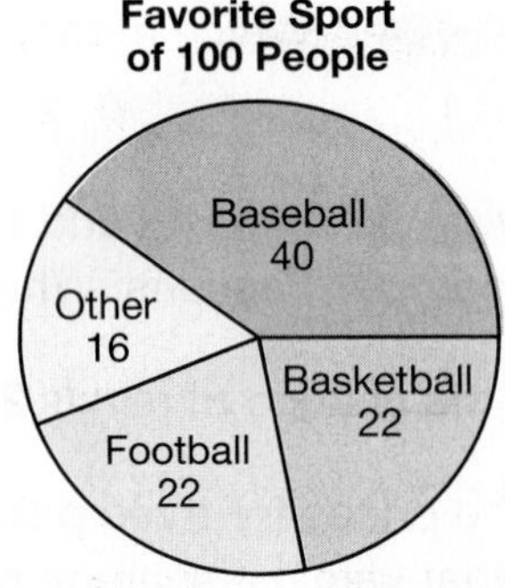

24. (40) How many more people favored baseball than favored football?

25. (40) What fraction of the people surveyed favored baseball?

26. (40) **Explain** Was any sport the favorite sport of the majority of the people surveyed? Write one or two sentences to explain your answer.

27. (40) **Connect** Since baseball was the favorite sport of 40 out of 100 people, it was the favorite sport of 40% of the people surveyed. What percent of the people answered that football was their favorite sport?

28. (41) What number is 40% of 200?

29. (19) Here we show 18 written as a product of prime numbers:

$$2 \cdot 3 \cdot 3$$

Write 20 as a product of prime numbers.

*** 30.** (Inv. 5) **Analyze** Judges awarded Sandra these scores for her performance on the vault:

9.1, 8.9, 9.0, 9.2, 9.2

What is the median score?

LESSON 63

Subtracting Mixed Numbers with Regrouping, Part 2

Power Up

Building Power

facts Power Up D

mental math

a. **Number Sense:** 5×140

b. **Number Sense:** $420 - 50$

c. **Calculation:** 4×63

d. **Calculation:** $8.50 − $2.50

e. **Number Sense:** Double $7\frac{1}{2}$.

f. **Number Sense:** $\frac{\$25}{100}$

g. **Measurement:** How many inches are in 6 feet?

h. **Calculation:** 5×10, $-\ 20$, $+\ 2$, $\div\ 4$, $+\ 1$, $\div\ 3$, $-\ 3$

problem solving Rhett chooses a marble at random from each of the four boxes below. From which box is he *most* likely to choose a blue marble?

A 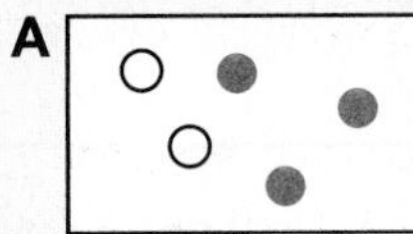
B 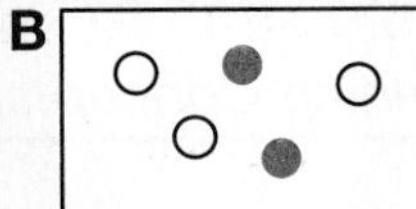
C 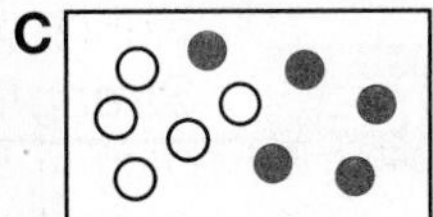
D 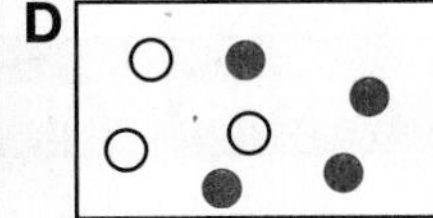

New Concept

Increasing Knowledge

Since Lesson 48 we have practiced subtracting mixed numbers with regrouping. In this lesson we will rename the fractions with common denominators before subtracting.

To subtract $1\frac{1}{2}$ from $3\frac{2}{3}$, we first rewrite the fractions with common denominators. Then we subtract the whole numbers and the fractions. If possible, we simplify.

$$\begin{array}{r} 3\frac{2}{3} \times \frac{2}{2} = 3\frac{4}{6} \\ -\ 1\frac{1}{2} \times \frac{3}{3} = 1\frac{3}{6} \\ \hline 2\frac{1}{6} \end{array}$$

When subtracting, it is sometimes necessary to regroup. We rewrite the fractions with common denominators before regrouping.

Example

Thinking Skill

Explain

Why did we regroup in this example but not in the previous problem?

Subtract: $5\frac{1}{2} - 1\frac{2}{3}$

Solution

We rewrite the fractions with common denominators. Before we can subtract, we must regroup. After we subtract, we simplify if possible.

$$\begin{array}{r} 5\frac{1}{2} \times \frac{3}{3} = \overset{4}{\cancel{5}}\frac{\overset{9}{\cancel{3}}}{6} \\ -\ 1\frac{2}{3} \times \frac{2}{2} = 1\frac{4}{6} \\ \hline 3\frac{5}{6} \end{array}$$

Justify Why can we use addition to check the answer?

Practice Set

Subtract:

a. $5\frac{1}{2} - 3\frac{1}{3}$ **b.** $4\frac{1}{4} - 2\frac{1}{3}$ **c.** $6\frac{1}{2} - 1\frac{3}{4}$

d. $7\frac{2}{3} - 3\frac{5}{6}$ **e.** $6\frac{1}{6} - 1\frac{1}{2}$ **f.** $4\frac{1}{3} - 1\frac{1}{2}$

g. $4\frac{5}{6} - 1\frac{1}{3}$ **h.** $6\frac{1}{2} - 3\frac{5}{6}$ **i.** $8\frac{2}{3} - 5\frac{3}{4}$

j. Formulate Write a word problem that involves subtracting mixed numbers.

Written Practice

Strengthening Concepts

* **1.** (12, 53) Connect What is the difference between the sum of 0.6 and 0.4 and the product of 0.6 and 0.4?

* **2.** (14) Analyze Mt. Whitney, the highest point in California, has an elevation of 14,494 feet *above* sea level. From there one can see Death Valley, which contains the lowest point in California, 282 feet *below* sea level. The floor of Death Valley is how many feet below the peak of Mt. Whitney?

* **3.** (10) Conclude It was 39° outside at 1 p.m. By 7 p.m. the temperature had dropped 11° and was below freezing. What was the temperature at 7 p.m.? On what scale is the temperature being measured?

* **4.** (62) Write the mixed number $4\frac{2}{3}$ as an improper fraction.

* **5.** (62) Explain Round $678.25 to the nearest ten dollars. Describe how you decided upon your answer.

6. (32) Explain What time is $2\frac{1}{2}$ hours after 10:15 a.m.? How did you find your answer?

7. (5) $(30 \times 15) \div (30 - 15)$

8. (56) Compare: $\frac{5}{8} \bigcirc \frac{2}{3}$

9. (43) $w - 3\frac{2}{3} = 1\frac{1}{2}$

10. (55) $\frac{6}{8} - \frac{3}{4}$

* **11.** $6\frac{1}{4} - 5\frac{5}{8}$
(63)

12. $\frac{3}{4} \times \frac{2}{5}$
(29)

13. $\frac{3}{4} \div \frac{2}{5}$
(54)

14. $(1 - 0.4)(1 + 0.4)$
(53)

15. How much money is 60% of $45?
(41)

16. $0.4 \div 8$
(45)

17. $8 \div 0.4$
(49)

18. **Predict** What is the next number in this sequence?
(10)

0.2, 0.4, 0.6, 0.8, ...

19. What is the tenth prime number?
(19)

* **20.** What is the perimeter of this rectangle?
(8, 59)

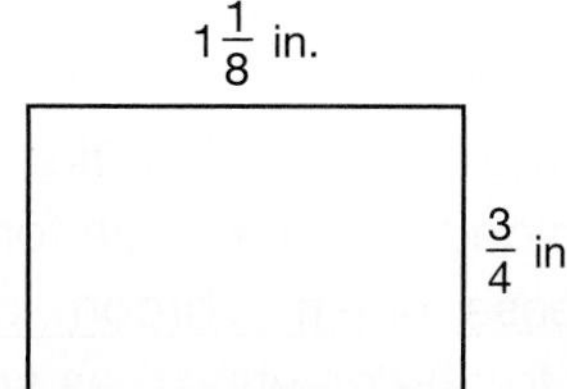

* **21.** A triangular prism has how many
(Inv. 6)

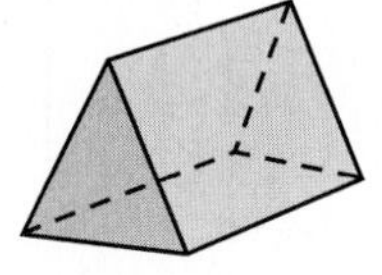

a. faces?

b. edges?

c. vertices?

22. Write $2\frac{1}{2}$ and $1\frac{1}{5}$ as improper fractions. Then multiply the improper fractions and simplify the product.
(16)

23. This rectangle is divided into two congruent regions. What is the area of the shaded region?
(31)

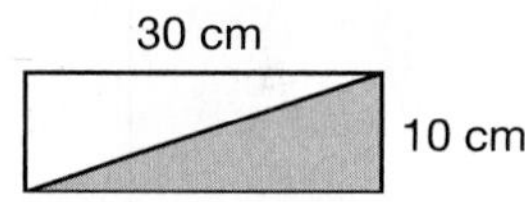

* **24.** A ton is 2000 pounds. How many pounds is $2\frac{1}{2}$ tons?
(15, 62)

25. **Connect** Which arrow could be pointing to 0.2 on this number line?
(50)

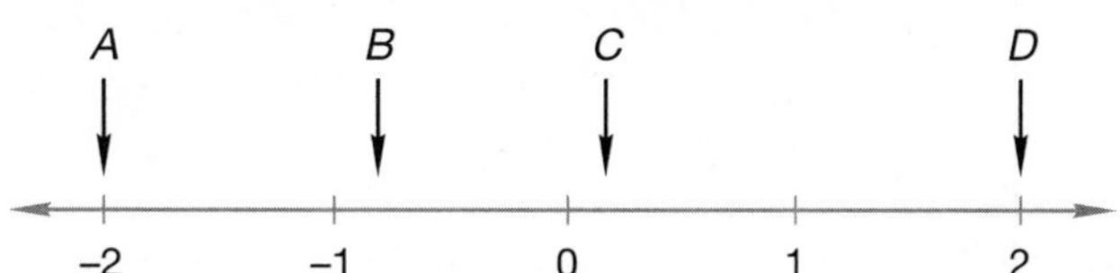

26. **Evaluate** The paper cup would not roll straight. One end was 7 cm in diameter, and the other end was 5 cm in diameter. In one roll of the cup,
(47)

a. how far would the larger end roll?

b. how far would the smaller end roll?

(Round each answer to the nearest centimeter. Use 3.14 for π.)

27. Jefferson got a hit 30% of the 240 times he went to bat during the season. Write 30% as a reduced fraction. Then find the number of hits Jefferson got during the season.
(29, 33)

28. Jena has run 11.5 miles of a 26.2-mile race. Find the remaining distance Jena has to run by solving this equation:
(43)

$$11.5 \text{ mi} + d = 26.2 \text{ mi}$$

29. The sales-tax rate was 7%. The two CDs cost $15.49 each. What was the total cost of the two CDs including tax?
(41)

* **30.** Rosa is mixing paint in ceramics class. She mixes $\frac{1}{2}$ teaspoon of yellow paint with $\frac{3}{4}$ teaspoon of red paint to make orange paint. How much orange paint does Rosa make?
(57)

Early Finishers
Real-World Application

The drama club had their first annual meeting this afternoon. The officers had decided to order pizza for all the new members this year. They ordered one cheese, one mushroom, and one tomato pizza. Due to the rain, the turnout for the meeting was small and there was a lot of pizza left. They had $\frac{1}{4}$ of the cheese pizza, $\frac{1}{2}$ of the mushroom pizza and $\frac{5}{12}$ of the tomato pizza left. How much leftover pizza did the drama club have?

LESSON 64

• Classifying Quadrilaterals

Power Up

Building Power

facts Power Up J

mental math

a. **Number Sense:** 5×240

b. **Number Sense:** $4500 + 450$

c. **Calculation:** 7×34

d. **Calculation:** $7.50 + $7.50

e. **Fractional Parts:** $\frac{1}{4}$ of $20.00

f. **Number Sense:** $\frac{\$75}{10}$

g. **Measurement:** How many meters are in 200 centimeters?

h. **Calculation:** 6×8, $\div 2$, $+ 1$, $\div 5$, $- 1$, $\times 4$, $\div 2$

problem solving

Emily has a blue folder, a green folder, and a red folder. She uses one folder each for her math, science, and history classes. She does not use her blue folder for math. Her green folder is not used for science. She does not use her red folder for history. If her red folder is not used for math, what folder does Emily use for each subject? Make a table to show your work.

New Concept

Increasing Knowledge

Math Language
The prefix *quadri-* means four. A quadrilateral is a polygon with four sides.

We learned in Lesson 60 that quadrilaterals are polygons with four sides. We can classify (sort) quadrilaterals by the characteristics of their sides and angles. The following table describes the various classifications of quadrilaterals:

Classifications of Quadrilaterals

Shape	Characteristic	Name
	No sides parallel	Trapezium
	One pair of parallel sides	Trapezoid
	Two pairs of parallel sides	Parallelogram
	Parallelogram with equal sides	Rhombus
	Parallelogram with right angles	Rectangle
	Rectangle with equal sides	Square

Notice that squares, rectangles, and rhombuses are all parallelograms. Also notice that a square is a special kind of rectangle, which is a special kind of parallelogram, which is a special kind of quadrilateral, which is a special kind of polygon. A square is also a special kind of rhombus.

Example 1

Thinking Skill

Justify

How can we rewrite the statement in example 1 so that it is true?

Is the following statement true or false?

All parallelograms are rectangles.

Solution

We are asked to decide whether every parallelogram is a rectangle. Since a rectangle is a special kind of parallelogram, some parallelograms are rectangles. However, some parallelograms are not rectangles. Since not all parallelograms are rectangles, the statement is **false.**

Example 2

Draw a pair of parallel lines. Then draw another pair of parallel lines. These lines should intersect the first pair but not be perpendicular to the first pair. What is the name for the quadrilateral that is formed by the intersecting lines?

Solution

We draw the first pair of parallel lines.

We draw the second pair of lines so that the lines are not perpendicular to the first pair.

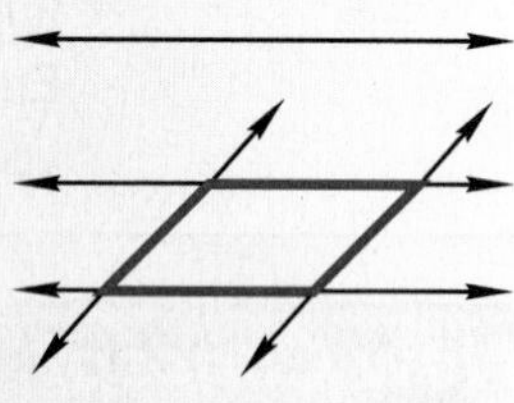

At right we have colored the segments that form the quadrilateral. The quadrilateral formed is a **parallelogram.**

Practice Set

a. What is a quadrilateral?

b. Describe the difference between a parallelogram and a trapezoid.

c. Model Draw a rhombus that is a square.

d. Model Draw a rhombus that is not a square.

e. Verify True or false: Some rectangles are squares.

f. Verify True or false: All squares are rectangles.

Written Practice

Strengthening Concepts

* **1.** (12, 53) When the sum of 1.3 and 1.2 is divided by the difference of 1.3 and 1.2, what is the quotient?

2. (13) William Shakespeare was born in 1564 and died in 1616. How many years did he live?

3. Duane kicked a 45-yard field goal. How many feet is 45 yards?
(15)

*** 4.** **Explain** Why is a square a regular quadrilateral?
(60)

*** 5.** A regular hexagon has a perimeter of 36 inches. How long is each side?
(60)

6. $\frac{1}{4} = \frac{?}{100}$
(42)

7. $\frac{8 \times 8}{8 + 8}$
(5)

*** 8.** $5\frac{2}{3} + 3\frac{3}{4}$
(59)

*** 9.** $\frac{1}{2} + \frac{2}{3} + \frac{1}{4}$
(61)

10. $\frac{9}{10} - \frac{1}{2}$
(57)

*** 11.** $6\frac{1}{2} - 2\frac{7}{8}$
(63)

12. Compare: $2 \times 0.4 \bigcirc 2 + 0.4$
(44)

13. 4.8×0.35
(39)

14. $1 \div 0.4$
(49)

15. How many $0.12 pencils can Mr. Velazquez buy for $4.80?
(15)

16. **Estimate** Round the product of 0.33 and 0.38 to the nearest hundredth.
(51)

17. Multiply the length by the width to find the area of this rectangle.
(31)

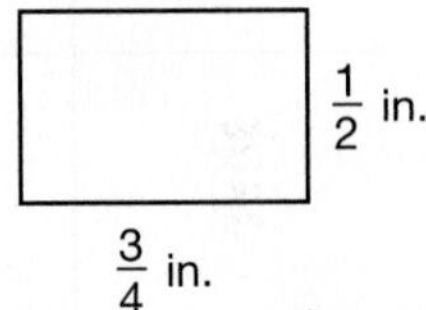

*** 18.** **Conclude** Is every rectangle a parallelogram?
(64)

19. **Analyze** What is the twelfth prime number?
(19)

20. The area of a square is 9 cm^2.
(38)

a. How long is each side of the square?

b. What is the perimeter of the square?

Refer to the box shown below to answer problems **21** and **22.**

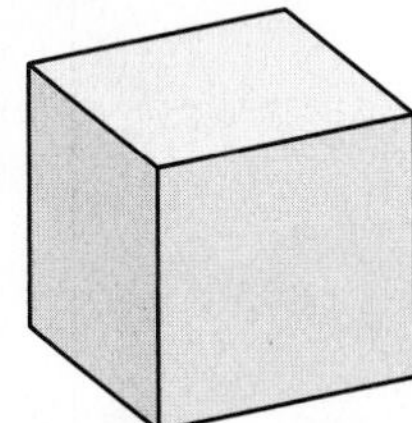

*** 21.** This box has how many faces? Draw a net to show how the box would look if you cut it apart and flattened it.
(Inv. 6)

*** 22.** If this box is a cube and each edge is 10 inches long, then
(Inv. 6)

a. what is the area of each face?

b. what is the total surface area of the cube?

23. There are 100 centimeters in a meter. How many centimeters equal 2.5 meters?
(15)

*** 24.** Write the mixed numbers $1\frac{1}{2}$ and $2\frac{1}{2}$ as improper fractions. Then multiply the improper fractions and simplify the product.
(62)

* **25.** (19) **Verify** The numbers 2, 3, 5, 7, and 11 are prime numbers. The numbers 4, 6, 8, 9, 10, and 12 are not prime numbers, but they can be formed by multiplying prime numbers.

$$4 = 2 \cdot 2$$

$$6 = 2 \cdot 3$$

$$8 = 2 \cdot 2 \cdot 2$$

Show how to form 9, 10, and 12 by multiplying prime numbers.

26. (33, 35) Write 75% as an unreduced fraction. Then write the fraction as a decimal number.

27. (54) Reduce: $\dfrac{2 \cdot 2 \cdot 2 \cdot 3 \cdot 3}{2 \cdot 2 \cdot 3 \cdot 5 \cdot 5}$

28. (43) **Analyze** Find the missing distance d in the equation below.

$$16.6 \text{ mi} + d = 26.2 \text{ mi}$$

Refer to the double-line graph below to answer problems **29** and **30**.

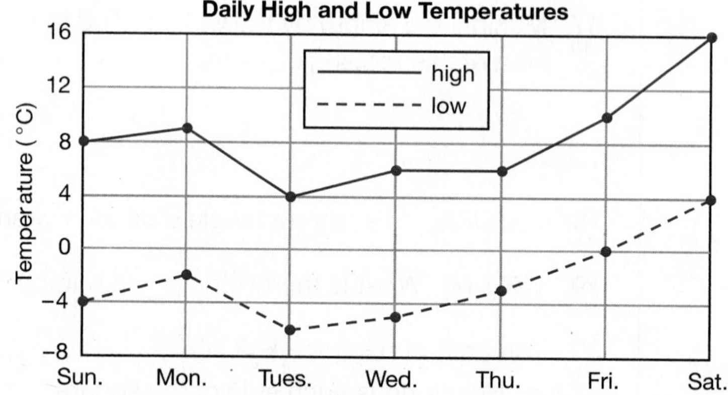

* **29.** (18) **a.** The difference between Tuesday's high and low temperatures was how many degrees?

b. The difference between the lowest temperature of the week and the highest temperature of the week was how many degrees?

* **30.** (18) **Predict** If the daily high temperature dropped 5 degrees the day after this graph was completed, what probably happened to the daily low temperature? Explain.

LESSON

65

• Prime Factorization
• Division by Primes
• Factor Trees

Power Up

Building Power

facts Power Up H

mental math

a. **Number Sense:** 5 × 60

b. **Number Sense:** 586 − 50

c. **Calculation:** 3 × 65

d. **Calculation:** \$20.00 − \$2.50

e. **Number Sense:** Double 75¢.

f. **Number Sense:** $\frac{\$75}{100}$

g. **Primes and Composites:** Name the prime numbers between 10 and 20.

h. **Calculation:** 9 × 9, − 1, ÷ 2, + 2, ÷ 6, + 3, ÷ 10

problem solving

Use the digits 6, 7, and 8 to complete this multiplication problem:

$$\begin{array}{r} 23\_ \\ \times \quad \_ \\ \hline 166\_ \end{array}$$

New Concepts

Increasing Knowledge

prime factorization

Every whole number greater than 1 is either a prime number or a **composite number.** A prime number has *only two* factors (1 and itself), while a composite number has *more than two* factors. As we studied in Lesson 19, the numbers 2, 3, 5, and 7 are prime numbers. The numbers 4, 6, 8, and 9 are composite numbers. All composite numbers can be formed by multiplying prime numbers together.

Thinking Skill

List

What are all the factors of 4, 6, 8, and 9?

$$4 = 2 \cdot 2$$

$$6 = 2 \cdot 3$$

$$8 = 2 \cdot 2 \cdot 2$$

$$9 = 3 \cdot 3$$

When we write a composite number as a product of its prime factors, we have written the **prime factorization** of the number. The prime factorizations of 4, 6, 8, and 9 are shown above. Notice that if we had written 8 as 2 · 4 instead of 2 · 2 · 2, we would not have completed the prime factorization of 8. Since the number 4 is not prime, we would complete prime factorization by "breaking" 4 into its prime factors of 2 and 2.

In this lesson we will show two methods for factoring a composite number, **division by primes** and **factor trees.** We will use both methods to factor the number 60.

division by primes

To factor a number using division by primes, we write the number in a division box and begin dividing by the smallest prime number that is a factor. The smallest prime number is 2. Since 60 is divisible by 2, we divide 60 by 2 to get 30.

$$\begin{array}{r} 30 \\ \mathbf{2}\overline{)60} \end{array}$$

Since 30 is also divisible by 2, we divide 30 by 2. The quotient is 15. Notice how we "stack" the divisions.

$$\begin{array}{r} 15 \\ \mathbf{2}\overline{)30} \\ \mathbf{2}\overline{)60} \end{array}$$

Although 15 is not divisible by 2, it is divisible by the next-smallest prime number, which is 3. Fifteen divided by 3 produces the quotient 5.

$$\begin{array}{r} 5 \\ \mathbf{3}\overline{)15} \\ \mathbf{2}\overline{)30} \\ \mathbf{2}\overline{)60} \end{array}$$

Five is a prime number. The only prime number that divides 5 is 5.

$$\begin{array}{r} 1 \\ \mathbf{5}\overline{)5} \\ \mathbf{3}\overline{)15} \\ \mathbf{2}\overline{)30} \\ \mathbf{2}\overline{)60} \end{array}$$

By dividing by prime numbers, we have found the prime factorization of 60.

$$60 = 2 \cdot 2 \cdot 3 \cdot 5$$

Example 1

Use division by primes to find the prime factorization of 36.

Solution

We begin by dividing 36 by its smallest prime-number factor, which is 2. We continue dividing by prime numbers until the quotient is 1.[1]

$$\begin{array}{r} \mathbf{1} \\ \mathbf{3}\overline{)\mathbf{3}} \\ \mathbf{3}\overline{)\mathbf{9}} \\ \mathbf{2}\overline{)\mathbf{18}} \\ \mathbf{2}\overline{)\mathbf{36}} \end{array}$$

$$\mathbf{36 = 2 \cdot 2 \cdot 3 \cdot 3}$$

[1] Some people prefer to divide only until the quotient is a prime number. When using that procedure, the final quotient is included in the prime factorization of the number.

factor trees

To make a factor tree for 60, we simply think of any two whole numbers whose product is 60. Since 6 × 10 equals 60, we can use 6 and 10 as the first two "branches" of the factor tree.

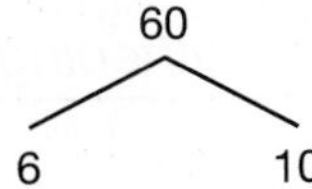

The numbers 6 and 10 are not prime numbers, so we continue the process by factoring 6 into 2 · 3 and by factoring 10 into 2 · 5.

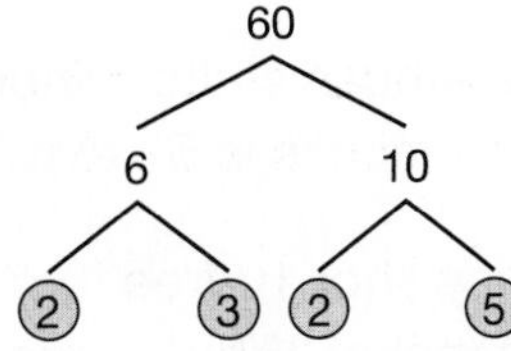

The circled numbers at the ends of the branches are all prime numbers. We have completed the factor tree. We will arrange the factors in order from least to greatest and write the prime factorization of 60.

$$60 = 2 \cdot 2 \cdot 3 \cdot 5$$

Example 2

Use a factor tree to find the prime factorization of 60. Use 4 and 15 as the first branches.

Solution

Thinking Skill

Connect

What other whole number pairs could we use as the first two branches of a factor tree for 60?

Some composite numbers can be divided into many different factor trees. However, when the factor tree is completed, the same prime numbers appear at the ends of the branches.

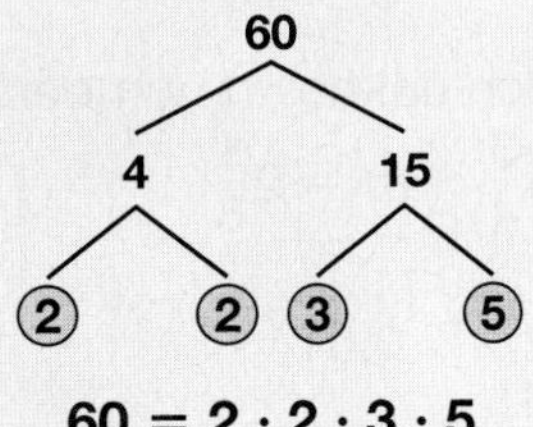

$$\mathbf{60 = 2 \cdot 2 \cdot 3 \cdot 5}$$

Practice Set

a. Classify Which of these numbers are composite numbers?

19, 20, 21, 22, 23

b. Write the prime factorization of each composite number in problem **a.**

c. Represent Use a factor tree to find the prime factorization of 36.

d. Use division by primes to find the prime factorization of 48.

e. Write 125 as a product of prime factors.

f. **Generalize** Write the prime factorization of 10, 100, 1000, and 10,000. What patterns do you see in the prime factorizations of these numbers?

Written Practice

Strengthening Concepts

1. (12) The total land area of the world is about fifty-seven million, five hundred six thousand square miles. Use digits to write that number of square miles.

2. (15) The African white rhinoceros can reach a height of about $5\frac{1}{2}$ feet. How many inches is $5\frac{1}{2}$ feet?

3. (29, 42) Jenny shot 10 free throws and made 6. What fraction of her shots did she make? What percent of her shots did she make?

*** 4.** (65) **Represent** Make a factor tree for 40. Then write the prime factorization of 40.

*** 5.** (65) **Classify** Which of these numbers is a composite number?

A 21 **B** 31 **C** 41

*** 6.** (62) Write $2\frac{2}{3}$ as an improper fraction. Then multiply the improper fraction by $\frac{3}{8}$. What is the product?

7. (58) Four of the ten marbles in the bag are red. If one marble is drawn from the bag, what ratio expresses the probability that the marble will not be red?

*** 8.** (61) $8\frac{1}{2} + 1\frac{1}{3} + 2\frac{1}{6}$

*** 9.** (61) $\frac{1}{12} + \frac{1}{6} + \frac{1}{2}$

Find each unknown number:

10. (43) $15\frac{3}{4} - m = 2\frac{1}{8}$

11. (42) $\frac{4}{25} = \frac{n}{100}$

12. (45) $12w = 0.0144$

13. (29) $\frac{3}{8} \times \frac{1}{3} = y$

14. (56) Compare: $\frac{1}{2} - \frac{1}{3} \bigcirc \frac{2}{3} - \frac{1}{2}$

15. (38) $1 - (0.2 + 0.48)$

16. (15, 41) **Explain** What is the total cost of two dozen erasers that are priced at 50¢ each if 8% sales tax is added? Describe a way to perform the calculation mentally.

17. (15) **Connect** The store manager put $20.00 worth of quarters in the change drawer. How many quarters are in $20.00?

*** 18.** (Inv. 6) A pyramid with a square base has how many

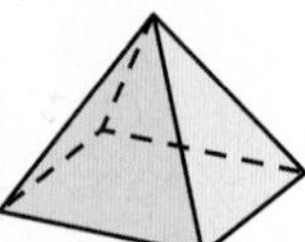

a. faces?

b. edges?

c. vertices?

* **19.** Use division by primes to find the prime factorization of 50.
(65)

* **20.** **Connect** What is the name of a six-sided polygon? How many vertices does it have?
(60)

* **21.** Write $3\frac{4}{7}$ as an improper fraction.
(62)

22. The area of a square is 36 square inches.
(38)

a. What is the length of each side?

b. What is the perimeter of the square?

23. Write 16% as a reduced fraction.
(33)

24. How many millimeters long is the line segment below?
(7)

cm 1 2 3 4 5 6 7

25. **Estimate** A meter is about $1\frac{1}{10}$ yards. About how many meters long is an automobile?
(7)

* **26.** Write the prime factorization of 375 and of 1000. What method did you use?
(65)

27. Reduce: $\frac{3 \cdot 5 \cdot 5 \cdot 5}{2 \cdot 2 \cdot 2 \cdot 5 \cdot 5 \cdot 5}$
(54)

28. **Estimate** The radius of the carousel is 15 feet. If the carousel turns around once, a person riding on the outer edge will travel how far? Round the answer to the nearest foot. (Use 3.14 for π.) Describe how to mentally check whether the answer is reasonable.
(47)

29. Eighty percent of the 20 answers were correct. How many answers were correct?
(29, 33)

30. **Verify** The prefix "rect-" in rectangle means "right." A rectangle is a "right-angle" shape. Why is every square also a rectangle?
(54)

LESSON 66

• Multiplying Mixed Numbers

Power Up

Building Power

facts Power Up J

mental math

a. **Number Sense:** 5×160

b. **Number Sense:** $376 + 99$

c. **Calculation:** 8×23

d. **Calculation:** $1.75 + $1.75

e. **Fractional Parts:** $\frac{1}{3}$ of $60.00

f. **Number Sense:** $\frac{\$30}{10}$

g. **Measurement:** Which is greater, 5 years or a decade?

h. **Calculation:** 8×8, $- 4$, $\div 2$, $+ 3$, $\div 3$, $+ 1$, $\div 6$, $\div 2$

problem solving

In this figure a square and a regular pentagon share a common side. The area of the square is 25 square centimeters. What is the perimeter of the pentagon?

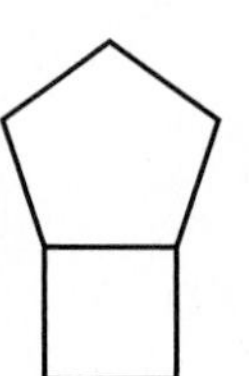

New Concept

Increasing Knowledge

Recall from Lesson 57 the three steps to solving an arithmetic problem with fractions.

Thinking Skill

Verify

How do we write a mixed number as a fraction?

Step 1: Put the problem into the correct shape (if it is not already).

Step 2: Perform the operation indicated.

Step 3: Simplify the answer if possible.

Remember that putting fractions into the correct shape for adding and subtracting means writing the fractions with common denominators. To multiply or divide fractions, we do not need to use common denominators. However, we must write the fractions in **fraction form.** This means we will write mixed numbers and whole numbers as improper fractions. We write a whole number as an improper fraction by making the whole number the numerator of a fraction with a denominator of 1.

Example 1

A length of fabric was cut into 4 equal sections. Each of 4 students received $2\frac{2}{3}$ yd of fabric. How much fabric was there before it was cut?

Solution

This is an equal groups problem. To find the original length of the fabric we multiply $2\frac{2}{3}$ yd by 4. First, we write $2\frac{2}{3}$ and 4 in fraction form.

$$\frac{8}{3} \times \frac{4}{1}$$

Second, we multiply the numerators to find the numerator of the product, and we multiply the denominators to find the denominator of the product.

$$\frac{8}{3} \times \frac{4}{1} = \frac{32}{3}$$

Third, we simplify the product by converting the improper fraction to a mixed number.

$$\frac{32}{3} = 10\frac{2}{3}$$

Before the fabric was cut it was $\mathbf{10\frac{2}{3}}$ **yd** long.

Evaluate How can we use estimation to check whether our answer is reasonable?

Example 2

Multiply: $\mathbf{2\frac{1}{2} \times 1\frac{1}{3}}$

Solution

First, we write the numbers in fraction form.

$$\frac{5}{2} \times \frac{4}{3}$$

Reading Math

Recall that the **terms** of a fraction are the numerator and the denominator.

Second, we multiply the terms of the fractions.

$$\frac{5}{2} \times \frac{4}{3} = \frac{20}{6}$$

Third, we simplify the product.

$$\frac{20}{6} = 3\frac{2}{6} = \mathbf{3\frac{1}{3}}$$

Sketching a rectangle on a grid is a way to check the reasonableness of a product of mixed numbers. To illustrate the product of $2\frac{1}{2}$ and $3\frac{1}{2}$, we use a grid that is at least 3 by 4 so that a $2\frac{1}{2}$ by $3\frac{1}{2}$ rectangle fits on the grid.

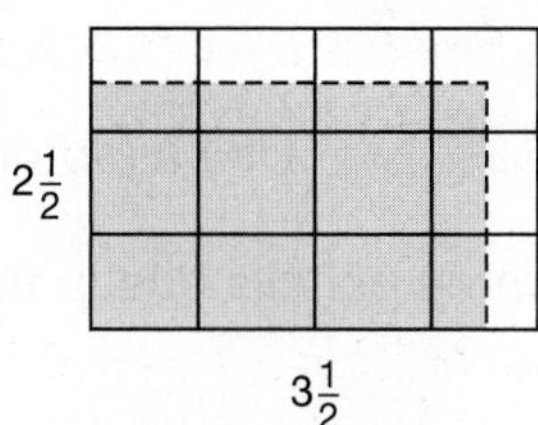

We sketch the rectangle and estimate the area. There are 6 full squares, 5 half squares, and a quarter square. Since 2 half squares equal a whole square, the area is about $8\frac{3}{4}$ square units.

Practice Set Multiply:

a. $1\frac{1}{2} \times \frac{2}{3}$ **b.** $1\frac{2}{3} \times \frac{3}{4}$ **c.** $1\frac{1}{2} \times 1\frac{2}{3}$

d. $1\frac{2}{3} \times 3$ **e.** $2\frac{1}{2} \times 2\frac{2}{3}$ **f.** $3 \times 1\frac{3}{4}$

g. $3\frac{1}{3} \times 1\frac{2}{3}$ **h.** $2\frac{3}{4} \times 2$ **i.** $2 \times 3\frac{1}{2}$

j. Check the reasonableness of the products in **e** and **h** by sketching rectangles on a grid.

k. **Formulate** Write and solve a word problem about multiplying a whole number and a mixed number.

Written Practice *Strengthening Concepts*

1. (29, 33) Fifty percent of the 60 questions on the test are multiple choice. Find the number of multiple-choice questions on the test.

2. (58) **Analyze** Twelve of the 30 students in the class are boys.

a. What is the ratio of boys to girls in the class?

b. If each student's name is placed in a hat and one name is drawn, what is the probability that it will be the name of a girl?

3. (15) **Analyze** Some railroad rails weigh 155 pounds per yard. How much would a 33-foot-long rail weigh?

* **4.** (66) $1\frac{1}{2} \times 2\frac{2}{3}$ * **5.** (66) $2\frac{2}{3} \times 2$

6. (18) The sum of five numbers is 200. What is the average of the numbers?

7. (5) $\frac{100 + 75}{100 - 75}$ **8.** (59) $1\frac{1}{5} + 3\frac{1}{2}$ * **9.** (61) $\frac{1}{3} + \frac{1}{6} + \frac{1}{12}$

* **10.** (63) $35\frac{1}{4} - 12\frac{1}{2}$ **11.** (29) $\frac{4}{5} \times \frac{1}{2}$ **12.** (54) $\frac{4}{5} \div \frac{1}{2}$

13. (45) $0.25 \div 5$ **14.** (49) $5 \div 0.25$

15. (39) What is the product of the answers to problems 13 and 14?

16. (54) **Verify** Which of the following is equal to $\frac{1}{2} + \frac{1}{2}$?

A $\frac{1}{2} - \frac{1}{2}$ **B** $\left(\frac{1}{2}\right)^2$ **C** $\frac{1}{2} \div \frac{1}{2}$

* **17.** (65) **Represent** Use a factor tree to find the prime factorization of 30.

18. (15) If three pencils cost a total of 75¢, how much would six pencils cost?

19. Seven and one half percent is equivalent to the decimal number
(41) 0.075. If the sales-tax rate is $7\frac{1}{2}\%$, what is the sales tax on a $10.00 purchase?

* **20.** **Analyze** One side of a regular pentagon measures 0.8 meter. What is
(60) the perimeter of the regular pentagon?

21. Twenty minutes is what fraction of an hour?
(29)

22. The temperature dropped from 12°C to −8°C. This was a drop of how
(14) many degrees?

The bar graph below shows the weights of different types of cereals packaged in the same size boxes. Refer to the graph to answer problems **23–25.**

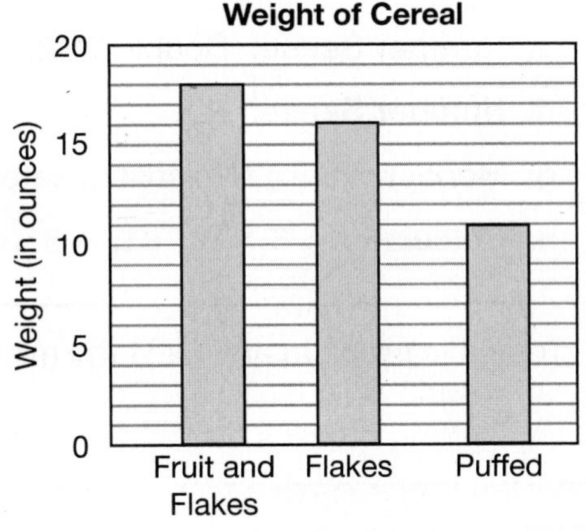

23. What is the range of the weights?
(Inv. 5)

24. What is the mean weight of the three types of cereal?
(Inv. 5)

* **25.** **Formulate** Write a comparison word problem that relates to the graph,
(13) and then answer the problem.

* **26.** **Connect** Use division by primes to find the prime factorization of 400.
(65)

27. **Analyze** Simon covered the floor of a square room with 144 square
(38) floor tiles. How many floor tiles were along each wall of the room?

28. **Estimate** The weight of a 1-kilogram object is about 2.2 pounds.
(46) A large man may weigh 100 kilograms. About how many pounds is that?

29. Reduce: $\dfrac{5 \cdot 5 \cdot 5 \cdot 7}{2 \cdot 2 \cdot 2 \cdot 5 \cdot 5 \cdot 5}$
(54)

* **30.** **Classify** Which of these polygons is not a regular polygon?
(60)

A (triangle) **B** (rectangle) **C** (pentagon) **D** (hexagon)

LESSON 67

Using Prime Factorization to Reduce Fractions

Power Up

Building Power

facts Power Up J

mental math

a. **Number Sense:** 5 × 260

b. **Number Sense:** 341 − 50

c. **Calculation:** 3 × 48

d. **Calculation:** \$9.25 − 75¢

e. **Number Sense:** Double \$1.25.

f. **Number Sense:** $\frac{\$30}{100}$

g. **Measurement:** Which is greater, 3 yards or 5 feet?

h. **Calculation:** 6 × 6, − 1, ÷ 5, × 2, + 1, ÷ 3, ÷ 2

problem solving Copy this problem and fill in the missing digits:

$$\begin{array}{r} \_\_\_,\_\_\_ \\ \times \quad 7 \\ \hline 999{,}999 \end{array}$$

New Concept

Increasing Knowledge

One way to **reduce** fractions with large terms is to factor the terms and then reduce the common factors. To reduce $\frac{125}{1000}$, we could begin by writing the prime factorizations of 125 and 1000.

$$\frac{125}{1000} = \frac{5 \cdot 5 \cdot 5}{2 \cdot 2 \cdot 2 \cdot 5 \cdot 5 \cdot 5}$$

We see three pairs of 5s that can be reduced. Each $\frac{5}{5}$ reduces to $\frac{1}{1}$.

$$\frac{\overset{1}{\cancel{5}} \cdot \overset{1}{\cancel{5}} \cdot \overset{1}{\cancel{5}}}{2 \cdot 2 \cdot 2 \cdot \underset{1}{\cancel{5}} \cdot \underset{1}{\cancel{5}} \cdot \underset{1}{\cancel{5}}} = \frac{1}{8}$$

We multiply the remaining factors and find that $\frac{125}{1000}$ reduces to $\frac{1}{8}$.

Thinking Skill

Explain

What are two strategies for finding the prime factorization of a number?

Example

Reduce: $\frac{375}{1000}$

Thinking Skill

Discuss

When is it helpful to use prime factorization to reduce a fraction?

Solution

We write the prime factorization of both the numerator and the denominator.

$$\frac{375}{1000} = \frac{3 \cdot 5 \cdot 5 \cdot 5}{2 \cdot 2 \cdot 2 \cdot 5 \cdot 5 \cdot 5} = \frac{3 \cdot \overset{1}{\cancel{5}} \cdot \overset{1}{\cancel{5}} \cdot \overset{1}{\cancel{5}}}{2 \cdot 2 \cdot 2 \cdot \underset{1}{\cancel{5}} \cdot \underset{1}{\cancel{5}} \cdot \underset{1}{\cancel{5}}} = \mathbf{\frac{3}{8}}$$

Then we reduce the common factors and multiply the remaining factors.

Practice Set Write the prime factorization of both the numerator and the denominator of each fraction. Then reduce each fraction.

a. $\frac{875}{1000}$

b. $\frac{48}{400}$

c. $\frac{125}{500}$

d. $\frac{36}{81}$

Written Practice *Strengthening Concepts*

* **1.** (66) Allison is making a large collage of a beach scene. She needs 2 yards of blue ribbon for the ocean, $\frac{1}{2}$ yard of yellow ribbon for the sun, and $\frac{3}{4}$ yard of green ribbon for the grass. Ribbon costs $2 a yard. How much money will Allison need for ribbon?

2. (13) **Estimate** A mile is 5280 feet. A nautical mile is about 6080 feet. A nautical mile is about how much longer than a mile?

3. (43) **Verify** Instead of dividing $1.50 by $0.05, Marcus formed an equivalent division problem by mentally multiplying both the dividend and the divisor by 100. Then he performed the equivalent division problem. What is the equivalent division problem Marcus formed, and what is the quotient?

Find each unknown number:

4. (3) $6 \text{ cm} + k = 11 \text{ cm}$

5. (43) $8g = 9.6$

6. (43) $\frac{7}{10} - w = \frac{1}{2}$

7. (42) $\frac{3}{5} = \frac{n}{100}$

* **8.** (60, 64) The perimeter of a quadrilateral is 172 inches. What is the average length of each side? Can we know for certain what type of quadrilateral this is? Why or why not?

9. (5) $100.00 − ($46.75 + $9.68)

10. (53) $(2 \times 0.3) - (0.2 \times 0.3)$

* **11.** (63) **Analyze** $4\frac{1}{4} - 2\frac{7}{8}$

* **12.** (38, 66) **Analyze** $2\frac{2}{3} \times \sqrt{9}$

13. (59) $3\frac{1}{3} + 2\frac{3}{4}$

* **14.** (66) $1\frac{1}{3} \times 2\frac{1}{4}$

15. (45) $1.44 \div 60$

16. (49) $6.00 ÷ $0.15

17. (15) Five dollars was divided evenly among 4 people. How much money did each receive?

18. (60) **Conclude** The area of a regular quadrilateral is 100 square inches. What is its perimeter? What is the name of the quadrilateral?

*** 19.** (67) Write the prime factorizations of 625 and of 1000. Then reduce $\frac{625}{1000}$.

*** 20.** (31, 66) What is the area of the rectangle shown below?

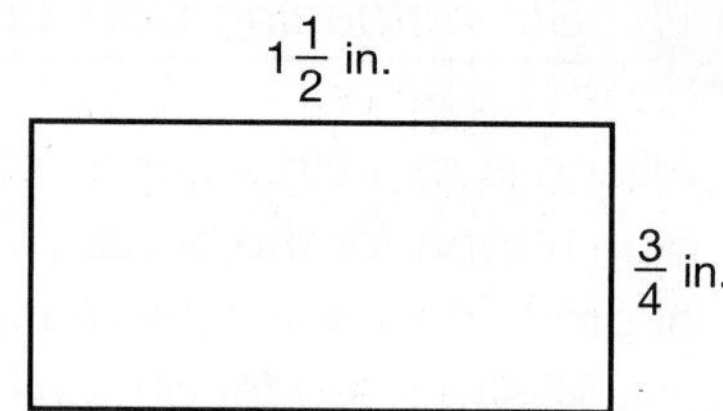

21. (29) Thirty-six of the 88 piano keys are black. What fraction of the piano keys are black?

*** 22.** (Inv. 6) **Represent** Draw a rectangular prism. Begin by drawing two congruent rectangles.

*** 23.** (30, 62) **Analyze** $1\frac{1}{2} \times \square = 1$

24. (15) There are 1000 meters in a kilometer. How many meters are in 2.5 kilometers?

25. (50) **Connect** Which arrow could be pointing to 0.1 on the number line?

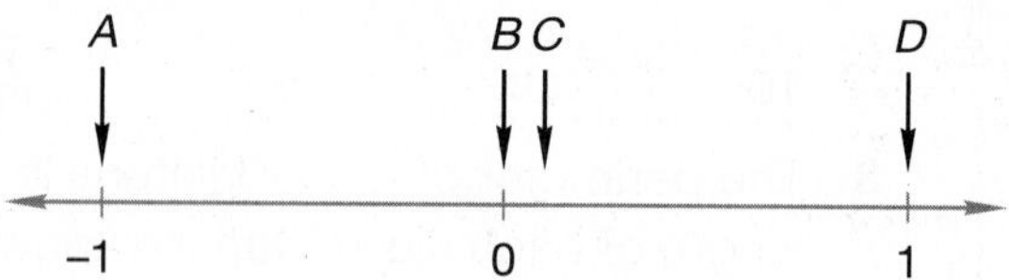

26. (47) **Estimate** If the tip of the minute hand is 6 inches from the center of the clock, how far does the tip travel in one hour? Round the answer to the nearest inch. (Use 3.14 for π.)

27. (Inv. 6) **Connect** A basketball is an example of what geometric solid?

28. (33, 35) Write 51% as a fraction. Then write the fraction as a decimal number.

*** 29.** (19, 58) **Represent** What is the probability of rolling a prime number with one toss of a number cube?

*** 30.** (64) **Conclude** This quadrilateral has one pair of parallel sides. What kind of quadrilateral is it?

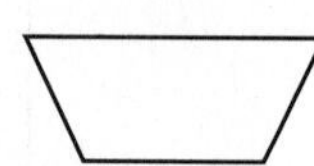

LESSON

68 • Dividing Mixed Numbers

Power Up

Building Power

facts Power Up I

mental math

a. **Number Sense:** 5 × 80

b. **Number Sense:** 275 + 1500

c. **Calculation:** 7 × 42

d. **Calculation:** $5.75 + 50¢

e. **Fractional Parts:** $\frac{1}{4}$ of $48.00

f. **Number Sense:** $\frac{\$120}{10}$

g. **Measurement:** Which is greater, 1 meter or 100 millimeters?

h. **Calculation:** 7 × 8, − 1, ÷ 5, × 2, − 1, ÷ 3, − 8

problem solving

Megan has many gray socks, white socks, and black socks in a drawer. In the dark she pulled out two socks that did not match. How many more socks does Megan need to pull from the drawer to be certain to have a matching pair?

New Concept

Increasing Knowledge

Recall the three steps to solving an arithmetic problem with fractions.

Step 1: Put the problem into the correct shape (if it is not already).

Step 2: Perform the operation indicated.

Step 3: Simplify the answer if possible.

In this lesson we will practice dividing mixed numbers. Recall from Lesson 66 that the correct shape for multiplying and dividing fractions is fraction form. So when dividing, we first write any mixed numbers or whole numbers as improper fractions.

Example 1

Shawna is pouring $2\frac{2}{3}$ cups of plant food into equal amounts to feed 4 plants. How much plant food is there for each plant?

Solution

Shawna is dividing $2\frac{2}{3}$ cups of plant food into four equal groups. We divide $2\frac{2}{3}$ by 4. We write the numbers as improper fractions.

$$\frac{8}{3} \div \frac{4}{1}$$

Math Language

Reciprocals are two numbers whose product is 1.

To divide, we find the number of 4s in 1. (That is, we find the reciprocal of 4.) Then we use the reciprocal of 4 to find the number of 4s in $\frac{8}{3}$.

$$1 \div \frac{4}{1} = \frac{1}{4}$$

$$\frac{8}{3} \times \frac{1}{4} = \frac{8}{12}$$

We simplify the answer.

$$\frac{8}{12} = \frac{2}{3}$$

There is $\frac{2}{3}$ **cup** of plant food for each plant. Notice that dividing a number by 4 is equivalent to finding $\frac{1}{4}$ of the number. Instead of dividing $2\frac{2}{3}$ by 4, we could have directly found $\frac{1}{4}$ of $2\frac{2}{3}$.

Example 2

Divide: $2\frac{2}{3} \div 1\frac{1}{2}$

Solution

Thinking Skill

Justify

Describe in your own words how to divide mixed numbers.

We write the mixed numbers as improper fractions.

$$\frac{8}{3} \div \frac{3}{2}$$

To divide, we find the number of $\frac{3}{2}$s in 1. (That is, we find the reciprocal of $\frac{3}{2}$.) Then we use the reciprocal of $\frac{3}{2}$ to find the number of $\frac{3}{2}$s in $\frac{8}{3}$.

$$1 \div \frac{3}{2} = \frac{2}{3}$$

$$\frac{8}{3} \times \frac{2}{3} = \frac{16}{9}$$

We simplify the improper fraction $\frac{16}{9}$ as shown below.

$$\frac{16}{9} = 1\frac{7}{9}$$

Practice Set

Find each product or quotient:

a. $1\frac{3}{5} \div 4$

b. $\frac{1}{4}$ of $1\frac{3}{5}$

c. $2\frac{2}{5} \div 3$

d. $\frac{1}{3}$ of $2\frac{2}{5}$

e. *Generalize* Why is dividing by 4 the same as multiplying by $\frac{1}{4}$?

f. $1\frac{2}{3} \div 2\frac{1}{2}$

g. $2\frac{1}{2} \div 1\frac{2}{3}$

h. $1\frac{1}{2} \div 1\frac{1}{2}$

i. $7 \div 1\frac{3}{4}$

j. Gabriel has $2\frac{1}{4}$ hours to finish three projects. If he divides his time equally, what fraction of an hour can he spend on each project?

Written Practice *Strengthening Concepts*

1. (12, 55) What is the difference between the sum of $\frac{1}{2}$ and $\frac{1}{4}$ and the product of $\frac{1}{2}$ and $\frac{1}{4}$?

2. (15) Bill ran a half mile in two minutes fifty-five seconds. How many seconds is that?

3. (15) The gauge of a railroad—the distance between the two tracks—is usually 4 feet $8\frac{1}{2}$ inches. How many inches is that?

* **4.** (68) $1\frac{1}{2} \div 2\frac{2}{3}$

* **5.** (68) $1\frac{1}{3} \div 4$

6. (18) In six games Yvonne scored a total of 108 points. How many points per game did she average?

* **7.** (67) Write the prime factorizations of 24 and 200. Then reduce $\frac{24}{200}$.

Find each unknown number:

* **8.** (59) $m - 5\frac{3}{8} = 1\frac{3}{16}$

9. (59) $3\frac{3}{5} + 2\frac{7}{10} = n$

10. (45) $25d = 0.375$

11. (42) $\frac{3}{4} = \frac{w}{100}$

* **12.** (63) $5\frac{1}{8} - 1\frac{1}{2}$

* **13.** (66) $3\frac{1}{3} \times 1\frac{1}{2}$

* **14.** (68) $3\frac{1}{3} \div 1\frac{1}{2}$

15. (31, 66) What is the area of a rectangle that is 4 inches long and $1\frac{3}{4}$ inches wide?

16. (53) $(3.2 + 1) - (0.6 \times 7)$

17. (49) $12.5 \div 0.4$

* **18.** (52) **Analyze** The product 3.2×10 equals which of the following?

A $32 \div 10$ **B** $320 \div 10$ **C** $0.32 \div 10$

19. (16) **Estimate** Find the sum of 6416, 5734, and 4912 to the nearest thousand.

20. (43) **Verify** Instead of dividing 800 by 24, Arturo formed an equivalent division problem by dividing both the dividend and the divisor by 8. Then he quickly found the quotient of the equivalent problem. What is the equivalent problem Arturo formed, and what is the quotient? Write the quotient as a mixed number.

21. (38) The perimeter of a square is 2.4 meters.

a. How long is each side of the square?

b. What is the area of the square?

22. (41) What is the tax on an $18,000 car if the tax rate is 8%?

23. (58) **Analyze** If the probability of an event occurring is 1 chance in a million, then what is the probability of the event not occurring?

24. (60) **Classify** Why is a circle not a polygon?

*** 25.** (66, 68) **Analyze** Compare: $\frac{1}{3} \times 4\frac{1}{2} \bigcirc 4\frac{1}{2} \div 3$

26. (17) **Estimate** Use a ruler to find the length of this line segment to the nearest eighth of an inch.

27. (28) **Conclude** Which angle in this figure is an obtuse angle?

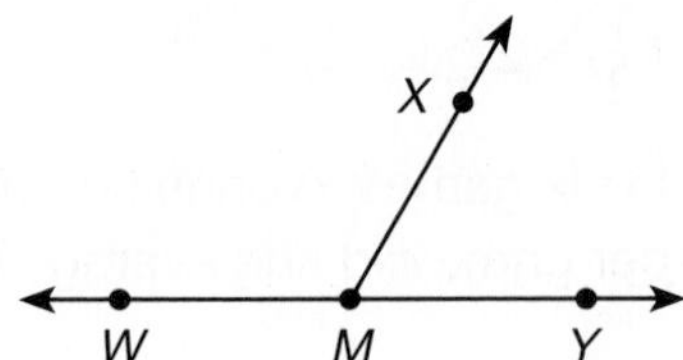

28. (33, 35) Write 3% as a fraction. Then write the fraction as a decimal number.

29. (Inv. 6) **Connect** A shoe box is an example of what geometric solid?

30. (32) Sunrise occurred at 6:20 a.m., and sunset occurred at 5:45 p.m. How many hours and minutes were there from sunrise to sunset?

Early Finishers

Choose A Strategy

Each 4 by 4 grid below is divided into 2 congruent sections.

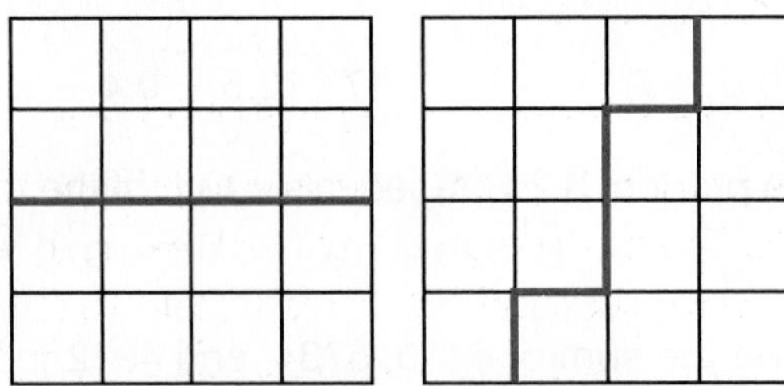

Find four ways to divide a 4 by 4 grid into 4 congruent sections.

LESSON

69

• Lengths of Segments
• Complementary and Supplementary Angles

Power Up — Building Power

facts Power Up J

mental math

a. **Number Sense:** 5×180

b. **Number Sense:** $530 - 50$

c. **Calculation:** 6×44

d. **Calculation:** $\$6.00 - \$1.75$

e. **Number Sense:** Double \$1.75.

f. **Number Sense:** $\frac{\$120}{100}$

g. **Measurement:** Which is greater, 36 inches or 1 yard?

h. **Calculation:** 6×5, $+ 2$, $\div 4$, $\times 3$, $\div 4$, $- 2$, $\div 2$, $\div 2$

problem solving Nathan used a one-yard length of string to form a rectangle that was twice as long as it was wide. What was the area that was enclosed by the string?

New Concepts — Increasing Knowledge

lengths of segments

Reading Math

We can use symbols to designate lines, rays, and segments:

$\overleftrightarrow{AB}$ = line AB

$\overrightarrow{AB}$ = ray AB

$\overline{AB}$ = segment AB

Letters are often used to designate points. Recall that we may use two points to identify a line, a ray, or a segment. Below we show a line that passes through points A and B. This line may be referred to as line AB or line BA. We may abbreviate line AB as $\overleftrightarrow{AB}$.

The ray that begins at point A and passes through point B is ray AB, which may be abbreviated $\overrightarrow{AB}$. The portion of line AB between and including points A and B is segment AB (or segment BA), which can be abbreviated $\overline{AB}$ (or $\overline{BA}$).

In the figure below we can identify three segments: $\overline{WX}$, $\overline{XY}$, and $\overline{WY}$. The length of $\overline{WX}$ plus the length of $\overline{XY}$ equals the length of $\overline{WY}$.

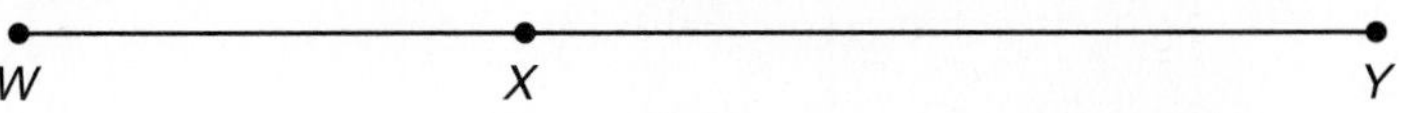

Example 1

In this figure the length of $\overline{LM}$ is 4 cm, and the length of $\overline{LN}$ is 9 cm. What is the length of $\overline{MN}$?

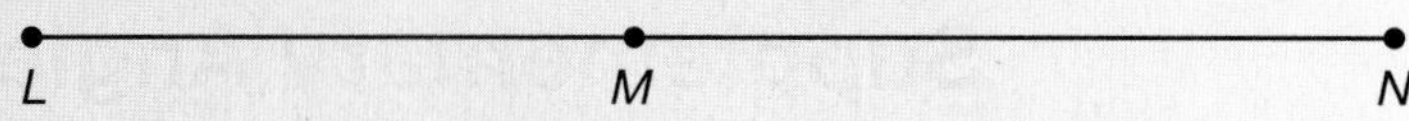

Solution

The length of $\overline{LM}$ plus the length of $\overline{MN}$ equals the length of $\overline{LN}$. With the information in the problem, we can write the equation shown below, where the letter l stands for the unknown length:

$$4 \text{ cm} + l = 9 \text{ cm}$$

Since 4 cm plus 5 cm equals 9 cm, we find that the length of $\overline{MN}$ is **5 cm.**

complementary and supplementary angles

Complementary angles are two angles whose measures total 90°. In the figure on the left below, $\angle PQR$ and $\angle RQS$ are complementary angles. In the figure on the right, $\angle A$ and $\angle B$ are complementary angles.

Thinking Skill

Verify

Can an angle that is complementary be an obtuse angle? Explain.

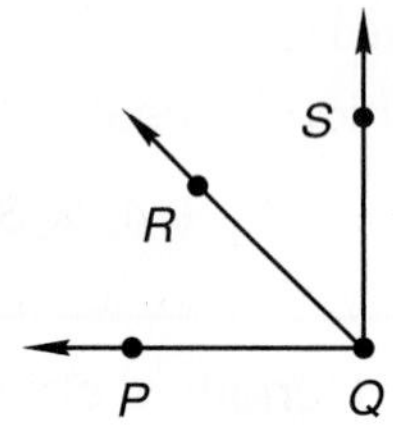

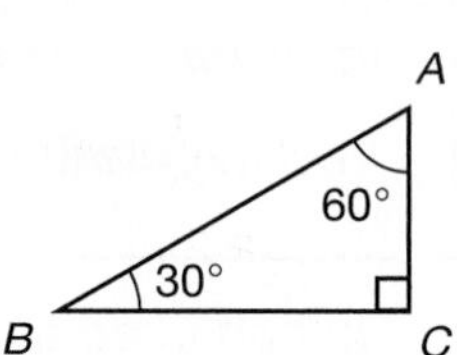

We say that $\angle A$ is the complement of $\angle B$ and that $\angle B$ is the complement of $\angle A$.

Supplementary angles are two angles whose measures total 180°. Below, $\angle 1$ and $\angle 2$ are supplementary, and $\angle A$ and $\angle B$ are supplementary. So $\angle A$ is the supplement of $\angle B$, and $\angle B$ is the supplement of $\angle A$.

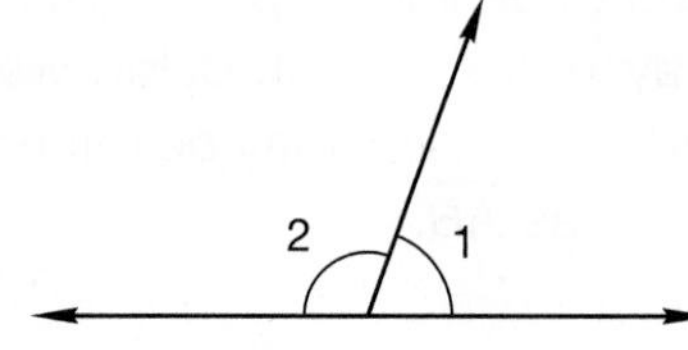

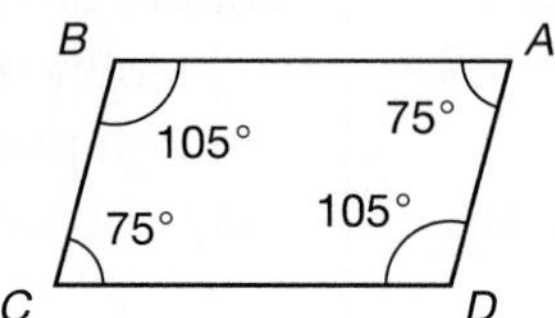

Example 2

In the figure at right, $\angle RWT$ is a right angle.

a. Which angle is the supplement of $\angle RWS$?

b. Which angle is the complement of $\angle RWS$?

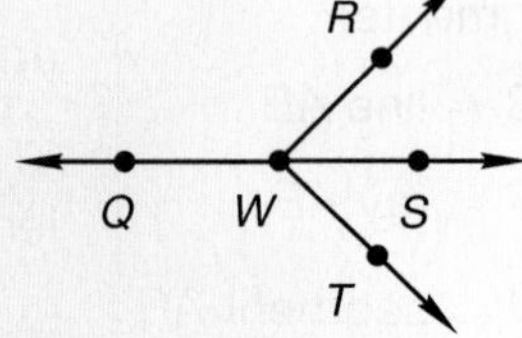

Solution

a. Supplementary angles total 180°. Angle QWS is 180° because it forms a line. So the angle that is the supplement of $\angle RWS$ is $\boldsymbol{\angle QWR}$ (or $\boldsymbol{\angle RWQ}$).

b. Complementary angles total 90°. Angle RWT is 90° because it is a right angle. So the complement of $\angle RWS$ is $\boldsymbol{\angle SWT}$ (or $\boldsymbol{\angle TWS}$).

Practice Set

a. In this figure the length of $\overline{AC}$ is 60 mm and the length of $\overline{BC}$ is 26 mm. Find the length of $\overline{AB}$.

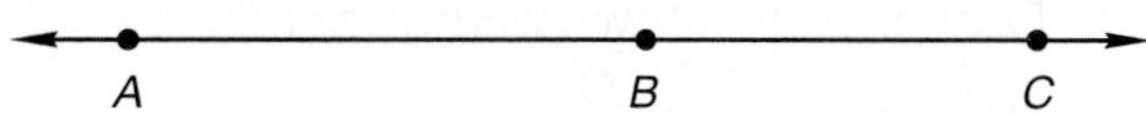

b. The complement of a 60° angle is an angle that measures how many degrees?

c. The supplement of a 60° angle is an angle that measures how many degrees?

d. Conclude If two angles are supplementary, can they both be acute? Why or why not?

e. Name two angles in the figure at right that appear to be supplementary.

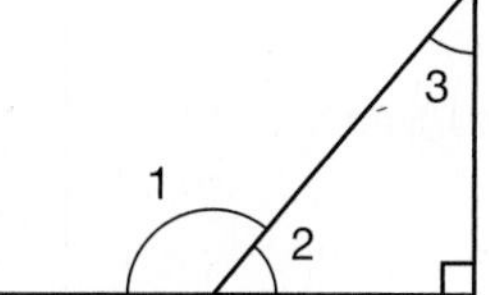

f. Name two angles that appear to be complementary.

Written Practice

Strengthening Concepts

*** 1.** (28, 64) Model Draw a pair of parallel lines. Then draw a second pair of parallel lines that are perpendicular to the first pair. Trace the quadrilateral that is formed by the intersecting pairs of lines. What kind of quadrilateral did you trace?

2. (54) Connect What is the quotient if the dividend is $\frac{1}{2}$ and the divisor is $\frac{1}{8}$?

3. (14) The highest weather temperature recorded was 136°F in Africa. The lowest was −129°F in Antarctica. How many degrees difference is there between these temperatures?

4. (15) Estimate A dollar bill is about 6 inches long. Placed end to end, about how many **feet** would 1000 dollar bills reach?

*** 5.** (67) Write the prime factorization of both the numerator and the denominator of this fraction. Then reduce the fraction.

$$\frac{45}{72}$$

*** 6.** (64) Conclude In quadrilateral *QRST,* which segment appears to be parallel to $\overline{RS}$?

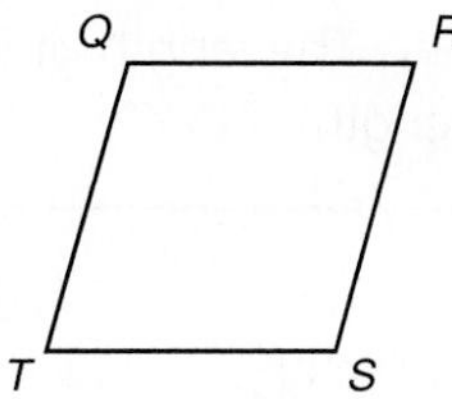

7. (13) In 10 days Juana saved $27.50. On average, how much did she save per day?

* **8.** (5) $\dfrac{1 \times 2 \times 3 \times 4 \times 5}{1 + 2 + 3 + 4 + 5}$

9. (61) $3\frac{1}{2} + 2\frac{3}{4} + 1\frac{5}{8}$

Find each unknown number:

10. (63) $m + 1\frac{3}{4} = 5\frac{3}{8}$

11. (56) $\frac{3}{4} - f = \frac{1}{3}$

12. (30) $\frac{2}{5}w = 1$

13. (42) $\frac{8}{25} = \frac{n}{100}$

* **14.** (68) $1\frac{2}{3} \div 2$

* **15.** (66) $2\frac{2}{3} \times 1\frac{1}{5}$

16. (49) $\dfrac{2.4}{0.08}$

17. (38) **a.** What is the perimeter of this square?

b. What is the area of this square?

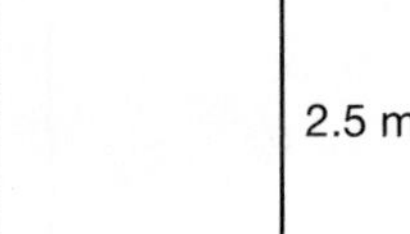

* **18.** (65) **Explain** How can you determine whether a counting number is a composite number?

* **19.** (65) **Represent** Make a factor tree to find the prime factorization of 250.

20. (60) A stop sign has the shape of an eight-sided polygon. What is the name of an eight-sided polygon?

21. (29) There were 15 boys and 12 girls in the class.

a. What fraction of the class was made up of girls?

b. What was the ratio of boys to girls in the class?

22. (43) **Verify** Instead of dividing $4\frac{1}{2}$ by $1\frac{1}{2}$, Carla doubled both numbers before dividing mentally. What was Carla's mental division problem and its quotient?

23. (30, 62) What is the reciprocal of $2\frac{1}{2}$?

24. (15) There are 1000 grams in 1 kilogram. How many grams are in 2.25 kilograms?

25. (7) How many **millimeters** long is the line below?

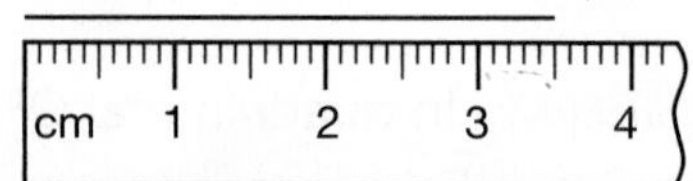

* **26.** (69) **Connect** The length of $\overline{WX}$ is 53 mm. The length of $\overline{XY}$ is 35 mm. What is the length of $\overline{WY}$?

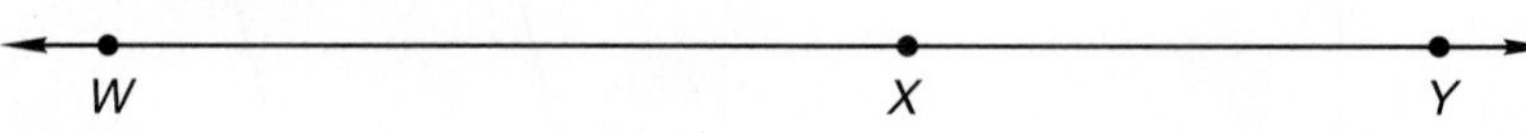

27. *(Inv. 6)* **Represent** Draw a cylinder.

28. *(50)* Arrange these numbers in order from least to greatest:

0.1, 1, −1, 0

29. *(33)* **Represent** Draw a circle and shade $\frac{1}{4}$ of it. What percent of the circle is shaded?

30. *(Inv. 6)* How many smaller cubes are in the large cube shown below?

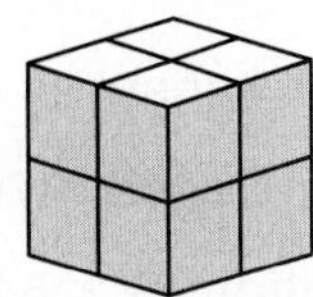

Early Finishers
Real-World Application

Taylor and her friends at school decided to make ribbons for their classmates to wear for spirit week. Taylor's mother offered to buy two rolls of ribbon. If each roll of ribbon is 25 yards in length and each ribbon is cut to $7\frac{1}{2}$ inches long, how many ribbons can Taylor and her friends make to give away at school? Show your work. Hint: 1 yard = 36 inches.

LESSON 70

Reducing Fractions Before Multiplying

Power Up

Building Power

facts

Power Up G

mental math

a. **Number Sense:** 5×280

b. **Number Sense:** $476 + 99$

c. **Calculation:** 3×54

d. **Calculation:** $4.50 + $1.75

e. **Fractional Parts:** $\frac{1}{3}$ of $90.00

f. **Number Sense:** $\frac{\$250}{10}$

g. **Geometry:** A square has a perimeter of 24 cm. What is the length of the sides of the square?

h. **Calculation:** 5×10, $\div 2$, $+ 5$, $\div 2$, $- 5$, $\div 10$, $- 1$

problem solving

The Crunch-O's cereal company makes two different cereal boxes. One is family size (12 in. high, 9 in. long, 2 in. wide) and the other is single-serving size (5 in. high, 3 in. long, 1 in. wide). Each of their boxes is made out of one piece of cardboard. To the right is a net of the family size box. Use this diagram to draw a net for the single-serving box.

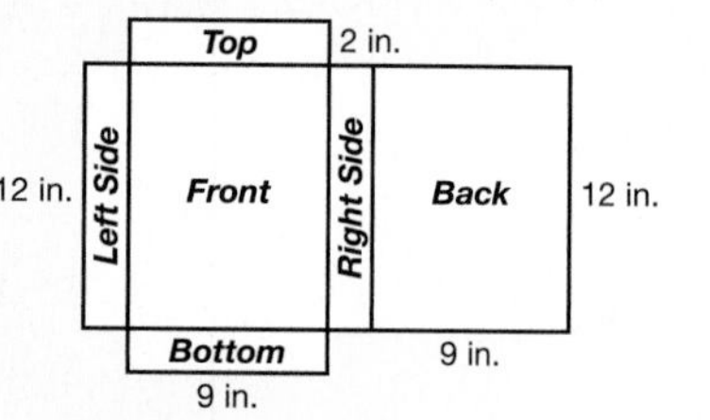

New Concept

Increasing Knowledge

Before two or more fractions are multiplied, we might be able to reduce the fraction terms, even if the reducing involves different fractions. For example, in the multiplication below we see that the number 3 appears as a numerator and as a denominator in different fractions.

$$\frac{\mathbf{3}}{5} \times \frac{2}{\mathbf{3}} = \frac{6}{15} \qquad \frac{6}{15} \text{ reduces to } \frac{2}{5}$$

We may reduce the common terms (the 3s) before multiplying. We reduce $\frac{3}{3}$ to $\frac{1}{1}$ by dividing both 3s by 3. Then we multiply the remaining terms.

$$\frac{\overset{1}{\cancel{3}}}{5} \times \frac{2}{\underset{1}{\cancel{3}}} = \frac{2}{5}$$

By reducing before we multiply, we avoid the need to reduce after we multiply. Reducing before multiplying is also known as **canceling.**

Thinking Skill

Connect

Sometimes we can reduce the numerators and the denominators of both fractions. Reduce the following:

$\frac{4}{9} \times \frac{3}{8}$

Example 1

Simplify: $\frac{5}{6} \times \frac{1}{5}$

Solution

We reduce before we multiply. Since 5 appears as a numerator and as a denominator, we reduce $\frac{5}{5}$ to $\frac{1}{1}$ by dividing both 5s by 5. Then we multiply the remaining terms.

$$\frac{\overset{1}{\cancel{5}}}{6} \times \frac{1}{\underset{1}{\cancel{5}}} = \mathbf{\frac{1}{6}}$$

Example 2

Simplify: $1\frac{1}{9} \times 1\frac{1}{5}$

Solution

First we write the numbers in fraction form.

$$\frac{10}{9} \times \frac{6}{5}$$

Thinking Skill

Discuss

Why might you want to reduce before you multiply?

We mentally pair 10 with 5 and 6 with 9.

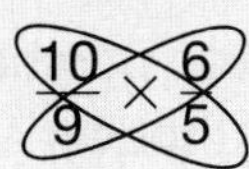

We reduce $\frac{10}{5}$ to $\frac{2}{1}$ by dividing both 10 and 5 by 5. We reduce $\frac{6}{9}$ to $\frac{2}{3}$ by dividing both 6 and 9 by 3.

$$\frac{\overset{2}{\cancel{10}}}{\underset{3}{\cancel{9}}} \times \frac{\overset{2}{\cancel{6}}}{\underset{1}{\cancel{5}}} = \frac{4}{3}$$

We multiply the remaining terms. Then we simplify the product.

$$\frac{4}{3} = \mathbf{1\frac{1}{3}}$$

Example 3

Simplify: $\frac{5}{6} \div \frac{5}{2}$

Solution

Thinking Skill

Justify

Explain how to divide any two fractions.

This is a division problem. We first find the number of $\frac{5}{2}$s in 1. The answer is the reciprocal of $\frac{5}{2}$. We then use the reciprocal of $\frac{5}{2}$ to find the number of $\frac{5}{2}$s in $\frac{5}{6}$.

$$1 \div \frac{5}{2} = \frac{2}{5}$$

$$\downarrow$$

$$\frac{5}{6} \times \frac{2}{5}$$

Now we have a multiplication problem. We cancel before we multiply.

$$\frac{\overset{1}{\cancel{5}}}{\underset{3}{\cancel{6}}} \times \frac{\overset{1}{\cancel{2}}}{\underset{1}{\cancel{5}}} = \mathbf{\frac{1}{3}}$$

Note: We may cancel the terms of fractions only when multiplying. A division problem must be rewritten as a multiplication problem before we may cancel the terms of the fractions. We do not cancel the terms of fractions in addition or subtraction problems.

Practice Set

Reduce before multiplying:

a. $\frac{3}{4} \cdot \frac{4}{5}$ **b.** $\frac{2}{3} \cdot \frac{3}{4}$ **c.** $\frac{8}{9} \cdot \frac{9}{10}$

Write in fraction form. Then reduce before multiplying.

d. $2\frac{1}{4} \times 4$ **e.** $1\frac{1}{2} \times 2\frac{2}{3}$ **f.** $3\frac{1}{3} \times 2\frac{1}{4}$

Rewrite each division problem as a multiplication problem. Then reduce before multiplying.

g. $\frac{2}{5} \div \frac{2}{3}$ **h.** $\frac{8}{9} \div \frac{2}{3}$ **i.** $\frac{9}{10} \div 1\frac{1}{5}$

Written Practice

Strengthening Concepts

1. (12) Alaska was purchased from Russia in 1867 for seven million, two hundred thousand dollars. Use digits to write that amount.

2. (54) **Connect** How many eighth notes equal a half note?

3. (43) **Verify** Instead of dividing $12\frac{1}{2}$ by $2\frac{1}{2}$, Shannon doubled both numbers and then divided. Write the division problem Shannon formed, as well as its quotient.

Reduce before multiplying:

***4.** (70) $\frac{5}{6} \cdot \frac{4}{5}$ ***5.** (70) $\frac{5}{6} \div \frac{5}{2}$ ***6.** (70) $\frac{9}{10} \cdot \frac{5}{6}$

7. (17) What number is halfway between $\frac{1}{2}$ and 1 on the number line?

8. (38) $\sqrt{100} + 10^2$ **9.** (59) $3\frac{2}{3} + 4\frac{5}{6}$

10. (63) $7\frac{1}{8} - 2\frac{1}{2}$ **11.** (38) $4.37 + 12.8 + 6$

12. (45) $0.46 \div 5$ **13.** (49) $60 \div 0.8$

14. (18) **Evaluate** What is the average of the three numbers marked by the arrows on this decimal number line? (First estimate whether the average will be more than 5 or less than 5.)

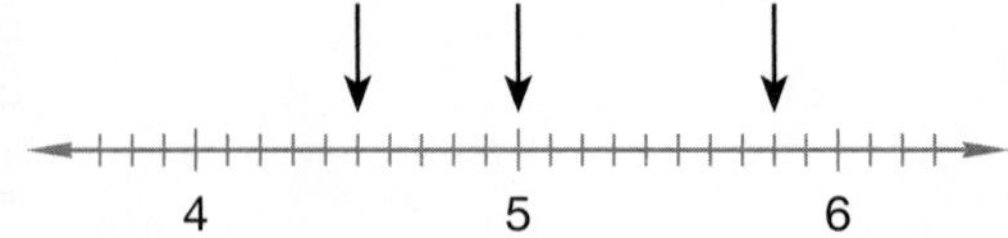

15. (49) **Verify** The division problem $1.5 \div 0.06$ is equivalent to which of the following?

A $15 \div 6$ **B** $150 \div 6$ **C** $150 \div 60$

16. (39) There are 1000 milliliters in 1 liter. How many milliliters are in 3.8 liters?

Find each unknown number:

17. (43) $\frac{2}{3} + n = 1$ **18.** (30) $\frac{2}{3}m = 1$ **19.** (56) $f - \frac{3}{4} = \frac{5}{6}$

20. (60, Inv. 6) A pyramid with a triangular base has how many

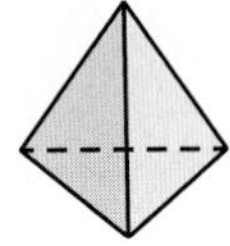

a. faces?

b. edges?

c. vertices?

Write the numbers in fraction form. Then reduce before multiplying.

*** 21.** (70) $1\frac{2}{3} \times 1\frac{1}{5}$ *** 22.** (70) $\frac{8}{9} \div 2\frac{2}{3}$

Refer to the line graph below to answer problems **23–25.**

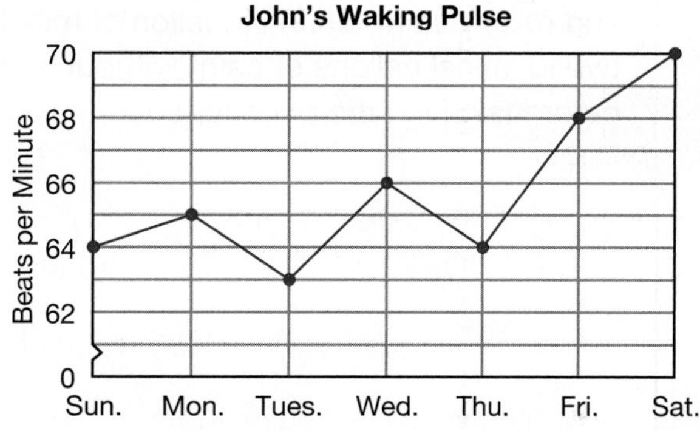

23. (18) When John woke on Saturday, his pulse was how many beats per minute more than it was on Tuesday?

24. (18) On Monday John took his pulse for 3 minutes before marking the graph. How many times did his heart beat in those 3 minutes?

25. (18) **Formulate** Write a question that relates to the graph and answer the question.

*** 26.** (67) **Analyze** Write the prime factorization of both the numerator and the denominator of this fraction. Then reduce the fraction.

$$\frac{72}{300}$$

Analyze In rectangle *ABCD* the length of $\overline{AB}$ is 2.5 cm, and the length of $\overline{BC}$ is 1.5 cm. Use this information and the figure below to answer problems **27–30.**

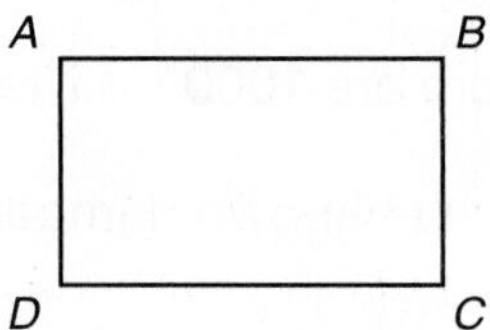

27. What is the perimeter of this rectangle?
(8)

28. What is the area of this rectangle?
(31)

* **29.** Name two segments perpendicular to $\overline{DC}$.
(64)

* **30.** If $\overline{BD}$ were drawn on the figure to divide the rectangle into two equal parts, what would be the area of each part?
(31)

Early Finishers
Real-World Application

Roland went to the local Super Store yesterday. He bought a new paint roller and roller pan for $8.97, a gallon of milk for $2.89, a magazine for $1.59, and two identical gallons of paint without marked prices. He paid a total of $47.83 before tax. Find the price for each gallon of paint.

Focus on

• The Coordinate Plane

By drawing two number lines perpendicular to each other and by extending the unit marks, we can create a grid called a **coordinate plane.**

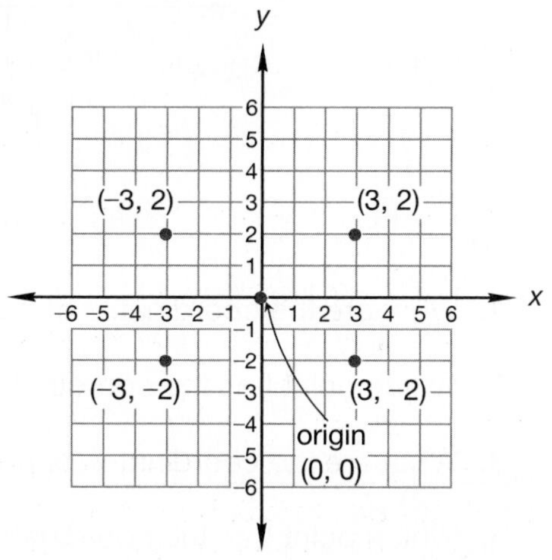

The point at which the number lines intersect is called the **origin.** The horizontal number line is called the ***x*-axis,** and the vertical number line is called the ***y*-axis.** We **graph** a point by marking a dot at the location of the point. We can name the location of any point on this coordinate plane with two numbers. The numbers that tell the location of a point are called the **coordinates** of the point.

Thinking Skill

Explain

On the coordinate plane, where will a point whose ordered pair contains two negative numbers be located?

The coordinates of a point are written as an **ordered pair** of numbers in parentheses; for example, (3, −2). The first number is the *x*-coordinate. It shows the horizontal (↔) direction and distance from the origin. The second number, the *y*-coordinate, shows the vertical (↕) direction and distance from the origin. The sign of the coordinate shows the direction. Positive coordinates are to the right or up, and negative coordinates are to the left or down.

Look at the coordinate plane above. To graph (3, −2), we begin at the origin and move three units to the right along the *x*-axis. From there we move down two units and mark a dot. We may label the point we graphed (3, −2).

On the coordinate plane, we also have graphed three other points and identified their coordinates. Notice that each pair of coordinates is different and designates a unique point:

(3, −2)

(3, 2)

(−3, 2)

(−3, −2)

Thinking Skill

Conclude

If you connected the points in alphabetical order (start with *AB* and end with *HA*), what type of polygon would you make?

Refer to the coordinate plane below to answer problems **1–6.**

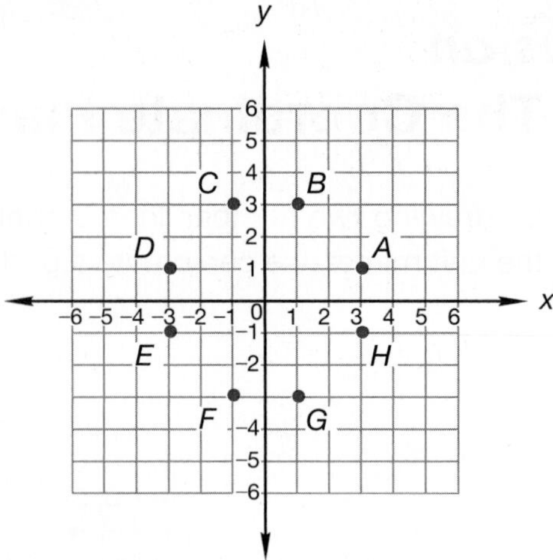

1. What are the coordinates of point *A*?
2. Which point has the coordinates (−1, 3)?
3. What are the coordinates of point *E*?
4. Which point has the coordinates (1, −3)?
5. What are the coordinates of point *D*?
6. Which point has the coordinates (3, −1)?

Visit www.SaxonPublishers.com/ActivitiesC1 *for a graphing calculator activity.*

The coordinate plane is useful in many fields of mathematics, including algebra and geometry.

In the next section of this investigation we will designate points on the plane as vertices of rectangles. Then we will calculate the perimeter and area of each rectangle.

Suppose we are told that the vertices of a rectangle are located at (3, 2), (−1, 2), (−1, −1), and (3, −1). We graph the points and then draw segments between the points to draw the rectangle.

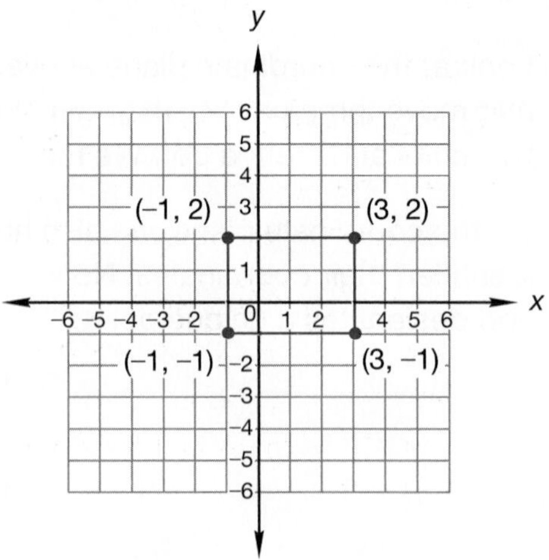

We see that the rectangle is four units long and three units wide. Adding the lengths of the four sides, we find that the perimeter is **14 units.** To find the area, we can count the unit squares within the rectangle. There are three rows of four squares, so the area of the rectangle is 3×4, which is **12 square units.**

Use graph paper or **Investigation Activity 15** to create a coordinate plane. Use the coordinate plane for the exercises that follow.

7. *Represent* The vertices of a rectangle are located at $(-2, -1)$, $(2, -1)$, $(2, 3)$, and $(-2, 3)$.

a. Graph the rectangle. What do we call this special type of rectangle?

b. What is the perimeter of the rectangle?

c. What is the area of the rectangle?

8. *Represent* The vertices of a rectangle are located at $(-4, 2)$, $(0, 2)$, $(0, 0)$, and $(-4, 0)$.

a. Graph the rectangle. Notice that one vertex is located at $(0, 0)$. What is the name for this point on the coordinate plane?

b. What is the perimeter of the rectangle?

c. What is the area of the rectangle?

9. Three vertices of a rectangle are located at $(3, 1)$, $(-2, 1)$, and $(-2, -3)$.

a. Graph the rectangle. What are the coordinates of the fourth vertex?

b. What is the perimeter of the rectangle?

c. What is the area of the rectangle?

As the following activity illustrates, we can use coordinates to give directions for making a drawing.

Activity

Drawing on the Coordinate Plane

Materials needed:

- 4 copies of **Investigation Activity 15**

10. *Verify* Christy made a drawing on a coordinate plane as shown on the next page. Then she wrote directions for making the drawing. Follow Christy's directions to make a similar drawing on your coordinate plane. The coordinates of the vertices are listed in order, as in a "dot-to-dot" drawing.

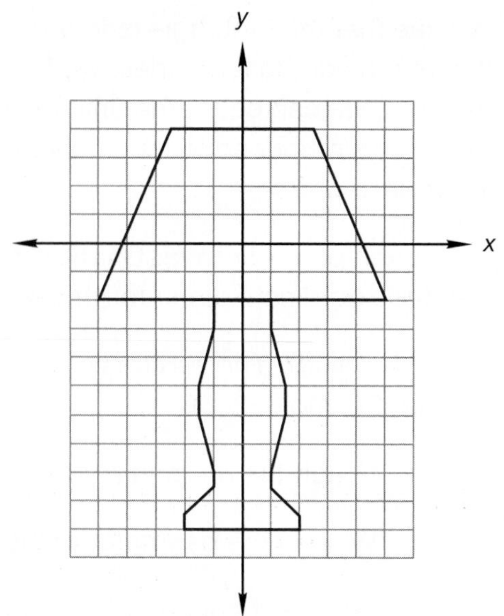

Christy's Directions

On your coordinate plane, draw segments to connect the following points in order:

a. $(-1, -2)$ **b.** $(-1, -3)$ **c.** $(-1\frac{1}{2}, -5)$ **d.** $(-1\frac{1}{2}, -6)$

e. $(-1, -8)$ **f.** $(-1, -8\frac{1}{2})$ **g.** $(-2, -9\frac{1}{2})$ **h.** $(-2, -10)$

i. $(2, -10)$ **j.** $(2, -9\frac{1}{2})$ **k.** $(1, -8\frac{1}{2})$ **l.** $(1, -8)$

m. $(1\frac{1}{2}, -6)$ **n.** $(1\frac{1}{2}, -5)$ **o.** $(1, -3)$ **p.** $(1, -2)$

Lift your pencil and restart:

a. $(-2\frac{1}{2}, 4)$ **b.** $(2\frac{1}{2}, 4)$ **c.** $(5, -2)$

d. $(-5, -2)$ **e.** $(-2\frac{1}{2}, 4)$

11. *Conclude* Carlos wrote the following directions for a drawing. Follow his directions to make the drawing on your own paper. Draw segments to connect the following points in order:

a. $(-9, 0)$ **b.** $(6, -1)$ **c.** $(8, 0)$

d. $(7, 1)$ **e.** $(6, \frac{1}{2})$ **f.** $(6, -1)$

g. $(9, -2\frac{1}{2})$ **h.** $(10, -2)$ **i.** $(7, 1)$

j. $(6, 1\frac{1}{2})$ **k.** $(-10\frac{1}{2}, 3)$ **l.** $(-11, 2)$

m. $(-10\frac{1}{2}, 0)$ **n.** $(-10, -1\frac{1}{2})$ **o.** $(9, -2\frac{1}{2})$

p. $(-3, -3\frac{1}{2})$ **q.** $(-7, -8)$ **r.** $(-10, -8)$

s. $(-9, -1\frac{1}{2})$

Lift your pencil and restart:

a. $(-10\frac{1}{2}, 0)$	**b.** $(-11, -\frac{1}{2})$	**c.** $(-12, \frac{1}{2})$
d. $(-11\frac{1}{2}, 1)$	**e.** $(-12, 1\frac{1}{2})$	**f.** $(-11\frac{1}{2}, 2)$
g. $(-12, 2\frac{1}{2})$	**h.** $(-11, 3\frac{1}{2})$	**i.** $(-10\frac{1}{2}, 3)$
j. $(-11\frac{1}{2}, 8)$	**k.** $(-9\frac{1}{2}, 8)$	**l.** $(-7, 3)$
m. $(-6, 2\frac{1}{2})$	**n.** $(-7, 3)$	**o.** $(-6, 5)$
p. $(-4, 5)$	**q.** $(-1, 2)$	

12. Model On a coordinate plane, make a straight-segment drawing. Then write directions for making the drawing by listing the coordinates of the vertices in "dot-to-dot" order. Trade directions with another student, and try to make each other's drawings.

extensions

a. Represent Use whole numbers, fractions, and mixed numbers to write the coordinates for each point.

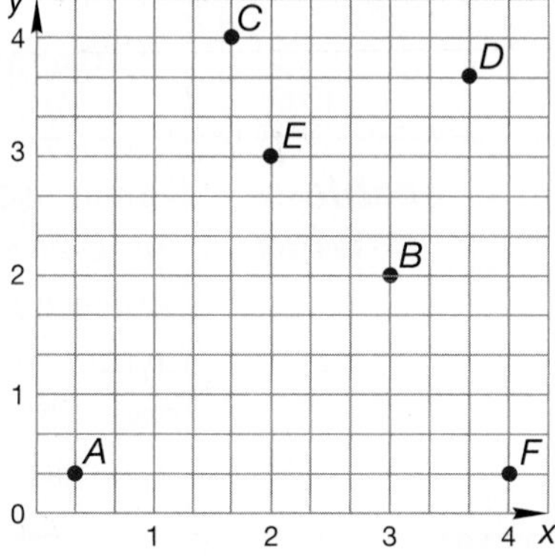

b. Represent Use the given coordinates to identify the point at each location.

(6, 6) (4.5, 3) (2.5, 1.5)
(3.5, 0) (3, 4.5) (0.5, 5.5)

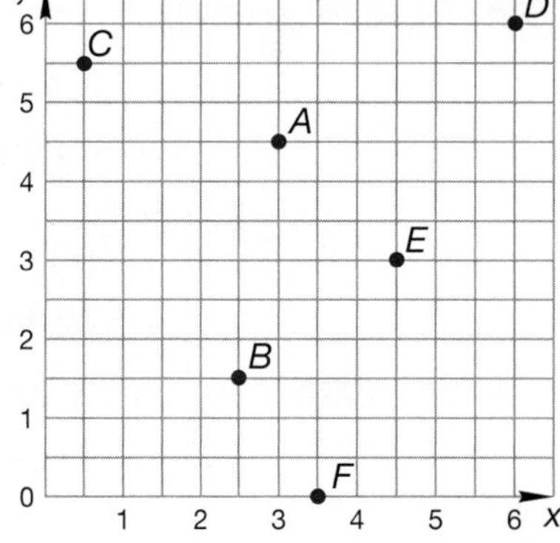

c. Generalize Graph these points on a coordinate graph. Then connect the points.

(4, 2), (6, 2), (2, 4), (8, 4), (4, 6), (6, 6)

- What polygon did you form?
- Name a set of points that when connected would form a hexagon inside the hexagon you drew.

LESSON 71

• Parallelograms

Power Up

Building Power

facts Power Up D

mental math

a. **Number Sense:** 5×480

b. **Number Sense:** $367 - 99$

c. **Calculation:** 8×43

d. **Calculation:** $10.00 − $8.75

e. **Number Sense:** Double $2.25.

f. **Number Sense:** $\frac{\$250}{100}$

g. **Geometry:** A square has an area of 25 in.^{2}. What is the length of the sides of the square?

h. **Calculation:** 8×9, $+ 3$, $\div 3$, $\times 2$, $- 10$, $\div 5$, $+ 3$, $\div 11$

problem solving

Griffin used 14 blocks to build this three-layer pyramid. How many blocks would he need to build a six-layer pyramid? How many blocks would he need for the bottom layer of a nine-layer pyramid?

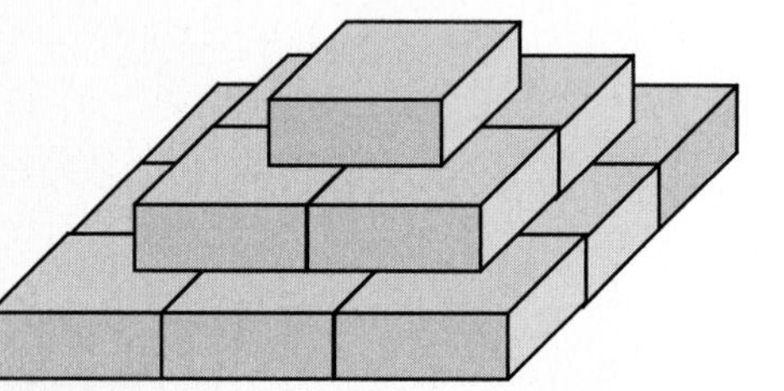

New Concept

Increasing Knowledge

In this lesson we will learn about various properties of parallelograms. The following example describes some angle properties of parallelograms.

Example 1

In parallelogram *ABCD,* the measure of angle *A* is 60°.

a. What is the measure of ∠*C*?

b. What is the measure of ∠*B*?

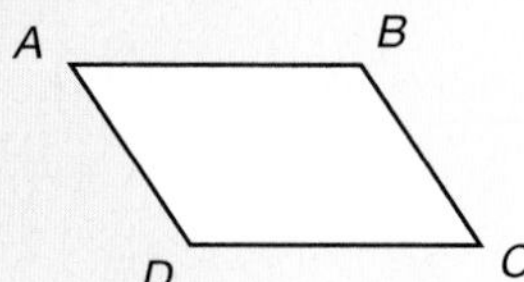

Reading Math

Give two other ways to name ∠*C*.

Solution

a. Angles *A* and *C* are opposite angles in that they are opposite to each other in the parallelogram. The opposite angles of a parallelogram have equal measures. So the measure of angle *C* equals the measure of angle *A*. Thus the measure of ∠*C* is **60°.**

Math Language

Remember that **supplementary angles** have a sum of 180°.

b. Angles A and B are adjacent angles in that they share a side. (Side AB is a side of $\angle A$ and a side of $\angle B$.) The adjacent angles of a parallelogram are supplementary. So $\angle A$ and $\angle B$ are supplementary, which means their measures total 180°. Since $\angle A$ measures 60°, $\angle B$ must measure **120°** for their sum to be 180°.

Model A flexible model of a parallelogram is useful for illustrating some properties of a parallelogram. A model can be constructed of brads and stiff tagboard or cardboard.

Lay two 8-in. strips of tagboard or cardboard over two parallel 10-in. strips as shown. Punch a hole at the center of the overlapping ends. Then fasten the corners with brads to hold the strips together.

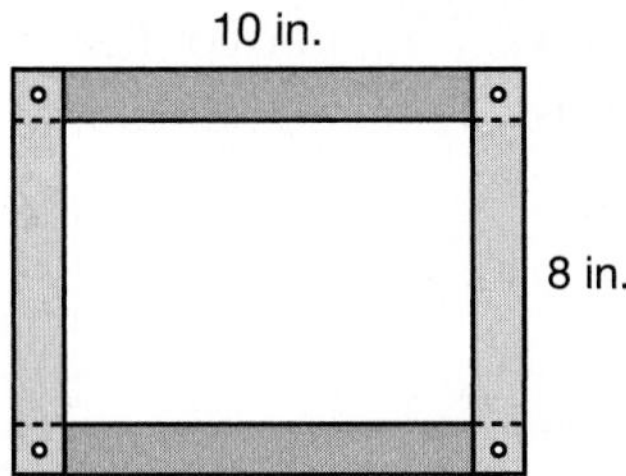

If we move the sides of the parallelogram back and forth, we see that opposite sides always remain parallel and equal in length. Though the angles change size, opposite angles remain equal and adjacent angles remain supplementary.

With this model we also can observe how the area of a parallelogram changes as the angles change. We hold the model with two hands and slide opposite sides in opposite directions. The maximum area occurs when the angles are 90°. The area reduces to zero as opposite sides come together.

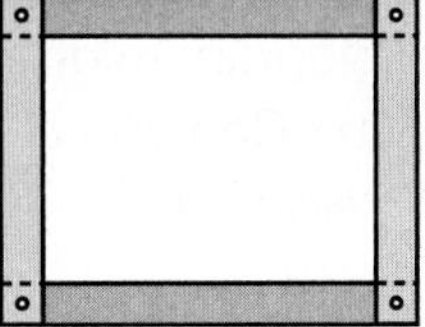
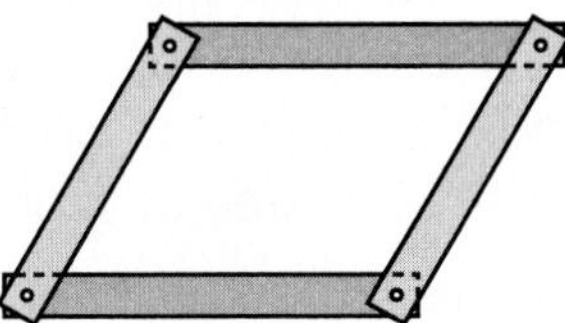
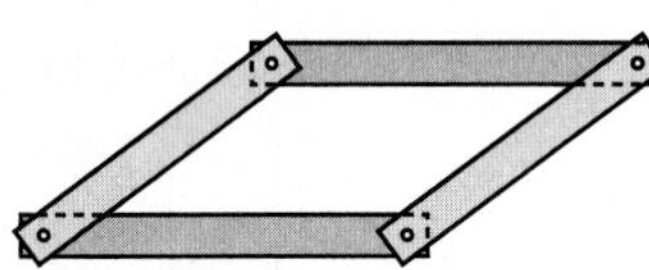

Discuss The area of a parallelogram changes as the angles change. Does the perimeter change?

The flexible model shows that parallelograms may have sides that are equal in length but areas that are different. To find the area of a parallelogram, we multiply two **perpendicular** measurements. We multiply the **base** by the **height** of the parallelogram.

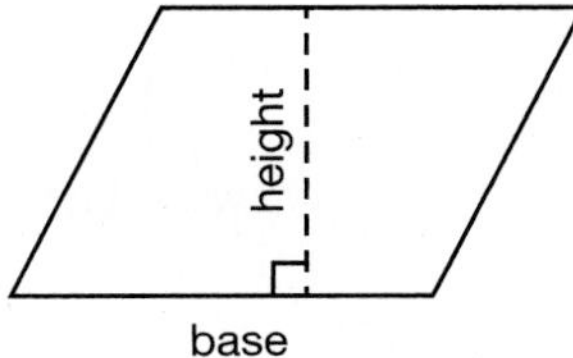

The base of a parallelogram is the length of one of the sides. The height of a parallelogram is the perpendicular distance from the base to the opposite side. The following activity will illustrate why the area of a parallelogram equals the base times the height.

Activity

Area of a Parallelogram

Materials needed:

- graph paper
- ruler
- pencil
- scissors

Represent Tracing over the lines on the graph paper, draw two parallel segments the same number of units long but shifted slightly as shown.

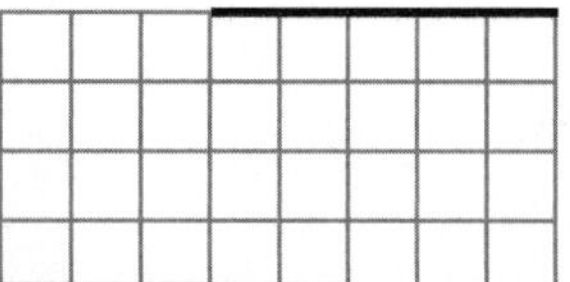

Then draw segments between the endpoints of the pair of parallel segments to complete the parallelogram.

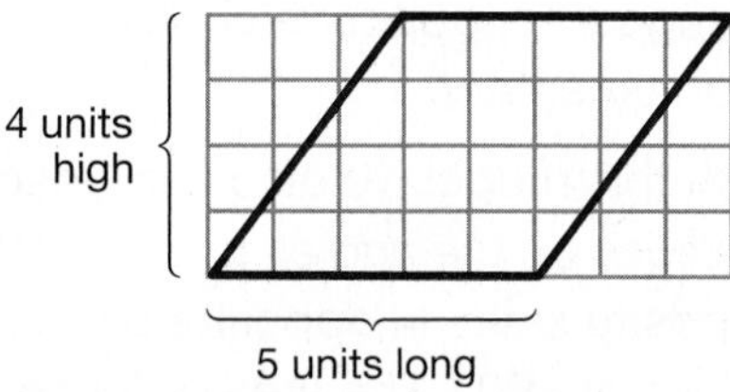

The base of the parallelogram we drew has a length of 5 units. The height of the parallelogram is 4 units. Your parallelogram might be different. How many units long and high is your parallelogram? Can you easily count the number of square units in the area of your parallelogram?

Model Use scissors to cut out your parallelogram.

Then select a line on the graph paper that is perpendicular to the first pair of parallel sides that you drew. Cut the parallelogram into two pieces along this line.

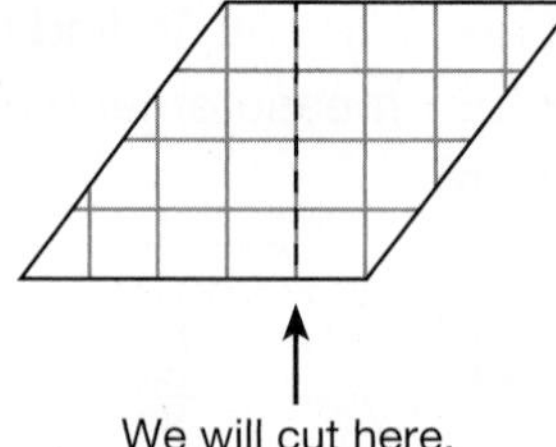

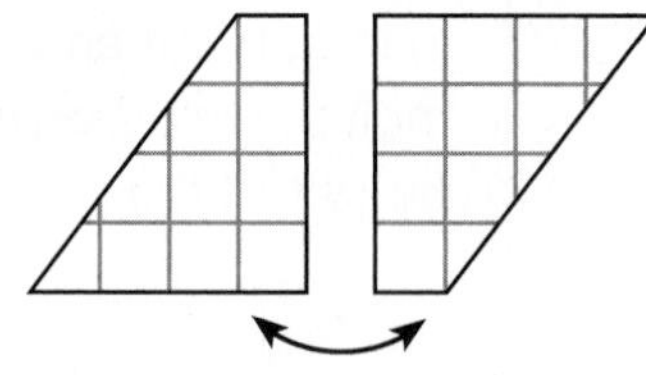

Rearrange the two pieces of the parallelogram to form a rectangle. What is the length and width of the rectangle? How many square units is the area of the rectangle?

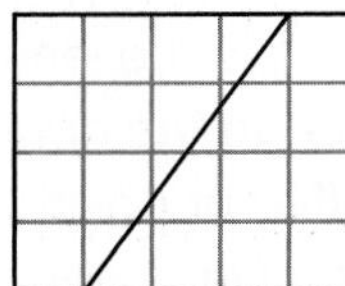

Our rectangle is 5 units long and 4 units wide. The area of the rectangle is 20 square units. So the area of the parallelogram is also 20 square units.

By making a perpendicular cut across the parallelogram and rearranging the pieces, we formed a rectangle having the same area as the parallelogram. The length and width of the rectangle equaled the base and height of the parallelogram. Therefore, by multiplying the base and height of a parallelogram, we can find its area.

Example 2

Find the area of this parallelogram:

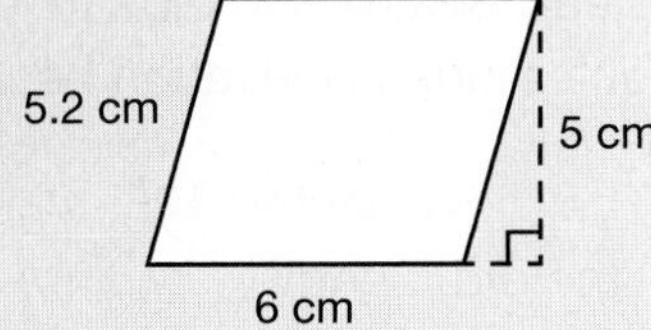

Solution

We multiply two perpendicular measurements, the base and the height. The height is often shown as a dashed line segment. The base is 6 cm. The height is 5 cm.

$$6 \text{ cm} \times 5 \text{ cm} = 30 \text{ sq. cm}$$

The area of the parallelogram is **30 sq. cm.**

Practice Set

Conclude Refer to parallelogram *QRST* to answer problems **a–d.**

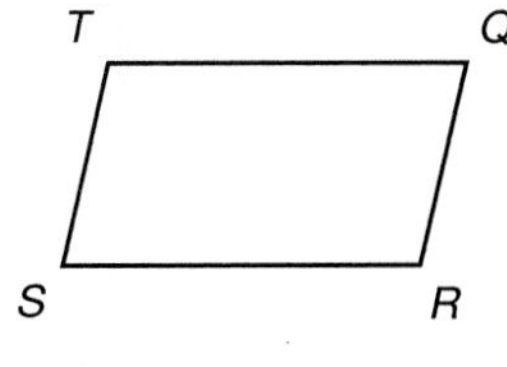

a. Which angle is opposite $\angle Q$?

b. Which angle is opposite $\angle T$?

c. Name two angles that are supplements of $\angle T$.

d. If the measure of $\angle R$ is 100°, what is the measure of $\angle Q$?

Calculate the perimeter and area of each parallelogram:

e.

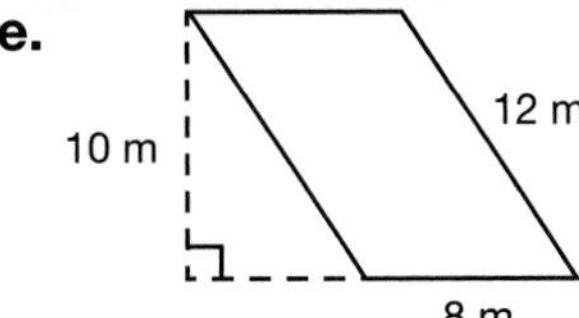

f.

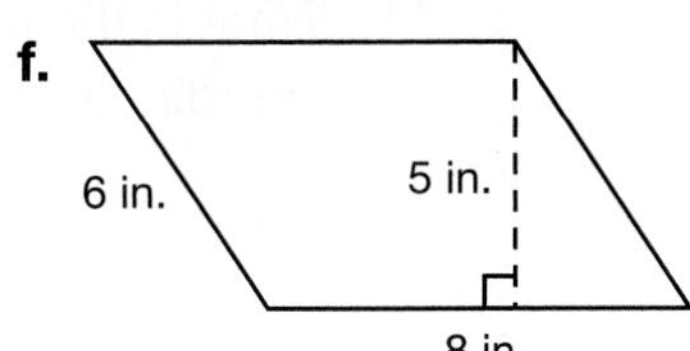

g. **Analyze** A formula for finding the area of a parallelogram is $A = bh$. This formula means

$$\text{Area} = \text{base} \times \text{height}$$

The base is the length of one side. The height is the perpendicular distance to the opposite side. Here we show the same parallelogram in two different positions, so the area of the parallelogram is the same in both drawings. What is the height in the figure on the right?

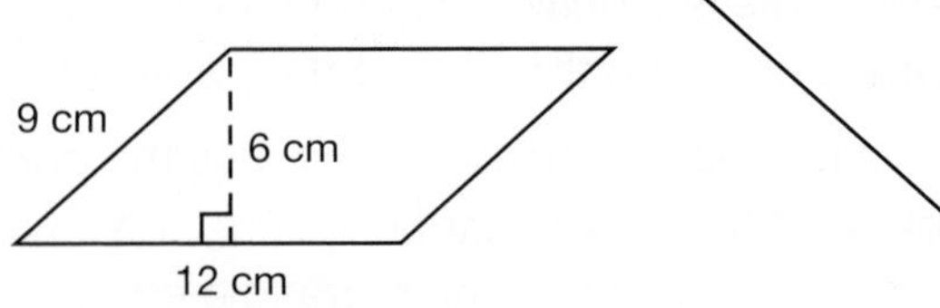

Written Practice

Strengthening Concepts

1. (30) What is the least common multiple of 6 and 10?

2. (14) **Analyze** The highest point on land is Mt. Everest, whose peak is 29,035 feet above sea level. The lowest point on land is the Dead Sea, which dips to 1371 feet below sea level. What is the difference in elevation between these two points?

3. (32) The movie lasted 105 minutes. If the movie started at 1:15 p.m., at what time did it end?

Analyze In problems **4–7,** reduce the fractions, if possible, before multiplying.

* **4.** (70) $\frac{2}{3} \cdot \frac{3}{8}$

* **5.** (70) $1\frac{1}{4} \cdot 2\frac{2}{3}$

* **6.** (70) $\frac{3}{4} \div \frac{3}{8}$

* **7.** (70) $4\frac{1}{2} \div 6$

8. (59) $6 + 3\frac{3}{4} + 2\frac{1}{2}$

9. (63) $5 - 3\frac{1}{8}$

10. (63) $5\frac{1}{4} - 1\frac{7}{8}$

11. (39) $(3.5)^2$

12. (2) $15\overline{)\$75.00}$

13. (53) $(1 + 0.6) \div (1 - 0.6)$

14. (41) Quan ordered a \$4.50 bowl of soup. The tax rate was $7\frac{1}{2}\%$ (which equals 0.075). He paid for the soup with a \$20 bill.

a. What was the tax on the bowl of soup?

b. What was the total price including tax?

c. How much money should Quan get back from his payment?

* **15.** (Inv. 7) What is the name for the point on the coordinate plane that has the coordinates (0, 0)?

* **16.** (Inv. 7) **Represent** Refer to the coordinate plane below to locate the points indicated.

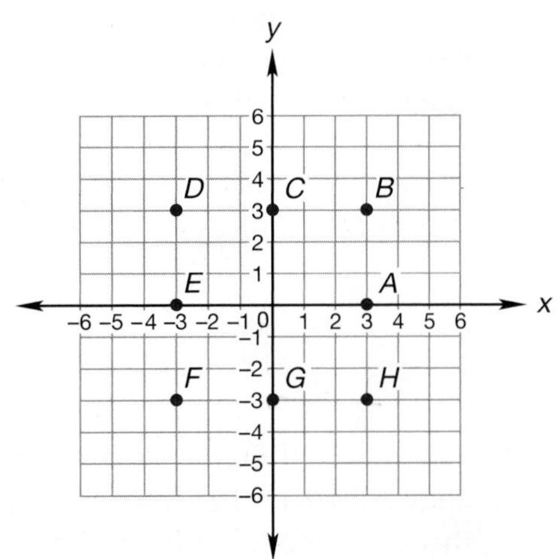

Name the points that have the following coordinates:

a. (−3, 3) **b.** (0, −3)

Identify the coordinates of the following points:

c. *H* **d.** *E*

Find each unknown number:

17. (49) $1.2f = 120$

18. (49) $\frac{120}{f} = 1.2$

* **19.** (67) Write the prime factorization of both the numerator and the denominator of this fraction. Then reduce the fraction.

$$\frac{64}{224}$$

20. (38) The perimeter of a square is 6.4 meters. What is its area?

21. (Inv. 2) **Analyze** What fraction of this circle is not shaded?

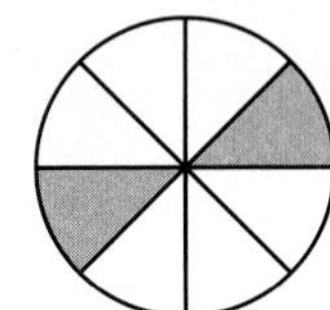

22. (47) **Explain** If the radius of this circle is 1 cm, what is the circumference of the circle? (Use 3.14 for π.) How did you find your answer?

23. (7) **Estimate** A centimeter is about as long as this segment:

———

About how many centimeters long is your little finger?

24. (10) **Connect** Water freezes at 32° Fahrenheit. The temperature shown on the thermometer is how many degrees Fahrenheit above the freezing point of water?

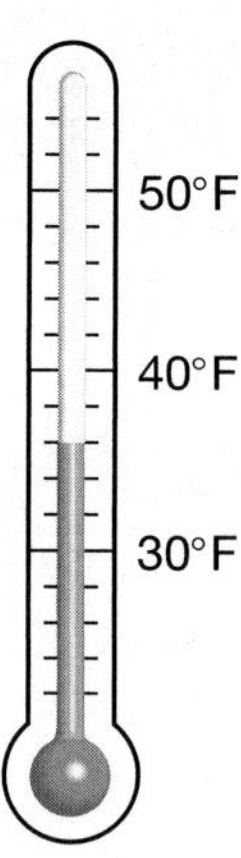

25. (29, 33) Ray watched TV for one hour. He determined that commercials were shown 20% of that hour. Write 20% as a reduced fraction. Then find the number of minutes that commercials were shown during the hour.

26. (Inv. 6) Name the geometric solid shown at right.

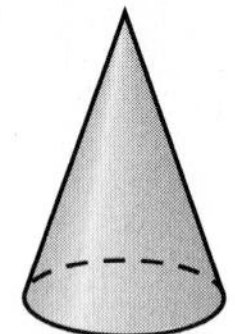

* **27.** (60) **Analyze** This square and regular triangle share a common side. The perimeter of the square is 24 cm. What is the perimeter of the triangle?

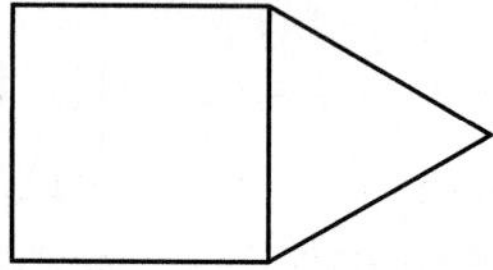

28. (31) Choose the appropriate unit for the area of your state.

A square inches **B** square yards **C** square miles

* **29.** (71) **a.** What is the perimeter of this parallelogram?

b. What is the area of this parallelogram?

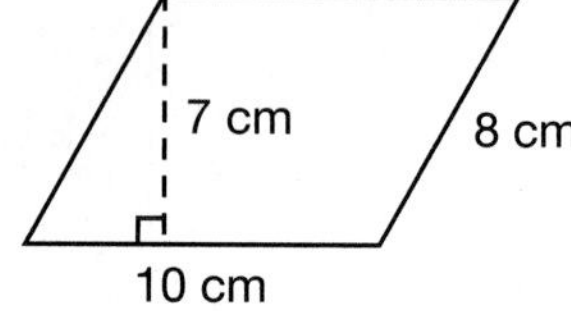

* **30.** (69) **Conclude** In this figure $\angle BMD$ is a right angle. Name two angles that are

a. supplementary.

b. complementary.

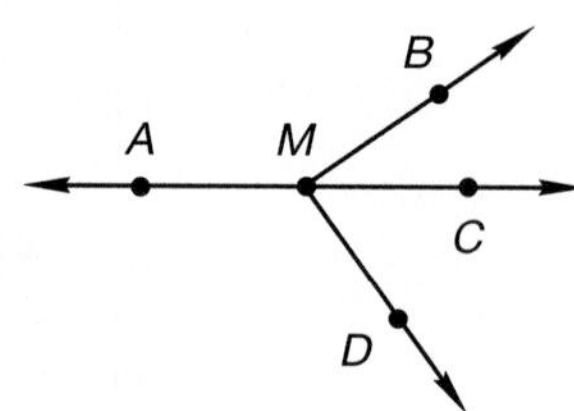

LESSON **72**

• Fractions Chart
• Multiplying Three Fractions

Power Up — *Building Power*

facts Power Up H

mental math

a. **Number Sense:** 3 × 125

b. **Number Sense:** 275 + 50

c. **Number Sense:** 3 × \$0.99

d. **Calculation:** \$20.00 − \$9.99

e. **Fractional Parts:** $\frac{1}{3}$ of \$6.60

f. **Decimals:** \$2.50 × 10

g. **Statistics:** Find the average 45, 33, and 60.

h. **Calculation:** 2 × 2, × 2, × 2, × 2, − 2, ÷ 2

problem solving Kioko was thinking of two numbers whose average was 24. If one of the numbers was half of 24, what was the other number?

New Concepts — *Increasing Knowledge*

fractions chart

We have learned three steps to take when performing pencil-and-paper arithmetic with fractions and mixed numbers:

Step 1: Write the problem in the correct **shape.**

Step 2: Perform the **operation.**

Step 3: **Simplify** the answer.

The letters S.O.S. can help us remember the steps as "shape," "operate," and "simplify." We summarize the S.O.S. rules we have learned in the following fractions chart.

Fractions Chart

	+ −	×	÷
1. Shape	Write fractions with common denominators.	Write numbers in fraction form.	
2. Operate	Add or subtract the numerators.	× Cancel.	÷ Find reciprocal of divisor, then cancel.
		Multiply numerators. Multiply denominators.	
3. Simplify	Reduce fractions. Convert improper fractions.		

- Below the + and − symbols we list the steps for adding or subtracting fractions.
- Below the × and ÷ symbols, we list the steps for multiplying or dividing fractions.

The "shape" step for addition and subtraction is the same; we write the fractions with common denominators. Likewise, the "shape" step for multiplication and division is the same; we write both numbers in fraction form.

Math Language
Recall that **canceling** means reducing before multiplying.

At the "operate" step, however, we separate multiplication and division. When multiplying fractions, we may reduce (cancel) before we multiply. Then we multiply the numerators to find the numerator of the product, and we multiply the denominators to find the denominator of the product. When dividing fractions, we first replace the divisor of the division problem with its reciprocal and change the division problem to a multiplication problem. We cancel terms, if possible, and then multiply.

The "simplify" step is the same for all four operations. We reduce answers when possible and convert answers that are improper fractions to mixed numbers.

multiplying three fractions

To multiply three or more fractions, we follow the same steps we take when multiplying two fractions:

Step 1: We write the numbers in fraction form.

Step 2: We cancel terms by reducing numerator-denominator pairs that have common factors. Then we multiply the remaining terms.

Step 3: We simplify if possible.

Example

Multiply: $\frac{2}{3} \times 1\frac{3}{5} \times \frac{3}{4}$

Solution

First we write $1\frac{3}{5}$ as the improper fraction $\frac{8}{5}$. Then we reduce where possible before multiplying. Multiplying the remaining terms, we find the product.

$$\frac{2}{\cancel{3}_1} \times \frac{\cancel{8}^2}{5} \times \frac{\cancel{3}^1}{\cancel{4}_1} = \mathbf{\frac{4}{5}}$$

Practice Set

a. Draw the fractions chart from this lesson.

b. Describe the three steps for adding fractions.

c. Describe the steps for dividing fractions.

Multiply:

d. $\frac{2}{3} \cdot \frac{4}{5} \cdot \frac{3}{8}$

e. $2\frac{1}{2} \times 1\frac{1}{10} \times 4$

Written Practice *Strengthening Concepts*

1. (18) What is the average of 4.2, 2.61, and 3.6?

2. (54) **Connect** Four tablespoons equals $\frac{1}{4}$ cup. How many tablespoons would equal one full cup?

3. (14) The temperature on the moon ranges from a high of about 130°C to a low of about -110°C. This is a difference of how many degrees?

4. (58) Four of the 12 marbles in the bag are blue. If one marble is taken from the bag, what is the probability that the marble is

a. blue? **b.** not blue?

c. What word names the relationship between the events in **a** and **b?**

5. (7, 47) The diameter of a circle is 1 meter. The circumference is how many centimeters? (Use 3.14 for π.)

6. (29) **Connect** What fraction of a dollar is a nickel?

Find each unknown number:

7. (56) $n - \frac{1}{2} = \frac{3}{5}$

8. (43) $1 - w = \frac{7}{12}$

9. (59) $w + 2\frac{1}{2} = 3\frac{1}{3}$

10. (43) $1 - w = 0.23$

11. (46) Write the standard decimal number for the following:

$$(6 \times 10) + \left(4 \times \frac{1}{10}\right) + \left(3 \times \frac{1}{100}\right)$$

12. (50) **Estimate** Which of these numbers is closest to 1?

A -1 **B** 0.1 **C** 10

13. (19) What is the largest prime number that is less than 100?

* **14** (64) **Classify** Which of these figures is not a parallelogram?

A

C 

B 

D 

15. (38) **Connect** A loop of string two feet around is formed to make a square.

a. How many inches long is each side of the square?

b. What is the area of the square in square inches?

*** 16.** (69) **Conclude** Figure *ABCD* is a rectangle.

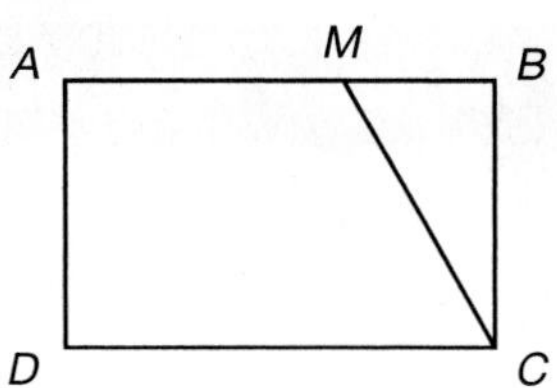

a. Name an angle complementary to $\angle DCM$.

b. Name an angle supplementary to $\angle AMC$.

Refer to this menu and the information that follows to answer problems **17–19.**

Menu

Grilled Chicken Sandwich	$3.49	Juice:	Small	$0.89
Green Salad	$3.29		Medium	$1.09
Pasta Salad	$2.89		Large	$1.29

From this menu the Johnsons ordered two grilled chicken sandwiches, one green salad, one small juice, and two medium juices.

17. (1) What was the total price of the Johnsons' order?

18. (41) If 7% tax is added to the bill, and if the Johnsons pay for the food with a $20 bill, how much money should they get back?

19. (1) **Formulate** Make up an order from the menu. Then calculate the bill, not including tax.

20. (47) If $A = lw$, and if l equals 2.5 and w equals 0.4, what does A equal?

*** 21.** (66) Write the prime factorization of both the numerator and the denominator of this fraction. Then reduce the fraction.

$$\frac{72}{120}$$

Refer to the coordinate plane below to answer problems **22** and **23.**

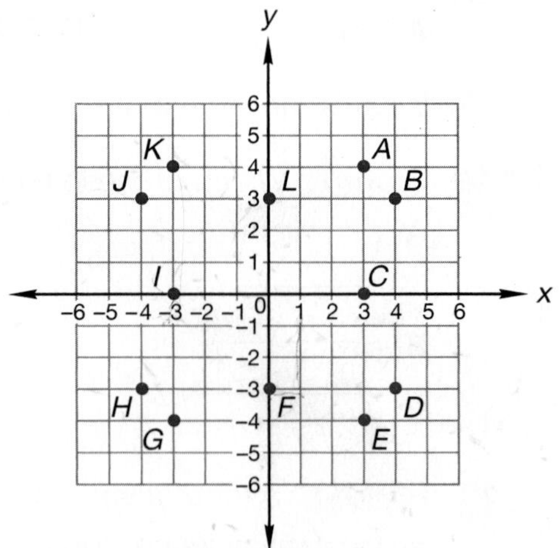

*** 22.** (Inv. 7) Identify the coordinates of the following points:

a. *K* **b.** *F*

*** 23.** (Inv. 7) Name the points that have the following coordinates:

a. (3, −4) **b.** (−3, 0)

*** 24.** (64) **Model** Draw a pair of parallel lines. Then draw a second pair of parallel lines perpendicular to the first pair of lines and about the same distance apart. Trace the quadrilateral that is formed by the intersecting lines. Is the quadrilateral a rectangle?

*** 25.** (72) $\frac{1}{2} \cdot \frac{5}{6} \cdot \frac{3}{5}$

*** 26.** (72) $3 \times 1\frac{1}{2} \times 2\frac{2}{3}$

27. (54) $\frac{3}{4} \div 2$

*** 28.** (68) $1\frac{1}{2} \div 1\frac{2}{3}$

29. (39) (0.12)(0.24)

30. (49) $0.6 \div 0.25$

Early Finishers
Real-World Application

LaDonna had errands to run and decided to park her car in front of a parking meter rather than drive from store to store. She calculated that she would spend about 20 minutes in the post office and 10 minutes at the hardware store. Then she would spend 5 minutes picking up her clothes from the cleaner's and another 30 minutes eating lunch.

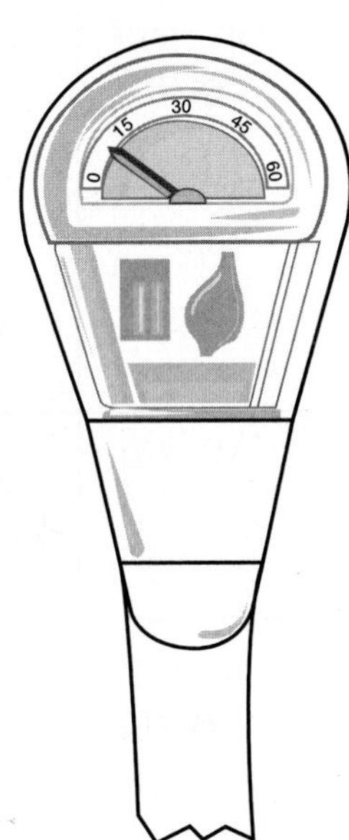

The sign on the meter read

$0.25 = 15 minutes

$0.10 = 6 minutes

$0.05 = 3 minutes

a. How much time will LaDonna spend to finish doing her errands?

b. If the meter has ten minutes left, how much money will she need to put into the meter?

LESSON 73

• Exponents
• Writing Decimal Numbers as Fractions, Part 2

Power Up

Building Power

facts Power Up J

mental math

a. **Calculation:** 4×112

b. **Number Sense:** $475 - 150$

c. **Calculation:** $4 \times \$0.99$

d. **Calculation:** $\$2.99 + \$1.99$

e. **Number Sense:** Double $3.50.

f. **Decimals:** $\$3.50 \div 10$

g. **Statistics:** Find the median of the set of numbers: 30, 61, 22, 46, 13

h. **Calculation:** 3×3, $\times 3$, $+ 3$, $\div 3$, $- 3$, $\times 3$

problem solving

How many different ways can Gunther spin a total of 6 if he spins each spinner once?

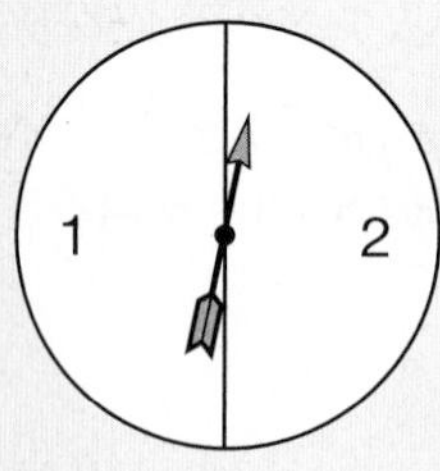

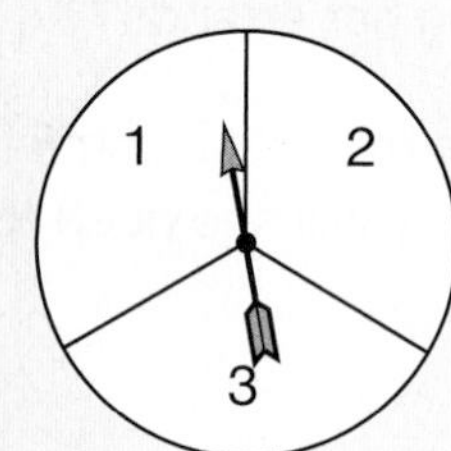

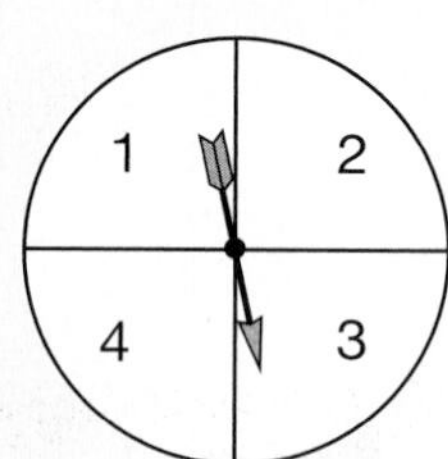

New Concepts

Increasing Knowledge

exponents

Math Language

Recall that an **exponent** is written as a small number on the upper right-hand side of the base number.

Since Lesson 38 we have used the exponent 2 to indicate that a number is multiplied by itself.

$$5^2 \text{ means } 5 \cdot 5$$

Exponents indicate repeated multiplication, so

$$5^3 \text{ means } 5 \cdot 5 \cdot 5$$

$$5^4 \text{ means } 5 \cdot 5 \cdot 5 \cdot 5$$

The exponent indicates how many times the base is used as a factor.

We read numbers with exponents as **powers.** Note that when the exponent is 2, we usually say "squared," and when the exponent is 3, we usually say "cubed." The following examples show how we read expressions with exponents:

Thinking Skill

Explain

How does the quantity *two to the third power* differ from the quantity *three to the second power?*

5^2	"five to the second power" or "five squared"
10^3	"ten to the third power" or "ten cubed"
3^4	"three to the fourth power"
2^5	"two to the fifth power"

Example 1

Compare: $3^4 \bigcirc 4^3$

Solution

We find the value of each expression.

3^4 means $3 \cdot 3 \cdot 3 \cdot 3$, which equals 81.

4^3 means $4 \cdot 4 \cdot 4$, which equals 64.

Since 81 is greater than 64, we find that 3^4 is greater than 4^3.

$$\mathbf{3^4 > 4^3}$$

Example 2

Write the prime factorization of 1000, using exponents to group factors.

Solution

Using a factor tree or division by primes, we find the prime factorization of 1000.

$$1000 = 2 \cdot 2 \cdot 2 \cdot 5 \cdot 5 \cdot 5$$

We group the three 2s and the three 5s with exponents.

$$1000 = \mathbf{2^3 \cdot 5^3}$$

Example 3

Simplify: $100 - 10^2$

Solution

We perform operations with exponents before we add, subtract, multiply, or divide. Ten squared is 100. So when we subtract 10^2 from 100, the difference is zero.

$$100 - 10^2$$

$$100 - 100 = \mathbf{0}$$

writing decimal numbers as fractions, part 2

We will review changing a decimal number to a fraction or mixed number. Recall from Lesson 35 that the number of places after the decimal point indicates the denominator of the decimal fraction (10 or 100 or 1000, etc.). The digits to the right of the decimal point make up the numerator of the fraction.

Example 4

Thinking Skill

Connect

What is the denominator of each of these decimal fractions: 0.2, 0.43, 0.658?

Write 0.5 as a common fraction.

Solution

We read 0.5 as "five tenths," which also names the fraction $\frac{5}{10}$. We reduce the fraction.

$$\frac{5}{10} = \mathbf{\frac{1}{2}}$$

Example 5

Write 3.75 as a mixed number.

Solution

Thinking Skill

Verify

How do you reduce $\frac{75}{100}$ to $\frac{3}{4}$?

The whole-number part of 3.75 is 3, and the fraction part is 0.75. Since 0.75 has two decimal places, the denominator is 100.

$$3.75 = 3\frac{75}{100}$$

We reduce the fraction.

$$3\frac{75}{100} = \mathbf{3\frac{3}{4}}$$

Practice Set

Find the value of each expression:

a. 10^4 **b.** $2^3 + 2^4$ **c.** $2^2 \cdot 5^2$

d. Write the prime factorization of 72 using exponents.

Write each decimal number as a fraction or mixed number:

e. 12.5 **f.** 1.25 **g.** 0.125

h. 0.05 **i.** 0.24 **j.** 10.2

Written Practice

Strengthening Concepts

1. (38) Formulate Tomas's temperature was 102°F. Normal body temperature is 98.6°F. How many degrees above normal was Tomas's temperature? Write an equation and solve the problem.

2. (11) Formulate Jill has read 42 pages of a 180-page book. How many pages are left for her to read? Write an equation and solve the problem.

3. (18) Formulate If Jill wants to finish the book in the next three days, then she should read an average of how many pages per day? Write an equation and solve the problem.

*** 4.** (73) Write 2.5 as a reduced mixed number.

*** 5.** (73) Write 0.35 as a reduced fraction.

6. (41) What is the total cost of a $12.60 item when $7\frac{1}{2}\%$ (0.075) sales tax is added?

*** 7.** (72) $\frac{3}{4} \times 2 \times 1\frac{1}{3}$

*** 8.** (73) **Analyze** $(100 - 10^2) \div 5^2$

9. (61) $3 + 2\frac{1}{3} + 1\frac{3}{4}$

10. (63) $5\frac{1}{6} - 3\frac{1}{2}$

*** 11.** (68) $\frac{3}{4} \div 1\frac{1}{2}$

12. (49) $7 \div 0.4$

*** 13.** (44, 73) Compare:

a. $5^2 \bigcirc 2^5$

b. $0.3 \bigcirc 0.125$

14. (47) The diameter of a quarter is about 2.4 cm.

a. What is the circumference of a quarter? (Use 3.14 for π.)

b. What is the ratio of the radius of the quarter to the diameter of the quarter?

Find each unknown number:

15. (45) $25m = 0.175$

16. (43) $1.2 + y + 4.25 = 7$

17. (34) Which digit is in the ten-thousands place in 123,456.78?

18. (56) Arrange these numbers in order from least to greatest:

$$1, \frac{1}{2}, \frac{1}{10}, \frac{1}{4}, 0$$

*** 19.** (73) Write the prime factorization of 200 using exponents.

20. (41) The store offered a 20% discount on all tools. The regular price of a hammer was $18.00.

a. How much money is 20% of $18.00?

b. What was the price of the hammer after the discount?

*** 21.** (69) **Connect** The length of $\overline{AB}$ is 16 mm. The length of $\overline{AC}$ is 50 mm. What is the length of $\overline{BC}$?

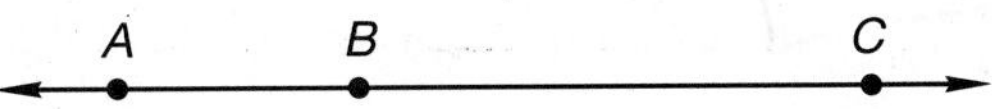

22. (31) One half of the area of this square is shaded. What is the area of the shaded region?

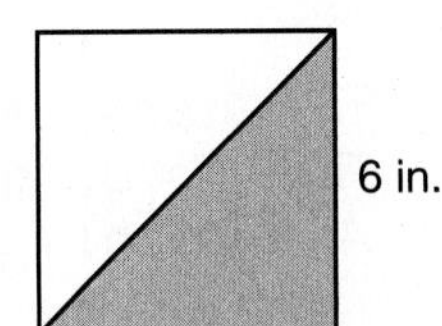

23. (64) **Verify** Is every square a rectangle?

*** 24.** (73) **Analyze** $\frac{2^2 + 2^3}{2}$

* **25.** (72) **Explain** The fractions chart from Lesson 72 says that the proper "shape" for multiplying fractions is "fraction form." What does that mean?

Refer to this coordinate plane to answer problems **26** and **27.**

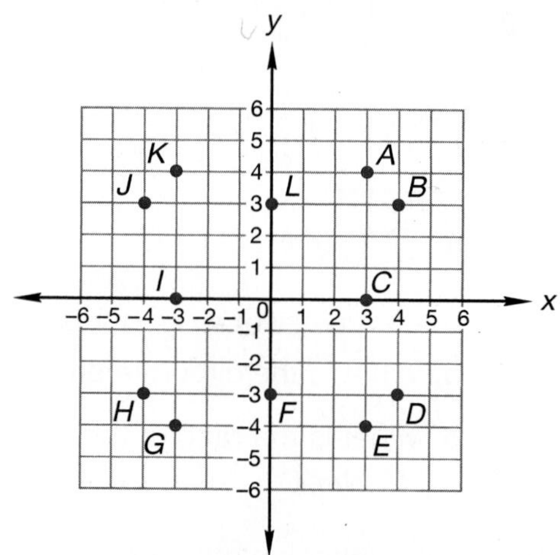

* **26.** (Inv. 7) Identify the coordinates of the following points:

a. *H* **b.** *L*

* **27.** (Inv. 7) Name the points that have the following coordinates:

a. (−4, 3) **b.** (3, 0)

* **28.** (73) If *s* equals 9, what does s^2 equal?

29. (Inv. 6) Name an every day object that has the same shape as each of these geometric solids:

a. cylinder **c.** sphere

b. rectangular prism **d.** cube

* **30.** (71) **Conclude** The measure of ∠*W* in parallelogram *WXYZ* is 75°.

a. What is the measure of ∠*X*?

b. What is the measure of ∠*Y*?

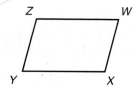

Early Finishers
Real-World Application

A teacher asked 23 students to close their eyes, then raise the hand up when they thought 60 seconds had elapsed. The results, in seconds, are shown below.

61 65 73 80 35 56 57 71 52 86 39 58
55 67 63 66 83 70 51 54 66 64 41

Which type of display—a stem-and-leaf plot or a line graph—is the most appropriate way to display this data? Draw your display and justify your choice.

LESSON 74

• Writing Fractions as Decimal Numbers
• Writing Ratios as Decimal Numbers

Power Up

Building Power

facts Power Up I

mental math

a. **Number Sense:** 3×230

b. **Number Sense:** $430 + 270$

c. **Calculation:** $5 \times \$0.99$

d. **Calculation:** $\$5.00 - \$1.98$

e. **Fractional Parts:** $\frac{1}{4}$ of $2.40

f. **Decimals:** $\$1.25 \times 10$

g. **Statistics:** Find the median of the set of numbers:

101, 26, 125, 84, 152

h. **Calculation:** 5×5, $- 5$, $\times 5$, $\div 2$, $+ 5$, $\div 5$

problem solving

A 60 in.-by-104 in. rectangular tablecloth was draped over a rectangular table. Eight inches of the 104-inch length of cloth hung over the left edge of the table, 3 inches over the back, 4 inches over the right edge, and 7 inches over the front.

In which directions (left, back, right, and/or forward) and by how many inches should the tablecloth be shifted so that equal amounts of cloth hang over opposite edges of the table? What are the dimensions of the table?

New Concepts

Increasing Knowledge

writing fractions as decimal numbers

We learned earlier that a fraction bar indicates division. So the fraction $\frac{1}{2}$ also means "1 divided by 2," which we can write as $2\overline{)1}$. By attaching a decimal point and zero, we can perform the division and write the quotient as a decimal number.

$$\frac{1}{2} \longrightarrow \begin{array}{r} 0.5 \\ 2\overline{)1.0} \\ \underline{1\,0} \\ 0 \end{array}$$

We find that $\frac{1}{2}$ equals the decimal number 0.5. To convert a fraction to a decimal number, we divide the numerator by the denominator.

Example 1

Convert $\frac{1}{4}$ to a decimal number.

Solution

The fraction $\frac{1}{4}$ means "1 divided by 4," which is $4\overline{)1}$. By attaching a decimal point and zeros, we can complete the division.

$$\begin{array}{r} \mathbf{0.25} \\ 4\overline{)1.00} \\ \underline{8} \\ 20 \\ \underline{20} \\ 0 \end{array}$$

Example 2

Use a calculator to convert $\frac{15}{16}$ to a decimal number.

Solution

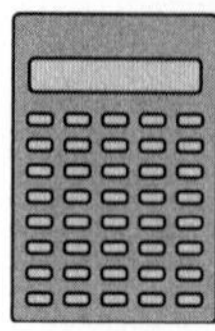

Begin by clearing the calculator. Then enter the fraction with these keystrokes.

After pressing the equal sign, the display shows the decimal equivalent of $\frac{15}{16}$:

0.9375

The answer is reasonable because both $\frac{15}{16}$ and 0.9375 are less than but close to 1.

Example 3

Write $7\frac{2}{5}$ as a decimal number.

Solution

The whole number part of $7\frac{2}{5}$ is 7, which we write to the left of the decimal point. We convert $\frac{2}{5}$ to a decimal by dividing 2 by 5.

$$\frac{2}{5} \longrightarrow \begin{array}{r} 0.4 \\ 5\overline{)2.0} \end{array}$$

Since $\frac{2}{5}$ equals 0.4, the mixed number $7\frac{2}{5}$ equals **7.4.**

Model Use a calculator to check the answer.

writing ratios as decimal numbers

Converting ratios to decimal numbers is similar to converting fractions to decimal numbers.

Example 4

A number cube is rolled once. Express the probability of rolling an even number as a decimal number.

Solution

Probabilities are often expressed as decimal numbers between 0 and 1. Since three of the six numbers on a number cube are even, the probability of rolling an even number is $\frac{3}{6}$, which equals $\frac{1}{2}$.

We convert $\frac{1}{2}$ to a decimal by dividing 1 by 2.

$$2\overline{)1.0}\quad 0.5$$

Thus the probability of rolling an even number with one roll of a number cube is **0.5.**

Practice Set

Convert each fraction or mixed number to a decimal number:

a. $\frac{3}{4}$ **b.** $4\frac{1}{5}$ **c.** $\frac{1}{8}$

d. $\frac{7}{20}$ **e.** $3\frac{3}{10}$ **f.** $\frac{7}{25}$

You may use a calculator to convert these numbers to decimal numbers:

g. $\frac{11}{16}$ **h.** $\frac{31}{32}$ **i.** $3\frac{24}{64}$

j. In a bag are three red marbles and two blue marbles. If Chad pulls one marble from the bag, what is the probability that the marble will be blue? Express the probability ratio as a fraction and as a decimal number.

Written Practice

Strengthening Concepts

* **1.** (73) Analyze What is the difference when five squared is subtracted from four cubed?

* **2.** (15) On LeAnne's map, 1 inch represents a distance of 10 miles. If Dallas, TX and Fort Worth, TX are 3 inches apart on the map, approximately how many miles apart are they?

* **3.** (74) Convert $2\frac{3}{4}$ to a decimal number.

* **4.** (58, 74) Tito spins the spinner once.

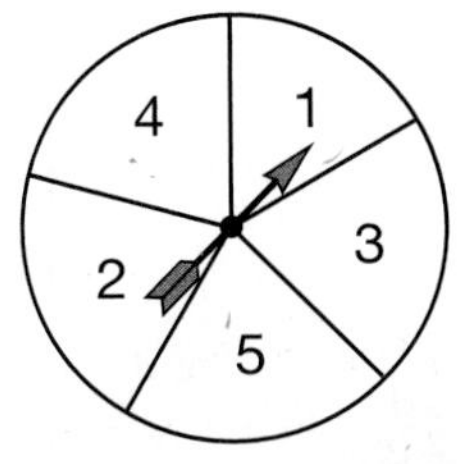

a. What is the sample space of the experiment?

b. What is the probability that he spins a number greater than 1? Express the probability ratio as a fraction and as a decimal.

* **5.** (73) Write 0.24 as a reduced fraction.

6. (13) **Formulate** Steve hit the baseball 400 feet. Lesley hit the golf ball 300 yards. How many feet farther did the golf ball travel than the baseball? After converting yards to feet, write an equation and solve the problem.

7. (2) If $A = bh$, and if b equals 12 and h equals 8, then what does A equal?

* **8.** (73) Compare: $3^2 \bigcirc 3 + 3$

9. (61) $\frac{1}{2} + \frac{2}{3} + \frac{1}{6}$

10. (63) $3\frac{1}{4} - 1\frac{7}{8}$

11. (72) $\frac{5}{8} \cdot \frac{3}{5} \cdot \frac{4}{5}$

12. (66) $3\frac{1}{3} \times 3$

13. (68) $\frac{3}{4} \div 1\frac{1}{2}$

14. (38) $(4 + 3.2) - 0.01$

15. (Inv. 6) **Represent** Draw a triangular prism.

16. (15) **Formulate** LaFonda bought a dozen golf balls for $10.44. What was the cost of each golf ball? Write an equation and solve the problem.

17. (16) Estimate the product of 81 and 38.

18. (18) In four days Jamar read 42 pages, 46 pages, 35 pages, and 57 pages. What was the average number of pages he read per day?

19. (30) What is the least common multiple of 6, 8, and 12?

20. (3) $24 + c + 96 = 150$

21. (67) Write the prime factorization of both the numerator and the denominator of this fraction. Then reduce the fraction.

$$\frac{40}{96}$$

* **22.** (47) **Analyze** If the perimeter of this square is 40 centimeters, then

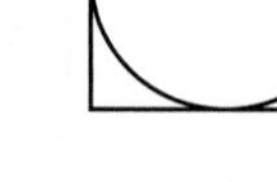

a. what is the diameter of the circle?

b. what is the circumference of the circle? (Use 3.14 for π.)

23. (29) Twenty-four of the three dozen cyclists rode mountain bikes. What fraction of the cyclists rode mountain bikes?

* **24.** (64) **Classify** Why are some rectangles not squares?

25. (17) **Connect** Which arrow could be pointing to $\frac{3}{4}$?

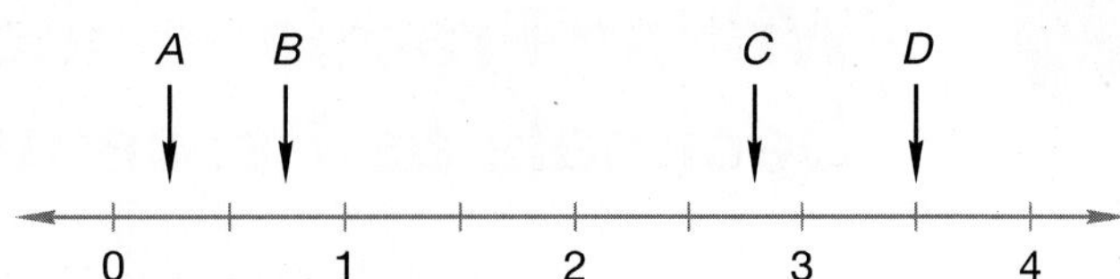

26. (64) **Conclude** In quadrilateral *PQRS,* which segment appears to be

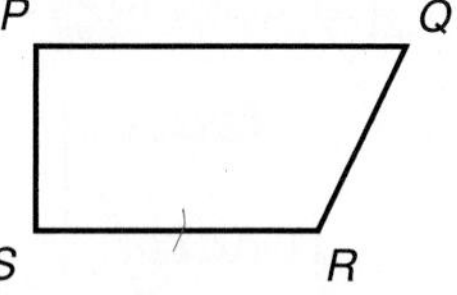

a. parallel to $\overline{PQ}$?

b. perpendicular to $\overline{PQ}$?

27. (Inv. 6) **Analyze** The figure at right shows a cube with edges 3 feet long.

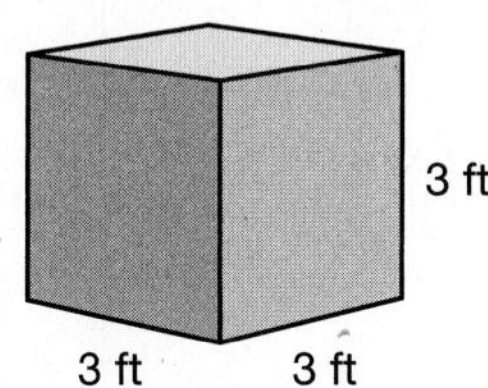

a. What is the area of each face of the cube?

b. What is the total surface area of the cube?

Refer to this coordinate plane to answer problems 28 and 29.

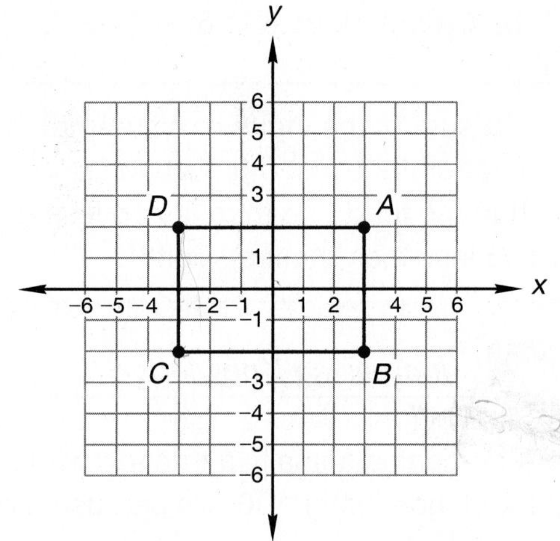

*** 28.** (Inv. 7) Identify the coordinates of the following points:

a. *C* **b.** origin

29. (64) **Connect** One pair of parallel segments in rectangle *ABCD* is $\overline{AB}$ and $\overline{DC}$. Name a second pair of parallel segments.

30. (41) Farmer Ruiz planted corn on 60% of his 300 acres. Find the number of acres planted with corn.

LESSON 75

• Writing Fractions and Decimals as Percents, Part 1

Power Up

Building Power

facts Power Up K

mental math

a. Calculation: 504×6

b. Number Sense: $625 - 250$

c. Calculation: $3 \times \$1.99$

d. Calculation: $\$2.50 + \$1.99$

e. Number Sense: Double $1.60.

f. Decimals: $\$12.50 \div 10$

g. Statistics: Find the median of the set of numbers:

28, 32, 44, 17, 15, 26

h. Calculation: 6×6, $- 6$, $\div 6$, $- 5$, $\times 2$, $+ 1$

problem solving The sum of the digits of a five-digit number is 25. What is the five digit number if the last digit is two less than the fourth, the fourth digit is two less than the third, the third is two less than the second, and the second digit is two less than the first digit?

New Concept

Increasing Knowledge

A percent is actually a fraction with a denominator of 100. Instead of writing the denominator 100, we can use a percent sign (%). So $\frac{25}{100}$ equals 25%.

Example 1

Write $\frac{3}{100}$ as a percent.

Solution

A percent is a fraction with a denominator of 100. Instead of writing the denominator, we write a percent sign. We write $\frac{3}{100}$ as **3%.**

Example 2

Write $\frac{3}{10}$ as a percent.

Solution

Thinking Skill

Explain

How do we write an equivalent fraction?

First we will write an equivalent fraction that has a denominator of 100.

$$\frac{3}{10} = \frac{?}{100}$$

We multiply $\frac{3}{10}$ by $\frac{10}{10}$.

$$\frac{3}{10} \cdot \frac{10}{10} = \frac{30}{100}$$

We write the fraction $\frac{30}{100}$ as **30%.**

Example 3

Of the 30 students who took the test, 15 earned an A. What percent of the students earned an A?

Solution

Fifteen of the 30 students earned an A. We write this as a fraction and reduce.

$$\frac{15}{30} = \frac{1}{2}$$

To write $\frac{1}{2}$ as a fraction with a denominator of 100, we multiply $\frac{1}{2}$ by $\frac{50}{50}$.

$$\frac{1}{2} \cdot \frac{50}{50} = \frac{50}{100}$$

The fraction $\frac{50}{100}$ equals **50%.**

Example 4

Write 0.12 as a percent.

Solution

The decimal number 0.12 is twelve hundredths.

$$0.12 = \frac{12}{100}$$

Twelve hundredths is equivalent to **12%.**

Example 5

Write 0.08 as a percent.

Solution

The decimal 0.08 is eight hundredths.

$$0.08 = \frac{8}{100}$$

Eight hundredths is equivalent to **8%.**

Example 6

Write 0.8 as a percent.

Solution

The decimal number 0.8 is eight tenths. If we place a zero in the hundredths place, the decimal is eighty hundredths.

$$0.8 = 0.80 = \frac{80}{100}$$

Eighty hundredths equals **80%.**

Notice that when a decimal number is converted to a percent, the decimal point is shifted two places to the right. In fact, shifting the decimal point two places to the right is a quick and useful way to write decimal numbers as percents.

Practice Set

Write each fraction as a percent:

a. $\frac{31}{100}$ **b.** $\frac{1}{100}$ **c.** $\frac{1}{10}$

d. $\frac{3}{50}$ **e.** $\frac{7}{25}$ **f.** $\frac{2}{5}$

g. Twelve of the 30 students earned a B on the test. What percent of the students earned a B?

h. Jorge correctly answered 18 of the 20 questions on the test. What percent of the questions did he answer correctly?

Write each decimal number as a percent:

i. 0.25 **j.** 0.3 **k.** 0.05

l. 1.0 **m.** 0.7 **n.** 0.15

Written Practice

Strengthening Concepts

1. *(30, 62)* **Connect** What is the reciprocal of two and three fifths?

2. *(32)* What time is one hour thirty-five minutes after 2:30 p.m.?

3. *(15)* A 1-pound box of candy cost $4.00. What was the cost per ounce (1 pound = 16 ounces)?

4. *(41)* **Estimate** Freda bought a sandwich for $4.00 and a drink for 94¢. Her grandson ordered a meal for $6.35. What was the total price of all three items when 8% sales tax was added? Explain how to use estimation to check whether your answer is reasonable.

5. *(58)* If the chance of rain is 50%, then what is the chance it will not rain?

6. **Represent** Draw a cube. How many edges does a cube have?
(Inv. 6)

* 7. a. Write $\frac{3}{4}$ as a decimal number.
(74, 75)

b. Write the answer to part **a** as a percent.

* 8. a. Write $\frac{3}{20}$ as a fraction with a denominator of 100.
(75)

b. Write $\frac{3}{20}$ as a percent.

* 9. Write 12% as a reduced fraction. Then write the fraction as a decimal number.
(33, 74)

Find each unknown number:

10. $\frac{7}{10} = \frac{n}{100}$
(42)

11. $5 - m = 3\frac{1}{8}$
(43)

12. $1 - w = 0.95$
(38, 43)

13. $m + 1\frac{2}{3} = 3\frac{1}{6}$
(59)

14. $\left(\frac{1}{2} + \frac{1}{3}\right) - \frac{1}{6}$
(57)

* 15. $3\frac{1}{2} \times 1\frac{1}{3} \times 1\frac{1}{2}$
(72)

16. $(0.43)(2.6)$
(39)

17. $0.26 \div 5$
(45)

* 18. Nathan correctly answered 17 of the 20 questions on the test. What percent of the questions did Nathan answer correctly?
(75)

19. **Estimate** The diameter of the big tractor tire was about 5 feet. As the tire rolled one full turn, the tire rolled about how many feet? Round the answer to the nearest foot. (Use 3.14 for π.)
(47)

20. **Connect** Which digit in 4.87 has the same place value as the 9 in 0.195?
(34)

21. Write the prime factorization of both the numerator and denominator of $\frac{18}{30}$. Then reduce the fraction.
(67)

22. What is the greatest common factor of 18 and 30?
(20)

23. If the product of two numbers is 1, then the two numbers are which of the following?
(30)

A equal **B** reciprocals **C** opposites **D** prime

24. **Verify** Why is every rectangle a quadrilateral?
(64)

25. If b equals 8 and h equals 6, what does $\frac{bh}{2}$ equal?
(47)

* 26. **Represent** Find the prime factorization of 400 using a factor tree. Then write the prime factorization of 400 using exponents.
(65, 73)

* **27.** (Inv. 7) **Represent** Draw a coordinate plane on graph paper. Then draw a rectangle with vertices located at (3, 1), (3, −1), (−1, 1), and (−1, −1).

28. (8, 31) Refer to the rectangle drawn in problem **27** to answer parts **a** and **b** below.

a. What is the perimeter of the rectangle?

b. What is the area of the rectangle?

* **29.** (71) **a.** What is the perimeter of this parallelogram?

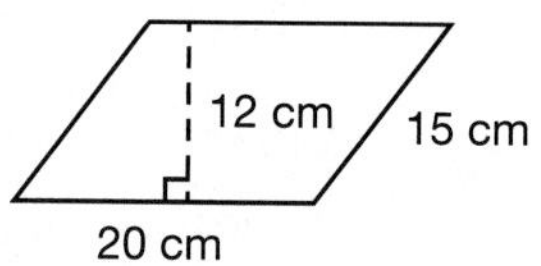

b. What is the area of this parallelogram?

30. (64) **Model** Draw two parallel segments of different lengths. Then form a quadrilateral by drawing two segments that connect the endpoints of the parallel segments. Is the quadrilateral a rectangle?

Early Finishers
Math and Science

The Moon is Earth's only natural satellite. The average distance from Earth to the Moon is approximately 620^2 kilometers. This distance is about 30 times the diameter of Earth.

a. Simplify 620^2 kilometers.

b. Find the diameter of Earth. Round your answer to the nearest kilometer.

LESSON

76

• Comparing Fractions by Converting to Decimal Form

Power Up

Building Power

facts Power Up G

mental math

a. **Calculation:** 4 × 208

b. **Calculation:** 380 + 155

c. **Calculation:** 4 × \$1.99

d. **Calculation:** \$10.00 − \$4.99

e. **Fractional Parts:** $\frac{1}{5}$ of \$4.50

f. **Decimals:** \$0.95 × 100

g. **Probability:** How many different four digit numbers can be made with the digits 6, 4, 2, 9 using each digit exactly once?

h. **Calculation:** 8 × 8, − 4, ÷ 2, + 2, ÷ 4, × 3, + 1, ÷ 5

problem solving

In the 4 × 200 m relay, Sarang ran first, then Gemmie, then Joyce, and finally Karla. Each girl ran her 200 meters 2 seconds faster than the previous runner. The team finished the race in exactly 1 minute and 50 seconds. How fast did each runner run her 200 meters?

New Concept

Increasing Knowledge

We have compared fractions by drawing pictures of fractions and by writing fractions with common denominators. Another way to compare fractions is to convert the fractions to decimal form.

Example 1

Compare these fractions. First convert each fraction to decimal form.

$$\frac{3}{5} \bigcirc \frac{5}{8}$$

Solution

We convert each fraction to a decimal number by dividing the numerator by the denominator.

$$\frac{3}{5} \longrightarrow 5\overline{)3.0}\ \text{quotient } 0.6 \qquad \frac{5}{8} \longrightarrow 8\overline{)5.000}\ \text{quotient } 0.625$$

We write both numbers with the same number of decimal places. Then we compare the two numbers.

Thinking Skill

Explain

How do we compare decimal numbers?

$$0.600 < 0.625$$

Since 0.6 is less than 0.625, we know that $\frac{3}{5}$ is less than $\frac{5}{8}$.

$$\mathbf{\frac{3}{5} < \frac{5}{8}}$$

Example 2

Compare: $\mathbf{\frac{3}{4} \bigcirc 0.7}$

Solution

First we write the fraction as a decimal.

$$\frac{3}{4} \longrightarrow 4\overline{)3.00} = 0.75$$

Then we compare the decimal numbers.

$$0.75 > 0.70$$

Since 0.75 is greater than 0.7, we know that $\frac{3}{4}$ is greater than 0.7.

$$\mathbf{\frac{3}{4} > 0.7}$$

Practice Set

Change the fractions to decimals to compare these numbers:

a. $\frac{3}{20} \bigcirc \frac{1}{8}$ **b.** $\frac{3}{8} \bigcirc \frac{2}{5}$ **c.** $\frac{15}{25} \bigcirc \frac{3}{5}$

d. $0.7 \bigcirc \frac{4}{5}$ **e.** $\frac{2}{5} \bigcirc 0.5$ **f.** $\frac{3}{8} \bigcirc 0.325$

Written Practice

Strengthening Concepts

* **1.** (73) Connect What is the product of ten squared and two cubed?

2. (50) Connect What number is halfway between 4.5 and 6.7?

3. (15) Formulate It is said that one year of a dog's life is the same as 7 years of a human's life. Using that thinking, a dog that is 13 years old is how many "human" years old? Write an equation and solve the problem.

* **4.** (76) Compare. First convert each fraction to decimal form.

$$\frac{2}{5} \bigcirc \frac{1}{4}$$

* **5.** (74, 75) **a.** What fraction of this circle is shaded?

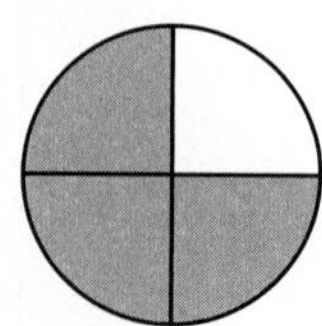

b. Convert the answer from part **a** to a decimal number.

c. What percent of this circle is shaded?

6. (7, 27) Choose the appropriate units for measuring the circumference of a juice glass.

A centimeters **B** meters **C** kilometers

*** 7.** (73, 74) **a.** Convert $2\frac{1}{2}$ to a decimal number.

b. Write 3.75 as a reduced mixed number.

*** 8.** (73, 75) **a.** Write 0.04 as a reduced fraction.

b. Write 0.04 as a percent.

9. (43) **Verify** Instead of dividing 200 by 18, Sam found half of each number and then divided. Show Sam's division problem and write the quotient as a mixed number.

10. (61) $6\frac{1}{3} + 3\frac{1}{4} + 2\frac{1}{2}$

11. (42) $\frac{4}{5} = \frac{?}{100}$

*** 12.** (72) **Analyze** $\left(2\frac{1}{2}\right)\left(3\frac{1}{3}\right)\left(1\frac{1}{5}\right)$

*** 13.** (68) $5 \div 2\frac{1}{2}$

Find each unknown number:

14. (43) $6.7 + 0.48 + n = 8$

15. (43) $12 - d = 4.75$

16. (39) 0.35×0.45

17. (38, 52) $4.3 \div 10^2$

18. (Inv. 5) Find the median of these numbers:

0.3, 0.25, 0.313, 0.2, 0.27

19. (16) **Estimate** Find the sum of 3926 and 5184 to the nearest thousand.

20. (19) **List** Name all the prime numbers between 40 and 50.

*** 21.** (75) Twelve of the 25 students in the class earned As on the test. What percent of the students earned As?

Refer to the triangle to answer problems **22** and **23.**

22. (8) What is the perimeter of this triangle?

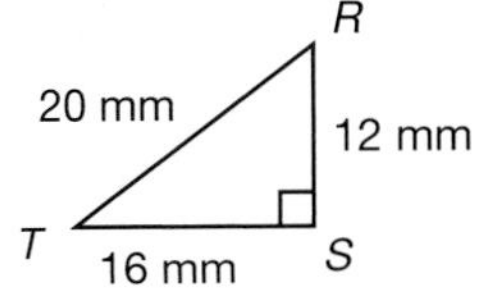

*** 23.** (69) **Analyze** Angles *T* and *R* are complementary. If the measure of $\angle R$ is 53°, then what is the measure of $\angle T$?

24. (7) **Estimate** About how many **millimeters** long is this line segment?

cm 1 2 3 4 5

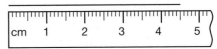

*** 25.** *(71)* This parallelogram is divided into two congruent triangles.

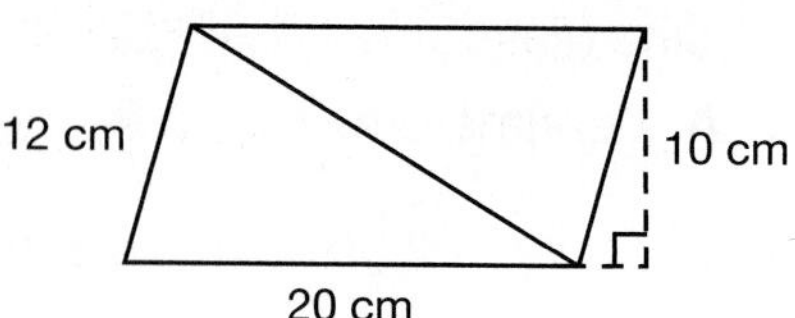

a. What is the area of the parallelogram?

b. What is the area of one of the triangles?

26. *(Inv. 6)* How many small cubes were used to form this rectangular prism?

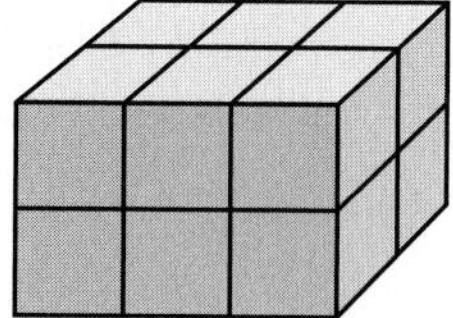

*** 27.** *(Inv. 7)* **Represent** Sketch a coordinate plane on graph paper. Graph point *A* (1, 2), point *B* (−3, −2), and point *C* (1, −2). Then draw segments to connect the three points. What type of polygon is figure *ABC*?

28. *(28)* **Conclude** In the figure drawn in problem 27,

a. which segment is perpendicular to $\overline{AC}$?

b. which angle is a right angle?

29. *(47)* If *b* equals 12 and *h* equals 9, what does $\frac{bh}{2}$ equal?

*** 30.** *(64)* **Model** Draw a pair of parallel lines. Draw a third line perpendicular to the parallel lines. Complete a quadrilateral by drawing a fourth line that intersects but is not perpendicular to the pair of parallel lines. Trace the quadrilateral that is formed. Is the quadrilateral a rectangle?

LESSON

77

• Finding Unstated Information in Fraction Problems

Power Up — *Building Power*

facts Power Up K

mental math

a. **Calculation:** 311 × 5

b. **Number Sense:** 565 − 250

c. **Calculation:** 5 × $1.99

d. **Calculation:** $7.50 + $1.99

e. **Number Sense:** Double 80¢.

f. **Decimals:** 6.5 ÷ 100

g. **Statistics:** Find the median of the set of numbers: 134, 147, 125, 149, 158, 185.

h. **Calculation:** 10 × 10, × 10, − 1, ÷ 9, − 11, ÷ 10

problem solving Kathleen read an average of 45 pages per day for four days. If she read a total of 123 pages during the first three days, how many pages did she read on the fourth day?

New Concept — *Increasing Knowledge*

Often fractional-parts statements contain more information than what is directly stated. Consider this fractional-parts statement:

Three fourths of the 28 students in the class are boys.

This sentence directly states information about the number of boys in the class. It also *indirectly* states information about the number of girls in the class. In this lesson we will practice finding several pieces of information from fractional-parts statements.

Example 1

Diagram this statement. Then answer the questions that follow.

Three fourths of the 28 students in the class are boys.

a. Into how many parts is the class divided?

b. How many students are in each part?

c. How many parts are boys?

d. How many boys are in the class?

e. **How many parts are girls?**

f. **How many girls are in the class?**

Solution

We draw a rectangle to represent the whole class. Since the statement uses fourths to describe a part of the class, we divide the rectangle into four parts. Dividing the total number of students by four, we find there are seven students in each part. We identify three of the four parts as boys and one of the four parts as girls. Now we answer the questions.

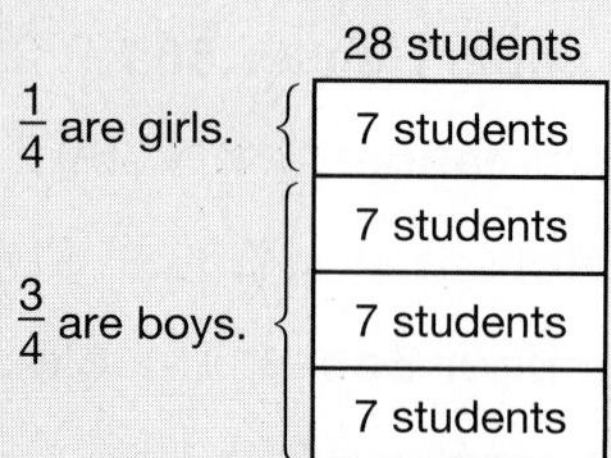

a. The denominator of the fraction indicates that the class is divided into **four parts** for the purpose of this statement. It is important to distinguish between the number of *parts* (as indicated by the denominator) and the number of *categories*. There are two categories of students implied by the statement—boys and girls.

b. In each of the four parts there are **seven students.**

c. The numerator of the fraction indicates that **three parts** are boys.

d. Since three parts are boys and since there are seven students in each part, we find that there are **21 boys** in the class.

e. Three of the four parts are boys, so only **one part** is girls.

f. There are seven students in each part. One part is girls, so there are **seven girls.**

Example 2

There are thirty marbles in a bag. If one marble is drawn from the bag, the probability of drawing red is $\frac{2}{5}$.

a. **How many marbles are red?**

b. **How many marbles are not red?**

c. **The complement of drawing a red marble is drawing a not red marble. What is the probability of drawing a not red marble?**

d. **What is the sum of the probabilities of drawing red and drawing not red?**

Solution

a. The probability of drawing red is $\frac{2}{5}$, so $\frac{2}{5}$ of the 30 marbles are red.

$$\frac{2}{5} \cdot 30 = 12$$

There are **12 red marbles.**

b. Twelve of the 30 marbles are red, so **18 marbles are not red.**

c. The probability of drawing a not red marble is $\frac{18}{30} = \mathbf{\frac{3}{5}}$.

d. The sum of the probabilities of an event and its complement is **1.**

$$\mathbf{\frac{2}{5} + \frac{3}{5} = 1}$$

Practice Set

Model Diagram this statement. Then answer the questions that follow.

Three eighths of the 40 little engines could climb the hill.

a. Into how many parts was the group divided?

b. How many engines were in each part?

c. How many parts could climb the hill?

d. How many engines could climb the hill?

e. How many parts could not climb the hill?

f. How many engines could not climb the hill?

Read the statement and then answer the questions that follow.

The face of a spinner is divided into 12 equal sectors. The probability of spinning red on one spin is $\frac{1}{4}$.

g. How many sectors are red?

h. How many sectors are not red?

i. What is the probability of spinning not red in one spin?

j. What is the sum of the probabilities of spinning red and not red? How are the events related?

Written Practice

Strengthening Concepts

1. (29) The weight of an object on the Moon is about $\frac{1}{6}$ of its weight on Earth. A person weighing 114 pounds on Earth would weigh about how much on the Moon?

*** 2.** (22) Estimate Estimate the weight of an object in your classroom such as a table or desk. Use the information in problem 1 to calculate what the approximate weight of the object would be on the moon. Round your answer to the nearest pound.

*** 3.** (29, 75) Mekhi was at bat 24 times and got 6 hits.

a. What fraction of the times at bat did Mekhi get a hit?

b. What percent of the times at bat did Mekhi get a hit?

* **4.** (77) **Model** Diagram this statement. Then answer the questions that follow.

There are 30 students in the class. Three fifths of them are boys.

a. Into how many parts is the class divided?

b. How many students are in each part?

c. How many boys are in the class?

d. How many girls are in the class?

* **5.** (74, 75) **a.** In the figure below, what fraction of the group is shaded?

b. Convert the fraction in part **a** to a decimal number.

c. What percent of the group is shaded?

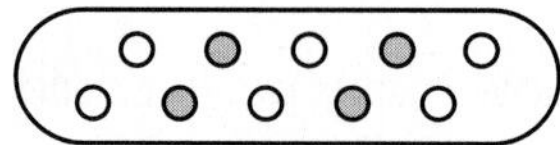

* **6.** (73) Write the decimal number 3.6 as a mixed number.

Find each unknown number:

7. (43) $3.6 + a = 4.15$

8. (30) $\frac{2}{5}x = 1$

9. (58) **Explain** If the chance of rain is 60%, is it more likely to rain or not to rain? Why?

* **10.** (75) Three fifths of a circle is what percent of a circle?

11. (10, 14) A temperature of -3°F is how many degrees below the freezing temperature of water?

* **12.** (73, 76) Compare:

a. $0.35 \bigcirc \frac{7}{20}$

b. $3^2 \bigcirc 2^3$

13. (57) $\frac{1}{2} + \frac{2}{3}$

14. (63) $3\frac{1}{5} - 1\frac{3}{5}$

15. (61) $\frac{1}{2} + \frac{3}{4} + \frac{7}{8}$

16. (66) $3 \times 1\frac{1}{3}$

17. (68) $3 \div 1\frac{1}{3}$

18. (68) $1\frac{1}{3} \div 3$

19. (8) What is the perimeter of this rectangle?

20. (31) What is the area of this rectangle?

1.5 cm

0.9 cm

* **21.** (65) Write the prime factorization of 1000 using exponents.

22. (41) Coats were on sale for 40% off. One coat was regularly priced at $80.

a. How much money would be taken off the regular price of the coat during the sale?

b. What would be the sale price of the coat?

23. (41) Patricia bought a coat that cost \$38.80. The sales-tax rate was 7%.

a. What was the tax on the purchase?

b. What was the total purchase price including tax?

24. (64) *Classify* Is every quadrilateral a polygon?

25. (32) What time is one hour fourteen minutes before noon?

26. (Inv. 2) *Estimate* What percent of this rectangle appears to be shaded?

A 20% **C** 60%

B 40% **D** 80%

* **27.** (Inv. 7) *Represent* Sketch a coordinate plane on graph paper. Graph point W (2, 3), point X (1, 0), point Y (−3, 0), and point Z (−2, 3). Then draw $\overline{WX}$, $\overline{XY}$, $\overline{YZ}$, and $\overline{ZW}$.

* **28.** (71) **a.** *Conclude* Which segment in problem 27 is parallel to $\overline{WX}$?

b. Which segment in problem 27 is parallel to $\overline{XY}$?

29. (67) Write the prime factorization of both the numerator and the denominator of this fraction. Then reduce the fraction.

$$\frac{210}{350}$$

30. (47, Inv. 6) **a.** *Connect* The moon has the shape of what geometric solid?

b. *Estimate* The diameter of the moon is about 2160 miles. Calculate the approximate circumference of the moon using 3.14 for π. Round the answer to the nearest ten miles.

Early Finishers
Choose A Strategy

Simplify each prime factorization below. Then identify which prime factorization does not belong in the group. Explain your reasoning.

$3^2 \times 2^3$ $2^2 \times 3^3 \times 5$ $2^2 \times 3^2 \times 5^2$ 3^5 $2^4 \times 7^2$

LESSON 78

• Capacity

Power Up

Building Power

facts Power Up J

mental math

a. **Calculation:** 4 × 325

b. **Number Sense:** 1500 + 275

c. **Calculation:** 3 × $2.99

d. **Calculation:** $20.00 − $2.99

e. **Fractional Parts:** $\frac{1}{3}$ of $2.40

f. **Decimals:** 1.75 × 100

g. **Statistics:** Find the median of the set of numbers: 384, 127, 388, 484, 488, 120.

h. **Calculation:** 9 × 11, + 1, ÷ 2, − 1, ÷ 7, − 2, × 5

problem solving Raul's PE class built a training circuit on a circular path behind their school. There are six light poles spaced evenly around the circuit, and it takes Raul 64 seconds to mow the path from the first pole to the third pole. At this rate, how long will it take Raul to mow once completely around the path?

New Concept

Increasing Knowledge

To measure quantities of liquid in the U.S. Customary System, we use the units gallons (gal), quarts (qt), pints (pt), cups (c), and ounces (oz). In the metric system we use liters (L) and milliliters (mL). The relationships between units within each system are shown in the following table:

Thinking Skill

Discuss

Name some real world situations where we use the word *quarter.*

Equivalence Table for Units of Liquid Measure

U.S. Customary System	Metric System
1 gallon = 4 quarts 1 quart = 2 pints 1 pint = 2 cups 1 cup = 8 ounces	1 liter = 1000 milliliters

Common container sizes based on the U.S. Customary System are illustrated below. These containers are named by their **capacity,** that is, by the amount of liquid they can contain. Notice that each container size has half the capacity of the next-largest container. Also, notice that a quart is one "quarter" of a gallon.

Math Language

Note that we use *ounces* to measure capacity and weight. However, these are two different units of measurement. When we measure capacity we often refer to *fluid ounces.*

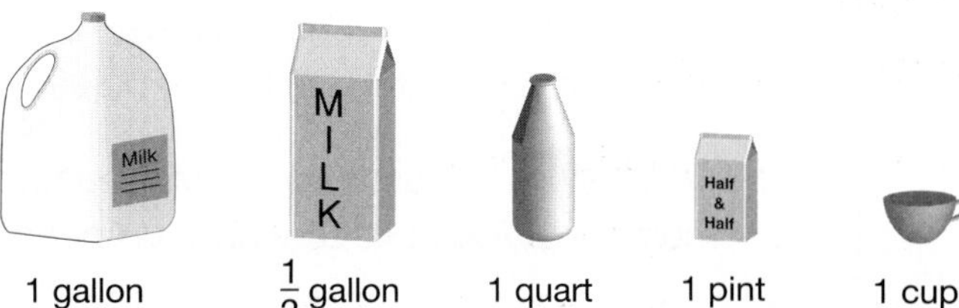

1 gallon $\frac{1}{2}$ gallon 1 quart 1 pint 1 cup

Food and beverage containers often have both U.S. Customary and metric capacities printed on the containers. Using the information found on 2-liter seltzer bottles and $\frac{1}{2}$-gallon milk cartons, we find that one liter is a little more than one quart. So the capacity of a 2-liter bottle is a little more than the capacity of a $\frac{1}{2}$-gallon container.

2-liter bottle $\frac{1}{2}$-gallon container

Example 1

A half gallon of milk is how many pints of milk?

Solution

Two pints equals a quart, and two quarts equals a half gallon. So a half gallon of milk is **4 pints.**

Example 2

Which has the greater capacity, a 12-ounce can or a 1-pint container?

Solution

A pint equals 16 ounces. **So a 1-pint container has more capacity than a 12-ounce can.**

Practice Set

a. What fraction of a gallon is a quart?

b. A 2-liter bottle has a capacity of how many milliliters?

c. A half gallon of orange juice will fill how many 8-ounce cups?

d. *Explain* The entire contents of a full 2-liter bottle are poured into an empty half-gallon carton. Will the half-gallon container overflow? Why or why not?

Written Practice — *Strengthening Concepts*

1. (12, 55) What is the difference when the product of $\frac{1}{2}$ and $\frac{1}{2}$ is subtracted from the sum of $\frac{1}{2}$ and $\frac{1}{2}$?

2. (7) The claws of a Siberian tiger are 10 centimeters long. How many millimeters is that?

3. (25) **Analyze** Sue was thinking of a number between 40 and 50 that is a multiple of 3 and 4. Of what number was she thinking?

*** 4.** (77) **Model** Diagram this statement. Then answer the questions that follow.

Four fifths of the 60 lights were on.

a. Into how many parts have the 60 lights been divided?

b. How many lights are in each part?

c. How many lights were on?

d. How many lights were off?

5. (65) **Classify** Which counting number is neither a prime number nor a composite number?

Find each unknown number:

6. (30) $\frac{4}{5}m = 1$

7. (43) $\frac{4}{5} + w = 1$

8. (43) $\frac{4}{5} \div x = 1$

9. (42, 43) $\frac{3}{4} = \frac{n}{100}$

*** 10.** (74, 75) **a.** What fraction of the rectangle below is shaded?

b. Write the answer to part **a** as a decimal number.

c. What percent of the rectangle is shaded?

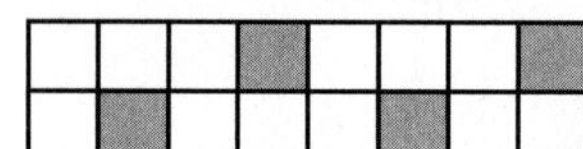

*** 11.** (73) Convert the decimal number 1.15 to a mixed number.

*** 12.** (73, 76) Compare:

a. $\frac{3}{5} \bigcirc 0.35$

b. $\sqrt{100} \bigcirc 1^4 + 2^3$

13. (57) $\frac{5}{6} - \frac{1}{2}$

14. (63) $4\frac{1}{4} - 3\frac{1}{3}$

15. (51) $\frac{1}{2} + \frac{2}{3} + \frac{5}{6}$

16. (66) $1\frac{1}{2} \times 2\frac{2}{3}$

17. (68) $1\frac{1}{2} \div 2\frac{2}{3}$

18. (68) $2\frac{2}{3} \div 1\frac{1}{2}$

19. (38) **a.** What is the perimeter of this square?

b. What is the area of this square?

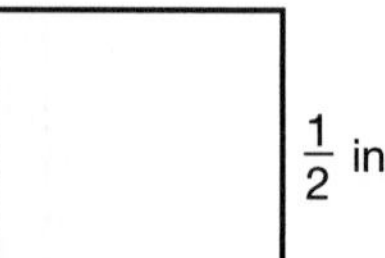

20. (64) **Verify** "The opposite sides of a rectangle are parallel." True or false?

* **21.** (73) What is the average of 3^3 and 5^2?

* **22.** (51, 74) The diameter of the small wheel was 7 inches. The circumference was about 22 inches. Write the ratio of the circumference to the diameter of the circle as a decimal number rounded to the nearest hundredth.

23. (66) How many inches is $2\frac{1}{2}$ feet?

24. (50) **Connect** Which arrow below could be pointing to 0.1?

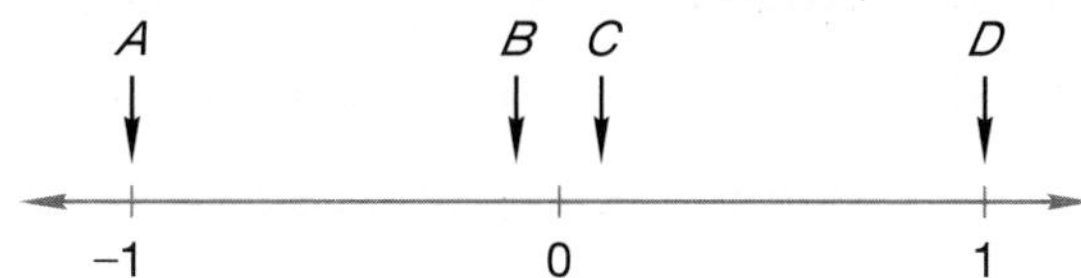

25. (64) **Represent** Draw a quadrilateral that is not a rectangle.

26. (65, 73) **Represent** Find the prime factorization of 900 by using a factor tree. Then write the prime factorization using exponents.

27. (Inv. 7) Three vertices of a rectangle have the coordinates (5, 3), (5, −1), and (−1, −1). What are the coordinates of the fourth vertex of the rectangle?

* **28.** (78) Refer to this table to answer **a** and **b**.

3 teaspoons = 1 tablespoon
16 tablespoons = 1 cup
2 cups = 1 pint
2 pints = 1 quart
4 quarts = 1 gallon

a. A teaspoon of soup is what fraction of a tablespoon of soup?

b. How many cups of milk is a gallon of milk?

* **29.** (78) **Estimate** A liter is closest in size to which of the following?

A pint **B** quart **C** $\frac{1}{2}$ gallon **D** gallon

* **30.** (78) In 1881 Clara Barton founded the American Red Cross, an organization that helps people during emergencies. The Red Cross organizes "blood drives" in which people can donate a pint of blood to help hospital patients who will undergo surgery. How many ounces is a pint?

LESSON 79

• Area of a Triangle

Power Up

Building Power

facts Power Up K

mental math

a. **Calculation:** 307×6

b. **Number Sense:** $1000 - 420$

c. **Calculation:** $4 \times \$2.99$

d. **Calculation:** $\$5.75 + \$2.99$

e. **Number Sense:** Double \$24.

f. **Decimals:** 0.125×100

g. **Measurement:** How many liters are in a kiloliter?

h. **Calculation:** $2 \times 2, \times 2, \times 2, - 1, \times 2, + 2, \div 2, \div 2$

problem solving

The restaurant serves four different soups and three different salads. How many different soup-and-salad combinations can diners order? Draw a diagram to support your answer.

New Concept

Increasing Knowledge

In this lesson we will demonstrate that the area of a triangle is half the area of a parallelogram with the same base and height.

Activity

Area of a Triangle

Materials needed:

- pencil and paper
- ruler
- scissors

Model Fold the paper in half, and draw a triangle on the folded paper.

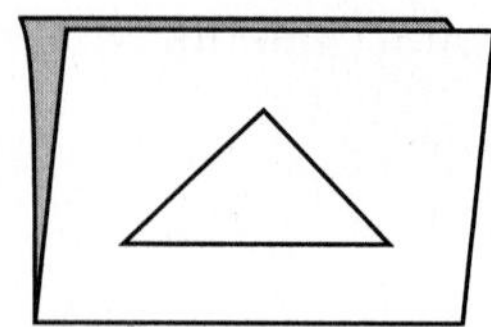

Math Language
Recall that **congruent** polygons have the same size and shape.

While the paper is folded, use your scissors to cut out the triangle so that you have two congruent triangles.

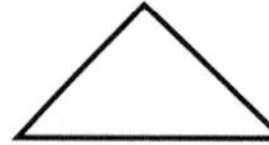 

Arrange the two triangles to form a parallelogram. What fraction of the area of the parallelogram is the area of one of the triangles?

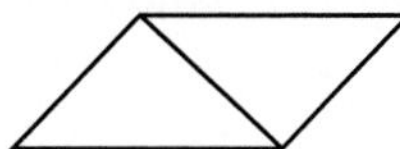

We find that whatever the shape of a triangle, its area is half the area of the parallelogram with the same base and height.

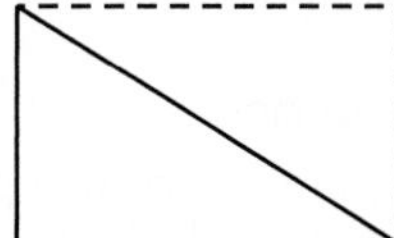

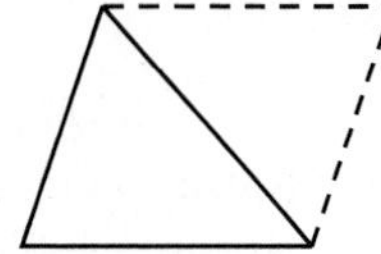

Recall that the area of a parallelogram can be found by multiplying its base by its height ($A = bh$). So the area of a triangle can be determined by finding half of the product of its base and height.

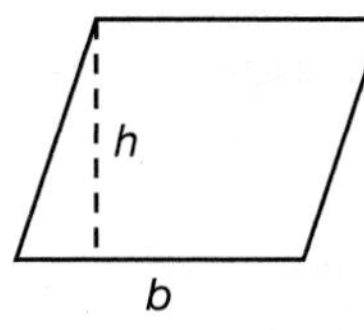

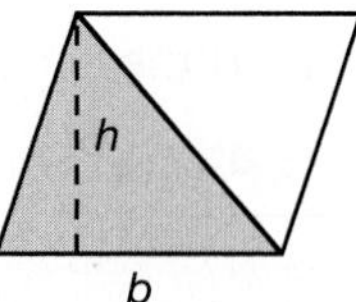

parallelogram
$A = bh$

triangle
$A = \frac{1}{2}bh$

Thinking Skill

Analyze

Why are multiplying by $\frac{1}{2}$ and dividing by 2 equivalent operations?

Since multiplying by $\frac{1}{2}$ and dividing by 2 are equivalent operations, the formula may also be written as

$$A = \frac{bh}{2}$$

In the following examples we will use both formulas stated above. From our calculations of the areas of rectangles and parallelograms, we remember that the base and the height are *perpendicular* measurements.

Example 1

Find the area of the triangle at right.

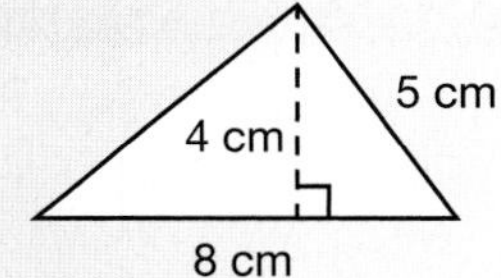

Solution

Reading Math

The area of a figure is expressed in square units. Read the abbreviation "cm²" as *square centimeters.*

The area of the triangle is half the product of the base and height. The height must be *perpendicular* to the base. The height in this case is 4 cm. Half the product of 8 cm × 4 cm is 16 cm^2.

$$A = \frac{1}{2}(8 \text{ cm})(4 \text{ cm})$$

$$A = \mathbf{16\ cm^2}$$

Example 2

Find the area of this right triangle:

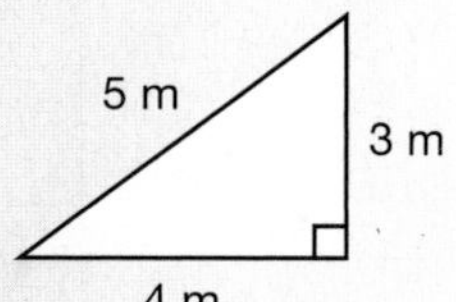

Solution

We find the area by multiplying the base by the height and then dividing by 2. All right triangles have two sides that are perpendicular, so we use the perpendicular sides as the base and height.

$$A = \frac{(4\text{ m})(3\text{ m})}{2}$$

$$A = \mathbf{6\ m^2}$$

Practice Set

Find the area of each triangle:

a.

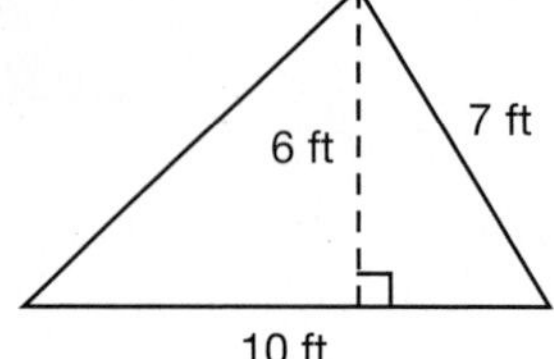

b.

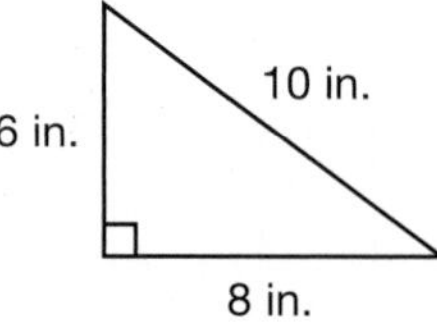

c.

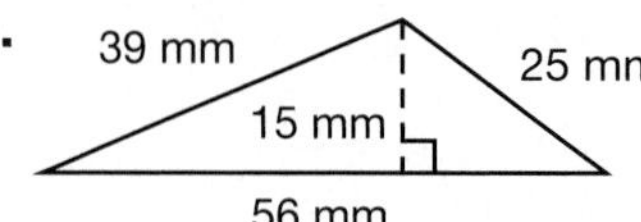

d.

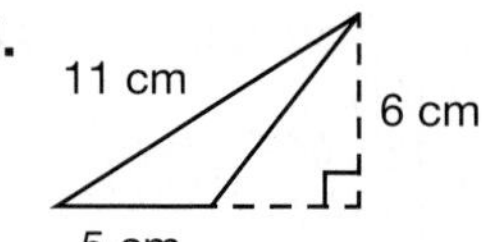

e. Predict If the height of the triangle in **c** is doubled to 30 mm, would the area double? Calculate to check your prediction.

Written Practice

Strengthening Concepts

1. (8) Explain If you know both the perimeter and the length of a rectangle, how can you determine the width of the rectangle?

*** 2.** (78) A 2-liter bottle contained 2 qt 3.6 oz of beverage. Use this information to compare a liter and a quart:

1 liter ◯ 1 quart

3. (11) Mr. Johnson was 38 years old when he started his job. He worked for 33 years. How old was he when he retired?

4. (64) Verify Answer "true" or "false" for each statement:

a. "Every rectangle is a square."

b. "Every rectangle is a parallelogram."

5. (23, 41) Ninety percent of 30 trees are birch trees.

a. How many trees are birch trees?

b. What is the ratio of birch trees to all other trees?

*** 6.** (75, 77) Eighteen of the twenty-four runners finished the race.

a. What fraction of the runners finished the race?

b. What fraction of the runners did not finish the race?

c. What percent of the runners did not finish the race?

*** 7.** (79) **Analyze** This parallelogram is divided into two congruent triangles. What is the area of each triangle?

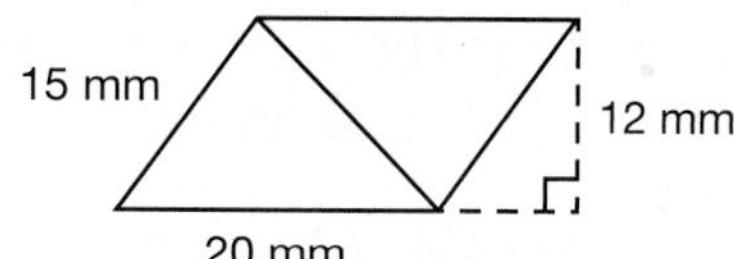

*** 8.** (73) $10^3 \div 10^2$

9. (38) $6.42 + 12.7 + 8$

10. (39) $1.2(0.12)$

11. (49) $64 \div 0.08$

12. (72) $3\frac{1}{3} \times \frac{1}{5} \times \frac{3}{4}$

13. (68) $2\frac{1}{2} \div 3$

Find each unknown number:

14. (43) $10 - q = 9.87$

15. (45) $24m = 0.288$

16. (63) $n - 2\frac{3}{4} = 3\frac{1}{3}$

17. (57) $w + \frac{1}{4} = \frac{5}{6}$

18. (38) The perimeter of a square is 80 cm. What is its area?

19. (46) Write the decimal number for the following:

$$(9 \times 10) + (6 \times 1) + \left(3 \times \frac{1}{100}\right)$$

*** 20.** (47) **Connect** Juana set the radius on the compass to 10 cm and drew a circle. What was the circumference of the circle? (Use 3.14 for π.)

21. (50) Which of these numbers is closest to zero?

A -2 **B** 0.2 **C** 1 **D** $\frac{1}{2}$

22. (51) **Estimate** Find the product of 6.7 and 7.3 by rounding each number to the nearest whole number before multiplying. Explain how you arrived at your answer.

*** 23.** (73) **Analyze** The expression 2^4 (two to the fourth power) is the prime factorization of 16. The expression 3^4 is the prime factorization of what number?

24. (18, 45) What number is halfway between 0.2 and 0.3?

25. (50) **Connect** To what decimal number is the arrow pointing on the number line below?

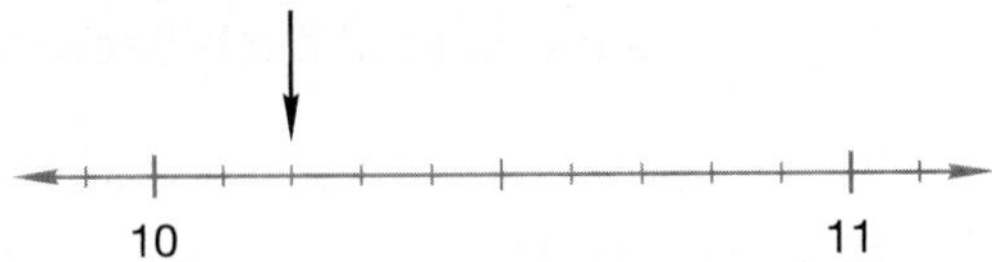

26. (64) Which quadrilateral has only one pair of parallel sides?

* **27.** (64, Inv. 7) **Analyze** The coordinates of the vertices of a quadrilateral are (–5, 5), (1, 5), (3, 1), and (–3, 1). What is the name for this kind of quadrilateral?

Analyze In the figure below, a square and a regular hexagon share a common side. The area of the square is 100 sq. cm. Use this information to answer problems **28** and **29.**

* **28.** (38) **a.** **Analyze** What is the length of each side of the square?

b. What is the perimeter of the square?

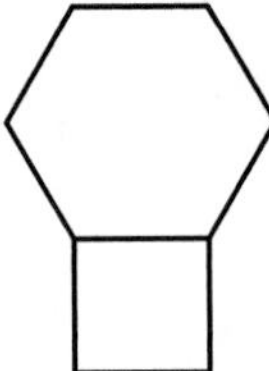

29. (8) **a.** What is the length of each side of the hexagon?

b. What is the perimeter of the hexagon?

30. (67) Write the prime factorization of both the numerator and the denominator of this fraction. Then reduce the fraction.

$$\frac{32}{48}$$

Early Finishers
Real-World Application

Mrs. Singh takes care of eight children. Half of the children drink four cups of milk a day. The other half drink two cups of milk a day. How many gallons of milk would Mrs. Singh have to purchase to have enough milk for the children for four days?

LESSON 80

Using a Constant Factor to Solve Ratio Problems

Power Up

Building Power

facts

Power Up I

mental math

a. **Calculation:** 4×315

b. **Number Sense:** $380 + 170$

c. **Calculation:** $5 \times \$2.99$

d. **Calculation:** $\$10.00 - \$7.99$

e. **Fractional Parts:** $\frac{1}{4}$ of \$4.80

f. **Decimals:** $37.5 \div 100$

g. **Measurement:** How many quarts are in a gallon?

h. **Calculation:** $5 \times 5, \times 5, - 25, \div 4, \div 5, - 5$

problem solving

A seven digit phone number consists of a three-digit prefix followed by four digits. How many different phone numbers are possible for a particular prefix?

New Concept

Increasing Knowledge

Consider the following ratio problem:

To make green paint, the ratio of blue paint to yellow paint is 3 to 2. For 6 ounces of yellow paint, how much blue paint is needed?

We see two uses for numbers in ratio problems. One use is to express a ratio. The other use is to express an actual count. A ratio box can help us sort the two uses by placing the ratio numbers in one column and the actual counts in another column. We write the items being compared along the left side of the rows.

	Ratio	Actual Count
Blue Paint	3	
Yellow Paint	2	6

We are told that the ratio of blue to yellow was 3 to 2. We place these numbers in the ratio column, assigning 3 to the blue paint row and 2 to the yellow paint row. We are given an actual count of 6 ounces of yellow paint, which we record in the box. We are asked to find the actual count of blue paint, so that portion of the ratio box is empty.

Ratio numbers and actual counts are related by **a constant factor.** If we multiply the terms of a ratio by the constant factor, we can find the actual count. Recall that a ratio is a reduced form of an actual count. If we can determine the factor by which the actual count was reduced to form the ratio, then we can recreate the actual count.

	Ratio	Actual Count
Blue Paint	3 × constant factor	?
Yellow Paint	2 × constant factor	6

We see that 2 can be multiplied by 3 to get 6. So 3 is the constant factor in this problem. That means we can multiply each ratio term by 3. Multiplying the ratio term 3 by the factor 3 gives us an actual count of 9 ounces of blue paint.

Example

Sadly, the ratio of flowers to weeds in the garden is 2 to 5. There are 30 flowers in the garden, about how many weeds are there?

Solution

We will begin by drawing a ratio box.

	Ratio	Actual Count
Flowers	2	30
Weeds	5	

To determine the constant factor, we study the row that has two numbers. In the "flowers" row we see the ratio number 2 and the actual count 30. If we divide 30 by 2, we find the factor, which is 15. Now we will use the factor to find the prediction of weeds in the garden.

$$\text{Ratio} \times \text{constant factor} = \text{actual count}$$

$$5 \times 15 = 75$$

There are about **75 weeds** in the garden.

Practice Set

Model Draw a ratio box and use a constant factor to solve each ratio problem:

a. The ratio of boys to girls in the cafeteria was 6 to 5. If there were 60 girls, how many boys were there?

b. The ratio of ants to flies at the picnic was 8 to 3. If there were 24 flies, predict how many ants were there?

Written Practice

Strengthening Concepts

1. *(Inv. 5)* What is the mean of 96, 49, 68, and 75? What is the range?

2. *(66)* The average depth of the ocean beyond the edges of the continents is $2\frac{1}{2}$ miles. How many feet is that? (1 mile = 5280 ft)

3. (15) **Formulate** The 168 girls who signed up for soccer were divided equally into 12 teams. How many players were on each team? Write an equation and solve the problem.

Analyze Parallelogram *ABCD* is divided into two congruent triangles. Segments *BA* and *CD* measure 3 in. Segments *AD* and *BC* measure 5 in. Segment *BD* measures 4 in. Refer to this figure to answer problems **4** and **5.**

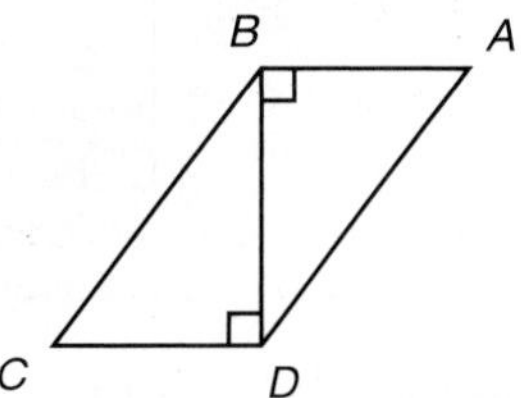

4. (71) **a.** What is the perimeter of the parallelogram?

b. What is the area of the parallelogram?

*** 5.** (79) **a.** What is the perimeter of each triangle?

b. What is the area of each triangle?

6. (64) **Conclude** This quadrilateral has one pair of parallel sides. What is the name of this kind of quadrilateral?

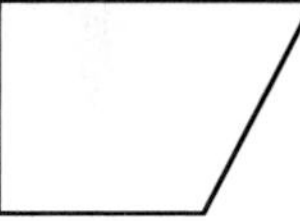

7. (64) **Verify** "All squares are rectangles." True or false?

*** 8.** (77) **Represent** If four fifths of the 30 students in the class were present, then how many students were absent?

*** 9.** (80) The ratio of dogs to cats in the neighborhood was 2 to 5. If there were 10 dogs, predict how many cats were there.

*** 10.** (75) Write as a percent:

a. $\frac{19}{20}$ **b.** 0.6

*** 11.** (74, 75) **Connect** **a.** What percent of the perimeter of a square is the length of one side?

b. What is the ratio of the side length of a square to its perimeter? Express the ratio as a fraction and as a decimal.

*** 12.** (76) Compare: **a.** $0.5 \bigcirc \frac{3}{4}$ **b.** 3 qts $\bigcirc$ 1 gal

*** 13.** (73) Write 4.4 as a reduced mixed number.

14. (74) Write $\frac{1}{8}$ as a decimal number.

15. (57) $\frac{5}{6} + \frac{1}{2}$ **16.** (57) $\frac{5}{8} - \frac{1}{4}$ **17.** (72) $2\frac{1}{2} \times 1\frac{1}{3} \times \frac{3}{5}$

Find each unknown number:

18. (43) $4 - a = 2.6$ **19.** (68) $3n = 1\frac{1}{2}$

20. (45) $5x = 0.36$ **21.** (49) $0.9y = 63$

22. (51) Round 0.4287 to the hundredths place.

Refer to the bar graph below to answer problems **23–25.**

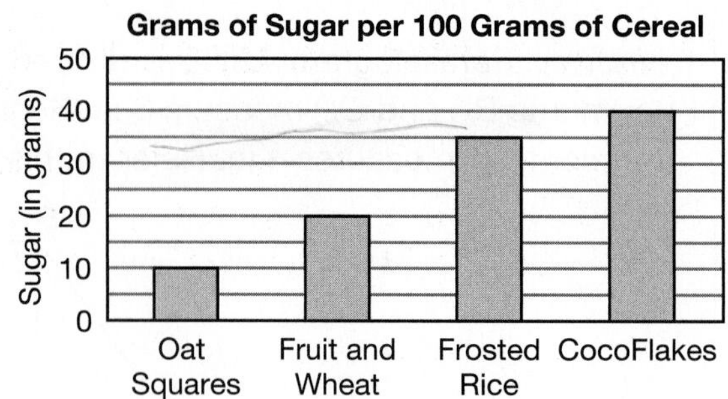

23. (Inv. 5) **Estimate** Frosted Rice contains about how many grams of sugar per 100 grams of cereal?

24. (Inv. 5) **Estimate** Fifty grams of CocoFlakes would contain about how many grams of sugar?

25. (Inv. 5) **Formulate** Write a problem about comparing that refers to the bar graph, and then answer the problem.

* **26.** (78) There was one quart of milk in the carton. Oscar poured one cup of milk on his cereal. How many cups of milk were left in the carton?

27. (Inv. 7) **Evaluate** Three vertices of a square are (3, 0), (3, 3), and (0, 3).

a. What are the coordinates of the fourth vertex of the square?

b. What is the area of the square?

28. (69) **Conclude** Which of these angles could be the complement of a 30° angle?

A ∠ **B** ∟ **C**

29. (47) If $A = \frac{1}{2}bh$, and if $b = 6$ and $h = 8$, then what does A equal?

30. (64) **Model** Draw a pair of parallel segments that are the same length. Form a quadrilateral by drawing two segments between the endpoints of the parallel segments. Is the quadrilateral a parallelogram?

INVESTIGATION 8

Focus on

• Geometric Construction of Bisectors

Since Lesson 27 we have used a compass to draw circles of various sizes. We can also use a compass together with a straightedge and a pencil to construct and divide various geometric figures.

Materials needed:

- Compass
- Ruler
- Pencil
- Several sheets of unlined paper
- Investigation Activity 17

Activity 1

Perpendicular Bisectors

Math Language

A **line** extends in opposite directions without end. A line cannot be bisected because it has no midpoint. A **segment** is a part of a line with two distinct endpoints.

Represent The first activity in this investigation is to **bisect** a segment. The word *bisect* means "to cut into two equal parts." We bisect a line segment when we draw a line (or segment) through the midpoint of the line segment. Below, segment *AB* is bisected by line *r* into two parts of equal length.

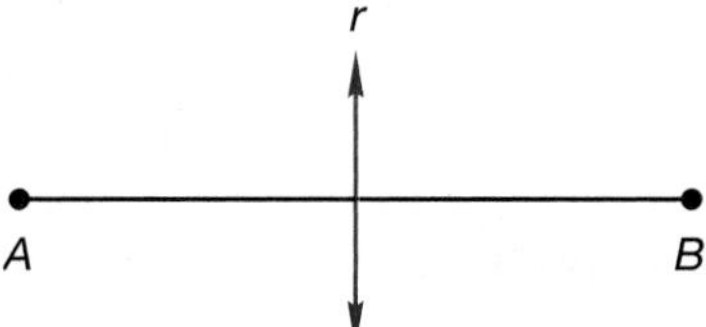

In section A of **Investigation Activity 17,** you will see segment *AB*. Follow these directions to bisect the segment.

Step 1: Set your compass so that the distance between the pivot point of the compass and the pencil point is more than half the length of the segment. Then place the pivot point of the compass on an endpoint of the segment and "swing an arc" on both sides of the segment, as illustrated.

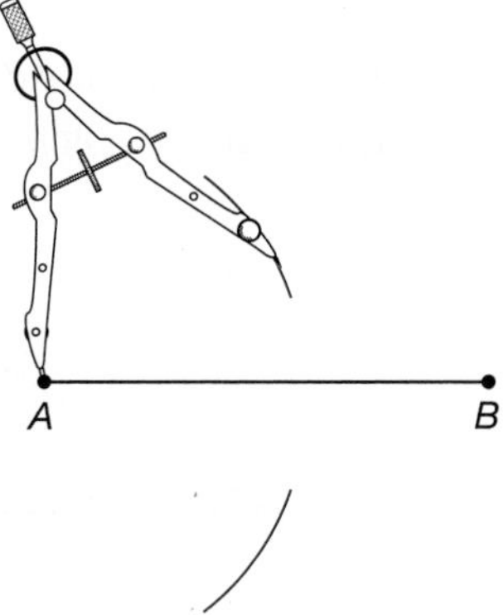

Step 2: Without resetting the radius of the compass, move the pivot point of the compass to the other endpoint of the segment. Swing an arc on both sides of the segment so that the arcs intersect as shown. (It may be necessary to return to the first endpoint to extend the first set of arcs until the arcs intersect on both sides of the segment.)

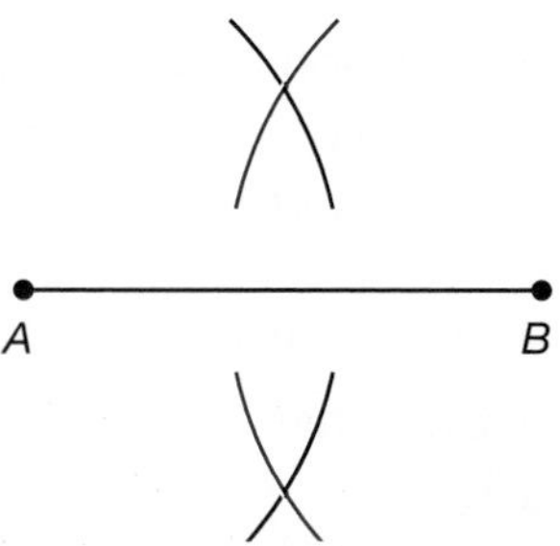

Thinking Skill

Explain

Why do we set our compass so that the distance from the pivot point to the pencil point is more than half the length of the segment?

Step 3: Draw a line through the two points where the arcs intersect.

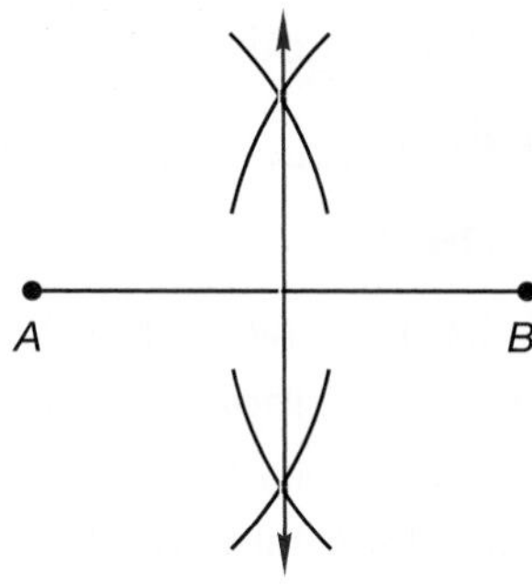

The line bisects the segment and is perpendicular to it. Thus the line is called a **perpendicular bisector** of the segment.

Check your work with a ruler. You should find that the perpendicular bisector has divided segment *AB* into two smaller segments of equal length.

Practice the procedure again by drawing your own line segment on a blank sheet of paper. Position the segment on the page so that there is enough area above and below the segment to draw the arcs you need to bisect the segment. Refer to the directions above if you need to refresh your memory.

Activity 2

Angle Bisectors

Math Language

A **ray** is part of a line with one endpoint. The **vertex** of an angle is the common endpoint of two rays that form the sides of the angle.

Represent In Section B of **Activity 17** an angle is shown. You will bisect the angle by drawing a ray halfway between the two sides of the angle.

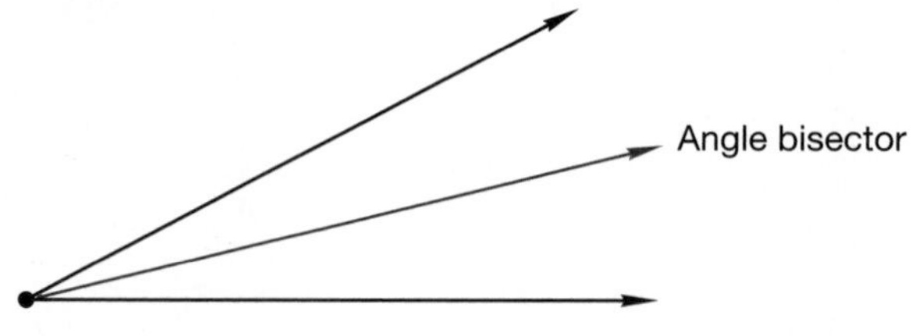

Follow these directions to bisect the angle.

Step 1: Place the pivot point of the compass on the vertex of the angle, and sweep an arc across both sides of the angle.

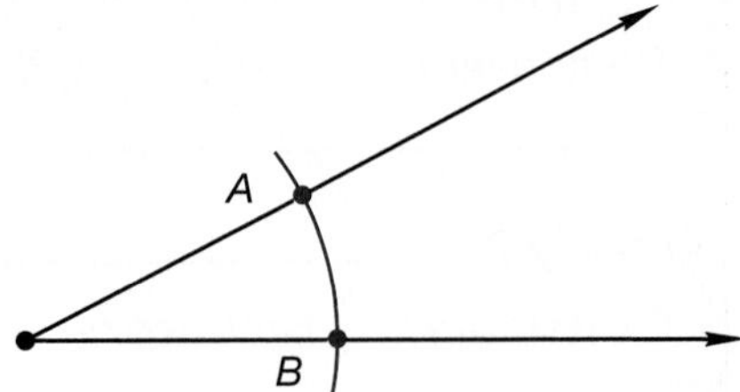

The arc intersects the sides of the angle at two points, which we have labeled *A* and *B*. Point *A* and point *B* are both the same distance from the vertex.

Step 2: Set the compass so that the distance between the pivot point of the compass and the pencil point is more than half the distance from point *A* to point *B*. Place the pivot point of the compass on point *A*, and sweep an arc as shown.

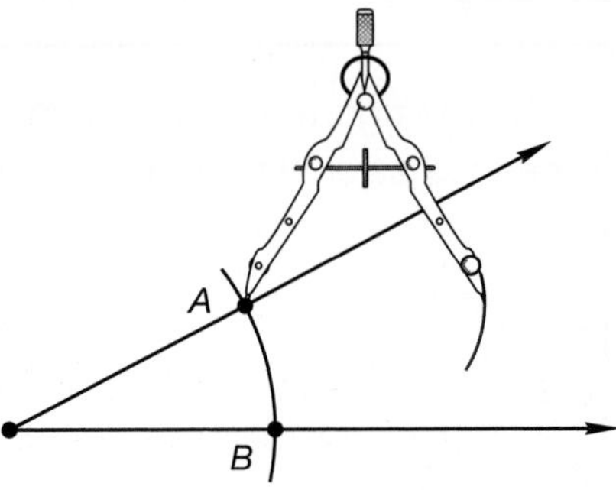

Step 3: Without resetting the radius of the compass, move the pivot point of the compass to point *B* and sweep an arc that intersects the arc drawn in step 2.

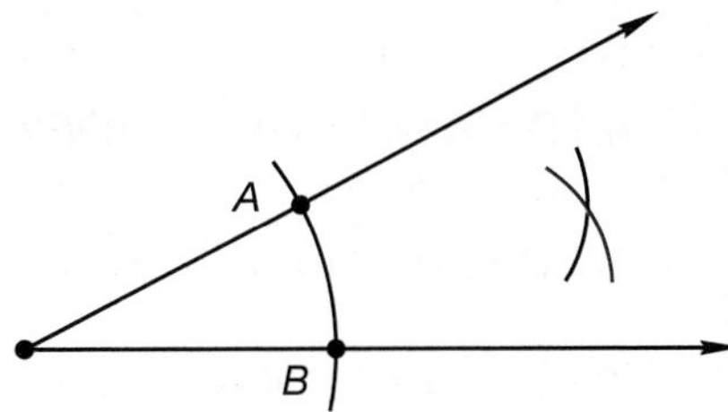

Step 4: Draw a ray from the vertex of the angle through the intersection of the arcs.

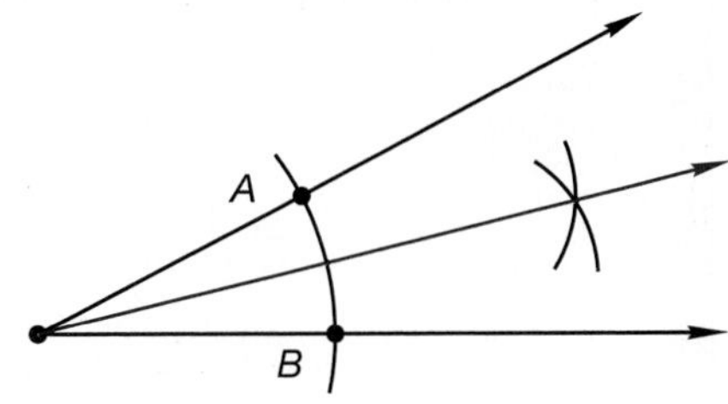

The ray is the **angle bisector** of the angle.

Use a protractor to check your work. You should find that the angle bisector has divided the angle into two smaller angles of equal measure.

Practice the procedure again by drawing an angle on a blank sheet of paper. Make the angle a different size from the one on the investigation activity. Then bisect the angle using the method presented in this activity.

Activity 3

Constructing Bisectors

In this activity you will make a page similar to **Investigation Activity 17** to give to another student. On an unlined sheet of paper, draw a line segment and draw an angle. The sizes of the segment and the angle should be different from the sizes of the line segment and angle on the activity page.

As your teacher directs, exchange papers. Using a compass and straightedge, construct the perpendicular bisector of the segment and the angle bisector of the angle on the sheet you are given. The arcs you draw in the construction should be visible so that your work can be checked.

extensions

Use the figure below and your protractor to answer questions **a–c.** Support each answer with angle measurements.

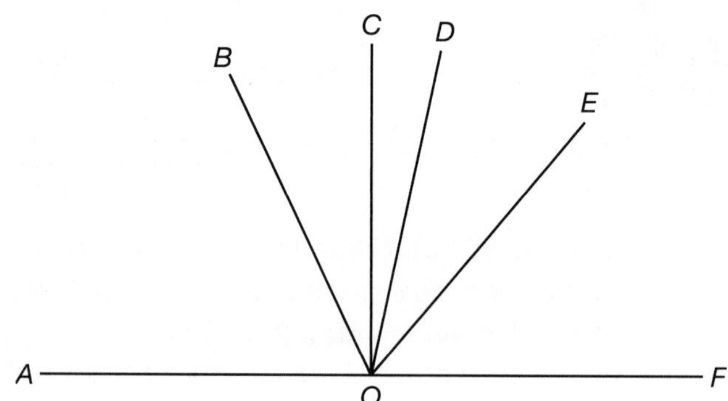

a. Does $\overrightarrow{OB}$ bisect $\angle AOE$?

b. Does $\overrightarrow{OD}$ bisect $\angle BOE$?

c. Use angle measures to classify these angles in the figure as acute, obtuse, or right.

$\angle AOE$	$\angle AOB$	$\angle BOD$
$\angle AOC$	$\angle EOF$	$\angle BOF$

LESSON 81

• Arithmetic with Units of Measure

Power Up

Building Power

facts Power Up K

mental math

a. **Calculation:** 311×7

b. **Number Sense:** $2000 - 1250$

c. **Calculation:** $4 \times \$9.99$

d. **Calculation:** $\$2.50 + \$9.99$

e. **Number Sense:** Double $5.50.

f. **Decimals:** 0.075×100

g. **Measurement:** How many milliliters are in 2 liters?

h. **Calculation:** 8×8, $+ 6$, $\div 2$, $+ 1$, $\div 6$, $\times 3$, $\div 2$

problem solving Alexis has 6 coins that total exactly $1.00. Name one coin she **must** have and one coin she cannot have.

New Concept

Increasing Knowledge

Recall that the operations of arithmetic are addition, subtraction, multiplication, and division. In this lesson we will practice adding, subtracting, multiplying, and dividing units of measure.

We may add or subtract measurements that have the same units. If the units are not the same, we first convert one or more measurements so that the units are the same. Then we add or subtract.

Example 1

Add: 2 ft + 12 in.

Solution

The units are not the same. Before we add, we either convert 2 feet to 24 inches or we convert 12 inches to 1 foot.

Convert to Inches	**Convert to Feet**
2 ft + 12 in.	2 ft + 12 in.
↓	↓
24 in. + 12 in. = **36 in.**	2 ft + 1 ft = **3 ft**

Either answer is correct, because 3 feet equals 36 inches.

Notice that in each equation in example 1, the units of the sum are the same as the units of the addends. The units do not change when we add or subtract measurements. However, the units *do* change when we multiply or divide measurements.

When we find the area of a figure, we multiply the lengths. Notice how the units change when we multiply.

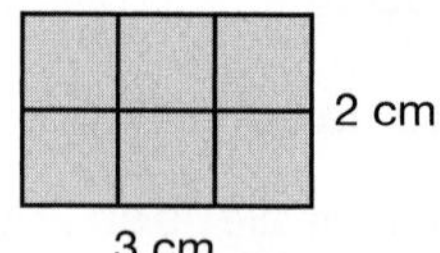

To find the area of this rectangle, we multiply 2 cm by 3 cm. The product has a different unit of measure than the factors.

Reading Math

Remember that the dot (·) indicates multiplication.

$$2 \text{ cm} \cdot 3 \text{ cm} = 6 \text{ sq. cm}$$

A centimeter and a square centimeter are two different kinds of units. A centimeter is used to measure length. It can be represented by a line segment.

1 cm

A square centimeter is used to measure area. It can be represented by a square that is 1 centimeter on each side.

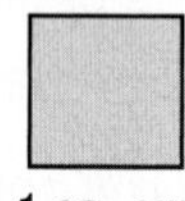

1 sq. cm

The unit of the product is a different unit because we multiplied the units of the factors. When we multiply 2 cm by 3 cm, we multiply both the numbers and the units.

$$2 \text{ cm} \cdot 3 \text{ cm} = \underbrace{2 \cdot 3}_{6} \ \underbrace{\text{cm} \cdot \text{cm}}_{\text{sq. cm}}$$

Instead of writing "sq. cm," we may use exponents to write "cm · cm" as "cm^2." Recall that we read cm^2 as "square centimeters."

$$2 \text{ cm} \cdot 3 \text{ cm} = \underbrace{2 \cdot 3}_{6} \ \underbrace{\text{cm} \cdot \text{cm}}_{\text{cm}^2}$$

Example 2

Multiply: 6 ft · 4 ft

Solution

We multiply the numbers. We also multiply the units.

$$6 \text{ ft} \cdot 4 \text{ ft} = \underbrace{6 \cdot 4}_{24} \ \underbrace{\text{ft} \cdot \text{ft}}_{\text{ft}^2}$$

The product is **24 ft^2,** which can also be written as "24 sq. ft."

Units also change when we divide measurements. For example, if we know both the area and the length of a rectangle, we can find the width of the rectangle by dividing.

Area = 21 cm²

7 cm

To find the width of this rectangle, we divide 21 cm² by 7 cm.

$$\frac{21\text{ cm}^2}{7\text{ cm}} = \frac{\overset{3}{\cancel{21}}}{\underset{1}{\cancel{7}}}\,\frac{\cancel{\text{cm}}\cdot\text{cm}}{\cancel{\text{cm}}}$$

We divide the numbers and write "cm²" as "cm · cm" in order to reduce the units. The quotient is 3 cm, which is the width of the rectangle.

Example 3

Divide: $\mathbf{\dfrac{25\text{ mi}^2}{5\text{ mi}}}$

Solution

To divide the units, we write "mi²" as "mi · mi" and reduce.

$$\frac{\overset{5}{\cancel{25}}}{\underset{1}{\cancel{5}}}\,\frac{\cancel{\text{mi}}\cdot\text{mi}}{\cancel{\text{mi}}}$$

The quotient is **5 mi.**

Sometimes when we divide measurements, the units will not reduce. When units will not reduce, we leave the units in fraction form. For example, if a car travels 300 miles in 6 hours, we can find the average speed of the car by dividing.

Reading Math

The word *per* means "for each" and is used in place of the division bar.

$$\frac{300\text{ mi}}{6\text{ hr}} = \frac{\overset{50}{\cancel{300}}}{\underset{1}{\cancel{6}}}\,\frac{\text{mi}}{\text{hr}}$$

The quotient is $50\frac{\text{mi}}{\text{hr}}$, which is 50 miles per hour (50 mph).

Notice that speed is a quotient of distance divided by time.

Example 4

Divide: $\mathbf{\dfrac{300\text{ mi}}{10\text{ gal}}}$

Solution

We divide the numbers. The units do not reduce.

$$\frac{300\text{ mi}}{10\text{ gal}} = \frac{\overset{30}{\cancel{300}}}{\underset{1}{\cancel{10}}}\,\frac{\text{mi}}{\text{gal}}$$

The quotient is $\mathbf{30\,\frac{mi}{gal}}$, which is 30 miles per gallon.

Practice Set Simplify:

a. 2 ft − 12 in. (Write the difference in inches.)

b. 2 ft × 4 ft **c.** $\frac{12 \text{ cm}^2}{3 \text{ cm}}$ **d.** $\frac{300 \text{ mi}}{5 \text{ hr}}$

Written Practice *Strengthening Concepts*

* **1.** (78) The Jones family had two gallons of milk before they ate breakfast. The family used two quarts of milk during breakfast. How many quarts of milk did the Jones family have after breakfast?

* **2.** (78) **Connect** One quart of milk is about 945 milliliters of milk. Use this information to compare a quart and a liter:

1 gallon ◯ 4 liters

3. (66) **Analyze** Carol cut $2\frac{1}{2}$ inches off her hair three times last year. How much longer would her hair have been at the end of the year if she had not cut it?

* **4.** (81) The plane flew 1200 miles in 3 hours. Divide the distance by the time to find the average speed of the plane.

5. (67) Write the prime factorization of both the numerator and the denominator of this fraction. Then reduce the fraction.

$$\frac{54}{135}$$

6. (29, 33) The basketball team scored 60% of its 80 points in the second half. Write 60% as a reduced fraction. Then find the number of points the team scored in the second half.

7. (71) What is the area of this parallelogram?

8. (71) What is the perimeter of this parallelogram?

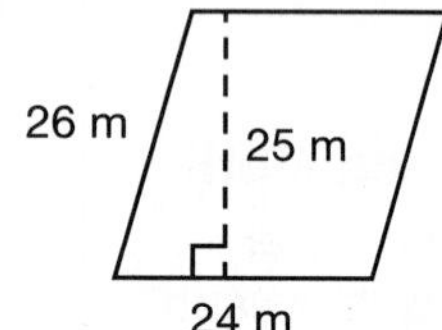

9. (64) **Verify** "Some rectangles are trapezoids." True or false? Why?

* **10.** (80) **Predict** The ratio of red marbles to blue marbles in the bag was 3 to 4. If 24 marbles were blue, how many were red?

* **11.** (76) Arrange these numbers in order from least to greatest:

$$\frac{1}{2}, \frac{1}{5}, 0.4$$

* **12.** (74, 75) **a.** What decimal number is equivalent to $\frac{4}{25}$?

b. What percent is equivalent to $\frac{4}{25}$?

13. (53) $(10 - 0.1) \times 0.1$

14. (53) $(0.4 + 3) \div 2$

15. (57) $\frac{5}{8} + \frac{3}{4}$

16. (63) $3 - 1\frac{1}{8}$

17. (63) $4\frac{1}{2} - 1\frac{3}{4}$

18. $\frac{5}{6} \cdot \frac{4}{5} \cdot \frac{3}{8}$ *(72)*

19. $4\frac{1}{2} \times 1\frac{1}{3}$ *(66)*

20. $3\frac{1}{3} \div 1\frac{2}{3}$ *(68)*

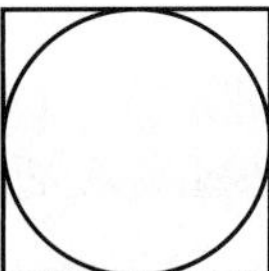

Analyze The perimeter of this square is two meters. Refer to this figure to answer problems **21** and **22.**

21. *(8)* How many centimeters long is each side of the square (1 meter = 100 centimeters)?

22. *(47)* **a.** What is the diameter of the circle?

b. What is the circumference of the circle? (Use 3.14 for π.)

23. *(41)* If the sales-tax rate is 6%, what is the tax on a $12.80 purchase?

24. *(32)* What time is two-and-one-half hours after 10:40 a.m.?

25. *(17)* Use a ruler to find the length of this line segment to the nearest sixteenth of an inch.

26. *(Inv. 7)* **Connect** What is the area of a quadrilateral with the vertices (0, 0), (4, 0), (6, 3), and (2, 3)?

27. *(Inv. 6)* What is the name of the geometric solid at right?

28. *(38)* If the area of a square is one square foot, what is the perimeter?

* **29.** *(81)* Simplify:

a. 2 yd + 3 ft (Write the sum in yards.)

b. 5 m × 3 m

c. $\frac{36 \text{ ft}^2}{6 \text{ ft}}$

d. $\frac{400 \text{ miles}}{20 \text{ gallons}}$

* **30.** *(64)* **Model** Draw a pair of parallel segments that are not the same length. Form a quadrilateral by drawing two segments between the endpoints of the parallel segments. What is the name of this type of quadrilateral?

LESSON 82

• Volume of a Rectangular Prism

Power Up

Building Power

facts Power Up L

mental math

a. Probability: What is the probability of rolling an even number on a number cube?

b. Number Sense: 284 − 150

c. Calculation: \$1.99 + \$2.99

d. Fractional Parts: $\frac{1}{3}$ of \$7.50

e. Decimals: 2.5 × 10

f. Number Sense: $\frac{800}{40}$

g. Measurement: How many ounces are in a cup?

h. Calculation: 10 × 10, − 1, ÷ 3, − 1, ÷ 4, + 1, ÷ 3

problem solving One-fifth of Ronnie's number is $\frac{1}{3}$. What is $\frac{3}{5}$ of Ronnie's number?

New Concept

Increasing Knowledge

The **volume** of a shape is the amount of space that the shape occupies. We measure volume by using units that take up space, called cubic units. The number of cubic units of space that the shape occupies is the volume measurement of that shape. We select units of appropriate size to describe a volume. For small volumes we can use cubic centimeters or cubic inches. For larger volumes we can use cubic feet and cubic meters.

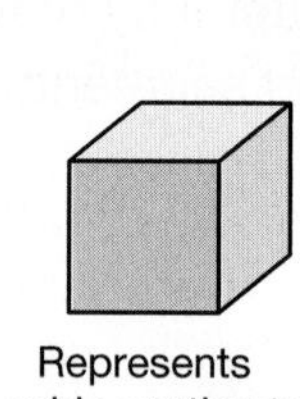

Represents
1 cubic centimeter

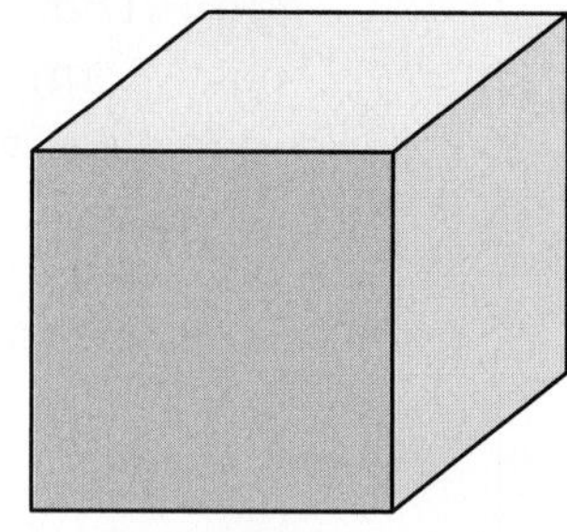

Represents
1 cubic inch

Math Language
The **base of a rectangular prism** is one of two parallel and congruent rectangular faces.

To calculate the volume of a rectangular prism, we can begin by finding the area of the base of the prism. Then we imagine building layers of cubes on the base up to the height of the prism.

Example 1

How many 1-inch cubes are needed to form this rectangular prism? (A 1-inch cube is a cube whose edges are 1 inch long.)

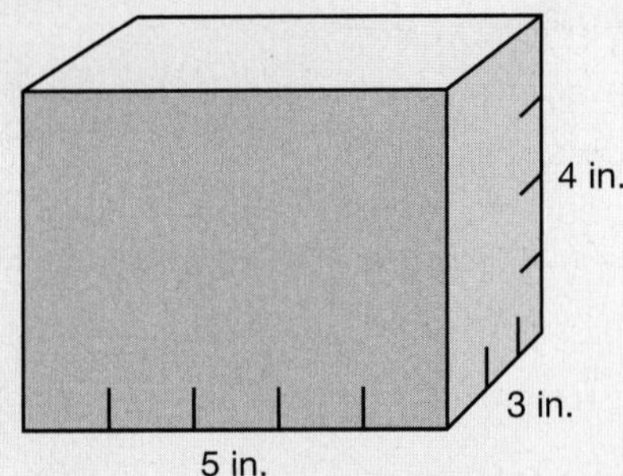

Solution

The area of the base is 5 inches times 3 inches, which equals 15 square inches. Thus 15 cubes are needed to make the bottom layer of the prism.

Reading Math

The formula for the area of the base is $A = l \times w$ or $A = lw$.

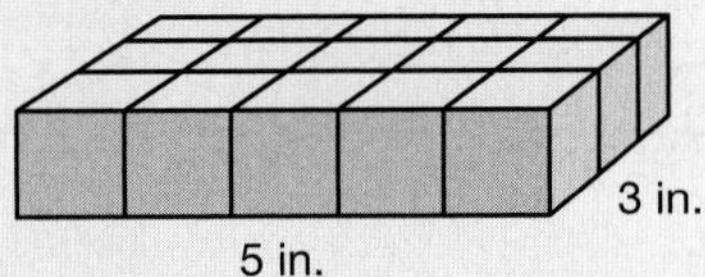

The prism is 4 inches high, so we will have 4 layers. The total number of cubes is 4 times 15, which is **60 cubes.**

Notice that in the example above we multiplied the length by the width to find the area of the base. Then we multiplied the area of the base by the height to find the volume.

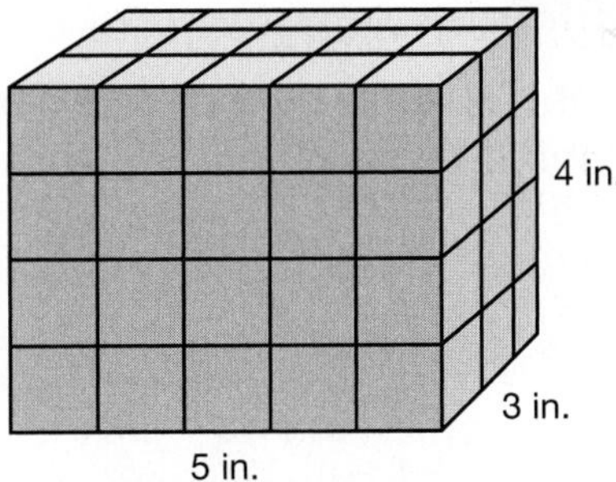

Thinking Skill

Discuss

Does the order in which we multiply the dimensions change the answer? Why or why not?

We can calculate the volume V of a rectangular prism by multiplying the three perpendicular dimensions of the prism: the length l, the width w, and the height h.

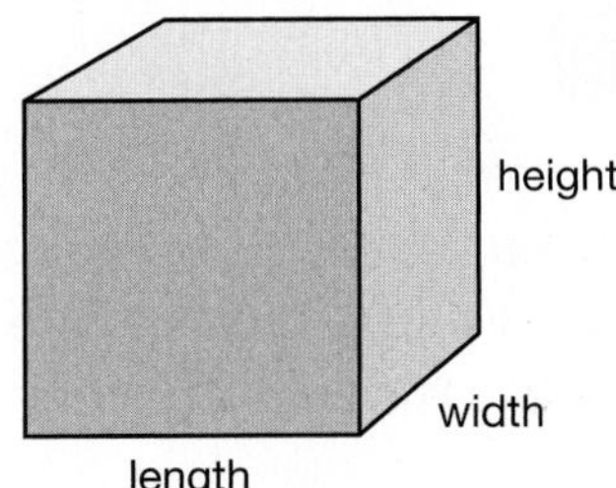

Thus the formula for finding the volume of a rectangular prism is

$$V = lwh$$

Example 2

What is the volume of a cube whose edges are 10 centimeters long?

Solution

Thinking Skill

Generalize

In addition to $V = lwh$, what other formulas could we use to find the volume of a cube?

The area of the base is 10 cm × 10 cm, or 100 sq. cm. Thus we could set 100 one-centimeter cubes on the bottom layer. There will be 10 layers, so it would take a total of 10 × 100, or 1000 cubes, to fill the cube. Thus the volume is **1000 cu. cm.**

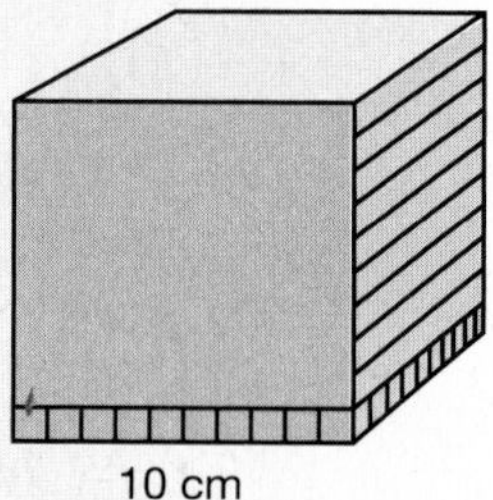

Example 3

Find the volume of a rectangular prism that is 4 feet long, 3 feet wide, and 2 feet high.

Solution

For l, w, and h we substitute 4 ft, 3 ft, and 2 ft. Then we multiply.

$$V = lwh$$

$$V = (4 \text{ ft})(3 \text{ ft})(2 \text{ ft})$$

$$V = \mathbf{24\ ft^3}$$

Notice that ft^3 means "cubic feet." We read 24 ft^3 as "24 cubic feet."

Example 4

Alison put small cubes together to build larger cubes. She made a table to record the number of cubes she used.

Length of Edge (Cubes Along Edge)	2	3	4	5
Volume (Number of Cubes Used)	8	27	64	125

Given this pattern, describe a rule that could be used to find the volume of a cube.

Solution

We can use a pattern to help us find a rule for determining the volume of a cube. In earlier examples we multiplied the length and width and height. For a cube these three measures are equal. We see that 2 × 2 × 2 is 8 and that 3 × 3 × 3 is 27. Thus cubing the edge of a cube gives us the volume, or $\boldsymbol{V = e^3}$**.**

Practice Set

a. How many 1-cm cubes would be needed to build a cube 4 cm on each edge?

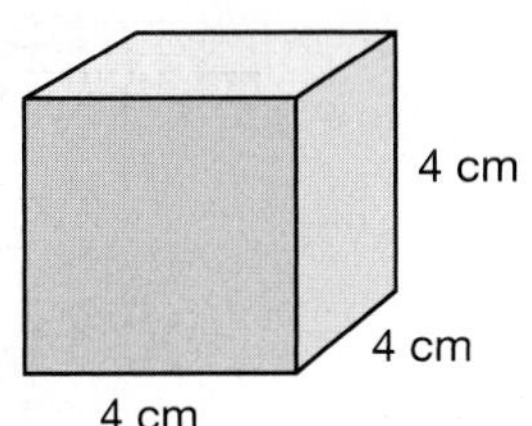

b. What is the volume of a rectangular box that is 5 feet long, 3 feet wide, and 2 feet tall?

c. **Analyze** The interior dimensions of a rectangular box are 10 inches by 6 inches by 4 inches. The box is to be filled with 1-inch cubes. How many cubes can fit on the bottom layer? How many cubes can fit in the box?

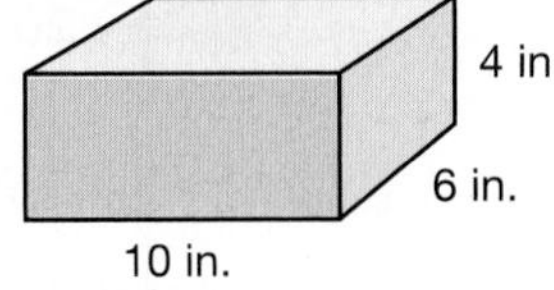

d. Choose the most appropriate unit to measure the volume of a refrigerator.

A cubic inches **B** cubic feet **C** cubic miles

Written Practice

Strengthening Concepts

1. (35) Write the number twenty-one and five hundredths.

2. (15) Tennis balls are sold in cans containing 3 balls. What would be the total cost of one dozen tennis balls if the price per can was $2.49?

3. (15) **Analyze** A cubit is about 18 inches. If Ruben was 4 cubits tall, about how many feet tall was he?

* **4.** (75) **a.** Write $\frac{7}{100}$ as a percent.

b. Write $\frac{7}{10}$ as a percent.

5. (33, 74) Write 90% as a reduced fraction. Then write the fraction as a decimal number.

6. (75) Of the 50 students who went on a trip, 23 wore a hat. What percent of the students wore a hat?

7. (75) Write $\frac{9}{25}$ as a percent.

8. (Inv. 6) **Connect** A box of cereal has the shape of what geometric solid?

Find each unknown number:

9. (63) $w - 3\frac{5}{6} = 2\frac{1}{3}$

10. (63) $3\frac{1}{4} - y = 1\frac{5}{8}$

11. (45) $6n = 0.12$

12. (49) $0.12m = 6$

13. (38) $5n = 10^2$

14. (68) $1\frac{1}{2}w = 6$

*** 15.** (75) **a.** What fraction of this group is shaded?

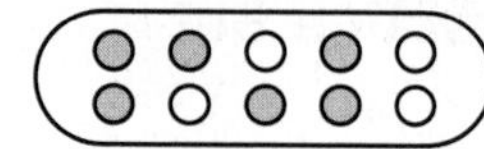

b. What percent of this group is shaded?

16. (53) $0.5 + (0.5 \div 0.5) + (0.5 \times 0.5)$

17. (61) $\frac{1}{2} + \frac{1}{5} + \frac{1}{10}$

18. (66) $1\frac{4}{5} \times 1\frac{2}{3}$

19. (34) Which digit in 6.3457 has the same place value as the 8 in 128.90?

20. (16) Estimate the product of 39 and 41.

21. (23, 58) In a bag there are 12 red marbles and 36 blue marbles.

a. What is the ratio of red marbles to blue marbles?

b. **Predict** If one marble is taken from the bag, what is the probability that the marble will be red? Express the probability ratio as a fraction and as a decimal.

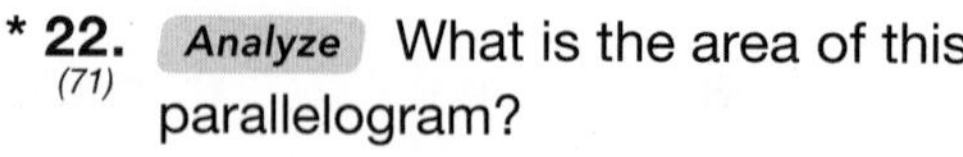

*** 22.** (71) **Analyze** What is the area of this parallelogram?

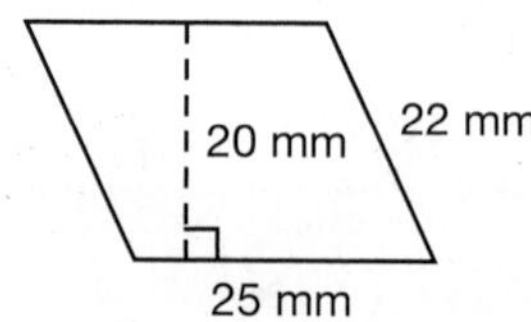

23. (71) What is the perimeter of this parallelogram?

24. (73) Write the prime factorization of 252 using exponents.

25. (60) **Verify** "Some triangles are quadrilaterals." True or false? Why?

*** 26.** (64, Inv. 7) **Model** A quadrilateral has vertices with the coordinates $(-2, -1)$, $(1, -1)$, $(3, 3)$, and $(-3, 3)$. Graph the quadrilateral on a coordinate plane. The figure is what type of quadrilateral?

*** 27.** (82) This cube is constructed of 1-inch cubes. What is the volume of the larger cube?

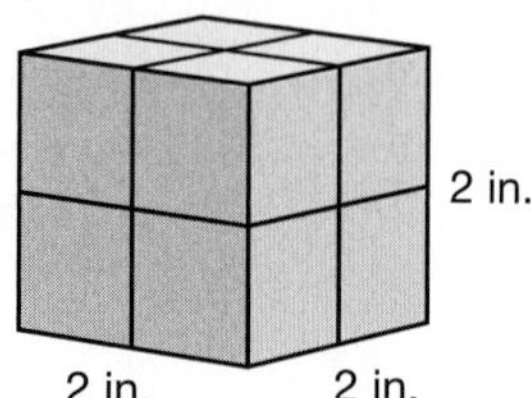

*** 28.** (81) Simplify:

a. 3 quarts + 2 pints (Write the sum in quarts.)

b. $\frac{49 \text{ m}^2}{7 \text{ m}}$

c. $\frac{400 \text{ miles}}{8 \text{ hours}}$

*** 29.** (75) Three of the dozen eggs were cracked. What percent of the eggs were cracked?

*** 30.** (78) **Estimate** A pint of milk weighs about a pound. About how many pounds does a gallon of milk weigh?

LESSON 83

• Proportions

Power Up

Building Power

facts Power Up I

mental math

a. Probability: What is the probability of rolling a number less than 3 on a number cube?

b. Number Sense: 1000 − 125

c. Calculation: 3 × $3.99

d. Number Sense: Double $3\frac{1}{2}$.

e. Decimals: 2.5 ÷ 100

f. Number Sense: 20 × 34

g. Measurement: How many milliliters are in 4 liters?

h. Calculation: 9 × 9, − 1, ÷ 2, + 2, ÷ 6, + 2, ÷ 3

problem solving

Compare the following two separate quantities:

$$1\frac{7}{8} + 2\frac{5}{6} \bigcirc 3\frac{4}{5} \qquad 6.142 \times 9.065 \bigcirc 54$$

Describe how you performed the comparisons.

New Concept

Increasing Knowledge

If peaches are on sale for 3 pounds for 4 dollars then the ratio $\frac{3}{4}$ expresses the relationship between the quantity and the price of peaches. Since the ratio is constant, we can buy 6 pounds for 8 dollars, 9 pounds for 12 dollars and so on. With two equal ratios we can write a proportion.

Peaches
3 lbs.
for
$4

Math Language
A **ratio** is a comparison of two numbers by division.

A **proportion** is a true statement that two ratios are equal. Here is an example of a proportion:

$$\frac{3}{4} = \frac{6}{8}$$

We read this proportion as "Three is to four as six is to eight." Two ratios that are not equivalent are not proportional.

Example 1

Which ratio forms a proportion with $\frac{2}{3}$?

A $\frac{2}{4}$ **B** $\frac{3}{4}$ **C** $\frac{4}{6}$ **D** $\frac{3}{2}$

Solution

Equivalent ratios form a proportion. Equivalent ratios also reduce to the same rate. Notice that $\frac{2}{4}$ reduces to $\frac{1}{2}$; that $\frac{3}{4}$ and $\frac{2}{3}$ are reduced, and that $\frac{4}{6}$ reduces to $\frac{2}{3}$. Thus the ratio equivalent to $\frac{2}{3}$ is **C.**

Verify How can we verify that $\frac{2}{3}$ and $\frac{4}{6}$ form a proportion?

Example 2

Write this proportion with digits: Four is to six as six is to nine.

Solution

We write "four is to six" as one ratio and "six is to nine" as the equivalent ratio. We are careful to write the numbers in the order stated.

$$\mathbf{\frac{4}{6} = \frac{6}{9}}$$

We can use proportions to solve a variety of problems. Proportion problems often involve finding an unknown term. The letter *a* represents an unknown term in this proportion:

$$\frac{3}{5} = \frac{6}{a}$$

Math Language

A **scale factor** is a number that relates corresponding sides of similar figures and corresponding terms in equivalent ratios.

One way to find an unknown term in a proportion is to determine the fractional name for 1 that can be multiplied by one ratio to form the equivalent ratio. The first terms in these ratios are 3 and 6. Since 3 times 2 equals 6, we find that the scale factor is 2. So we multiply $\frac{3}{5}$ by $\frac{2}{2}$ to form the equivalent ratio.

$$\frac{3}{5} \cdot \frac{2}{2} = \frac{6}{10}$$

We find that *a* represents the number 10.

Example 3

Complete this proportion: Two is to six as what number is to 30?

Solution

We write the terms of the proportion in the stated order, using a letter to represent the unknown number.

$$\frac{2}{6} = \frac{n}{30}$$

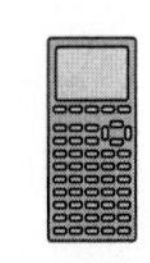

Visit www.SaxonPublishers.com/ActivitiesC1 *for a graphing calculator activity.*

We are not given both first terms, but we are given both second terms, 6 and 30. The scale factor is 5, since 6 times 5 equals 30. We multiply $\frac{2}{6}$ by $\frac{5}{5}$ to complete the proportion.

$$\frac{2}{6} \times \frac{5}{5} = \frac{10}{30}$$

The unknown term of the proportion is **10.**

$$\frac{2}{6} = \frac{10}{30}$$

Evaluate How can we check the answer?

Practice Set

a. Which ratio forms a proportion with $\frac{5}{2}$?

A $\frac{3}{2}$ **B** $\frac{4}{10}$ **C** $\frac{15}{6}$ **D** $\frac{5}{20}$

b. Write this proportion with digits: Six is to eight as nine is to twelve.

c. Write and complete this proportion: Four is to three as twelve is to what number? How did you find your answer?

d. **Explain** Write and complete this proportion: Six is to nine as what number is to thirty-six? How can you check your answer?

Written Practice

Strengthening Concepts

1. (12, 53) What is the product when the sum of 0.2 and 0.2 is multiplied by the difference of 0.2 and 0.2?

2. (66) **Analyze** Arabian camels travel about 3 times as fast as Bactrian camels. If Bactrian camels travel at $1\frac{1}{2}$ miles per hour, at how many miles per hour do Arabian camels travel?

3. (32) **Connect** Mark was paid at a rate of $4 per hour for cleaning up a neighbor's yard. If he worked from 1:45 p.m. to 4:45 p.m., how much was he paid?

4. (33) Write 55% as a reduced fraction.

5. (75) **a.** Write $\frac{9}{100}$ as a percent.

b. Write $\frac{9}{10}$ as a percent.

6. (75) The whole class was present. What percent of the class was present?

7. (29, 75) **Connect** A century is 100 years. A decade is 10 years.

a. What fraction of a century is a decade?

b. What percent of a century is a decade?

8. (73, 75) **a.** Write 0.48 as a reduced fraction.

b. Write 0.48 as a percent.

9. (74) Write $\frac{7}{8}$ as a decimal number.

10. (48) $\left(1\frac{1}{3} + 1\frac{1}{6}\right) - 1\frac{2}{3}$

11. (72) $1\frac{1}{2} \times 3 \times 1\frac{1}{9}$

12. (68) $4\frac{2}{3} \div 1\frac{1}{6}$

13. (38) $0.1 + (1 - 0.01)$

* **14.** (83) **Analyze** Write and complete this proportion: Three is to four as nine is to what number?

15. (15) **Verify** Which ratio below forms a proportion with $\frac{3}{5}$?

A $\frac{3}{10}$ **B** $\frac{6}{15}$ **C** $\frac{12}{20}$ **D** $\frac{5}{3}$

16. (32) Write the standard numeral for the following:

$$(8 \times 10{,}000) + (4 \times 100) + (2 \times 10)$$

17. (7, 73) **a.** Compare: $2^4 \bigcirc 4^2$ **b.** 1 km ◯ 1 mi

18. (67) Write the prime factorization of both the numerator and the denominator of this fraction. Then reduce the fraction.

$$\frac{24}{32}$$

* **19.** (20, 83) **a.** **Analyze** What is the greatest common factor of 24 and 32?

b. Which two fractions reduce to the same number?

$$\frac{24}{32} \quad \frac{9}{16} \quad \frac{9}{12} \quad \frac{12}{18}$$

20. (47) **Estimate** If the diameter of a ceiling fan is 4 ft, then the tip of one of the blades on the fan moves about how far during one full turn? Choose the closest answer.

A 8 ft **B** 12 ft **C** $12\frac{1}{2}$ ft **D** 13 ft

21. (8) What is the perimeter of this trapezoid?

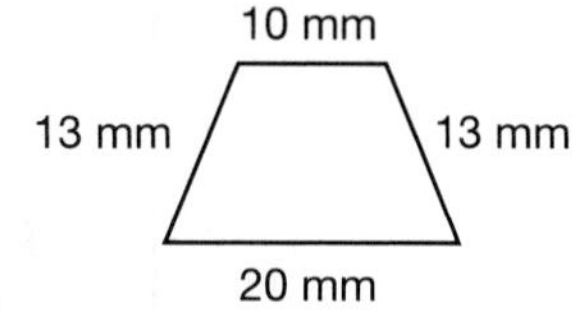

* **22.** (82) A cube has edges 3.1 cm long. What is a good estimate for the volume of the cube?

* **23.** (71, 79) **a.** What is the area of this parallelogram?

b. What is the area of the shaded triangle?

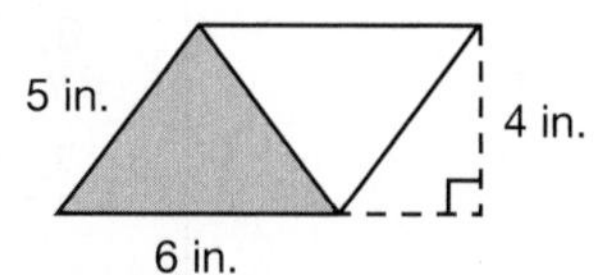

* **24.** (77) One fourth of the 120 students took wood shop. How many students did not take wood shop?

25. (7) How many millimeters is 2.5 centimeters?

*** 26.** *(Inv. 6)* **a.** What is the name of this geometric solid?

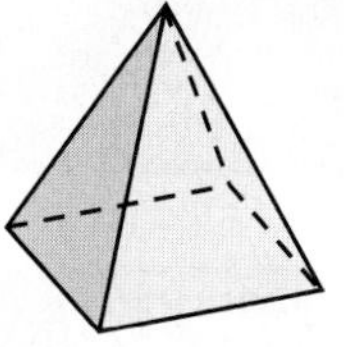

b. Sketch a net of this solid.

*** 27.** *(81)* Simplify:

a. 3 quarts + 2 pints (Write the sum in pints.)

b. $\dfrac{64\ \text{cm}^2}{8\ \text{cm}}$ **c.** $\dfrac{60\ \text{students}}{3\ \text{teachers}}$

*** 28.** *(75)* DeShawn delivers newspapers to 20 of the 25 houses on North Street. What percent of the houses on North Street does DeShawn deliver papers to?

29. *(28, 60)* **Represent** Draw a triangle that has two perpendicular sides.

*** 30.** *(15, 83)* **Analyze** The ratio of dimes to nickels in Pilar's change box is $\frac{2}{3}$. Pilar has \$0.75 in nickels.

a. How many nickels does Pilar have?

b. How many dimes does Pilar have? (*Hint:* Write and complete a proportion using the ratio given above and your answer to **a.**)

c. In all, how much money does Pilar have in her change box?

Early Finishers
Real-World Application

Use a protractor to measure each angle listed. Then classify each angle as acute, right, obtuse, or straight.

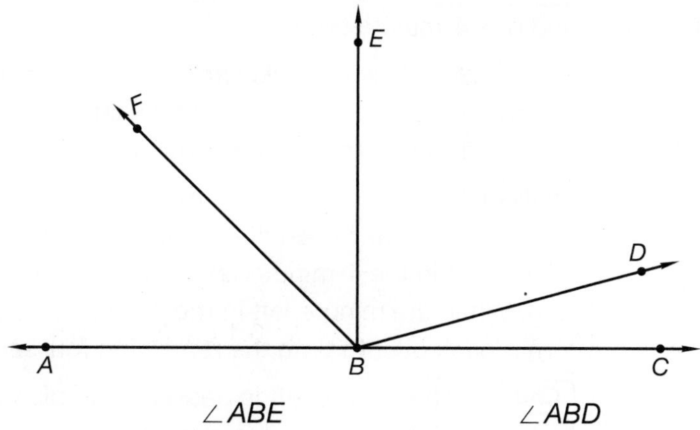

∠*ABF*	∠*ABE*	∠*ABD*
∠*ABC*	∠*FBE*	∠*FBD*
∠*CBD*	∠*CBE*	∠*CBF*
∠*DBE*		

LESSON

84

• Order of Operations, Part 2

Power Up

Building Power

facts Power Up K

mental math

a. **Probability:** What is the probability of rolling a number greater than 2 on a number cube?

b. **Calculation:** 980 − 136

c. **Calculation:** \$5.99 + \$2.99

d. **Fractional Parts:** $\frac{1}{4}$ of \$10.00

e. **Decimals:** 7.5 × 100

f. **Number Sense:** $\frac{480}{20}$

g. **Measurement:** How many ounces are in a pint?

h. **Calculation:** 8 × 8, − 4, ÷ 3, + 4, ÷ 4, + 2, ÷4

problem solving

Katrina's hourglass sand timer runs for exactly three minutes. Jessi's timer runs for exactly four minutes. The two girls want to play a game where they each get one five-minute turn. Explain how the girls can use their timers to mark off exactly five minutes.

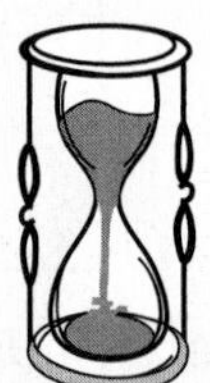

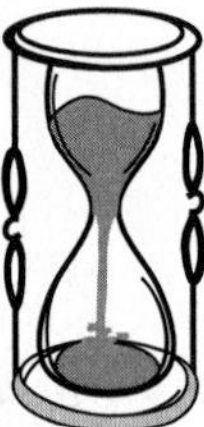

Understand We need to time a 5-minute turn. We have one 3-minute timer and one 4-minute timer.

Plan We will *work backwards* and *use logical reasoning* to find the answer. We can use the 4-minute timer by itself to mark off four minutes, so we will look for a way to mark off one minute using both timers.

Solve We see that we can mark off one minute by turning both timers over at the same time. When the 3-minute timer is empty, there is exactly one minute left in the 4-minute timer. At this point the player should begin her turn. When the minute left in the 4-minute timer runs out, we can immediately turn it back over to time the remaining four minutes.

Check The minute left in Jessi's timer plus the four minutes when it is turned back over totals five minutes.

New Concept — *Increasing Knowledge*

Recall that the four operations of arithmetic are addition, subtraction, multiplication, and division. When more than one type of operation occurs in the same expression, we perform the operations in the order described below.

Order of Operations

1. Perform operations within parentheses.
2. Multiply and divide from left to right.
3. Add and subtract from left to right

Example 1

Simplify: $2 \cdot 8 + 2 \cdot 6$

Solution

Multiplication and addition occur in this expression. We multiply first.

$$\underbrace{2 \times 8}_{16} + \underbrace{2 \times 6}_{12}$$

Then we add.

$$16 + 12 = \mathbf{28}$$

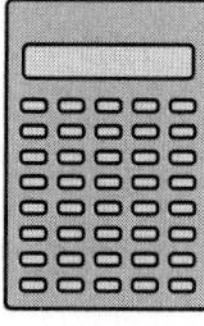

Some calculators are designed to recognize the standard order of operations and some are not. If a variety of calculator models are available in the classroom, you can test their design by using the expression from example 1. Enter these keystrokes:

"Algebraic logic" calculators should display the following after the equal sign is pressed:

28.

Example 2

Simplify: $0.5 + 0.5 \div 0.5 - 0.5 \times 0.5$

Solution

First we multiply and divide from left to right.

$$0.5 + \underbrace{0.5 \div 0.5}_{1} - \underbrace{0.5 \times 0.5}_{0.25}$$

Then we add and subtract from left to right.

$$0.5 + 1 - 0.25 = \mathbf{1.25}$$

Example 3

Simplify: 2(8 + 6)

Solution

First we perform the operation within the parentheses.

$$2(8 + 6)$$
$$2(14)$$

Then we multiply.

$$2(14) = \mathbf{28}$$

Practice Set

Simplify:

a. $5 + 5 \times 5 - 5 \div 5$

b. $32 + 1.8(20)$

c. $5 + 4 \times 3 \div 2 - 1$

d. $2(10) + 2(6)$

e. $3 + 3 \times 3 - 3 \div 3$

f. $2(10 + 6)$

Written Practice

Strengthening Concepts

1. (23, 65) **Classify** What is the ratio of prime numbers to composite numbers in this list?

2, 3, 4, 5, 6, 7, 8, 9, 10

* **2.** (78) Bianca poured four cups of milk from a full half-gallon container. How many cups of milk were left in the container?

* **3.** (84) **Analyze** $6 + 6 \times 6 - 6 \div 6$

4. (33, 74) Write 30% as a reduced fraction. Then write the fraction as a decimal number.

Find the area of each triangle:

* **5.** (79)

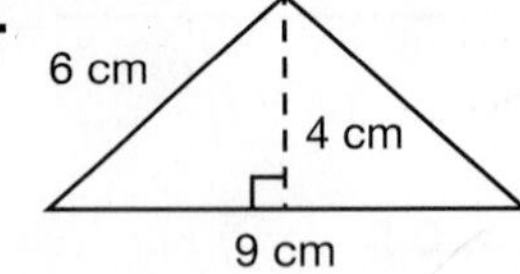

* **6.** (79)

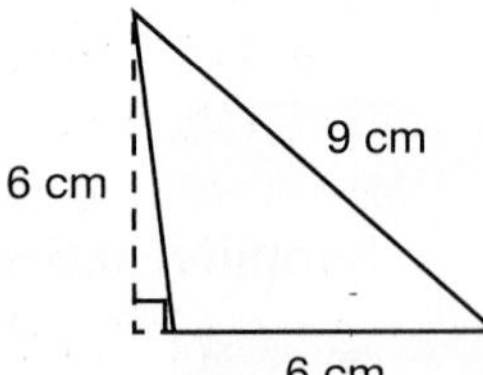

* **7.** (74, 75) **a.** Write $\frac{1}{20}$ as a decimal number.

b. Write $\frac{1}{20}$ as a percent.

8. (64) **Verify** "Some parallelograms are rectangles." True or false? Why?

9. What is the area of this parallelogram?
(71)

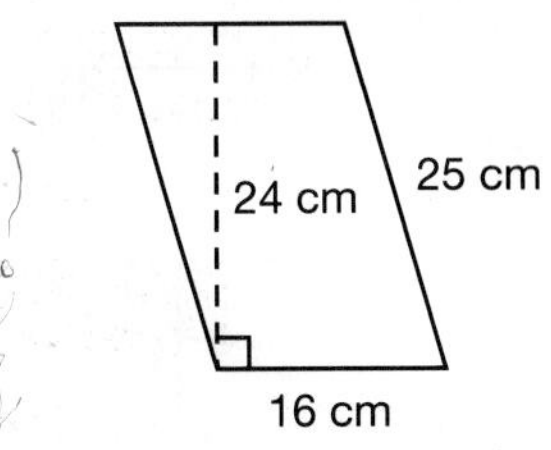

10. What is the perimeter of this parallelogram?
(71)

11. $\left(3\frac{1}{8} + 2\frac{1}{4}\right) - 1\frac{1}{2}$
(48)

12. $\frac{5}{6} \times 2\frac{2}{3} \times 3$
(72)

13. $8\frac{1}{3} \div 100$
(68)

14. $(4 - 3.2) \div 10$
(53)

15. $0.5 \times 0.5 + 0.5 \div 0.5$
(84)

16. $8 \div 0.04$
(49)

17. Which digit is in the hundredths place in 12.345678?
(34)

18. **Explain** How do you round $5\frac{1}{8}$ to the nearest whole number?
(51)

19. **Analyze** Write the prime factorization of 700 using exponents.
(73)

20. Two ratios form a proportion if the ratios reduce to the same fraction. Which two ratios below form a proportion?
(83)

$$\frac{15}{12} \quad \frac{15}{9} \quad \frac{25}{10} \quad \frac{35}{21}$$

21. **Connect** The perimeter of a square is 1 meter. How many centimeters long is each side?
(8)

* **22.** Fong scored 9 of the team's 45 points.
(29, 75)

a. What fraction of the team's points did Fong score?

b. What percent of the team's points did she score?

23. What time is 5 hours 30 minutes after 9:30 p.m.?
(32)

* **24.** **Analyze** Write and complete this proportion: Six is to four as what number is to eight?
(83)

25. **Conclude** Figure *ABCD* is a parallelogram. Its opposite angles ($\angle A$ and $\angle C$, $\angle B$ and $\angle D$) are congruent. Its adjacent angles (such as $\angle A$ and $\angle B$) are supplementary. If angle *A* measures 70°, what are the measures of $\angle B$, $\angle C$, and $\angle D$?
(69)

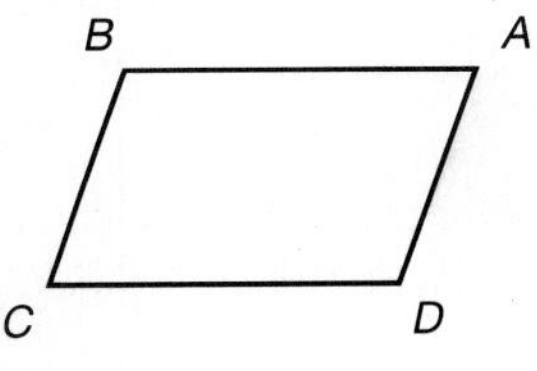

* **26.** If each small cube has a volume of 1 cm^3, what is the volume of this rectangular prism?
(82)

Simplify:

* **27.** 2 ft + 24 in. (Write the sum in inches.)
(81)

* **28.** (81) **a.** $\frac{100 \text{ cm}^2}{10 \text{ cm}}$ **b.** $\frac{180 \text{ pages}}{4 \text{ days}}$

29. (Inv. 7) *Model* A triangle has vertices at the coordinates (4, 4) and (4, 0) and at the origin. Draw the triangle on graph paper. Notice that inside the triangle are some full squares and some half squares.

a. How many full squares are in the triangle?

b. How many half squares are in the triangle?

30. (23) *Analyze* This year Moises has read 24 books. Sixteen of the books were non-fiction and the rest were fiction. What is the ratio of fiction to non-fiction books Moises has read this year?

Early Finishers
Real-World Application

Alejandra owns a triangular plot of land. She hopes to buy another triangular section adjacent to the one she owns. Use the figure below to find the area of the land Alejandra owns and the area of the land she hopes to buy.

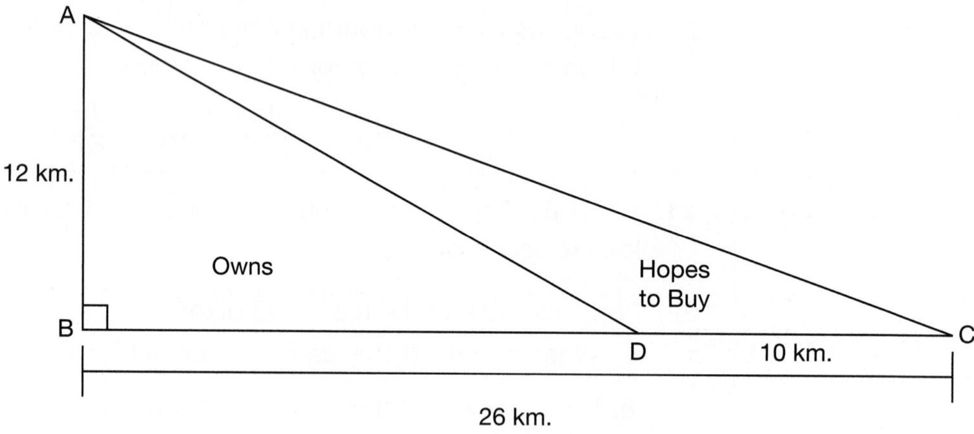

LESSON 85

Using Cross Products to Solve Proportions

Power Up

Building Power

facts Power Up L

mental math

a. **Number Sense:** 50×50

b. **Number Sense:** $1000 - 625$

c. **Calculation:** $4 \times \$3.99$

d. **Number Sense:** Double \$1.25.

e. **Decimals:** $7.5 \div 10$

f. **Number Sense:** 20×35

g. **Measurement:** How many liters are in 3000 milliliters?

h. **Calculation:** 7×7, $+ 1$, $\div 2$, $- 1$, $\div 2$, $\times 5$, $\div 2$

problem solving

Copy this problem and fill in the missing digits. No two digits in the problem may be alike.

$$\begin{array}{r} \_\,\_\,\_ \\ \times \quad 7 \\ \hline 9\,\_\,\_ \end{array}$$

New Concept

Increasing Knowledge

We have compared fractions by writing the fractions with common denominators. A variation of this method is to determine whether two fractions have equal **cross products.** If the cross products are equal, then the fractions are equal. The cross products of two fractions are found by cross multiplication, as we show below.

$$8 \times 3 = 24 \qquad \frac{3}{4} \times \frac{6}{8} \qquad 4 \times 6 = 24$$

Both cross products are 24. Since the cross products are equal, we can conclude that the fractions are equal.

Equal fractions have equal cross products.

Example 1

Use cross products to determine whether $\frac{3}{5}$ and $\frac{4}{7}$ are equal.

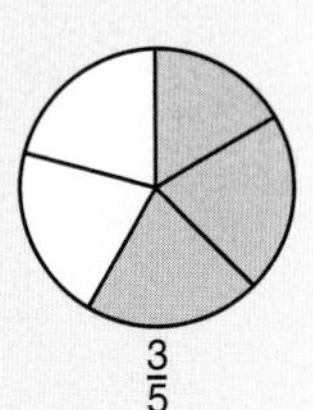

$\frac{3}{5}$ $\frac{4}{7}$

Solution

To find the cross products, we multiply the numerator of each fraction by the denominator of the other fraction. We write the cross product above the numerator that is multiplied.

$$\overset{21}{\nwarrow}\ \frac{3}{5} \times \frac{4}{7}\ \overset{20}{\nearrow}$$

The cross products are not equal, so **the fractions are not equal.** The greater cross product is above the greater fraction. So $\frac{3}{5}$ is greater than $\frac{4}{7}$.

When we find the cross products of two fractions, we are simply renaming the fractions with common denominators. The common denominator is the product of the two denominators and is usually not written. Look again at the two fractions we compared:

$$\frac{3}{5} \qquad \frac{4}{7}$$

The denominators are 5 and 7.

If we multiply $\frac{3}{5}$ by $\frac{7}{7}$ and multiply $\frac{4}{7}$ by $\frac{5}{5}$, we form two fractions that have common denominators.

$$\frac{3}{5} \times \frac{7}{7} = \frac{21}{35} \qquad \frac{4}{7} \times \frac{5}{5} = \frac{20}{35}$$

The numerators of the renamed fractions are 21 and 20, which are the cross products of the fractions. So when we compare cross products, we are actually comparing the numerators of the renamed fractions.

Example 2

Math Language

A **proportion** is a statement that shows two ratios are equal.

Do these two ratios form a proportion?

$$\mathbf{\frac{8}{12}, \frac{12}{18}}$$

Solution

If the cross products of two ratios are equal, then the ratios are equal and therefore form a proportion. To find the cross products of the ratios above, we multiply 8 by 18 and 12 by 12.

$$\overset{144}{\nwarrow}\ \frac{8}{12} \times \frac{12}{18}\ \overset{144}{\nearrow}$$

The cross products are 144 and 144, so **the ratios form a proportion.**

$$\frac{8}{12} = \frac{12}{18}$$

Since equivalent ratios have equal cross products, we can use cross products to find an unknown term in a proportion. By cross multiplying, we form an equation. Then we solve the equation to find the unknown term of the proportion.

Example 3

Use cross products to complete this proportion: $\frac{6}{9} = \frac{10}{m}$

Solution

The cross products of a proportion are equal. So 6 times *m* equals 9 times 10, which is 90.

$$\frac{6}{9} = \frac{10}{m}$$

$$6m = 9 \cdot 10$$

We solve this equation:

$$6m = 90$$

$$m = 15$$

The unknown term is 15. We complete the proportion.

$$\mathbf{\frac{6}{9} = \frac{10}{15}}$$

Example 4

Use cross products to find the unknown term in this proportion: Fifteen is to twenty-one as what number is to seventy?

Solution

We write the ratios in the order stated.

$$\frac{15}{21} = \frac{w}{70}$$

The cross products of a proportion are equal.

$$15 \cdot 70 = 21w$$

To find the unknown term, we divide $15 \cdot 70$ by 21. Notice that we can reduce as follows:

$$\frac{\overset{5}{\cancel{15}} \cdot \overset{10}{\cancel{70}}}{\underset{1}{\cancel{\underset{7}{\cancel{21}}}}} = w$$

The unknown term is **50.**

Practice Set

Use cross products to determine whether each pair of ratios forms a proportion:

a. $\frac{6}{9}, \frac{7}{11}$

b. $\frac{6}{8}, \frac{9}{12}$

Use cross products to complete each proportion:

c. $\frac{6}{10} = \frac{9}{x}$

d. $\frac{12}{16} = \frac{y}{20}$

e. Use cross products to find the unknown term in this proportion: 10 is to 15 as 30 is to what number?

Written Practice *Strengthening Concepts*

* **1.** Twenty-one of the 25 books Aretha has are about crafts. What percent of the books are about crafts?
(75)

2. By the time the blizzard was over, the temperature had dropped from 17°F to − 6°F. This was a drop of how many degrees?
(14)

3. The cost to place a collect call was $1.50 for the first minute plus $1.00 for each additional minute. What was the cost of a 5-minute phone call?
(12)

* **4.** The ratio of runners to walkers at the 10K fund-raiser was 5 to 7. If there were 350 runners, how many walkers were there?
(80)

* **5.** **Conclude** The two acute angles in $\triangle ABC$ are complementary. If the measure of $\angle B$ is 55°, what is the measure of $\angle A$?
(69)

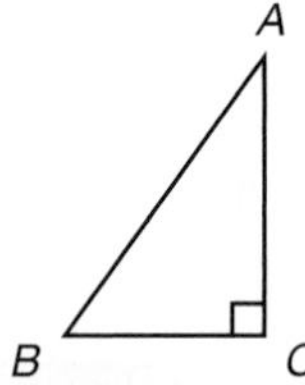

6. Athletic shoes are on sale for 20% off. Toni wants to buy a pair of running shoes that are regularly priced at $55.
(41)

a. How much money will be subtracted from the regular price if she buys the shoes on sale?

b. What will be the sale price of the shoes?

7. Freddy bought a pair of shoes for a sale price of $39.60. The sales-tax rate was 8%.
(41)

a. What was the sales tax on the purchase?

b. What was the total price including tax?

* **8.** **a.** Write $\frac{1}{25}$ as a decimal number.
(74, 75)

b. Write $\frac{1}{25}$ as a percent.

* **9.** Use cross products to determine whether this pair of ratios forms a proportion:
(85)

$$\frac{5}{11}, \frac{6}{13}$$

* **10.** **Explain** Use cross products to find the unknown term in this proportion: 4 is to 6 as 10 is to what number? Describe how you found your answer.
(85)

11. $10 \div 2\frac{1}{2}$
(68)

12. $6.5 - (4 - 0.32)$
(38)

13. $(6.25)(1.6)$
(39)

14. $0.06 \div 12$
(45)

Find each unknown number:

15. (59) $2\frac{1}{2} + x = 3\frac{1}{4}$

16. (48) $4\frac{1}{8} - y = 1\frac{1}{2}$

* **17.** (85) $\frac{9}{12} = \frac{n}{20}$

* **18.** (76) Arrange in order from least to greatest:

$\frac{1}{2}$, 0.4, 30%

19. (23) In a school with 300 students and 15 teachers, what is the student-teacher ratio?

20. (58, 65) If a number cube is rolled once, what is the probability that it will stop with a composite number on top?

21. (77) One fourth of 32 students have pets. How many students do not have pets?

22. (Inv. 7, 71) **Connect** What is the area of a parallelogram that has vertices with the coordinates (0, 0), (4, 0), (5, 3), and (1, 3)?

* **23.** (84) $2 + 2 \times 2 - 2 \div 2$

24. (32) Alejandro started the 10-kilometer race at 8:22 a.m. He finished the race at 9:09 a.m. How long did it take him to run the race?

Reading Math

The symbol $\approx$ means "is approximately equal to."

25. (15) **Estimate** Refer to the table below to answer this question: Ten kilometers is about how many miles? Round the answer to the nearest mile.

1 meter $\approx$ 1.093 yards
1 kilometer $\approx$ 0.621 mile

* **26.** (82) **Analyze** Lindsey packed boxes that were 1 foot long, 1 foot wide, and 1 foot tall into a larger box that was 5 feet long, 4 feet wide, and 3 feet tall.

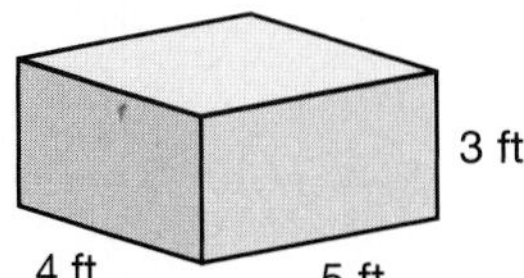

a. How many boxes could be packed on the bottom layer of the larger box?

b. Altogether, how many small boxes could be packed in the larger box?

* **27.** (81) Simplify:

a. 2 ft + 24 in. (Write the sum in feet.)

b. 3 yd · 3 yd

28. (75, 78) **Connect** A quart is what percent of a gallon?

Analyze This figure shows a square inside a circle, which is itself inside a larger square. Refer to this figure to answer problems **29** and **30.**

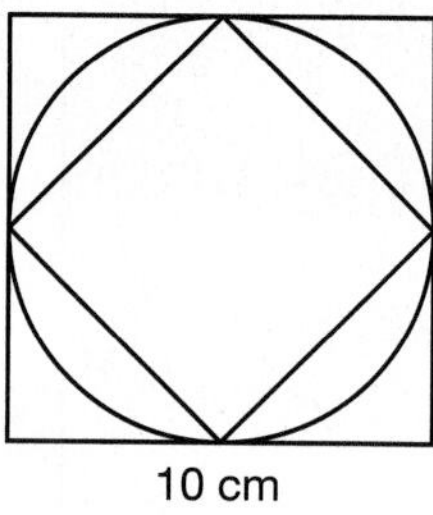

29. (31) The area of the smaller square is half the area of the larger square.

a. What is the area of the larger square?

b. What is the area of the smaller square?

30. (31) **Predict** Based on your answers to the questions in problem 29, make an educated guess as to the area of the circle. Explain your reasoning.

Early Finishers
Real-World Application

One day Mr. Holmes brought home a $50\frac{1}{4}$ ounce tin of maple syrup. Mrs. Holmes knew they could never finish that much maple syrup. Mr. Holmes explained to her that he was going to share the maple syrup with two neighbors. Mr. Holmes brought back $26\frac{1}{4}$ ounces of maple syrup after sharing with both neighbors. How much maple syrup did each of Mr. Holmes' neighbors receive if they received equal amounts of syrup?

LESSON 86

• Area of a Circle

Power Up

Building Power

facts Power Up G

mental math

a. Number Sense: 60 · 60

b. Number Sense: 850 − 170

c. Calculation: \$8.99 + \$4.99

d. Fractional Parts: $\frac{1}{5}$ of \$2.50

e. Decimals: 0.08 × 100

f. Number Sense: $\frac{360}{120}$

g. Measurement: How many cups are in a pint?

h. Calculation: 6 × 6, − 6, ÷ 2, − 1, ÷ 2, × 8, − 1, ÷ 5

problem solving Thomasita was thinking of a number less than 90 that she says when counting by sixes and when counting by fives, but not when counting by fours. Of what number was she thinking?

New Concept

Increasing Knowledge

We can estimate the area of a circle drawn on a grid by counting the number of square units enclosed by the figure.

Example 1

This circle is drawn on a grid.

a. How many units is the radius of the circle?

b. Estimate the area of the circle.

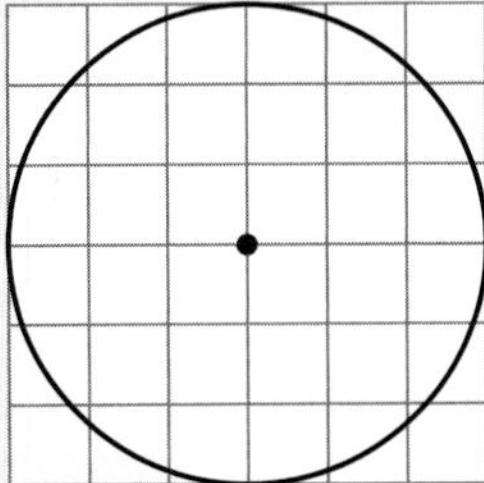

Solution

a. To find the radius of the circle, we may either find the diameter of the circle and divide by 2, or we may locate the center of the circle and count units to the circle. We find that the radius is **3 units.**

Thinking Skill

Discuss

What is another way we could count the squares and find the area of the circle?

b. To estimate the area of the circle, we count the square units enclosed by the circle. We show the circle again, this time shading the squares that lie completely or mostly within the circle. We have also marked with dots the squares that have about half their area inside the circle.

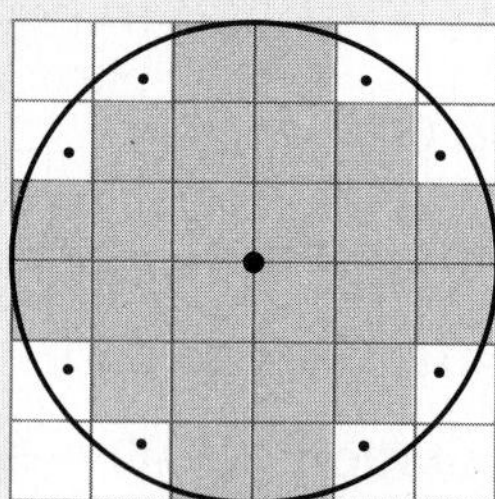

We count 24 squares that lie completely or mostly within the circle. We count 8 "half squares." Since $\frac{1}{2}$ of 8 is 4, we add 4 square units to 24 square units to get an estimate of **28 square units** for the area of the circle.

Finding the exact area of a circle involves the number π. To find the area of a circle, we first find the area of a square built on the radius of the circle. The circle below has a radius of 10 mm, so the area of the square is 100 mm^2. Notice that four of these squares would cover more than the area of the circle. However, the area of three of these squares is less than the area of the circle.

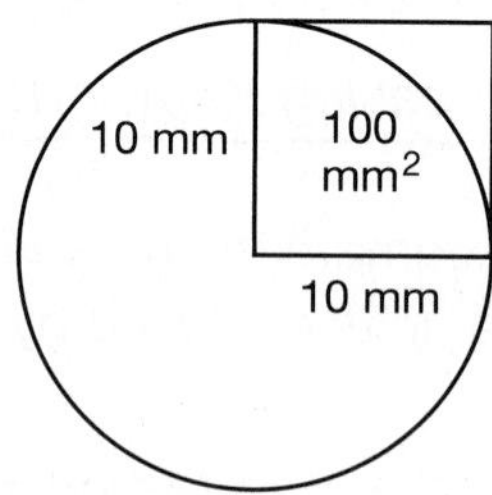

The area of the circle is exactly equal to π times the area of one of these squares. To find the area of this circle, we multiply the area of the square by π. We will continue to use 3.14 for the approximation of π.

$$3.14 \times 100 \text{ mm}^2 = 314 \text{ mm}^2$$

The area of the circle is approximately 314 mm^2.

Example 2

The radius of a circle is 3 cm. What is the area of the circle? (Use 3.14 for π. Round the answer to the nearest square centimeter.)

Solution

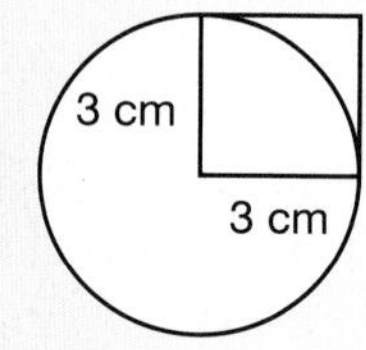

We will find the area of a square whose sides equal the radius. Then we multiply that area by 3.14.

Area of square: $3 \text{ cm} \times 3 \text{ cm} = 9 \text{ cm}^2$

Area of circle: $(3.14)(9 \text{ cm}^2) = 28.26 \text{ cm}^2$

We round 28.26 cm^2 to the nearest whole number of square centimeters and find that the area of the circle is approximately **28 cm^2.**

The area of any circle is π times the area of a square built on a radius of the circle. The following formula uses A for the area of a circle and r for the radius of the circle to relate the area of a circle to its radius:

$$A = \pi r^2$$

Practice Set

a. The radius of this circle is 4 units. Estimate the area of the circle by counting the squares within the circle.

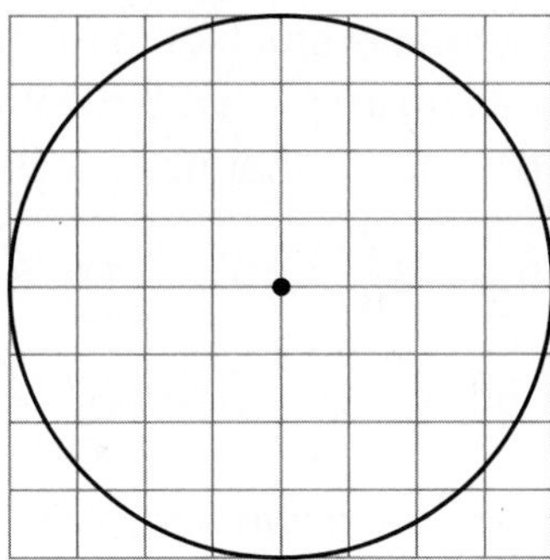

In problems **b–e,** use 3.14 for π.

b. Calculate the area of a circle with a radius of 4 cm.

c. Calculate the area of a circle with a radius of 2 feet.

d. Calculate the area of a circle with a diameter of 2 feet.

e. Calculate the area of a circle with a diameter of 10 inches.

Written Practice *Strengthening Concepts*

1. (49) What is the quotient when the decimal number ten and six tenths is divided by four hundredths?

2. (32) The time in Los Angeles is 3 hours earlier than the time in New York. If it is 1:15 p.m. in New York, what time is it in Los Angeles?

3. (12) Geraldine paid with a \$10 bill for 1 dozen keychains that cost 75¢ each. How much should she get back in change?

* 4. (84) **Analyze** $32 + 1.8(50)$

* 5. (82) **Analyze** If each block has a volume of one cubic inch, what is the volume of this tower?

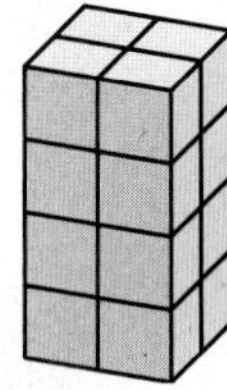

6. (80) **Predict** The ratio of hardbacks to paperbacks in the school library was 5 to 2. If there were 600 hardbacks, how many paperbacks were there?

7. (75) Nate missed three of the 20 questions on the test. What percent of the questions did he miss?

8. (41) **Analyze** The credit card company charges 1.5% (0.015) interest on the unpaid balance each month. If Mr. Jones has an unpaid balance of $2000, how much interest does he need to pay this month?

9. (74, 75) **a.** Write $\frac{4}{5}$ as a decimal number.

b. Write $\frac{4}{5}$ as a percent.

10. (58) Serena is stuck on a multiple-choice question that has four choices. She has no idea what the correct answer is, so she just guesses. What is the probability that her guess is correct?

11. (59) $5\frac{1}{2} + 3\frac{7}{8}$ **12.** (63) $3\frac{1}{4} - \frac{5}{8}$ **13.** (66) $\left(4\frac{1}{2}\right)\left(\frac{2}{3}\right)$

14. (68) $12\frac{1}{2} \div 100$ **15.** (68) $5 \div 1\frac{1}{2}$ **16.** (29) $\frac{5}{6}$ of $30

Find each unknown number:

17. (43) $4.72 + 12 + n = 50.4$ **18.** (3) $\$10 - m = \$9.87$

19. (45) $3n = 0.48$ **20.** (85) $\frac{w}{8} = \frac{25}{20}$

21. (38) **Predict** What are the next three terms in this sequence of perfect squares?

1, 4, 9, 16, 25, 36, 49, 64, 81, 100, ______, ______, ______, . . .

*** 22.** (71, 79) **Analyze** This parallelogram is divided into two congruent triangles.

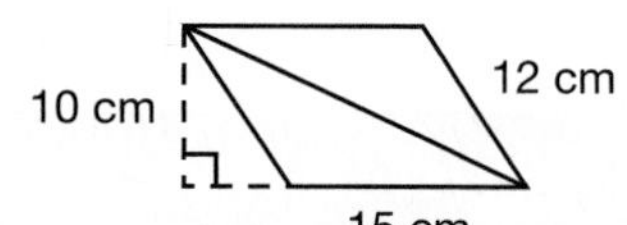

a. What is the area of the parallelogram?

b. What is the area of each triangle?

*** 23.** (86) Sydney drew a circle with a radius of 10 cm. What was the approximate area of the circle? (Use 3.14 for π.)

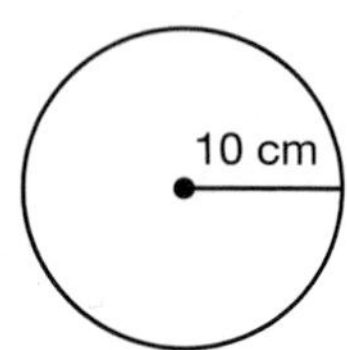

24. (31) Choose the appropriate unit for the area of a garage.

A square inches **B** square feet **C** square miles

25. (83, 85) **Connect** Which two ratios form a proportion? How do you know?

$$\frac{9}{12} \quad \frac{8}{14} \quad \frac{12}{21} \quad \frac{20}{36}$$

26. (86) The wheel of the covered wagon turned around once in about 12 feet. The diameter of the wheel was about

A 6 feet **B** 4 feet **C** 3 feet **D** 24 feet

27. (81) Anabel drove her car 348 miles in 6 hours. Divide the distance by the time to find the average speed of the car.

28. (69, 71) **Conclude** The opposite angles of a parallelogram are congruent. The adjacent angles are supplementary. If $\angle X$ measures 110°, then what are the measures of $\angle Y$ and $\angle Z$?

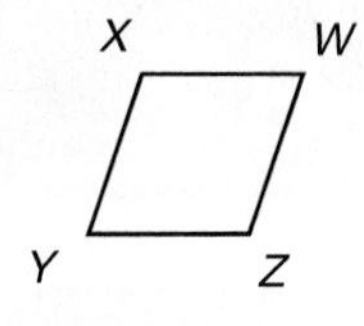

29. (47) **Estimate** The diameter of each wheel on the lawn mower is 10 inches. How far must the lawn mower be pushed in order for each wheel to complete one full turn? Round the answer to the nearest inch. (Use 3.14 for π.)

* **30.** (Inv. 7) **Connect** The coordinates of three vertices of a parallelogram are $(-3, 3)$, $(2, 3)$, and $(4, -1)$. What are the coordinates of the fourth vertex?

Early Finishers
Math and Science

The Great Frigates are large birds with long, slender wings. Frigates are great flyers and have one of the greatest wingspan to weight ratios of all birds. If a 3-pound Great Frigate bird has a wingspan of 6 feet, what would be the approximate wingspan of a 4-pound Great Frigate bird? Assume that the ratio of wingspan to weight is fairly constant.

LESSON 87

• Finding Unknown Factors

Power Up — *Building Power*

facts Power Up D

mental math

a. **Number Sense:** 70 · 70

b. **Number Sense:** 1000 − 375

c. **Calculation:** 5 × $4.99

d. **Number Sense:** Double $0.85.

e. **Decimals:** 62.5 ÷ 100

f. **Number Sense:** 20 × 45

g. **Algebra:** If $n = 2$, what does $2n$ equal?

h. **Calculation:** 5 × 5, − 5, × 5, ÷ 2, − 1, ÷ 7, × 3, − 1, ÷ 2

problem solving

In his 1859 autobiography, Abraham Lincoln wrote, "Of course when I came of age I did not know much. Still somehow, I could read, write, and cipher to the Rule of Three." When Lincoln wrote these words, to "cipher to the rule of three" was what students called setting up a proportion: "3 is to 12 as 5 is to __." In this book we use equations to solve proportions like this one: $\frac{3}{12} = \frac{5}{x}$. We find that the answer is 20 using both methods.

Cipher to the rule of three the numbers 2, 6, and 7.

New Concept — *Increasing Knowledge*

Since Lesson 4 we have practiced solving unknown factor problems. In this lesson we will solve problems in which the unknown factor is a mixed number or a decimal number. Remember that we can find an unknown factor by dividing the product by the known factor.

Example 1

Solve: $5n = 21$

Solution

Thinking Skill

Verify

Why can we divide the product by the known factor?

To find an unknown factor, we divide the product by the known factor.

$$\begin{array}{r} 4\frac{1}{5} \\ 5\overline{)21} \\ \underline{20} \\ 1 \end{array}$$

$$n = \mathbf{4\frac{1}{5}}$$

Note: We will write the answer as a mixed number unless there are decimal numbers in the problem.

Example 2

Solve: 0.6*m* = 0.048

Solution

Thinking Skill

Justify

Write the steps needed to solve example 2.

Again, we find the unknown factor by dividing the product by the known factor. Since there are decimal numbers in the problem, we write our answer as a decimal number.

$$\begin{array}{r} 0.08 \\ 06.\overline{)00.48} \\ \underline{48} \\ 0 \end{array}$$

$m =$ **0.08**

Example 3

Solve: 45 = 4*x*

Solution

This problem might seem "backward" because the multiplication is on the right-hand side. However, an equal sign is not directional. It simply states that the quantities on either side of the sign are equal. In this case, the product is 45 and the known factor is 4. We divide 45 by 4 to find the unknown factor.

$$\begin{array}{r} 11\frac{1}{4} \\ 4\overline{)45} \\ \underline{4} \\ 05 \\ \underline{4} \\ 1 \end{array}$$

$x =$ $\mathbf{11\frac{1}{4}}$

Practice Set

Solve:

a. $6w = 21$ **b.** $50 = 3f$ **c.** $5n = 36$

d. $0.3t = 0.24$ **e.** $8m = 3.2$ **f.** $0.8 = 0.5x$

Written Practice

Strengthening Concepts

1. (4) If the divisor is 12 and the quotient is 24, what is the dividend?

2. (29) The brachiosaurus, one of the largest dinosaurs, weighed only $\frac{1}{4}$ as much as a blue whale. A blue whale can weigh 140 tons. How much could a brachiosaurus have weighed?

3. (23) **Analyze** Fourteen of the 32 students in the class are boys. What is the ratio of boys to girls in the class?

Find each unknown number:

*** 4.** (87) $0.3m = 0.27$

*** 5.** (87) $31 = 5n$

*** 6.** (38) **Analyze** $3n = 6^2$

*** 7.** (87) $4n = 0.35$

8. (73) Write 0.25 as a fraction and add it to $3\frac{1}{4}$. What is the sum?

9. (74) Write $\frac{3}{5}$ as a decimal and add it to 6.5. What is the sum?

10. (74, 75) Write $\frac{1}{50}$ as a decimal number and as a percent.

11. (63) $12\frac{1}{5} - 3\frac{4}{5}$ **12.** (66) $6\frac{2}{3} \times 1\frac{1}{5}$ **13.** (68) $11\frac{1}{9} \div 100$

14. (38) $4.75 + 12.6 + 10$ * **15.** (84) $35 - (0.35 \times 100)$

* **16.** (85) $4 + 4 \times 4 - 4 \div 4$

17. (35) Write the decimal numeral twelve and five hundredths.

* **18.** (82) Find the volume of this rectangular prism.

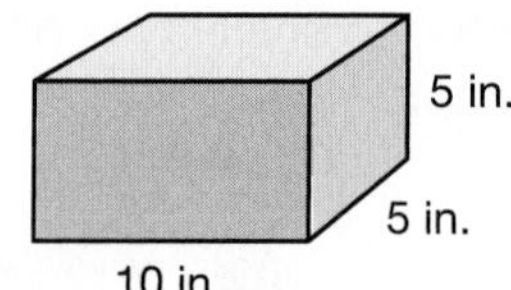

19. (47) **Evaluate** If a equals 15, then what number does $2a - 5$ equal?

20. (71) What is the area of this parallelogram?

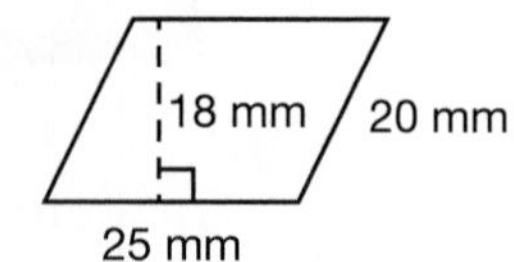

21. (8) What is the perimeter of this parallelogram?

22. (64) **Verify** "All rectangles are parallelograms." True or false?

* **23.** (77) Charles spent $\frac{1}{10}$ of his 100 shillings. How many shillings does he still have?

24. (14) The temperature rose from −18°F to 19°F. How many degrees did the temperature increase?

25. (7) How many **centimeters** long is the line below?

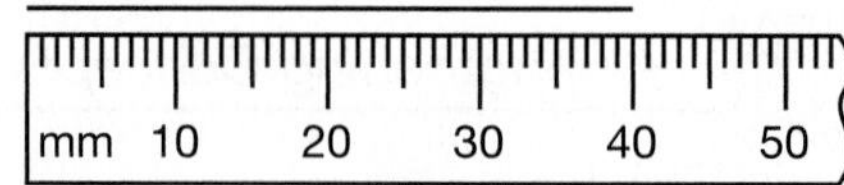

26. (78) Johann poured 500 mL of water from a full 2-liter container. How many milliliters of water were left in the container?

27. (Inv. 8) Name this geometric solid.

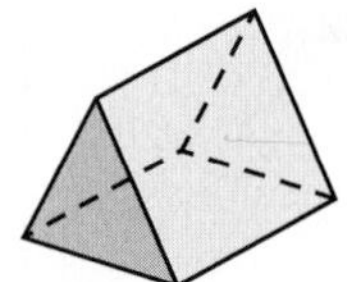

* **28.** (81) Simplify:

  a. 2 meters + 100 centimeters (Write the answer in meters.)

  b. 2 m · 4 m

* **29.** (84) **Analyze** Solve this proportion: $\frac{12}{m} = \frac{18}{9}$

30. (Inv. 7) **Connect** What is the perimeter of a rectangle with vertices at $(-4, -4)$, $(-4, 4)$, $(4, 4)$, and $(4, -4)$?

Early Finishers
Real-World Application

Frida and her family went on a summer vacation to Chicago, Illinois from Boston, Massachusetts. Her family drove 986 miles in 17 hours.

a. How many miles per hour did Frida's family average on their drive to Chicago?

b. If the family's car averages 29 miles per gallon, how many gallons of gas did they use on their trip?

LESSON 88

Using Proportions to Solve Ratio Word Problems

Power Up — *Building Power*

facts Power Up L

mental math

a. **Number Sense:** 80 · 80

b. **Number Sense:** 720 − 150

c. **Calculation:** $1.98 + $1.98

d. **Fractional Parts:** $\frac{1}{10}$ of $5.00

e. **Decimals:** 0.15 × 100

f. **Number Sense:** $\frac{750}{250}$

g. **Measurement:** How many milliliters are in 5 liters?

h. **Calculation:** 4 × 4, − 1, × 2, + 3, ÷ 3, − 1, × 10, − 1, ÷ 9

problem solving

Kim hit a target like the one shown 6 times, earning a total score of 20. Find two sets of scores Kim could have earned.

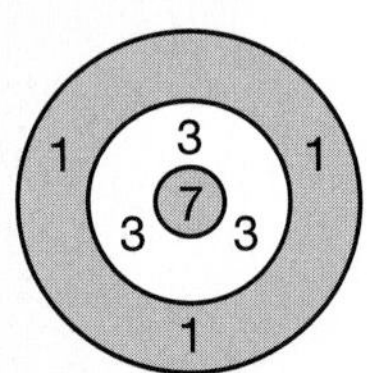

New Concept — *Increasing Knowledge*

Proportions can be used to solve many types of word problems. In this lesson we will use proportions to solve ratio word problems such as those in the following examples.

Example 1

The ratio of salamanders to frogs was 5 to 7. If there were 20 salamanders, how many frogs were there?

Solution

In this problem there are two kinds of numbers: ratio numbers and actual-count numbers. The ratio numbers are 5 and 7. The number 20 is an actual count of the salamanders. We will arrange these numbers in two columns and two rows to form a ratio box.

	Ratio	Actual Count
Salamanders	5	20
Frogs	7	*f*

We were not given the actual count of frogs, so we use the letter f to stand for the actual number of frogs.

Instead of using scale factors in this lesson, we will practice using proportions. We use the positions of numbers in the ratio box to write a proportion. By solving the proportion, we find the actual number of frogs.

	Ratio	Actual Count
Salamanders	5	20
Frogs	7	f

$$\frac{5}{7} = \frac{20}{f}$$

We can solve the proportion in two ways. We can multiply $\frac{5}{7}$ by $\frac{4}{4}$, or we can use cross products. Here we show the solution using cross products:

Thinking Skill

Justify

What steps do we follow to find the value of f?

$$\frac{5}{7} = \frac{20}{f}$$

$$5f = 7 \cdot 20$$

$$5f = \frac{7 \cdot 20}{5}$$

$$f = 28$$

We find that there were **28 frogs.**

Example 2

If 3 sacks of concrete will make 12 square feet of sidewalk, predict how many sacks of concrete are needed to make 40 square feet of sidewalk?

Solution

Thinking Skill

Discuss

Two methods for solving proportions are using cross products and using a constant factor. Under what circumstance is one method preferable to the other?

We are given the ratio 3 sacks to 12 square feet and the actual count of square feet of sidewalk needed.

	Ratio	Actual Count
Sacks	3	n
Sq. ft	12	40

$$\frac{3}{12} = \frac{n}{40}$$

The constant factor from the first ratio to the second is not obvious, so we use cross products.

$$\frac{3}{12} = \frac{n}{40}$$

$$12 \cdot n = 3 \cdot 40$$

$$n = \frac{3 \cdot 40}{12}$$

$$n = 10$$

We find that **10 sacks** of concrete are needed.

Practice Set

Model For each problem, draw a ratio box. Then solve the problem using proportions.

a. The ratio of DVDs to CDs was 5 to 4. If there were 60 CDs, how many DVDs were there?

b. **Explain** At the softball game, the ratio of fans for the home team to the fans for the away team is 5 to 3. If there are 30 fans for the home team, how many fans for the away team are there? How can you check your answer?

Written Practice

Strengthening Concepts

1. (75) Mavis scored 12 of the team's 20 points. What percent of the team's points did Mavis score?

2. (50) One fourth of an inch of snow fell every hour during the storm. How many hours did the storm last if the total accumulation of snow was 4 inches?

* **3.** (87) Eamon wants to buy a new baseball glove that costs $50. He has $14 and he earns $6 per hour cleaning yards. How many hours must he work to have enough money to buy the glove?

* **4.** (79) **Analyze** Find the area of this triangle.

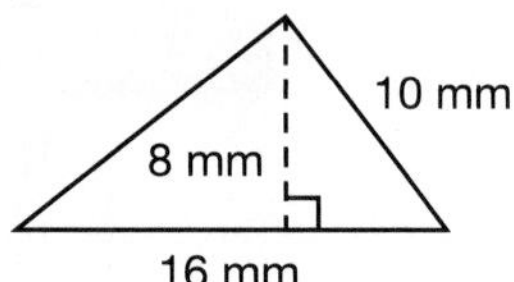

* **5.** (88) **Model** Draw a ratio box for this problem. Then solve the problem using a proportion.

The ratio of adults to students on the field trip is 3 to 5. If there are 15 students on the field trip, how many adults are there?

* **6.** (82) What is the volume of this rectangular prism?

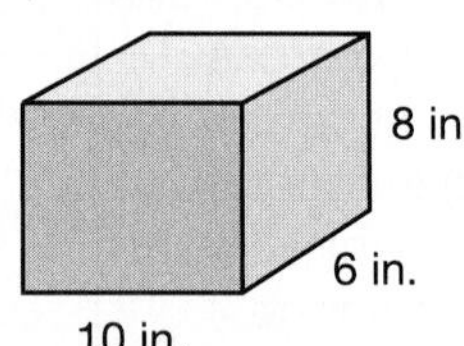

Find each unknown factor:

* **7.** (87) $10w = 25$

* **8.** (87) $20 = 9m$

* **9.** (8, 79) **a.** What is the perimeter of this triangle?

b. What is the area of this triangle?

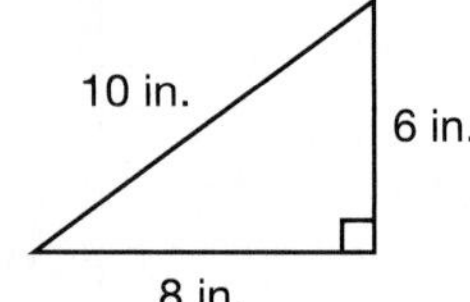

10. (33, 75) Write 5% as a

a. decimal number.

b. fraction.

11. (74) Write $\frac{2}{5}$ as a decimal, and multiply it by 2.5. What is the product?

12. Compare: $\frac{2}{3} + \frac{3}{2} \bigcirc \frac{2}{3} \cdot \frac{3}{2}$
(29, 56)

13. $\frac{1}{3} \times \frac{100}{1}$
(70)

14. $6 \div 1\frac{1}{2}$
(68)

15. $12 \div 0.25$
(49)

16. 0.025×100
(46)

17. If the tax rate is 7%, what is the tax on a \$24.90 purchase?
(41)

18. The prime factorization of what number is $2^2 \cdot 3^2 \cdot 5^2$?
(73)

19. **Classify** Which of these is a composite number?
(65)

A 61 **B** 71 **C** 81 **D** 101

20. Round the decimal number one and twenty-three hundredths to the nearest tenth.
(51)

21. Albert baked 5 dozen muffins and gave away $\frac{7}{12}$ of them. How many muffins were left?
(77)

* **22.** $6 \times 3 - 6 \div 3$
(84)

23. How many milliliters is 4 liters?
(78)

24. **Model** Draw a line segment $2\frac{1}{4}$ inches long. Label the endpoints *A* and *C*. Then make a dot at the midpoint of $\overline{AC}$ (the point halfway between points *A* and *C*), and label the dot *B*. What are the lengths of $\overline{AB}$ and $\overline{BC}$?
(69)

25. On a coordinate plane draw a rectangle with vertices at (−2, −2), (4, −2), (4, 2), and (−2, 2). What is the area of the rectangle?
(Inv. 7)

26. What is the ratio of the length to the width of the rectangle in problem 25?
(23)

* **27.** **Explain** How do you calculate the area of a triangle?
(79)

28. In the figure at right, angles *ADB* and *BDC* are supplementary.
(69)

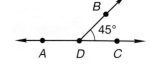

a. What is m∠*ADB*?

b. What is the ratio of m∠*BDC* to m∠*ADB*? Write the answer as a reduced fraction.

* **29.** **Analyze** Nathan drew a circle with a radius of 10 cm. Then he drew a square around the circle.
(86)

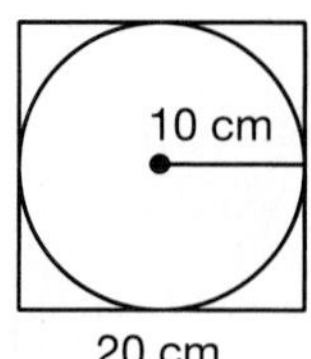

a. What was the area of the square?

b. What was the area of the circle? (Use 3.14 for π.)

* **30.** **Formulate** Write a word problem that can be solved using the proportion $\frac{6}{8} = \frac{w}{100}$. Solve the problem.
(85)

LESSON 89

Estimating Square Roots

Power Up

Building Power

facts Power Up K

mental math

a. **Number Sense:** $90 \cdot 90$

b. **Number Sense:** $1000 - 405$

c. **Calculation:** $6 \times \$7.99$

d. **Number Sense:** Double \$27.00

e. **Decimals:** $87.5 \div 100$

f. **Number Sense:** 20×36

g. **Geometry:** A rectangular solid is 5 in. × 3 in. × 3 in. What is the volume of the solid?

h. **Calculation:** 3×3, $+ 2$, $\times 5$, $- 5$, $\times 2$, $\div 10$, $+ 5$, $\div 5$

problem solving

A loop of string was arranged to form a square with sides 9 inches long. If the same loop of string is arranged to form an equilateral triangle, how long will each side be? If a regular hexagon is formed, how long will each side be?

New Concept

Increasing Knowledge

We have practiced finding square roots of perfect squares from 1 to 100. In this lesson we will find the square roots of perfect squares greater than 100. We will also use a guess-and-check method to estimate the square roots of numbers that are not perfect squares. As we practice, our guesses will improve and we will begin to see clues to help us estimate.

Example 1

Simplify: $\sqrt{400}$

Solution

We need to find a number that, when multiplied by itself, has a product of 400.

$$\square \times \square = 400$$

We know that $\sqrt{400}$ is more than 10, because 10×10 equals 100. We also know that $\sqrt{400}$ is much less than 100, because 100×100 equals 10,000. Since $\sqrt{4}$ equals 2, the 4 in $\sqrt{400}$ hints that we should try 20.

$$20 \times 20 = 400$$

We find that $\sqrt{400}$ equals **20.**

Example 2

Simplify: $\sqrt{625}$

Solution

In example 1 we found that $\sqrt{400}$ equals 20. Since $\sqrt{625}$ is greater than $\sqrt{400}$, we know that $\sqrt{625}$ is greater than 20. We find that $\sqrt{625}$ is less than 30, because 30 × 30 equals 900. Since the last digit is 5, perhaps $\sqrt{625}$ is 25. We multiply to find out.

$$\begin{array}{r} 25 \\ \times\ 25 \\ \hline 125 \\ 50 \\ \hline 625 \end{array}$$

We find that $\sqrt{625}$ equals **25.**

We have practiced finding the square roots of numbers that are perfect squares. Now we will practice estimating the square root of numbers that are not perfect squares.

Example 3

Math Language
Consecutive whole numbers are two numbers we count in sequence, such as 4 and 5, or 15 and 16.

Between which two consecutive whole numbers is $\sqrt{20}$?

Solution

Notice that we are not asked to find the square root of 20. To find the whole numbers on either side of $\sqrt{20}$, we can first think of the perfect squares that are on either side of 20. Here we show the first few perfect squares, starting with 1.

1, 4, 9, 16, 25, 36, 49

We see that 20 is between the perfect squares 16 and 25. So $\sqrt{20}$ is between $\sqrt{16}$, and $\sqrt{25}$.

$$\sqrt{16},\ \sqrt{20},\ \sqrt{25}$$

Since $\sqrt{16}$ is 4 and $\sqrt{25}$ is 5, we see that $\sqrt{20}$ is between **4** and **5.**

Using the reasoning in example 3, we know there must be some number between 4 and 5 that is the square root of 20. We try 4.5.

$$4.5 \times 4.5 = 20.25$$

We see that 4.5 is too large, so we try 4.4.

$$4.4 \times 4.4 = 19.36$$

We see that 4.4 is too small. So $\sqrt{20}$ is greater than 4.4 but less than 4.5. (It is closer to 4.5.) If we continued this process, we would never find a decimal number or fraction that exactly equals $\sqrt{20}$. This is because $\sqrt{20}$ belongs to a number family called the **irrational numbers.**

Reading Math

Recall that the wavy equal sign means "is approximately equal to."

Irrational numbers cannot be expressed exactly as a ratio (that is, as a fraction or decimal). We can only use fractions or decimals to express the *approximate* value of an irrational number.

$$\sqrt{20} \approx 4.5$$

The square root of 20 is approximately equal to 4.5.

Example 4

Use a calculator to approximate the value of $\sqrt{20}$ to two decimal places.

Solution

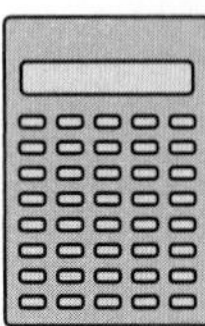

We clear the calculator and then enter √ 2 0 (or 2 0 √).[1] The display will show 4.472135955. The actual value of $\sqrt{20}$ contains an infinite number of decimal places. The display approximates $\sqrt{20}$ to nine or so decimal places (depending on the model). We are asked to show two decimal places, so we round the displayed number to **4.47.**

Practice Set

Find each square root:

a. $\sqrt{169}$ **b.** $\sqrt{484}$ **c.** $\sqrt{961}$

Each of these square roots is between which two consecutive whole numbers? Find the answer without using a calculator.

d. $\sqrt{2}$ **e.** $\sqrt{15}$ **f.** $\sqrt{40}$

g. $\sqrt{60}$ **h.** $\sqrt{70}$ **i.** $\sqrt{80}$

Estimate Use a calculator to approximate each square root to two decimal places:

j. $\sqrt{3}$ **k.** $\sqrt{10}$ **l.** $\sqrt{50}$

Written Practice *Strengthening Concepts*

1. (12, 72) What is the difference when the product of $\frac{1}{2}$ and $\frac{1}{2}$ is subtracted from the sum of $\frac{1}{4}$ and $\frac{1}{4}$?

2. (78) A dairy cow can give 4 gallons of milk per day. How many cups of milk is that (1 gallon = 4 quarts; 1 quart = 4 cups)?

3. (29) The recipe called for $\frac{3}{4}$ cup of sugar. If the recipe is doubled, how much sugar should be used?

*** 4.** (88) **Model** Draw a ratio box for this problem. Then solve the problem using a proportion.

The recipe called for sugar and flour in the ratio of 2 to 9. If the chef used 18 pounds of flour, how many pounds of sugar were needed?

[1] The order of keystrokes depends on the model of calculator. See the instructions for your calculator if the keystroke sequences described in this lesson do not work for you.

* **5.** (89) **Classify** Which of these numbers is greater than 6 but less than 7?

A $\sqrt{6.5}$ **B** $\sqrt{67}$ **C** $\sqrt{45}$ **D** $\sqrt{76}$

* **6.** (87) Express the unknown factor as a mixed number:

$$7n = 30$$

7. (47) Amanda used a compass to draw a circle with a radius of 4 inches.

a. What is the diameter of the circle?

b. What is the circumference of the circle?

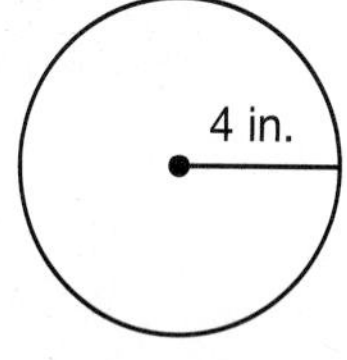

Use 3.14 for π.

* **8.** (86) In problem **7** what is the area of the circle Amanda drew?

* **9.** (79) What is the area of the triangle at right?

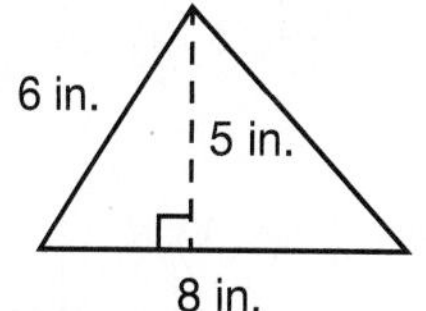

10. (71) **a.** What is the area of this parallelogram?

b. What is the perimeter of this parallelogram?

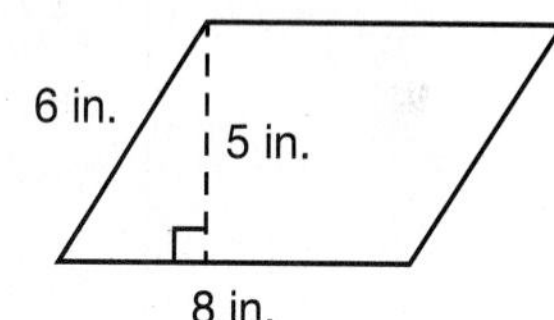

11. (63, 73) Write 0.5 as a fraction and subtract it from $3\frac{1}{4}$. What is the difference?

12. (74) Write $\frac{3}{4}$ as a decimal, and multiply it by 0.6. What is the product?

* **13.** (84) **Analyze** $2 \times 15 + 2 \times 12$

* **14.** (89) **Analyze** $\sqrt{900}$

15. (2) $\$6 \div 8$

16. (66) $1\frac{3}{5} \times 10 \times \frac{1}{4}$

17. (68) $37\frac{1}{2} \div 100$

18. (68) $3 \div 7\frac{1}{2}$

19. (34) What is the place value of the 7 in 987,654.321?

20. (35) Write the decimal number five hundred ten and five hundredths.

21. (3) $30 + 60 + m = 180$

22. (72) **Analyze** Half of the students are girls. Half of the girls have brown hair. Half of the brown-haired girls wear their hair long. Of the 32 students, how many are girls with long, brown hair?

Refer to the pictograph below to answer problems **23–25.**

Books Read This Year

Johnny	
Mary	
Pat	

represents 4 books.

23. *(Inv. 5)* How many books has Johnny read?

24. *(Inv. 5)* Mary has read how many more books than Pat?

25. *(Inv. 5)* **Formulate** Write a question that relates to this graph and answer the question.

* **26.** *(85)* **Analyze** Solve this proportion: $\frac{12}{8} = \frac{21}{m}$

27. *(58, 74)* The face of this spinner is divided into 12 congruent regions. If the spinner is spun once, what is the probability that it will stop on a 3? Express the probability ratio as a fraction and as a decimal number rounded to the nearest hundredth.

28. *(28, 69)* **Conclude** If two angles are complementary, and if one angle is acute, then the other angle is what kind of angle?

A acute **B** right **C** obtuse

* **29.** *(81)* Simplify:

a. 100 cm + 100 cm (Write the answer in meters.)

b. $\frac{(5\text{ in.})(8\text{ in.})}{2}$

30. *(82)* If each small block has a volume of 1 cubic inch, then what is the volume of this cube?

LESSON

90

• Measuring Turns

Power Up

Building Power

facts

Power Up J

mental math

a. Power/Roots: $\sqrt{100}$

b. Calculation: 781 − 35

c. Calculation: \$1.98 + \$2.98

d. Fractional Parts: $\frac{1}{3}$ of \$24.00

e. Decimals: 0.375 × 100

f. Number Sense: $\frac{1200}{300}$

g. Geometry: A cube has a height of 4 cm. What is the volume of the cube?

h. Calculation: 2 × 2, × 2, × 2, − 1, × 2, + 2, ÷ 4, ÷ 4

problem solving

One state used a license plate that included one letter followed by five digits. How many different license plates could be made that started with the letter A?

New Concept

Increasing Knowledge

Math Language

Clockwise means "moving in the direction of the hands of a clock." *Counterclockwise* means "moving in the opposite direction of the hands of a clock."

Every hour the minute hand of a clock completes one full turn in a clockwise direction. How many degrees does the minute hand turn in an hour?

Turns can be measured in degrees. A full turn is a 360° turn. So the minute hand turns 360° in one hour.

If you turn 360°, you will end up facing the same direction you were facing before you turned. A half turn is half of 360°, which is 180°. If you turn 180°, you will end up facing opposite the direction you were facing before you turned.

If you are facing north and turn 90°, you will end up facing either east or west, depending on the direction in which you turned. To avoid confusion, we often specify the direction of a turn as well as the measure of the turn. Sometimes the direction is described as being to the right or to the left. Other times it is described as clockwise or counterclockwise.

Example 1

Leila was traveling north. At the light she turned 90° to the left and traveled one block to the next intersection. At the intersection she turned 90° to the left. What direction was Leila then traveling?

Solution

A picture may help us answer the question. Leila was traveling north when she turned 90° to the left. After that first turn Leila was traveling west. When she turned 90° to the left a second time, she began traveling to the **south.** Notice that the two turns in the same direction (left) total 180°. So we would expect that after the two turns Leila was heading in the direction opposite to her starting direction.

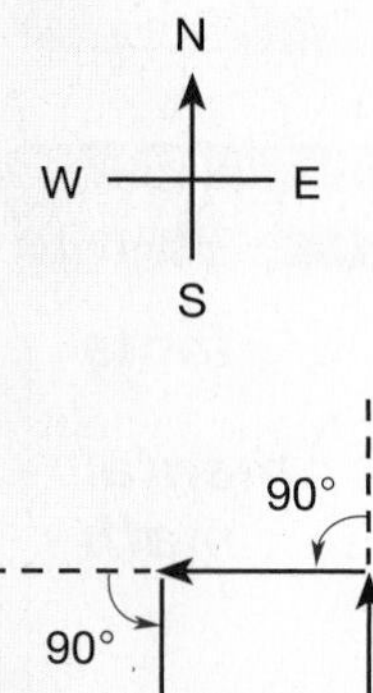

Example 2

Andy and Barney were both facing north. Andy made a quarter turn (90°) clockwise to face east, while Barney turned counterclockwise until he faced east. How many degrees did Barney turn?

Solution

We will draw the two turns. Andy made a quarter turn clockwise. We see that Barney made a three-quarter turn counterclockwise. We can calculate the number of degrees in three quarters of a turn by finding $\frac{3}{4}$ of 360°.

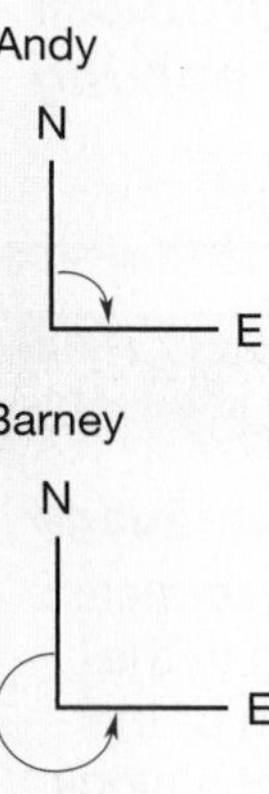

$$\frac{3}{4} \times 360° = 270°$$

Another way to find the number of degrees is to recognize that each quarter turn is 90°. So three quarters is three times 90°.

$$3 \times 90° = 270°$$

Barney turned **270°** counterclockwise.

Example 3

As Elizabeth ran each lap around the park, she made six turns to the left (and no turns to the right). What was the average number of degrees of each turn?

Solution

Thinking Skill

Discuss

Does the number of laps Elizabeth ran affect the answer? Why or why not?

We are not given the measure of any of the turns, but we do know that Elizabeth made six turns to the left to get completely around the park. That is, after six turns she once again faced the same direction she faced before the first turn. So after six turns she had turned a total of 360°. We find the average number of degrees in each turn by dividing 360° by 6.

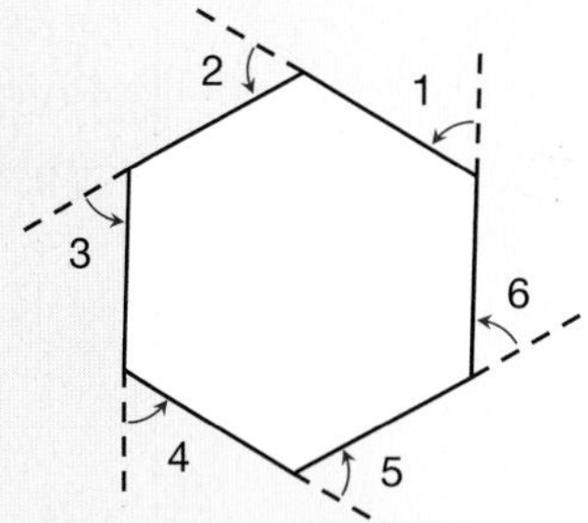

$$360° \div 6 = 60°$$

Each of Elizabeth's turns averaged **60°.**

Practice Set

a. **Analyze** Jose was heading south on his bike. When he reached Sycamore, he turned 90° to the right. Then at Highland he turned 90° to the right, and at Elkins he turned 90° to the right again. Assuming each street was straight, in which direction was Jose heading on Elkins?

b. Kiara made one full turn counterclockwise. Mary made two full turns clockwise. How many degrees did Mary turn?

c. **Model** David ran three laps around the park. On each lap he made five turns to the left and no turns to the right. What was the average number of degrees in each of David's turns? Draw a picture of the problem.

Written Practice

Strengthening Concepts

1. (Inv. 5) What is the mean of 4.2, 4.8, and 5.1?

2. (32) The movie is 120 minutes long. If it begins at 7:15 p.m., when will it be over?

3. (75) Fifteen of the 25 students in Room 20 are boys. What percent of the students in Room 20 are boys?

4. (Inv. 6) This triangular prism has how many more edges than vertices?

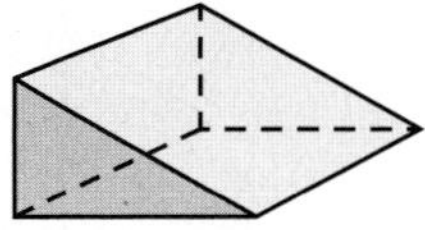

*** 5.** (86) The teacher cut a 12-inch diameter circle from a sheet of construction paper.

a. What was the radius of the circle?

b. What was the area of the circle? (Use 3.14 for π.)

6. (64) **Explain** Write a description of a trapezoid.

7. (14, 17) Arrange these numbers in order from least to greatest:

$$1, -2, 0, -4, \frac{1}{2}$$

* **8.** (87) **Analyze** Express the unknown factor as a mixed number:

$$25n = 70$$

Refer to the triangle to answer questions **9–11.**

* **9.** (79) What is the area of this triangle?

10. (8) What is the perimeter of this triangle?

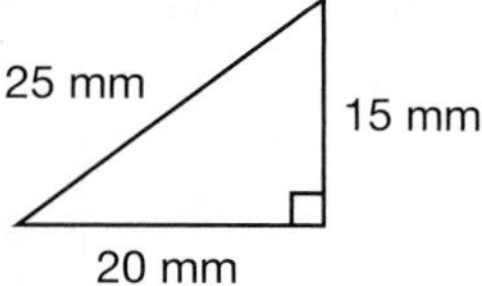

* **11.** (23, 74) What is the ratio of the length of the shortest side to the length of the longest side? Express the ratio as a fraction and as a decimal.

12. (63, 73) **Connect** Write 6.25 as a mixed number. Then subtract $\frac{5}{8}$ from the mixed number. What is the difference?

* **13.** (90) **Analyze** Ali was facing north. Then he turned to his left 180°. What direction was he facing after he turned?

14. (41) Write 28% as a reduced fraction.

* **15.** (85) **Analyze** $\frac{n}{12} = \frac{20}{30}$

16. (52) $0.625 \div 10$

17. (49) $\frac{25}{0.8}$

18. (59) $3\frac{3}{8} + 3\frac{3}{4}$

19. (48) $5\frac{1}{8} - 1\frac{7}{8}$

20. (72) $6\frac{2}{3} \times \frac{3}{10} \times 4$

21. (77) One third of the two dozen knights were on horseback. How many knights were not on horseback?

* **22.** (18) **Evaluate** Weights totaling 38 ounces were placed on the left side of this scale, while weights totaling 26 ounces were placed on the right side of the scale. How many ounces of weights should be moved from the left side to the right side to balance the scale? (*Hint:* Find the average of the weights on the two sides of the scale.)

23. (82) The cube at right is made up of smaller cubes that each have a volume of 1 cubic centimeter. What is the volume of the larger cube?

24. (51) Round forty-eight hundredths to the nearest tenth.

* **25.** (89) $\sqrt{144} - \sqrt{121}$

26. (23) The ratio of dogs to cats in the neighborhood is 6 to 5. What is the ratio of cats to dogs?

27. (84) $10 + 10 \times 10 - 10 \div 10$

28. (78) **Analyze** The Thompsons drink a gallon of milk every two days. There are four people in the Thompson family. Each person drinks an average of how many pints of milk per day? Explain your thinking.

* **29.** (81) Simplify:

a. 10 cm + 100 mm (Write the answer in millimeters.)

b. 300 books ÷ 30 students

30. (Inv. 7) **Model** On a coordinate plane draw a segment from point *A* (−3, −1) to point *B* (5, −1). What are the coordinates of the point that is halfway between points *A* and *B*?

Early Finishers
Real-World Application

The students in Mrs. Fitzgerald's cooking class will make buttermilk biscuits using the recipe below. If 56 students each work with a partner to make one batch of biscuits, how much of each ingredient will Mrs. Fitzgerald need to buy? Hint: Keep each item in the unit measure given.

$1\frac{3}{4}$ cups of all-purpose flour

1 teaspoon of baking soda

1 stick of butter

$1\frac{1}{4}$ cups of milk

INVESTIGATION 9

Focus on

• Experimental Probability

In Lesson 58 we determined probabilities for the outcomes of experiments without actually performing the experiments. For example, in the case of rolling a number cube, the sample space of the experiment is {1, 2, 3, 4, 5, 6} and all six outcomes are equally likely. Each outcome therefore has a probability of $\frac{1}{6}$. In the spinner example we assume that the likelihood of the spinner landing in a particular sector is proportional to the area of the sector. Thus, if the area of sector A is twice the area of sector B, the probability of the spinner landing in sector A is twice the probability of the spinner landing in sector B.

Probability that is calculated by performing "mental experiments" (as we have been doing since Lesson 58) is called **theoretical probability.** Probabilities associated with many real-world situations, though, cannot be determined by theory. Instead, we must perform the experiment repeatedly or collect data from a sample experiment. Probability determined in this way is called **experimental probability.**

A survey is one type of probability experiment. Suppose a pizza company is going to sell individual pizzas at a football game. Three types of pizzas will be offered: cheese, tomato, and mushroom. The company wants to know how many of each type of pizza to prepare, so it surveys a representative sample of 500 customers.

The company finds that 175 of these customers would order cheese pizzas, 225 would order tomato pizzas, and 100 would order mushroom pizzas. To estimate the probability that a particular pizza will be ordered, the company uses **relative frequency.** This means they divide the frequency (the number in each category) by the total (in this case, 500).

	Frequency	Relative Frequency
Cheese	175	$\frac{175}{500} = 0.35$
Tomato	225	$\frac{225}{500} = 0.45$
Mushroom	100	$\frac{100}{500} = 0.20$

Notice that the sum of the three relative frequencies is 1. This means that the entire sample is represented. We can change the relative frequencies from decimals to percents.

0.35 ⟶ 35%   0.45 ⟶ 45%   0.20 ⟶ 20%

Recall from Lesson 58 that we use the term *chance* to describe a probability expressed as a percent. So the company makes the following estimates about any given sale. The chance that a cheese pizza will be ordered is **35%.** The chance that a tomato pizza will be ordered is **45%.** The chance that a mushroom pizza will be ordered is **20%.** The company plans to make 3000 pizzas for the football game, so about 20% of the 3000 pizzas should be mushroom. How many pizzas will be mushroom?

Now we will apply these ideas to another survey. Suppose a small town has only four markets: Bob's Market, The Corner Grocery, Express Grocery, and Fine Foods. A representative sample of 80 adults was surveyed. Each person chose his or her favorite market: 30 chose Bob's Market, 12 chose Corner Grocery, 14 chose Express Grocery, and 24 chose Fine Foods.

1. *Represent* Present the data in a relative frequency table similar to the one for pizza.
2. Estimate the probability that in this town an adult's favorite market is Express Grocery. Write your answer as a decimal.
3. Estimate the probability that in this town an adult's favorite market is Bob's Market. Write your answer as a fraction in reduced form.
4. Estimate the chance that in this town an adult's favorite market is Fine Foods. Write your answer as a percent.
5. Suppose the town has 4000 adult residents. The Corner Grocery is the favorite market of about how many adults in the town?

Thinking Skill

Discuss

How can the frequency table you made in problem 1 help us find the answer to problem 5?

A survey is just one way of conducting a probability experiment. In the following activity we will perform an experiment that involves drawing two marbles out of a bag. By performing the experiment repeatedly and recording the results, we gather information that helps us determine the probability of various outcomes.

Activity

Probability Experiment

Materials needed:

- 6 marbles (4 green and 2 white)
- Small, opaque bag from which to draw the marbles
- Pencil and paper

The purpose of this experiment is to determine the probability that two marbles drawn from the bag at the same time will be green. We will create a relative frequency table to answer the question.

To estimate the probability, put 4 green marbles and 2 white marbles in a bag. Pair up with another student, and work through problems **6–8** together.

Thinking Skill

Predict

Which outcome do you think will occur most often?

6. Choose one student to draw from the bag and the other to record results. Shake the bag; then remove two objects at the same time. Record the result by marking a tally in a table like the one below. Replace the marbles and repeat this process until you have performed the experiment exactly 25 times.

Outcome	Tally
Both green	
Both white	
One of each	

7. Use your tally table to make a relative frequency table. (Divide each row's tally by 25 and express the quotient as a decimal.)

8. Estimate the probability that both marbles drawn will be green. Write your answer as a reduced fraction and as a decimal.

If, for example, you drew two green marbles 11 times out of 25 draws, your best estimate of the probability of drawing two green marbles would be

$$\frac{11}{25} = \frac{44}{100} = 0.44$$

But this is only an estimate. The more times you draw, the more likely it is that the estimate will be close to the theoretical probability. It is better to repeat the experiment 500 times than 25 times. Thus, combining your results with other students' results is likely to produce a better estimate. To combine results, add everyone's tallies together; then calculate the new frequency.

extensions

a. *Represent* Ask 10 other students the following question: "What is your favorite sport: baseball, football, soccer, or basketball?" Record each response. Create a relative frequency table of your results. Share the results of the survey with your class.

b. *Analyze* In groups, conduct an experiment by drawing two counters out of a bag containing 3 green counters and 3 white counters. Each group should perform the experiment 30 times. Record each group's tallies in a frequency table like the one shown on the next page.

	Both Green		Both White		One of Each	
	Tally	Rel. Freq.	Tally	Rel. Freq.	Tally	Rel. Freq.
Group 1						
Group 2						
Group 3						
Group 4						
Group 5						
Group 6						
Whole Class						

Calculate the relative frequency for each group by dividing the tallies by 30 (the number of times each group performed the experiment). Then combine the results from all the groups. To combine the results, add the tallies in each column and write the totals in the last row of the table. Then divide each of these totals by the *total* number of times the experiment was performed (equal to the number of groups times 30). The resulting quotients are the whole-class relative frequencies for each event. Discuss your findings.

On the basis of their own data, which groups would guess that the probabilities were less than the "Whole class" data indicate? Which groups would guess that the probabilities were greater than the whole class's data indicate?

c. Choose a partner and roll two number cubes 100 times. Each time, observe the sum of the upturned faces, and fill out a relative frequency table like the one below. The sample space of this experiment has 11 outcomes.

Predict Are the outcomes equally likely? If not, which outcomes are more likely and which are less likely?

Sum	2	3	4	5	6	7	8	9	10	11	12
Frequency											
Relative Frequency											

After the experiment and calculation, estimate the probability that the sum of a roll will be 8. Estimate the probability that the sum will be at least 10. Estimate the probability that the sum will be odd.

LESSON 91

• Geometric Formulas

Power Up

Building Power

facts Power Up H

mental math

a. **Power/Roots:** $5^2 + \sqrt{100}$

b. **Number Sense:** 1000 − 875

c. **Calculation:** \$6.99 × 5

d. **Number Sense:** Double \$125.00.

e. **Decimals:** 12.5 ÷ 100

f. **Number Sense:** 20 × 42

g. **Measurement:** How many pints are in a quart?

h. **Calculation:** 3 × 4, ÷ 2, × 3, + 2, × 2, + 2, ÷ 2, ÷ 3

problem solving

If Sam can read 20 pages in 30 minutes, how long will it take Sam to read 200 pages?

New Concept

Increasing Knowledge

We have found the area of a rectangle by multiplying the length of the rectangle by its width. This procedure can be described with the following formula:

$$A = lw$$

The letter A stands for the area of the rectangle. The letters l and w stand for the length and width of the rectangle. Written side by side, lw means that we multiply the length by the width. The table below lists formulas for the perimeter and area of squares, rectangles, parallelograms, and triangles.

Figure	Perimeter	Area
Square	$P = 4s$	$A = s^2$
Rectangle	$P = 2l + 2w$	$A = lw$
Parallelogram	$P = 2b + 2s$	$A = bh$
Triangle	$P = s_1 + s_2 + s_3$	$A = \frac{1}{2}bh$

The letters P and A are abbreviations for *perimeter* and *area.* Other abbreviations are illustrated below:

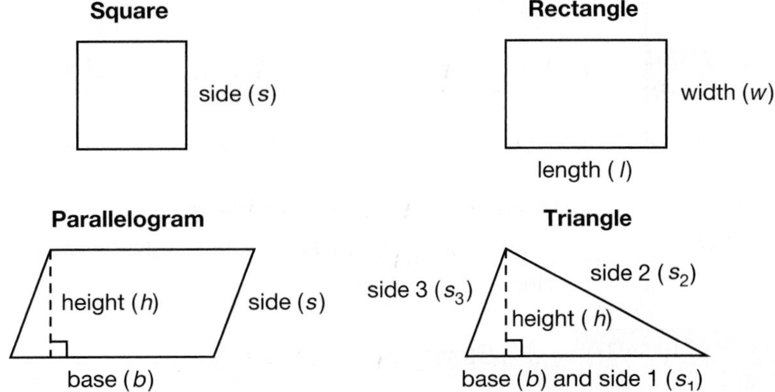

Since squares and rectangles are also parallelograms, the formulas for the perimeter and area of parallelograms may also be used for squares and rectangles.

To use a formula, we substitute each known measure in place of the appropriate letter in the formula. When substituting a number in place of a letter, it is a good practice to write the number in parentheses.

Example

Write the formula for the perimeter of a rectangle. Then substitute 8 cm for the length and 5 cm for the width. Solve the equation to find *P*.

Solution

The formula for the perimeter of a rectangle is

$$P = 2l + 2w$$

We rewrite the equation, substituting 8 cm for l and 5 cm for w. We write these measurements in parentheses.

$$P = 2(8 \text{ cm}) + 2(5 \text{ cm})$$

We multiply 2 by 8 cm and 2 by 5 cm.

$$P = 16 \text{ cm} + 10 \text{ cm}$$

Now we add 16 cm and 10 cm.

$$P = 26 \text{ cm}$$

The perimeter of the rectangle is **26 cm.**

We summarize the steps below to show how your work should look.

$$P = 2l + 2w$$

$$P = 2(8 \text{ cm}) + 2(5 \text{ cm})$$

$$P = 16 \text{ cm} + 10 \text{ cm}$$

$$P = 26 \text{ cm}$$

Practice Set

a. Write the formula for the area of a rectangle. Then substitute 8 cm for the length and 5 cm for the width. Solve the equation to find the area of the rectangle.

b. Write the formula for the perimeter of a parallelogram. Then substitute 10 cm for the base and 6 cm for the side. Solve the equation to find the perimeter of the parallelogram.

c. **Estimate** Look around the room for a rectangular or triangular shape. Estimate its dimensions and write a word problem about its perimeter or area. Then solve the problem.

Written Practice

Strengthening Concepts

1. (23, 65) What is the ratio of prime numbers to composite numbers in this list?

10, 11, 12, 13, 14, 15, 16, 17, 18, 19, 20, 21

2. (32) Sunrise was at 6:15 a.m. and sunset was at 5:45 p.m. How many hours and minutes were there from sunrise to sunset?

* **3.** (88) **Model** Draw a ratio box for this problem. Then solve the problem using a proportion.

When the good news was announced many leaped for joy and others just smiled. The ratio of leapers to smilers was 3 to 2. If 12 leaped for joy, how many just smiled?

4. (Inv. 6) A rectangular prism has how many more faces than a triangular prism?

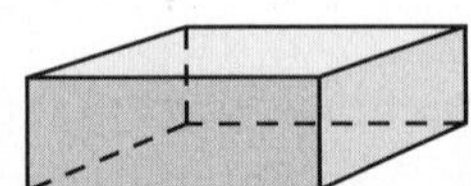

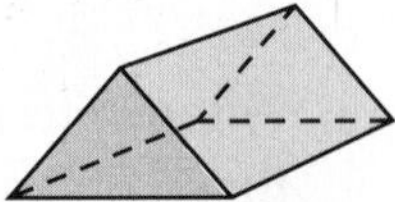

* **5.** (91) Write the formula for the area of a parallelogram as given in this lesson. Then substitute 15 cm for the base and 4 cm for the height. Solve the equation to find the area of the parallelogram.

* **6.** (90) **Connect** How many degrees does the minute hand of a clock turn in 45 minutes?

7. (Inv. 6) A pyramid with a triangular base is shown at right.

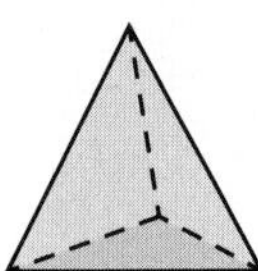

a. How many faces does it have?

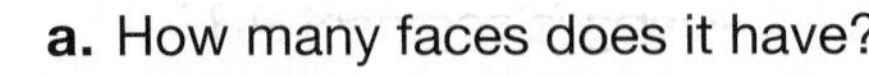

b. How many edges does it have?

c. How many vertices does it have?

8. (71) **a.** What is the perimeter of this parallelogram?

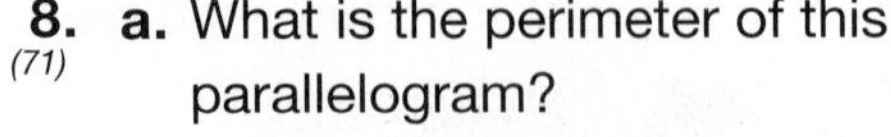

b. What is the area of this parallelogram?

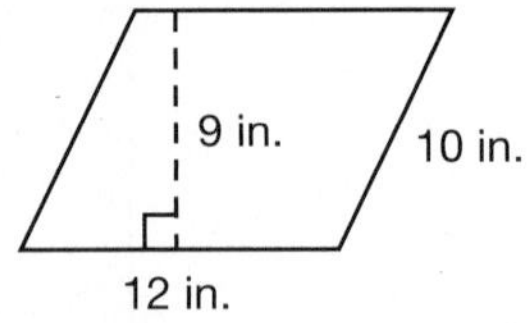

9. (74, 75) **a.** Write $\frac{7}{20}$ as a decimal number.

b. Write $\frac{7}{20}$ as a percent.

* **10.** (91) **Connect** The distance between bases in a softball diamond is 60 feet. What is the shortest distance a player could run to score a home run?

11. (59) $6\frac{2}{3} + 1\frac{3}{4}$

12. (63) $5 - 1\frac{2}{5}$

13. (63) $4\frac{1}{4} - 3\frac{5}{8}$

14. (72) $3 \times \frac{3}{4} \times 2\frac{2}{3}$

15. (68) $6\frac{2}{3} \div 100$

16. (68) $2\frac{1}{2} \div 3\frac{3}{4}$

17. (75) Compare: $\frac{9}{20}$ ◯ 50%

18. (75) **a.** What fraction of this group is shaded?

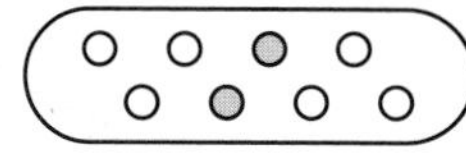

b. What percent of this group is shaded?

19. (77) If $\frac{5}{6}$ of the 300 seeds sprouted, how many seeds did not sprout?

* **20.** (87) $6y = 10$

* **21.** (85) $\frac{w}{20} = \frac{12}{15}$

22. (79) What is the area of this triangle?

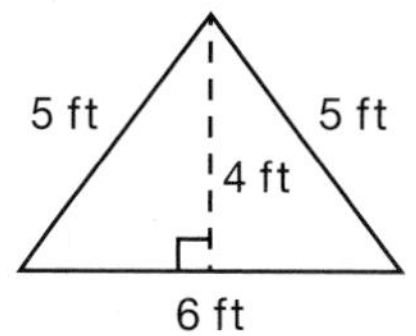

23. (82) The illustration below shows a cube with edges 1 foot long. (Thus, the edges are also 12 inches long.) What is the volume of the cube in cubic inches?

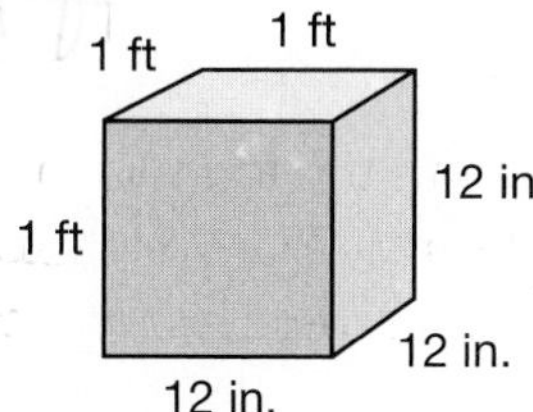

24. (73) Write the prime factorization of 225 using exponents.

25. (69) **Connect** The length of segment *AC* is 56 mm. The length of segment *BC* is 26 mm. How long is segment *AB*?

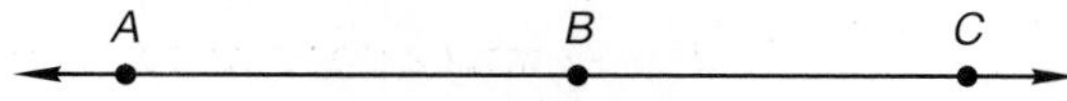

* **26.** (89) **Estimate** On a number line, $\sqrt{60}$ is between which two consecutive whole numbers?

* **27.** (89) Which whole number equals $\sqrt{225}$?

28. (Inv. 7) **Model** A square has vertices at the coordinates (2, 0), (0, −2), (−2, 0), and (0, 2). Graph the points on graph paper, and draw segments from point to point in the order given. To complete the square, draw a fourth segment from (0, 2) to (2, 0).

*** 29.** (Inv. 7) **Evaluate** The square in problem 28 encloses some whole squares and some half squares on the graph paper.

a. How many whole squares are enclosed by the square?

b. How many half squares are enclosed by the square?

c. Counting two half squares as a whole square, calculate the area of the entire square.

*** 30.** (Inv. 9) **Analyze** John will toss a coin three times. What is the probability that the coin will land heads up all three times? Express the probability ratio as a fraction and as a decimal.

Early Finishers
Real-World Application

The square practice field where Raul plays baseball has an area of 820 square yards. Before practice, Raul runs once around the bases to warm up.

a. Estimate the distance Raul runs to warm up. Explain how you made your estimate.

b. Next week, Raul has to run around the bases twice to warm up for practice. If he practices 3 days next week, estimate the total distance he will run to warm up.

LESSON 92

• Expanded Notation with Exponents
• Order of Operations with Exponents
• Powers of Fractions

Power Up

Building Power

facts Power Up L

mental math

a. **Number Sense:** 30 · 50

b. **Number Sense:** 486 + 50

c. **Percent:** 50% of 24

d. **Calculation:** \$20.00 − \$14.75

e. **Decimals:** 100 × 1.25

f. **Number Sense:** $\frac{600}{30}$

g. **Algebra:** If $n = 3$, what does $5n$ equal?

h. **Calculation:** $\sqrt{36}$, + 4, × 3, + 2, ÷ 4, + 1, $\sqrt{\ }$[1]

problem solving The basketball team's points-per-game average is 88 after its first four games. How many points does the team need to score during its fifth game to have a points-per-game average of 90?

New Concepts

Increasing Knowledge

expanded notation with exponents

In Lesson 32 we began writing whole numbers in **expanded notation.** Here we show 365 in expanded notation:

$$365 = (3 \times 100) + (6 \times 10) + (5 \times 1)$$

When writing numbers in expanded notation, we may write the powers of 10 with exponents.

$$365 = (3 \times 10^2) + (6 \times 10^1) + (5 \times 10^0)$$

Notice that 10^0 equals 1. The table below shows whole-number place values using powers of 10:

Trillions			Billions			Millions			Thousands			Ones		
hundreds	tens	ones	hundreds	tens	ones	hundreds	tens	ones	hundreds	tens	ones	hundreds	tens	ones
10^{14}	10^{13}	10^{12}	10^{11}	10^{10}	10^9	10^8	10^7	10^6	10^5	10^4	10^3	10^2	10^1	10^0

[1] Read $\sqrt{\ }$ as "find the square root."

Example 1

The speed of light is about 186,000 miles per second. Write 186,000 in expanded notation using exponents.

Solution

We write each nonzero digit (1, 8, and 6) multiplied by its place value.

$$\mathbf{186{,}000 = (1 \times 10^5) + (8 \times 10^4) + (6 \times 10^3)}$$

order of operations with exponents

In the order of operations, we simplify expressions with exponents or roots before we multiply or divide.

Order of Operations

1. Simplify within parentheses.
2. Simplify powers and roots.
3. Multiply and divide from left to right.
4. Add and subtract from left to right.

Some students remember the order of operations by using this memory aid:

Please
Excuse
My **D**ear
Aunt **S**ally

The first letter of each word is meant to remind us of the order of operations.

Parentheses
Exponents
Multiplication **D**ivision
Addition **S**ubtraction

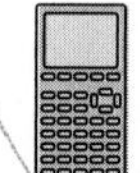

Visit www.SaxonPublishers.com/ActivitiesC1 *for a graphing calculator activity.*

Example 2

Simplify: $\mathbf{5 - (8 + 8) \div \sqrt{16} + 3^2 \times 2}$

Solution

We follow the order of operations.

$5 - (8 + 8) \div \sqrt{16} + 3^2 \times 2$	original problem
$5 - 16 \div \sqrt{16} + 3^2 \times 2$	simplified in parentheses
$5 - 16 \div 4 + 9 \times 2$	simplified powers and roots
$5 - 4 + 18$	multiplied and divided
19	added and subtracted

powers of fractions

We may use exponents with fractions and with decimals. With fractions, parentheses help clarify that an exponent applies to the whole fraction, not just its numerator.

$$\left(\frac{1}{2}\right)^3 \text{ means } \frac{1}{2} \cdot \frac{1}{2} \cdot \frac{1}{2}$$

$$(0.1)^2 \text{ means } 0.1 \times 0.1$$

Example 3

Simplify: $\left(\frac{2}{3}\right)^2$

Solution

We write $\frac{2}{3}$ as a factor twice and then multiply.

$$\frac{2}{3} \cdot \frac{2}{3} = \mathbf{\frac{4}{9}}$$

Practice Set

a. Write 2,500,000 in expanded notation using exponents.

b. Write this number in standard notation:

$$(5 \times 10^9) + (2 \times 10^8)$$

Simplify:

c. $10 + 2^3 \times 3 - (7 + 2) \div \sqrt{9}$

d. $\left(\frac{1}{2}\right)^3$ **e.** $(0.1)^2$ **f.** $\left(1\frac{1}{2}\right)^2$

g. $(2 + 3)^2 - (2^2 + 3^2)$

Written Practice

Strengthening Concepts

1. (58) **Explain** The weather forecast stated that the chance of rain for Wednesday is 40%. Does this forecast mean that it is more likely to rain or not to rain? Why?

2. (58, 74) A set of 36 shape cards contains an equal number of cards with hexagons, squares, circles, and triangles. What is the probability of drawing a square from this set of cards? Express the probability ratio as a fraction and as a decimal.

3. (18) **Connect** If the sum of three numbers is 144, what is the average of the three numbers?

4. (60) **Verify** "All quadrilaterals are polygons." True or false?

*** 5.** (89) $\sqrt{441}$

*** 6.** (92) $2 \cdot 3^2 - \sqrt{9} + (3 - 1)^3$

*** 7.** (91) Write the formula for the perimeter of a rectangle. Then substitute 12 in. for the length and 6 in. for the width. Solve the equation to find the perimeter of the rectangle.

8. (44) Arrange these numbers in order from least to greatest:

$$1, 0, 0.1, -1$$

9. (77) If $\frac{5}{6}$ of the 30 members were present, how many members were absent?

10. (70) Reduce before multiplying or dividing: $\dfrac{(24)(36)}{48}$

* **11.** (92) $\dfrac{\sqrt{100}}{\sqrt{25}}$

12. (59) $12\frac{5}{6} + 15\frac{1}{3}$

13. (38) $100 - 9.9$

14. (29) $\frac{4}{7} \times 100$

15. (42) $\frac{5}{8} = \frac{w}{48}$

16. (39) $0.25 \times \$4.60$

* **17.** (86) **Estimate** The diameter of a circular saucepan is 6 inches. What is the area of the circular base of the pan? Round the answer to the nearest square inch. (Use 3.14 for π.)

18. (74) Write $3\frac{3}{4}$ as a decimal number and subtract that number from 7.4.

19. (75) What percent of the first ten letters of the alphabet are vowels?

* **20.** (90) **Connect** Bobby rode his bike north. At Grand Avenue he turned left 90°. When he reached Arden Road, he turned left 90°. In what direction was Bobby riding on Arden Road?

21. (51) **Estimate** Find the product of 6.95 and 12.1 to the nearest whole number.

* **22.** (85) **Analyze** Write and solve a proportion for this statement: 16 is to 10 as what number is to 25?

23. (79) What is the area of the triangle below?

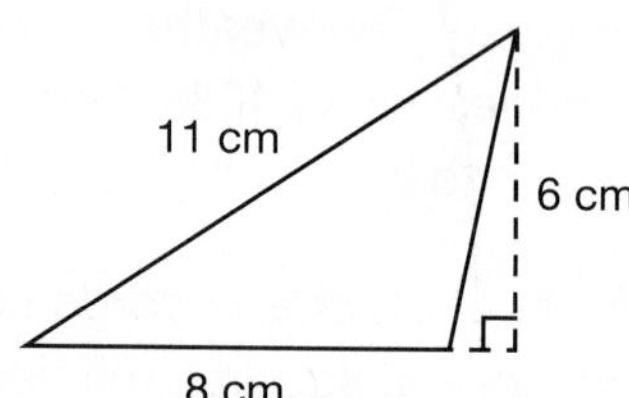

24. (Inv. 6) This figure is a rectangular prism.

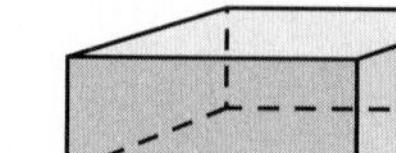

a. How many faces does it have?

b. How many edges does it have?

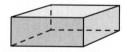

25. (17) **Predict** Each term in this sequence is $\frac{1}{16}$ more than the previous term. What are the next four terms in the sequence?

$$\frac{1}{16}, \frac{1}{8}, \frac{3}{16}, \frac{1}{4}, \_\_\_\_, \_\_\_\_, \_\_\_\_, \_\_\_\_, \ldots$$

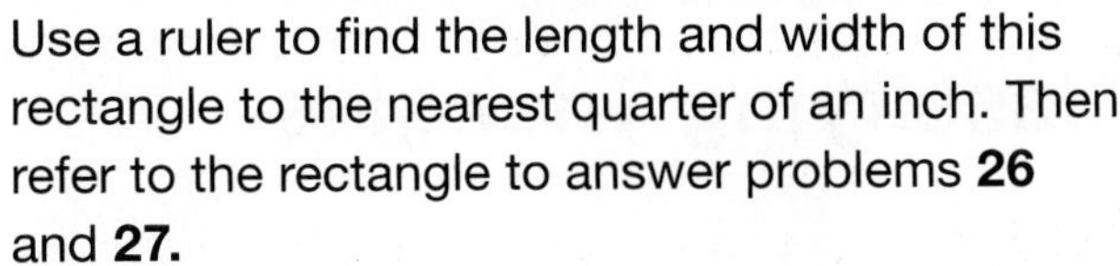

Use a ruler to find the length and width of this rectangle to the nearest quarter of an inch. Then refer to the rectangle to answer problems **26** and **27.**

* **26.** *(91)* What is the perimeter of the rectangle?

* **27.** *(91)* What is the area of the rectangle?

28. *(Inv. 7)* **Connect** The coordinates of the vertices of a parallelogram are (4, 3), (−2, 3), (0, −2), and (−6, −2). What is the area of the parallelogram?

* **29.** *(81)* Simplify:

a. (12 cm)(8 cm)

b. $\frac{36 \text{ ft}^2}{4 \text{ ft}}$

30. *(78)* Fernando poured water from one-pint bottles into a three-gallon bucket. How many pints of water could the bucket hold?

Early Finishers
Math and Geography

There are close to 4 million people living on the island of Puerto Rico, making it one of the most densely populated islands in the world. If the population density is approximately 1,000 people per square mile, how many people live in a 3.5 square mile area? Write and solve a proportion to answer this question.

LESSON 93

• Classifying Triangles

Power Up

Building Power

facts Power Up I

mental math

a. **Number Sense:** $40 \cdot 60$

b. **Number Sense:** $234 - 50$

c. **Percent:** 25% of 24

d. **Calculation:** \$5.99 + \$2.47

e. **Decimals:** $1.2 \div 100$

f. **Number Sense:** 30×25

g. **Algebra:** If $x = 5$, what does $5x$ equal?

h. **Calculation:** 8×9, $+ 3$, $\div 3$, $\sqrt{\ }$, $\times 6$, $+ 3$, $\div 3$, $- 10$

problem solving

Benjamin put 2 purple marbles, 7 red marbles, and 1 brown marble in a bag and shook the bag. If he reaches in and chooses a marble without looking, what is the probability that he chooses a red marble? A purple or brown marble? What is the probability of *not* choosing a red marble? If Benjamin does choose a red marble, but gives it away, what is the probability he will choose another red marble?

New Concept

Increasing Knowledge

Thinking Skill

Generalize

Explain in your own words how the number of equal sides of a triangle compares to the number of equal angles it has.

All three-sided polygons are triangles, but not all triangles are alike. We distinguish between different types of triangles by using the lengths of their sides and the measures of their angles. We will first classify triangles based on the lengths of their sides.

Triangles Classified by Their Sides

Name	Example	Description
Equilateral triangle		All three sides are equal in length.
Isosceles triangle		At least two of the three sides are equal in length.
Scalene triangle		All three sides have different lengths.

An **equilateral triangle** has three equal sides and three equal angles.

An **isosceles triangle** has at least two equal sides and two equal angles.

A **scalene triangle** has three unequal sides and three unequal angles.

Thinking Skill

Justify

Is an equilateral triangle also an isosceles triangle? Why or why not?

Next, we consider triangles classified by their angles. In Lesson 28 we learned the names of three different kinds of angles: **acute, right,** and **obtuse.** We can also use these words to describe triangles.

Triangles Classified by Their Angles

Name	Example	Description
Acute triangle		All three angles are acute.
Right triangle		One angle is a right angle.
Obtuse triangle		One angle is an obtuse angle.

Each angle of an equilateral triangle measures 60°, so an equilateral triangle is also an acute triangle. An isosceles triangle may be an acute triangle, a right triangle, or an obtuse triangle. A scalene triangle may also be an acute triangle, a right triangle, or an obtuse triangle.

Practice Set

a. One side of an equilateral triangle measures 15 cm. What is the perimeter of the triangle?

b. *Verify* "An equilateral triangle is also an acute triangle." True or false?

c. *Verify* "All acute triangles are equilateral triangles." True or false?

d. Two sides of a triangle measure 3 inches and 4 inches. If the perimeter is 10 inches, what type of triangle is it?

e. *Verify* "Every right triangle is a scalene triangle." True or false?

Written Practice

Strengthening Concepts

* **1.** (88) *Model* Draw a ratio box for this problem. Then solve the problem using a proportion.

The ratio of the length to the width of the rectangular lot was 5 to 2. If the lot was 60 ft wide, how long was the lot?

* **2.** (Inv. 9) Mitch does not know the correct answer to two multiple-choice questions. The choices are A, B, C, and D. If Mitch just guesses, what is the probability that Mitch will guess both answers correctly?

3. (18) If the sum of four numbers is 144, what is the average of the four numbers?

4. (82) The rectangular prism shown below is constructed of 1-cubic-centimeter blocks. What is the volume of the prism?

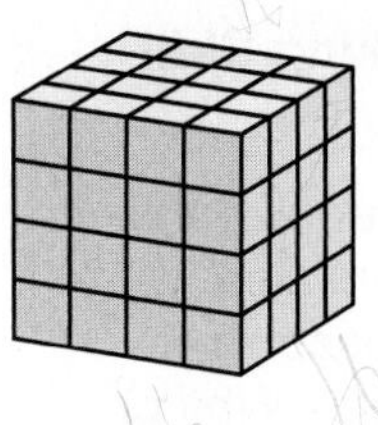

5. (74, 75) Write $\frac{9}{25}$ as a decimal number and as a percent.

6. (74) Write $3\frac{1}{5}$ as a decimal number and add it to 3.5. What is the sum?

7. (41) What number is 45% of 80?

8. (73) $(0.3)^3$

9. (73) $\left(2\frac{1}{2}\right)^2$

* **10.** (92) $\sqrt{9} \cdot \sqrt{100}$

11. (77) Twenty of the two dozen members voted yes. What fraction of the members voted yes?

12. (23) **Analyze** If the rest of the members in problem 11 voted no, then what was the ratio of "no" votes to "yes" votes?

Find each unknown number:

13. (63) $w + 4\frac{3}{4} = 9\frac{1}{3}$

* **14.** (85) $\frac{6}{5} = \frac{m}{30}$

* **15.** (93) **Conclude** In what type of triangle are all three sides the same length?

16. (42) What mixed number is $\frac{3}{8}$ of 100?

* **17.** (92) **Analyze** $10 + 6^2 \div 3 - \sqrt{9} \times 3$

18. (Inv. 6) A triangular prism has how many faces?

19. (78) How many quarts of milk is $2\frac{1}{2}$ gallons of milk?

20. (65, 73) **Represent** Use a factor tree to find the prime factors of 800. Then write the prime factorization of 800 using exponents.

21. (51) Round the decimal number one hundred twenty-five thousandths to the nearest tenth.

* **22.** (87) $0.08n = \$1.20$

23. (79) The diagonal segment through this rectangle divides the rectangle into two congruent right triangles. What is the area of one of the triangles?

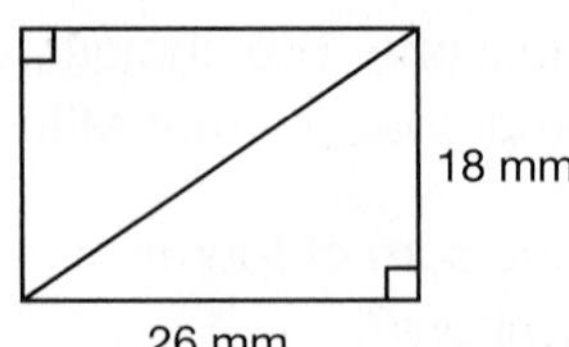

24. (75) Write $\frac{17}{20}$ as a percent.

* **25.** (89) **Estimate** On this number line the arrow could be pointing to which of the following?

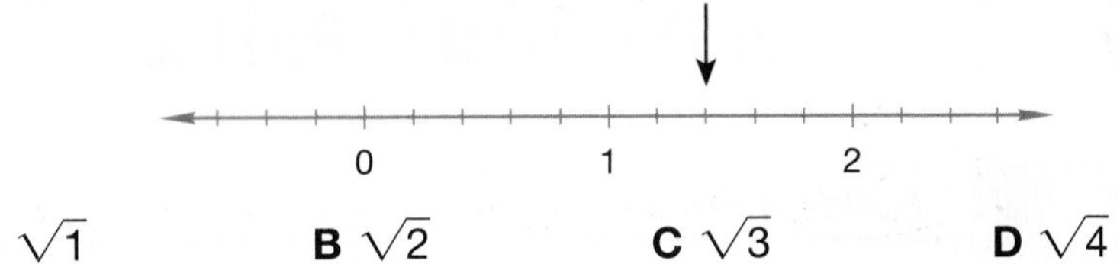

A $\sqrt{1}$ **B** $\sqrt{2}$ **C** $\sqrt{3}$ **D** $\sqrt{4}$

* **26.** (92) Write this number in standard notation:

$$(7 \times 10^9) + (2 \times 10^8) + (5 \times 10^7)$$

27. (58) **a.** What is the probability of rolling a 6 with a single roll of a number cube?

b. What is the probability of rolling a number less than 6 with a single roll of a number cube?

c. Name the event and its complement. Then describe the relationship between the two probabilities.

28. (64, Inv. 7) **Represent** The coordinates of the four vertices of a quadrilateral are (−3, −2), (0, 2), (3, 2), and (5, −2). What is the name for this type of quadrilateral?

* **29.** (91) **Explain** The formula for the area of a triangle is

$$A = \frac{bh}{2}$$

If the base measures 20 cm and the height measures 15 cm, then what is the area of the triangle? Explain your thinking.

30. (10, 17) **Generalize** Write the rule for this sequence. Then write the next four numbers.

$$\frac{1}{16}, \frac{1}{8}, \frac{3}{16}, \frac{1}{4}, \frac{5}{16}, \frac{3}{8}, \_\_\_, \_\_\_, \_\_\_, \_\_\_, \ldots$$

Early Finishers

Real-World Application

A teacher emptied a 1.5 oz snack-sized box of raisins into a dish. The teacher then asked for volunteers to estimate the number of raisins in the dish. Twelve volunteers gave the following estimates.

84 100 50 75 66 75 70 90 85 77 91 80

a. Which type of display—a circle graph or a stem-and-leaf plot—is the most appropriate way to display this data? Draw your display and justify your choice.

b. The dish contained exactly 85 raisins. How many volunteers made a reasonable estimate? Give a reason to support your answer. (Hint: You might think of how far off an estimate is in terms of a percent of 85.)

LESSON 94

Writing Fractions and Decimals as Percents, Part 2

Power Up

Building Power

facts Power Up K

mental math

a. **Number Sense:** $50 \cdot 70$

b. **Number Sense:** $572 + 150$

c. **Percent:** 50% of 80

d. **Calculation:** $10.00 − $6.36

e. **Decimals:** 100×0.02

f. **Number Sense:** $\frac{640}{20}$

g. **Algebra:** If $r = 6$, what does $9r$ equal?

h. **Calculation:** 4×5, $+ 1$, $\div 3$, $\times 8$, $- 1$, $\div 5$, $\times 4$, $- 2$, $\div 2$

problem solving What are the next four numbers in this sequence: $\frac{1}{12}, \frac{1}{6}, \frac{1}{4}, \frac{1}{3}, \ldots$

New Concept

Increasing Knowledge

Since Lesson 75 we have practiced changing a fraction or decimal to a percent by writing an equivalent fraction with a denominator of 100.

$$\frac{3}{5} = \frac{60}{100} = 60\%$$

$$0.4 = 0.40 = \frac{40}{100} = 40\%$$

In this lesson we will practice another method of changing a fraction to a percent. Since 100% equals 1, we can multiply a fraction by 100% to form an equivalent number. Here we multiply $\frac{3}{5}$ by 100%:

$$\frac{3}{5} \times \frac{100\%}{1} = \frac{300\%}{5}$$

Then we simplify and find that $\frac{3}{5}$ equals 60%.

$$\frac{300\%}{5} = 60\%$$

Thinking Skill

Discuss

How can you use mental math to change a decimal to a percent?

We can use the same procedure to change decimals to percents. Here we multiply 0.375 by 100%.

$$0.375 \times 100\% = 37.5\%$$

To change a number to a percent, multiply the number by 100%.

Example 1

Change $\frac{1}{3}$ to a percent.

Solution

We multiply $\frac{1}{3}$ by 100%.

$$\frac{1}{3} \times \frac{100\%}{1} = \frac{100\%}{3}$$

To simplify, we divide 100% by 3 and write the quotient as a mixed number.

$$\begin{array}{r} \mathbf{33\tfrac{1}{3}\%} \\ 3\overline{)100\%} \\ \underline{9} \\ 10 \\ \underline{9} \\ 1 \end{array}$$

Example 2

Write 1.2 as a percent.

Solution

We multiply 1.2 by 100%.

$$1.2 \times 100\% = \mathbf{120\%}$$

In some applications a percent may be greater than 100%. If the number we are changing to a percent is greater than 1, then the percent is greater than 100%.

Example 3

Write $2\frac{1}{4}$ as a percent.

Solution

We show two methods below.

Method 1: We split the whole number and fraction. The mixed number $2\frac{1}{4}$ means "$2 + \frac{1}{4}$." We change each part to a percent and then add.

$$2 + \frac{1}{4}$$

$$\swarrow \qquad \downarrow$$

$$200\% + 25\% = \mathbf{225\%}$$

Method 2: We change the mixed number to an improper fraction. The mixed number $2\frac{1}{4}$ equals the improper fraction $\frac{9}{4}$. We then change $\frac{9}{4}$ to a percent.

$$\frac{9}{\cancel{4}_{1}} \times \frac{\overset{25}{\cancel{100}}\%}{1} = \mathbf{225\%}$$

Example 4

Write $2\frac{1}{6}$ as a percent.

Solution

Method 1 shown in example 3 is quick, if we can recall the percent equivalent of a fraction. Method 2 is easier if the percent equivalent does not readily come to mind. We will use method 2 in this example. We write the mixed number $2\frac{1}{6}$ as the improper fraction $\frac{13}{6}$ and multiply by 100%.

$$\frac{13}{6} \times \frac{100\%}{1} = \frac{1300\%}{6}$$

Now we divide 1300% by 6 and write the quotient as a mixed number.

$$\frac{1300\%}{6} = \mathbf{216\frac{2}{3}\%}$$

Example 5

Twenty of the thirty students on the bus were girls. What percent of the students on the bus were girls?

Solution

We first find the fraction of the students that were girls. Then we convert the fraction to a percent.

Girls were $\frac{20}{30}$ $\left(\text{or } \frac{2}{3}\right)$ of the students on the bus. Now we multiply the fraction by 100%, which is the percent name for 1. We can use either $\frac{20}{30}$ or $\frac{2}{3}$, as we show below.

$$\frac{20}{\underset{3}{\cancel{30}}} \times \overset{10}{\cancel{100}}\% = \frac{200\%}{3} = 66\frac{2}{3}\%$$

or

$$\frac{2}{3} \times 100\% = \frac{200\%}{3} = 66\frac{2}{3}\%$$

We find that $\mathbf{66\frac{2}{3}\%}$ of the students on the bus were girls.

Practice Set

Change each decimal number to a percent by multiplying by 100%:

a. 0.5 **b.** 0.06 **c.** 0.125

d. 0.45 **e.** 1.3 **f.** 0.025

g. 0.09 **h.** 1.25 **i.** 0.625

Change each fraction or mixed number to a percent by multiplying by 100%:

j. $\frac{2}{3}$ **k.** $\frac{1}{6}$ **l.** $\frac{1}{8}$

m. $1\frac{1}{4}$ **n.** $2\frac{4}{5}$ **o.** $1\frac{1}{3}$

p. What percent of this rectangle is shaded?

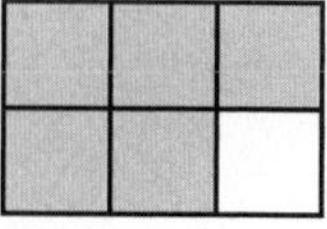

q. Connect What percent of a yard is a foot?

Written Practice
Strengthening Concepts

* **1.** (94) **Analyze** Ten of the thirty students on the bus were boys. What percent of the students on the bus were boys?

2. (18) **Connect** On the Celsius scale water freezes at 0°C and boils at 100°C. What temperature is halfway between the freezing and boiling temperatures of water?

3. (69) **Connect** If the length of segment *AB* is $\frac{1}{3}$ the length of segment *AC*, and if segment *AC* is 12 cm long, then how long is segment *BC*?

4. (94) What percent of this group is shaded?

* **5.** (94) Change $1\frac{2}{3}$ to a percent by multiplying $1\frac{2}{3}$ by 100%.

* **6.** (94) Change 1.5 to a percent by multiplying 1.5 by 100%.

7. (74) $6.4 - 6\frac{1}{4}$ (Begin by writing $6\frac{1}{4}$ as a decimal number.)

* **8.** (92) $10^4 - 10^3$

9. (70) How much is $\frac{3}{4}$ of 360?

Tommy placed a cylindrical can of spaghetti sauce on the counter. He measured the diameter of the can and found that it was about 8 cm. Use this information to answer problems **10** and **11**.

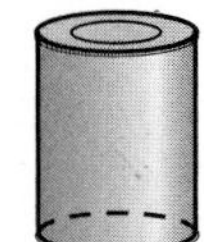

Use 3.14 for π.

10. (47) The label wraps around the circumference of the can. How long does the label need to be?

* **11.** (86) **Analyze** How many square centimeters of countertop does the can occupy?

12. (61) $3\frac{1}{2} + 1\frac{3}{4} + 4\frac{5}{8}$

13. (72) $\frac{9}{10} \cdot \frac{5}{6} \cdot \frac{8}{9}$

* **14.** (92) Write 250,000 in expanded notation using exponents.

15. (1) \$8.47 + 95¢ + \$12

16. (51) $37.5 \div 100$

17. (85) $\frac{3}{7} = \frac{21}{x}$

18. (68) $33\frac{1}{3} \div 100$

19. (41) If ninety percent of the answers were correct, then what percent were incorrect?

20. (35) Write the decimal number one hundred twenty and three hundredths.

21. (76) Arrange these numbers in order from least to greatest:

$$-2.5, \frac{2}{5}, \frac{5}{2}, -5.2$$

22. (Inv. 6) **Conclude** A pyramid with a square base has how many edges?

23. (71) What is the area of this parallelogram?

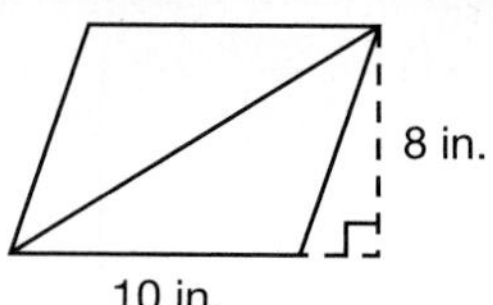

* **24.** (93) **Classify** The parallelogram in problem **23** is divided into two congruent triangles. Both triangles may be described as which of the following?

A acute **B** right **C** obtuse

25. (14) During the year, the temperature ranged from −37°F in winter to 103°F in summer. How many degrees was the range of temperature for the year?

26. (Inv. 7, 79) **Model** The coordinates of the three vertices of a triangle are (0, 0), (0, −4), and (−4, 0). Graph the triangle and find its area.

27. (Inv. 5) Margie's first nine test scores are shown below.

21, 25, 22, 19, 22, 24, 20, 22, 24

a. What is the mode of these scores?

b. What is the median of these scores?

* **28.** (92) $2^3 + \sqrt{25} \times 3 - 4^2 \div \sqrt{4}$

29. (78) Sandra filled the aquarium with 24 quarts of water. How many gallons of water did Sandra pour into the aquarium?

30. (58, 74) A bag contains lettered tiles, two for each letter of the alphabet. What is the probability of drawing a tile with the letter A? Express the probability ratio as a fraction and as a decimal rounded to the nearest hundredth.

LESSON 95

Reducing Rates Before Multiplying

Power Up

Building Power

facts Power Up G

mental math

a. **Number Sense:** 60 · 80

b. **Number Sense:** 437 − 150

c. **Percent:** 25% of 80

d. **Calculation:** \$3.99 + \$4.28

e. **Decimals:** 17.5 ÷ 100

f. **Number Sense:** 30 × 55

g. **Algebra:** If $w = 10$, what does $7w$ equal?

h. **Calculation:** 6 × 8, + 1, $\sqrt{\ }$, × 5, + 1, $\sqrt{\ }$, × 3, ÷ 2, $\sqrt{\ }$

problem solving

Between the prime number 2 and its double, 4, there is a prime number

2 ③ 4

Is there at least one prime number between every prime number and its double?

New Concept

Increasing Knowledge

Since Lesson 70 we have practiced reducing fractions before multiplying. This is sometimes called *canceling.*

$$\frac{\overset{1}{\cancel{3}}}{\underset{2}{\cancel{4}}} \cdot \frac{\overset{1}{\cancel{2}}}{\underset{1}{\cancel{3}}} \cdot \frac{\overset{1}{\cancel{5}}}{\underset{2}{\cancel{6}}} = \frac{1}{4}$$

We can cancel **units** before multiplying just as we cancel numbers.

$$\frac{4 \text{ miles}}{1 \cancel{\text{ hour}}} \times \frac{2 \cancel{\text{ hours}}}{1} = \frac{8 \text{ miles}}{1} = 8 \text{ miles}$$

Since rates are ratios of two measures, multiplying and dividing rates involves multiplying and dividing units.

Example 1

Multiply 55 miles per hour by six hours.

Solution

Math Language
Recall that a ratio is a comparison of two numbers by division.

We write the rate 55 miles per hour as the ratio 55 miles over 1 hour, because "per" indicates division. We write six hours as the ratio 6 hours over 1.

$$\frac{55 \text{ miles}}{1 \text{ hour}} \times \frac{6 \text{ hours}}{1}$$

The unit "hour" appears above and below the division line, so we can cancel hours.

$$\frac{55 \text{ miles}}{1 \cancel{\text{hour}}} \times \frac{6 \cancel{\text{hours}}}{1} = \textbf{330 miles}$$

Connect Can you think of a word problem to fit this equation?

Example 2

Multiply 5 feet by 12 inches per foot.

Solution

We write ratios of 5 feet over 1 and 12 inches over 1 foot. We then cancel units and multiply.

$$\frac{5 \cancel{\text{feet}}}{1} \cdot \frac{12 \text{ inches}}{1 \cancel{\text{foot}}} = \textbf{60 inches}$$

Practice Set

When possible, cancel numbers and units before multiplying:

a. $\frac{3 \text{ dollars}}{1 \text{ hour}} \times \frac{8 \text{ hours}}{1}$

b. $\frac{6 \text{ baskets}}{10 \text{ shots}} \times \frac{100 \text{ shots}}{1}$

Reading Math
The abbreviation "kwh" stands for *kilowatt hours,* a rate used to measure energy.

c. $\frac{10 \text{ cents}}{1 \text{ kwh}} \times \frac{26.3 \text{ kwh}}{1}$

d. $\frac{160 \text{ km}}{2 \text{ hours}} \cdot \frac{10 \text{ hours}}{1}$

e. Multiply 18 teachers by 29 students per teacher.

f. Multiply 2.3 meters by 100 centimeters per meter.

g. Solve this problem by multiplying two ratios: How far will the train travel in 6 hours at 45 miles per hour?

Written Practice

Strengthening Concepts

1. (41) What is the total price of a $45.79 item when 7% sales tax is added to the price?

*** 2.** (95) Jeff is 1.67 meters tall. How many centimeters tall is Jeff ? (Multiply 1.67 meters by 100 centimeters per meter.)

3. (77) **Analyze** If $\frac{5}{8}$ of the 40 seeds sprouted, how many seeds did not sprout?

4. (46) Write this number in standard notation:

$$(5 \times 100) + (6 \times 10) + \left(7 \times \frac{1}{10}\right) + \left(3 \times \frac{1}{100}\right)$$

*** 5.** (94) Change $\frac{1}{6}$ to its percent equivalent by multiplying $\frac{1}{6}$ by 100%.

*** 6.** (94) **Analyze** What is the percent equivalent of 2.5?

7. (41) How much money is 30% of $12.00?

*** 8.** (90) **Connect** The minute hand of a clock turns 180° in how many minutes?

9. (27) **Evaluate** The circumference of the front tire on Elizabeth's bike is 6 feet. How many complete turns does the front wheel make as Elizabeth rides down her 30-foot driveway?

10. (82) Chad built this stack of one-cubic-foot boxes. What is the volume of the stack?

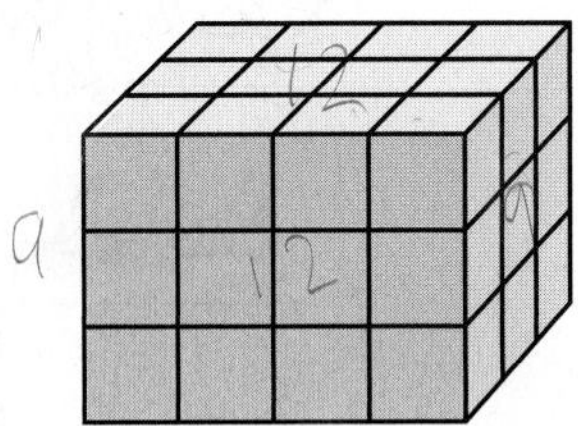

11. (57) $\frac{3}{4} + \frac{3}{5}$

12. (3) $18\frac{1}{8} - 12\frac{1}{2}$

13. (72) $3\frac{3}{4} \times 2\frac{2}{3} \times 1\frac{1}{10}$

14. (92) $\frac{2^5}{2^3}$

15. (68) How many fourths are in $2\frac{1}{2}$?

16. (38) $12 + 8.75 + 6.8$

17. (38, 39) $(1.5)^2$

18. (74) $6\frac{2}{5} \div 0.8$ (decimal answer)

19. (51) **Estimate** Find the sum of $6\frac{1}{4}$, 4.95, and 8.21 by rounding each number to the nearest whole number before adding. Explain how you arrived at your answer.

*** 20.** (86) **Analyze** The diameter of a round tabletop is 60 inches.

a. What is the radius of the tabletop?

b. What is the area of the tabletop? (Use 3.14 for π.)

21. (75) Arrange these numbers in order from least to greatest:

$$\frac{1}{4}, 4\%, 0.4$$

Find each unknown number.

22. (43) $y + 3.4 = 5$

23. (85) $\frac{4}{8} = \frac{x}{12}$

24. (Inv. 6, 82) A cube has edges that are 6 cm long.

a. What is the area of each face of the cube?

b. What is the volume of the cube?

c. What is the surface area of the cube?

25. (69) Connect $\overline{AB}$ is 24 mm long. $\overline{AC}$ is 42 mm long. How long is $\overline{BC}$?

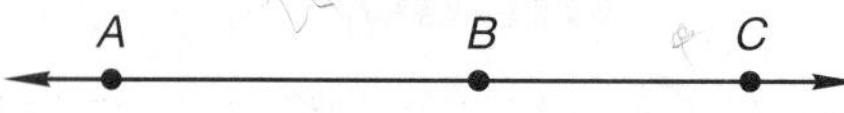

*** 26.** (89, 92) $6^2 \div \sqrt{9} + 2 \times 2^3 - \sqrt{400}$

27. (78) What is the ratio of a pint of water to a quart of water?

*** 28.** (91) The formula for the area of a parallelogram is $A = bh$. If the base of a parallelogram is 1.2 m and the height is 0.9 m, what is the area of the parallelogram? How can estimation help you check your answer?

*** 29.** (95) Multiply 2.5 liters by 1000 milliliters per liter.

$$\frac{2.5 \text{ liters}}{1} \times \frac{1000 \text{ milliliters}}{1 \text{ liter}}$$

30. (58, 74) If this spinner is spun once, what is the probability that the arrow will end up pointing to an even number? Express the probability ratio as a fraction and as a decimal.

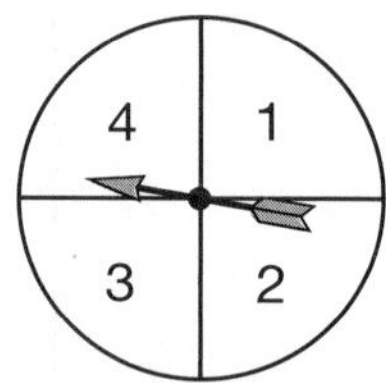

Early Finishers
Real-World Application

The local university football stadium seats 60,000 fans, and average attendance at home games is 48,500. It has been determined that an average fan consumes 2.25 beverages per game.

a. If each beverage is served in a cup, about how many cups are used during an average game? Express your answer in scientific notation.

b. Next week is the homecoming game, which is always sold out. A box of cups contains 1×10^3 cups. How many boxes of cups will be needed for the game?

LESSON

96

• Functions
• Graphing Functions

Power Up

Building Power

facts

Power Up D

mental math

a. **Number Sense:** 70 · 90

b. **Number Sense:** 364 + 250

c. **Percent:** 50% of 60

d. **Calculation:** \$5.00 − \$0.89

e. **Decimals:** 100 × 0.015

f. **Number Sense:** $\frac{750}{30}$

g. **Measurement:** How many pints are in 2 quarts?

h. **Calculation:** 6 × 6, − 1, ÷ 5, × 8, − 1, ÷ 11, × 8, × 2, + 1, $\sqrt{}$

problem solving

Copy this factor tree and fill in the missing numbers:

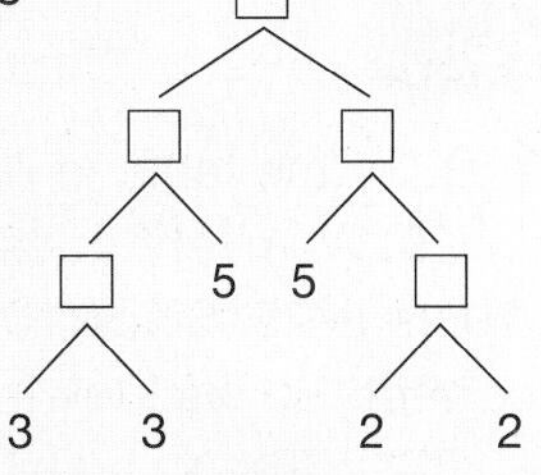

New Concepts

Increasing Knowledge

functions

We know that the surface area of a rectangular prism is the sum of the areas of its sides. A cube is a special rectangular prism with six square faces. If we know the area of one side of a cube, then we can find the surface area of that cube. We can make a table to show the surface areas of cubes based on the area of one side of the cube.

Area of Each Side of a Cube (cm^2)	Surface Area of the Cube (cm^2)
4	24
9	54
16	96
25	150

Discuss Use the data in the table to help you create a formula for the surface area of a cube. Let A be the area of each side of the cube and S be the surface area of the cube.

$$(S = 6A)$$

Your formula is an example of a **function.** A function is a rule for using one number (an input) to calculate another number (an output). In this function, side area is the input and surface area is the output. Because the surface area of a cube depends on the area of each side, we say that the surface area of a cube is a function of the area of a side. If we know the area of one side of a cube, we can apply the function's rule (formula) to find the surface area of the cube.

Example 1

Find the rule for this function. Then use the rule to find the value of *m* when *l* is *7*.

l	*m*
5	20
7	
10	25
15	30

Solution

We study the table to discover the function rule. We see that when l is 5, m is 20. We might guess that the rule is to multiply l by 4. However, when l is 10, m is 25. Since 10×4 does not equal 25, we know that this guess is incorrect. So we look for another rule.

We notice that 20 is 15 more than 5 and that 25 is 15 more than 10. Perhaps the rule is to add 15 to l. We see that the values in the bottom row of the table ($l = 15$ and $m = 30$) fit this rule. So the rule is, **to find *m*, add 15 to *l*.** To find m when l is *7,* we add 15 to 7.

$$7 + 15 = 22$$

The missing number in the table is **22.**

Instead of using the letter m at the top of the table, we could have written the rule. In the table at right, $l + 15$ has replaced m. This means we add 15 to the value of l. We show this type of table in the next example.

l	*l* + 15
5	20
7	
10	25
15	30

Example 2

Find the missing number in this function table:

x	2	3	4
$3x - 2$	4	7	

Solution

This table is arranged horizontally. The rule of the function is stated in the table: multiply the value of x by 3, then subtract 2. To find the missing number in the table, we apply the rule of the function when x is 4.

$$3x - 2$$

$$3(4) - 2 = 10$$

We find that the missing number is **10.**

graphing functions

Many functions can be graphed on a coordinate plane. Here we show a function table that relates the perimeter of a square to the length of one of its sides. On the coordinate plane we have graphed the number pairs that appear in the table. The coordinate plane's horizontal axis shows the length of a side, and its vertical axis shows the perimeter.

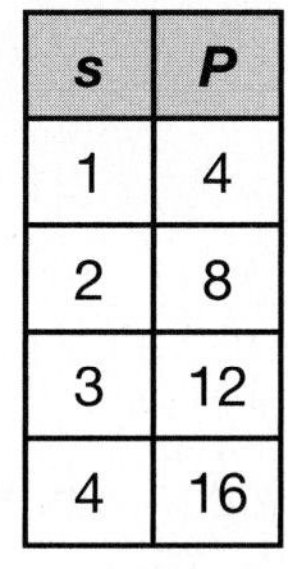

s	P
1	4
2	8
3	12
4	16

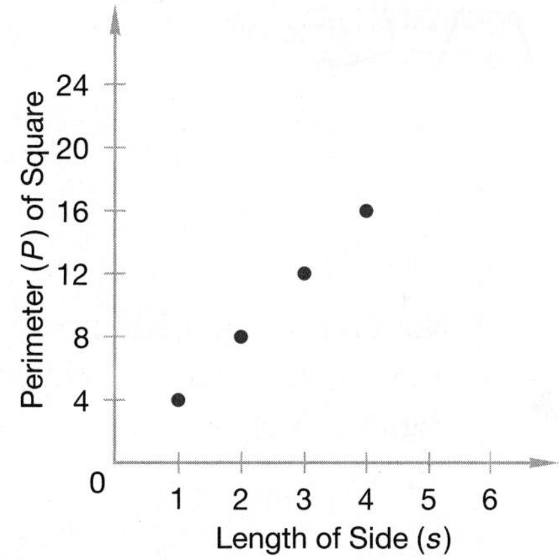

We have used different scales on the two axes so that the graph is not too steep. The graphed points show the side length and perimeter of four squares with side lengths of 1, 2, 3, and 4 units. Notice that the graphed points are aligned. Of course, we could graph many more points and represent squares with side lengths of 100 units or more. We could also graph points for squares with side lengths of 0.01 or less. In fact, we can graph points for any side length whatsoever! Such a graph would look like a ray, as shown on the next page.

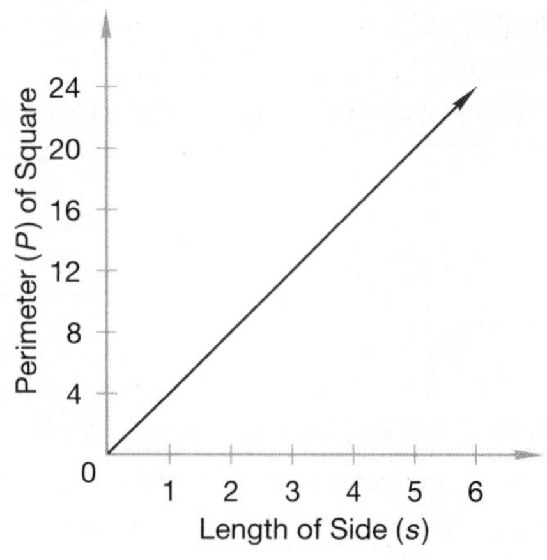

Example 3

The perimeter of an equilateral triangle is a function of the length of its sides. Make a table for this function using side lengths of 1, 2, 3, and 4 units. Then graph the ordered pairs on a coordinate plane. Extend a ray through the points to represent the function for all equilateral triangles.

Solution

Thinking Skill

Generalize

What is the rule for this table?

We create a table of ordered pairs. The letter *s* stands for the length of a side, and *P* stands for the perimeter.

s	*P*
1	3
2	6
3	9
4	12

Now we graph these points on a coordinate plane with one axis for perimeter and the other axis for side length. Then we draw a ray from the origin through these points.

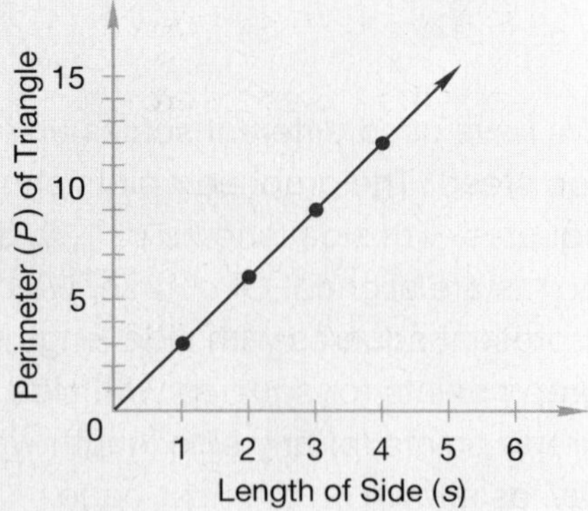

Every point along the ray represents the side length and perimeter of an equilateral triangle.

Practice Set

Generalize Find the missing number in each function table:

a.

x	y
3	1
5	3
6	4
10	

b.

a	b
3	8
5	10
7	12
	15

c.

x	3	6	8
$3x + 1$	10	19	

d.

x	3	4	7
$3x - 1$	8		20

e. **Model** The chemist mixed a solution that weighed 2 pounds per quart. Create a table of ordered pairs for this function for 1, 2, 3, and 4 quarts. Then graph the points on a coordinate plane, using the horizontal axis for quarts and the vertical axis for pounds. Would it be appropriate to draw a ray through the points? Why or why not?

Written Practice

Strengthening Concepts

1. (12, 53) When the sum of 2.0 and 2.0 is subtracted from the product of 2.0 and 2.0, what is the difference?

2. (49) A 4.2-kilogram object weighs the same as how many objects that each weigh 0.42 kilogram?

3. (18) If the average of 8 numbers is 12, what is the sum of the 8 numbers?

4. (64) **Conclude** What is the name of a quadrilateral that has one pair of sides that are parallel and one pair of sides that are not parallel?

*** 5.** (94) **a.** Write 0.15 as a percent.

b. Write 1.5 as a percent.

*** 6.** (94) Write $\frac{5}{6}$ as a percent.

7. (41, 76) **Classify** Three of the numbers below are equivalent. Which one is not equivalent to the others?

A 1 **B** 100% **C** 0.1 **D** $\frac{100}{100}$

8. (73) 11^3

9. (70) How much is $\frac{5}{6}$ of 360?

*** 10.** (89) **Estimate** Between which two consecutive whole numbers is $\sqrt{89}$?

*** 11.** (90) **Analyze** Silvester ran around the field, turning at each of the three backstops. What was the average number of degrees he turned at each of the three corners?

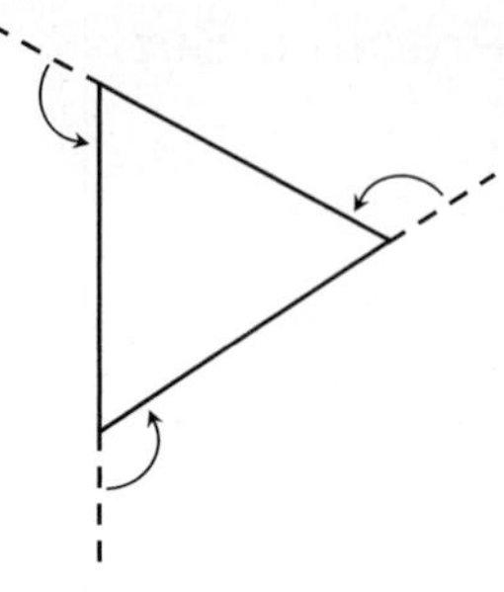

*** 12.** (96) **Generalize** Find the missing number in this function table.

x	4	7	13	15
2x − 1	7	13		29

13. (67) Factor and reduce: $\frac{(45)(54)}{81}$

14. (49) $\frac{30}{0.08}$

15. (68) $16\frac{2}{3} \div 100$

16. (61) $2\frac{1}{2} + 3\frac{1}{3} + 4\frac{1}{6}$

17. (72) $6 \times 5\frac{1}{3} \times \frac{3}{8}$

18. (22) $\frac{2}{5}$ of \$12.00

19. (39) $0.12 \times \$6.50$

20. (74) $5.3 - 3\frac{3}{4}$ (decimal answer)

21. (23) What is the ratio of the number of cents in a dime to the number of cents in a quarter?

Find each unknown number:

*** 22.** (87) $4n = 6 \cdot 14$

*** 23.** (87) $0.3n = 12$

24. (7) **Model** Draw a segment $1\frac{3}{4}$ inches long. Label the endpoints *R* and *T*. Then find and mark the midpoint of $\overline{RT}$. Label the midpoint *S*. What are the lengths of $\overline{RS}$ and $\overline{ST}$?

*** 25.** (85) Solve this proportion: $\frac{6}{9} = \frac{36}{w}$

*** 26.** (95) Multiply 4 hours by 6 dollars per hour:

$$\frac{4 \text{ hours}}{1} \times \frac{6 \text{ dollars}}{1 \text{ hours}}$$

27. (Inv. 7, 71) **Connect** The coordinates of the vertices of a parallelogram are (0, 0), (6, 0), (4, 4), and (–2, 4). What is the area of the parallelogram?

28. (78) **Estimate** The saying "A pint's a pound the world around" refers to the fact that a pint of water weighs about one pound. About how many pounds does a gallon of water weigh?

*** 29.** (92) **Analyze** $3^2 + 2^3 - \sqrt{4} \times 5 + 6^2 \div \sqrt{16}$

30. (58, 74) What is the probability of rolling a prime number with one roll of a number cube? Express the ratio as a fraction and as a decimal.

LESSON 97

• Transversals

Power Up

Building Power

facts

Power Up L

mental math

a. **Number Sense:** 20 · 50

b. **Number Sense:** 517 − 250

c. **Percent:** 25% of 60

d. **Calculation:** \$7.99 + \$7.58

e. **Decimals:** 0.1 ÷ 100

f. **Number Sense:** 20 × 75

g. **Measurement:** How many liters are in 1000 milliliters?

h. **Calculation:** 5 × 9, − 1, ÷ 2, − 1, ÷ 3, × 10, + 2, ÷ 9, − 2, ÷ 2

problem solving

Chad and his friends played three games that are scored from 1–100. His lowest score was 70 and his highest score is 100. What is Chad's lowest possible three-game average? What is his highest possible three-game average?

New Concept

Increasing Knowledge

A line that intersects two or more other lines is a **transversal.** In this drawing, line *r* is a transversal of lines *s* and *t.*

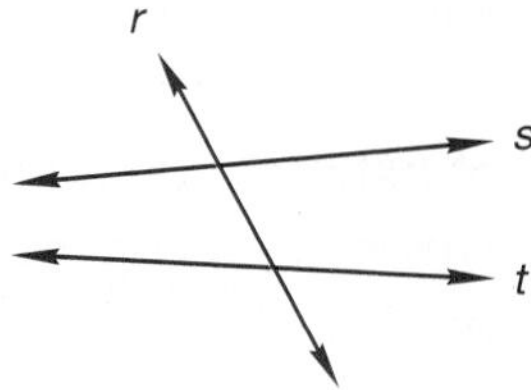

Math Language

Parallel lines are lines in the same plane that do not intersect and are always the same distance apart.

In the drawing, lines *s* and *t* are not parallel. However, in this lesson we will focus on the effects of a transversal intersecting parallel lines.

Below we show parallel lines *m* and *n* intersected by transversal *p*. Notice that eight angles are formed. In this figure there are four obtuse angles (numbered 1, 3, 5, and 7) and four acute angles (numbered 2, 4, 6, and 8).

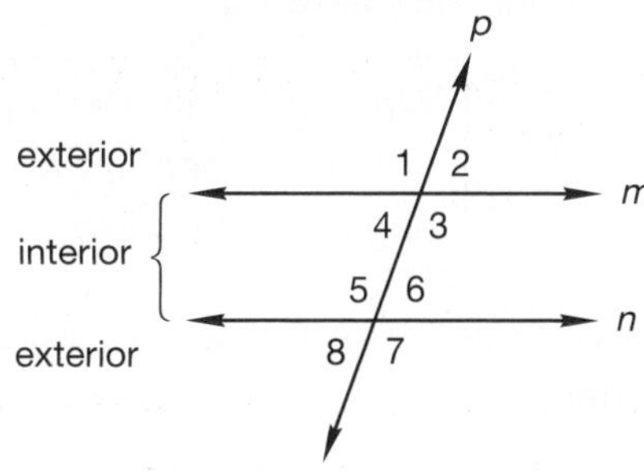

Thinking Skill

Verify

Why does every pair of supplementary angles in the diagram contain one obtuse and one acute angle?

Notice that obtuse angle 1, and acute angle 2, together form a straight line. These angles are **supplementary,** which means their measures total 180°. So if ∠1 measures 110°, then ∠2 measures 70°. Also notice that ∠2 and ∠3 are supplementary. If ∠2 measures 70°, then ∠3 measures 110°. Likewise, ∠3 and ∠4 are supplementary, so ∠4 would measure 70°.

There are names to describe some of the angle pairs. For example, we say that ∠1 and ∠5 are **corresponding angles** because they are in the same relative positions. Notice that ∠1 is the "upper left angle" from line *m*, while ∠5 is the "upper left angle" from line *n*.

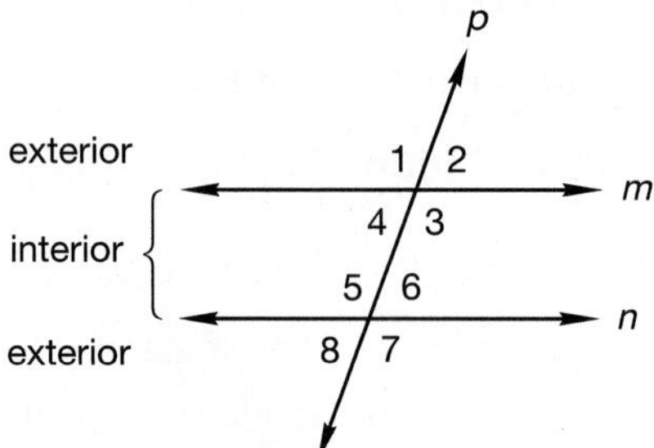

Which angle corresponds to ∠2?

Which angle corresponds to ∠7?

Since lines *m* and *n* are parallel, line *p* intersects line *m* at the same angle as it intersects line *n*. So the corresponding angles are congruent. Thus, if we know that ∠1 measures 110°, we can conclude that ∠5 also measures 110°.

The angles between the parallel lines (numbered 3, 4, 5, and 6 in the figure on previous page) are **interior angles.** Angle 3 and ∠5 are on opposite sides of the transversal and are called **alternate interior angles.**

Name another pair of alternate interior angles.

Alternate interior angles are congruent if the lines intersected by the transversal are parallel. So if ∠5 measures 110°, then ∠3 also measures 110°.

Angles not between the parallel lines are **exterior angles.** Angle 1 and ∠7, which are on opposite sides of the transversal, are **alternate exterior angles.**

Name another pair of alternate exterior angles.

Alternate exterior angles formed by a transversal intersecting parallel lines are congruent. So if the measure of ∠1 is 110°, then the measure of ∠7 is also 110°.

While we practice the terms for describing angle pairs, it is useful to remember the following.

> When a transversal intersects parallel lines, all acute angles formed are equal in measure, and all obtuse angles formed are equal in measure.

Thus any acute angle formed will be supplementary to any obtuse angle formed.

Example

Transversal *w* intersects parallel lines *x* and *y*.

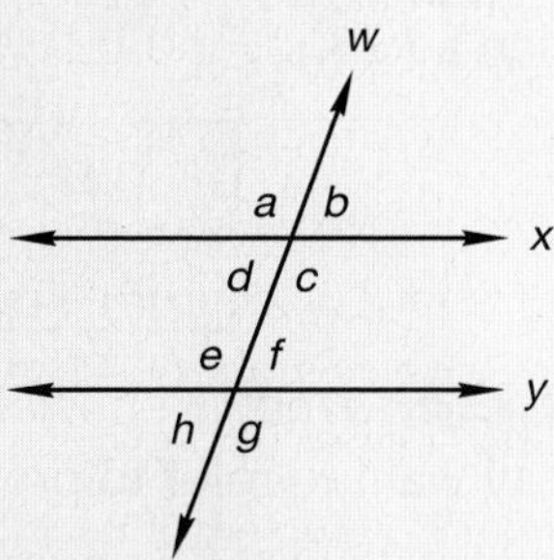

a. Name the pairs of corresponding angles.

b. Name the pairs of alternate interior angles.

c. Name the pairs of alternate exterior angles.

d. If the measure of $\angle a$ is 115°, then what are the measures of $\angle e$ and $\angle f$?

Solution

a. $\angle a$ and $\angle e$, $\angle b$ and $\angle f$, $\angle c$ and $\angle g$, $\angle d$ and $\angle h$

b. $\angle d$ and $\angle f$, $\angle c$ and $\angle e$

c. $\angle a$ and $\angle g$, $\angle b$ and $\angle h$

d. If $\angle a$ measures 115°, then $\angle e$ also measures **115°** and $\angle f$ measures **65°.**

Practice Set

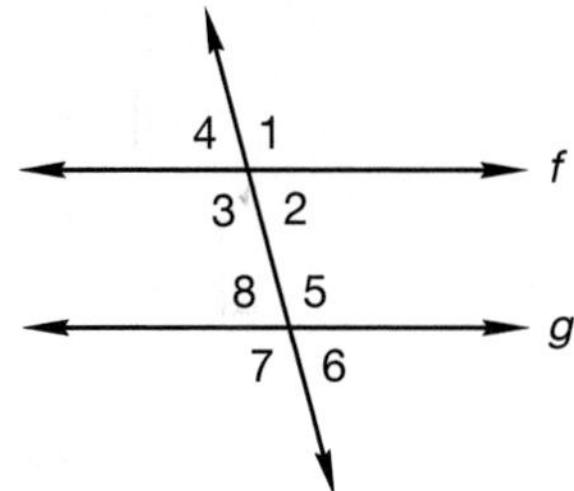

a. Which line in the figure at right is a transversal?

b. Which angle is an alternate interior angle to $\angle 3$?

c. Which angle corresponds to $\angle 8$?

d. Which angle is an alternate exterior angle to $\angle 7$?

e. *Conclude* If the measure of $\angle 1$ is 105°, what is the measure of each of the other angles in the figure?

Written Practice

Strengthening Concepts

1. (49) How many quarter-pound hamburgers can be made from 100 pounds of ground beef ?

2. (18) *Connect* On the Fahrenheit scale water freezes at 32°F and boils at 212°F. What temperature is halfway between the freezing and boiling temperatures of water?

*** 3.** (96) This function table shows the relationship between temperatures measured in degrees Celsius and degrees Fahrenheit. (To find the Fahrenheit temperature, multiply the temperature in Celsius by 1.8, then add 32.) Find the missing number in the table.

C	**0**	**10**	**20**	**30**
$1.8C + 32$	32	50	68	

Predict What is special about the result when $C = 100$? (*Hint:* You may want to refer to problem **2.**)

4. (76) Compare: $\frac{5}{8} \bigcirc 0.675$

*** 5.** (94) Write $2\frac{1}{4}$ as a percent.

*** 6.** (94) Write $1\frac{2}{5}$ as a percent.

*** 7.** (94) Write 0.7 as a percent.

*** 8.** (94) Write $\frac{7}{8}$ as a percent.

9. (73) Use division by primes to find the prime factors of 320. Then write the prime factorization of 320 using exponents.

*** 10.** (90) In one minute the second hand of a clock turns 360°. How many degrees does the minute hand of a clock turn in one minute?

*** 11.** (90) **Analyze** Jason likes to ride his skateboard around Parallelogram Park. If he made four turns on each trip around the park, what was the average number of degrees in each turn?

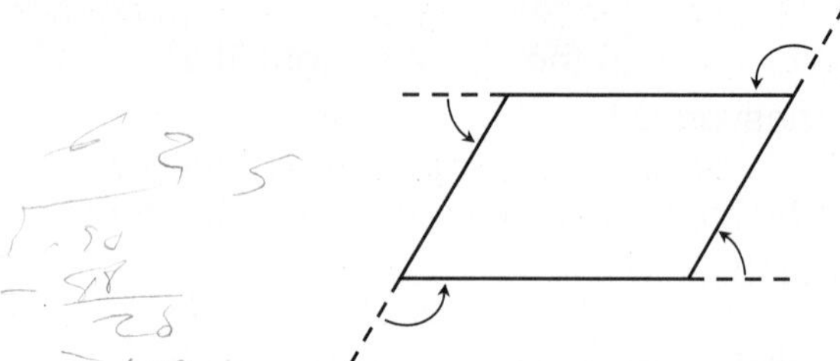

12. (59) $6\frac{3}{4} + 5\frac{7}{8}$

13. (63) $6\frac{1}{3} - 2\frac{1}{2}$

14. (68) $2\frac{1}{2} \div 100$

15. (38) $6.93 + 8.429 + 12$

16. (53) $(1 - 0.1)(1 \div 0.1)$

17. (74) $4.2 + \frac{7}{8}$ (decimal answer)

18. (73) Jovita bought $3\frac{1}{3}$ cubic yards of mulch for the garden. She will need 2.5 cubic yards for the flowerbeds. How much mulch is left for Jovita to use for her vegetable garden? Write your answer as a fraction.

19. (41) **Analyze** If 80% of the 30 students passed the test, how many students did not pass?

20. (50) Compare: $\frac{1}{2} \div \frac{1}{3} \bigcirc \frac{1}{3} \div \frac{1}{2}$

21. (10) **Predict** What is the next number in this sequence?

$$\ldots, 1000, 100, 10, 1, \ldots$$

Find each unknown number:

22. (3) $a + 60 + 70 = 180$

23. (85) $\frac{7}{4} = \frac{w}{44}$

24. (79) The perimeter of this square is 48 in. What is the area of one of the triangles?

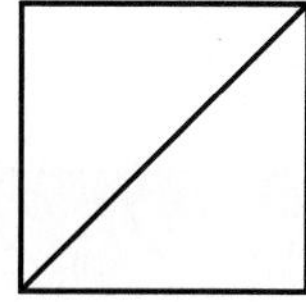

Refer to the table below to answer problems **25–27.**

Mark's Personal Running Records

Distance	Time (minutes:seconds)
$\frac{1}{4}$ mile	0:58
$\frac{1}{2}$ mile	2:12
1 mile	5:00

25. (32) If Mark set his 1-mile record by keeping a steady pace, then what was his $\frac{1}{2}$-mile time during the 1-mile run?

26. (32) **Conclude** What is a reasonable expectation for the time it would take Mark to run 2 miles?

A 9:30 **B** 11:00 **C** 15:00

27. (32) **Formulate** Write a question that relates to this table and answer the question.

* **28.** (97) Transversal *t* intersects parallel lines *r* and *s*. Angle 2 measures 78°.

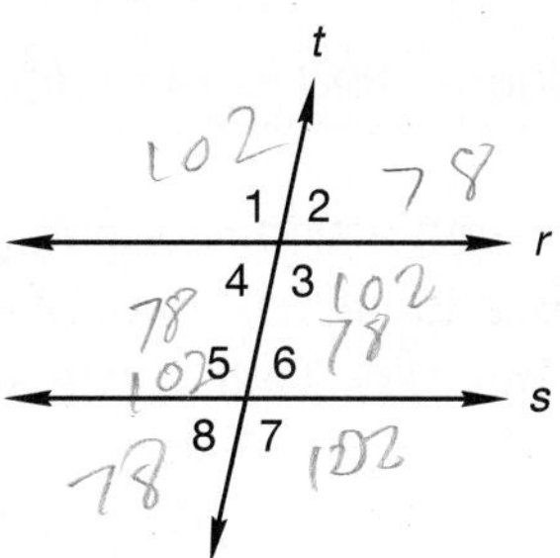

a. **Analyze** Which angle corresponds to ∠2?

b. Find the measures of ∠5 and ∠8.

* **29.** (92) $10^2 - \sqrt{49} - (10 + 8) \div 3^2$

30. (58) What is the probability of rolling a composite number with one roll of a number cube?

LESSON 98

Sum of the Angle Measures of Triangles and Quadrilaterals

Power Up

Building Power

facts Power Up J

mental math

a. **Number Sense:** 40 · 50

b. **Number Sense:** 293 + 450

c. **Percent:** 50% of 48

d. **Calculation:** \$20.00 − \$18.72

e. **Decimals:** 12.5 × 100

f. **Number Sense:** $\frac{360}{40}$

g. **Measurement:** How many cups are in 2 pints?

h. **Calculation:** 8 × 8, − 1, ÷ 9, × 4, + 2, ÷ 2, + 1, $\sqrt{\ }$, $\sqrt{\ }$

problem solving

If two people shake hands, there is one handshake. If three people shake hands, there are three handshakes. If four people shake hands with one another, we can picture the number of handshakes by drawing four dots (for people) and connecting the dots with segments (for handshakes). Then we count the segments (six). Use this method to count the number of handshakes that will take place between Bill, Phil, Jill, Lil, and Wil.

New Concept

Increasing Knowledge

If we extend a side of a polygon, we form an **exterior angle.** In this figure ∠1 is an exterior angle, and ∠2 is an **interior angle.** Notice that these angles are supplementary. That is, the sum of their measures is 180°.

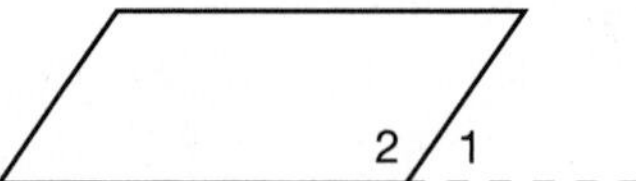

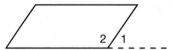

Thinking Skill

Verify

Act out the turns Elizabeth made to verify the number of degrees.

Recall from Lesson 90 that a full turn measures 360°. So if Elizabeth makes three turns to get around a park, she has turned a total of 360°. Likewise, if she makes four turns to get around a park, she has also turned 360°.

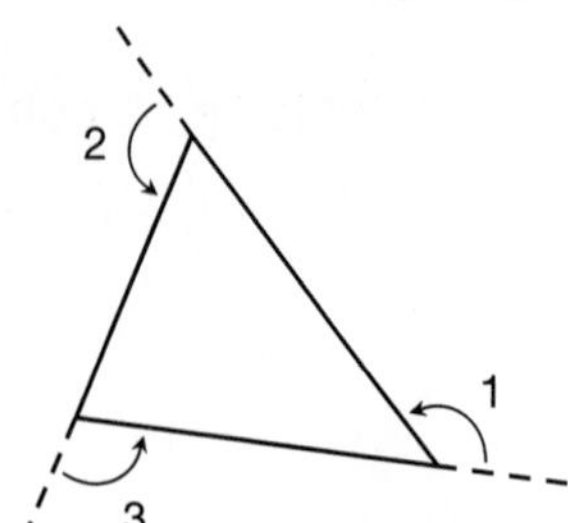

The sum of the measures of angles 1, 2, and 3 is 360°.

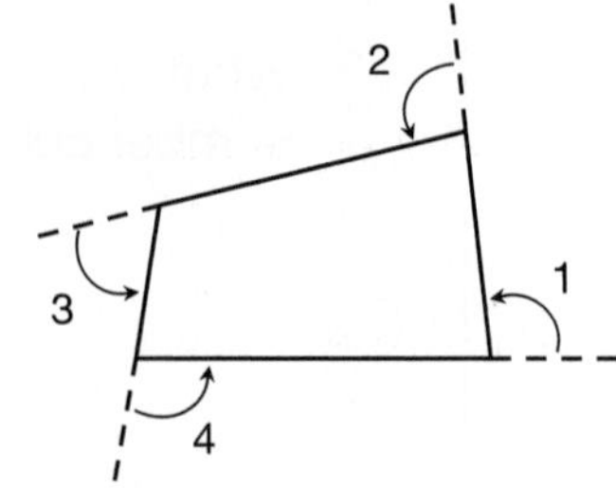

The sum of the measures of angles 1, 2, 3, and 4 is 360°.

If Elizabeth makes three turns to get around the park, then each turn averages 120°.

$$\frac{360^\circ}{3 \text{ turns}} = 120^\circ \text{ per turn}$$

If she makes four turns to get around the park, then each turn averages 90°.

$$\frac{360^\circ}{4 \text{ turns}} = 90^\circ \text{ per turn}$$

Recall that these turns correspond to exterior angles of the polygons and that the exterior and interior angles at a turn are supplementary. Since the exterior angles of a triangle average 120°, the interior angles must average 60°. A triangle has three interior angles, so the sum of the interior angles is 180° ($3 \times 60^\circ = 180^\circ$).

The sum of the interior angles of a triangle is 180°.

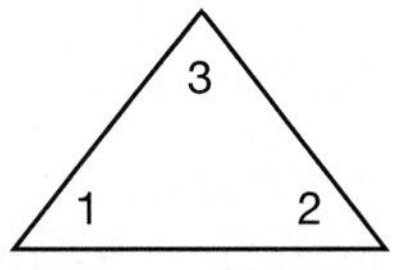

The sum of angles 1, 2, and 3 is 180°.

Since the exterior angles of a quadrilateral average 90°, the interior angles must average 90°. So the sum of the four interior angles of a quadrilateral is 360° ($4 \times 90^\circ = 360^\circ$).

The sum of the interior angles of a quadrilateral is 360°.

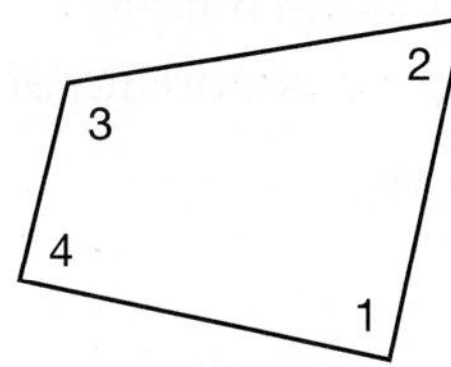

The sum of angles 1, 2, 3, and 4 is 360°.

Example 1

What is m∠*A* in △ *ABC*?

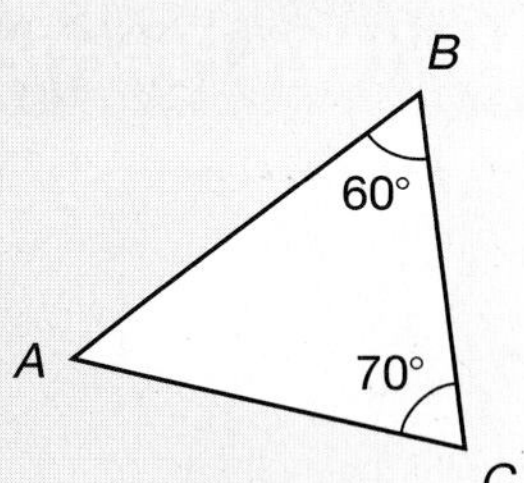

Solution

The measures of the interior angles of a triangle total 180°.

$$m\angle A + 60^\circ + 70^\circ = 180^\circ$$

Since the measures of ∠*B* and ∠*C* total 130°, m∠*A* is **50°**.

Example 2

What is m∠T in quadrilateral $QRST$?

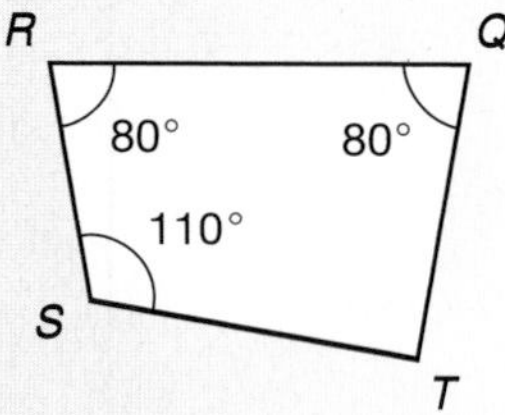

Solution

The measures of the interior angles of a quadrilateral total 360°.

$$m\angle T + 80° + 80° + 110° = 360°$$

The measures of ∠Q, ∠R, and ∠S total 270°. So m∠T is **90°.**

Practice Set

Quadrilateral *ABCD* is divided into two triangles by segment *AC*. Use for problems **a–c.**

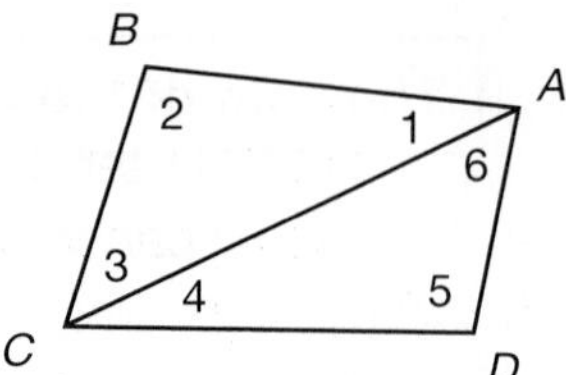

a. What is the sum of m∠1, m∠2, and m∠3?

b. What is the sum of m∠4, m∠5, and m∠6?

c. Generalize What is the sum of the measures of the four interior angles of the quadrilateral?

d. What is m∠P in △PQR?

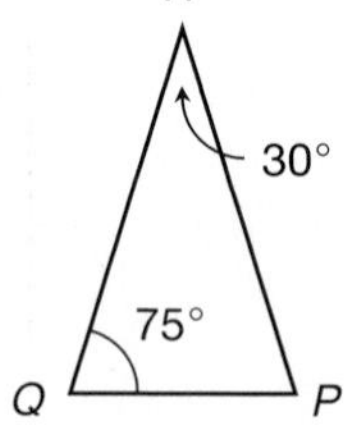

e. What is the measure of each interior angle of a regular quadrilateral?

f. Model Elizabeth made five left turns as she ran around the park. Draw a sketch that shows the turns in her run around the park. Then find the average number of degrees in each turn.

Written Practice *Strengthening Concepts*

1. (12, 72) When the sum of $\frac{1}{2}$ and $\frac{1}{4}$ is divided by the product of $\frac{1}{2}$ and $\frac{1}{4}$, what is the quotient?

* **2.** (95) Analyze Jenny is $5\frac{1}{2}$ feet tall. She is how many inches tall?

3. (77) If $\frac{4}{5}$ of the 200 runners finished the race, how many runners did not finish the race?

* **4.** Lines p and q are parallel.
(97)

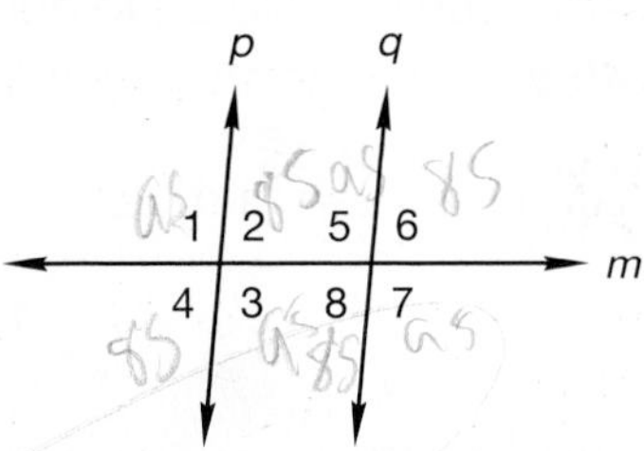

a. Which angle is an alternate interior angle to $\angle 2$?

b. If $\angle 2$ measures 85°, what are the measures of $\angle 6$ and $\angle 7$?

* **5.** **Analyze** The circumference of the earth is about 25,000 miles. Write that distance in expanded notation using exponents.
(92)

6. **Estimate** Use a ruler to measure the diameter of a quarter to the nearest sixteenth of an inch. How can you use that information to find the radius and the circumference of the quarter?
(17, 27)

7. **Connect** Which of these bicycle wheel parts is the best model of the circumference of the wheel?
(27)

A spoke **B** axle **C** tire

8. **Predict** As this sequence continues, each term equals the sum of the two previous terms. What is the next term in this sequence?
(10)

1, 1, 2, 3, 5, 8, 13, . . .

9. If there is a 20% chance of rain, what is the probability that it will not rain?
(58)

* **10.** Write $1\frac{1}{3}$ as a percent.
(94)

* **11.** **Analyze** $0.08w = \$0.60$
(87)

12. $\dfrac{1 - 0.001}{0.03}$
(49)

13. $\dfrac{3\frac{1}{3}}{100}$
(68)

14. If the volume of each small block is one cubic inch, what is the volume of this rectangular prism?
(82)

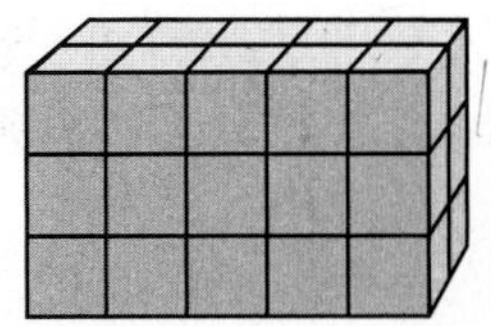

15. $6\frac{1}{2} + 4.95$ (decimal)
(74)

16. $2\frac{1}{6} - 1.5$ (fraction)
(73)

17. If a shirt costs \$19.79 and the sales-tax rate is 6%, what is the total price including tax? Explain how you can check your answer using estimation.
(41)

* **18.** What fraction of a foot is 3 inches?
(95)

19. What percent of a meter is 3 centimeters?
(75)

20. (88) The ratio of children to adults in the theater was 5 to 3. If there were 45 children, how many adults were there?

21. (14, 17) Arrange these numbers in order from least to greatest:

$$1, -1, 0, \frac{1}{2}, -\frac{1}{2}$$

22. (64) **Classify** These two triangles together form a quadrilateral with only one pair of parallel sides. What type of quadrilateral is formed?

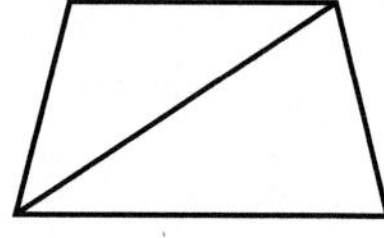

23. (60) **Conclude** Do the triangles in this quadrilateral appear to be congruent or not congruent?

* **24.** (98) **a.** **Analyze** What is the measure of $\angle A$ in $\triangle ABC$?

b. **Analyze** What is the measure of the exterior angle marked x?

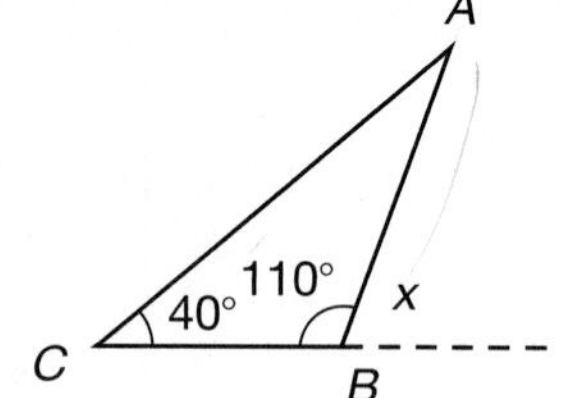

25. (33, 74) Write 40% as a

a. simplified fraction.

b. simplified decimal number.

26. (86) The diameter of this circle is 20 mm. What is the area of the circle? (Use 3.14 for π.)

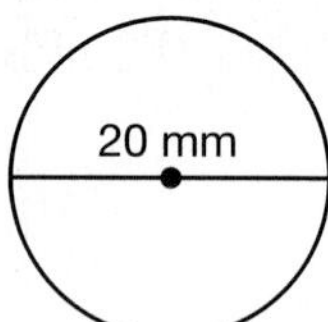

* **27.** (92) $2^3 + \sqrt{81} \div 3^2 + \left(\frac{1}{2}\right)^2$

* **28.** (95) Multiply 120 inches by 1 foot per 12 inches.

$$\frac{120 \text{ in.}}{1} \times \frac{1 \text{ ft}}{12 \text{ in.}}$$

29. (23, 58) A bag contains 20 red marbles and 15 blue marbles.

a. What is the ratio of red marbles to blue marbles?

b. If one marble is drawn from the bag, what is the probability that the marble will be blue?

* **30.** (93) **Conclude** An architect drew a set of plans for a house. In the plans, the roof is supported by a triangular framework. When the house is built, two sides of the framework will be 19 feet long and the base will be 33 feet long. Classified by side length, what type of triangle will be formed?

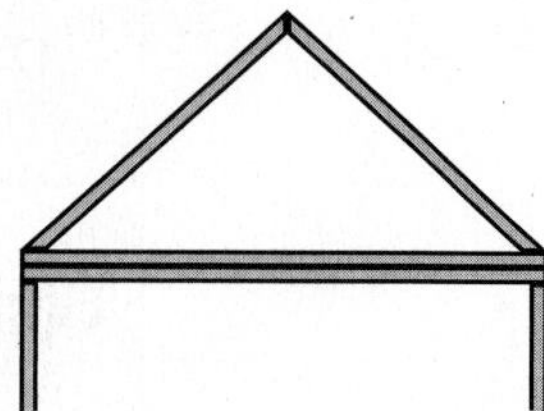

LESSON 99

Fraction-Decimal-Percent Equivalents

Power Up

Building Power

facts Power Up K

mental math

a. **Number Sense:** 60 · 50

b. **Number Sense:** 741 − 450

c. **Percent:** 25% of 48

d. **Calculation:** \$12.99 + \$4.75

e. **Decimals:** 37.5 ÷ 100

f. **Number Sense:** 30 × 15

g. **Measurement:** Which is greater 1 liter or 1000 milliliters?

h. **Calculation:** 7 × 7, + 1, ÷ 2, $\sqrt{\ }$, × 4, − 2, ÷ 3, × 5, + 3, ÷ 3

problem solving If the last page of a section of large newspaper is page 36, what is the fewest number of sheets of paper that could be in that section?

New Concept

Increasing Knowledge

Fractions, decimals, and percents are three ways to express parts of a whole. An important skill is being able to change from one form to another. This lesson asks you to complete tables that show equivalent fractions, decimals, and percents.

Example

Complete the table.

Fraction	Decimal	Percent
$\frac{1}{2}$	a.	b.
c.	0.3	d.
e.	f.	40%

Solution

The numbers in each row should be equivalent. For $\frac{1}{2}$ we write a decimal and a percent. For 0.3 we write a fraction and a percent. For 40% we write a fraction and a decimal.

a. $\frac{1}{2} = 2\overline{)1.0}$ **0.5**

b. $\frac{1}{2} \times \frac{100\%}{1} =$ **50%**

c. $0.3 = \mathbf{\frac{3}{10}}$

d. $0.3 \times 100\% =$ **30%**

e. $40\% = \frac{40}{100} = \mathbf{\frac{2}{5}}$

f. $40\% = 0.40 =$ **0.4**

Practice Set

Connect Complete the table.

Fraction	Decimal	Percent
$\frac{3}{5}$	**a.**	**b.**
c.	0.8	**d.**
e.	**f.**	20%
$\frac{3}{4}$	**g.**	**h.**
i.	0.12	**j.**
k.	**l.**	5%

Written Practice

Strengthening Concepts

1. (68) Analyze A foot-long ribbon can be cut into how many $1\frac{1}{2}$-inch lengths?

2. (Inv. 6) A can of beans is the shape of what geometric solid?

3. (77) Analyze If $\frac{3}{8}$ of the group voted yes and $\frac{3}{8}$ voted no, then what fraction of the group did not vote?

4. (29, 75) Connect Nine months is

a. what fraction of a year?

b. what percent of a year?

5. (82) One-cubic-foot boxes were stacked as shown. What was the volume of the stack of boxes?

*** 6.** (90) Analyze Tom was facing east. Then he turned counterclockwise 270°. After the turn, what direction was Tom facing?

7. (75) If $\frac{1}{5}$ of the pie was eaten, what percent of the pie was left?

*** 8.** (94) Write the percent form of $\frac{1}{7}$.

9. (74) $6\frac{3}{4} - 6.2$ (decimal answer)

10. (5) $5 \cdot 4 \cdot 3 \cdot 2 \cdot 1 \cdot 0$

11. (49) $\frac{4.5}{0.18}$

*** 12.** (89) $\sqrt{1600}$

13. $\sqrt{64} + 5^2 - \sqrt{25} \times (2 + 3)$
(92)

14. **Analyze** Solve this proportion: $\frac{15}{20} = \frac{24}{n}$
(85)

15. $12\frac{1}{2} \times 1\frac{3}{5} \times 5$
(72)

16. $(4.2 \times 0.05) \div 7$
(53)

17. If the sales-tax rate is 7%, what is the tax on a $111.11 purchase?
(41)

18. **Analyze** The table shows the percent of the population aged 25–64 with some senior high school education. The figures are for the year 2001. Use the table to answer **a–c.**
(Inv. 5)

Country	Percent
Peru	44%
Iceland	57%
Poland	46%
Italy	43%
Greece	51%
Chile	46%
Luxembourg	53%

a. Find the mode of the data.

b. If the data were arranged from least to greatest, which country or countries would have the middle score?

c. What is the term used for the answer to problem **b?** Will this quantity always be the same as the mode in every set of data? Explain.

19. Write the prime factorization of 900 using exponents.
(73)

20. Think of two different prime numbers, and write them on your paper. Then write the greatest common factor (GCF) of the two prime numbers.
(20)

21. **Explain** The perimeter of a square is 2 meters. How many centimeters long is each side? Explain your thinking.
(7, 8)

* 22. a. What is the area of this triangle?
(79, 93)

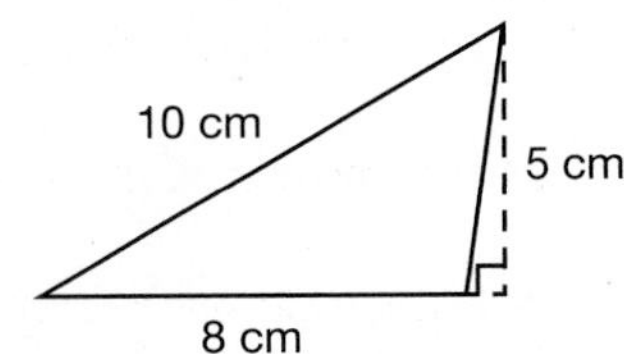

b. **Classify** Is this an acute, right, or obtuse triangle?

* **23.** (98) **a.** What is the measure of $\angle B$ in quadrilateral *ABCD*?

b. What is the measure of the exterior angle at *D*?

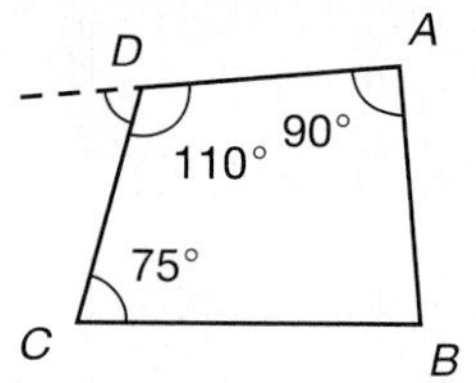

Complete the table to answer problems **24–26.**

	Fraction	Decimal	Percent
* **24.** (99)	**a.**	0.6	**b.**
* **25.** (99)	**a.**	**b.**	15%
* **26.** (99)	$\frac{3}{10}$	**a.**	**b.**

27. (17) **Model** Draw $\overline{AC}$ $1\frac{1}{4}$ inches long. Find and mark the midpoint of $\overline{AC}$, and label the midpoint *B*. What are the lengths of $\overline{AB}$ and $\overline{BC}$?

28. (58) There are 32 cards in a bag. Eight of the cards have letters written on them. What is the chance of drawing a card with a letter written on it?

29. (78) Compare: 1 gallon ◯ 4 liters

* **30.** (27, 96) **Generalize** This function table shows the relationship between the radius (*r*) and diameter (*d*) of a circle. The radius is the input and the diameter is the output. Describe the rule and find the missing number.

r	*d*
1.2	2.4
0.7	1.4
	5
15	30

Early Finishers
Real-World Application

Jesse displays trophies on 4 shelves in the family room. Two of the 6 trophies on each shelf are for soccer. How many trophies are NOT for soccer?

Write one equation and use it to solve the problem.

LESSON

100

• Algebraic Addition of Integers

Power Up

Building Power

facts Power Up I

mental math

a. **Number Sense:** $50 \cdot 80$

b. **Number Sense:** $380 + 550$

c. **Percent:** 50% of 100

d. **Calculation:** \$40.00 − \$21.89

e. **Decimals:** 0.8×100

f. **Number Sense:** $\frac{750}{25}$

g. **Measurement:** How many pints are in 2 quarts?

h. **Calculation:** $5 + 5$, $\times 10$, $- 1$, $\div 9$, $+ 1$, $\div 3$, $\times 7$, $+ 2$, $\div 2$

problem solving

How many different triangles of any size are in this figure?

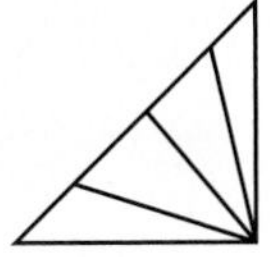

New Concept

Increasing Knowledge

Math Language

Integers consist of the counting numbers (1, 2, 3, ...), the negative counting numbers (−1, −2, −3, ...), and 0. All numbers that fall between these numbers are not integers.

Thinking Skill

Analyze

How is a thermometer like a number line? How is it different?

In this lesson we will practice adding integers.

The dots on this number line mark the integers from negative five to positive five (−5 to +5).

If we consider a rise in temperature of five degrees as a positive five (+5) and a fall in temperature of five degrees as a negative five (−5), we can use the scale on a thermometer to keep track of the addition.

Imagine that the temperature is 0°F. If the temperature falls five degrees (−5) and then falls another five degrees (−5), the resulting temperature is ten degrees below zero (−10°F). When we add two negative numbers, the sum is negative.

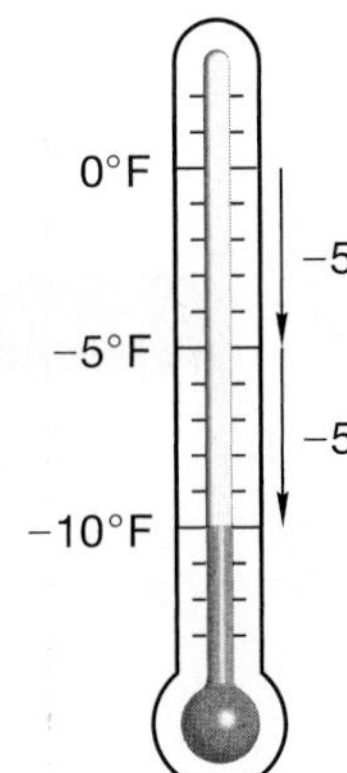

$$-5 + -5 = -10$$

Math Language

Opposites are numbers that can be written with the same digits but with opposite signs. They are the same distance, in opposite directions, from zero on the number line.

Imagine a different situation. We will again start with a temperature of 0°F. First the temperature falls five degrees (−5). Then the temperature rises five degrees (+5). This brings the temperature back to 0°F. The numbers −5 and +5 are opposites. When we add opposites, the sum is zero.

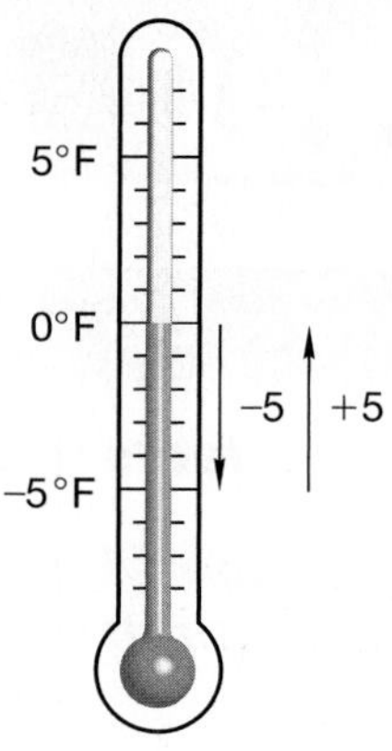

$$-5 + +5 = 0$$

Starting from 0°F, if the temperature rises five degrees (+5) and then falls ten degrees (−10), the temperature will fall through zero to −5°F. The sum is less than zero because the temperature fell more than it rose.

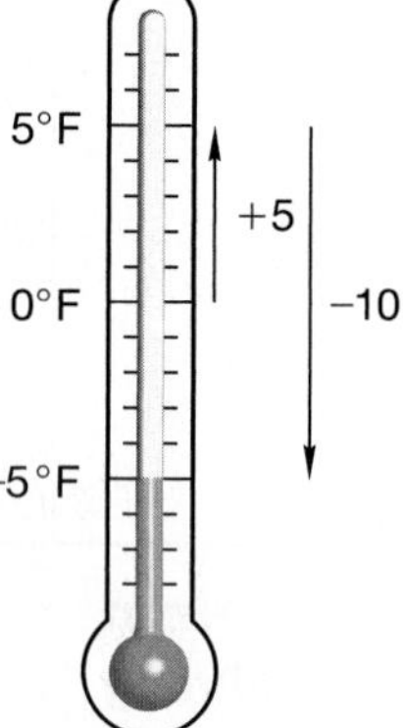

$$+5 + -10 = -5$$

Example 1

Add: +8 + −5

Solution

We will illustrate this addition on a number line. We begin at zero and move eight units in the positive direction (to the right). From +8 we move five units in the negative direction (to the left) to +3.

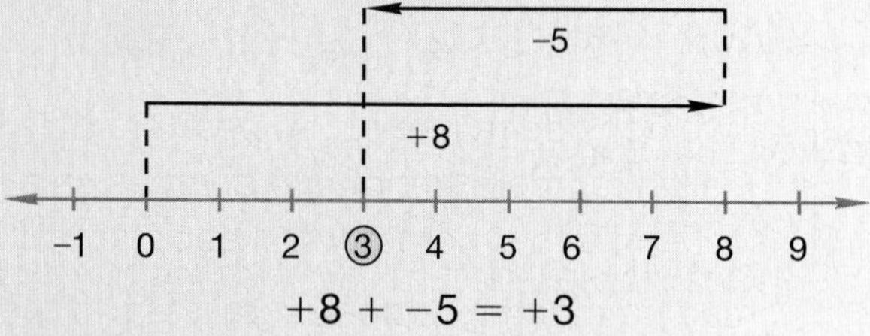

$$+8 + -5 = +3$$

The sum is **+3,** which we write as 3.

Example 2

Add: −5 + −3

Solution

Again using a number line, we start at zero and move in the negative direction, or to the left, five units to -5. From -5 we continue moving left three units to -8.

Thinking Skill

Generalize

When two negative integers are added, is the sum negative or positive?

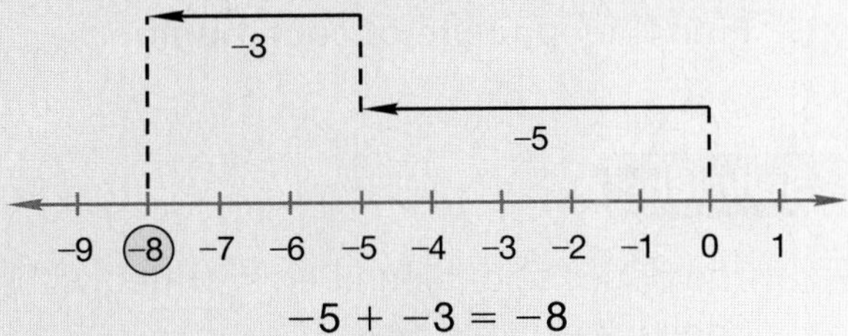

$$-5 + -3 = -8$$

The sum is **−8.**

Example 3

Add: −6 + +6

Solution

We start at zero and move six units to the left. Then we move six units to the right, returning to **zero.**

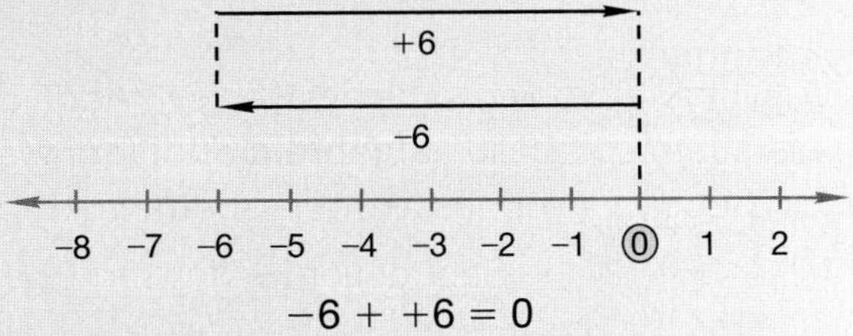

$$-6 + +6 = 0$$

Example 4

Add: (+6) + (−6)

Solution

Sometimes positive and negative numbers are written with parentheses. The parentheses help us see that the positive or negative sign is the sign of the number and not an addition or subtraction operation.

$$(+6) + (-6) = \mathbf{0}$$

Negative 6 and positive 6 are **opposites.** Opposites are numbers that can be written with the same digits but with opposite signs. The opposite of 3 is -3, and the opposite of -5 is 5 (which can be written as $+5$).

On a number line, we can see that any two opposites lie equal distances from zero. However, they lie on opposite sides of zero from each other.

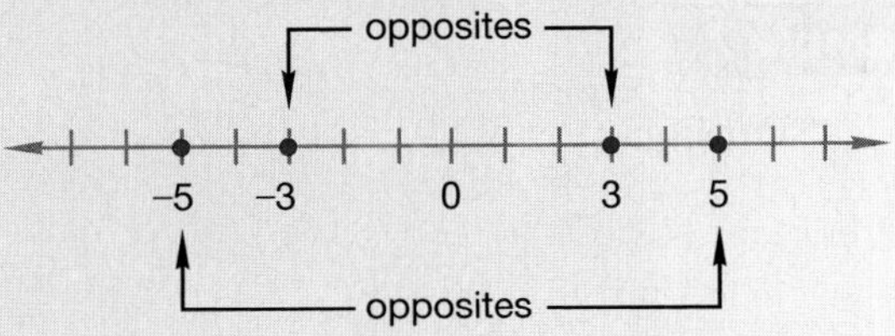

If opposites are added, the sum is zero.

$$-3 + +3 = 0 \qquad -5 + +5 = 0$$

Example 5

Find the opposite of each number:

a. -7 **b.** 10

Solution

The opposite of a number is written with the same digits but with the opposite sign.

a. The opposite of -7 is **+7,** which is usually written as 7.

b. The opposite of 10 (which is positive) is **−10.**

Using opposites allows us to change any subtraction problem into an addition problem. Consider this subtraction problem:

$$10 - 6$$

Instead of subtracting 6 from 10, we can add the opposite of 6 to 10. The opposite of 6 is -6.

$$10 + -6$$

In both problems the answer is 4. Adding the opposite of a number to subtract is called **algebraic addition.** We change subtraction to addition by adding the opposite of the subtrahend.

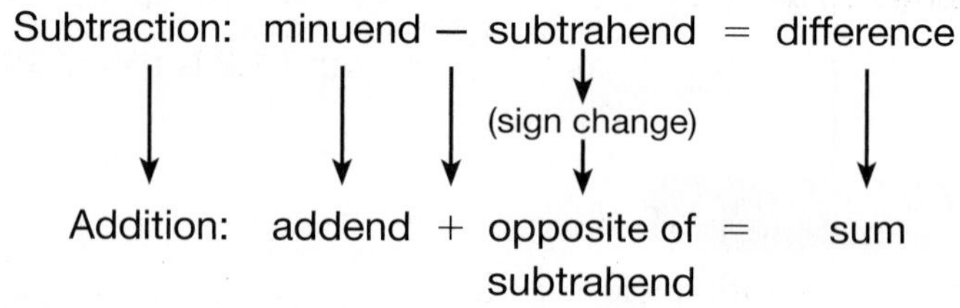

Example 6

Simplify: −10 − −6

Solution

This problem directs us to subtract a negative six from negative ten. Instead, we may add the opposite of negative six to negative ten.

$$-10 - -6$$

$$\downarrow \quad \downarrow$$

$$-10 + +6 = \mathbf{-4}$$

Example 7

Simplify: (−3) − (+5)

Solution

Instead of subtracting a positive five, we add a negative five.

$$(-3) - (+5)$$
$$\downarrow \quad \downarrow$$
$$(-3) + (-5) = \mathbf{-8}$$

Practice Set

Model Find each sum. Draw a number line to show the addition for problems **a** and **b.** Solve problems **c–h** mentally.

a. $-3 + +4$ **b.** $-3 + -4$

c. $-3 + +3$ **d.** $+4 + -3$

e. $(+3) + (-4)$ **f.** $(+10) + (-5)$

g. $(-10) + (-5)$ **h.** $(-10) + (+5)$

Find the opposite of each number:

i. -8 **j.** 4 **k.** 0

Solve each subtraction problem using algebraic addition:

l. $-3 - -4$ **m.** $-4 - +2$

n. $(+3) - (-6)$ **o.** $(-2) - (-4)$

Written Practice

Strengthening Concepts

1. (39) If 0.6 is the divisor and 1.2 is the quotient, what is the dividend?

2. (12) If a number is twelve less than fifty, then it is how much more than twenty?

3. (18) If the sum of four numbers is 14.8, what is the average of the four numbers?

*** 4.** (100) **Model** Illustrate this problem on a number line:

$$-3 + +5$$

*** 5.** (100) Find each sum mentally:

a. $-4 + +4$ **b.** $-2 + -3$

c. $-5 + +3$ **d.** $+5 + -10$

*** 6.** (100) Solve each subtraction problem using algebraic addition:

a. $-2 - -5$ **b.** $-3 - -3$

c. $+2 - -3$ **d.** $-2 - +3$

*** 7.** (93, 98) **Analyze** What is the measure of each angle of an equilateral triangle?

8. (71) Quadrilateral *ABCD* is a parallelogram. If angle *A* measures 70°, what are the measures of angles *B, C,* and *D?*

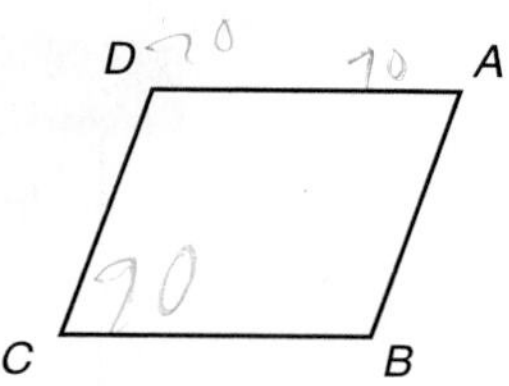

9. (58) **a.** If the spinner is spun once, what is the probability that it will stop in a sector with a number 2? How do you know your answer is correct?

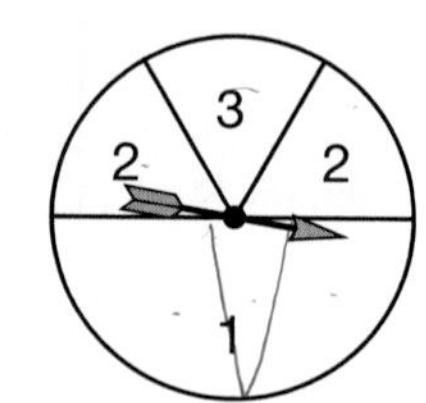

b. Estimate If the spinner is spun 30 times, about how many times would it be expected to stop in the sector with the number 3?

10. (82) Find the volume of the rectangular prism at right.

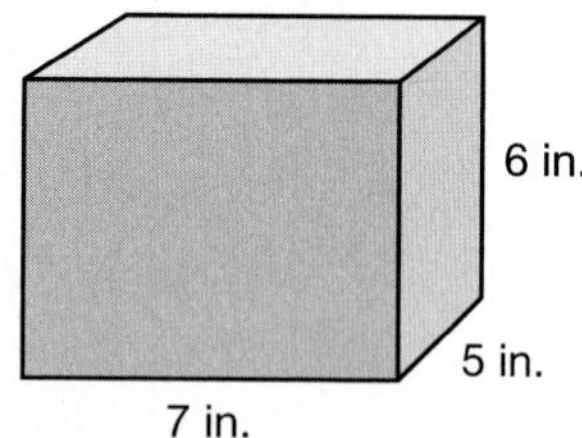

11. (23) Twelve of the 27 students in the class are boys. What is the ratio of girls to boys in the class?

* **12.** (92) Analyze $10^2 + (5^2 - 11) \div \sqrt{49} - 3^3$

* **13.** (94) The fraction $\frac{2}{3}$ is equal to what percent?

14. (33) If 20% of the students brought their lunch to school, then what fraction of the students did not bring their lunch to school?

* **15.** (92) $\frac{4^2}{2^4}$

16. (59) $5\frac{7}{8} + 4\frac{3}{4}$

17. (68) $1\frac{1}{2} \div 2\frac{1}{2}$

18. (38) $5 - (3.2 + 0.4)$

19. (80) Estimate If the diameter of a circular plastic swimming pool is 6 feet, then the area of the bottom of the pool is about how many square feet? Round to the nearest square foot. (Use 3.14 for π.)

20. (82) Explain We use squares to measure the area of a rectangle. Why do we use cubes instead of squares to measure the volume of a rectangular prism?

21. (85) Solve this proportion: $\frac{9}{12} = \frac{15}{x}$

Rectangle *ABCD* is 8 cm long and 6 cm wide. Segment *AC* is 10 cm long. Use this information to answer problems **22** and **23.**

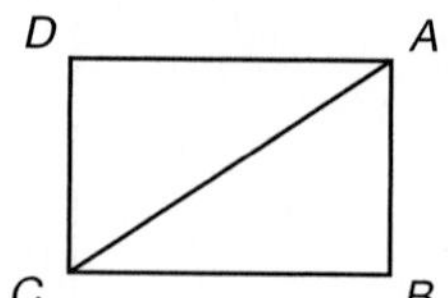

22. (79) What is the area of triangle *ABC?*

23. (8) What is the perimeter of triangle *ABC?*

24. Measure the diameter of a nickel to the nearest millimeter.
(27)

*** 25.** **Estimate** Calculate the circumference of a nickel. Round to the nearest millimeter. (Use 3.14 for π.)
(47)

26. A bag contains 12 marbles. Eight of the marbles are red and 4 are blue. If you draw a marble from the bag without looking, what is the probability that the marble will be blue? Express the probability ratio as a fraction and as a decimal rounded to the nearest hundredth.
(58, 74)

Connect Complete the table to answer problems **27–29.**

	Fraction	Decimal	Percent
*** 27.** *(99)*	$\frac{9}{10}$	**a.**	**b.**
*** 28.** *(99)*	**a.**	1.5	**b.**
*** 29.** *(99)*	**a.**	**b.**	4%

30. A full one-gallon container of milk was used to fill two one-pint containers. How many quarts of milk were left in the one-gallon container?
(78)

Early Finishers
Choose A Strategy

These three prime factorizations represent numbers that are powers of 10. Simplify each prime factorization.

$2^2 \times 5^2$ $\quad\quad$ $2^4 \times 5^4$ $\quad\quad$ $2^5 \times 5^5$

Use exponents to write the prime factorization of another number that is a power of 10.

INVESTIGATION 10

Focus on

• Compound Experiments

Some experiments whose outcomes are determined by chance contain more than one part. Such experiments are called **compound experiments.** In this investigation we will consider compound experiments that consist of two parts performed in order. Here are three experiments:

1. A spinner with sectors A, B, and C is spun; then a marble is drawn from a bag that contains 4 blue marbles and 2 white marbles.
2. A marble is drawn from a bag with 4 blue marbles and 2 white marbles; then, without the first marble being replaced, a second marble is drawn.
3. A number cube is rolled; then a coin is flipped.

The second experiment is actually a way to look at drawing two marbles from the bag at once. We estimated probabilities for this compound experiment in Investigation 9.

A **tree diagram** can help us visualize the sample space for a compound experiment. Here is a tree diagram for compound experiment 1:

Math Language

Recall that the list of all possible outcomes in an experiment is called a *sample space.* A tree diagram is one way to represent the sample space for an experiment.

Spinner	Marble	Compound Outcome
A	blue	A, blue
	white	A, white
B	blue	B, blue
	white	B, white
C	blue	C, blue
	white	C, white

Each *branch* of the tree corresponds to a possible outcome. There are three possible spinner outcomes. For each spinner outcome, there are two possible marble outcomes. To find the total number of **compound outcomes,** we multiply the number of branches in the first part of the experiment by the number of branches in the second part of the experiment. There are $3 \times 2 = 6$ branches, so there are six possible compound outcomes. In the column titled "Compound Outcome," we list the outcome for each branch. "A, blue" means that the spinner stopped on A, then the marble drawn was blue. Although there are six different outcomes, not all the outcomes are equally likely. We need to determine the probability of each part of the experiment in order to find the probability for each compound outcome. To do this, we will use the multiplication principle for compound probability.

The probability of a compound outcome is the product of the probabilities of each part of the outcome.

We will use this principle to calculate the probability of the first branch of experiment 1, the spinner-marble experiment, which corresponds to the compound outcome "A, blue."

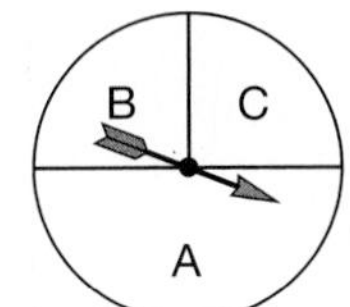

The first part of the outcome is that the spinner stops in sector A. The probability of this outcome is $\frac{1}{2}$, since sector A occupies half the area of the circle.

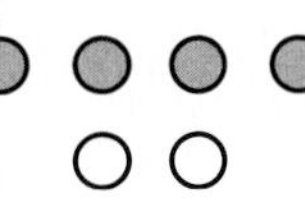

The second part of the outcome is that a blue marble is drawn from the bag. Since four of the six marbles are blue, the probability of this outcome is $\frac{4}{6}$, which simplifies to $\frac{2}{3}$.

To find the probability of the compound outcome, we multiply the probabilities of each part.

The probability of "A, blue" is $\frac{1}{2} \cdot \frac{2}{3}$, which equals $\frac{1}{3}$.

Notice that although "A, blue" is one of six possible outcomes, the probability of "A, blue" is greater than $\frac{1}{6}$. This is because "A" is the most likely of the three possible spinner outcomes, and "blue" is the more likely of the two possible marble outcomes.

Thinking Skill

Predict

What do you expect the sum of the probabilities to be?

For problems **1–6,** copy the table below and calculate the probability of each possible outcome. For the last row, find the sum of the probabilities of the six possible outcomes.

Outcome	Probability
A, blue	$\frac{1}{2} \cdot \frac{2}{3} = \frac{1}{3}$
A, white	**1.**
B, blue	**2.**
B, white	**3.**
C, blue	**4.**
C, white	**5.**
sum of probabilities	**6.**

For problems **7–9** we will consider compound experiment 2, which involves two draws from a bag of marbles that contains four blue marbles and two white marbles. The first part of the experiment is that one marble is drawn from the bag and is not replaced. The second part is that a second marble is drawn from the marbles remaining in the bag.

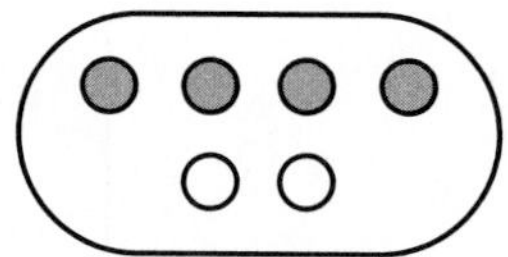

7. *Model* Copy and complete this tree diagram showing all possible outcomes of the compound experiment:

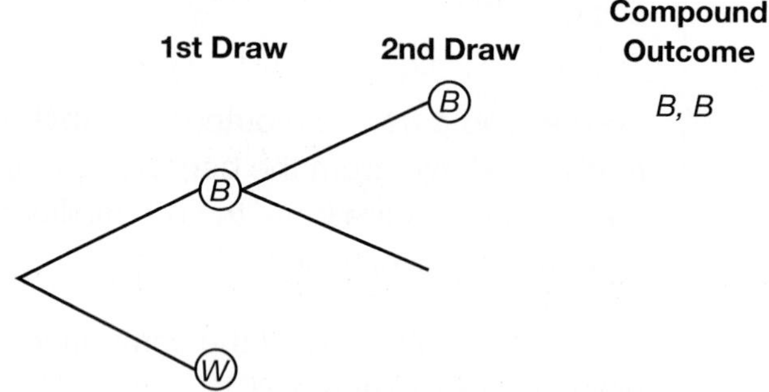

We will calculate the probability of the outcome blue, blue (B, B). On the first draw four of the six marbles are blue, so the probability of blue is $\frac{4}{6}$, which equals $\frac{2}{3}$.

If the first marble drawn is blue, then three blue marbles and two white marbles remain (see picture at right). So the probability of drawing a blue marble on the second draw is $\frac{3}{5}$. Therefore, the probability of the outcome blue, blue is $\frac{2}{3} \cdot \frac{3}{5} = \frac{2}{5}$.

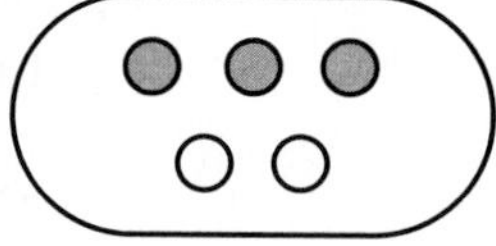

8. *Represent* Copy and complete this table to show the probability of each remaining possible outcome and the sum of the probabilties of all outcomes. Remember that the first draw changes the collection of marbles in the bag for the second draw.

Outcome	Probability
blue, blue	$\frac{4}{6} \cdot \frac{3}{5} = \frac{2}{5}$
sum of probabilities	

9. Suppose we draw three marbles from the bag, one at a time and without replacement. What is the probability of drawing three white marbles? What is the probability of drawing three blue marbles?

For problems **10–14,** consider a compound experiment in which a nickel is flipped and then a quarter is flipped.

10. Represent Create a tree diagram that shows all of the possible outcomes of the compound experiment.

11. Represent Make a table that shows the probability of each possible outcome.

Use the table you made in problem **11** to answer problems **12–14.**

12. What is the probability that one of the coins shows "heads" and the other coin shows "tails"?

13. What is the probability that at least one of the coins shows "heads"?

14. What is the probability that the nickel shows "heads" and the quarter shows "tails"?

extensions

Analyze For extensions **a** and **b,** consider experiment 2 in which a bag contains 4 blue marbles and 2 white marbles. One marble is drawn from the bag and not replaced, and then a second marble is drawn.

a. Find the probability that the two marbles drawn from the bag are different colors.

b. Find the probability that the two marbles drawn from the bag are the same color.

Analyze For extensions **c** and **d,** consider experiment 1 involving spinning the spinner and then drawing a marble.

c. The complement of "A, blue" is "not A, blue". Find the probability that the compound outcome will *not* be "A, blue."

d. Find the probability that the compound outcome will *not* include "A" and will *not* include "blue."

e. The probabilities in exercises **c** and **d** are different. Explain why.

For extensions **f** and **g,** consider the compound experiment consisting of rolling a number cube and then flipping a quarter.

f. Represent Draw a tree diagram to show the sample space for the experiment.

g. Find the probability of each compound outcome.

LESSON

101 • Ratio Problems Involving Totals

Power Up

Building Power

facts Power Up H

mental math

a. **Number Sense:** 20 · 300

b. **Number Sense:** 920 − 550

c. **Percent:** 25% of 100

d. **Calculation:** $18.99 + $5.30

e. **Decimals:** 3.75 ÷ 100

f. **Number Sense:** 40 × 25

g. **Measurement:** Which is greater: 1 liter or 500 milliliters?

h. **Calculation:** Find half of 100, − 1, $\sqrt{\ }$, × 5, + 1, $\sqrt{\ }$, × 3, + 2, ÷ 2

problem solving

The numbers in these boxes form number patterns. What one number should be placed in both empty boxes to complete the patterns?

1	2	3
2	4	
3		9

New Concept

Increasing Knowledge

In some ratio problems a total is used as part of the calculation. Consider this problem:

The ratio of boys to girls in a class was 5 to 4. If there were 27 students in the class, how many girls were there?

Thinking Skill

Model

Use red and yellow counters or buttons to model the problem.

We begin by drawing a ratio box. In addition to the categories of boys and girls, we make a third row for the total number of students. We will use the letters *b* and *g* to represent the actual counts of boys and girls.

	Ratio	Actual Count
Boys	5	*b*
Girls	4	*g*
Total	9	27

Math Language

A **proportion** is a statement that shows two ratios are equal.

In the ratio column we add the ratio numbers for boys and girls and get the ratio number 9 for the total. We were given 27 as the actual count of students. We will use two of the three rows from the ratio box to write a proportion. **We use the row we want to complete and the row that is already complete.** Since we are asked to find the actual number of girls, we will use the "girls" row. And since we know both "total" numbers, we will also use the "total" row. We solve the proportion below.

	Ratio	Actual Count
Boys	5	b
Girls	4	g
Total	9	27

$$\frac{4}{9} = \frac{g}{27}$$

$$9g = 4 \cdot 27$$

$$g = 12$$

We find that there were 12 girls in the class. If we had wanted to find the number of boys, we would have used the "boys" row along with the "total" row to write a proportion.

Example

The ratio of football players to band members on the football field was 2 to 5. Altogether, there were 175 football players and band members on the football field. How many football players were on the field?

Solution

We use the information in the problem to make a table. We include a row for the total. The ratio number for the total is 7.

	Ratio	Actual Count
Football Players	2	f
Band Members	5	b
Total	7	175

Next we write a proportion using two rows of the table. We are asked to find the number of football players, so we use the "football players" row. We know both totals, so we also use the "total" row. Then we solve the proportion.

	Ratio	Actual Count
Football Players	2	f
Band Members	5	b
Total	7	175

$$\frac{2}{7} = \frac{f}{175}$$

$$7f = 2 \cdot 175$$

$$f = 50$$

We find that there were **50 football players** on the field.

Practice Set | **Represent** Use ratio boxes to solve problems **a** and **b.**

a. Sparrows and crows perched on the wire in the ratio of 5 to 3. If the total number of sparrows and crows on the wire was 72, how many were crows?

b. Raisins and nuts were mixed by weight in a ratio of 2 to 3. If 60 ounces of mix were prepared, how many ounces of raisins were used?

c. **Model** Using 20 red and 20 yellow color tiles (or 20 shaded and unshaded circles) create a ratio of 3 to 2. How many of each color (or shading) do you have?

Written Practice *Strengthening Concepts*

* **1.** (101) **Represent** Draw a ratio box for this problem. Then solve the problem using a proportion.

The ratio of boys to girls in the class was 3 to 2. If there were 30 students in the class, how many girls were there?

2. (Inv. 6) **Connect** A shoe box is the shape of what geometric solid?

3. (18) **Analyze** If the average of six numbers is 12, what is the sum of the six numbers?

4. (27, 68) If the diameter of a circle is $1\frac{1}{2}$ inches, what is the radius of the circle?

* **5.** (95) What is the cost of 2.6 pounds of meat priced at $1.65 per pound?

6. (69) Suppose $\overline{AC}$ is 12 cm long. If $\overline{AB}$ is $\frac{1}{4}$ the length of $\overline{AC}$, then how long is $\overline{BC}$?

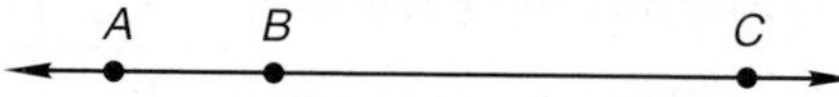

* **7.** (100) Find each sum mentally:

a. $-3 + -4$ **b.** $+5 + -5$

c. $-6 + +3$ **d.** $+6 + -3$

* **8.** (100) Solve each subtraction problem using algebraic addition:

a. $-3 - -4$ **b.** $+5 - -5$

c. $-6 - +3$ **d.** $-6 - -6$

e. **Generalize** Describe how to change a subtraction problem into an addition problem.

* **9.** (Inv. 10) **Explain** Two coins are tossed.

a. What is the probability that both coins will land heads up?

b. What is the probability that one of the coins will be heads and the other tails?

Complete the table to answer problems **10–12.**

	Fraction	Decimal	Percent
* **10.** (99)	$\frac{3}{4}$	**a.**	**b.**
* **11.** (99)	**a.**	1.6	**b.**
* **12.** (99)	**a.**	**b.**	5%

13. (66) $1\frac{1}{2} \times 4$

14. (68) $6 \div 1\frac{1}{2}$

15. (92) $(0.4)^2 \div 2^3$

Find each unknown number:

16. (43) $x + 2\frac{1}{2} = 5$

17. (42) $\frac{8}{5} = \frac{40}{x}$

18. (49) $0.06n = \$0.15$

19. (87) $6n = 21 \cdot 4$

20. (82) **Connect** Nia's garage is 20 feet long, 20 feet wide, and 8 feet high.

a. How many 1-by-1-by-1-foot boxes can she fit on the floor (bottom layer) of her garage?

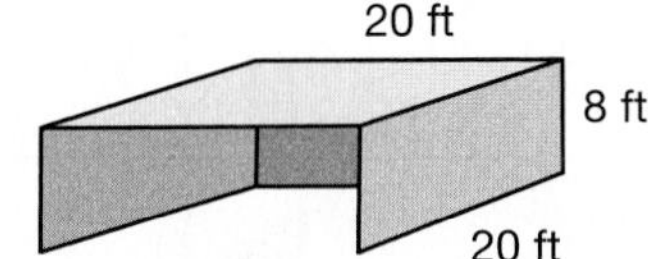

b. Altogether, how many boxes can Nia fit in her garage if she stacks the boxes 8 feet high?

21. (47) **Estimate** If a roll of tape has a diameter of $2\frac{1}{2}$ inches, then removing one full turn of tape yields about how many inches? Choose the closest answer.

A $2\frac{1}{2}$ in. **B** 5 in. **C** $7\frac{3}{4}$ in. **D** $9\frac{1}{4}$ in.

22. (92) $9^2 - \sqrt{9} \times 10 - 2^4 \times 2$

Use the figure to answer problems **23** and **24.**

23. (60) Together, these three triangles form what kind of polygon?

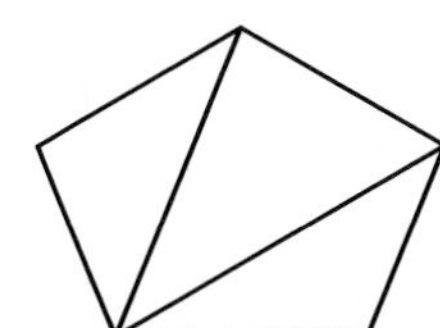

* **24.** (98) **Generalize** What is the sum of the measures of the angles of each triangle?

* **25.** (14, 100) At 6 a.m. the temperature was −8°F. By noon the temperature was 15°F. The temperature had risen how many degrees?

26. (50) **Connect** To what decimal number is the arrow pointing on the number line below?

27. (38, 58) What is the probability of rolling a perfect square with one roll of a number cube?

28. (Inv. 7, 79) **Connect** What is the area of a triangle with vertices located at (4, 0), (0, −3), and (0, 0)?

* **29.** (95) **Explain** How can you convert 18 feet to yards?

30. (78) If a gallon of milk costs $3.80, what is the cost per quart?

Early Finishers
Math and Science

The surface of the Dead Sea is approximately 408 meters below sea level. Its greatest depth is 330 meters. In contrast, Mt. Everest reaches a height of 8,850 meters. What is the difference in elevation between the summit of Mt. Everest and the bottom of the Dead Sea? Show your work.

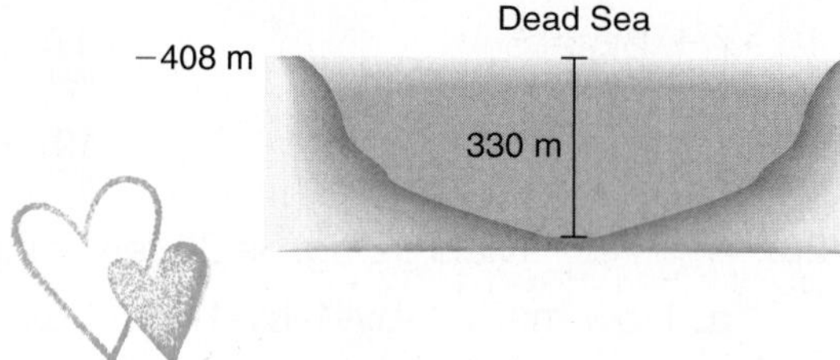

LESSON

102

• Mass and Weight

Power Up

Building Power

facts Power Up M

mental math

a. **Number Sense:** 30 · 400

b. **Number Sense:** 462 + 150

c. **Percent:** 50% of 40

d. **Calculation:** \$100.00 − \$47.50

e. **Decimals:** 0.06 × 100

f. **Number Sense:** 50 × 15

g. **Measurement:** How many pints are in 4 quarts?

h. **Calculation:** 12 + 12, + 1, $\sqrt{\ }$, × 3, + 1, $\sqrt{\ }$, × 2, + 2, × 5

problem solving

"Casting out nines" is a technique for checking long multiplication. To cast out nines, we sum the digits of each number from left to right and "cast out" (subtract) 9 from the resulting sums. For instance:

6,749	6 + 7 + 4 + 9 = 26	26 − 9 = 17; 17 − 9 = **8**
× 85	8 + 5 = 13	13 − 9 = **4**
573,665	5 + 7 + 3 + 6 + 6 + 5 = 32	32 − 9 = 23; 23 − 9 = 14; 14 − 9 = **5**

To verify the product is correct, we multiply the 8 and the 4 (8 × 4 = 32), add the resulting digits (3 + 2 = 5), and compare the result to the product after casting out nines. The number 573,665 results in 5 after casting out nines, so the product is most likely correct. If the numbers had been different, we would know that our original product was incorrect. Matching results after casting out nines does not always guarantee that our product is correct, but the technique catches most random errors.

Check 1234 × 56 = 69,106 by casting out nines.

New Concept

Increasing Knowledge

Physical objects are composed of matter. The amount of matter in an object is its **mass.** In the metric system we measure the mass of objects in milligrams (mg), grams (g), and kilograms (kg).

Math Language

The prefix **kilo-** means *one thousand*. The prefix **milli-** means *one thousandth*. Remembering what the prefixes mean helps us convert units.

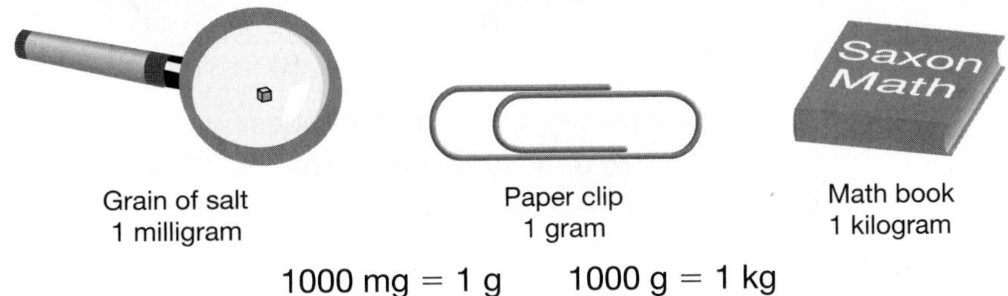

Grain of salt 1 milligram

Paper clip 1 gram

Math book 1 kilogram

1000 mg = 1 g   1000 g = 1 kg

A particular object has the same mass on Earth as it has on the moon, in orbit, or anywhere else in the universe. In other words, the mass of an object does not change with changes in the force of gravity. However, the **weight** of an object does change with changes in the force of gravity. For example, astronauts who are in orbit feel no gravitational force, so they experience weightlessness. An astronaut who weighs 154 pounds on Earth weighs zero pounds in weightless conditions. Although the weight of the astronaut has changed, his or her mass has not changed.

In the U.S. Customary System we measure the weight of objects in ounces (oz), pounds (lb), or tons (tn). On Earth an object with a mass of 1 kilogram weighs about 2.2 pounds.

Envelope and letter
1 ounce

Shoe
1 pound

Small car
1 ton

16 ounces = 1 pound   2000 pounds = 1 ton

Example 1

Thinking Skill

Predict

When you convert an amount from a larger unit to a smaller unit, will the result be more units or fewer units?

Two kilograms is how many grams?

Solution

One kilogram is 1000 grams. So 2 kilograms equals **2000 grams.**

Some measures are given using a mix of units. For example, Sam might finish a facts practice test in 2 minutes 34 seconds. His sister may have weighed 7 pounds 12 ounces when she was born. The following example shows how to add and subtract measures in pounds and ounces.

Example 2

a. Add:

$$\begin{array}{r} \textbf{7 lb 12 oz} \\ \underline{\textbf{+ 2 lb \ \ 6 oz}} \end{array}$$

b. Subtract:

$$\begin{array}{r} \textbf{9 lb 10 oz} \\ \underline{\textbf{− 7 lb 12 oz}} \end{array}$$

Solution

a. The sum of 12 oz and 6 oz is 18 oz, which is 1 lb 2 oz. We record the 2 oz and then add the pound to 7 lb and 2 lb.

$$\begin{array}{r} \overset{1}{} \\ \text{7 lb 12 oz} \\ \underline{\text{+ 2 lb \ \ 6 oz}} \\ \textbf{10 lb \ \ 2 oz} \end{array}$$

b. Before we can subtract ounces, we convert 9 pounds to 8 pounds plus 16 ounces. We combine the 16 ounces and the 10 ounces to get 26 ounces. Then we subtract.

$$\begin{array}{r} \overset{8}{\cancel{9}}\text{ lb } \overset{26}{\cancel{10}}\text{ oz} \\ -\ 7\text{ lb } 12\text{ oz} \\ \hline \mathbf{1\text{ lb } 14\text{ oz}} \end{array}$$

Practice Set

a. Half of a kilogram is how many grams?

b. The mass of a liter of water is 1 kilogram. So the mass of 2 liters of beverage is about how many grams?

c. $\begin{array}{r} 5\text{ lb } 10\text{ oz} \\ +\ 1\text{ lb } \ \ 9\text{ oz} \\ \hline \end{array}$

d. $\begin{array}{r} 9\text{ lb } \ \ 8\text{ oz} \\ -\ 6\text{ lb } 10\text{ oz} \\ \hline \end{array}$

e. A half-ton pickup truck can haul a half-ton load. Half of a ton is how many pounds?

Written Practice

Strengthening Concepts

On his first six tests, Chris had scores of 90%, 92%, 96%, 92%, 84%, and 92%. Use this information to answer problems **1** and **2.**

1. *(Inv. 5)* **a.** Which score occurred most frequently? That is, what is the mode of the scores?

b. The difference between Chris's highest score and his lowest score is how many percentage points? That is, what is the range of the scores?

2. *(18)* What was Chris's average score for the six tests? That is, what is the mean of the scores?

* **3.** *(87)* In basketball there are one-point baskets, two-point baskets, and three-point baskets. If a team scored 96 points and made 18 one-point baskets and 6 three-point baskets, how many two-point baskets did the team make? Explain how you found your answer.

* **4.** *(83)* **Analyze** Which ratio forms a proportion with $\frac{4}{7}$?

A $\frac{7}{4}$ **B** $\frac{14}{17}$ **C** $\frac{12}{21}$ **D** $\frac{2}{3}$

* **5.** *(85)* Complete this proportion: Four is to five as what number is to twenty?

6. *(50)* Arrange these numbers in order from least to greatest:

$$-1, 1, 0.1, -0.1, 0$$

7. *(92)* The product of $10^3 \cdot 10^2$ equals which of the following?

A 10^9 **B** 10^6 **C** 10^5 **D** 10

8. (86) The area of the square in this figure is 100 mm^2.

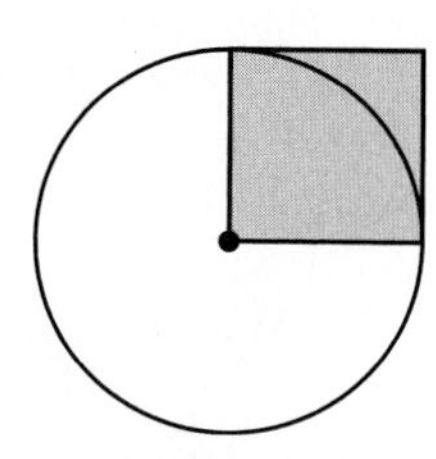

a. What is the radius of the circle?

b. What is the diameter of the circle?

c. What is the area of the circle? (Use 3.14 for π.)

Connect Complete the table to answer problems **9–11.**

	Fraction	Decimal	Percent
*** 9.** (99)	$\frac{4}{25}$	**a.**	**b.**
*** 10.** (99)	**a.**	0.01	**b.**
*** 11.** (99)	**a.**	**b.**	90%

12. (61) $1\frac{2}{3} + 3\frac{1}{2} + 4\frac{1}{6}$

13. (72) $\frac{5}{6} \times \frac{3}{10} \times 4$

14. (68) $6\frac{1}{4} \div 100$

15. (38) $6.437 + 12.8 + 7$

16. (74) **Estimate** Convert $\frac{1}{7}$ to a decimal number by dividing 1 by 7. Stop dividing after three decimal places, and round your answer to two decimal places.

17. (60) An octagon has how many more sides than a pentagon?

18. (92) $4 \times 5^2 - 50 \div \sqrt{4} + (3^2 - 2^3)$

*** 19.** (Inv. 10) **Analyze** Sector 2 on this spinner is a 90° sector. If the spinner is spun twice, what is the probability that it will stop in sector 2 both times?

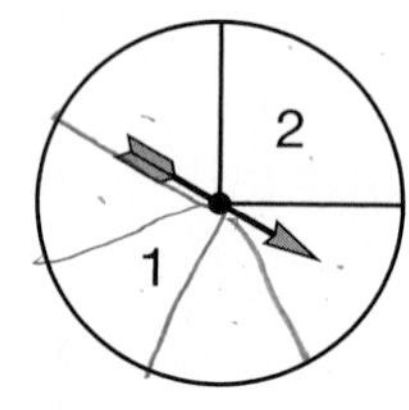

20. (58) If the spinner is spun 100 times, about how many times would it be expected to stop in sector 1?

21. (82) How many 1 inch cubes would be needed to build this larger cube?

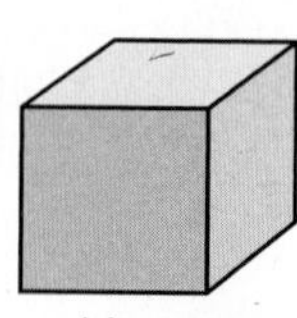

22. (18) The average of four numbers is 5. What is their sum?

*** 23.** (102) **Connect** When Andy was born, he weighed 8 pounds 4 ounces. Three weeks later he weighed 10 pounds 1 ounce. How many pounds and ounces had he gained in three weeks?

*** 24.** (97) Lines *s* and *t* are parallel.

a. Which angle is an alternate interior angle to ∠5?

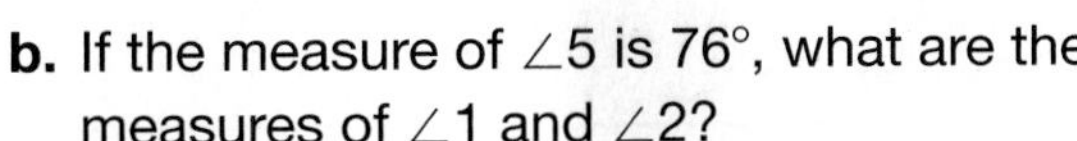

b. If the measure of ∠5 is 76°, what are the measures of ∠1 and ∠2?

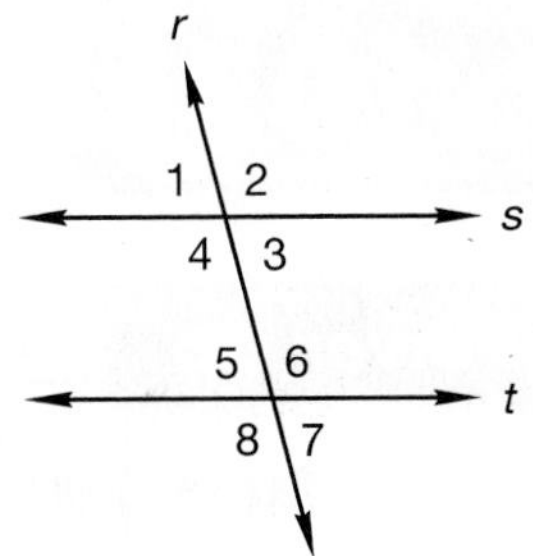

*** 25.** (96) **Generalize** Find the missing number in this function table:

x	1	2	4	5
3x − 5	−2	1	7	

26. (8) What is the perimeter of this hexagon? Dimensions are in centimeters.

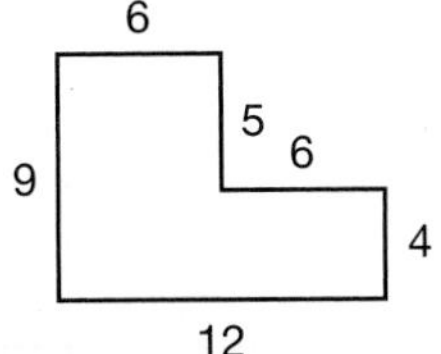

27. (71, 79) **a.** What is the area of the parallelogram at right?

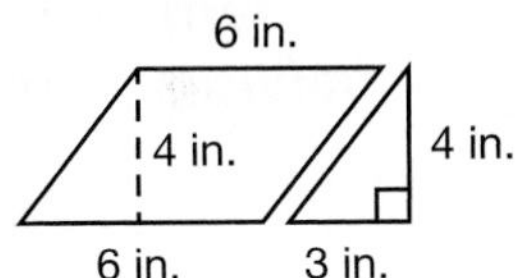

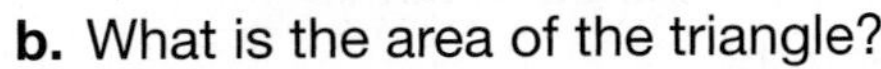

b. What is the area of the triangle?

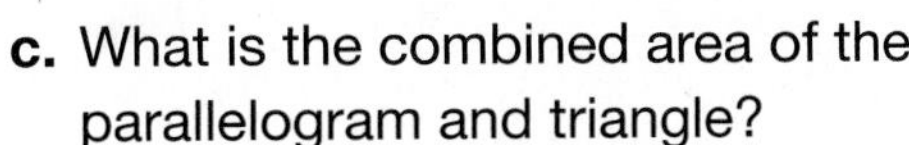

c. What is the combined area of the parallelogram and triangle?

*** 28.** (102) How many milligrams is half of a gram?

29. (Inv. 7) **Model** The coordinates of the endpoints of a line segment are (3, −1) and (3, 5). The midpoint of the segment is the point halfway between the endpoints. What are the coordinates of the midpoint?

30. (47) **Estimate** Tania took 10 steps to walk across the tetherball circle and 31 steps to walk around the tetherball circle. Use this information to find the approximate number of diameters in the circumference of the tetherball circle.

LESSON 103

• Perimeter of Complex Shapes

Power Up — *Building Power*

facts Power Up M

mental math

a. **Number Sense:** $50 \cdot 60$

b. **Number Sense:** $543 - 250$

c. **Percent:** 25% of 40

d. **Calculation:** $5.65 + $3.99

e. **Decimals:** $87.5 \div 100$

f. **Number Sense:** $\frac{500}{20}$

g. **Measurement:** How many milliliters are in 10 liters?

h. **Calculation:** 6×6, $- 1$, $\div 5$, $\times 6$, $- 2$, $\div 5$, $\times 4$, $- 2$, $\times 3$

problem solving

Here are the front, top, and side views of an object. Draw a three-dimensional view of the object from the perspective of the upper right front.

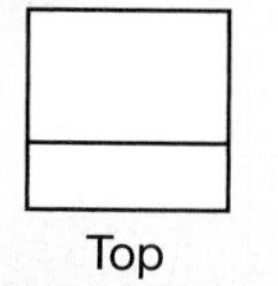

New Concept — *Increasing Knowledge*

Thinking Skill

Conclude

Why is the shape described as complex?

In this lesson we will practice finding the perimeters of complex shapes. The figure below is an example of a complex shape. Notice that the lengths of two of the sides are not given. We will first find the lengths of these sides; then we will find the perimeter of the shape. (In this book, assume that corners that look square are square.)

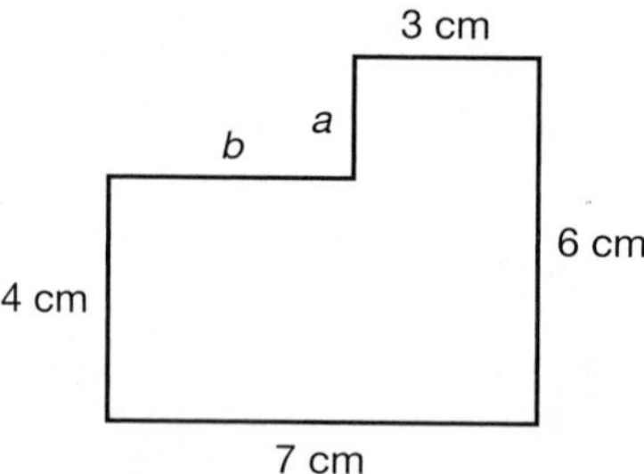

We see that the figure is 7 cm long. The sides marked *b* and 3 cm together equal 7 cm. So *b* must be 4 cm.

$$b + 3 \text{ cm} = 7 \text{ cm}$$

$$b = 4 \text{ cm}$$

The width of the figure is 6 cm. The sides marked 4 cm and a together equal 6 cm. So a must equal 2 cm.

$$4 \text{ cm} + a = 6 \text{ cm}$$

$$a = 2 \text{ cm}$$

We have found that b is 4 cm and a is 2 cm.

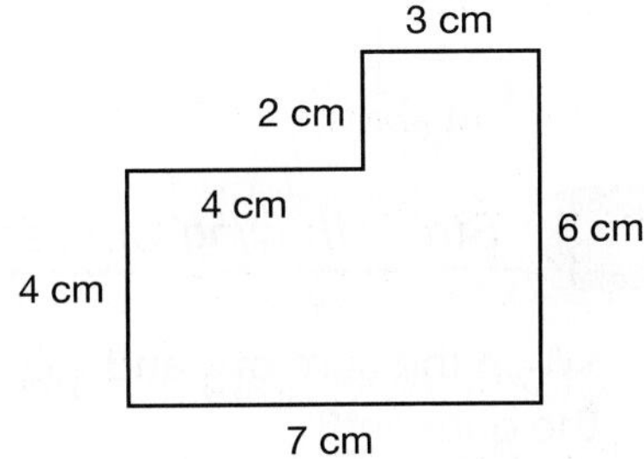

We add the lengths of all the sides and find that the perimeter is 26 cm.

$$6 \text{ cm} + 7 \text{ cm} + 4 \text{ cm} + 4 \text{ cm} + 2 \text{ cm} + 3 \text{ cm} = 26 \text{ cm}$$

Example

Find the perimeter of this figure.

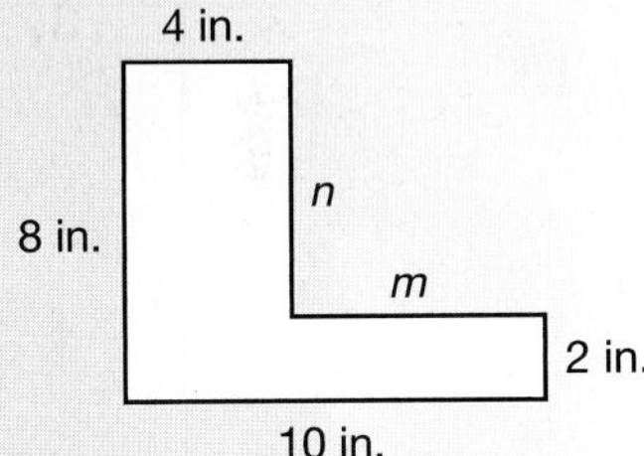

Solution

To find the perimeter, we add the lengths of the six sides. The lengths of two sides are not given in the illustration. We will write two equations to find the lengths of these sides. The length of the figure is 10 inches. The sides parallel to the 10-inch side have lengths of 4 inches and m inches. Their combined length is 10 inches, so m must equal 6 inches.

$$4 \text{ in.} + m = 10 \text{ in.}$$

$$m = 6 \text{ in.}$$

The width of the figure is 8 inches. The sides parallel to the 8-inch side have lengths of n inches and 2 inches. Their combined measures equal 8 inches, so n must equal 6 inches.

$$n + 2 \text{ in.} = 8 \text{ in.}$$

$$n = 6 \text{ in.}$$

We add the lengths of the six sides to find the perimeter of the complex shape.

$$10 \text{ in.} + 8 \text{ in.} + 4 \text{ in.} + 6 \text{ in.} + 6 \text{ in.} + 2 \text{ in.} = \textbf{36 in.}$$

Practice Set

Find the perimeter of each complex shape:

a.

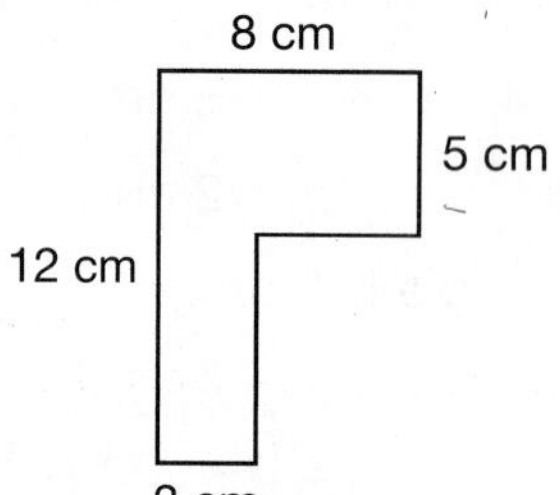

b.

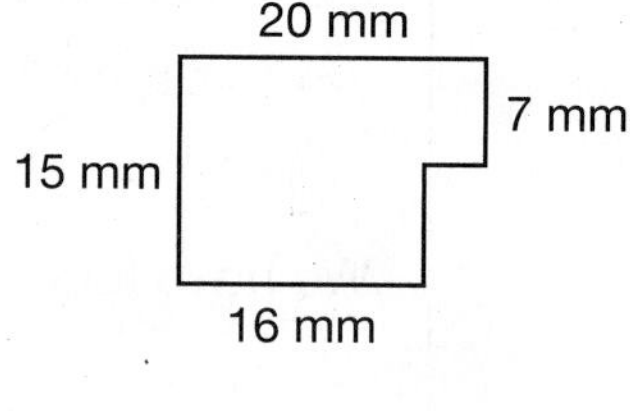

Written Practice

Strengthening Concepts

1. (12, 72) When the sum of $\frac{1}{2}$ and $\frac{1}{3}$ is divided by the product of $\frac{1}{2}$ and $\frac{1}{3}$, what is the quotient?

2. (18) The average age of three men is 24 years.

a. What is the sum of their ages?

b. If two of the men are 22 years old, how old is the third?

3. (38) A string one yard long is formed into the shape of a square.

a. How many inches long is each side of the square?

b. How many square inches is the area of the square?

4. (85) Complete this proportion: Five is to three as thirty is to what number?

5. (23) Mr. Cho has 30 books. Fourteen of the books are mysteries. What is the ratio of mysteries to non-mysteries?

*** 6.** (101) **Analyze** In another class of 33 students, the ratio of boys to girls is 4 to 7. How many girls are in that class?

*** 7.** (92) $100 \div 10^2 + 3 \times (2^3 - \sqrt{16})$

*** 8.** (58, 102) Robert complained that he had a "ton" of homework.

a. How many pounds is a ton?

b. **Conclude** What is the probability that Robert would literally have a ton of homework?

Connect Complete the table to answer problems **9–11.**

	Fraction	Decimal	Percent
*** 9.** (99)	$\frac{1}{100}$	**a.**	**b.**
*** 10.** (99)	**a.**	0.4	**b.**
*** 11.** (99)	**a.**	**b.**	8%

12. (68) $10\frac{1}{2} \div 3\frac{1}{2}$

13. (53) $(6 + 2.4) \div 0.04$

Find each unknown number:

14. (61) $7\frac{1}{2} + 6\frac{3}{4} + n = 15\frac{3}{8}$

15. (63) $x - 1\frac{3}{4} = 7\frac{1}{2}$

16. (43) **Verify** Instead of dividing $10\frac{1}{2}$ by $3\frac{1}{2}$, Guadalupe doubled both numbers before dividing. What was Guadalupe's division problem and its quotient?

17. (47) **Estimate** Mariabella used a tape measure to find the circumference and the diameter of a plate. The circumference was about 35 inches, and the diameter was about 11 inches. Find the approximate number of diameters in the circumference. Round to the nearest tenth.

18. (92) Write twenty million, five hundred thousand in expanded notation using exponents.

19. (19) **List** Name the prime numbers between 40 and 50.

* **20.** (100) **Analyze** Calculate mentally:

a. $-3 + -8$

b. $-3 - -8$

c. $-8 + +3$

d. $-8 - +3$

* **21.** (98) **Conclude** In $\triangle ABC$ the measure of $\angle A$ is 40°. Angles B and C are congruent. What is the measure of $\angle C$?

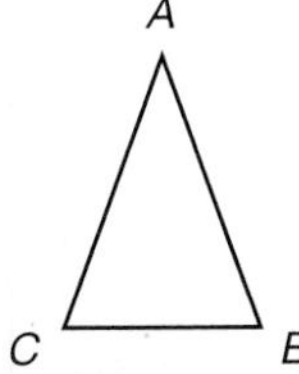

* **22.** (8, 79) **a.** What is the perimeter of this triangle?

b. What is the area of this triangle?

c. What is the ratio of the length of the 20 mm side to the length of the longest side? Express the ratio as a fraction and as a decimal.

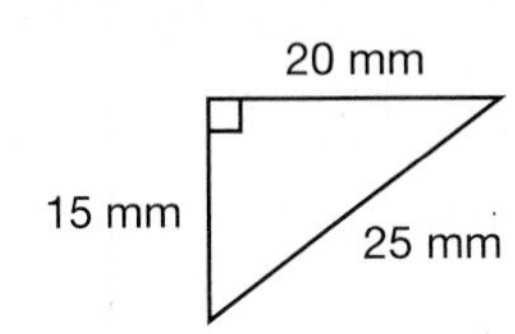

23. (82) **Analyze** The Simpsons rented a trailer that was 8 feet long and 5 feet wide. If they load the trailer with 1-by-1-by-1-foot boxes to a height of 3 feet, how many boxes can be loaded onto the trailer?

24. (58) What is the probability of drawing the queen of spades from a normal deck of 52 cards?

*** 25.** (10, 100) **a.** Connect What temperature is shown on this thermometer?

b. If the temperature rises 12°F, what will the temperature be?

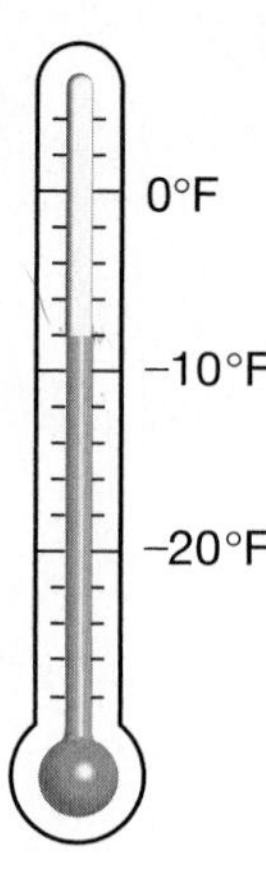

*** 26.** (103) Find the perimeter of the figure below.

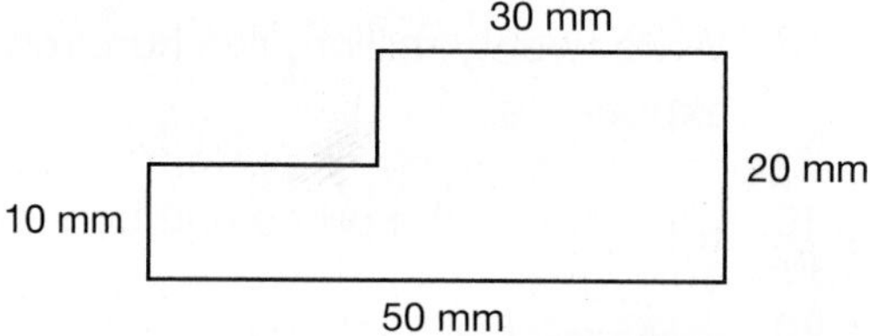

27. (37) **a.** Analyze What is the area of the shaded rectangle?

b. What is the area of the unshaded rectangle?

c. What is the combined area of the two rectangles?

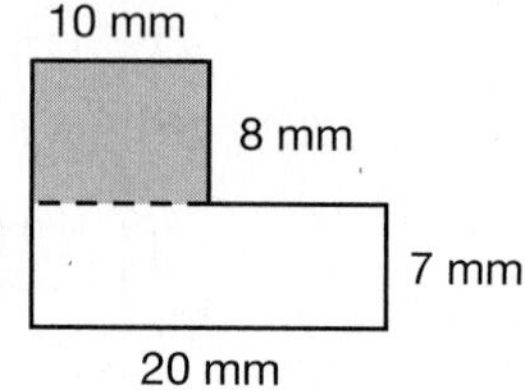

28. (Inv. 7) Connect What are the coordinates of the point halfway between (−3, −2) and (5, −2)?

*** 29.** (78, 102) Estimate A pint of milk weighs about 16 ounces. About how many pounds does a half gallon of milk weigh?

30. (90) Evaluate Ruben walked around a building whose perimeter was shaped like a regular pentagon.

a. At each corner of the building, Ruben turned about how many degrees?

b. What is the measure of each interior angle of the regular pentagon?

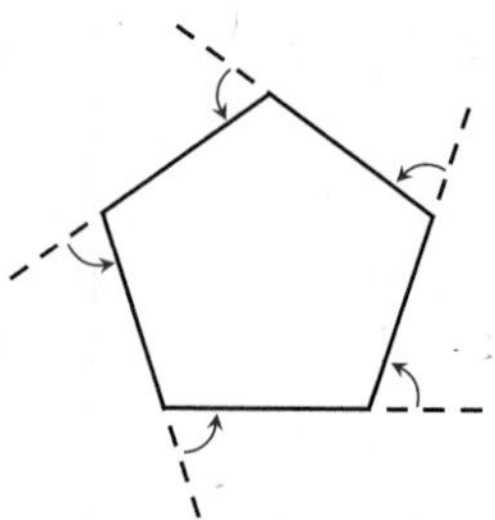

LESSON

104 • Algebraic Addition Activity

Power Up

Building Power

Note: Because the New Concept in this lesson takes about half of a class period, today's Power Up has been omitted.

problem solving

Sonya, Sid, and Sinead met at the gym on Monday. Sonya goes to the gym every two days. The next day she will be at the gym is Wednesday. Sid goes to the gym every three days. The next day Sid will be at the gym is Thursday. Sinead goes to the gym every four days. She will next be at the gym on Friday. What will be the next day that Sonya, Sidney, and Sinead are at the gym on the same day?

New Concept

Increasing Knowledge

One model for the addition of signed numbers is the number line. Another model for the addition of signed numbers is the electrical-charge model, which is used in the Sign Game. In this model, signed numbers are represented by positive and negative charges that can neutralize each other when they are added. The game is played with sketches, as shown here. The first two levels may be played with two color counters.

Thinking Skill

Discuss

What addition equation shows that a positive charge and a negative charge neutralize each other?

Activity

Sign Game

In the Sign Game pairs of positive and negative charges become neutral. After determining the neutral pairs we count the signs that remain and then write our answer. There are four skill levels to the game. Be sure you are successful at one level before moving to the next level.

level 1

Positive and negative signs are placed randomly on a "screen." When the game begins positive and negative pairs are neutralized so we cross out the signs as shown. (Appropriate sound effects strengthen the experience!) (If using counters, remove all pairs of counters that have different colors.)

Before

After

Two positives remain.

After marking positive-negative pairs we count the remaining positives or negatives. In the example shown above, two positives remain. With counters, two counters of one color remain. See whether you can determine what will remain on the three practice screens below:

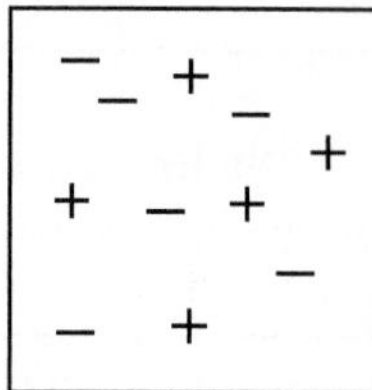

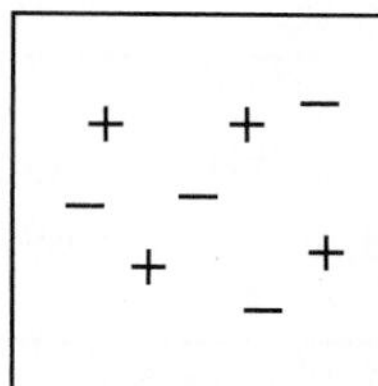

– – – – – – –

level 2 Positives and negatives are displayed in counted clusters or stacked counters. The suggested strategy is to combine the same signs first. So $+3$ combines with $+1$ to form $+4$, and -5 combines with -2 to form -7. Then determine how many of which charge (sign) remain.

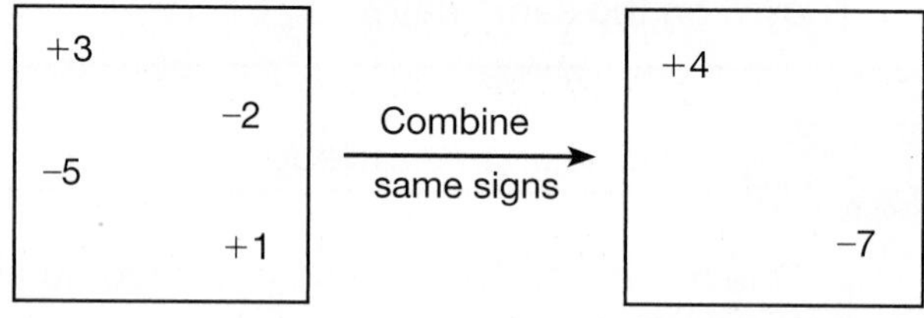

Three negatives, or -3, remain.

There were three more negatives than positives, so -3 remain. With counters, stacks of equal height and different colors are removed. Only one color (or no counters) remains. See whether you can determine how many of which charge will remain for the three practice screens below:

$+5$, -6, -4, $+3$, $+4$

$+3$, $+4$, $+1$, $+2$

$+1$, -3, $+4$, -2

level 3 Positive and negative clusters can be displayed with two signs, one sign, or no sign. Clusters appear "in disguise" by taking on an additional sign or by dropping a sign. The first step is to remove the disguise. A cluster with no sign, with "$- -$," or with "$+ +$" is a positive cluster. A cluster with "$+ -$" or with "$- +$" is a negative cluster. If a cluster has a "shield" (parentheses), look through the shield to see the sign. With counters, for "$- -$" invert a negative to a positive, and for "$- +$" invert a positive to a negative.

Reading Math

Symbols

A negative sign indicates the opposite of a number. -3 means "the opposite of 3." Likewise, $-(-3)$ means "the opposite of -3," which is 3. $-(+3)$ means "the opposite of $+3$," which is -3.

Examples of Positives	Examples of Negatives
$-(-3) = +3$	$-(+2) = -2$
$--2 = +2$	$+(-3) = -3$
$4 = +4$	$+-1 = -1$
$++1 = +1$	$-+4 = -4$

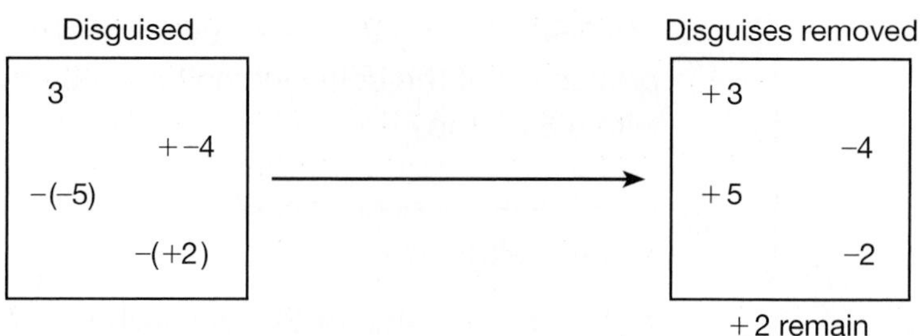

See whether you can determine how many of which charge remain for the following practice screens:

−−3 +(−5) −+6 +(+4)	+(−3) −2 −(+4)	−(+6) −−3 +4 −(+2) +−6

level 4 Extend Level 3 to a line of clusters without using a screen.

$$-3 + (-4) - (-5) - (+2) + (+6)$$

Use the following steps to find the answer:

Step 1: Remove the disguises: $-3 - 4 + 5 - 2 + 6$

Step 2: Group forces: $-9 + 11$

Step 3: Find what remains: $+2$

Practice Set

Simplify:

a. $-2 + -3 - -4 + -5$

b. $-3 + (+2) - (+5) - (-6)$

c. $+3 + -4 - +6 + +7 - -1$

d. $2 + (-3) - (-9) - (+7) + (+1)$

e. $3 - -5 + -4 - +2 + +8$

f. $(-10) - (+20) - (-30) + (-40)$

Written Practice *Strengthening Concepts*

1. (Inv. 6) *Conclude* A pyramid with a square base has how many more edges than vertices?

*** 2.** (102) Becki weighed 7 lb 8 oz when she was born and 12 lb 6 oz at 3 months. How many pounds and ounces did Becki gain in 3 months?

3. (23) There are 6 fish and 10 snails in the aquarium. What is the ratio of fish to snails?

*** 4.** (101) A team's win-loss ratio was 3 to 2. If the team had played 20 games without a tie, how many games had it won?

* **5.** (Inv. 10) **Analyze** If Molly tosses a coin and rolls a number cube, what is the probability of the coin landing heads up and the number cube stopping with a 6 on top?

6. (71) **a.** What is the perimeter of this parallelogram?

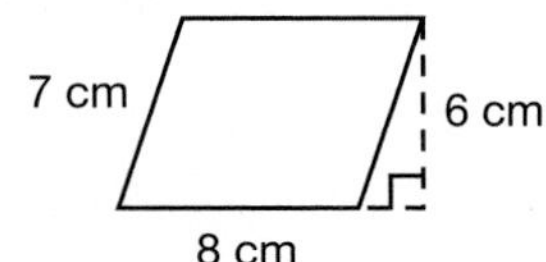

b. What is the area of this parallelogram?

7. (71) **Conclude** If each acute angle of a parallelogram measures 59°, then what is the measure of each obtuse angle?

8. (86) **Estimate** The center of this circle is the origin. The circle passes through (2, 0).

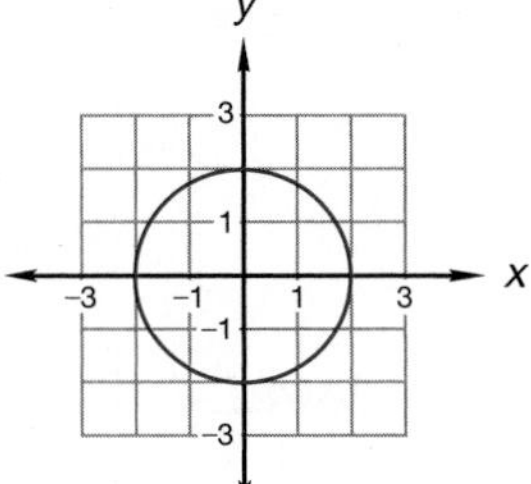

a. Estimate the area of the circle in square units by counting squares.

b. Calculate the area of the circle by using 3.14 for π.

9. (83) Which ratio forms a proportion with $\frac{2}{3}$?

A $\frac{2}{4}$ **B** $\frac{3}{4}$ **C** $\frac{4}{6}$ **D** $\frac{3}{2}$

10. (85) Complete this proportion: $\frac{6}{8} = \frac{a}{12}$

* **11.** (103) What is the perimeter of the hexagon at right? Dimensions are in centimeters.

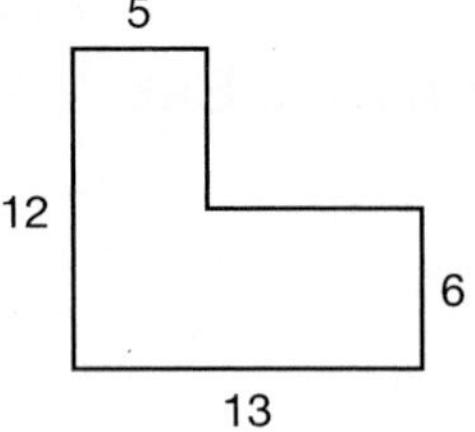

Connect Complete the table to answer problems **12–14.**

	Fraction	Decimal	Percent
* **12.** (99)	$\frac{3}{20}$	**a.**	**b.**
* **13.** (99)	**a.**	1.2	**b.**
* **14.** (99)	**a.**	**b.**	10%

15. (41) Sharon bought a notebook for 40% off the regular price of $6.95. What was the sale price of the notebook?

16. (89) Between which two consecutive whole numbers is $\sqrt{200}$?

* **17.** (92, Inv. 10) **Analyze** Compare:

$\left(\frac{1}{2}\right)^3$ ◯ the probability of 3 consecutive "heads" coin tosses

18. (49) **Estimate** Divide 0.624 by 0.05 and round the quotient to the nearest whole number.

19. The average of three numbers is 20. What is the sum of the three numbers?
(18)

20. Write the prime factorization of 450 using exponents.
(73)

* **21.** $-3 + -5 - -4 - +2$
(104)

* **22.** $3^4 + 5^2 \times 4 - \sqrt{100} \times 2^3$
(92)

23. How many blocks 1 inch on each edge would it take to fill a shoe box that is 12 inches long, 6 inches wide, and 5 inches tall?
(82)

24. Three fourths of the 60 athletes played in the game. How many athletes did not play?
(77)

* **25.** The distance a car travels can be found by multiplying the **speed** of the car by the amount of **time** the car travels at that speed. How far would a car travel in 4 hours at 88 kilometers per hour?
(95)

$$\frac{88 \text{ km}}{1 \text{ hr}} \times \frac{4 \text{ hr}}{1}$$

26. **Analyze** Use the figure on the right to answer **a–c**.
(31)

5 cm
8 cm
12 cm
6 cm

a. What is the area of the shaded rectangle?

b. What is the area of the unshaded rectangle?

c. What is the combined area of the two rectangles?

27. **Estimate** Colby measured the circumference and diameter of four circles. Then he divided the circumference by the diameter of each circle to find the number of diameters in a circumference. Here are his answers:
(18, 51)

3.12, 3.2, 3.15, 3.1

Find the average of Colby's answers. Round the average to the nearest hundredth.

28. **Explain** Hector was thinking of a two-digit counting number, and he asked Simon to guess the number. Describe how you can find the probability that Simon will guess correctly on the first try.
(58)

29. **Connect** The coordinates of three vertices of a triangle are (3, 5), (−1, 5), and (−1, −3). What is the area of the triangle?
(Inv. 7, 79)

30. $\frac{2 \text{ gal}}{1} \times \frac{4 \text{ qt}}{1 \text{ gal}} \times \frac{2 \text{ pt}}{1 \text{ qt}}$
(95)

LESSON 105

Using Proportions to Solve Percent Problems

Power Up

Building Power

facts Power Up M

mental math

a. **Number Sense:** 200 · 40

b. **Number Sense:** 567 − 150

c. **Percent:** 50% of 200

d. **Calculation:** \$17.20 + \$2.99

e. **Decimals:** 7.5 ÷ 100

f. **Number Sense:** $\frac{440}{20}$

g. **Measurement:** How many quarts are in 2 gallons?

h. **Calculation:** 6 × 8, + 1, $\sqrt{\ }$, × 3, − 1, ÷ 2, × 10, − 1, ÷ 9

problem solving

Copy this problem and fill in the missing digits:

$$\frac{\square}{4} + \frac{\square}{6} = \frac{11}{12}$$

New Concept

Increasing Knowledge

We know that a percent can be expressed as a fraction with a denominator of 100.

$$30\% = \frac{30}{100}$$

A percent can also be regarded as a ratio in which 100 represents the total number in the group, as we show in the following example.

Math Language

Percent means per hundred. 30% means 30 out of 100

Example 1

Thirty percent of the cars in the parade are antique cars. If 12 vehicles are antique cars, how many vehicles are in the parade in all?

Solution

We construct a ratio box. The ratio numbers we are given are 30 and 100. We know from the word *percent* that 100 represents the ratio total. The actual count we are given is 12. Our categories are "Antiques" and "not Antiques."

	Percent	Actual Count
Antiques	30	12
Not Antiques		
Total	100	

Since the ratio total is 100, we calculate that the ratio number for "not Antiques" is 70. We use n to stand for "not Antiques" and t for "total" in the actual-count column. We use two rows from the table to write a proportion. Since we know both numbers in the "Antiques" row, we use the numbers in the "Antiques" row for the proportion. Since we want to find the total number of students, we also use the numbers from the "total" row. We will then solve the proportion using cross products.

	Percent	Actual Count
Antiques	30	12
Not Antiques	70	n
Total	100	t

$$\frac{30}{100} = \frac{12}{t}$$

$$30t = 12 \cdot 100$$

$$t = \frac{\overset{4}{\cancel{12}} \cdot \overset{10}{\cancel{100}}}{\underset{\underset{1}{\cancel{3}}}{\cancel{30}}}$$

$$t = 40$$

We find that a total of **40 vehicles** were in the parade.

In the above problem we did not need to use the 70% who were "not Antiques." In the next example we will need to use the "not" percent in order to solve the problem.

Example 2

Only 40% of the team members played in the game. If 24 team members did not play, then how many did play?

Solution

We construct a ratio box. The categories are "played," "did not play," and "total." Since 40% played, we calculate that 60% did not play. We are asked for the actual count of those who played. So we use the "played" row and the "did not play" row (because we know both numbers in that row) to write the proportion.

	Percent	Actual Count
Played	40	p
Did Not Play	60	24
Total	100	t

$$\frac{40}{60} = \frac{p}{24}$$

$$60p = 40 \cdot 24$$

$$p = \frac{\overset{4}{\cancel{40}} \cdot \overset{4}{\cancel{24}}}{\underset{\underset{1}{\cancel{6}}}{\cancel{60}}}$$

$$p = 16$$

We find that **16 team members** played in the game.

Example 3

Buying the shoes on sale, Nathan paid $45.60, which was 60% of the full price. What was the full price of the shoes?

Solution

Nathan paid 60% instead of 100%, so he saved 40% of the full price. We are given what Nathan paid. We are asked for the full price, which is the 100% price.

	Percent	Actual Count
Paid	60%	$45.60
Saved	40%	s
Full Price	100%	f

$$\frac{60}{100} = \frac{45.60}{f}$$

$$60f = 4560$$

$$f = 76$$

Full price for the shoes was **$76.**

Practice Set

Model Solve these percent problems using proportions. Make a ratio box for each problem.

a. Forty percent of the cameras in a store are digital cameras. If 24 cameras are not digital, how many cameras are in the store in all?

b. Seventy percent of the team members played in the game. If 21 team members played, how many team members did not play?

c. *Model* Referring to problem **b,** what proportion would we use to find the number of members on the team?

d. Joan walked 0.6 miles in 10 minutes. How far can she walk in 25 minutes at that rate? Write and solve a proportion to find the answer.

e. *Formulate* Create and solve your own percent problem using the method shown in this lesson.

Written Practice *Strengthening Concepts*

* **1.** How far would a car travel in $2\frac{1}{2}$ hours at 50 miles per hour?
(95)

$$\frac{50 \text{ mi}}{1 \text{ hr}} \times \frac{2\frac{1}{2} \text{ hr}}{1}$$

2. **Connect** A map of Texas is drawn to a scale of 1 inch = 50 miles. Houston and San Antonio are 4 inches apart on the map. What is the actual distance between Houston and San Antonio?
(95)

3. The ratio of humpback whales to orcas was 2 to 1. If there were 900 humpback whales, how many orcas were there?
(88)

4. When Robert measured a half-gallon box of frozen yogurt, he found it had the dimensions shown in the illustration. What was the volume of the box in cubic inches?
(82)

* **5.** Calculate mentally:
(100, 104)

a. $+10 + -10$ **b.** $-10 - -10$ **c.** $+6 + -5 - -4$

* **6.** **Estimate** On Earth a 1-kilogram object weighs about 2.2 pounds. A rock weighs 50 kilograms. About how many pounds does the rock weigh?
(102)

* **7.** Sonia has only dimes and nickels in her coin jar; they are in a ratio of 3 to 5. If she has 120 coins in the jar, how many are dimes?
(101)

* **8.** **Analyze** The airline sold 25% of the seats on the plane at a discount. If 45 seats were sold at a discount, how many seats were on the plane? How do you know your answer is correct?
(105)

Connect Complete the table to answer problems **9–11.**

	Fraction	Decimal	Percent
* **9.** (99)	$\frac{3}{50}$	**a.**	**b.**
* **10.** (99)	**a.**	0.04	**b.**
* **11.** (99)	**a.**	**b.**	150%

12. $4\frac{1}{12} + 5\frac{1}{6} + 2\frac{1}{4}$
(61)

13. $\frac{4}{5} \times 3\frac{1}{3} \times 3$
(72)

14. 0.125×80
(39)

15. $(1 + 0.5) \div (1 - 0.5)$
(53)

16. Solve: $\frac{c}{12} = \frac{3}{4}$
(85)

17. What is the total cost of an $8.75 purchase after 8% sales tax is added?
(41)

18. Write the decimal number one hundred five and five hundredths.
(35)

*** 19.** **Conclude** The measure of $\angle A$ in quadrilateral *ABCD* is 115°. What are the measures of $\angle B$ and $\angle C$?
(98)

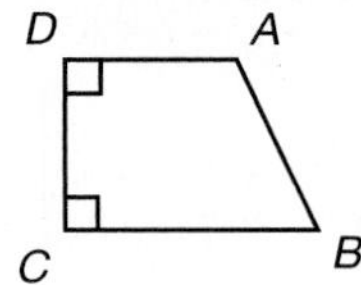

20. Write the prime factorization of 500 using exponents.
(73)

21. **Estimate** A quart is a little less than a liter, so a gallon is a little less than how many liters?
(78)

22. Diane will spin the spinner twice. What is the probability that it will stop in sector 2 both times?
(Inv. 10)

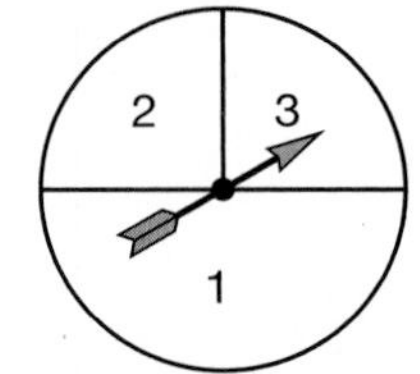

23. The perimeter of this isosceles triangle is 18 cm. What is the length of its longest side?
(8)

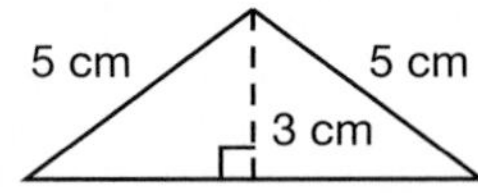

24. What is the area of the triangle in problem **23?**
(79)

25. The temperature was −5°F at 6:00 a.m. By noon the temperature had risen 12 degrees. What was the noontime temperature?
(14, 100)

26. The weather report stated that the chance of rain is 30%. Use a decimal number to express the probability that it will not rain.
(58)

***27.** Find the perimeter of this figure. Dimensions are in inches.
(103)

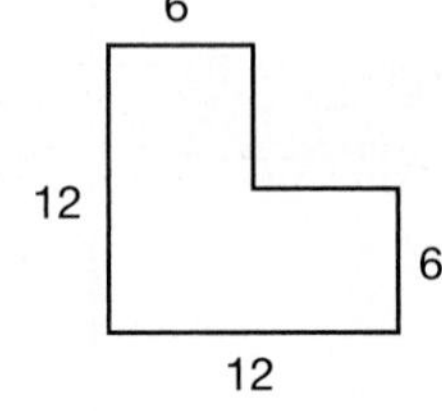

Math Language
Recall that a **function** is a rule for using one number to calculate another number.

***28.** **Generalize** Study this function table and describe the rule that helps you find *y* if you know *x*.
(96)

x	$\frac{1}{2}$	1	$1\frac{1}{2}$	2
y	$1\frac{1}{2}$	3	$4\frac{1}{2}$	6

29. A room is 15 feet long and 12 feet wide.
(7, 31)

a. The room is how many **yards** long and wide?

b. What is the area of the room in square yards?

***30.** **Explain** Ned rolled a die and it turned up 6. If he rolls the die again, what is the probability that it will turn up 6?
(58)

LESSON 106

• Two-Step Equations

Power Up

Building Power

facts Power Up K

mental math

a. **Number Sense:** $40 \cdot 600$

b. **Number Sense:** $429 + 350$

c. **Percent:** 25% of 200

d. **Calculation:** \$60.00 − \$59.45

e. **Decimals:** 1.2×100

f. **Number Sense:** 60×12

g. **Measurement:** Which is greater, 2000 milliliters or 1 liter?

h. **Calculation:** Square 5, − 1, ÷ 4, × 5, + 2, ÷ 4, × 3, + 1, $\sqrt{\ }$

problem solving

Every whole number can be expressed as the sum of, *at most,* four square numbers. In the diagram, we see that 12 is made up of one 3 × 3 square and three 1 × 1 squares. The number sentence that represents the diagram is 12 = 9 + 1 + 1 + 1.

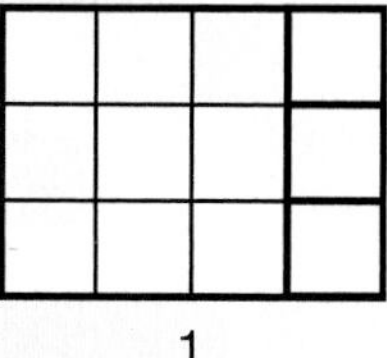

Diagram how 15, 18, and 20 are composed of four smaller squares, at most, and then write an equation for each diagram. (*Hint:* Diagrams do not have to be perfect rectangles.)

New Concept

Increasing Knowledge

Since Lessons 3 and 4 we have solved one-step equations in which we look for an unknown number in addition, subtraction, multiplication, or division. In this lesson we will begin solving two-step equations that involve more than one operation.

Example 1

Solve: $3n - 1 = 20$

Solution

Thinking Skill

Discuss

In this two-step equation, what two steps do we use to find the solution?

Let us think about what this equation means. When 1 is subtracted from $3n$, the result is 20. So $3n$ equals 21.

$$3n = 21$$

Since $3n$ means "3 times n" and $3n$ equals 21, we know that n equals 7.

$$\boldsymbol{n = 7}$$

We show our work this way:

$$3n - 1 = 20$$
$$3n = 21$$
$$n = 7$$

We check our answer this way:

$$3(7) - 1 = 20$$
$$21 - 1 = 20$$
$$20 = 20$$

Example 1 describes one method for solving equations. It is a useful method for solving equations by inspection—by mentally calculating the solution. However, there is an algebraic method that is helpful for solving more complicated equations. The method uses inverse operations to isolate the variable—to get the variable by itself on one side of the equal sign. We show this method in example 2.

Example 2

Solve $3n - 1 = 20$ using inverse operations.

Solution

We focus our attention on the side of the equation with the variable. We see that n is multiplied by 3 and 1 is subtracted from that product. We will undo the subtraction by adding, and we will undo the multiplication by dividing in order to isolate the variable.

Step:	Justification:
$3n - 1 = 20$	Given equation
$3n - 1 + 1 = 20 + 1$	Added 1 to both sides of the equation
$3n = 21$	Simplified both sides
$\frac{3n}{3} = \frac{21}{3}$	Divided both sides by 3
$\mathbf{n = 7}$	Simplified both sides

We performed two operations (addition and division) to the left side of the equation to isolate the variable. Notice that we also performed the same two operations on the right side to keep the equation balanced at each step.

Practice Set

Solve each equation showing the steps of the solution. Then check your answer.

a. $3n + 1 = 16$

b. $2x - 1 = 9$

c. $3y - 2 = 22$

d. $5m + 3 = 33$

e. $4w - 1 = 35$

f. $7a + 4 = 25$

Written Practice *Strengthening Concepts*

1. (18) **Evaluate** The average of three numbers is 20. If the greatest is 28 and the least is 15, what is the third number?

* **2.** (95) A map is drawn to the scale of 1 inch = 10 miles. How many miles apart are two points that are $2\frac{1}{2}$ inches apart on the map?

$$\frac{2\frac{1}{2}\text{ in.}}{1} \times \frac{10\text{ mi}}{1\text{ in.}}$$

3. (29) What number is one fourth of 360?

4. (75) **Connect** What percent of a quarter is a nickel?

5. (58, 74) **Analyze** Anita places a set of number cards 1 through 30 in a bag. She draws out a card. What is the probability that the number will have the digit 1 in it? Express the probability as a fraction and as a decimal.

Solve and check:

* **6.** (106) $8x + 1 = 25$

* **7.** (106) $3w - 5 = 25$

* **8.** (100, 104) Calculate mentally:

a. $-15 + +20$

b. $-15 - +20$

c. $(-3) + (-2) - (-1)$

* **9.** (102) A sign in the elevator says that the maximum load is 4000 pounds. How many tons is 4000 pounds?

10. (78) **Connect** One gallon minus one quart equals how many pints?

11. (88) The ratio of kangaroos to koalas was 9 to 5. If there were 414 kangaroos, how many koalas were there?

Connect Complete the table to answer problems **12–14.**

	Fraction	Decimal	Percent
12. (99)	$\frac{1}{8}$	**a.**	**b.**
13. (99)	**a.**	1.8	**b.**
14. (99)	**a.**	**b.**	3%

15. (63) $8\frac{1}{3} - 3\frac{1}{2}$

16. (68) $2\frac{1}{2} \div 100$

17. (49) $0.014 \div 0.5$

18. (92) Write the standard notation for the following:

$$(6 \times 10^4) + (9 \times 10^2) + (7 \times 10^0)$$

19. (73) **Evaluate** The prime factorization of one hundred is $2^2 \cdot 5^2$. The prime factorization of one thousand is $2^3 \cdot 5^3$. Write the prime factorization of one million using exponents.

20. (63) A 1-foot ruler broke into two pieces so that one piece was $5\frac{1}{4}$ inches long. How long was the other piece?

* **21.** (92) $6 + 3^2(5 - \sqrt{4})$

22. (82) **Connect** If each small block has a volume of one cubic centimeter, what is the volume of this rectangular prism?

23. (75) Three inches is what percent of a foot?

24. (31) Use the figure on the right to answer **a–c.**

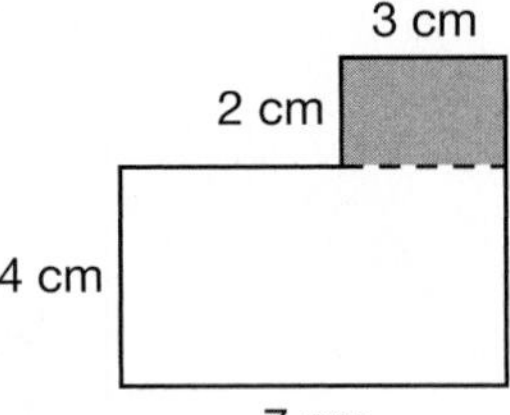

a. What is the area of the shaded rectangle?

b. What is the area of the unshaded rectangle?

c. What is the combined area of the two rectangles?

* **25.** (103) **Analyze** What is the perimeter of the hexagon in problem 24?

26. (47) **Estimate** The diameter of each tire on Jan's bike is two feet. The circumference of each tire is closest to which of the following? (Use 3.14 for π.)

A 6 ft **B** 6 ft 3 in. **C** 6 ft 8 in. **D** 7 ft

* **27.** (79) What is the area of this triangle?

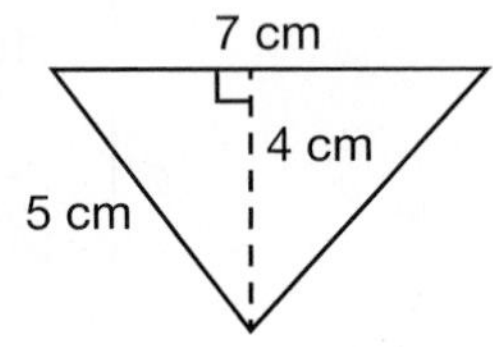

This table shows the number of miles Celina rode her bike each day during the week. Use this information to answer problems **28–30.**

28. (Inv. 5) If the data were rearranged in order of distance (with 3 miles listed first and 10 miles listed last), then which distance would be in the middle of the list?

29. (18) What was the average number of miles Celina rode each day?

* **30.** (13) **Formulate** Write a comparison question that relates to the table, and then answer the question.

Miles of Bike Riding for the Week

Day	Miles
Sunday	7
Monday	3
Tuesday	6
Wednesday	10
Thursday	5
Friday	4
Saturday	7

LESSON 107

Area of Complex Shapes

Power Up

Building Power

facts Power Up M

mental math

a. **Number Sense:** $100 \cdot 100$

b. **Number Sense:** $376 - 150$

c. **Percent:** 10% of 200

d. **Calculation:** \$12.89 + \$9.99

e. **Decimals:** $6.0 \div 100$

f. **Number Sense:** $\frac{360}{60}$

g. **Measurement:** How many pints are in a gallon?

h. **Calculation:** 10×6, $+ 4$, $\sqrt{\ }$, $\times 3$, $+ 1$, $\sqrt{\ }$, $\times 7$, $+ 1$, $\sqrt{\ }$

problem solving Laura has nickels, dimes, and quarters in her pocket. She has half as many dimes as nickels and half as many quarters as dimes. If Laura has four dimes, then how much money does she have in her pocket?

New Concept

Increasing Knowledge

In Lesson 103 we found the perimeter of complex shapes. In this lesson we will practice finding the area of complex shapes. One way to find the area of a complex shape is to divide the shape into two or more parts, find the area of each part, and then add the areas. Think of how the shape below could be divided into two rectangles.

Example

Find the area of this figure.

Thinking Skill

Classify

Based on the number of sides and their lengths, what is the geometric name of this figure?

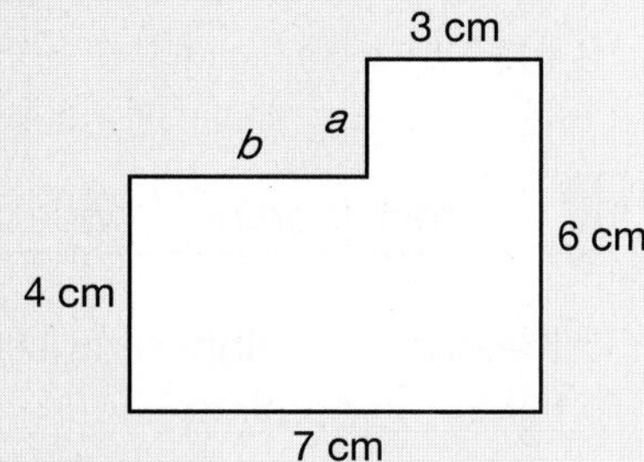

Solution

We will show two ways to divide this shape into two rectangles. We use the skills we learned in Lesson 103 to find that side *a* is 2 cm and side *b* is 4 cm. We extend side *b* with a dashed line segment to divide the figure into two rectangles.

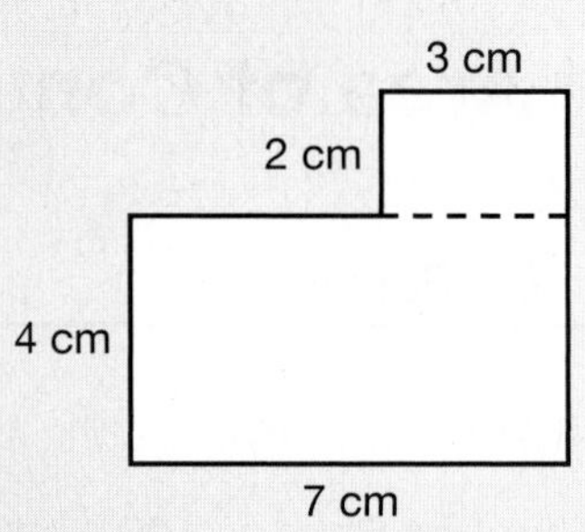

The length and width of the smaller rectangle are 3 cm and 2 cm, so its area is 6 cm^2. The larger rectangle is 7 cm by 4 cm, so its area is 28 cm^2. We find the combined area of the two rectangles by adding.

$$6 \text{ cm}^2 + 28 \text{ cm}^2 = \mathbf{34 \text{ cm}^2}$$

A second way to divide the figure into two rectangles is to extend side *a.*

Thinking Skill

Define

What does it mean when we say that the shape of a figure is *complex*?

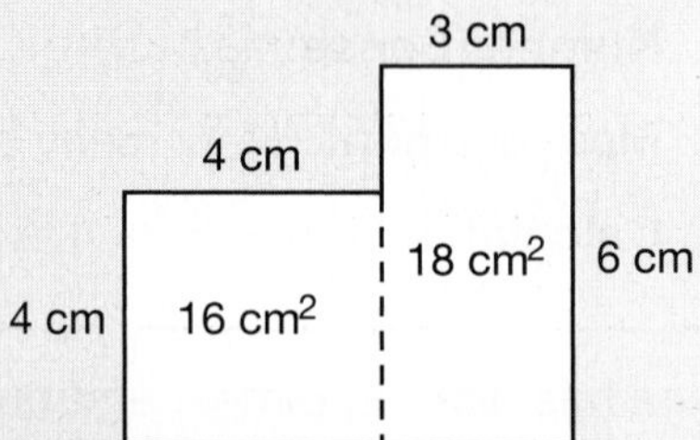

Extending side *a* forms a 4-cm by 4-cm rectangle and a 3-cm by 6-cm rectangle. Again we find the combined area of the two rectangles by adding.

$$16 \text{ cm}^2 + 18 \text{ cm}^2 = 34 \text{ cm}^2$$

Either way we divide the figure, we find that its area is 34 cm^2.

Practice Set

a. *Model* Draw two ways to divide this figure into two rectangles. Then find the area of the figure each way.

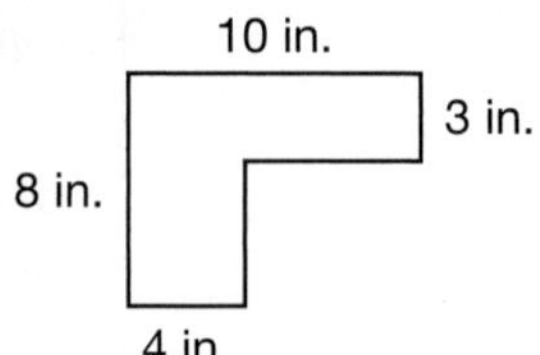

b. This trapezoid can be divided into a rectangle and a triangle. Find the area of the trapezoid.

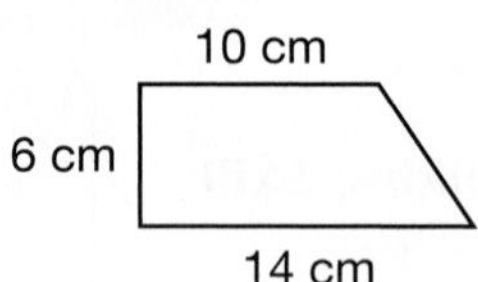

Written Practice *Strengthening Concepts*

1. (49) If the divisor is eight tenths and the dividend is forty-eight hundredths, what is the quotient?

2. (95) The plans for the clubhouse were drawn so that 1 inch equals 2 feet. In the plans the clubhouse was 4 inches tall. The actual clubhouse will be how tall?

3. (23) If 600 roses and 800 tulips were sold, what was the ratio of tulips sold to roses sold?

4. (8, 60) **Conclude** What percent of the perimeter of a regular pentagon is the length of one side?

*** 5.** (102) **Analyze** The mass of a dollar bill is about one gram. A gram is what fraction of a kilogram?

*** 6.** (100, 104) Calculate mentally:

a. $+15 + -10$

b. $-15 - -10$

c. $(+3) + (-5) - (-2) - (+4)$

7. (92) $10^3 - (10^2 - \sqrt{100}) - 10^3 \div 100$

*** 8.** (85, 105) **Analyze** Complete this proportion: $\frac{6}{u} = \frac{8}{1.2}$

Connect Complete the table to answer problems **9–11.**

	Fraction	Decimal	Percent
9. (99)	$1\frac{1}{10}$	**a.**	**b.**
10. (99)	**a.**	0.45	**b.**
11. (99)	**a.**	**b.**	80%

12. (61) $5\frac{3}{8} + 4\frac{1}{4} + 3\frac{1}{2}$

13. (72) $\frac{8}{3} \cdot \frac{5}{12} \cdot \frac{9}{10}$

14. (38) $64.8 + 8.42 + 24$

15. (93, 98) **Conclude** If one acute angle of a right triangle measures 55°, then what is the measure of the other acute angle?

16. (78) How many ounces is one half of a pint?

*** 17.** (106) Solve and check: $3m + 8 = 44$

18. (92) Write one hundred ten million in expanded notation using exponents.

19. (20) What is the greatest common factor of 30 and 45?

*** 20.** (29, 31) A square with sides 1 inch long is divided into $\frac{1}{2}$-by-$\frac{1}{4}$-inch rectangles.

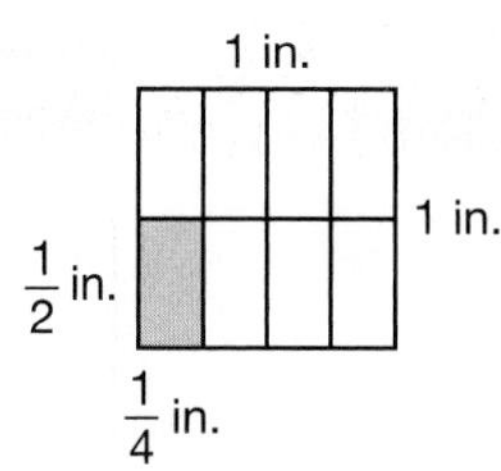

a. What is the area of each $\frac{1}{2}$-by-$\frac{1}{4}$-in. rectangle?

b. What fraction of the square is shaded?

21. (82) How many blocks that are 1 foot long on each edge would be needed to fill a cubical box with edges 1 yard long?

22. (49) $0.3n = \$6.39$

* **23.** What is the perimeter of this hexagon?
(103)

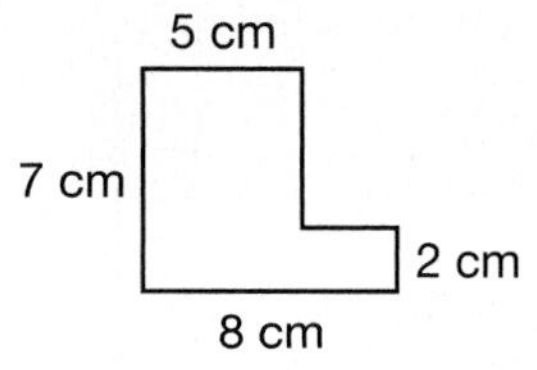

* **24.** Divide the hexagon at right into two rectangles. What is the combined area of the two rectangles?
(107)

* **25.** This trapezoid has been divided into two triangles. Find the area of the trapezoid by adding the areas of the two triangles.
(107)

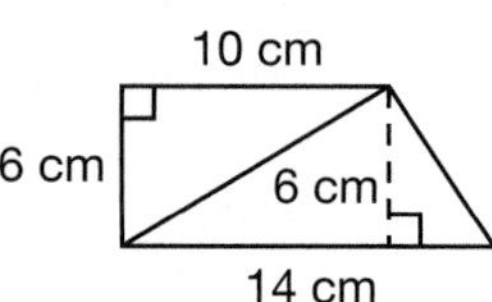

The table shows the age of the first nine American presidents at the time they were inaugurated. Use the information for problems **26–27.**

President	Age in Years at Inauguration
George Washington	57
John Adams	61
Thomas Jefferson	57
James Madison	57
James Monroe	58
John Quincy Adams	57
Andrew Jackson	61
Martin Van Buren	54
William Henry Harrison	68

26. **a.** What age appears most frequently?
(Inv. 5)

b. When the ages are arranged in order from least to greatest, what age appears in the middle?

c. **Estimate** Find the average age at inauguration of the first nine presidents. Round your answer to the nearest whole year.

d. **Connect** Name the mathematical term for the answers to **a, b,** and **c.**

* **27.** **Represent** Choose an appropriate graph and display the data in the table.
(Inv. 5)

* **28.**
(102)

$$\begin{array}{r} 12\text{ lb } 3\text{ oz} \\ -\ 8\text{ lb } 7\text{ oz} \\ \hline \end{array}$$

29. **Model** Use a compass to draw a circle that has a diameter of 10 cm.
(47)

a. What is the radius of the circle?

b. Calculate the circumference of the circle. (Use 3.14 for π.)

30. $\frac{10\text{ gallons}}{1} \times \frac{31.5\text{ miles}}{1\text{ gallon}}$
(95)

LESSON 108

• Transformations

Power Up

Building Power

facts Power Up M

mental math

a. **Number Sense:** $70 \cdot 70$

b. **Number Sense:** $296 - 150$

c. **Percent:** 25% of $20

d. **Calculation:** $8.23 + $8.99

e. **Number Sense:** $75 \div 100$

f. **Number Sense:** $\frac{800}{40}$

g. **Measurement:** Which is greater 2 liters or 3000 milliliters?

h. **Calculation:** 8×8, $- 4$, $\div 2$, $+ 5$, $\div 5$, $\times 8$, $- 1$, $\div 5$, $\times 2$, $- 1$, $\div 3$

problem solving

If two people shake hands, there is one handshake. If three people shake hands, there are three handshakes. From this table can you predict the number of handshakes with 6 people? Draw a diagram or act it out to confirm your prediction.

Number in Group	Number of Handshakes
2	1
3	3
4	6
5	10
6	

New Concept

Increasing Knowledge

Math Language
Two figures are **congruent** if one figure has the same shape and size as the other figure.

One way to determine whether two figures are congruent is to position one figure "on top of" the other. The two triangles below are congruent. As we will see below, triangle *ABC* can be positioned "on top of" triangle *XYZ,* illustrating that it is congruent to triangle *XYZ.*

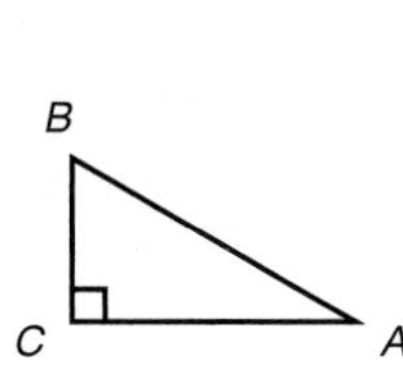

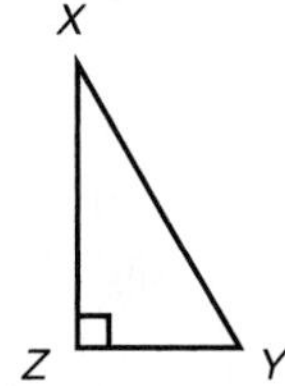

To position triangle *ABC* on triangle *XYZ,* we make three different kinds of moves. First, we **rotate** (turn) triangle *ABC* 90° counterclockwise.

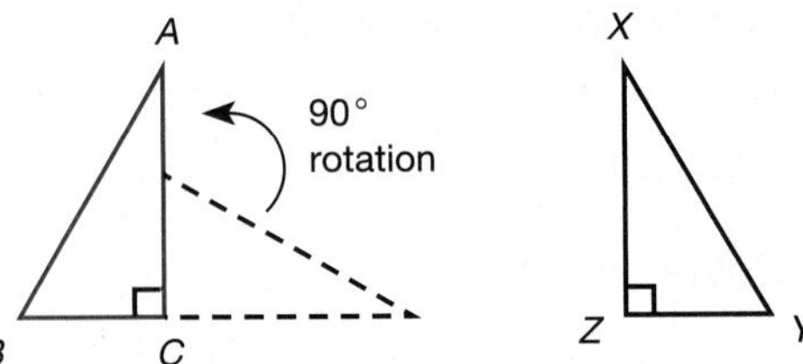

Second, we **translate** (slide) triangle *ABC* to the right so that side *AC* aligns with side *XZ.*

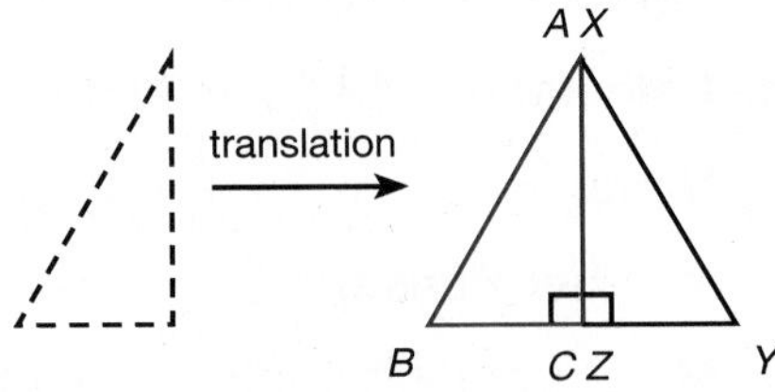

Third, we **reflect** (flip) triangle *ABC* so that angle *B* is positioned on top of angle *Y.*

Thinking Skill

Generalize

If a triangle is rotated, translated, or reflected, will the resulting triangle be congruent to the original triangle? Explain.

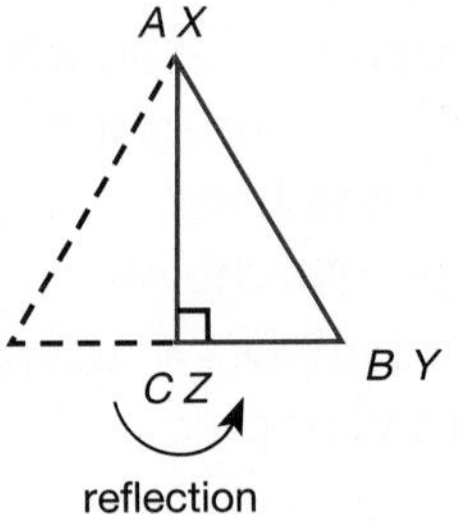

The three different kinds of moves we made are called **transformations.** We list them in the following table:

Transformations

Name	Movement
Rotation	turning a figure about a certain point
Translation	sliding a figure in one direction without turning the figure
Reflection	reflecting a figure as in a mirror or "flipping" a figure over a certain line

Activity

Transformations

Materials needed:

- scissors
- pencil and paper

Follow these steps to cut out a pair of congruent triangles with a partner or small group.

Step 1: Fold a piece of paper in half.

Step 2: Draw a triangle on the folded paper.

Step 3: While the paper is folded, cut out the triangle so that two triangles are cut out at the same time.

Have one partner (or group member) place the two triangles on a desk or table so that the triangles are apart and in different orientations. Let the other partner (or group member) move one of the triangles until it is positioned on top of the other triangle. The moves permitted are rotation, translation, and reflection. Take the moves one at a time and describe them as you go. After successfully aligning the triangles, switch roles and repeat the procedure. Allow each student one or two opportunities to perform and describe a transformation.

Practice Set

Conclude For problems **a–e,** what transformation(s) could be used to position triangle I on triangle II? For exercise **f** triangle *ABC* is reflected across the *y*-axis. Write the coordinates of the vertices of $\triangle ABC$ and the coordinates of the vertices of its reflection $\triangle A'B'C'$ (Read, "*A* prime, *B* prime, *C* prime.") Explain all answers.

a.

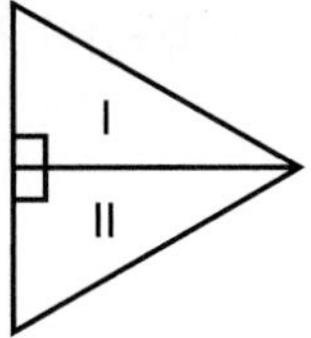

b.

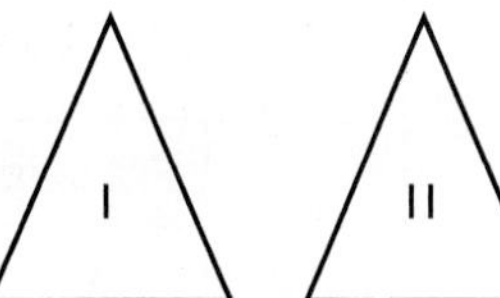

c.

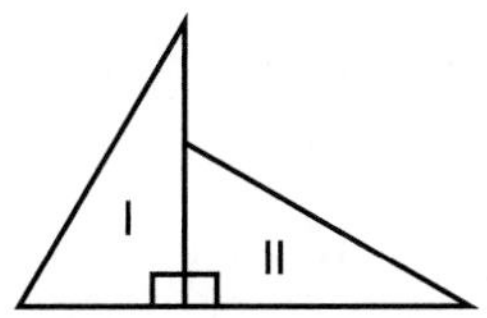

d.

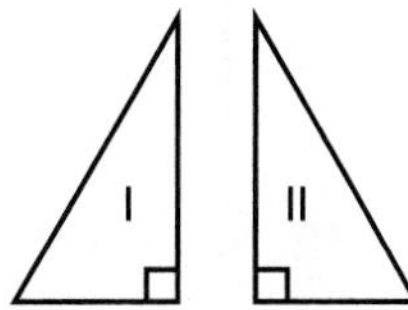

e.

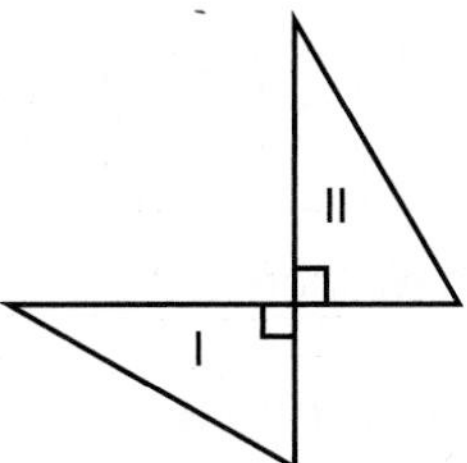

f.

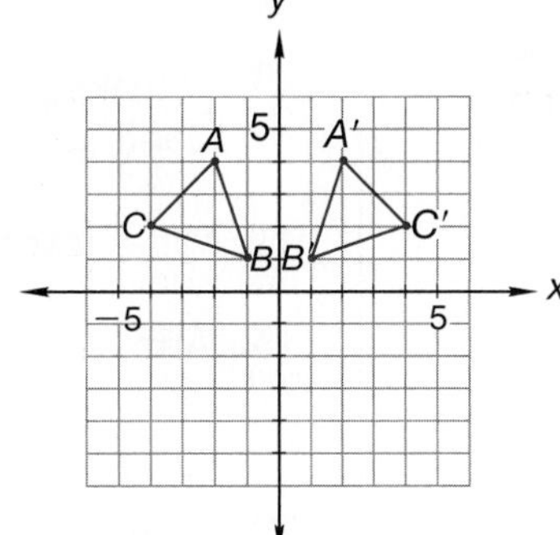

Written Practice

Strengthening Concepts

1. (10) What is the sum of the first five positive even numbers?

2. (88) **Analyze** The team's win-loss ratio is 4 to 3. If the team has won 12 games, how many games has the team lost?

*** 3.** (105) **Justify** Five students were absent today. The teacher reported that 80% of the students were present. Find the number of students who were present and justify your answer.

4. (86) **Estimate** Kaliska joined the band and got a new drum. Its diameter is 12 inches. What is the area of the top of the drum? Round your answer to the nearest square inch. (Use 3.14 for π.)

5. (77) Three eighths of the 48 band members played woodwinds. How many woodwind players were in the band?

6. (30) What is the least common multiple (LCM) of 6, 8, and 12?

*** 7.** (108) **Verify** Triangles I and II are congruent. Describe the transformations that would position triangle I on triangle II.

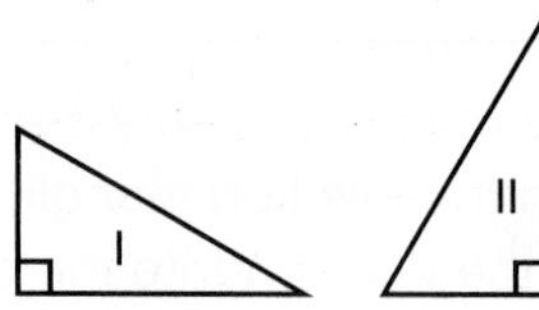

*** 8.** (85, 105) Complete this proportion: $\frac{0.7}{20} = \frac{n}{100}$

Connect Complete the table to answer problems **9–11.**

	Fraction	Decimal	Percent
9. (99)	$1\frac{2}{5}$	**a.**	**b.**
10. (99)	**a.**	0.24	**b.**
11. (99)	**a.**	**b.**	35%

12. (63) $4\frac{3}{4} + \left(2\frac{1}{4} - \frac{7}{8}\right)$

13. (38) $1\frac{1}{5} \div \left(2 \div 1\frac{2}{3}\right)$

14. (38) $6.2 + (9 - 2.79)$

*** 15.** (104) $-3 + +7 + -8 - -1$

16. (41) **Estimate** Find 6% of $2.89. Round the product to the nearest cent.

17. (95) What fraction of a meter is a millimeter?

18. (44) Arrange these numbers in order from least to greatest:

0.3, 0.31, 0.305

19. (82) If each edge of a cube is 10 centimeters long, then its volume is how many cubic centimeters?

20. (92) $2^5 - 5^2 + \sqrt{25} \times 2$

*** 21.** (106) Solve and check: $8a - 4 = 60$

22. (28, Inv. 3) **Conclude** Acute angle a is one third of a right angle. What is the measure of angle a?

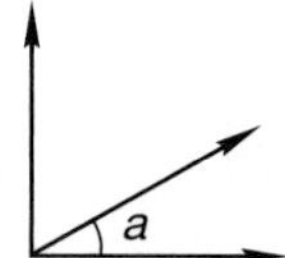

Refer to the figure at right to answer problems **23** and **24.** Dimensions are in millimeters.

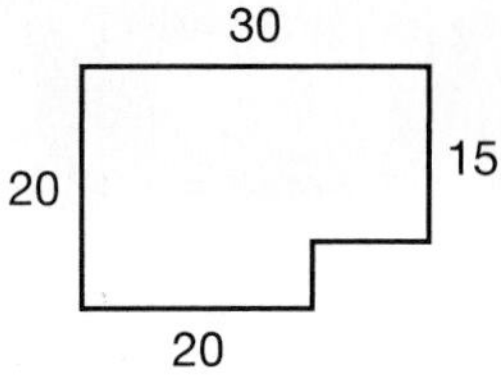

* **23.** What is the perimeter of this polygon?
(103)

* **24.** What is the area of this polygon?
(107)

25. **Estimate** A pint of water weighs about one pound. About how much does a two-gallon bucket of water weigh? (Disregard the weight of the bucket.)
(78)

* **26.** The parallel sides of this trapezoid are 10 mm apart. The trapezoid is divided into two triangles. What is the area of the trapezoid?
(107)

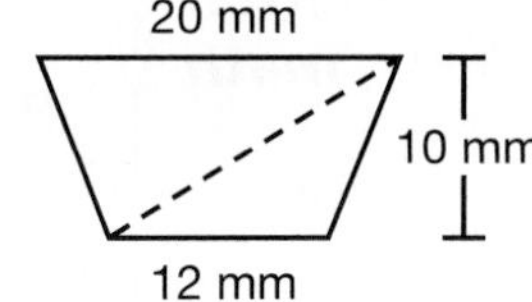

27. The cubic container shown can contain one liter of water. One liter is how many milliliters?
(78)

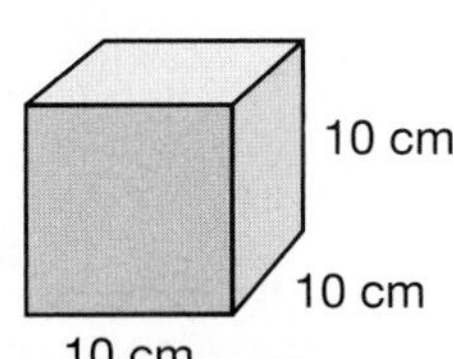

* **28.** **Analyze** A bag contains 6 red marbles and 4 blue marbles. If Delia draws one marble from the bag and then draws another marble without replacing the first, what is the probability that both marbles will be red?
(Inv. 10)

29. One and one half kilometers is how many meters?
(95)

* **30.** **Model** On a coordinate plane draw triangle *RST* with these vertices: *R* (−1, 4), *S* (−3, 1), *T* (−1, 1). Then draw its reflection across the *y*-axis. Name the reflection $\triangle R'S'T'$. What are the coordinates of the vertices of $\triangle R'S'T'$?
(Inv. 7, 108)

LESSON 109

• Corresponding Parts
• Similar Figures

Power Up

Building Power

facts Power Up N

mental math

a. Number Sense: $400 \cdot 30$

b. Number Sense: $687 + 250$

c. Percent: 10% of \$20

d. Calculation: \$10.00 − \$6.87

e. Decimals: 0.5×100

f. Number Sense: 70×300

g. Measurement: How many cups are in a quart?

h. Calculation: Square 7, + 1, ÷ 2, × 3, − 3, ÷ 8, $\sqrt{}$

problem solving One state uses a license plate that contains two letters followed by four digits. How many different license plates are possible if all of the letters and numbers are used?

New Concepts

Increasing Knowledge

corresponding parts

The two triangles below are congruent. Each triangle has three angles and three sides. The angles and sides of triangle *ABC* **correspond** to the angles and sides of triangle *XYZ*.

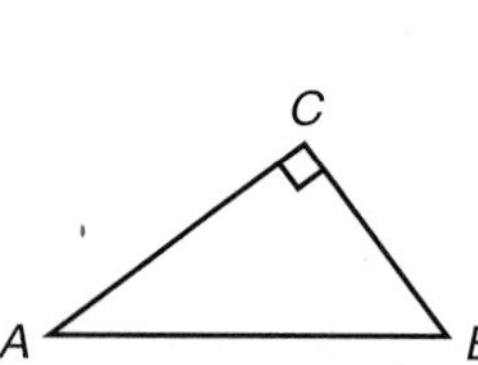

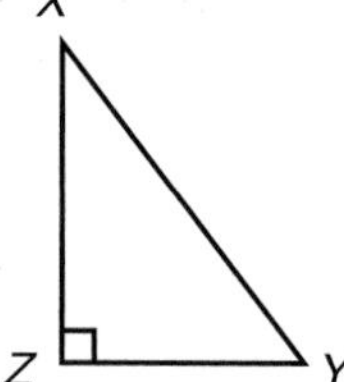

By rotating, translating, and reflecting triangle *ABC*, we could position it on top of triangle *XYZ*. Then their **corresponding parts** would be in the same place.

$\angle A$ corresponds to $\angle X$.

$\angle B$ corresponds to $\angle Y$.

$\angle C$ corresponds to $\angle Z$.

$\overline{AB}$ corresponds to $\overline{XY}$.

$\overline{BC}$ corresponds to $\overline{YZ}$.

$\overline{AC}$ corresponds to $\overline{XZ}$.

If two figures are congruent, their corresponding parts are congruent. So the measures of the corresponding parts are equal.

Example 1

These triangles are congruent. What is the perimeter of each?

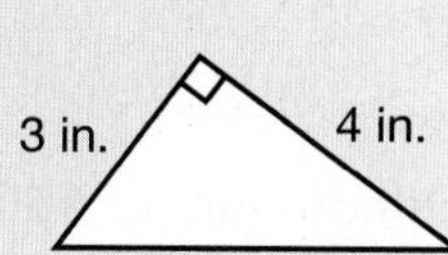

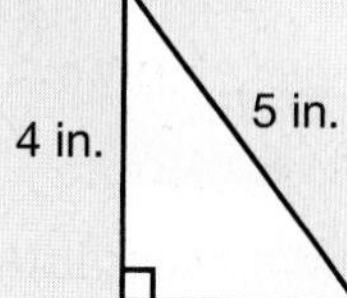

Solution

We will rotate the triangle on the left so that the corresponding parts are easier to see.

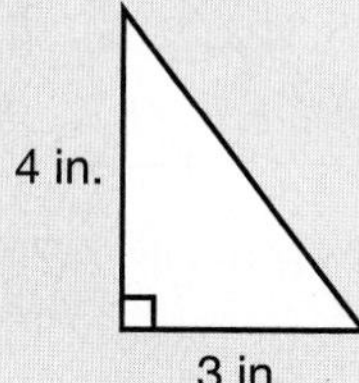

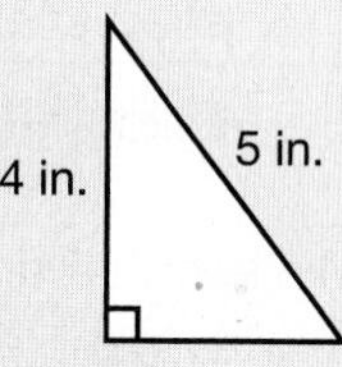

Now we can more easily see that the unmarked side on the left-hand triangle corresponds to the 5-inch side on the right-hand triangle. Since the triangles are congruent, the measures of the corresponding parts are equal. So each triangle has sides that measure 3 inches, 4 inches, and 5 inches. Adding, we find that the perimeter of each triangle is **12 inches.**

$$3 \text{ in.} + 4 \text{ in.} + 5 \text{ in.} = 12 \text{ in.}$$

similar figures

Figures that have the same shape but are not necessarily the same size are **similar.** Three of these four triangles are similar:

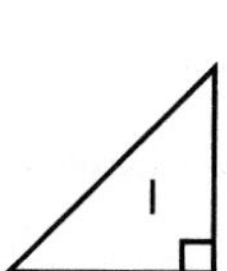

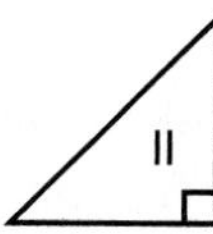

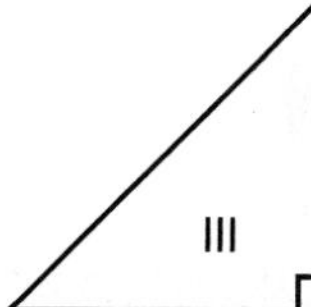

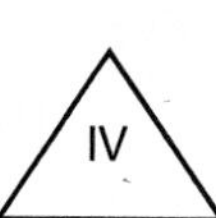

Triangles I and II are similar. They are also congruent. Remember, congruent figures have the same shape *and* size. Triangle III is similar to triangles I and II. It has the same shape but not the same size as triangles I and II. Notice that the corresponding angles of similar figures have the same measure. Triangle IV is not similar to the other triangles.

Analyze Can we reduce or enlarge Triangle IV to make it match the other triangles in the diagram? Explain.

Example 2

The two triangles below are similar. What is the measure of angle *A*?

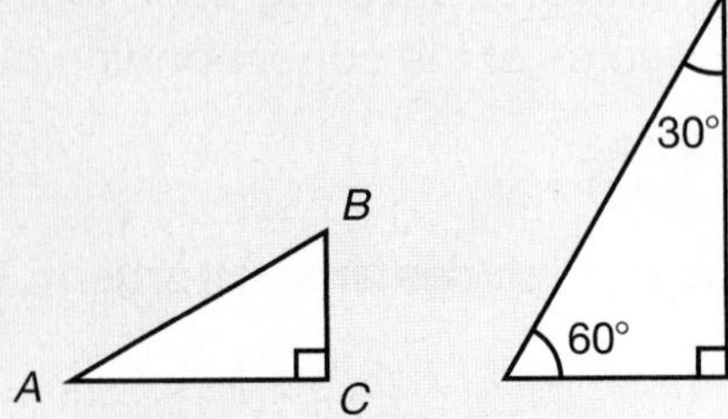

Solution

We will rotate and reflect triangle *ABC* so that the corresponding angles are easier to see.

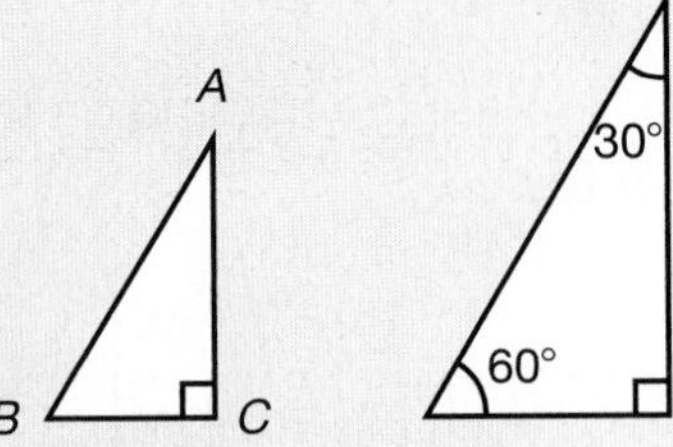

We see that angle *A* in triangle *ABC* corresponds to the 30° angle in the similar triangle. Since corresponding angles of similar triangles have the same measure, the measure of angle *A* is **30°.**

Here are two important facts about similar polygons.

1. The corresponding angles of similar polygons are congruent.
2. The corresponding sides of similar polygons are proportional.

The first fact means that even though the sides of similar polygons might not have matching lengths, the corresponding angles do match. The second fact means that similar figures are related by a scale factor. The scale factor is a number. Multiplying the side length of a polygon by the scale factor gives the side length of the corresponding side of the similar polygon.

Example 3

The two rectangles below are similar. What is the ratio of corresponding sides? By what scale factor is rectangle *ABCD* larger than rectangle *EFGH*?

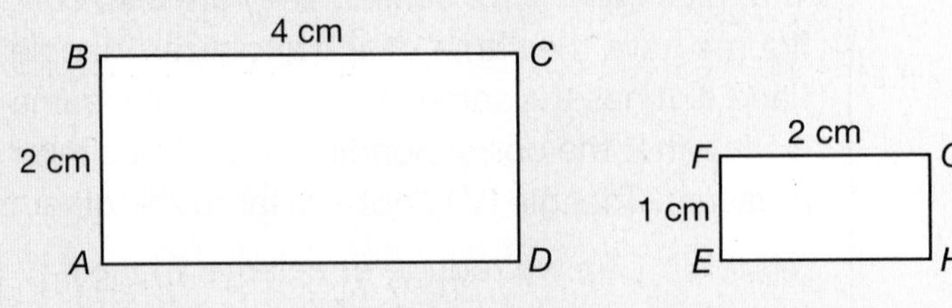

Solution

First, we find the ratios of corresponding sides.

Side *AB*: Side *EF* = $\frac{2}{1}$

Side BC: Side $FG = \frac{4}{2} = \frac{2}{1}$

In all similar polygons, such as these two rectangles, the ratios of corresponding sides are **equal**.

The sides of rectangle *ABCD* are 2 times larger than the sides of rectangle *EFGH.* So rectangle *ABCD* is larger than rectangle *EFGH* by a scale factor of **2**.

Practice Set

a. *Verify* "All squares are similar." True or false?

b. *Verify* "All similar triangles are congruent." True or false?

c. *Verify* "If two polygons are similar, then their corresponding angles are equal in measure." True or false?

d. These two triangles are congruent. Which side of triangle *PQR* is the same length as $\overline{AB}$?

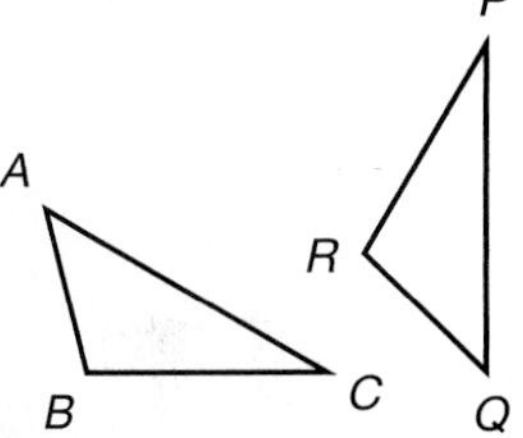

e. *Classify* Which of these two triangles appear to be similar?

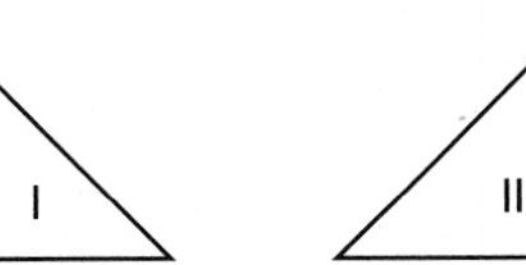

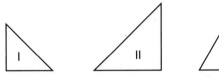

f. These two pentagons are similar. The scale factor for corresponding sides is 3. How long is segment *AE*? How long is segment *IJ*?

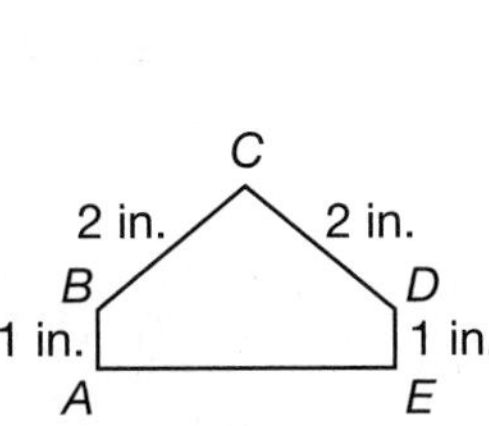

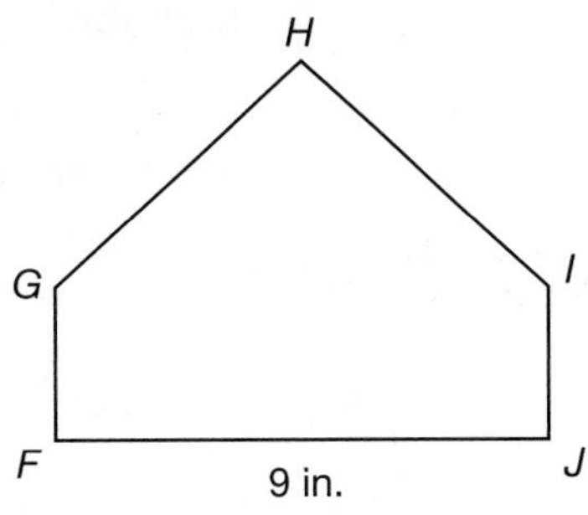

Written Practice

Strengthening Concepts

1. The first three prime numbers are 2, 3, and 5. Their product is 30. What is the product of the next three prime numbers?
(19)

2. On the map 2 cm equals 1 km. What is the actual length of a street that is 10 cm long on the map?
(95)

3. Between 8 p.m. and 9 p.m. the station broadcasts 8 minutes of commercials. What was the ratio of commercial time to noncommercial time during that hour?
(23)

4. (93) **a.** *Classify* Which of the following triangles appears to have a right angle as one of its angles?

b. In which triangle do all three sides appear to be the same length?

A 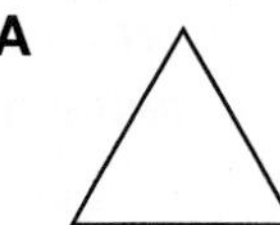 **B** 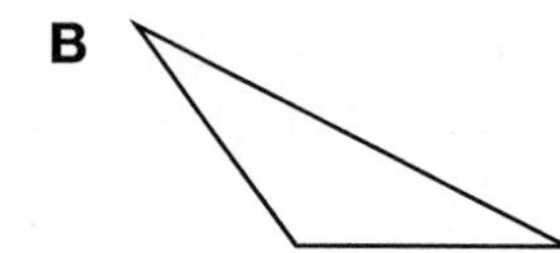 **C** 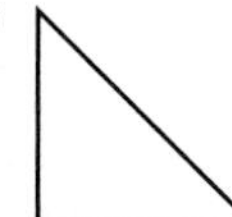

*** 5.** (28, 98) *Conclude* If the two acute angles of a right triangle are congruent, then what is the measure of each acute angle?

6. (58) Ms. Hernandez is assigning each student in her class one of the fifty U.S. states on which to write a report. What is the probability that Manuela will be assigned one of the 5 states that has coastline on the Pacific Ocean? Express the probability ratio as a fraction and as a decimal.

Solve:

*** 7.** (106) $7w - 3 = 60$

*** 8.** (85, 105) $\frac{8}{n} = \frac{4}{2.5}$

Connect Complete the table to answer problems **9–11.**

	Fraction	Decimal	Percent
9. (99)	$\frac{5}{8}$	**a.**	**b.**
10. (99)	**a.**	1.25	**b.**
11. (99)	**a.**	**b.**	70%

12. (58) **a.** If the spinner is spun once, what is the probability that it will stop on a number less than 4?

b. If the spinner is spun 100 times, how many times would it be expected to stop on a prime number?

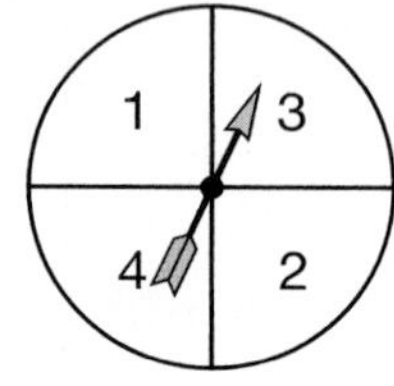

13. (95) Convert 200 centimeters to meters by completing this multiplication:

$$\frac{200 \text{ cm}}{1} \cdot \frac{1 \text{ m}}{100 \text{ cm}}$$

14. (38) $(6.2 + 9) - 2.79$

15. (92) $10^3 \div 10^2 - 10^1$

*** 16.** (96) *Represent* Create a function table for *x* and *y*. In your table, record four pairs of numbers that follow this rule:

y is twice *x*

17. (74) Write the fraction $\frac{2}{3}$ as a decimal number rounded to the hundredths place.

18. (82) The Zamoras rent a storage room that is 10 feet wide, 12 feet long, and 8 feet high. How many cube-shaped boxes 1 foot on each edge can the Zamoras store in the room?

* **19.** (109) These two rectangles are similar.

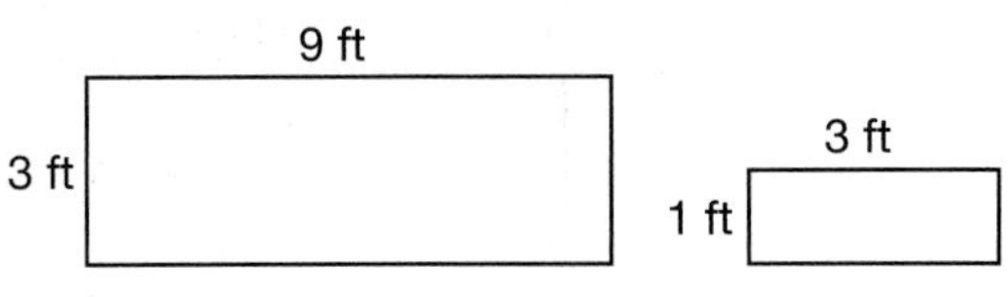

a. What is the scale factor from the smaller rectangle to the larger rectangle?

b. What is the scale factor from the larger rectangle to the smaller rectangle?

20. (49) $0.12m = \$4.20$

* **21.** (100) Calculate mentally:

a. $+7 + -8$ **b.** $-7 + +8$

c. $-7 - +8$ **d.** $-7 - -8$

Triangles I and II below are congruent. Refer to the triangles to answer problems **22** and **23.**

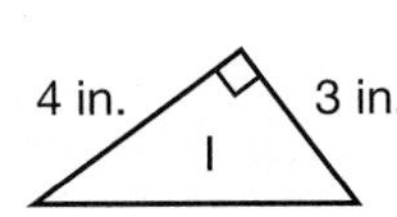

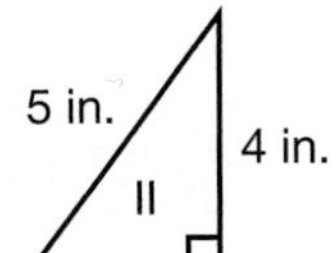

* **22.** (109) What is the area of each triangle?

* **23.** (108) **Conclude** Name the transformations that would position triangle I on triangle II.

* **24.** (107) This trapezoid has been divided into a rectangle and a triangle. What is the area of the trapezoid?

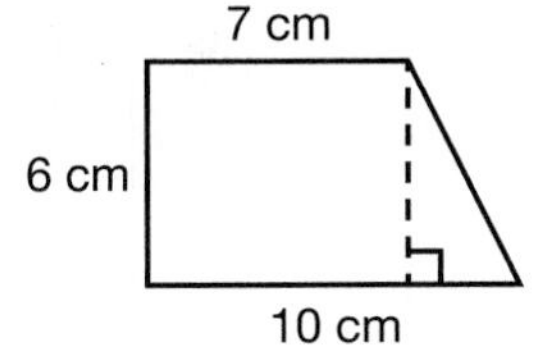

25. (47) **Estimate** A soup label must be long enough to wrap around a can. If the diameter of the can is 7 cm, then the label must be at least how long? Round up to the nearest whole number.

* **26.** (109) **Analyze** The triangles below are similar. What is the measure of angle A?

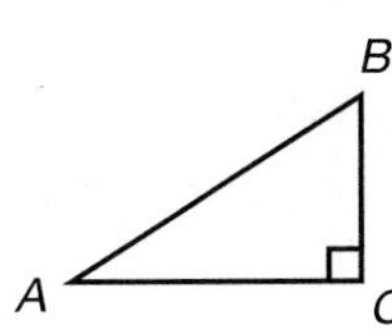

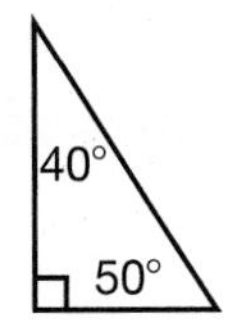

27. (88) The ratio of almonds to cashews in the mix was 9 to 2. Horace counted 36 cashews in all. How many almonds were there?

28. (7, 50) Write the length of the segment below in

a. millimeters.

b. centimeters.

cm 1 2 3 4

29. (68) $6\frac{2}{3} \div 100$

30. (92) Compare: $\left(\frac{1}{10}\right)^2 \bigcirc 0.01$

Early Finishers
Real-World Application

Issam wants to make a rock garden in his backyard. He bought $\frac{1}{5}$ ton of gravel and $\frac{1}{2}$ ton of rocks. Issam is not sure if the gravel and the rocks will fit on the trailer he rented. If the trailer can carry a maximum load of 1250 pounds, can Issam take all the gravel and rocks in one trip? Note: 1 ton = 2000 pounds. Support your answer.

LESSON 110

• Symmetry

Power Up

Building Power

facts Power Up M

mental math

a. **Number Sense:** 90 · 90

b. **Number Sense:** 726 − 250

c. **Percent:** 50% of $50

d. **Calculation:** $7.62 + $3.98

e. **Decimals:** 8 ÷ 100

f. **Number Sense:** $\frac{350}{50}$

g. **Geometry:** A circle has a diameter of 4 yd. What is the circumference of the circle?

h. **Calculation:** Square 10, − 1, ÷ 9, × 3, − 1, ÷ 4, × 7, + 4, ÷ 3

problem solving

Copy the problem and fill in the missing digits. Use only zeros or ones in the spaces.

```
       9_
 __)____
    99
    --
     __
     __
      _
```

New Concept

Increasing Knowledge

A figure has line **symmetry** if it can be divided in half so that the halves are mirror images of each other. We can observe symmetry in nature. For example, butterflies, leaves, and most types of fish are symmetrical. In many respects our bodies are also symmetrical. Manufactured items such as lamps, chairs, and kitchen sinks are sometimes designed with symmetry.

Two-dimensional figures can also be symmetrical. A two-dimensional figure is symmetrical if a line can divide the figure into two mirror images. Line *r* divides the triangle below into two mirror images. Thus the triangle is symmetrical, and line *r* is called a **line of symmetry.**

Example 1

This rectangle has how many lines of symmetry?

Solution

There are two ways to divide the rectangle into mirror images:

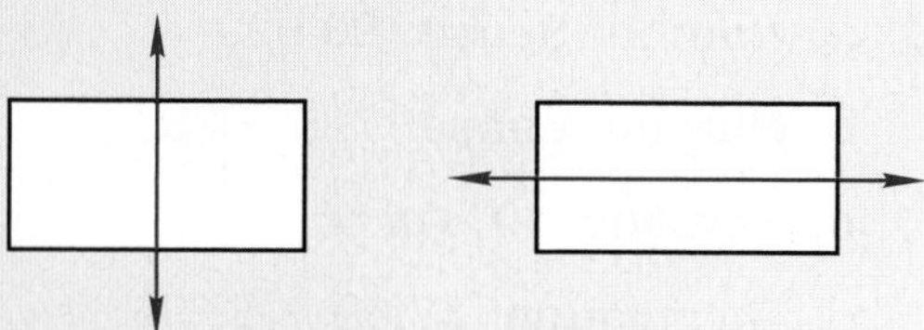

We see that the rectangle has **2 lines of symmetry.**

Thinking Skill

Analyze

How can we be sure that there are no more lines of symmetry?

Example 2

Which of these triangles does not appear to be symmetrical?

A

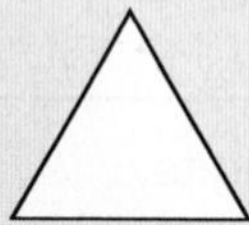

B

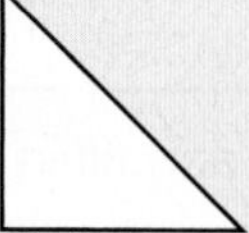

C

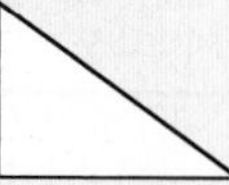

Solution

We check each triangle to see whether we can find a line of symmetry. In choice **A** all three sides of the triangle are the same length. We can find three lines of symmetry in the triangle.

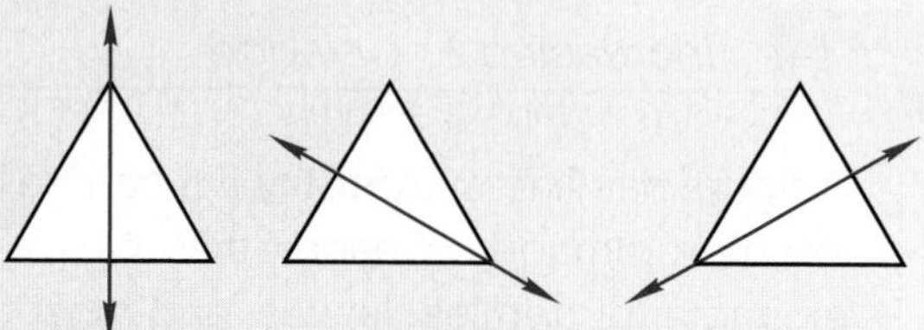

In choice **B** two sides of the triangle are the same length. The triangle in choice **B** has one line of symmetry.

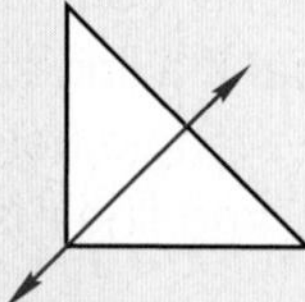

In choice **C** each side of the triangle is a different length. The triangle has no line of symmetry, so the triangle is not symmetrical. The answer is **C.**

A figure has **rotational symmetry** if the image of the figure re-appears in the same position as it is turned *less than* a full turn. For example the image of a square re-appears in the same position as it is turned 90°, 180°, and 270°.

original position 45° turn **90° turn** 150° turn **180° turn** 210° turn **270° turn**

Example 3

Which of these figures have rotational symmetry? Choose all correct answers.

A B C D

Solution

Turning your book might help you see which shapes have rotational symmetry. If the figures in choices A and B are rotated 180°, the images of the figures re-appear, so **choices A and B have rotational symmetry.** The figures in choices C and D re-appear only after a full turn.

Practice Set

a. *Model* Draw four squares. Then draw a different line of symmetry for each square.

b. *Classify* All but one of these letters can be drawn to have a line of symmetry. Which of these letters does not have a line of symmetry?

A **B** **C** **D** **E** **F**

c. Which two of these letters have rotational symmetry? (*Hint:* Rotating your book might help you find the answer.)

L **M** **N** **O** **P** **Q**

Written Practice

Strengthening Concepts

1. (2) When the greatest four-digit number is divided by the greatest two-digit number, what is the quotient?

2. (88) The ratio of the length to the width of the Alamo is about 5 to 3. If the width of the Alamo is approximately 63 ft, about how long is the Alamo?

3. (82) A box of crackers in the shape of a square prism had a length, width, and height of 4 inches, 4 inches, and 10 inches respectively. How many cubic inches was the volume of the box?

Crackers
10 in.
4 in.
4 in.

4. (90) **Analyze** A full turn is 360°. How many degrees is $\frac{1}{6}$ of a turn?

Refer to these triangles to answer problems **5–7:**

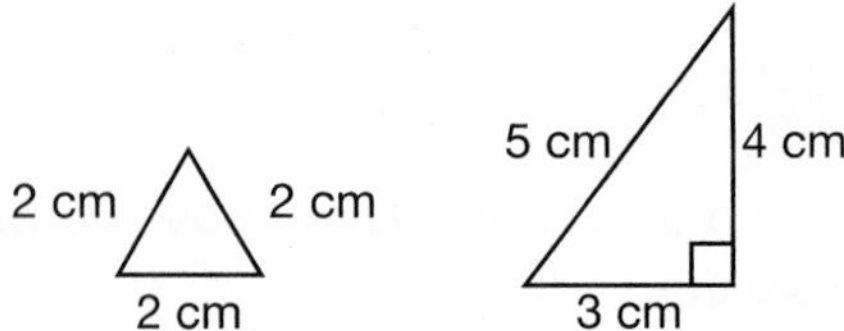

*** 5.** (110) **Explain** Does the equilateral triangle have rotational and line symmetry? Use words or diagrams to explain how you know.

*** 6.** (93, 109) Sketch a triangle similar to the equilateral triangle. Make the scale factor from the equilateral triangle to your sketch 3. What is the perimeter of the triangle you sketched?

7. (93) What is the area of the right triangle?

*** 8.** (105) **Model** Draw a ratio box for this problem. Then solve the problem using a proportion.

Ms. Mendez is sorting her photographs. She notes that 12 of the photos are black and white and that 40% of the photos are color. How many photos does she have?

Connect Complete the table to answer problems **9–11.**

	Fraction	Decimal	Percent
9. (99)	$2\frac{3}{4}$	**a.**	**b.**
10. (99)	**a.**	1.1	**b.**
11. (99)	**a.**	**b.**	64%

12. (61) $24\frac{1}{6} + 23\frac{1}{3} + 22\frac{1}{2}$

13. (68) $\left(1\frac{1}{5} \div 2\right) \div 1\frac{2}{3}$

14. (38) $9 - (6.2 + 2.79)$

15. (49) $0.36m = \$63.00$

16. (51) Find 6.5% of $24.89 by multiplying 0.065 by $24.89. Round the product to the nearest cent.

17. (51) **Estimate** Round the quotient to the nearest thousandth:

$$0.065 \div 4$$

18. (73) Write the prime factorization of 1000 using exponents.

19. (109) **Verify** "All squares are similar." True or false?

20. $3^3 - 3^2 \div 3 - 3 \times 3$
(92)

*** 21.** What is the perimeter of this polygon?
(103)

*** 22.** What is the area of this polygon?
(107)

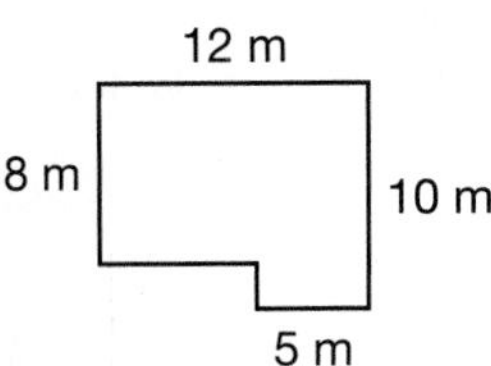

Triangles I and II are congruent. Refer to these triangles to answer problems **23** and **24.**

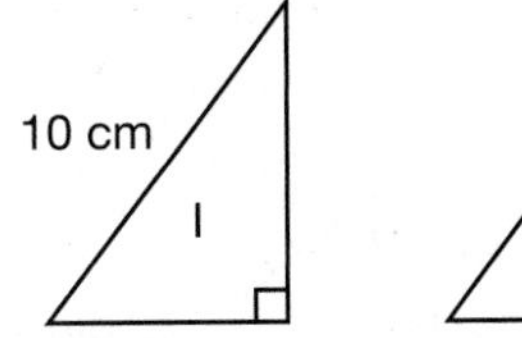

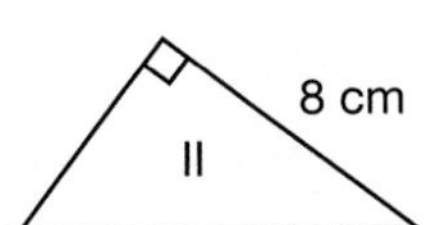

*** 23.** **Conclude** Name the transformations that would position triangle I on triangle II.
(108)

*** 24.** The perimeter of each triangle is 24 cm. What is the length of the shortest side of each triangle?
(8, 109)

25. **Estimate** The first Ferris wheel was built in 1893 for the world's fair in Chicago. The diameter of the Ferris wheel was 250 ft. Find the circumference of the original Ferris wheel to the nearest hundred feet.
(47)

26. Use a ruler to draw $\overline{AB}$ $1\frac{3}{4}$ inches long. Then draw a dot at the midpoint of $\overline{AB}$, and label the point *M*. How long is $\overline{AM}$?
(7)

27. **Model** Use a compass to draw a circle on a coordinate plane. Make the center of the circle the origin, and make the radius five units. At which two points does the circle cross the *x*-axis?
(27, Inv. 7)

28. What is the area of the circle in problem **27?** (Use 3.14 for π.)
(86)

*** 29.** $-3 + -4 - -5 - +7$
(104)

*** 30.** If Freddy tosses a coin four times, what is the probability that the coin will turn up heads, tails, heads, tails in that order?
(Inv. 10)

Focus on
Scale Factor: Scale Drawings and Models

Recall from Lesson 109 that the dimensions of similar figures are related by a scale factor, as shown below.

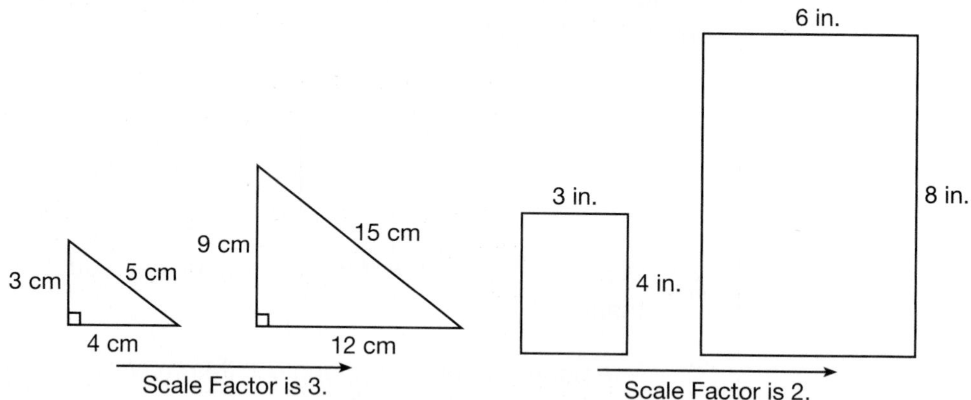

Similar figures are often used by manufacturers to design products and by architects to design buildings. Architects create **scale drawings** to guide the construction of a building. Sometimes, a **scale model** of the building is also constructed to show the appearance of the finished project.

Scale drawings, such as architectural plans, are two-dimensional representations of larger objects. Scale models, such as model cars and action figures, are three-dimensional representations of larger objects. In some cases, however, a scale drawing or model represents an object smaller than the model itself. For example, we might want to construct a large model of a bee in order to more easily portray its anatomy.

In scale drawings and models the **legend** gives the relationship between a unit of length in the drawing and the actual measurement that the unit represents. The drawing below shows the floorplan of Angela's studio apartment. The legend for this scale drawing is $\frac{1}{2}$ inch = 5 feet.

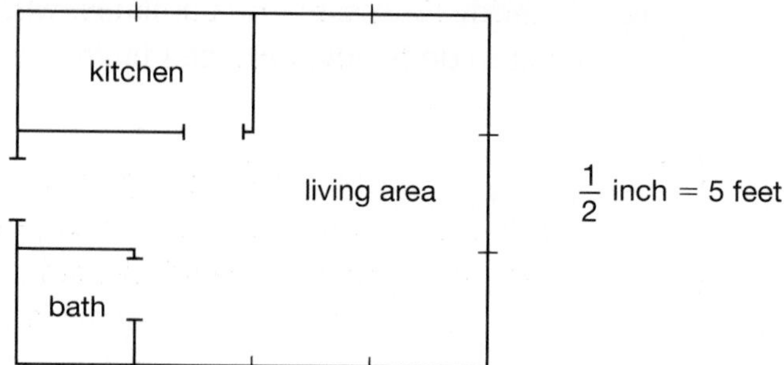

If we measure the scale drawing above, we find that it is 2 inches long and $1\frac{1}{2}$ inches wide. Using these measurements, we can determine the actual dimensions of Angela's apartment. On the next page we show some relationships that are based on the scale drawing's legend.

$$\begin{aligned} \tfrac{1}{2}\text{ inch} &= 5\text{ feet (given)} \\ 1\text{ inch} &= 10\text{ feet} \\ 1\tfrac{1}{2}\text{ inches} &= 15\text{ feet} \\ 2\text{ inches} &= 20\text{ feet} \end{aligned}$$

Since the scale drawing is 2 inches long by $1\frac{1}{2}$ inches wide, we find that Angela's apartment is 20 feet long by 15 feet wide.

1. **Connect** What are the actual length and width of Angela's kitchen?

2. In the scale drawing each doorway measures $\frac{1}{4}$ inch wide. Since $\frac{1}{4}$ inch is half of $\frac{1}{2}$ inch, what is the actual width of each doorway in Angela's apartment?

3. **Connect** A dollhouse was built as a scale model of an actual house using 1 inch to represent 1.5 feet. What are the dimensions of a room in the actual house if the corresponding dollhouse room measures 8 in. by 10 in.?

4. A scale model of an airplane is built using 1 inch to represent 2 feet. The wingspan of the model airplane is 24 inches. What is the wingspan of the actual airplane in feet?

The lengths of corresponding parts of scale drawings or models and the objects they represent are proportional. Since the relationships are proportional, we can use a ratio box to organize the numbers and a proportion to find the unknown.

Connect To answer problems **5–8,** use a ratio box and write a proportion. Then solve for the unknown measurement either by using cross products or by writing an equivalent ratio. Make one column of the ratio box for the model and the other column for the actual object.

5. A scale model of a sports car is 7 inches long. The car itself is 14 feet long. If the model is 3 inches wide, how wide is the actual car?

6. **Explain** For the sports car in problem **5,** suppose the actual height is 4 feet. What is the height of the model? How do you know your answer is correct?

7. **Analyze** The femur is the large bone that runs from the knee to the hip. In a scale drawing of a human skeleton the length of the femur measures 3 cm, and the full skeleton measures 12 cm. If the drawing represents a 6-ft-tall person, what is the actual length of the person's femur?

8. The humerus is the bone that runs from the elbow to the shoulder. Suppose the humerus of a 6-ft-tall person is 1 ft long. How long should the humerus be on the scale drawing of the skeleton in problem **7?**

9. A scale drawing of a room addition that measures 28 ft by 16 ft is shown below. The scale drawing measures 7 cm by 4 cm.

a. **Connect** Complete this legend for the scale drawing:

1 cm = ______ ft

b. **Estimate** What is the actual length and width of the bathroom, rounded to the nearest foot.

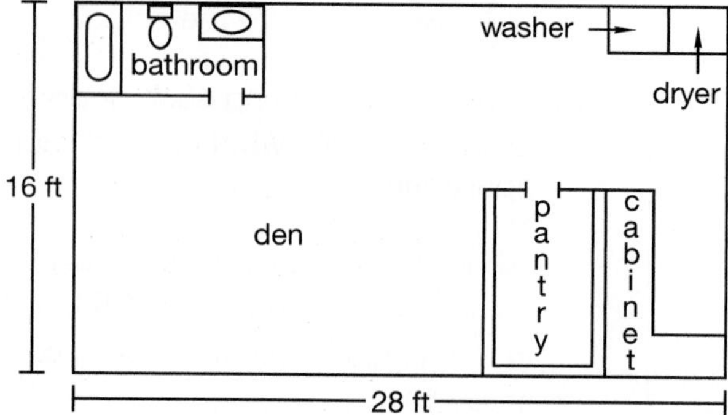

10. A natural history museum contains a 44-inch-long scale model of a *Stegosaurus* dinosaur. The actual length of the *Stegosaurus* was 22 feet. What should be the legend for the scale model of the dinosaur?

1 inch = ______ feet

Maps, blueprints, and models are called *renderings.* If a rendering is smaller than the actual object it represents, then the dimensions of the rendering are a fraction of the dimensions of the actual object. This fraction is called the **scale** of the rendering.

To determine the scale of a rendering, we form a fraction using corresponding dimensions and the same units for both. Then we reduce.

$$\text{scale} = \frac{\text{dimension of rendering}}{\text{dimension of object}}$$

In the case of the *Stegosaurus* in problem **10,** the corresponding lengths are 44 inches and 22 feet. Before reducing the fraction, we will convert 22 feet to 264 inches. Then we write a fraction, using the length of the model as the numerator and the dinosaur's actual length as the denominator.

$$\text{scale} = \frac{44 \text{ inches}}{22 \text{ feet}} = \frac{44 \text{ inches}}{264 \text{ inches}} = \frac{1}{6}$$

So the model is a $\frac{1}{6}$ scale model. The reciprocal of the scale is the **scale factor.** So the scale factor from the model to the actual *Stegosaurus* is 6. This means we can multiply any dimension of the model by 6 to determine the corresponding dimension of the actual object.

11. What is the scale of the model car in problem **5?** What is the scale factor?

12. A scale may be written as a ratio that uses a colon. For example, we can write the scale of the *Stegosaurus* model as 1:6. Suppose that a toy company makes action figures of sports stars using a scale of 1:10. How many inches tall will a figure of a 6-ft-8-in. basketball player be?

13. In a scale drawing of a wall mural, the scale factor is 6. If the scale drawing is 3 feet long by 1.5 feet wide, what are the dimensions of the actual mural?

extensions

a. Model Make a scale drawing of your bedroom's floor plan, where 1 in. = 3 ft. Include in your drawing the locations of doors and windows as well as major pieces of furniture. What is the scale factor you used?

b. Model Cut out and assemble the pieces from Activity 21 and 22 to make a scale model of the *Freedom 7* spacecraft. This spacecraft was piloted by Alan B. Shepard, the first American to go into space. With *Freedom 7* sitting atop a rocket, Shepard blasted off from Cape Canaveral, Florida, on May 5, 1961. Because he did not orbit (circle) the earth, the trip lasted only 15 minutes before splashdown in the Atlantic Ocean. Shepard was one of six astronauts to fly a Mercury spacecraft like *Freedom 7.* Each Mercury spacecraft could carry only one astronaut, because the rockets available in the early 1960s were not powerful enough to lift heavier loads.

Your completed *Freedom 7* model will have a scale of 1:24. After you have constructed the model, measure its length. Use this information to determine the length of an actual Mercury spacecraft.

c. Represent On a coordinate plane draw a square with vertices at (2, 2), (4, 2), (4, 4), and (2, 4). Then apply a scale factor of 2 to the square so that the dimensions of the square double but the point (3, 3) remains the center of the larger square. Draw the larger square on the same coordinate plane. What are the coordinates of its vertices?

d. Represent Using plastic straws, scissors, and string, make a scale model of a triangle whose sides are 3 ft, 4 ft, and 5 ft. For the model, let $\frac{3}{4}$ in. = 1 foot. What is the scale factor? How long are the sides of the model?

e. Model Draw these shapes on grid paper and measure the angles. Write the measure by the angle and then mark each angle as acute, obtuse, or right.

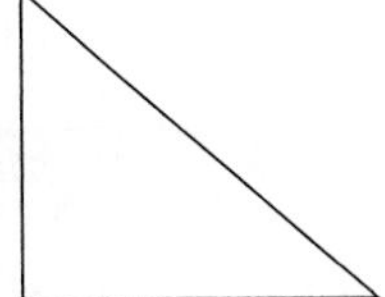

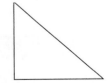

Now use the grid paper to draw each figure increased by a scale factor of 2. Then develop a mathematical argument proving or disproving that the angle classification of the images are the same as the given figures.

LESSON

111

• Applications Using Division

Power Up

Building Power

facts

Power Up N

mental math

a. **Fractional Parts:** $\frac{2}{3}$ of 24

b. **Calculation:** 7×35

c. **Percent:** 50% of $48

d. **Calculation:** $10.00 − $8.59

e. **Decimals:** 0.5×100

f. **Number Sense:** $\frac{1600}{400}$

g. **Geometry:** A rectangular solid is 10 ft. × 6 ft. × 4 ft. What is the volume of the solid?

h. **Calculation:** 8×8, -1, $\div 9$, $\times 4$, $+2$, $\div 2$, $\div 3$, $\times 5$

problem solving

Twenty students in a homeroom class are signing up for fine arts electives. So far, 5 students have signed up for band, 6 have signed up for drama, and 12 signed up for art. There are 3 students who signed up for both drama and art, 2 students who signed up for both band and art, and 1 student who signed up for all three. How many students have not yet registered for an elective?

New Concept

Increasing Knowledge

When a division problem has a remainder, there are several ways to write the answer: with a remainder, as a mixed number, or as a decimal number.

$$4\overline{)15}\ \text{3 R 3} \qquad 4\overline{)15}\ 3\tfrac{3}{4} \qquad 4\overline{)15.00}\ 3.75$$

How a division answer should be written depends upon the question to be answered. In real-world applications we sometimes need to round an answer up, and we sometimes need to round an answer down. The quotient of 15 ÷ 4 rounds up to 4 and rounds down to 3.

Example 1

One hundred students are to be assigned to 3 classrooms. How many students should be in each class so that the numbers are as balanced as possible?

Solution

Dividing 100 by 3 gives us 33 R 1. Assigning 33 students per class totals 99 students. We add the remaining student to one of the classes, giving that class 34 students. We write the answer **33, 33,** and **34.**

Example 2

Matinee movie tickets cost \$8. Jim has \$30. How many tickets can he buy?

Solution

We divide 30 dollars by 8 dollars per ticket. The quotient is $3\frac{3}{4}$ tickets.

$$\frac{30 \text{ dollars}}{8 \text{ dollars per ticket}} = 3\frac{3}{4} \text{ tickets}$$

Jim cannot buy $\frac{3}{4}$ of a ticket, so we round down to the nearest whole number. Jim can buy **3 tickets.**

Example 3

Fifteen children need a ride to the fair. Each car can transport 4 children. How many cars are needed to transport 15 children?

Solution

We divide 15 children by 4 children per car. The quotient is $3\frac{3}{4}$ cars.

$$\frac{15 \text{ children}}{4 \text{ children per car}} = 3\frac{3}{4} \text{ cars}$$

Three cars are not enough. Four cars will be needed. One of the cars will be $\frac{3}{4}$ full. We round $3\frac{3}{4}$ cars up to **4 cars.**

Example 4

Dale cut a 10-foot board into four equal lengths. How long was each of the four boards?

Solution

We divide 10 feet by 4.

$$\frac{10 \text{ ft}}{4} = 2\frac{2}{4} \text{ ft} = 2\frac{1}{2} \text{ ft}$$

Each board was $\mathbf{2\frac{1}{2}}$ **ft** long.

Example 5

Kimberly is on the school swim team. At practice she swam the 50 m freestyle three times. Her times were 37.53 seconds, 36.90 seconds, and 36.63 seconds. What was the mean of her three times?

Solution

To find the mean we add the three times and divide by 3.

$$\begin{array}{r} 37.53 \\ 36.90 \\ \underline{36.63} \\ 111.06 \end{array} \qquad 3\overline{)111.06}^{\;37.02}$$

Kimberly's mean time was **37.02 seconds.**

Practice Set

a. **Infer** Ninety students were assigned to four classrooms as equally as possible. How many students were in each of the four classrooms?

b. Movie tickets cost \$9.50. Aluna has \$30.00. How many movie tickets can she buy?

c. **Infer** Twenty-eight children need a ride to the fair. Each van can carry six children. How many vans are needed?

d. Corinne folded an $8\frac{1}{2}$ in. by 11 in. piece of paper in half. Then she folded the paper in half again as shown. After the two folds, what are the dimensions of the rectangle that is formed? How can you check your answer?

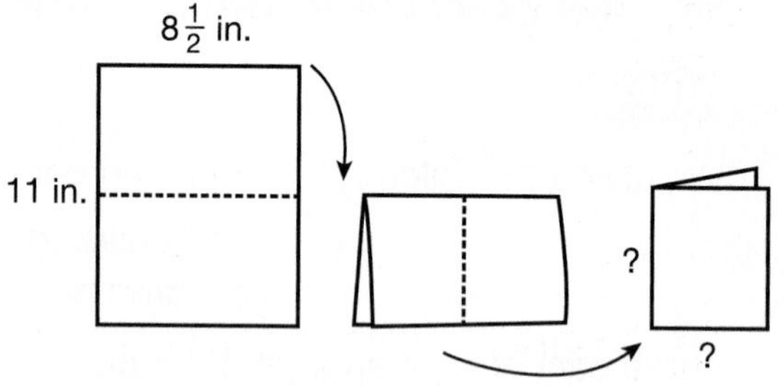

e. Kevin ordered four books at the book fair for summer reading. The books cost \$6.95, \$7.95, \$6.45, and \$8.85. Find the average (mean) price of the books.

Written Practice *Strengthening Concepts*

1. (111) Eighty students will be assigned to three classrooms. How many students should be in each class so that the numbers are as balanced as possible? (Write the numbers.)

2. (111) Four friends went out to lunch. Their bill was \$45. If the friends divide the bill equally, how much will each friend pay?

3. (49) Shauna bought a sheet of 39¢ stamps at the post office for \$15.60. How many stamps were in the sheet?

4. (82) Eight cubes were used to build this 2-by-2-by-2 cube. How many cubes are needed to build a cube that has three cubes along each edge?

5. (92) Write the standard notation for the following:

$$(5 \times 10^3) + (4 \times 10^1) + (3 \times 10^0)$$

6. (7) **Estimate** Use a centimeter ruler and an inch ruler to answer this question. Twelve inches is closest to how many centimeters? Round the answer to the nearest centimeter.

*** 7.** (Inv. 11) Create a scale drawing of a room at home or of your classroom. Choose a scale that allows the drawings to fit on one sheet of paper. Write the legend on the scale drawing.

*** 8.** (98) **Conclude** If two angles of a triangle measure 70° and 80°, then what is the measure of the third angle?

Connect Complete the table to answer problems **9–11.**

	Fraction	Decimal	Percent
9. (99)	$\frac{11}{20}$	**a.**	**b.**
10. (99)	**a.**	1.5	**b.**
11. (99)	**a.**	**b.**	1%

*** 12.** (100, 104) **Analyze** Calculate mentally:

a. $-6 + -12$ **b.** $-6 - -12$

c. $-12 + +6$ **d.** $-12 - +6$

13. (68) $6\frac{1}{4} \div 100$

14. (49) $0.3m = \$4.41$

*** 15.** (105) **Analyze** Kim scored 15 points, which was 30% of the team's total. How many points did the team score in all?

*** 16.** (111) Andrea received the following scores in a gymnastic event.

6.7	7.6	6.6	6.7	6.5	6.7	6.8

The highest score and the lowest score are not counted. What is the average of the remaining scores?

17. (13) **Formulate** Refer to problem **16** to write a comparison question about the scores Andrea received from the judges. Then answer the question.

*** 18.** (107) What is the area of the quadrilateral below?

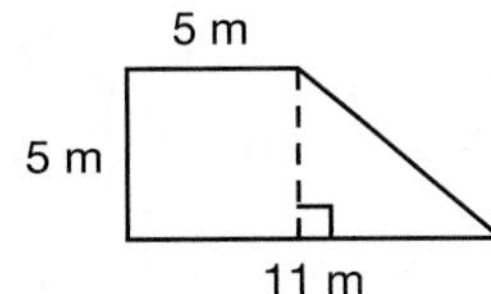

19. (Inv. 6) What is the ratio of vertices to edges on a pyramid with a square base?

* **20.** (110) **Conclude** Line r is called a line of symmetry because it divides the equilateral triangle into two mirror images. Which other line is also a line of symmetry?

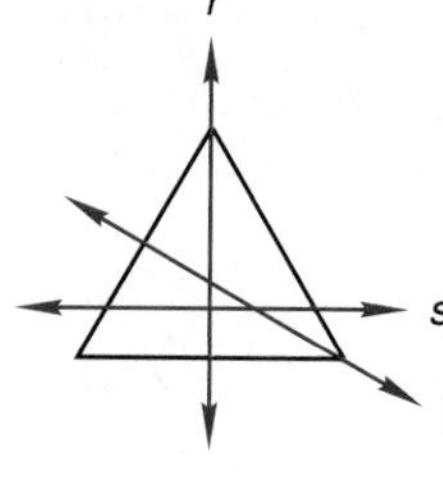

* **21.** (106) Solve and check: $3m + 1 = 100$

22. (73) Write the prime factorization of 600 using exponents.

23. (Inv. 6) **Conclude** You need to make a three-dimensional model of a soup can using paper and tape. What 3 two-dimensional shapes do you need to cut to make the model?

24. (41) The price of an item is 89¢. The sales-tax rate is 7%. What is the total for the item, including tax?

25. (58) The probability of winning a prize in the drawing is one in a million. What is the probability of not winning a prize in the drawing?

Conclude Triangles ABC and CDA are congruent. Refer to this figure to answer problems **26** and **27.**

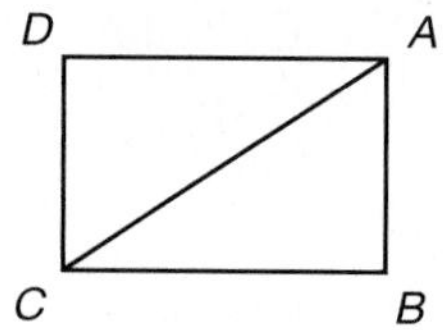

* **26.** (109) Which angle in triangle ABC corresponds to angle D in triangle CDA?

* **27.** (108) **Connect** Which transformations would position triangle CDA on triangle ABC?

28. (47) Malik used a compass to draw a circle with a radius of 5 centimeters. What was the circumference of the circle? (Use 3.14 for π.)

29. (85) Solve this proportion: $\frac{10}{16} = \frac{25}{y}$

30. (95) **Evaluate** The formula $d = rt$ shows that the distance traveled (d) equals the rate (r) times the time (t) spent traveling at that rate. (Here, *rate* means "speed.") This function table shows the relationship between distance and time when the rate is 50.

t	1	2	3	4
d	50	100	150	200

Find the value of d in $d = rt$ when r is $\frac{50 \text{ mi}}{1 \text{ hr}}$ and t is 5 hr.

LESSON 112

• Multiplying and Dividing Integers

Power Up

Building Power

facts Power Up M

mental math

a. **Fractional Parts:** $\frac{3}{4}$ of 24

b. **Calculation:** 6×48

c. **Percent:** 25% of \$48

d. **Calculation:** \$4.98 + \$2.49

e. **Decimals:** $0.5 \div 10$

f. **Number Sense:** $500 \cdot 30$

g. **Geometry:** A cube has a volume of 27 in. What is the length of the sides of the cube?

h. **Calculation:** 11×4, $+ 1$, $\div 5$, $\sqrt{\ }$, $\times 4$, $- 2$, $\times 5$, $- 1$, $\sqrt{\ }$

problem solving

When all the cards from a 52-card deck are dealt to three players, each player receives 17 cards, and there is one extra card. Dean invented a new deck of cards so that any number of players up to 6 can play and there will be no extra cards. How many cards are in Dean's deck if the number is less than 100?

New Concept

Increasing Knowledge

We know that when we multiply two positive numbers the product is positive.

$$(+3)(+4) = +12$$

positive × positive = positive

Reading Math

$(+3)(+4)$ is the same as 3×4 or $3 \cdot 4$.

Notice that when we write $(+3)(+4)$ there is no + or − sign between the sets of parentheses.

When we multiply a positive number and a negative number, the product is negative. We show an example on this number line by multiplying 3 and −4.

$$3 \times -4 \text{ means } (-4) + (-4) + (-4)$$

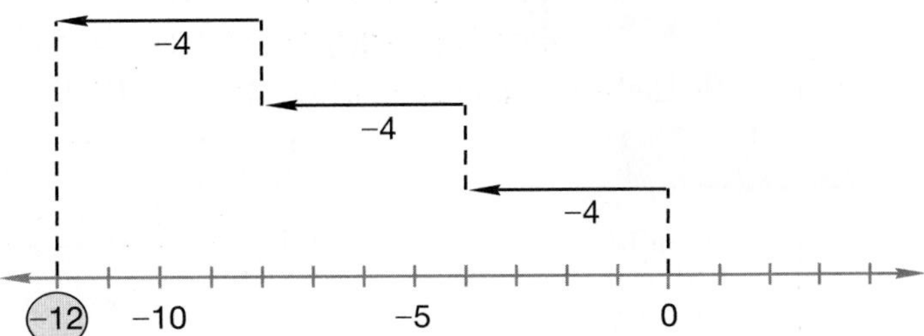

We write the multiplication this way:

$$(+3)(-4) = -12$$

Positive three times *negative* four equals *negative* 12.

positive × negative = negative

When we multiply two negative numbers, the product is positive. Consider this sequence of equations:

1. Three times 4 is 12. $3 \times 4 = 12$
2. "Three times the opposite of 4" is "the opposite of 12." $3 \times -4 = -12$
3. The opposite of "3 times the opposite of 4" is the opposite of "the opposite of 12." $-3 \times -4 = +12$

negative × negative = positive

Recall that we can rearrange the numbers of a multiplication fact to make two division facts.

Thinking Skill

Discuss

How is multiplying integers similar to multiplying whole numbers? How is it different?

Multiplication Facts	Division Facts	
$(+3)(+4) = +12$	$\frac{+12}{+3} = +4$	$\frac{+12}{+4} = +3$
$(+3)(-4) = -12$	$\frac{-12}{+3} = -4$	$\frac{-12}{-4} = +3$
$(-3)(-4) = +12$	$\frac{+12}{-3} = -4$	$\frac{+12}{-4} = -3$

Studying these nine facts, we can summarize the results in two rules:

1. If the two numbers in a multiplication or division problem have the **same sign,** the answer is positive.
2. If the two numbers in a multiplication or division problem have **different signs,** the answer is negative.

Example

Calculate mentally:

a. $(+8)(+4)$ **b.** $(+8) \div (+4)$

c. $(+8)(-4)$ **d.** $(+8) \div (-4)$

e. $(-8)(+4)$ **f.** $(-8) \div (+4)$

g. $(-8)(-4)$ **h.** $(-8) \div (-4)$

Solution

a. +32 **b.** +2

c. −32 **d.** −2

e. −32 **f.** −2

g. +32 **h.** +2

Practice Set

Predict First predict which problems will have a positive answer and which will have a negative answer. Then simplify each problem.

a. $(-5)(+4)$ **b.** $(-5)(-4)$

c. $(+5)(+4)$ **d.** $(+5)(-4)$

e. $\frac{+12}{-2}$ **f.** $\frac{+12}{+2}$ **g.** $\frac{-12}{+2}$ **h.** $\frac{-12}{-2}$

Written Practice

Strengthening Concepts

* **1.** (111) Two hundred students are traveling by bus on a field trip. The maximum number of students allowed on each bus is 84. How many buses are needed for the trip?

* **2.** (Inv. 11) Estimate The wingspan of a jumbo jet is about 210 feet. The wingspan of a model of a jumbo jet measures 25.2 inches.

a. What is the approximate scale of the model?

b. The model is 28 inches long. To the nearest foot, how long is the jumbo jet?

* **3.** (112) Calculate mentally:

a. $(-2)(-6)$ **b.** $\frac{+6}{-2}$

c. $\frac{-6}{-6}$ **d.** $(-2)(+6)$

* **4.** (100) Calculate mentally:

a. $-2 + -6$ **b.** $-2 - -6$

c. $+2 + -6$ **d.** $+2 - -6$

* **5.** (105) Analyze The chef chopped 27 carrots, which was 90% of the carrots in the bag. How many carrots remained?

6. (92) Write twenty million, five hundred ten thousand in expanded notation using exponents.

7. (41) Find 8% of \$3.65 and round the product to the nearest cent.

8. (92) $\left(\frac{1}{2}\right)^2 + \frac{1}{8} \div \frac{1}{2}$

Connect Complete the table to answer problems **9–11.**

	Fraction	Decimal	Percent
9. (99)	$1\frac{4}{5}$	**a.**	**b.**
10. (99)	**a.**	0.6	**b.**
11. (99)	**a.**	**b.**	2%

Solve:

12. (63) $5\frac{1}{2} - m = 2\frac{5}{6}$

* **13.** (85, 105) $\frac{6}{10} = \frac{0.9}{n}$

*** 14.** (106) $9x - 7 = 92$

15. (49) $0.05w = 8$

16. (15) All eight books in the stack are the same weight. Three books weigh a total of six pounds.

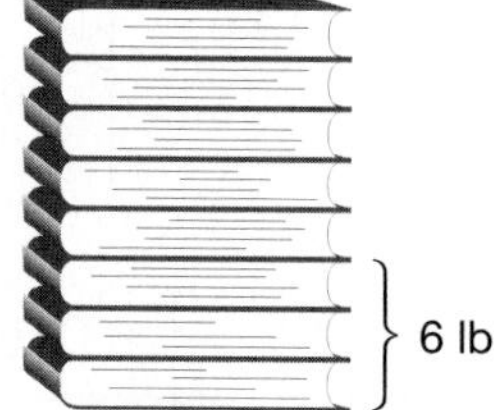

a. How much does each book weigh?

b. How much do all eight books weigh?

17. (91) Find the volume of a rectangular prism using the formula $V = lwh$ when the length is 8 cm, the width is 5 cm, and the height is 2 cm.

18. (95) How many millimeters is 1.2 meters (1 m = 1000 mm)?

*** 19.** (103) What is the perimeter of the polygon at right? (Dimensions are in millimeters.)

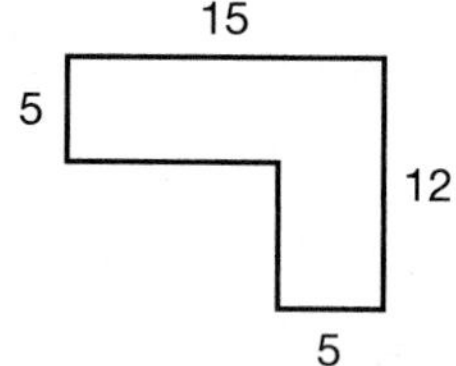

*** 20.** (107) What is the area of the polygon in problem 19?

21. (Inv. 6) **Verify** If the pattern shown below were cut out and folded on the dotted lines, would it form a cube, a pyramid, or a cylinder? Explain how you know.

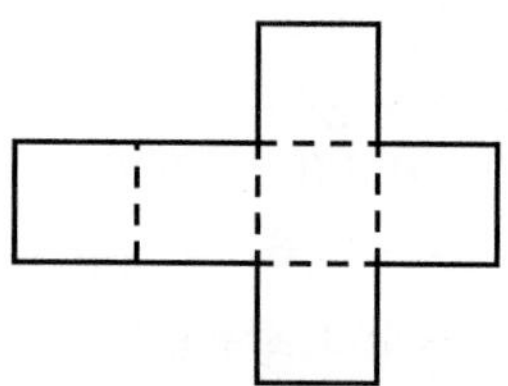

22. (65) **Classify** Which one of these numbers is not a composite number?

A 34 **B** 35 **C** 36 **D** 37

23. (47) **Estimate** Debbie wants to decorate a cylindrical wastebasket by wrapping it with wallpaper. The diameter of the wastebasket is 12 inches. The length of the wallpaper should be at least how many inches? Round up to the next inch.

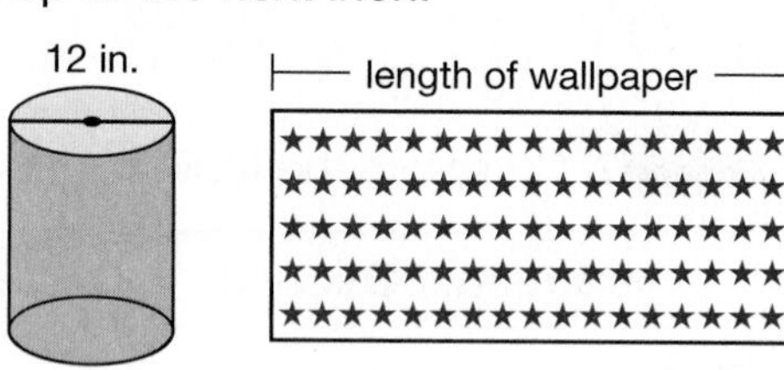

*** 24.** (110) **a.** **Conclude** Which one of these letters has two lines of symmetry?

H A V E

b. Which letter has rotational symmetry?

25. (17) **Connect** Which arrow is pointing to $-\frac{1}{2}$?

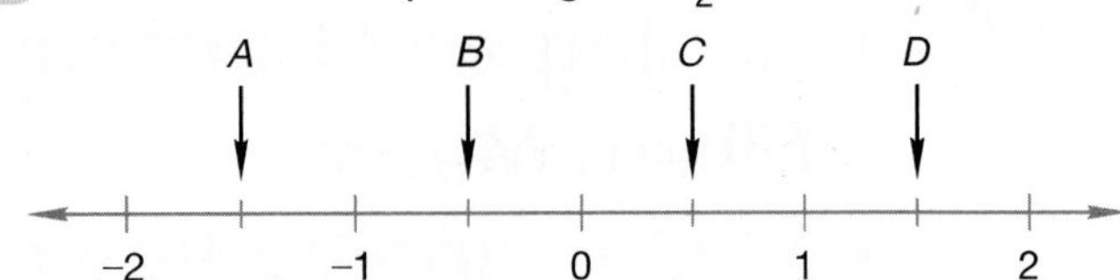

26. (101) The ratio of nonfiction to fiction books on Shawna's bookshelf is 2 to 3. If the total number of books on her shelf is 30, how many nonfiction books are there?

27. (Inv. 7) **Connect** What are the coordinates of the point that is halfway between $(-2, -3)$ and $(6, -3)$?

28. (Inv. 10) A set of 40 number cards contains one each of the counting numbers from 1 through 40. A multiple of 10 is drawn from the set and is not replaced. A second card is drawn from the remaining 39 cards. What is the probability that the second card will be a multiple of 10?

* **29.** (107) Combine the areas of the two triangles to find the area of this trapezoid.

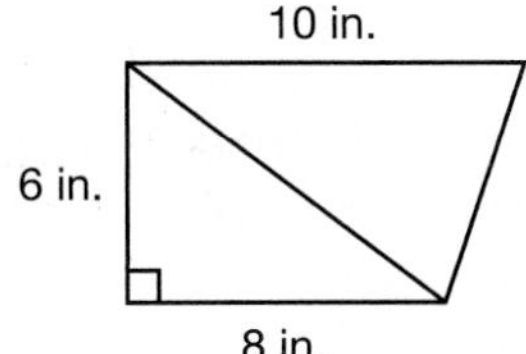

30. (92) $3^2 + 2 \times 5^2 - 50 \div \sqrt{25}$

Early Finishers
Real-World Application

Hannah's mother gave her $20 for her birthday. Her aunt gave her $25. She spent one-fifth of her birthday money at the bookstore. How much did Hannah spend at the bookstore?

Write one equation and use it to solve the problem.

LESSON

113

• Adding and Subtracting Mixed Measures
• Multiplying by Powers of Ten

Power Up

Building Power

facts Power Up N

mental math

a. **Fractional Parts:** $\frac{3}{10}$ of 40

b. **Calculation:** 4×38

c. **Percent:** 25% of $200

d. **Calculation:** $100.00 − $9.50

e. **Decimals:** $0.12 \div 10$

f. **Number Sense:** $\frac{2000}{500}$

g. **Geometry:** A circle has a diameter of 10 mm. What is the circumference of the circle?

h. **Calculation:** $6 \times 8, + 2, \times 2, - 1, \div 3, - 1, \div 4, + 2, \div 10, - 1$

problem solving

A hexagon can be divided into four triangles by three diagonals drawn from a single vertex. How many triangles can a dodecagon be divided into using diagonals drawn from one vertex?

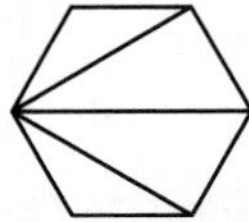

New Concepts

Increasing Knowledge

adding and subtracting mixed measures

Measurements that include more than one unit of measurement are mixed measures. If we say that a movie is an hour and 40 minutes long, we have used a mixed measure that includes hours and minutes. When adding or subtracting mixed measures, we may need to convert from one unit to another unit. In Lesson 102 we added and subtracted mixed measures involving pounds and ounces. In this lesson we will consider other mixed measures.

Example 1

The hike from the trailhead to the waterfall took 1 hr 50 min. The return trip took 1 hr 40 min. Altogether, how many hours and minutes long was the hike?

Solution

We add 50 minutes and 40 minutes to get 90 minutes which equals 1 hour 30 minutes.

$$\begin{array}{r} 1 \text{ hr } 50 \text{ min} \\ + \ 1 \text{ hr } 40 \text{ min} \\ \hline 90 \text{ min} \end{array} \quad \text{(which is 1 hr 30 min)}$$

We change 90 minutes to 1 hour 30 minutes. We write "30 minutes" in the minutes column and add the 1 hour to the hours column. Then we add the hours.

$$\begin{array}{r} \overset{1}{1} \text{ hr } 50 \text{ min} \\ + \ 1 \text{ hr } 40 \text{ min} \\ \hline 3 \text{ hr } \underset{30}{\cancel{90}} \text{ min} \end{array}$$

The hike took **3 hours 30 minutes.**

Example 2

To measure his vertical leap, Tyrone first reaches as high as he can against a wall. He reaches 6 ft 9 in. Then he put chalk on his fingertips, and jumping as high as he can, he slaps the wall. The top of the chalk mark is 8 ft 7 in. How high off the ground did Tyrone leap?

Solution

Thinking Skills

Justify

Why did we need to rename 8 ft before we subtracted?

We find the difference between the two measures. Before we subtract inches, we rename 8 feet 7 inches. The 12 inches combine with the 7 inches to make 19 inches. Then we subtract.

$$\begin{array}{r} \overset{7}{\cancel{8}} \text{ ft } \overset{19}{\cancel{7}} \text{ in.} \\ - \ 6 \text{ ft } 9 \text{ in.} \\ \hline 1 \text{ ft } 10 \text{ in.} \end{array}$$

Tyrone leaped **1 ft 10 in.** 

multiplying by powers of ten

Math Language

In **standard notation** all the digits in a number are shown. When multiplying by powers of ten, only place values containing non-zero digits are shown.

We can multiply by **powers** of ten very easily. Multiplying by powers of ten does not change the digits, only the place value of the digits. We can change the place value by moving the decimal point the number of places shown by the exponent. To write 1.2×10^3 in standard notation, we simply move the decimal point three places to the right and fill the empty places with zeros.

$$1.2 \times 10^3 = 1200. = 1200$$

Example 3

Write 6.2×10^2 in standard notation.

Solution

To multiply by a power of ten, simply move the decimal point the number of places shown by the exponent. In this case, we move the decimal point two places to the right.

$$6.2 \times 10^2 = 620. = \mathbf{620}$$

Sometimes powers of ten are named with words instead of numbers. For example, we might read that the population of Hong Kong is about 6.8 million people. The number 6.8 million means $6.8 \times 1{,}000{,}000$. We can write this number by shifting the decimal point of 6.8 six places to the right, which gives us 6,800,000.

Example 4

Write $\frac{1}{2}$ billion in standard notation.

Solution

The expression $\frac{1}{2}$ billion means "one half of one billion." First we write $\frac{1}{2}$ as the decimal number 0.5. Then we multiply by one billion, which shifts the decimal point nine places.

$$\frac{1}{2}\text{ billion} = 0.5 \times 1{,}000{,}000{,}000 = \mathbf{500{,}000{,}000}$$

Connect How can we write 500,000,000 using powers of ten?

Practice Set

Find each sum or difference:

a. 6 ft 5 in. + 4 ft 8 in.

b. 3 hr 15 min − 1 hr 40 min

Write the standard notation for each of the following numbers. Change fractions and mixed numbers to decimal numbers before multiplying.

c. 1.2×10^4

d. 1.5 million

e. $2\frac{1}{2}$ billion

f. $\frac{1}{4}$ million

Written Practice

Strengthening Concepts

*** 1.** (111) **Analyze** For cleaning the yard, four teenagers were paid a total of $75.00. If they divide the money equally, how much money will each teenager receive?

2. (7) **Estimate** Which of the following is the best estimate of the length of a bicycle?

A 0.5 m **B** 2 m **C** 6 m **D** 36 m

3. (58) If the chance of rain is 80%, what is the probability that it will not rain? Express the answer as a decimal.

*** 4.** (101) The ratio of students who walk to school to students who ride a bus to school is 5 to 3. If there are 120 students, how many students walk to school?

*** 5.** (113) **Analyze** Write 4.5×10^6 as a standard numeral.

*** 6.** Calculate mentally:
(112)

a. $(-12)(+3)$ **b.** $(-12)(-3)$

c. $\frac{-12}{+3}$ **d.** $\frac{-12}{-3}$

*** 7.** Calculate mentally:
(100)

a. $-12 + -3$ **b.** $-12 - -3$

c. $+3 + -12$ **d.** $+3 - -12$

8. *(76)* **Explain** Describe a method for arranging these fractions from least to greatest:

$$\frac{3}{4}, \frac{3}{5}, \frac{4}{5}$$

Connect Complete the table to answer problems **9–11.**

	Fraction	Decimal	Percent
9. *(99)*	$\frac{1}{50}$	**a.**	**b.**
10. *(99)*	**a.**	1.75	**b.**
11. *(99)*	**a.**	**b.**	25%

12. *(63)* $12\frac{1}{4}$ in. $- 3\frac{5}{8}$ in.

13. *(66)* $3\frac{1}{3}$ ft $\times 2\frac{1}{4}$ ft

14. *(81)* (3 cm)(3 cm)(3 cm)

15. *(81)* 0.6 m $\times$ 0.5 m

16. *(92)* $5^2 + 2^5$

*** 17.** *(107)* **Justify** Find the area of this trapezoid. Show and explain your work to justify your answer.

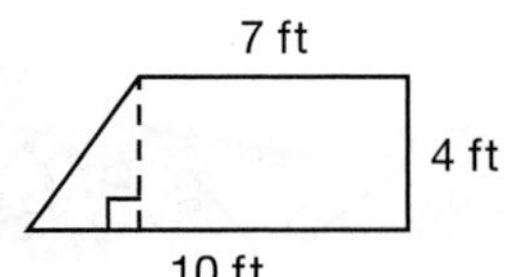

*** 18.** *(113)* 2 feet 3 inches − 1 foot 9 inches

*** 19.** *(110)* **a.** **Conclude** Which line in this figure is not a line of symmetry?

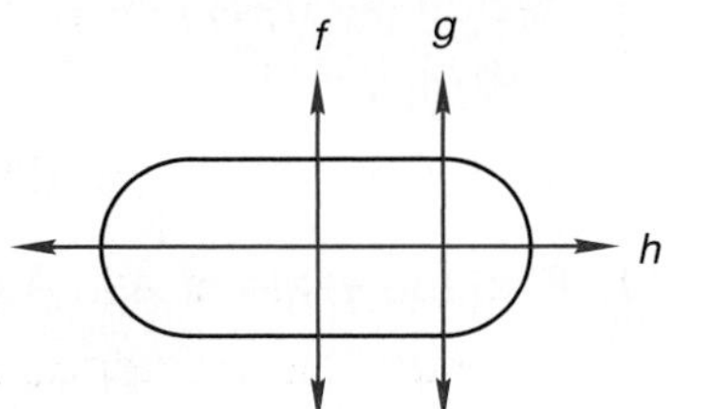

b. Does the figure have rotational symmetry? (Disregard lines *f*, *g*, and *h*.) Explain your answer.

20. *(82)* How many cubes one centimeter on each edge would be needed to fill this box?

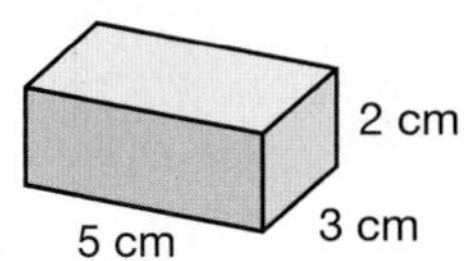

21. (88) Elizabeth worked for three days and earned $240. At that rate, how much would she earn in ten days?

22. (65) **Analyze** Seventy is the product of which three prime numbers?

23. (12) Saturn is about 900 million miles from the Sun. Write that distance in standard notation.

Math Language

A **rhombus** is a parallelogram in which all four sides are equal in length.

* **24.** (71) Use the rhombus at the right for problems **a–c**.

a. What is the perimeter of this rhombus?

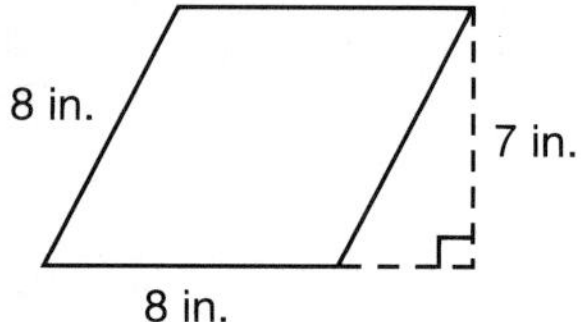

b. What is the area of this rhombus?

c. **Analyze** If an acute angle of this rhombus measures 61°, then what is the measure of each obtuse angle?

17

25. (88) The ratio of quarters to dimes in Keiko's savings jar is 5 to 8. If there were 120 quarters, how many dimes were there?

26. (Inv. 7, 79) **a.** **Connect** The coordinates of the three vertices of a triangle are (0, 0), (0, 4), and (4, 4). What is the area of the triangle?

b. If the triangle were reflected across the y-axis, what would be the coordinates of the vertices of the reflection?

The following list shows the ages of the children attending a luncheon. Use this information to answer problems **27** and **28.**

8, 9, 8, 8, 7, 9, 12, 12, 11, 16

27. (Inv. 5) What was the median age of the children attending the luncheon?

28. (Inv. 5) What was the mean age of the children at the luncheon?

29. (47) The diameter of a playground ball is 10 inches. What is the circumference of the ball? (Use 3.14 for π.) How can estimation help you determine if your answer is reasonable?

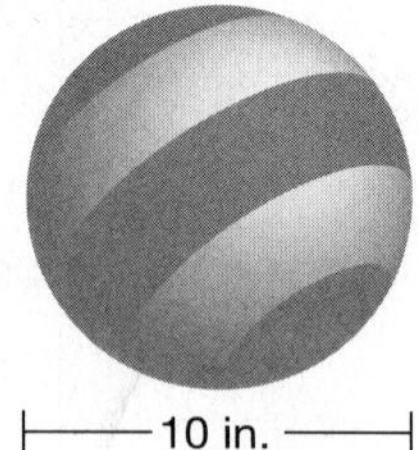

30. (91) Find the value of A in $A = s^2$ when s is 10 m.

LESSON 114

• Unit Multipliers

Power Up

Building Power

facts Power Up M

mental math

a. **Fractional Parts:** $\frac{7}{10}$ of 40

b. **Number Sense:** 6×480

c. **Percent:** 10% of \$500

d. **Calculation:** \$4.99 + 65¢

e. **Decimals:** 0.125×1000

f. **Number Sense:** $40 \cdot 900$

g. **Geometry:** You need to fill a show box with sand. What do you need to measure to find how much sand you will need?

h. **Calculation:** 5×7, $+ 1$, $\div 4$, $\sqrt{\ }$, $\times 7$, $- 1$, $\times 3$, $- 10$, $\times 2$, $\sqrt{\ }$

problem solving

On a balanced scale are a 25-gram mass, a 100-gram mass, and five identical blocks marked *x*, which are distributed as shown. What is the mass of each block marked *x*? Write an equation illustrated by this balanced scale.

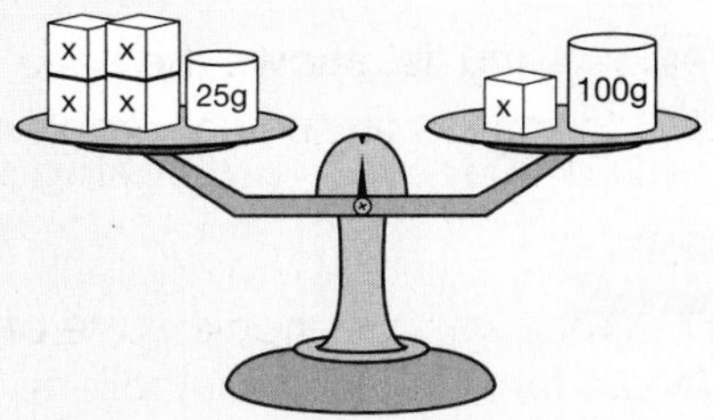

New Concept

Increasing Knowledge

A **unit multiplier** is a fraction that equals 1 and that is written with two different units of measure. Recall that when the numerator and denominator of a fraction are equal (and are not zero), the fraction equals 1. Since 1 foot equals 12 inches, we can form two unit multipliers with the measures 1 foot and 12 inches.

$$\frac{1 \text{ ft}}{12 \text{ in.}} \qquad \frac{12 \text{ in.}}{1 \text{ ft}}$$

Each of these fractions equals 1 because the numerator and denominator of each fraction are equal.

We can use unit multipliers to help us convert from one unit of measure to another. If we want to convert 60 inches to feet, we can multiply 60 inches by the unit multiplier $\frac{1 \text{ ft}}{12 \text{ in.}}$.

$$\frac{\overset{5}{\cancel{60}}\ \cancel{\text{in.}}}{1} \times \frac{1 \text{ ft}}{\underset{1}{\cancel{12}}\ \cancel{\text{in.}}} = 5 \text{ ft}$$

Example 1

a. Write two unit multipliers using these equivalent measures:

$$3 \text{ ft} = 1 \text{ yd}$$

b. Which unit multiplier would you use to convert 30 yards to feet?

Solution

a. We use the equivalent measures to write two fractions equal to 1.

$$\frac{3 \text{ ft}}{1 \text{ yd}} \qquad \frac{1 \text{ yd}}{3 \text{ ft}}$$

b. We want the units we are changing **from** to appear in the denominator and the units we are changing **to** to appear in the numerator. To convert 30 yards to feet, we use the unit multiplier that has yards in the denominator and feet in the numerator.

$$\frac{30 \text{ yd}}{1} \times \frac{3 \text{ ft}}{1 \text{ yd}}$$

Here we show the work. Notice that the yards "cancel," and the product is expressed in feet.

$$\frac{30 \cancel{\text{yd}}}{1} \times \frac{3 \text{ ft}}{1 \cancel{\text{yd}}} = 90 \text{ ft}$$

Example 2

Convert 30 feet to yards using a unit multiplier.

Solution

We can form two unit multipliers.

$$\frac{1 \text{ yd}}{3 \text{ ft}} \text{ and } \frac{3 \text{ ft}}{1 \text{ yd}}$$

We are asked to convert from feet to yards, so we use the unit multiplier that has feet in the denominator and yards in the numerator.

$$\frac{\overset{10}{\cancel{30}} \cancel{\text{ft}}}{1} \times \frac{1 \text{ yd}}{\cancel{3} \cancel{\text{ft}}} = 10 \text{ yd}$$

Thirty feet converts to **10 yards.**

Practice Set

a. Write two unit multipliers for these equivalent measures:

$$1 \text{ gal} = 4 \text{ qt}$$

b. Which unit multiplier from problem **a** would you use to convert 12 gallons to quarts?

c. Write two unit multipliers for these equivalent measures:

$$1 \text{ m} = 100 \text{ cm}$$

d. Which unit multiplier from problem **c** would you use to convert 200 centimeters to meters?

e. Use a unit multiplier to convert 12 quarts to gallons.

f. Use a unit multiplier to convert 200 meters to centimeters.

g. Use a unit multiplier to convert 60 feet to yards (1 yd = 3 ft).

Written Practice *Strengthening Concepts*

* **1.** (111) Tickets to the matinee are $6 each. How many tickets can Maela buy with $20?

2. (88) **Analyze** Maria ran four laps of the track at an even pace. If it took 6 minutes to run the first three laps, how long did it take to run all four laps?

3. (77) Fifteen of the 25 members played in the game. What fraction of the members did not play?

4. (77) Two fifths of the 160 acres were planted with alfalfa. How many acres were not planted with alfalfa?

5. (12) Which digit in 94,763,581 is in the ten-thousands place?

* **6.** (114) **a.** Write two unit multipliers for these equivalent measures:

1 gallon = 4 quarts

b. Which of the two unit multipliers from part **a** would you use to convert 8 gallons to quarts? Why?

7. (51) **Estimate** What is the sum of $36.43, $41.92, and $26.70 to the nearest dollar.

8. (92) $4 + 4^2 \div \sqrt{4} - \frac{4}{4}$

9. (61) $3\frac{1}{4}$ in. + $2\frac{1}{2}$ in. + $4\frac{5}{8}$ in.

Connect Complete the table to answer problems **10–12.**

	Fraction	Decimal	Percent
10. (99)	$\frac{1}{8}$	**a.**	**b.**
11. (99)	**a.**	0.9	**b.**
12. (99)	**a.**	**b.**	60%

13. (73) $3.25 \div \frac{2}{3}$ (fraction answer)

Solve:

* **14.** (106) $3m - 10 = 80$

* **15.** (85, 105) $\frac{3}{2} = \frac{1.8}{m}$

* **16.** Calculate mentally:
(112)

a. $(-5)(-20)$ **b.** $(-5)(+20)$

c. $\frac{-20}{+5}$ **d.** $\frac{-20}{-5}$

* **17.** The distance between San Francisco and Los Angeles is about 387 miles. If Takara leaves San Francisco and travels 6 hours at an average speed of 55 miles per hour, will she reach Los Angeles? How far will she travel?
(95)

* **18.** **Analyze** What is the area of this polygon?
(107)

* **19.** What is the perimeter of this polygon?
(103)

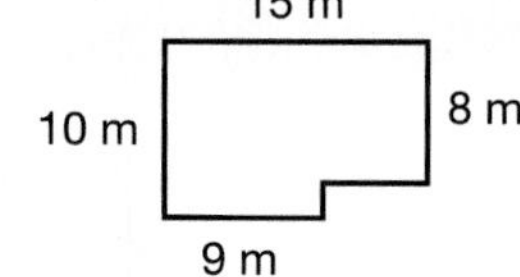

20. Calculate mentally:
(100)

a. $-5 + -20$ **b.** $-20 - -5$

c. $-5 - -5$ **d.** $+5 - -20$

* **21.** **Conclude** Transversal t intersects parallel lines q and r. Angle 1 is half the measure of a right angle.
(97)

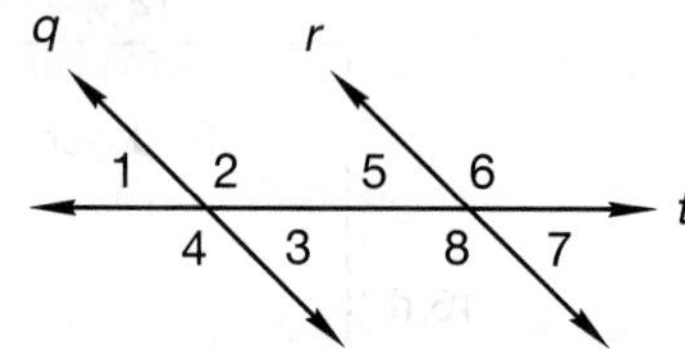

a. Which angle corresponds to $\angle 1$?

b. What is the measure of each obtuse angle?

* **22.** Fifty people responded to the survey, a number that represented 5% of the surveys mailed. How many surveys were mailed? Explain how you found your answer.
(105)

23. Think of two different prime numbers, and write them on your paper. Then find the least common multiple (LCM) of the two prime numbers.
(19, 30)

* **24.** Write 1.5×10^6 as a standard number.
(113)

25. A classroom that is 30 feet long, 30 feet wide, and 10 feet high has a volume of how many cubic feet?
(82)

* **26.** Convert 8 quarts to gallons using a unit multiplier.
(114)

27. **Analyze** A circle was drawn on a coordinate plane. The coordinates of the center of the circle were (1, 1). One point on the circle was (1, –3).
(Inv. 7, 86)

a. What was the radius of the circle?

b. What was the area of the circle? (Use 3.14 for π.)

28. (Inv. 5) During one season, the highest number of points scored in one game by the local college basketball team was 95 points. During that same season, the range of the team's scores was 35 points. What was the team's lowest score?

* **29.** (113) 4 ft 3 in. − 2 ft 9 in.

30. (96) **Generalize** Study this function table and describe a rule for finding *A* when *s* is known. Explain how you know.

s	*A*
1	1
2	4
3	9
4	16

+ 0

Early Finishers
Real-World Application

Robin wants to install crown molding in her two upstairs bedrooms. Crown molding lines the perimeter of a room, covering the joint formed by the wall and the ceiling. Use the dimensions on the floor plan below to answer **a** and **b.**

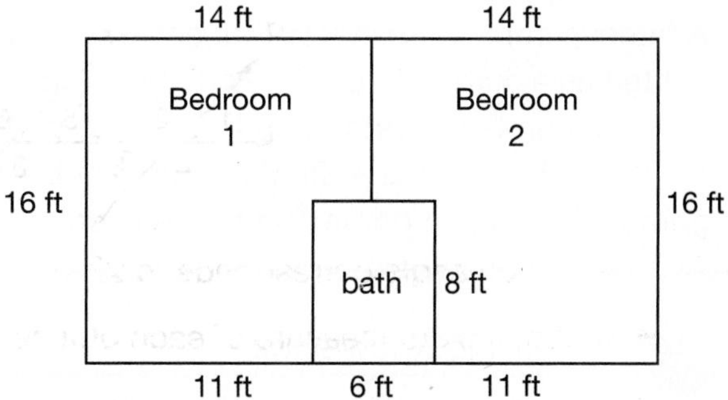

a. How many feet of crown molding will Robin need for both bedrooms?

b. If crown molding is sold in 8 ft. sections, how many sections will Robin need?

LESSON

115 • Writing Percents as Fractions, Part 2

Power Up

Building Power

facts Power Up *N*

mental math

a. **Fractional Parts:** $\frac{3}{4}$ of 16

b. **Calculation:** 9×507

c. **Percent:** 10% of \$2.50

d. **Calculation:** \$10.00 − \$9.59

e. **Decimals:** $0.5 \div 100$

f. **Number Sense:** $\frac{2400}{300}$

g. **Probability:** What is the probability of rolling an odd number on a number cube?

h. **Calculation:** 10×9, $-\ 10$, $\div\ 2$, $+\ 2$, $\div\ 6$, $\times\ 10$, $+\ 2$, $\div\ 9$, $-\ 9$

problem solving

A famous conjecture states that any even number greater than two can be written as a sum of two prime numbers ($12 = 5 + 7$). Another states that any odd number greater than five can be written as the sum of three prime numbers ($11 = 7 + 2 + 2$). Write the numbers 10, 15, and 20 as the sums of primes. (The same prime number may be used more than once in a sum.)

New Concept

Increasing Knowledge

Thinking Skill

Justify

How do we simplify $\frac{50}{100}$ to lowest terms?

Recall that a percent is a fraction with a denominator of 100. We can write a percent in fraction form by removing the percent sign and writing the denominator 100.

$$50\% = \frac{50}{100}$$

We then simplify the fraction to lowest terms. If the percent includes a fraction, we actually divide by 100 to simplify the fraction.

$$33\tfrac{1}{3}\% = \frac{33\frac{1}{3}}{100}$$

In this case we divide $33\frac{1}{3}$ by 100. We have performed division problems similar to this in the problem sets.

$$33\frac{1}{3} \div 100 = \frac{\overset{1}{\cancel{100}}}{3} \times \frac{1}{\underset{1}{\cancel{100}}} = \frac{1}{3}$$

We see that $33\frac{1}{3}\%$ equals $\frac{1}{3}$.

Example

Convert $3\frac{1}{3}\%$ to a fraction.

Solution

We remove the percent sign and write the denominator 100.

$$3\frac{1}{3}\% = \frac{3\frac{1}{3}}{100}$$

We perform the division.

$$3\frac{1}{3} \div 100 = \frac{\overset{1}{\cancel{10}}}{3} \times \frac{1}{\underset{10}{\cancel{100}}} = \frac{1}{30}$$

We find that $3\frac{1}{3}\%$ equals $\mathbf{\frac{1}{30}}$.

Practice Set

a. Convert $66\frac{2}{3}\%$ to a fraction.

b. Convert $6\frac{2}{3}\%$ to a fraction.

c. Convert $12\frac{1}{2}\%$ to a fraction.

d. Write $14\frac{2}{7}\%$ as a fraction.

e. Write $83\frac{1}{3}\%$ as a fraction.

Written Practice

Strengthening Concepts

1. (41) What is the total cost of a $12.60 item plus 7% sales tax?

*** 2.** (115) Convert $16\frac{2}{3}\%$ to a fraction.

*** 3.** (105) **Model** Draw a ratio box for this problem. Then solve the problem using a proportion.

Ines missed three questions on the test but answered 90% of the questions correctly. How many questions were on the test?

4. (95) Sound travels about 331 meters per second in air. How far will it travel in 60 seconds?

$$\frac{331\text{ m}}{1\text{ s}} \cdot \frac{60\text{ s}}{1}$$

5. (92) Write the standard number for $(5 \times 10^4) + (6 \times 10^2)$.

6. (27, 37) If the radius of a circle is seventy-five hundredths of a meter, what is the diameter?

7. (66) **Estimate** Round the product of $3\frac{2}{3}$ and $2\frac{2}{3}$ to the nearest whole number.

*** 8.** (Inv. 10) In a bag are three red marbles and three white marbles. If two marbles are taken from the bag at the same time, what is the probability that both marbles will be red?

Connect Complete the table to answer problems **9** and **10.**

	Fraction	Decimal	Percent
9. (99)	$2\frac{2}{5}$	**a.**	**b.**
10. (99)	**a.**	0.85	**b.**

Solve:

* **11.** (106) $7x - 3 = 39$

12. (85) $\frac{x}{7} = \frac{35}{5}$

* **13.** (112) Calculate mentally:

a. $(-3)(-15)$

b. $\frac{-15}{+3}$

c. $\frac{-15}{-3}$

d. $(+3)(-15)$

* **14.** (104) $-6 + -7 + +5 - -8$

15. (49) $0.12 \div (12 \div 0.4)$

16. (74) Write $\frac{22}{7}$ as a decimal rounded to the hundredths place.

17. (89) What whole number multiplied by itself equals 10,000?

* **18.** (107) What is the area of this hexagon?

19. (103) What is the perimeter of this hexagon?

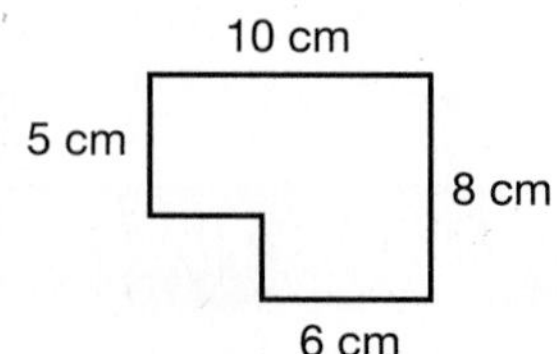

20. (82) What is the volume of this cube?

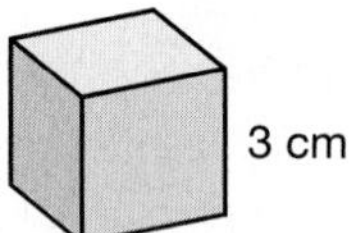

21. (Inv. 5) What is the mode of the number of days in the twelve months of the year?

22. (88) **Analyze** If seven of the containers can hold a total of 84 ounces, then how many ounces can 10 containers hold?

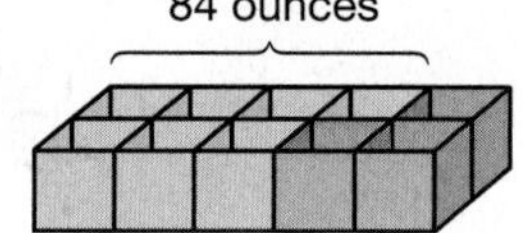

* **23.** (113) Write the standard number for $4\frac{1}{2}$ million.

24. (16) Round 58,697,284 to the nearest million.

25. (50) **Connect** Which arrow is pointing to 0.4?

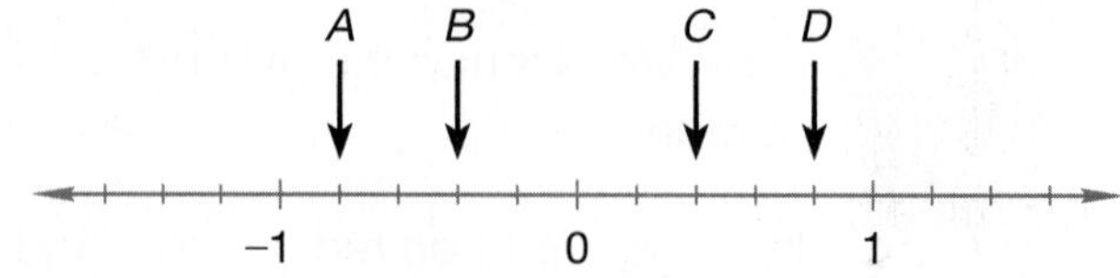

* **26.** (102, 113) When Rosita was born, she weighed 7 pounds 9 ounces. Two months later she weighed 9 pounds 7 ounces. How much weight did she gain in two months?

27. (Inv. 7, 71) **Connect** The coordinates of the vertices of a parallelogram are (0, 0), (5, 0), (6, 3), and (1, 3). What is the area of the parallelogram?

28. (44) Which is the greatest weight?

A 6.24 lb **B** 6.4 lb **C** 6.345 lb

* **29.** (113)

$$\begin{array}{r} 2 \text{ gal } 2 \text{ qt } 1 \text{ pt} \\ + \; 2 \text{ gal } 2 \text{ qt } 1 \text{ pt} \\ \hline \end{array}$$

30. (81) **Analyze** Gilbert started the trip with a full tank of gas. He drove 323.4 miles and then refilled the tank with 14.2 gallons of gas. How can Gilbert calculate the average number of miles he traveled on each gallon of gas?

Early Finishers
Real-World Application

A few students from the local high school decide to survey the types of vehicles in the parking lot. Their results are as follows: 210 cars, 125 trucks, and 14 motorcycles.

a. Find the simplified ratios for cars to trucks, cars to motorcycles, and trucks to motorcycles.

b. Find the fraction, decimal, and percent for the ratio of cars to the total number of vehicles. Round to nearest thousandth and tenth of a percent.

LESSON 116

• Compound Interest

Power Up

Building Power

facts Power Up M

mental math

a. **Fractional Parts:** $\frac{3}{8}$ of 16

b. **Number Sense:** 4×560

c. **Percent:** 50% of $2.50

d. **Calculation:** $8.98 + 49¢

e. **Decimals:** 0.375×100

f. **Number Sense:** $50 \cdot 400$

g. **Statistics:** Find the mode and range of the set of numbers: 84, 27, 91, 84, 22, 72, 27, 84.

h. **Calculation:** 11×6, $-\ 2$, $\sqrt{\ }$, $\times\ 3$, $+\ 1$, $\sqrt{\ }$, $\times\ 10$, $-\ 1$, $\sqrt{\ }$

problem solving

Four identical blocks marked *x,* a 250-gram mass, and a 500-gram mass were balanced on a scale as shown. Write an equation to represent this balanced scale, and find the mass of each block marked *x.*

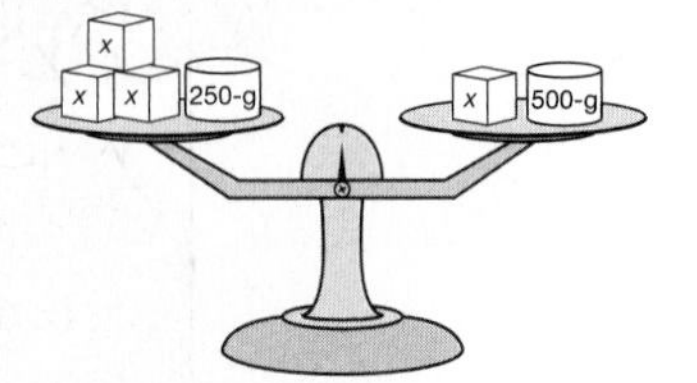

New Concept

Increasing Knowledge

When you deposit money in a bank, the bank uses a portion of that money to make loans and investments that earn money for the bank. To attract deposits, banks offer to pay **interest,** a percentage of the money deposited. The amount deposited is called the **principal.**

There is a difference between **simple interest** and **compound interest.** Simple interest is paid on the principal only and not paid on any accumulated interest. For instance, if you deposited $100 in an account that pays 3% simple interest, you would be paid 3% of $100 ($3) each year your $100 was on deposit. If you take your money out after three years, you would have a total of $109.

Simple Interest

$100.00	principal
$3.00	first-year interest
$3.00	second-year interest
+ $3.00	third-year interest
$109.00	total

Most interest-bearing accounts, however, are compound-interest accounts. In a compound-interest account, interest is paid on accumulated interest as well as on the principal. If you deposited $100 in an account with 3% annual percentage rate, the amount of interest you would be paid each year increases if the earned interest is left in the account. After three years you would have a total of $109.27.

Compound Interest

$100.00	principal
$3.00	first-year interest (3% of $100.00)
$103.00	total after one year
$3.09	second-year interest (3% of $103.00)
$106.09	total after two years
$3.18	third-year interest (3% of $106.09)
$109.27	total after three years

Visit www.SaxonPublishers.com/ActivitiesC1 *for a graphing calculator activity.*

Example

Mrs. Vasquez opened a $2000 retirement account that has grown at a rate of 10% a year for three years. What is the current value of her account?

Solution

We calculate the total amount of money in the account at the end of each year by adding 10% to the value of the account at the beginning of that year.

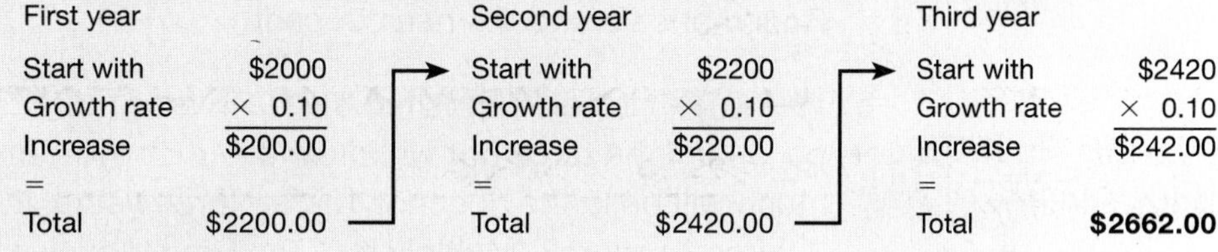

First year
Start with $2000
Growth rate × 0.10
Increase $200.00
=
Total $2200.00
Second year
Start with $2200
Growth rate × 0.10
Increase $220.00
=
Total $2420.00
Third year
Start with $2420
Growth rate × 0.10
Increase $242.00
=
Total **$2662.00**

Notice that the account grew by a larger number of dollars each year, even though the growth rate stayed the same. This increase occurred because the starting amount increased year by year. The effect of compounding becomes more dramatic as the number of years increases.

In our solution above we multiplied each starting amount by 10% (0.10) to find the amount of increase, then we added the increase to the starting amount. Instead of multiplying by 10% and adding, we can multiply by 110% (1.10) to find the value of the account after each year.

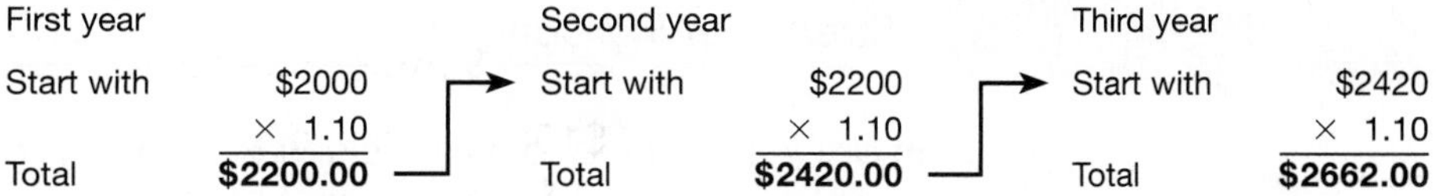

First year
Start with $2000
× 1.10
Total **$2200.00**
Second year
Start with $2200
× 1.10
Total **$2420.00**
Third year
Start with $2420
× 1.10
Total **$2662.00**

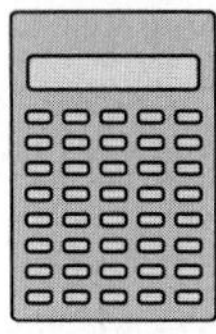

We will use this second method with a calculator to find the amount of money in Mrs. Vasquez's account after one, two, and three years. We use 1.1 for 110% and follow this keystroke sequence:[1]

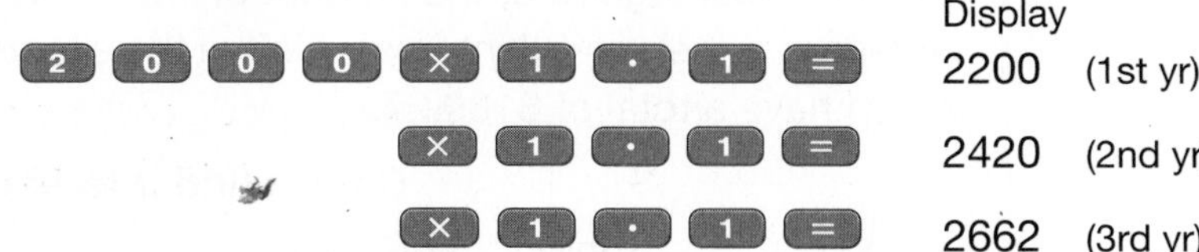

Math Language
Abbreviations on memory keys may vary from one calculator to another. We will use →M for "enter memory" and MR for "memory recall."

We can use the calculator memory to reduce the number of keystrokes. Instead of entering 1.1 for every year, we enter 1.1 into the memory with these keystrokes:

1 . 1 →M

Now we can use the "memory recall" key instead of 1.1 to perform the calculations. We find the amount of money in Mrs. Vasquez's account after one, two, and three years with this sequence of keystrokes:

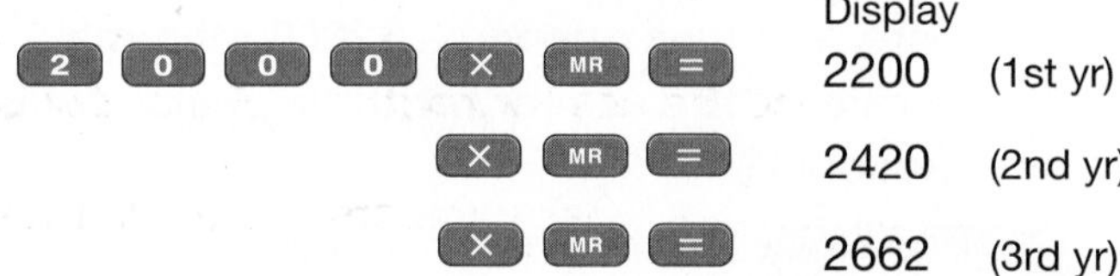

Practice Set

a. *Estimate* After the third year, $2662 was in Mrs. Vasquez's account. If the account continues to grow 10% annually, how much money will be in the account **1.** after the tenth year and **2.** after the twentieth year? Round answers to the nearest cent.

b. *Estimate* Nelson deposited $2000 in an account that pays 4% interest per year. If he does not withdraw any money from the account, how much will be in the account **1.** after three years, **2.** after 10 years, and **3.** after 20 years? (Multiply by 1.04. Round answers to the nearest cent.)

c. How much more money will be in Mrs. Vasquez's account than in Nelson's account **1.** after three years, **2.** after 10 years, and **3.** after 20 years?

Connect Find an advertisement that gives a bank's or other saving institution's interest rate for savings accounts. Write a problem based on the advertisement. Be sure to provide the answer to your problem.

Written Practice *Strengthening Concepts*

1. (79, 93) John drew a right triangle with sides 6 inches, 8 inches, and 10 inches long. What was the area of the triangle?

2. (95) If 5 feet of ribbon costs $1.20, then 10 feet of ribbon would cost how much?

[1] If this keystroke sequence does not produce the indicated result, consult the manual for the calculator for the appropriate keystroke sequence.

3. (22, 75) **a.** Six is what fraction of 15?

b. Six is what percent of 15?

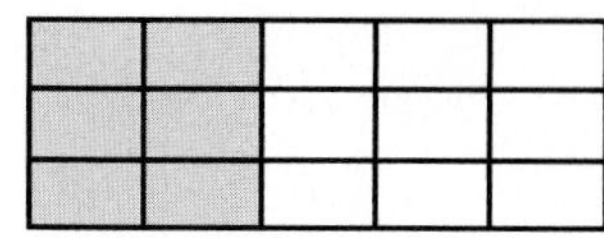

4. (58) The multiple-choice question has four choices. Raidon knows that one of the choices must be correct, but he has no idea which one. If Raidon simply guesses, what is the chance that he will guess the correct answer?

5. (77) If $\frac{2}{5}$ of the 30 students in the class buy lunch in the school cafeteria, what is the ratio of students who buy lunch in the school cafeteria to students who bring their lunch from home?

*** 6.** (113) **Connect** Write 1.2×10^9 as a standard number.

7. (95) **Evaluate** The cost (c) of apples is related to its price per pound (p) and its weight (w) by this formula:

$$c = pw$$

Find the cost when p is $\frac{\$1.25}{1 \text{ pound}}$ and w is 5 pounds.

8. (44) Arrange these numbers in order from least to greatest:

9.9, 9.95, 9.925, 9.09

Connect Complete the table to answer problems **9** and **10.**

	Fraction	Decimal	Percent
9. (99)	$3\frac{3}{8}$	**a.**	**b.**
10. (99)	**a.**	**b.**	15%

Solve and check:

*** 11.** (106) $9x + 17 = 80$

*** 12.** (85, 105) $\frac{x}{3} = \frac{1.6}{1.2}$

*** 13.** (104) $-6 + -4 - +3 - -8$

14. (74) $6 + 3\frac{3}{4} + 4.6$ (decimal answer)

*** 15.** (114) The Gateway Arch in St. Louis, Missouri, is approximately 210 yards tall. How tall is it in feet?

16. (73) Use division by primes to find the prime factors of 648. Then write the prime factorization of 648 using exponents.

17. (15) If a 32-ounce box of cereal costs \$3.84, what is the cost per ounce?

* **18.** (107) Find the area of this trapezoid by combining the area of the rectangle and the area of the triangle.

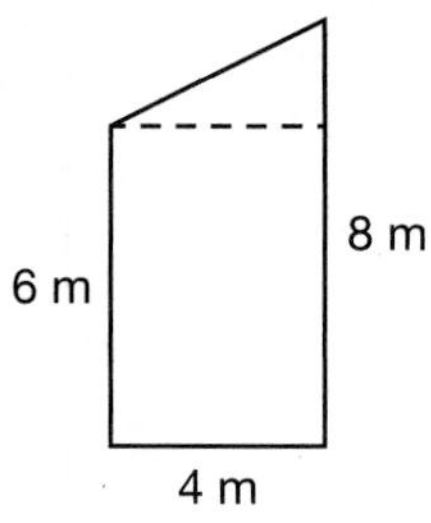

19. (47, 86) The radius of a circle is 10 cm. Use 3.14 for π to calculate the

a. circumference of the circle.

b. area of the circle.

20. (82) **Conclude** The volume of the pyramid is $\frac{1}{3}$ the volume of the cube. What is the volume of the pyramid?

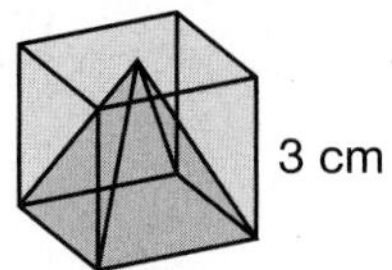

21. (49) Solve: $0.6y = 54$

* **22.** (112) Calculate mentally:

a. $(-8)(-2)$ **b.** $(+8)(-2)$

c. $\frac{+8}{-2}$ **d.** $\frac{-8}{-2}$

23. (111) **Analyze** The 306 students were assigned to ten rooms so that there were 30 or 31 students in each room. How many rooms had exactly 30 students?

24. (98) Two angles of a triangle measure 40° and 110°.

a. What is the measure of the third angle?

b. **Represent** Make a rough sketch of the triangle.

25. (58) **Estimate** If Anthony spins the spinner 60 times, about how many times should he expect the arrow to stop in sector 3?

A 60 times **B** 40 times

C 20 times **D** 10 times

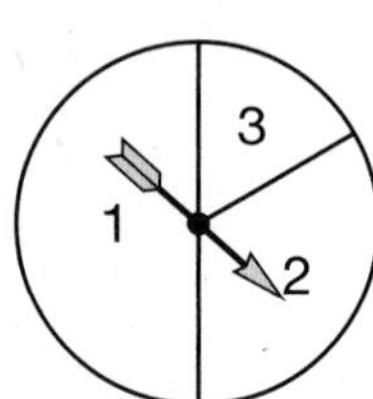

* **26.** (8, 38) **Analyze** An equilateral triangle and a square share a common side. If the area of the square is 100 mm^2, then what is the perimeter of the equilateral triangle?

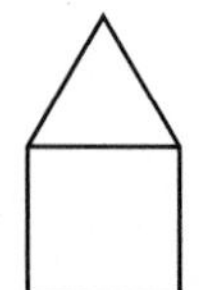

* **27.** (115) Write $11\frac{1}{9}\%$ as a reduced fraction.

28. (Inv. 5) The heights of the five starters on the basketball team are listed below. Find the mean, median, and range of these measures.

181 cm, 177 cm, 189 cm, 158 cm, 195 cm

* **29.** (102, 113) Keisha bought two bunches of bananas. The smaller bunch weighed 2 lb 12 oz. The larger bunch weighed 3 lb 8 oz. What was the total weight of the two bunches of bananas?

* **30.** (93, 110) *Classify* Which type of triangle has no lines of symmetry?

A equilateral **B** isosceles **C** scalene

Early Finishers
Real-World Application

Gerard plays basketball on his high school team. This season he scored 372 of his team's 1488 points.

a. What percent of the points were scored by Gerard's teammates?

b. Gerard's team played 12 games. How many points did Gerard average per game?

LESSON 117

• Finding a Whole When a Fraction Is Known

Power Up — *Building Power*

facts Power Up N

mental math

a. **Fractional Parts:** $\frac{7}{8}$ of 16

b. **Number Sense:** 3×760

c. **Percent:** 25% of $80

d. **Percent:** 5 is what % of 20?

e. **Decimals:** 0.6×40

f. **Number Sense:** $60 \cdot 700$

g. **Statistics:** Find the mode and range of the set of numbers: 99, 101, 34, 44, 120, 34, 43.

h. **Calculation:** 8×8, -1, $\div 7$, $\sqrt{\ }$, $\times 4$, $\div 2$, $\times 3$, $\div 2$

problem solving Three numbers add to 180. The second number is twice the first number, and the third number is three times the first number. Create a visual representation of the equation, and then find the three numbers.

New Concept — *Increasing Knowledge*

Consider the following fractional-parts problem:

Two fifths of the students in the class are boys. If there are ten boys in the class, how many students are in the class?

Thinking Skill

Discuss

What information in the problem suggests that we divide the rectangle into 5 parts?

A diagram can help us understand and solve this problem. We have drawn a rectangle to represent the whole class. The problem states that two fifths are boys, so we divide the rectangle into five parts. Two of the parts are boys, so the remaining three parts must be girls.

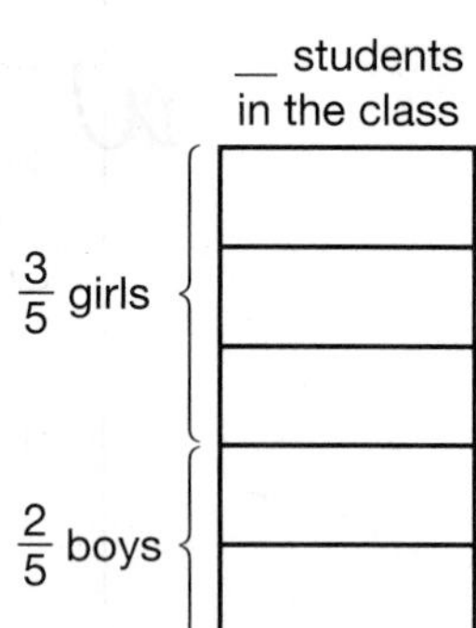

We are also told that there are ten boys in the class. In our diagram ten boys make up two of the parts. Since ten divided by two is five, there are five students in each part. All five parts together represent the total number of students, so there are 25 students in all. We complete the diagram.

25 students in the class	
$\frac{3}{5}$ girls	5
	5
	5
$\frac{2}{5}$ boys	5
	5

Example 1

Three eighths of the townspeople voted. If 120 of the townspeople voted, how many people live in the town?

Solution

We are told that $\frac{3}{8}$ of the town voted, so we divide the whole into eight parts and mark off three of the parts. We are told that these three parts total 120 people. Since the three parts total 120, each part must be 40 ($120 \div 3 = 40$). Each part is 40, so all eight parts must be 8 times 40, which is **320 people.**

___ people live in the town.	
$\frac{3}{8}$ voted.	40
	40
	40
$\frac{5}{8}$ did not vote.	40
	40
	40
	40
	40

Example 2

Six is $\frac{2}{3}$ of what number?

Solution

A larger number has been divided into three parts. Six is the total of two of the three parts. So each part equals three, and all three parts together equal **9.**

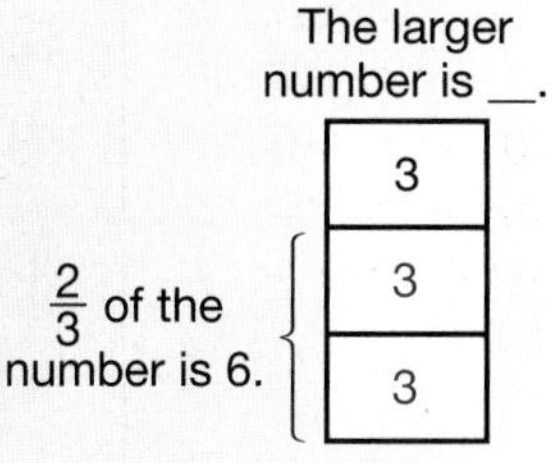

Practice Set

Model Solve. Draw a diagram for problems **a–c.**

a. Eight is $\frac{1}{5}$ of what number?

b. Eight is $\frac{2}{5}$ of what number?

c. Nine is $\frac{3}{4}$ of what number?

d. Sixty is $\frac{3}{8}$ of what number?

e. Three fifths of the students in the class were girls. If there were 18 girls in the class, how many students were in the class altogether?

Written Practice

Strengthening Concepts

*** 1.** (117) Three fifths of the townspeople voted. If 120 of the townspeople voted, how many people live in the town?

*** 2.** (111) **Analyze** If 130 children are separated as equally as possible into four groups, how many will be in each group? (Write four numbers, one for each of the four groups.)

3. (32, 95) If the parking lot charges \$1.25 per half hour, what is the cost of parking a car from 11:15 a.m. to 2:45 p.m.?

4. (86) **Analyze** If the area of the square is 400 m^2, then what is the area of the circle? (Use 3.14 for π.)

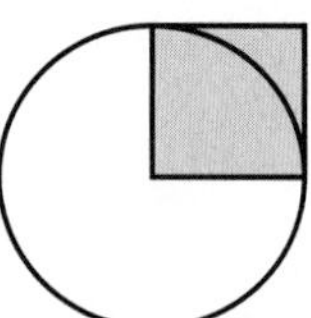

5. (94) Only 4 of the 50 states have names that begin with the letter A. What percent of the states have names that begin with letters other than A?

6. (Inv. 7, 79) **Connect** The coordinates of the vertices of a triangle are (3, 6), (5, 0), and (0, 0). What is the area of the triangle?

7. (35) Write one hundred five thousandths as a decimal number.

8. (51) **Estimate** Round the quotient of \$7.00 ÷ 9 to the nearest cent.

9. (99) Arrange in order from least to greatest:

$$81\%, \frac{4}{5}, 0.815$$

Solve and check:

10. (106) $6x - 12 = 60$

11. (85) $\frac{9}{15} = \frac{m}{25}$

12. (117) Six is $\frac{2}{5}$ of what number?

13. (63) $\left(5 - 1\frac{2}{3}\right) - 1\frac{1}{2}$

14. (68) $2\frac{2}{5} \div 1\frac{1}{2}$

15. (39) 0.625×2.4

16. (104) $-5 + -5 + -5$

17. (73) The prime factorization of 24 is $2 \cdot 2 \cdot 2 \cdot 3$, which we can write as $2^3 \cdot 3$. Write the prime factorization of 36 using exponents.

18. (41) What is the total price of a $12.50 item plus 6% sales tax?

*** 19.** (107) What is the area of this pentagon?

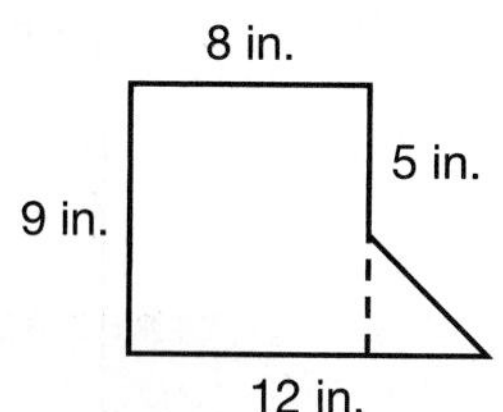

*** 20.** (113) Write the standard numeral for 6×10^5.

*** 21.** (112) **Analyze** Calculate mentally:

a. $\frac{-20}{-4}$ **b.** $\frac{-36}{6}$

c. $(-3)(8)$ **d.** $(-4)(-9)$

22. (82) If each small cube has a volume of one cubic inch, then what is the volume of this rectangular solid?

23. (93) **Represent** Draw a triangle in which each angle measures less than 90°. What type of triangle did you draw?

24. (Inv. 5) **Estimate** The mean, median, and mode of student scores on a test were 89, 87, and 92 respectively. About half of the students scored what score or higher?

*** 25.** (116) **Analyze** A bank offers an annual percentage rate (APR) of 6.5%.

a. By what decimal number do we multiply a deposit to find the total amount in an account after one year at this rate?

b. Maria deposited $1000 into an account at this rate. How much money was in the account after three years? (Assume that the account earns compound interest.)

*** 26.** (Inv. 10) **Analyze** If the spinner is spun twice, what is the probability that the arrow will stop on a number greater than 1 on both spins?

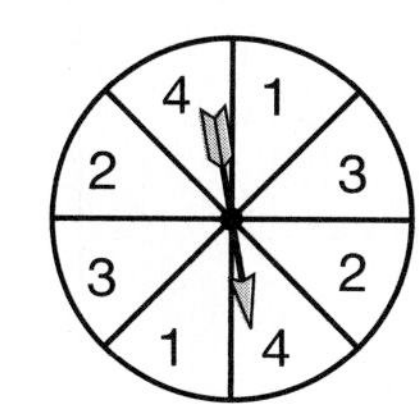

*** 27.** (108) **Conclude** By rotation and translation, these two congruent triangles can be arranged to form a:

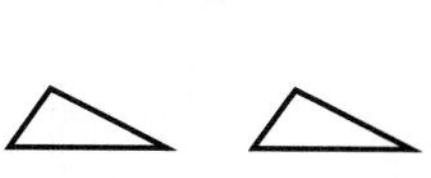

A square **B** parallelogram **C** octagon

*** 28.** (101) **Model** Draw a ratio box for this problem. Then solve the problem using a proportion.

The ratio of cattle to horses on the ranch was 15 to 2. The combined number of cattle and horses was 1020. How many horses were on the ranch?

* **29.** $\sqrt{100} + 3^2 \times 5 - \sqrt{81} \div 3$
(92)

* **30.** **a.** Conclude Which of these figures has the greatest number of lines of symmetry?
(110)

A △ **B** □ **C** ○

b. Which of these figures has rotational symmetry?

Early Finishers
Real-World Application

You and your friends want to rent go-karts this Saturday. Go-Kart Track A rents go-karts for a flat fee of $10 per driver plus $7 per hour. Go-Kart Track B rents go-karts for a flat fee of $7 per driver plus $8 per hour.

a. If you and your friends plan to stay for 2 hours, which track has the better deal? Show all your work.

b. Which track is the better value if you stay for four hours?

LESSON

118 • Estimating Area

Power Up

Building Power

facts Power Up M

mental math

a. Fractional Parts: 6 is $\frac{1}{3}$ of what number?

b. Fractional Parts: $\frac{2}{3}$ of 15

c. Percent: 40% of $20

d. Percent: 10 is what % of 40?

e. Decimals: 0.3×20

f. Number Sense: $300 \cdot 300$

g. Statistics: Find the mode and range of the set of numbers: 567, 899, 576, 345, 899, 907.

h. Calculation: 10×10, $-\ 10$, $\div\ 2$, $-\ 1$, $\div\ 4$, $\times\ 3$, $-\ 1$, $\div\ 4$

problem solving

Carlos reads 5 pages in 4 minutes, and Tom reads 4 pages in 5 minutes. If they both begin reading 200-page books at the same time and do not stop until they are done, how many minutes before Tom finishes will Carlos finish?

New Concept

Increasing Knowledge

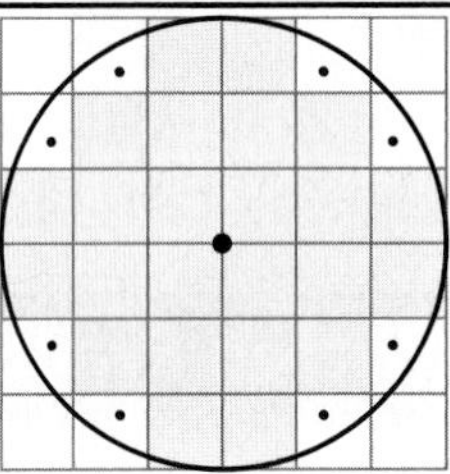

In Lesson 86 we used a grid to estimate the area of a circle. Recall that we counted squares with most of their area within the circle as whole units. We counted squares with about half of their area in the circle as half units. In the figure at right, we have marked these "half squares" with dots.

We can use a grid to estimate the areas of shapes whose areas would otherwise be difficult to calculate.

Example

A one-acre grid is placed over an aerial photograph of a lake. Estimate the surface area of the lake in acres.

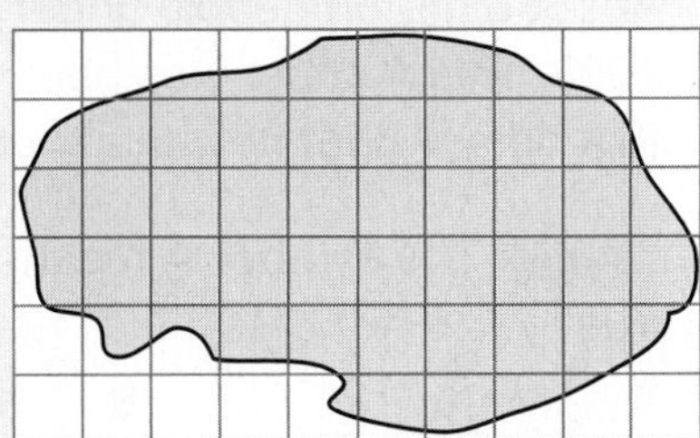

Solution

Each square on the grid represents an area of one acre. The curve is the shoreline of the lake. We count each square that is entirely or mostly within the curve as one acre. We count as half acres those squares that are about halfway within the curve. (Those squares are marked with dots in the figure below.) We ignore bits of squares within the curve because we assume that they balance out the bits of squares we counted that lay outside the curve.

Thinking Skill

Explain

Why do you think area in this example is expressed in acres, rather than in square acres?

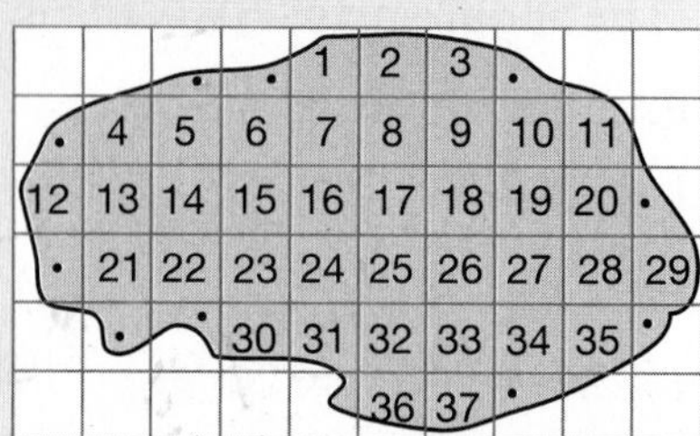

We count 37 entire or nearly entire squares within the shoreline. We also count ten squares with about half of their area within the shoreline. Ten half squares is equivalent to five whole squares. So we estimate the surface area of the lake to be about **42 acres.**

Justify How can we check to see if the answer is reasonable?

Practice Set

Estimate the area of the paw print shown below. Describe the method you used to find your answer.

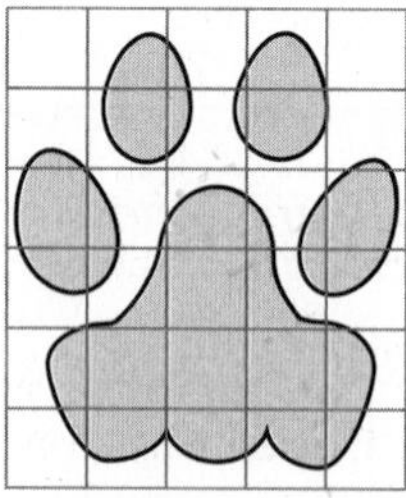

Written Practice

Strengthening Concepts

1. (111) Tabari is giving out baseball cards. Seven students sit in a circle. If he goes around the circle giving out 52 of his baseball cards, how many students will get 8 cards?

2. (7) *Estimate* About how long is a new pencil?

A 1.8 cm **B** 18 cm **C** 180 cm

3. (117) *Estimate* Texas is the second most populous state in the United States. About 6 million people under the age of 18 lived in Texas in the year 2000. This number was about $\frac{3}{10}$ of the total population of the state at that time. About how many people lived in Texas in 2000?

4. (76) *Verify* The symbol $\neq$ means "is not equal to." Which statement is true?

A $\frac{3}{4} \neq \frac{9}{12}$ **B** $\frac{3}{4} \neq \frac{9}{16}$ **C** $\frac{3}{4} \neq 0.75$

5. What is the total price, including 7% tax, of a \$14.49 item?
(41)

6. As Elsa peered out her window she saw 48 trucks, 84 cars, and 12 motorcycles go by her home. What was the ratio of trucks to cars that Elsa saw?
(23)

7. What is the mean of 17, 24, 27, and 28?
(Inv. 5)

8. Arrange in order from least to greatest:
(74)

$$6.1, \sqrt{36}, 6\tfrac{1}{4}$$

*** 9.** **Analyze** Nine cookies were left in the package. That was $\frac{3}{10}$ of the original number of cookies. How many were in the package originally?
(117)

10. **Explain** Buz measured the circumference of the trunk of the old oak tree. How can Buz calculate the approximate diameter of the tree?
(47)

*** 11.** Twelve is $\frac{3}{4}$ of what number?
(117)

12. $2\frac{2}{3} + \left(5\frac{1}{3} - 2\frac{1}{2}\right)$
(72)

13. $6\frac{2}{3} \div 4\frac{1}{6}$
(68)

14. $4\frac{1}{4} + 3.2$ (decimal answer)
(74)

15. $1 - (0.1)^2$
(92)

16. $\sqrt{441}$
(89)

*** 17.** On Earth a kilogram is about 2.2 pounds. Use a unit multiplier to convert 2.2 pounds to ounces. (Round to the nearest ounce.)
(114)

18. The quadrilateral at right is a parallelogram.
(71)

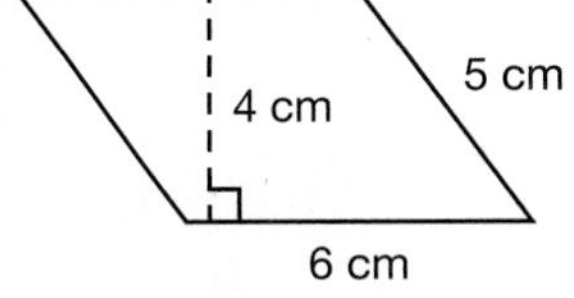

a. What is the area of the parallelogram?

b. If each obtuse angle measures 127°, then what is the measure of each acute angle?

19. What is the perimeter of this hexagon?
(103)

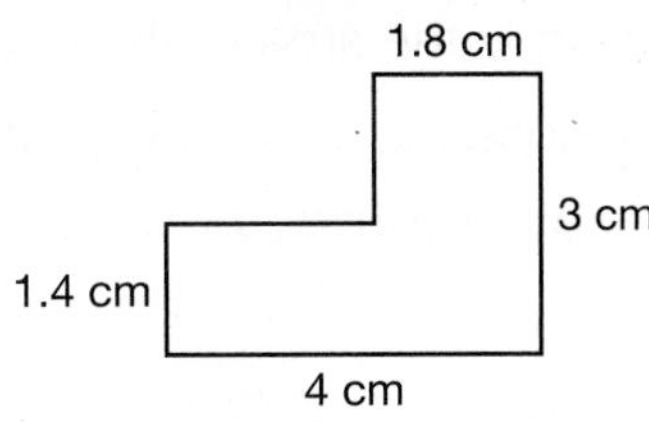

*** 20.** **Conclude** We show two lines of symmetry for this square. A square has a total of how many lines of symmetry?
(110)

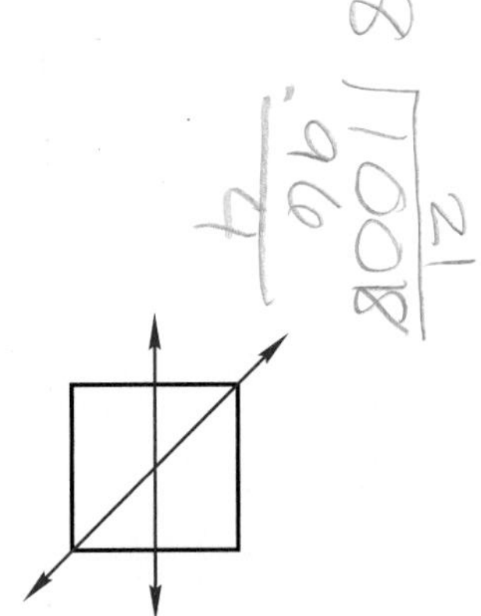

*** 21.** **Estimate** Each edge of a cube measures 4.11 feet. What is a good estimate of the cube's volume?
(82)

22. Complete the proportion: $\frac{f}{12} = \frac{12}{16}$
(85)

*** 23.** (96) **Generalize** Find the rule for this function. Then use the rule to find the missing number.

x	y
2	10
3	15
5	25
	40

*** 24.** (113) Write the standard number for 1.25×10^4.

*** 25.** (104) $-5 + +2 - +3 - -4 + -1$

*** 26.** (116) **Analyze** Esmerelda deposited $4000 in an account that pays $2\frac{1}{2}$% interest compounded annually. How much money was in the account after two years?

*** 27.** (115) Convert $7\frac{1}{2}$% to a fraction.

28. (14) At noon the temperature was −3°F. By sunset the temperature had dropped another five degrees. What was the temperature at sunset?

29. (Inv. 10) There were three red marbles, three white marbles, and three blue marbles in a bag. Luis drew a white marble out of the bag and held it. If he draws another marble out of the bag, what is the probability that the second marble also will be white?

*** 30.** (118) **Estimate** Humberto is designing a garden. The area of the outer garden, square *ABCD* is 16 units^2.

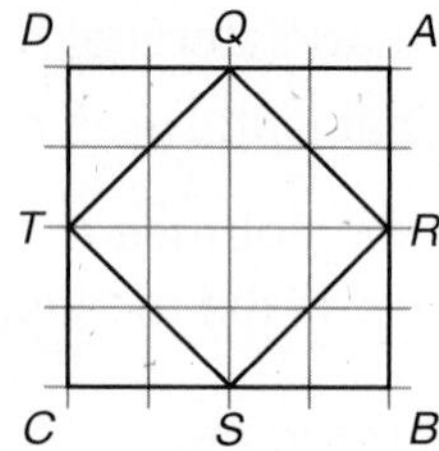

a. What is the area of the inner garden square *QRST?*

b. Choose the appropriate unit for the area of Humberto's garden.

A square inches **B** square feet **C** square miles

LESSON **119**

Finding a Whole When a Percent Is Known

Power Up — *Building Power*

facts Power Up N

mental math

a. **Fractional Parts:** $\frac{2}{3}$ of what number is 6?

b. **Fractional Parts:** $\frac{7}{10}$ of 60

c. **Percent:** 70% of \$50

d. **Percent:** 3 is what % of 6?

e. **Decimals:** 0.8×70

f. **Number Sense:** $\frac{5000}{25}$

g. **Statistics:** Find the mode and range of the set of numbers: 78, 89, 34, 89, 56, 89, 56.

h. **Calculation:** $\frac{1}{2}$ of 50, $\sqrt{\ }$, $\times$ 6, + 2, ÷ 4, $\times$ 3, + 1, $\times$ 4, $\sqrt{\ }$

problem solving

Michelle's grandfather taught her this math method for converting kilometers to miles: "Divide the kilometers by 8, and then multiply by 5." Michelle's grandmother told her to, "Just multiply the kilometers by 0.6." Use both methods to convert 80 km to miles and compare the results. If one kilometer is closer to 0.62 miles, whose method produces the more accurate answer?

New Concept — *Increasing Knowledge*

We have solved problems like the following using a ratio box. In this lesson we will practice writing equations to help us solve these problems.

Thirty percent of the football fans in the stadium are waving team banners. There are 150 football fans waving banners. How many football fans are in the stadium in all?

The statement above tells us that 30% of the fans are waving banners and that the number is 150. We will write an equation using *t* to stand for the total number of fans.

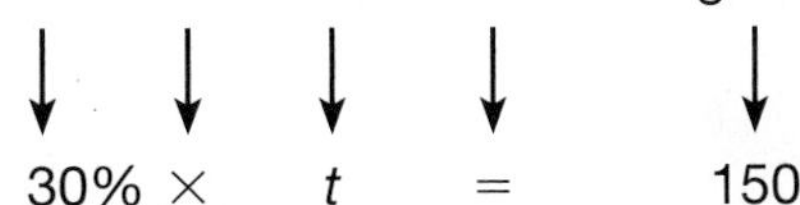

Now we change 30% to a fraction or to a decimal. For this problem we choose to write 30% as a decimal.

$$0.3t = 150$$

Now we find t by dividing 150 by three tenths.

$$0.3\overline{)150.0} \rightarrow 03.\overline{)1500.} \quad 500.$$

We find that there were 500 fans in all. We can use a model to represent the problem.

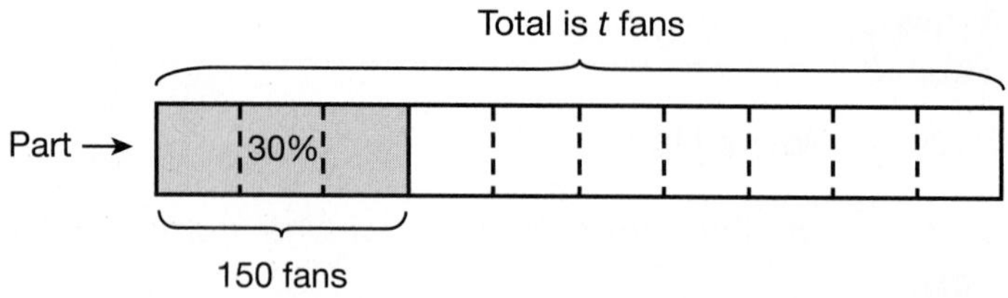

Example 1

Thirty percent of what number is 120? Find the answer by writing and solving an equation. Model the problem with a sketch.

Solution

To translate the question into an equation, we translate the word *of* into a multiplication sign and the word *is* into an equal sign. For the words *what number* we write the letter n.

Thirty percent of what number is 120?

$$30\% \times n = 120$$

Thinking Skill

Explain

How can we check the answer?

We may choose to change 30% to a fraction or a decimal number. We choose the decimal form.

$$0.3n = 120$$

Now we find n by dividing 120 by 0.3.

$$03.\overline{)1200.} \quad 400.$$

Thirty percent of **400** is 120.

We were given a part (30% is 120) and were asked for the whole. Since 30% is $\frac{3}{10}$ we divide 100% into ten divisions instead of 100.

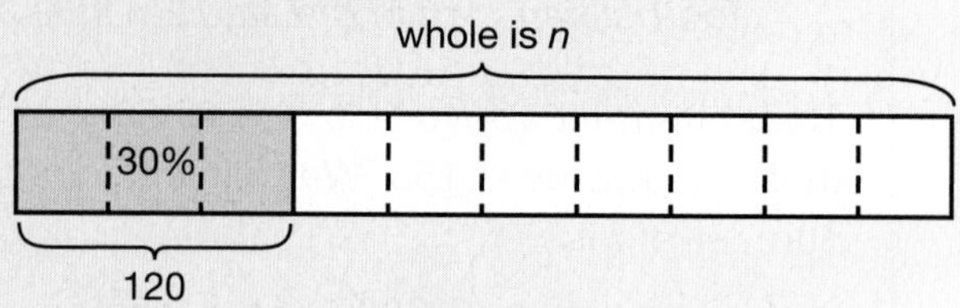

Example 2

Sixteen is 25% of what number? Solve with an equation and model with a sketch.

Solution

We translate the question into an equation, using an equal sign for *is*, a multiplication sign for *of*, and a letter for *what number*.

Sixteen is 25% of what number?

$$16 = 25\% \times n$$

Because of the way the question was asked, the numbers are on opposite sides of the equal sign as compared to example 1. We can solve the equation in this form, or we can rearrange the equation. Either form of the equation may be used.

$$16 = 25\% \times n$$

$$25\% \times n = 16$$

Thinking Skill

Discuss

Why did we use a fraction to solve this problem rather than a decimal?

We will use the first form of the equation and change 25% to the fraction $\frac{1}{4}$.

$$16 = 25\% \times n$$

$$16 = \frac{1}{4}n$$

We find n by dividing 16 by $\frac{1}{4}$.

$$16 \div \frac{1}{4} = \frac{16}{1} \times \frac{4}{1} = 64$$

Sixteen is 25% of **64.**

We are given a part and asked for the whole. Since 25% is $\frac{1}{4}$ we model 100% with four sections of 25%.

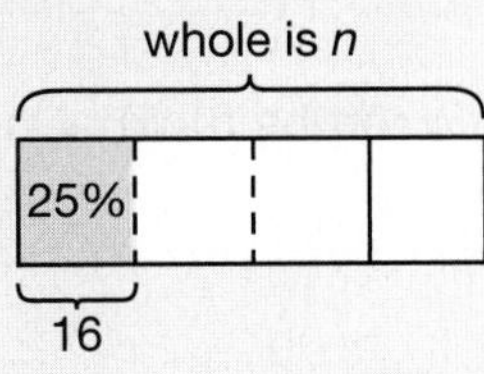

Practice Set

Formulate Translate each question into an equation and solve:

a. Twenty percent of what number is 120?

b. Fifty percent of what number is 30?

c. Twenty-five percent of what number is 12?

d. Twenty is 10% of what number?

e. Twelve is 100% of what number?

f. Fifteen is 15% of what number?

g. Write and solve a word problem for the equation below.

$$15\% \times n = 12$$

Written Practice — *Strengthening Concepts*

1. (25) Divide 555 by 12 and write the quotient

a. with a remainder.

b. as a mixed number.

2. (37) The six gymnasts scored 9.75, 9.8, 9.9, 9.4, 9.9, and 9.95. The lowest score was not counted. What was the sum of the five highest scores?

*** 3.** (105) **Analyze** Cantara said that the six trumpet players made up 10% of the band. The band had how many members?

*** 4.** (117) Eight is $\frac{2}{3}$ of what number?

5. (92) Write the standard number for the following:

$$(1 \times 10^5) + (8 \times 10^4) + (6 \times 10^3)$$

*** 6.** (Inv. 11) On Rob's scale drawing, each inch represents 8 feet. One of the rooms in his drawing is $2\frac{1}{2}$ inches long. How long is the actual room?

7. (98) Two angles of a triangle each measure 45°.

a. **Generalize** What is the measure of the third angle?

b. **Represent** Make a rough sketch of the triangle.

*** 8.** (115) Convert $8\frac{1}{3}\%$ to a fraction.

9. (75) Nine dollars is what percent of $12?

*** 10.** (119) **Analyze** Twenty percent of what number is 12?

*** 11.** (117) Three tenths of what number is 9?

12. (104) $(-5) - (+6) + (-7)$

13. (112) $(-15)(-6)$

14. (29) Reduce: $\frac{60}{84}$

15. (63) $2\frac{1}{2} - 1\frac{2}{3}$

16. (101) **Analyze** Stephen competes in a two-event race made up of biking and running. The ratio of the length of the distance run to the length of the bike ride is 2 to 5. If the distance run was 10 kilometers, then what was the total length of the two-event race?

17. (79) The area of the shaded triangle is 2.8 cm^2. What is the area of the parallelogram?

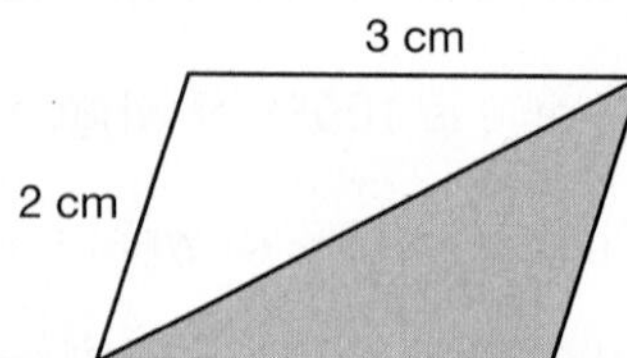

18. (82) The figurine was packed in a box that was 10 in. long, 3 in. wide, and 4 in. deep. What was the volume of the box?

19. (110) A rectangle that is not a square has a total of how many lines of symmetry?

20. (Inv. 6) **Conclude** If this shape were cut out and folded on the dotted lines, would it form a cube, a pyramid, or a cone?

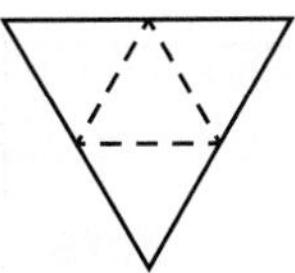

21. (106) $3m - 5 = 25$

* **22.** (96) **Generalize** Write the rule for this function as an equation. Then use the rule to find the missing number.

x	y
3	12
4	16
6	24
	32

23. (102) How many pounds is 10 tons?

24. (64) **Classify** Which of these polygons is not a quadrilateral?

A parallelogram **B** pentagon **C** trapezoid

25. (86) Compare: area of the square ◯ area of the circle

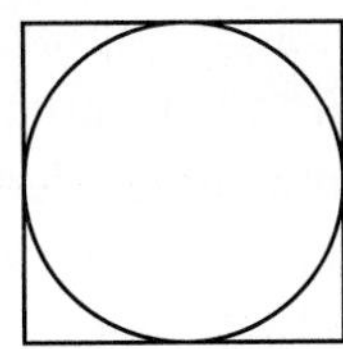

26. (Inv. 7) **Evaluate** The coordinates of three points that are on the same line are $(-3, -2)$, $(0, 0)$, and $(x, 4)$. What number should replace x in the third set of coordinates?

27. (58) Robert flipped a coin. It landed heads up. He flipped the coin a second time. It landed heads up. If he flips the coin a third time, what is the probability that it will land heads up?

28. (Inv. 10) James is going to flip a coin three times. What is the probability that the coin will land heads up all three times?

29. (38, 86) The diameter of the circle is 10 cm.

a. What is the area of the square?

b. What is the area of the circle?

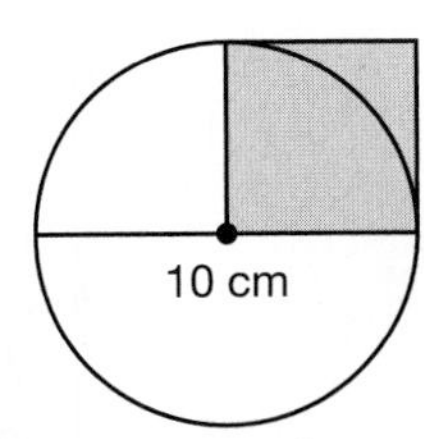

30. (Inv. 5) What is the mode and the range of this set of numbers?

4, 7, 6, 4, 5, 3, 2, 6, 7, 9, 7, 4, 10, 7, 9

LESSON 120

Volume of a Cylinder

Power Up

Building Power

facts Power Up M

mental math

a. **Fractional Parts:** $\frac{2}{3}$ of 27

b. **Percent:** 80% of \$60

c. **Fractional Parts:** $\frac{3}{4}$ of what number is 9?

d. **Percent:** 5 is what % of 5?

e. **Decimals:** 0.8×400

f. **Number Sense:** $10 \cdot 20 \cdot 30$

g. **Statistics:** Find the mode and range of the set of numbers: 908, 234, 980, 243, 908, 567.

h. **Calculation:** $\frac{1}{4}$ of 24, $\times$ 5, + 5, $\div$ 7, $\times$ 8, + 2, $\div$ 7, + 1, $\div$ 7

problem solving

To remind himself of where he buried his treasure, a pirate made this map. He made two of the statements true and one statement false to confuse his enemies in case they captured the map. How will the pirate know where he buried his treasure?

New Concept

Increasing Knowledge

Imagine pressing a quarter down into a block of soft clay.

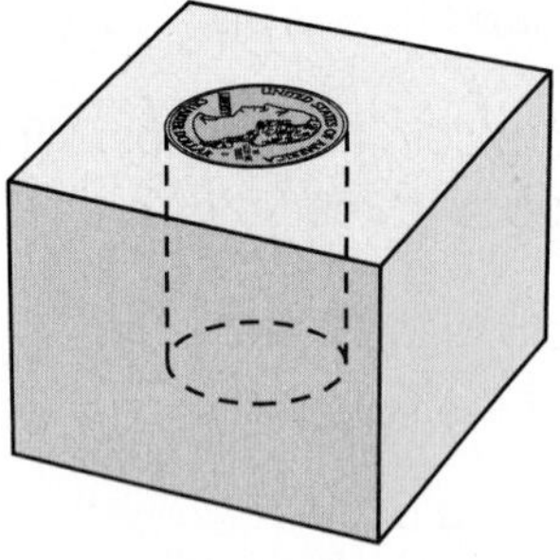

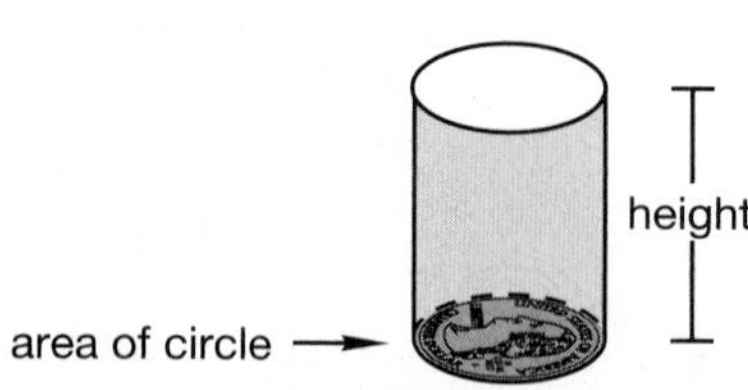

As the quarter is pressed into the block, it creates a hole in the clay. The quarter sweeps out a cylinder as it moves through the clay. We can calculate the volume of the cylinder by multiplying the area of the circular face of the quarter by the distance it moved through the clay. The distance the quarter moved is the **height** of the cylinder.

Thinking Skill

What is the formula for the area of a circle?

Example

The diameter of this cylinder is 20 cm. Its height is 10 cm. What is its volume?

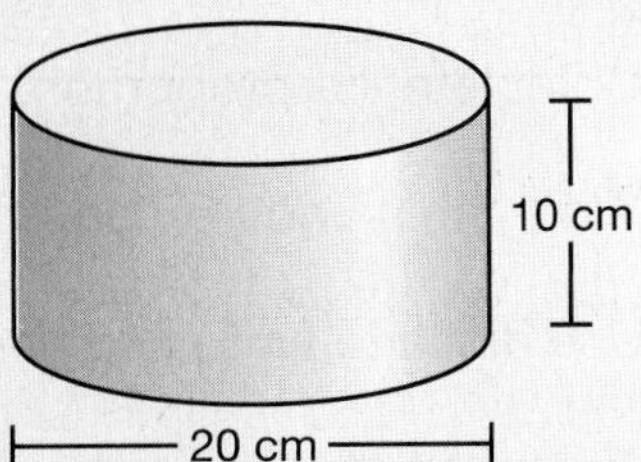

Solution

Thinking Skill

Explain

Why is the volume of this cylinder an approximation and not a precise number?

To calculate the volume of a cylinder, we find the area of a circular end of the cylinder and multiply that area by the height of the cylinder—the distance between the circular ends.

Since the diameter of the cylinder is 20 cm, the radius is 10 cm. A square with a side the length of the radius has an area of 100 cm^2. So the area of the circle is about 3.14 times 100 cm^2, which is 314 cm^2.

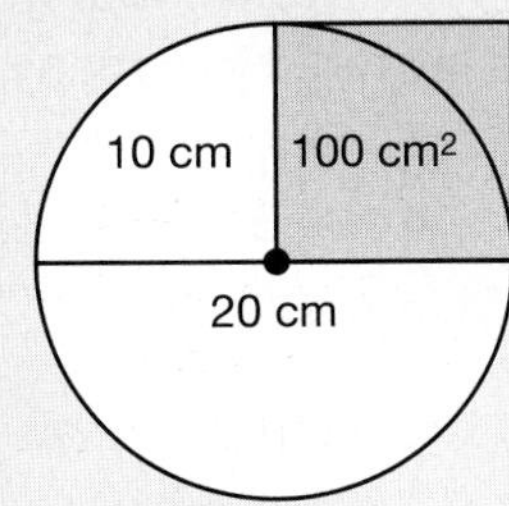

Now we multiply the area of the circular end of the cylinder by the height of the cylinder.

$$314 \text{ cm}^2 \times 10 \text{ cm} = 3140 \text{ cm}^3$$

We find that the volume of the cylinder is approximately **3140 cm^3.**

Practice Set

A large can of soup has a diameter of about 8 cm and a height of about 12 cm. The volume of the can is about how many cubic centimeters? Round your answer to the nearest hundred cubic centimeters.

Written Practice

Strengthening Concepts

1. Write the prime factorization of 750 using exponents.
(73)

2. *Estimate* About how long is your little finger?
(7)
A 0.5 mm **B** 5 mm **C** 50 mm **D** 500 mm

3. (88) Analyze If 3 parts is 24 grams, how much is 8 parts?

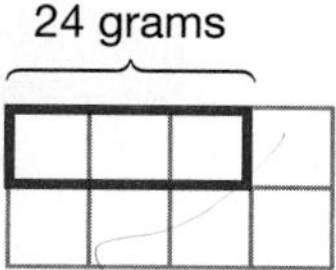

4. (85) Complete the proportion: $\frac{3}{24} = \frac{8}{w}$

5. (92) Write the standard number for $(7 \times 10^3) + (4 \times 10^0)$.

6. (12) Use digits to write two hundred five million, fifty-six thousand.

7. (Inv. 5) The mean of four numbers is 25. Three of the numbers are 17, 23, and 25.

a. What is the fourth number?

b. What is the range of the four numbers?

8. (100, 112) Calculate mentally:

a. $-6 - -4$ **b.** $-10 + -15$ **c.** $(-10)(-10)$

*** 9.** (115) Write $16\frac{2}{3}\%$ as a reduced fraction.

*** 10.** (117) Analyze Twenty-four guests came to the party. This was $\frac{4}{5}$ of those who were invited. How many guests were invited?

11. (61) $1\frac{1}{3} + 3\frac{3}{4} + 1\frac{1}{6}$

12. (72) $\frac{5}{6} \times 3 \times 2\frac{2}{3}$

13. (38) $5.62 + 0.8 + 4$

14. (49) $0.08 \div (1 \div 0.4)$

15. (104) $(-2) + (-2) + (-2)$

16. (89) $\sqrt{2500} + \sqrt{25}$

*** 17.** (114) At $1.12 per pound, what is the price per ounce (1 pound = 16 ounces)?

18. (47) The children held hands and stood in a circle. The diameter of the circle was 10 m. What was the circumference of the circle? (Use 3.14 for π.)

19. (38) Analyze If the area of a square is 36 cm^2, what is the perimeter of the square?

20. (82) If each small cube has a volume of 1 cm^3, then what is the volume of this rectangular solid?

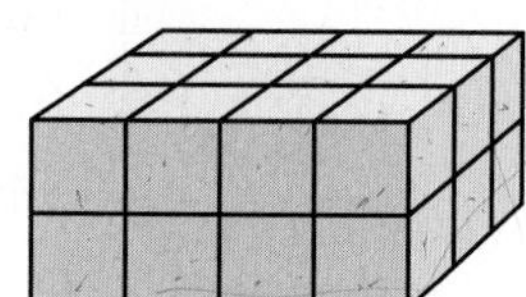

21. (105) Sixty percent of the votes were cast for Shayla. If Shayla received 18 votes, how many votes were cast in all?

22. (Inv. 10) Evaluate Kareem has a spinner marked *A, B, C, D*. Each letter fills one fourth of the face of his spinner. If he spins the spinner three times, what is the probability he will spin *A* three times in a row?

23. (58) If the spinner from problem 22 is spun twenty times, how many times is the spinner likely to land on *C*?

24. (100) **a.** $(-8) - (+7)$ **b.** $(-8) - (-7)$

25. (104) $+3 + -5 - -7 - +9 + +11 + -7$

26. (Inv. 7, 79) **Connect** The three vertices of a triangle have the coordinates (0, 0), (−8, 0), and (−8, −8). What is the area of the triangle?

27. (Inv. 10) Kaya tossed a coin and it landed heads up. What is the probability that her next two tosses of the coin will also land heads up?

* **28.** (120) The inside diameter of a mug is 8 cm. The height of the mug is 7 cm. What is the capacity of the mug in cubic centimeters? (Think of the capacity of the mug as the volume of a cylinder with the given dimensions.)

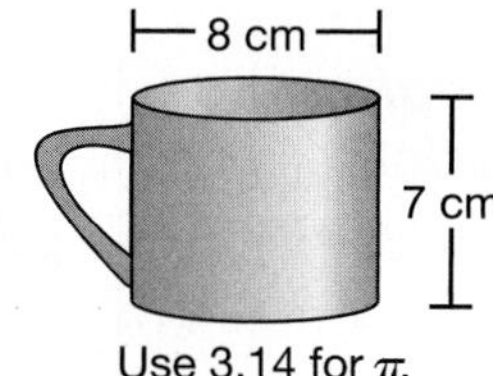

Use 3.14 for π.

29. (78) **Estimate** A cubic centimeter of liquid is a milliliter of liquid. The mug in problem 28 will hold how many milliliters of hot chocolate? Round to the nearest ten milliliters.

* **30.** (105) **Model** Draw a ratio box for this problem. Then solve the problem using a proportion.

Ricardo correctly answered 90% of the trivia questions. If he incorrectly answered four questions, how many questions did he answer correctly?

Early Finishers
Choose A Strategy

Lisa plans to use 20 tiles for a border around a square picture frame. She wants a two-color symmetrical design that has a 2 to 3 ratio of white tiles to gray tiles. What could her design look like?

Focus on

• Volume of Prisms, Pyramids, Cylinders, and Cones

• Surface Area of Prisms and Cylinders

volume of prisms and pyramids

As we learned in Investigation 6, a **prism** is a polyhedron with two congruent, parallel bases. A **pyramid** is a three-dimensional object with a polygon as its base and triangular faces that meet at a vertex.

The height of a prism is the perpendicular distance from the prism's base to its opposite face. The height of a pyramid is the perpendicular distance from the pyramid's base to its vertex.

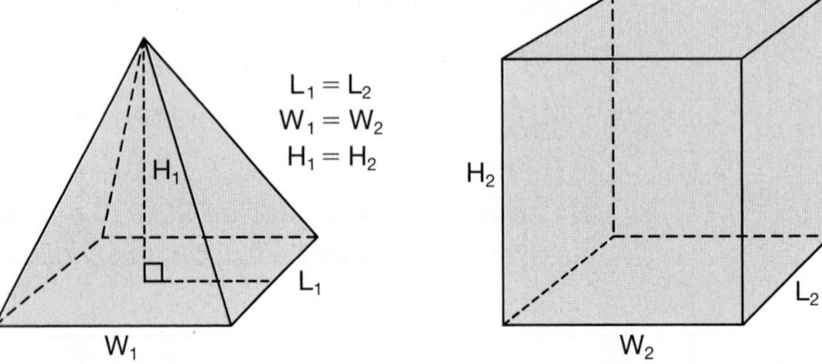

In the figure above, the base of the pyramid is congruent to the bases of the prism, so $L_1 = L_2$ and $W_1 = W_2$. Also, the height of the two solids is the same so $H_1 = H_2$.

Thus, the fundamental difference between the two solids is that the pyramid has a vertex rather than a second base. As a result, its volume is smaller. We can see this clearly when we compare the Relational GeoSolids of the two figures.

We know from Lesson 82 that the formula for finding the volume of a prism is:

$$V = lwh$$

Since *lw* gives us the area of the base, we can also write the formula as:

$$V = \text{area of } B \times h$$

We can use this formula to derive (develop) the formula for the volume of a pyramid. To find the volume of a pyramid, we first find the volume of a similar prism (cube).

By drawing segments from one vertex of a cube to four other vertices, we can see how a cube can be divided into pyramids. The base of the cube is the base of one pyramid. Its right face is the base of a second pyramid and its back face is the base of the third pyramid.

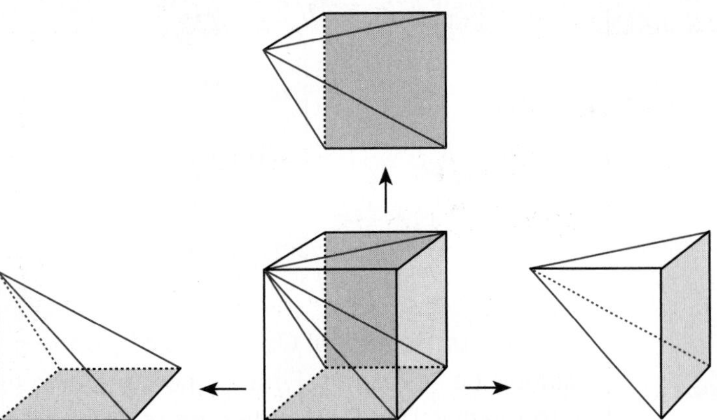

We see that the cube is divided into three congruent pyramids indicating that the volume of each pyramid is $\frac{1}{3}$ the volume of the cube.

To find the volume of one pyramid, we find the volume of $\frac{1}{3}$ of a prism with the same base and height.

$$V \text{ of a pyramid} = \frac{1}{3} \text{ area of } B \times \text{height} = \frac{1}{3}(B \times h)$$

Generalize Use the derived formula to find the volume of each of the following pyramids.

1.

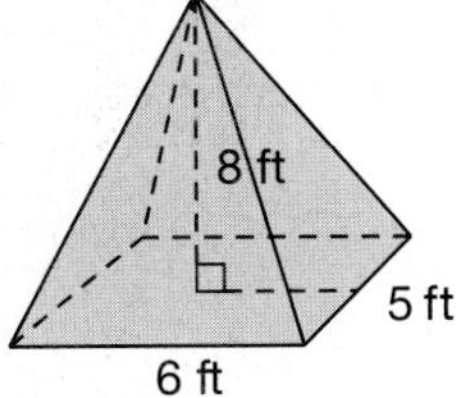

2.

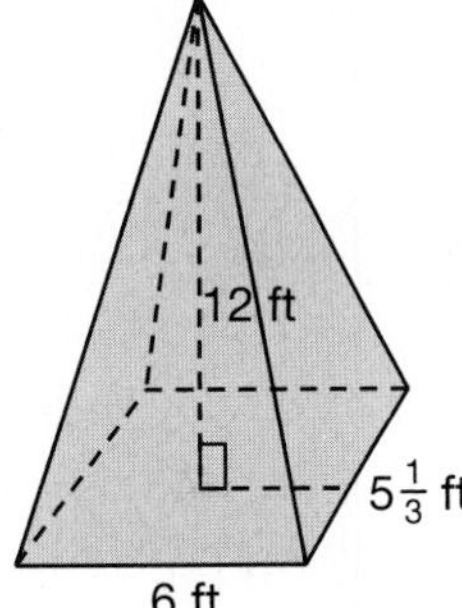

3.

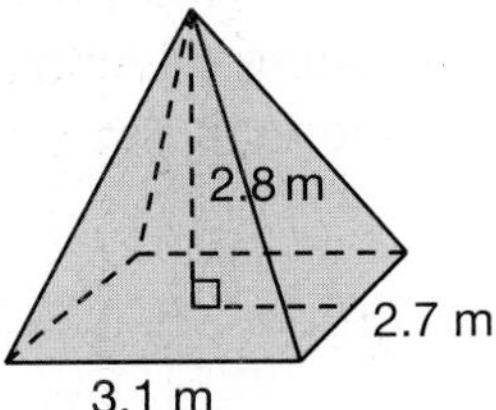

volume of cylinders and cones

A **cylinder** is a solid with two circular bases that are opposite and parallel to each other. Its face is curved. A **cone** is a solid with one circular base and a single vertex. Its face is curved. In the figure below, the base of the cone is congruent to the bases of the cylinder, and the height of the two solids is the same. Thus, the fundamental difference between the two solids is that the cone has a vertex rather than a second base. As a result, its volume is smaller.

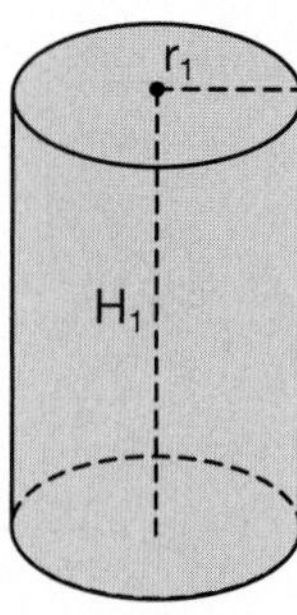

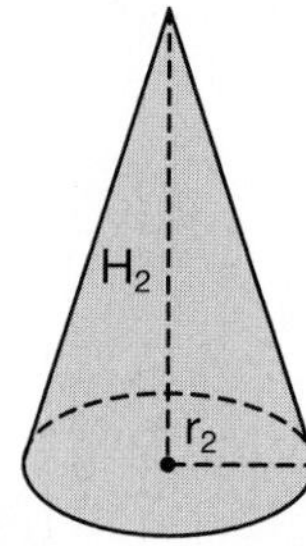

Using your Relational GeoSolids of a cone and a cylinder, demonstrate the difference in volume. First fill the cone with rice or salt. Then empty the cone into the cylinder. Repeat two more times to show that the volume of the cylinder is three times the volume of the cone.

In Lesson 120, we learned that the volume of a cylinder can be found by multiplying the area of the circular end and multiplying the result by the height of the cylinder. We can express this process as a formula:

$$V \text{ of a cylinder} = \pi \cdot r^2 \times h, \text{ or } \pi \cdot r^2h$$

Look at the figures below. Apply what we learned about the volumes of a prism and a pyramid to make a reasonable statement about the volumes of a cylinder and a cone.

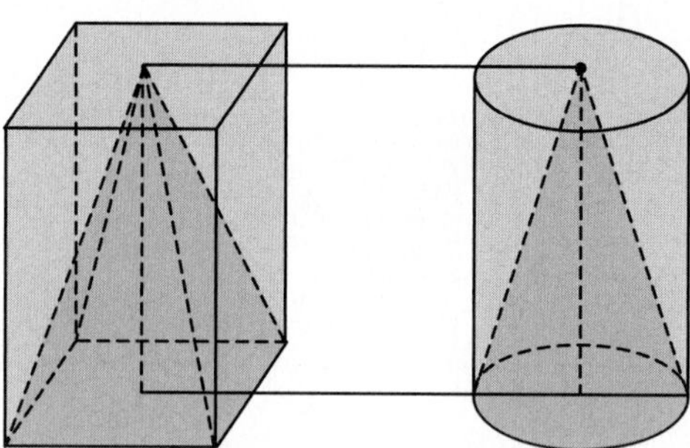

Recall that the volume of a pyramid is $\frac{1}{3}$ the volume of a rectangular prism with the same base and height. The same relationship is true for cones and cylinders. That is, the volume of the cone is $\frac{1}{3}$ the volume of a cylinder with the same height and base area.

$$V \text{ of a cone} = \frac{1}{3} \text{ area of } B \times \text{height} = \frac{1}{3}(B \times h)$$

When we insert the formula for *B,* the area of the base of the cylinder, we get:

$$V \text{ of a cone} = \frac{1}{3}(\pi r^2 h)$$

Generalize Use the appropriate formulas to find the volume of each of the following:

4. Leave π as π.

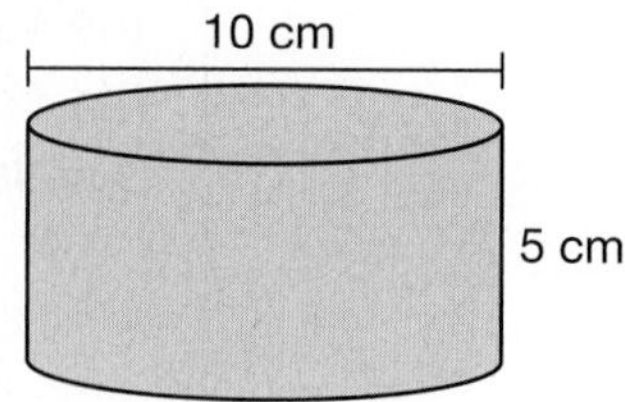

5. Use $\frac{22}{7}$ for π. Estimate to find the answer.

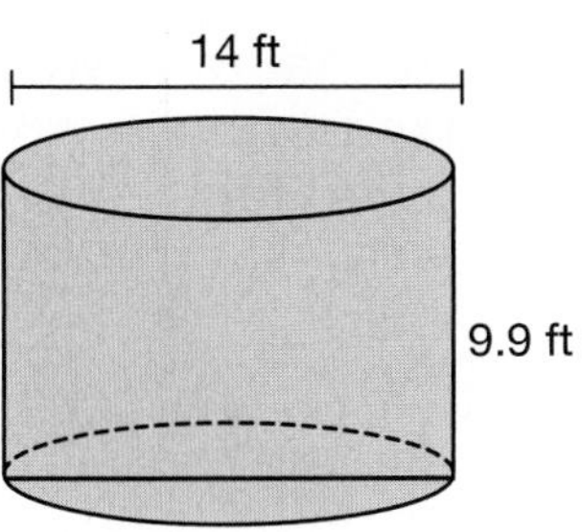

6. Leave π as π.

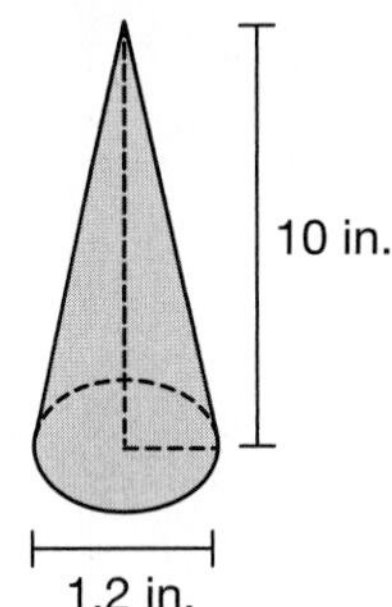

surface area of a prism

The **surface area** of a prism is equal to the sum of the areas of its surfaces. In **Investigation 6,** we found that we could use a net to help us find surface area.

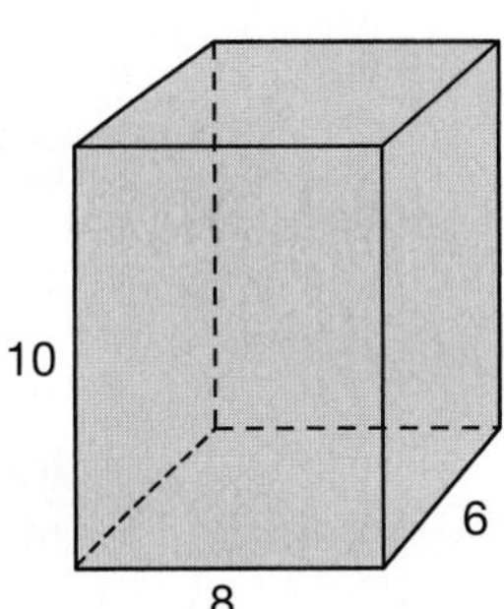

If we compare the rectangular prism in the Relational GeoSolids to the figures on the previous page, we see that we can find the surface area by adding the area of the six sides.

Area of two faces (top and bottom) = $(6 \times 8) + (6 \times 8)$

Area of two faces (sides) = $(6 \times 10) + (6 \times 10)$

Area of two faces (front and back) = $(8 \times 10) + (8 \times 10)$

Thus the total surface area of the prism is 376 in.^{2}

From this, we can develop a formula for the surface area of a prism:

$$SA = 2lw + 2lh + 2wh$$

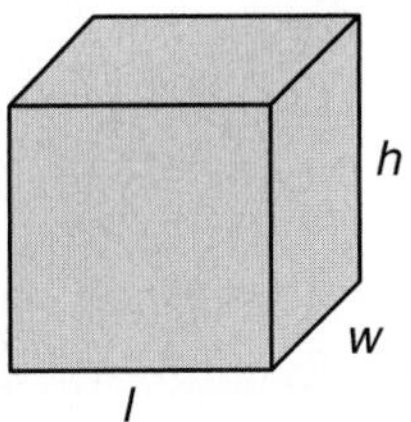

Generalize Find the surface area of the following rectangular prisms.

7.

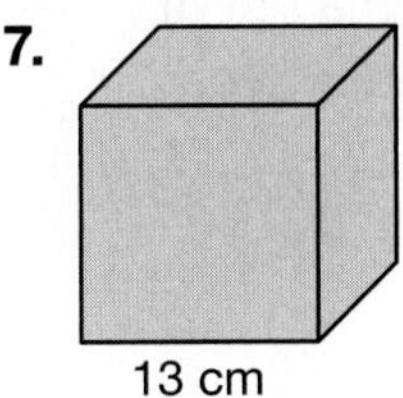

8.

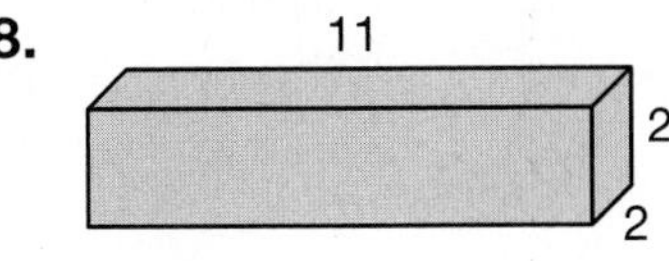

surface area of a cylinder

We can think of the surface area of a cylinder as having three parts—the area of the two bases and the area of its face, its **lateral surface area.** A net of the cylinder makes this easy to see.

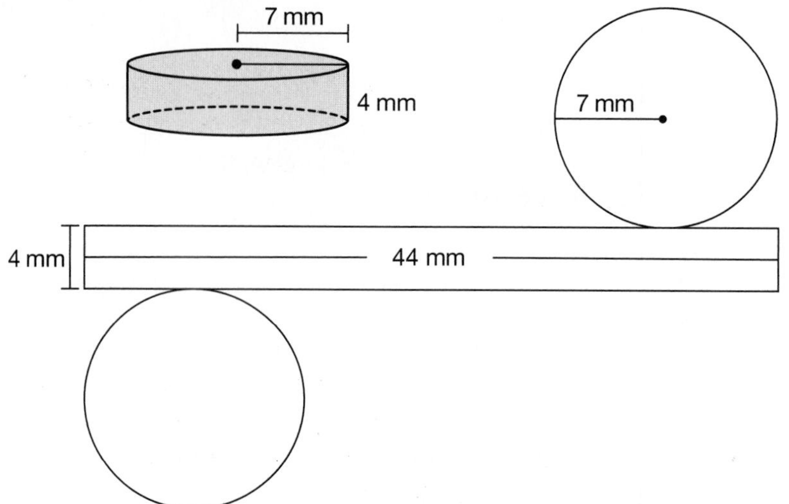

To calculate the area of the circular bases, use the formula for the area of a circle, $A = \pi r^2$. Since there are two bases, multiply the formula by 2.

$$\text{Area} = 2\pi r^2$$

To calculate the lateral surface area, which is a rectangle, use the formula for the area of a rectangle *lw.* In this case, the length (*l*) is the circumference of the circle, or $2\pi r$. The width (*w*) is height of the cylinder. We'll use $\frac{22}{7}$ for π.

We can calculate the surface area of the cylinder above as follows:

$$SA = 2\pi r^2 + 2\pi r\text{h}$$

$$SA \approx 2 \cdot \left(\frac{22}{7}\right) \cdot 7^2 + 2 \cdot \left(\frac{22}{7}\right) \cdot 7 \cdot 4$$

$$SA \approx 2 \cdot \left(\frac{22}{7}\right) \cdot 7^2 + 2 \cdot \left(\frac{22}{7}\right) \cdot 7 \cdot 4$$

$$SA \approx 308 + 176$$

$$SA \approx 484 \text{ mm}^2$$

The surface area of the cylinder is about 484 mm^2.

Applications

9. Martin is installing a 10-feet-tall cylindrical tank to collect rainwater. He wants to know how much water the tank can hold. If the radius of the tank is 2 feet, what is the approximate volume of the tank? To wrap the *entire* tank with insulation, Martin needs to find the total surface area. Draw a net of the cylinder and estimate the surface area. Leave π as π.

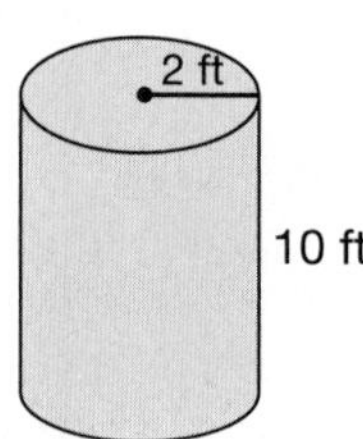

10. *Estimate* Lydia is making coffee for dinner guests and wants to know how much ground coffee her new filter will hold. Estimate the volume of a cone-shaped filter with a diameter of 14 cm and a height of 9 cm. Use $\frac{22}{7}$ for π.

11. A cone is inscribed in a right cylinder as shown. What is the volume of the cone? What is the surface area of the cylinder? Leave π as π.

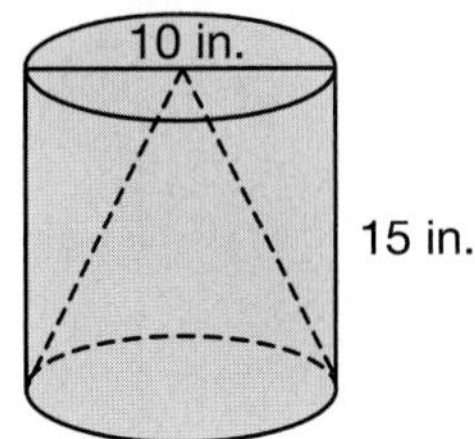

12. Jenna's piano teacher gave her a pyramid-shaped metronome to count time. The metronome's base measures 4 inches by $3\frac{1}{2}$ inches. Calculate the metronome's volume if its height is 9 in.

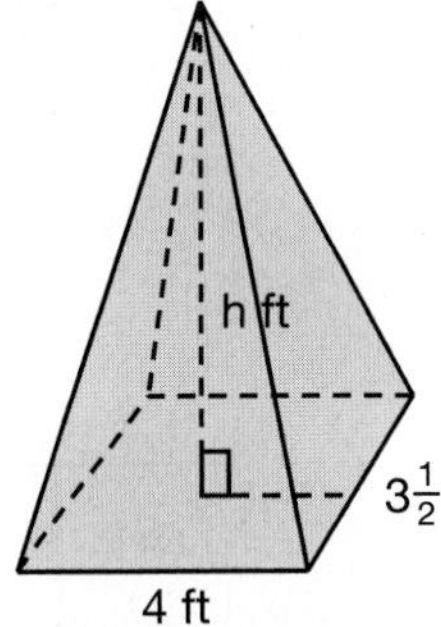

13. *Estimate* Geoff and Sasha drew a sketch of a skateboard ramp they plan to build using scrap wood. To determine how much wood they need to build the ramp, which is shaped like a right triangular prism, estimate the total surface area.

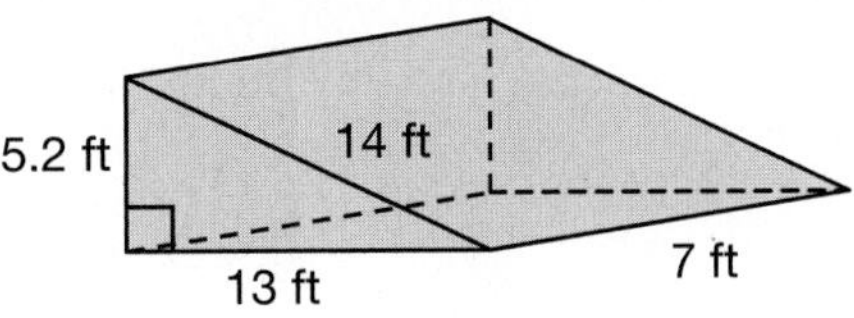

extensions

a. *Represent* Draw a rectangular prism with the same base and height as the pyramid shown. Calculate the volume of the prism. Units are in meters.

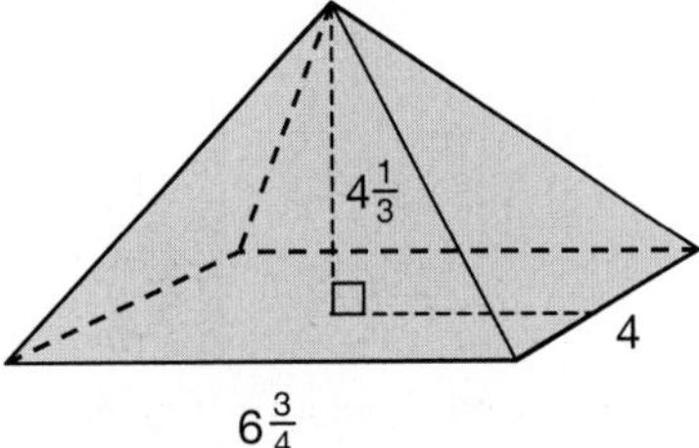

b. *Represent* Sketch a cone inside the pyramid shown with the same height. The diameter of the cone's base equals the width of the pyramid's base. Find the volume of the cone. Leave π as π. Discuss which is greater, the volume of the cone or the volume of the pyramid.

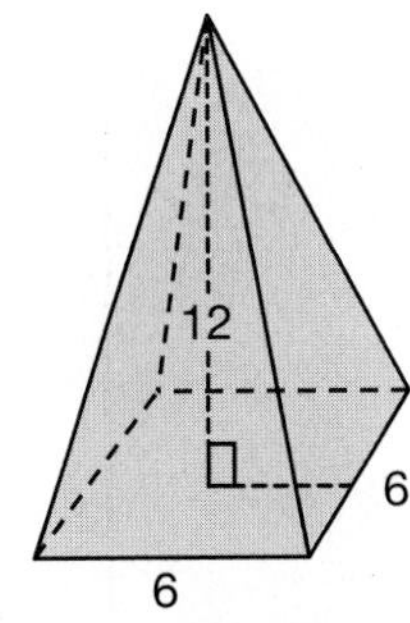

c. Find the surface area of a cube that has a volume of 27 $in.^3$

d. Find the surface area of a cube with an edge of 4 cm.

e. The heights of a rectangular prism with a square base and a cylinder are equal, and the diameter of the cylinder is equal to one edge of the prism's square base. Develop a mathematical argument to prove that surface areas of the two figures are not equal. (*Hint:* Use what you know about the areas of circles and squares to prove your answer.)

A

acute angle
ángulo agudo
(28)

An angle whose measure is more than 0° and less than 90°.

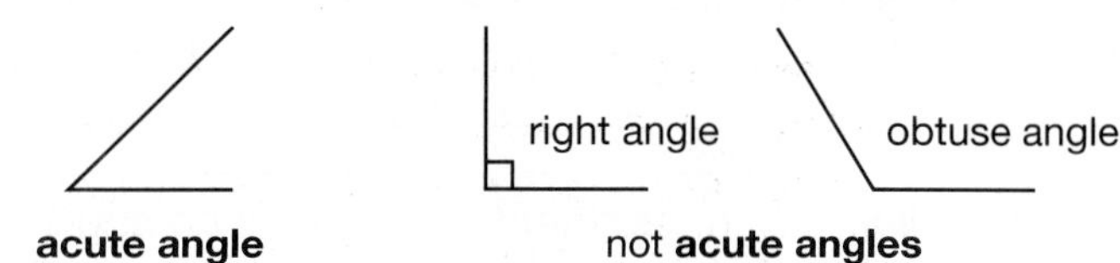

*An **acute angle** is smaller than both a right angle and an obtuse angle.*

---

acute triangle
triángulo acutángulo
(93)

A triangle whose largest angle measures less than 90°.

obtuse triangle

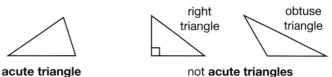

acute triangle not **acute triangles**

---

addend
sumando
(1)

One of two or more numbers that are added to find a sum.

*7 + 3 = 10 The **addends** in this problem are 7 and 3.*

---

algebraic addition
suma algebraica
(100)

The combining of positive and negative numbers to form a sum.

*We use **algebraic addition** to find the sum of −3, +2, and −11:*
$(-3) + (+2) + (-11) = -12$

---

alternate exterior angles
ángulos alternos externos
(97)

A special pair of angles formed when a transversal intersects two lines. Alternate exterior angles lie on opposite sides of the transversal and are outside the two intersected lines.

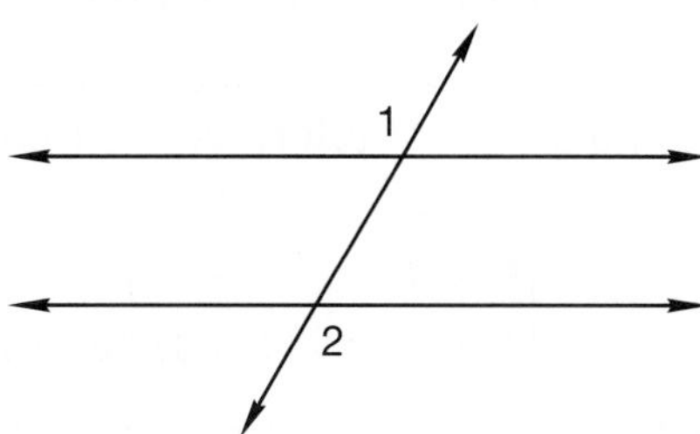

*∠1 and ∠2 are **alternate exterior angles.** When a transversal intersects parallel lines, as in this figure, **alternate exterior angles** have the same measure.*

---

alternate interior angles
ángulos alternos internos
(97)

A special pair of angles formed when a transversal intersects two lines. Alternate interior angles lie on opposite sides of the transversal and are inside the two intersected lines.

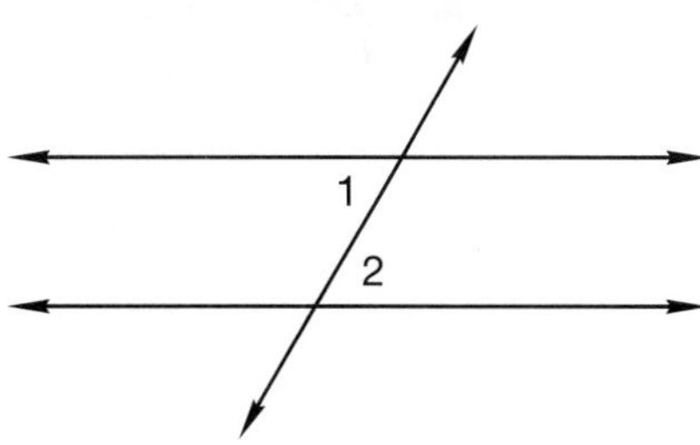

*∠1 and ∠2 are **alternate interior angles.** When a tr… intersects parallel lines, as in this figure, **alternate int…** **angles** have the same measure.*

Sa

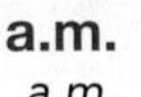

a.m. *a.m.* *(32)*
The period of time from midnight to just before noon.

I get up at 7 ***a.m.*** *I get up at 7 o'clock in the morning.*

---

angle(s) *ángulo(s)* *(28)*
The opening that is formed when two lines, rays, or segments intersect.

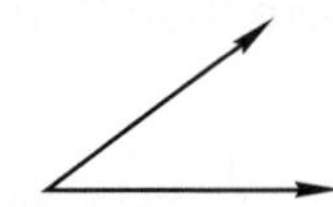

These rays form an ***angle.***

---

angle bisector *bisectriz* *(Inv. 8)*
A line, ray, or segment that divides an angle into two congruent parts.

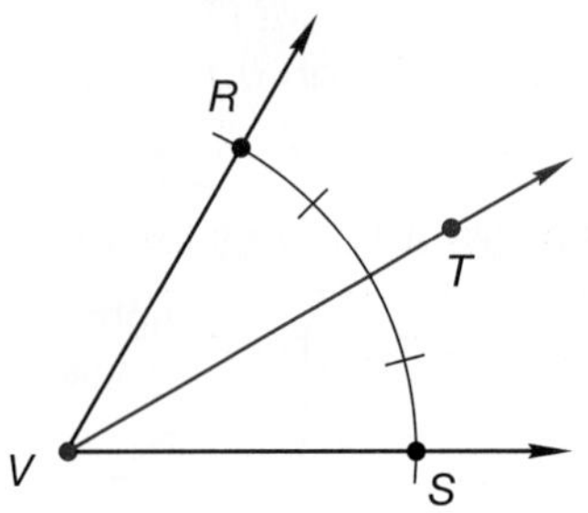

$\overrightarrow{VT}$ *is an* ***angle bisector.***
It divides $\angle RVS$ *in half.*

---

area *área* *(31)*
The number of square units needed to cover a surface.

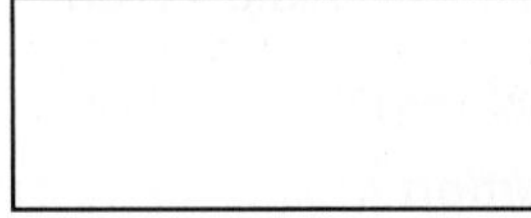

The ***area*** *of this rectangle is 10 square inches.*

Associative Property of Addition *propiedad asociativa de la suma* *(5)*
The grouping of addends does not affect their sum. In symbolic form, $a + (b + c) = (a + b) + c$. Unlike addition, subtraction is not associative.

$(8 + 4) + 2 = 8 + (4 + 2)$	$(8 - 4) - 2 \neq 8 - (4 - 2)$
Addition is ***associative.***	*Subtraction is not* ***associative.***

Associative Property of Multiplication *propiedad asociativa de la multiplicación* *(5)*
The grouping of factors does not affect their product. In symbolic form, $a \times (b \times c) = (a \times b) \times c$. Unlike multiplication, division is not associative.

$(8 \times 4) \times 2 = 8 \times (4 \times 2)$	$(8 \div 4) \div 2 \neq 8 \div (4 \div 2)$
Multiplication is ***associative.***	*Division is not* ***associative.***

---

average *promedio* *(18)*
The number found when the sum of two or more numbers is divided by the number of addends in the sum; also called *mean.*

To find the ***average*** *of the numbers 5, 6, and 10, first add.*

$$5 + 6 + 10 = 21$$

Then, since there were three addends, divide the sum by 3.

$$21 \div 3 = 7$$

The ***average*** *of 5, 6, and 10 is 7.*

B

bar graph(s) *gráfica(s) de barras* *(Inv. 1)*

Displays numerical information with shaded rectangles or bars.

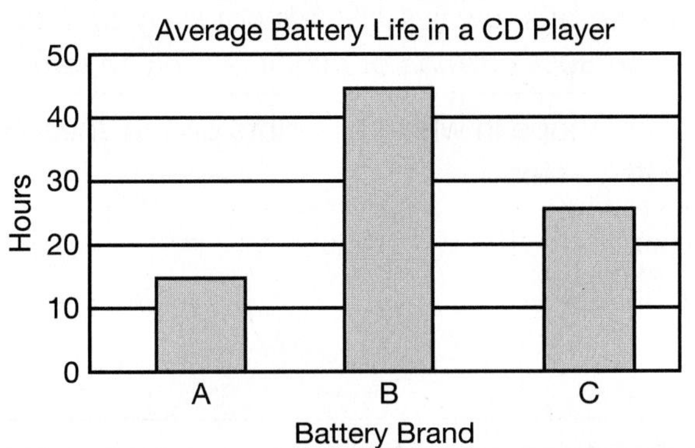

*This **bar graph** shows data for three different brands of batteries.*

base *base* *(71)*

1. A designated side or face of a geometric figure.

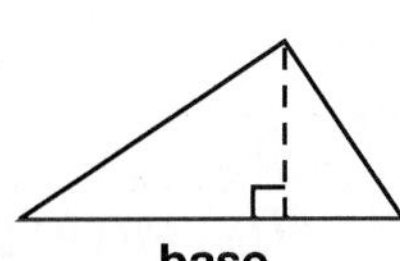

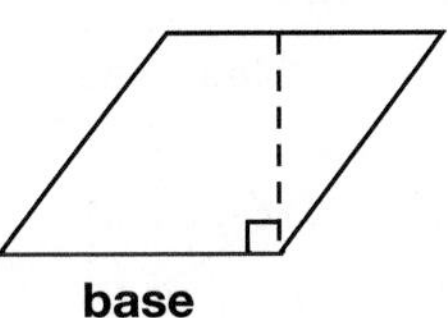

base **base** **base**

2. The lower number in an exponential expression.

base ⟶ 5^3 ⟵ *exponent*

5^3 means $5 \times 5 \times 5$, and its value is 125.

bimodal *bimodal* *(Inv. 5)*

Having two modes.

*The numbers 5 and 7 are the modes of the data at right. This set of data is **bimodal.***

5, 1, 44, 5, 7, 13, 9, 7

bisect *bisecar* *(Inv. 8)*

To divide a segment or angle into two equal halves.

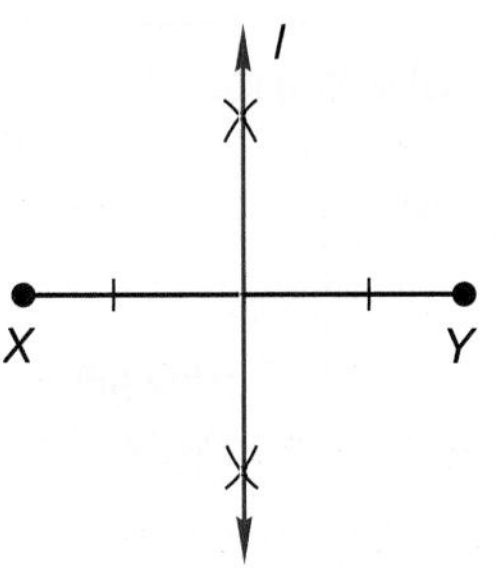

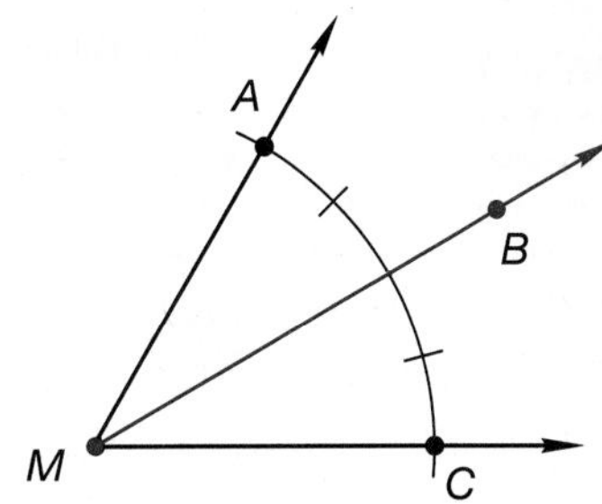

*Line l **bisects** $\overline{XY}$.*

*Ray MB **bisects** ∠AMC.*

C

capacity *capacidad* *(78)*

The amount of liquid a container can hold.

*Cups, gallons, and liters are units of **capacity.***

Celsius scale *escala Celsius* *(10)*

A scale used on some thermometers to measure temperature.

*On the **Celsius scale,** water freezes at 0°C and boils at 100°C.*

chance
posibilidad
(58)

A way of expressing the likelihood of an event; the probability of an event expressed as a percentage.

*The **chance** of snow is 10%. It is not likely to snow.*
*There is an 80% **chance** of rain. It is likely to rain.*

circle
círculo
(27)

A closed, curved shape in which all points on the shape are the same distance from its center.

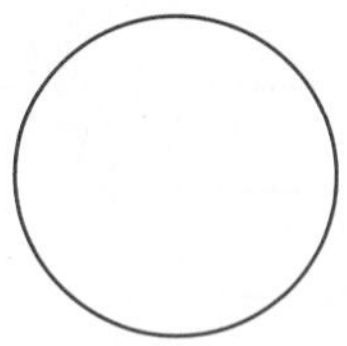

circle

circle graph
gráfica circular
(40)

A method of displaying data, often used to show information about percentages or parts of a whole. A circle graph is made of a circle divided into sectors.

Class Test Grades

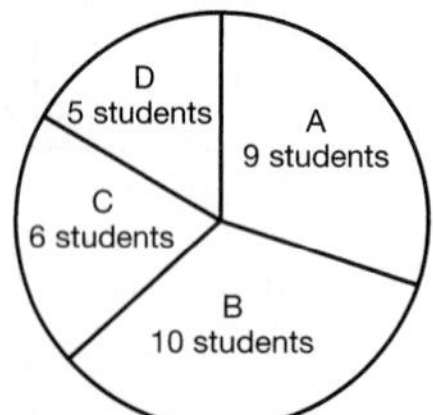

*This **circle graph** shows data for a class's test grades.*

circumference
circunferencia
(27)

The perimeter of a circle.

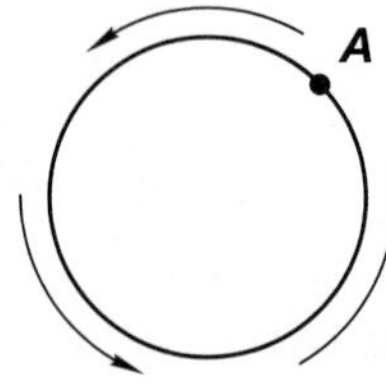

*If the distance from point A around to point A is 3 inches, then the **circumference** of the circle is 3 inches.*

closed-option survey
encuesta de opción cerrada
(Inv. 1)

A survey in which the possible responses are limited.

closed-option survey

common denominator
denominador común
(55)

A number that is the denominator of two or more fractions.

*The fractions $\frac{2}{5}$ and $\frac{3}{5}$ have **common denominators.***

Commutative Property of Addition
propiedad conmutativa de la suma
(1)

Changing the order of addends does not affect their sum. In symbolic form, $a + b = b + a$. Unlike addition, subtraction is not commutative.

$8 + 2 = 2 + 8$	$8 - 2 \neq 2 - 8$
*Addition is **commutative.***	*Subtraction is not **commutative.***

Commutative Property of Multiplication
propiedad conmutativa de la multiplicación
(3)

Changing the order of factors does not affect their product. In symbolic form, $a \times b = b \times a$. Unlike multiplication, division is not commutative.

$8 \times 2 = 2 \times 8$	$8 \div 2 \neq 2 \div 8$
Multiplication is ***commutative.***	*Division is not* ***commutative.***

compass
compás
(27)

A tool used to draw circles and arcs.

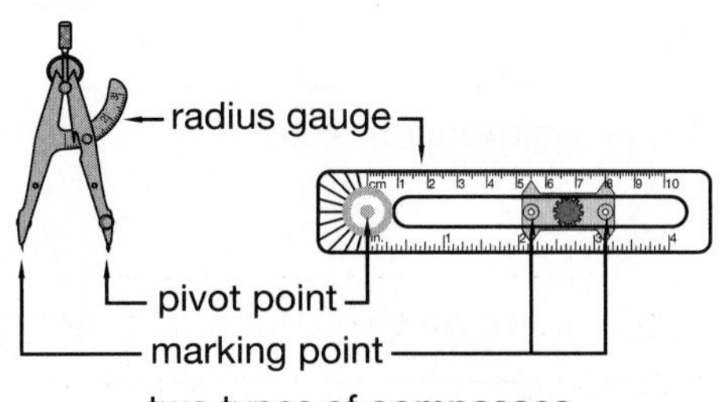

two types of **compasses**

complementary angles
ángulos complementarios
(69)

Two angles whose sum is 90°.

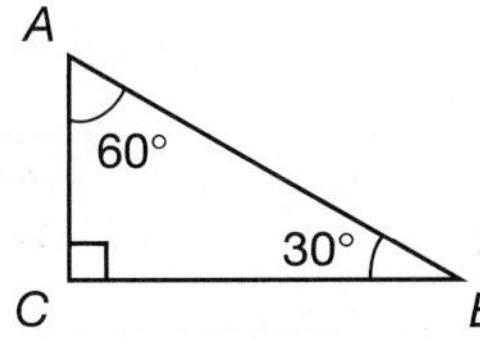

∠A and ∠B are ***complementary angles.***

complement of an event
complemento de un evento
(58)

The opposite of an event. The complement of event B is "not B." The probability of an event and the probability of its complement add up to 1.

composite number
número compuesto
(65)

A counting number greater than 1 that is divisible by a number other than itself and 1. Every composite number has three or more factors.

9 is divisible by 1, 3, and 9. It is ***composite.***
11 is divisible by 1 and 11. It is not ***composite.***

compound experiments
experimentos compuestos
(Inv. 10)

Experiments that contain more than one part performed in order.

compound interest
interés compuesto
(116)

Interest that pays on previously earned interest.

Compound Interest

$100.00	*principal*
+ $6.00	*first-year interest (6% of $100)*
$106.00	*total after one year*
+ $6.36	*second-year interest (6% of $106)*
$112.36	*total after two years*

Simple Interest

$100.00	*principal*
$6.00	*first-year interest (6% of $100)*
+ $6.00	*second-year interest (6% of $100)*
$112.00	*total after two years*

compound outcomes
resultados compuestos
(Inv. 10)

The outcomes to a compound experiment.

concentric circles
círculos concéntricos
(27)

Two or more circles with a common center.

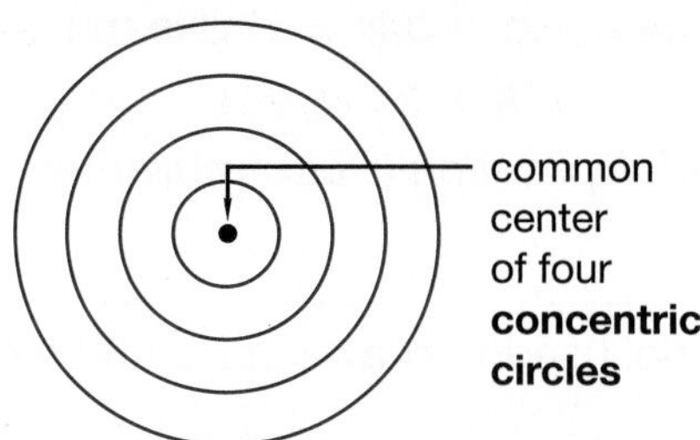

cone
cono
(Inv. 6)

A three-dimensional solid with a circular base and a single vertex.

congruent
congruente
(60)

Having the same size and shape.

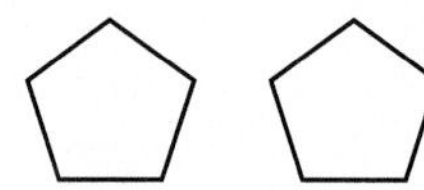

*These polygons are **congruent.** They have the same size and shape.*

coordinate(s)
coordenada(s)
(Inv. 7)

1. A number used to locate a point on a number line.

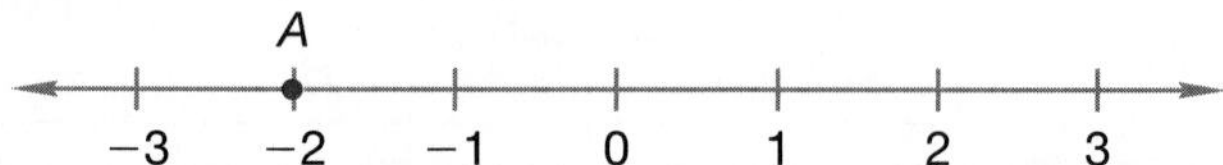

*The **coordinate** of point A is −2.*

2. An ordered pair of numbers used to locate a point in a coordinate plane.

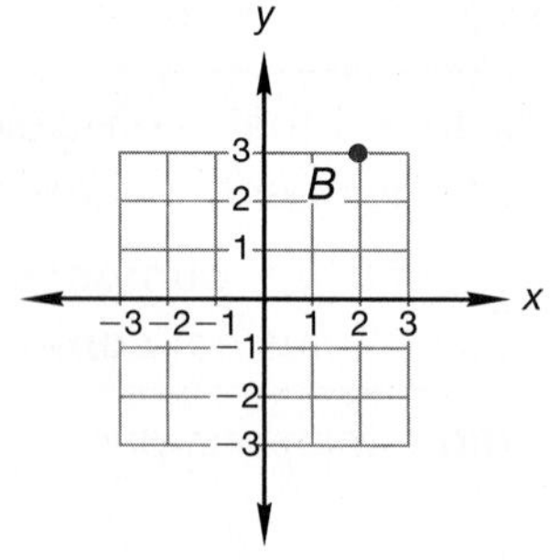

*The **coordinates** of point B are (2, 3). The x-coordinate is listed first, the y-coordinate second.*

coordinate plane
plano coordenado
(Inv. 7)

A grid on which any point can be identified by an ordered pair of numbers.

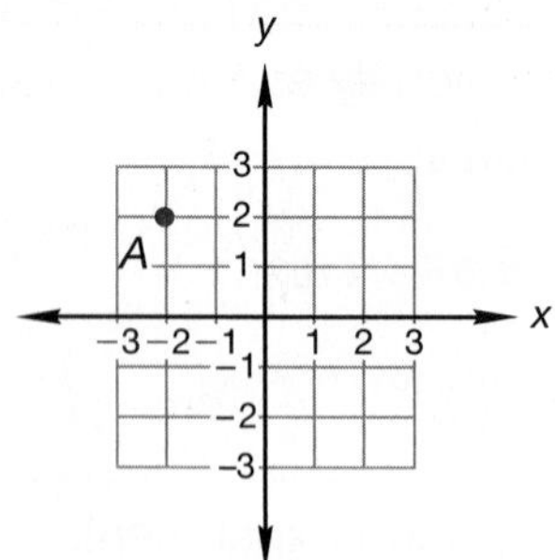

*Point A is located at (−2, 2) on this **coordinate plane.***

corresponding angles
ángulos correspondientes
(97)

A special pair of angles formed when a transversal intersects two lines. Corresponding angles lie on the same side of the transversal and are in the same position relative to the two intersected lines.

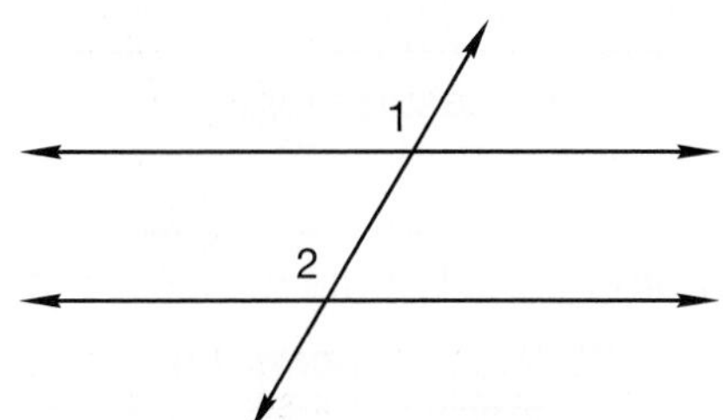

*∠1 and ∠2 are **corresponding angles.** When a transversal intersects parallel lines, as in this figure, **corresponding angles** have the same measure.*

corresponding parts
partes correspondientes
(109)

Sides or angles that occupy the same relative positions in similar polygons.

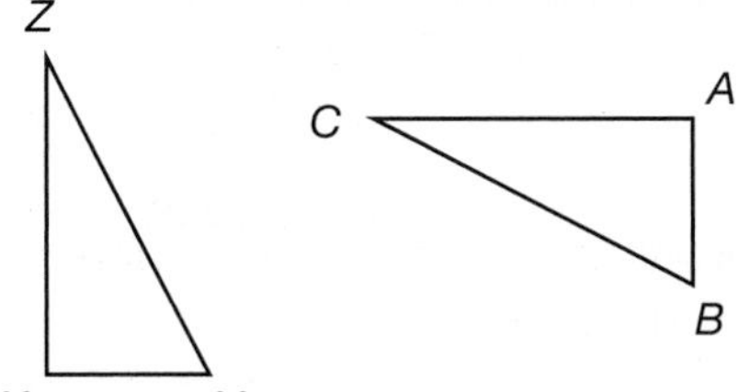

*$\overline{BC}$ **corresponds** to $\overline{YZ}$.*
*∠A **corresponds** to ∠X.*

counting numbers
números de conteo
(9)

The numbers used to count; the members of the set {1, 2, 3, 4, 5, ...}. Also called *natural numbers.*

*1, 24, and 108 are **counting numbers.***
*−2, 3.14, 0, and $2\frac{7}{9}$ are not **counting numbers.***

cross products
productos cruzados
(85)

The product of the numerator of one fraction and the denominator of another.

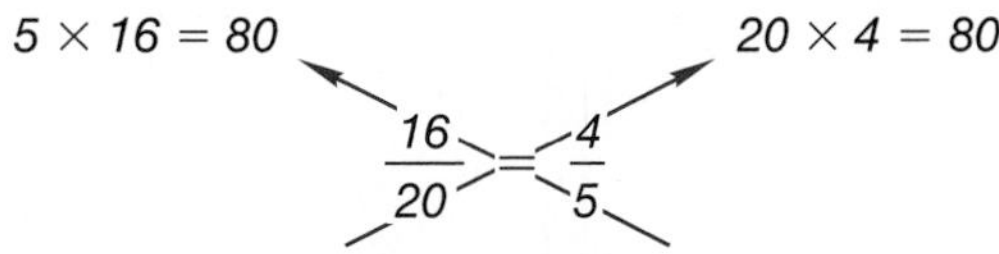

*The **cross products** of these two fractions are equal.*

cube
cubo
(Inv. 6)

A three-dimensional solid with six square faces. Adjacent faces are perpendicular and opposite faces are parallel.

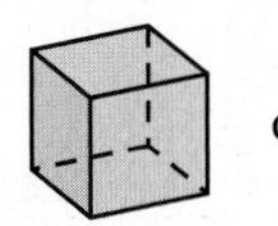

cube

cylinder
cilindro
(Inv. 6)

A three-dimensional solid with two circular bases that are opposite and parallel to each other.

cylinder

D

data
datos
(Inv. 4)

Information that is gathered and organized in a way that conclusions can be drawn from it.

---

data points
puntos de datos
(Inv. 5)

Individual measurements or numbers in a set of data.

---

decimal number
número decimal
(34)

A numeral that contains a decimal point.

*23.94 is a **decimal number** because it contains a decimal point.*

---

decimal places
cifras decimales
(34)

Places to the right of a decimal point.

*5.47 has two **decimal places.***
*6.3 has one **decimal place.***
*8 has no **decimal places.***

---

decimal point
punto decimal
(34)

The symbol in a decimal number used as a reference point for place value.

34.15
↑
decimal point

---

degree (°)
grado
(Inv. 3)

1. A unit for measuring angles.

360°

*There are 90 **degrees** (90°) in a right angle.*

*There are 360 **degrees** (360°) in a circle.*

2. A unit for measuring temperature.

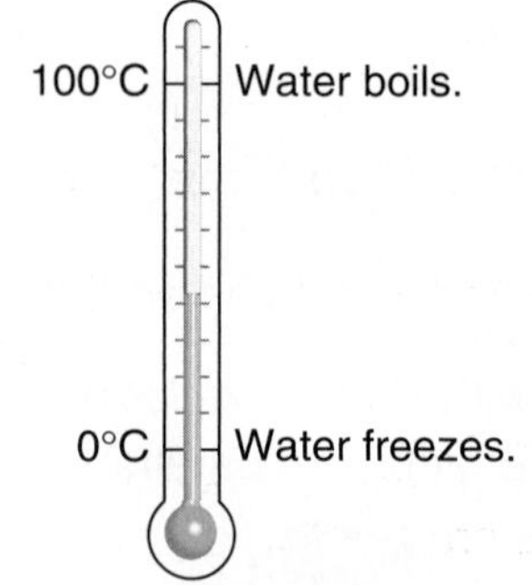

*There are 100 **degrees** between the freezing and boiling points of water on the Celsius scale.*

---

denominator
denominador
(6)

The bottom term of a fraction.

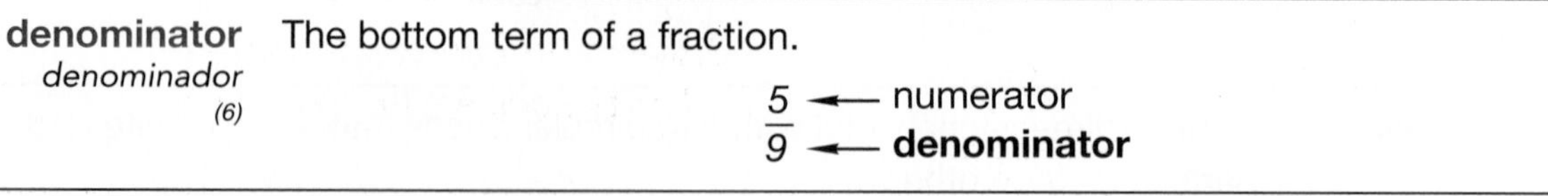

$\frac{5}{9}$ ← numerator
← **denominator**

---

diameter
diámetro
(27)

The distance across a circle through its center.

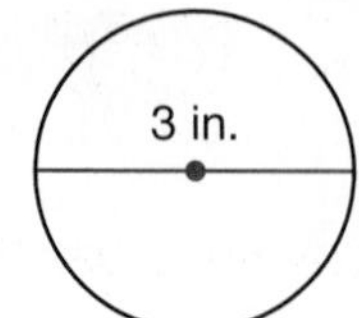

*The **diameter** of this circle is 3 inches.*

difference The result of subtraction.
diferencia
(1)

$12 - 8 = 4$ *The **difference** in this problem is 4.*

digit Any of the symbols used to write numbers: 0, 1, 2, 3, 4, 5, 6, 7, 8, 9.
dígito
(2)

*The last **digit** in the number 7862 is 2.*

dividend A number that is divided.
dividendo
(2)

$12 \div 3 = 4$ $3\overline{)12}$ (quotient 4) $\frac{12}{3} = 4$ *The **dividend** is 12 in each of these problems.*

divisible Able to be divided by a whole number without a remainder.
divisible
(19)

$4\overline{)20}$ (quotient 5) *The number 20 is **divisible** by 4, since 20 ÷ 4 has no remainder.*

$3\overline{)20}$ (quotient 6 R 2) *The number 20 is not **divisible** by 3, since 20 ÷ 3 has a remainder.*

divisor **1.** A number by which another number is divided.
divisor
(2)

$12 \div 3 = 4$ $3\overline{)12}$ (quotient 4) $\frac{12}{3} = 4$ *The **divisor** is 3 in each of these problems.*

2. A factor of a number.

*2 and 5 are **divisors** of 10.*

E

edge A line segment formed where two faces of a polyhedron intersect.
arista
(Inv. 6)

*One **edge** of this cube is colored blue. A cube has 12 **edges.***

endpoint A point at which a segment ends.
extremo
(7)

A B

*Points A and B are the **endpoints** of segment AB.*

equation A statement that uses the symbol "=" to show that two quantities are equal.
ecuación
(3)

$x = 3$ $3 + 7 = 10$ — **equations**

$4 + 1$ $x < 7$ — not **equations**

equilateral triangle A triangle in which all sides are the same length.
triángulo equilátero
(93)

*This is an **equilateral triangle.** All of its sides are the same length.*

equivalent fractions Different fractions that name the same amount.
fracciones equivalentes
(42)

$\frac{1}{2}$ = $\frac{2}{4}$

*$\frac{1}{2}$ and $\frac{2}{4}$ are **equivalent fractions.***

estimate *estimar* (16)
To determine an approximate value.

*We **estimate** that the sum of 199 and 205 is about 400.*

evaluate *evaluar* (73)
To find the value of an expression.

*To **evaluate** $a + b$ for $a = 7$ and $b = 13$, we replace a with 7 and b with 13:*

$$7 + 13 = 20$$

even numbers *números pares* (10)
Numbers that can be divided by 2 without a remainder; the members of the set {..., −4, −2, 0, 2, 4, ...}.

***Even numbers** have 0, 2, 4, 6, or 8 in the ones place.*

event *evento* (58)
Outcome(s) resulting from an experiment or situation.

- *Events that are certain to occur have a probability of 1.*
- *Events that are certain not to occur have a probability of zero.*
- *Events that are uncertain have probabilities that fall anywhere between zero and one.*

expanded notation *notación expandida* (32)
A way of writing a number as the sum of the products of the digits and the place values of the digits.

*In **expanded notation** 6753 is written*

$$(6 \times 1000) + (7 \times 100) + (5 \times 10) + (3 \times 1).$$

experimental probability *probabilidad experimental* (Inv. 9)
The probability of an event occurring as determined by experimentation.

*If we roll a number cube 100 times and get 22 threes, the **experimental probability** of rolling three is $\frac{22}{100}$, or $\frac{11}{50}$.*

exponent *exponente* (38)
The upper number in an exponential expression; it shows how many times the base is to be used as a factor.

base → 5^3 ← **exponent**

5^3 means $5 \times 5 \times 5$, and its value is 125.

exponential expression *expresión exponencial* (73)
An expression that indicates that the base is to be used as a factor the number of times shown by the exponent.

$$4^3 = 4 \times 4 \times 4 = 64$$

*The **exponential expression** 4^3 is evaluated by using 4 as a factor 3 times. Its value is 64.*

expression *expresión*
A combination of numbers and/or variables by operations, but not including an equal or inequality sign.

	equation	inequality
$3x + 2y\ (x - 1)^2$	$y = 3x - 1$	$x < 4$
expressions	not **expressions**	

exterior angle *ángulo externo* (98)
In a polygon, the supplementary angle of an interior angle.

exterior angle

F

face *cara* *(Inv. 6)*
A flat surface of a geometric solid.

*One **face** of the cube is shaded.*
*A cube has six **faces.***

fact family *familia de operaciones* *(1)*
A group of three numbers related by addition and subtraction or by multiplication and division.

*The numbers 3, 4, and 7 are a **fact family.** They make these four facts:*

$3 + 4 = 7 \quad 4 + 3 = 7 \quad 7 - 3 = 4 \quad 7 - 4 = 3$

factor *factor* *(2)*
1. Noun: One of two or more numbers that are multiplied.

   $3 \times 5 = 15$ *The **factors** in this problem are 3 and 5.*

2. Noun: A whole number that divides another whole number without a remainder.

   *The numbers 3 and 5 are **factors** of 15.*

3. Verb: To write as a product of factors.

   *We can **factor** the number 15 by writing it as 3×5.*

factor tree *árbol de factores* *(65)*
A method of finding all the prime factors of a number.

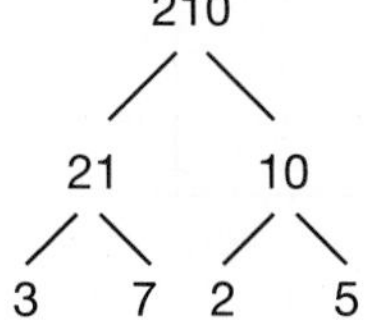

*The numbers on each branch of this **factor tree** are factors of the number 210. Each number at the end of a branch is a prime factor of 210.*

The prime factors of 210 are 2, 3, 5 and 7.

Fahrenheit scale *escala Fahrenheit* *(10)*
A scale used on some thermometers to measure temperature.

*On the **Fahrenheit scale,** water freezes at 32°F and boils at 212°F.*

fraction(s) *fracción* *(6)*
A number that names part of a whole.

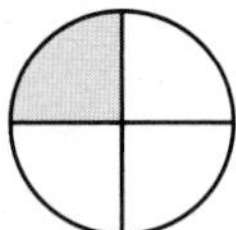

$\frac{1}{4}$ *of the circle is shaded.*

$\frac{1}{4}$ *is a **fraction.***

frequency table *tabla de frecuencias* *(Inv. 1)*
Displays the number of times a value occurs in data.

Daily Temperature Highs in October

Temperature (°F)	Tally	Frequency
81–85	𝍸 𝍸 𝍸 |	16
76–80	𝍸 ||||	9
71–75	𝍸	5
65–70	|	1

*This **frequency table** shows data for temperatures in October.*

function
función
(96)

A rule for using one number (an input) to calculate another number (an output). Each input produces only one output.

$y = 3x$

x	y
3	9
5	15
7	21
10	30

There is exactly one resulting number for every number we multiply by 3. Thus, $y = 3x$ is a ***function.***

G

geometric solid
sólido geométrico
(Inv. 6)

A three-dimensional geometric figure.

geometric solids | not **geometric solids**

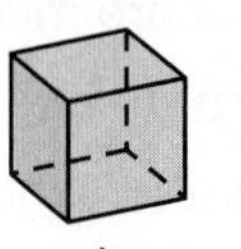

cube cylinder

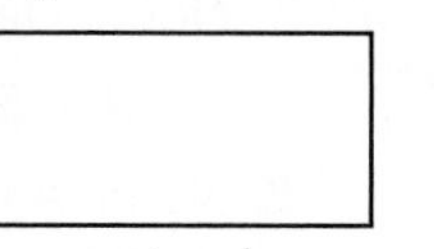

circle rectangle hexagon

graph
gráfica
(Inv. 7)

1. Noun: A diagram, such as a bar graph, a circle graph (pie chart), or a line graph, that displays quantitative information.

Rainy Days

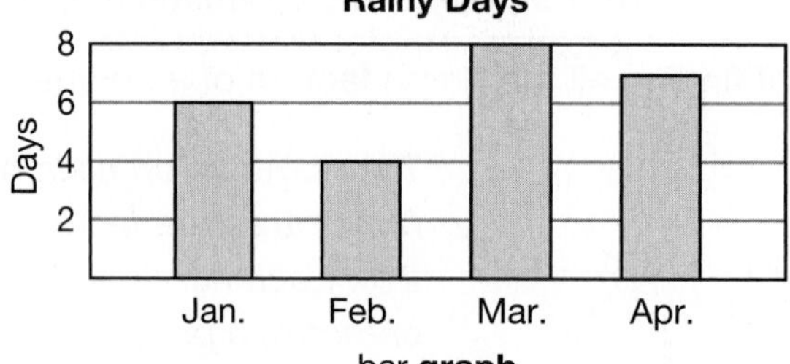

bar **graph**

Hair Colors of Students

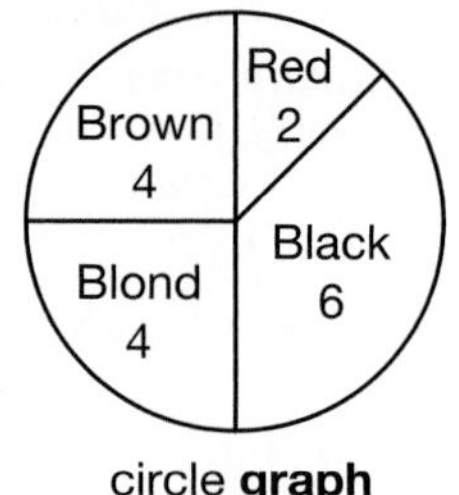

circle **graph**

2. Noun: A point, line, or curve on a coordinate plane.

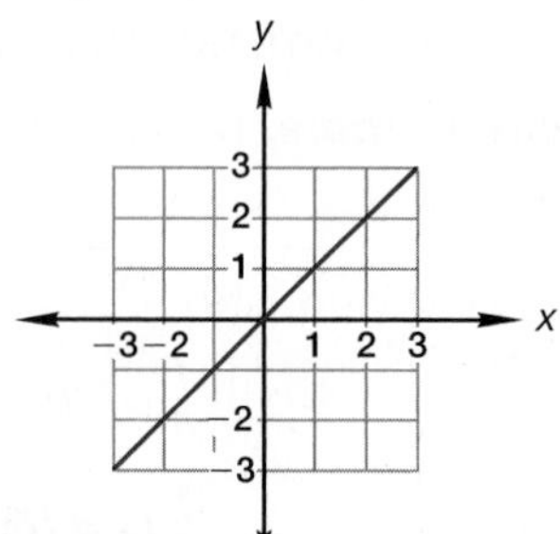

The ***graph*** *of the equation $y = x$*

3. Verb: To draw a point, line, or curve on a coordinate plane.

greatest common factor (GCF)
máximo común divisor (MCD)
(20)

The largest whole number that is a factor of two or more given numbers.

The factors of 12 are 1, 2, 3, 4, 6, and 12.
The factors of 18 are 1, 2, 3, 6, 9, and 18.
The ***greatest common factor*** *of 12 and 18 is 6.*

H

height *altura* *(71)* The perpendicular distance from the base to the opposite side of a parallelogram or trapezoid; from the base to the opposite face of a prism or cylinder; or from the base to the opposite vertex of a triangle, pyramid, or cone.

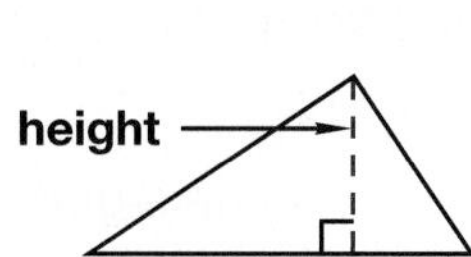

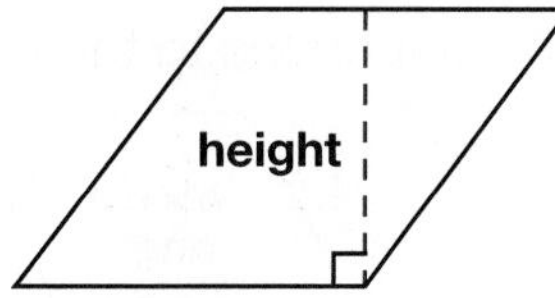

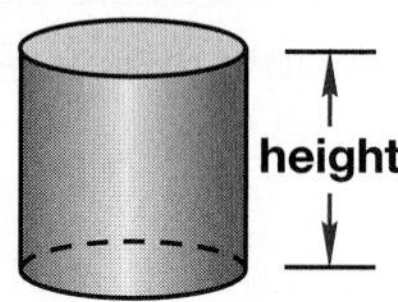

histogram *histograma* *(Inv. 1)* A method of displaying a range of data. A histogram is a special type of bar graph that displays data in intervals of equal size with no space between bars.

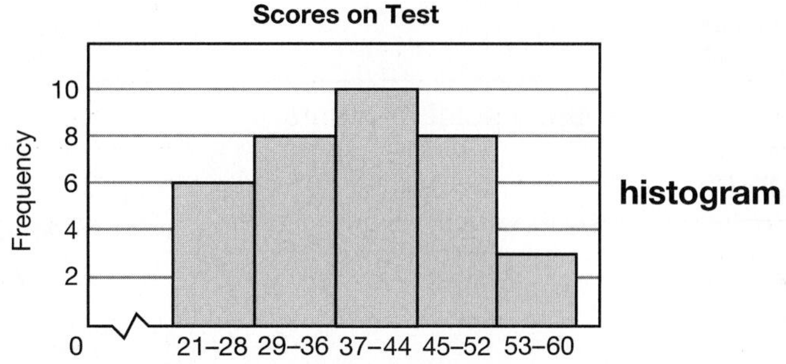

horizontal *horizontal* *(18)* Parallel to the horizon; perpendicular to vertical.

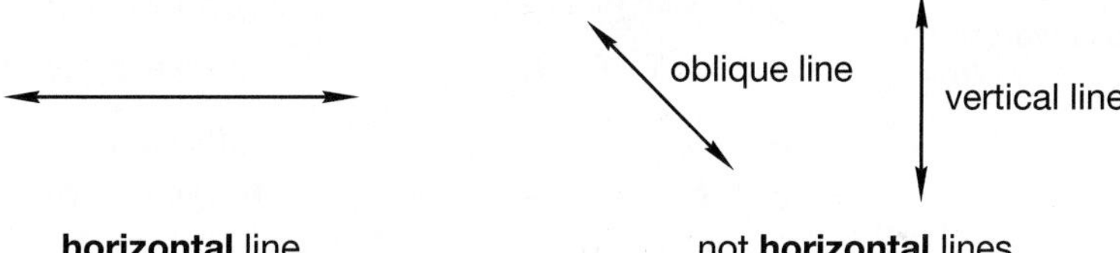

horizontal line

not **horizontal** lines

I

Identity Property of Addition *propiedad de identidad de la suma* *(1)* The sum of any number and 0 is equal to the initial number. In symbolic form, $a + 0 = a$. The number 0 is referred to as the *additive identity.*

*The **Identity Property of Addition** is shown by this statement:*

$$13 + 0 = 13$$

Identity Property of Multiplication *propiedad de identidad de la multiplicación* *(2, 3)* The product of any number and 1 is equal to the initial number. In symbolic form, $a \times 1 = a$. The number 1 is referred to as the *multiplicative identity.*

*The **Identity Property of Multiplication** is shown by this statement:*

$$94 \times 1 = 94$$

improper fraction *fracción impropia* *(Inv. 2)* A fraction with a numerator equal to or greater than the denominator.

$\frac{12}{12}$, $\frac{57}{3}$, *and* $2\frac{15}{2}$ *are **improper fractions.***

*All **improper fractions** are greater than or equal to 1.*

integers *números positivos, negativos y el cero* *(14)* The set of counting numbers, their opposites, and zero; the members of the set $\{\ldots, -2, -1, 0, 1, 2, \ldots\}$.

*-57 and 4 are **integers.** $\frac{15}{8}$ and -0.98 are not **integers.***

interest *interés* *(116)*
An amount added to a loan, account, or fund, usually based on a percentage of the principal.

If we borrow $500.00 from the bank and repay the bank $575.00 for the loan, the ***interest*** *on the loan is $575.00 − $500.00 = $75.00.*

---

interior angle *ángul interno* *(98)*
An angle that opens to the inside of a polygon.

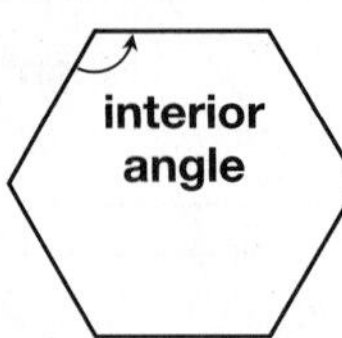

This hexagon has six ***interior angles.***

International System *Sistema internacional* *(7)*
See **metric system.**

---

intersect *intersecar* *(28)*
To share a point or points.

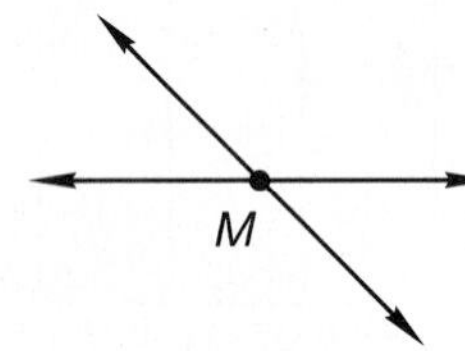

These two lines ***intersect.*** *They share the point M.*

---

inverse operations *operaciones inversas* *(1)*
Operations that "undo" one another.

$a + b - b = a$ $a - b + b = a$	*Addition and subtraction are* ***inverse operations.***
$a \times b \div b = a \quad (b \neq 0)$ $a \div b \times b = a \quad (b \neq 0)$	*Multiplication and division are* ***inverse operations.***
$\sqrt{a^2} = a \quad (a \geq 0)$ $(\sqrt{a})^2 = a \quad (a \geq 0)$	*Squaring and finding square roots are* ***inverse operations.***

---

irrational numbers *números irracionales* *(89)*
Numbers that cannot be expressed as a ratio of two integers. Their decimal expansions are nonending and nonrepeating.

π and $\sqrt{3}$ are ***irrational numbers.***

---

isosceles triangle *triángulo isósceles* *(93)*
A triangle with at least two sides of equal length.

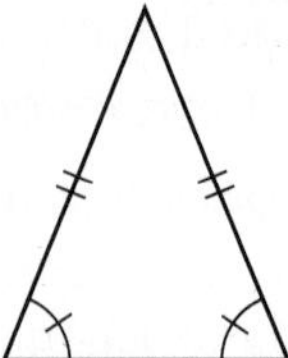

Two of the sides of this ***isosceles triangle*** *have equal lengths.*

L

---

least common multiple (LCM) *mínimo común múltiplo (mcm)* *(30)*
The smallest whole number that is a multiple of two or more given numbers.

Multiples of 6 are 6, 12, 18, 24, 30, 36, ...
Multiples of 8 are 8, 16, 24, 32, 40, 48, ...
The ***least common multiple*** *of 6 and 8 is 24.*

legend *rótulo* *(Inv. 11)* A notation on a map, graph, or diagram that describes the meaning of the symbols and/or the scale used.

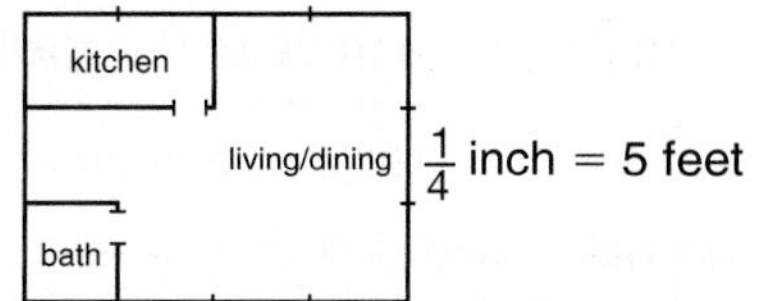

*The **legend** of this scale drawing shows that $\frac{1}{4}$ inch represents 5 feet.*

---

line *línea* *(7)* A straight collection of points extending in opposite directions without end.

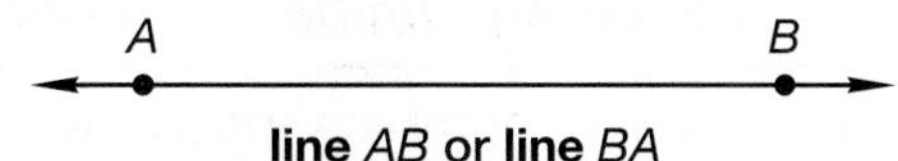

line *AB* or **line** *BA*

---

line graph *gráfica lineal* *(18)* A method of displaying numerical information as points connected by line segments.

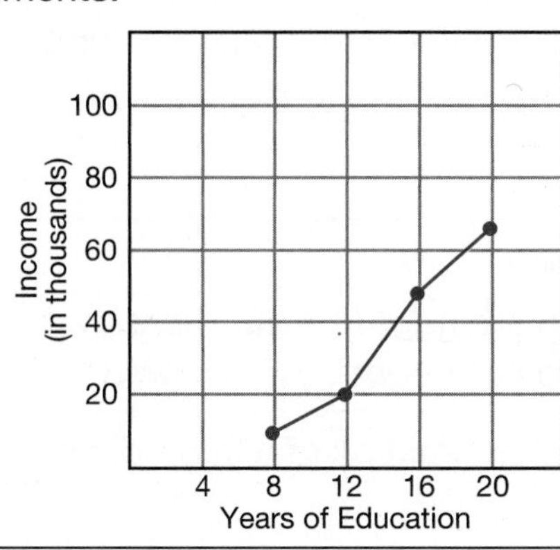

*This **line graph** has a horizontal axis that shows the number of completed years of education and a vertical axis that shows the average yearly income.*

---

line of symmetry *línea de simetría* *(110)* A line that divides a figure into two halves that are mirror images of each other.

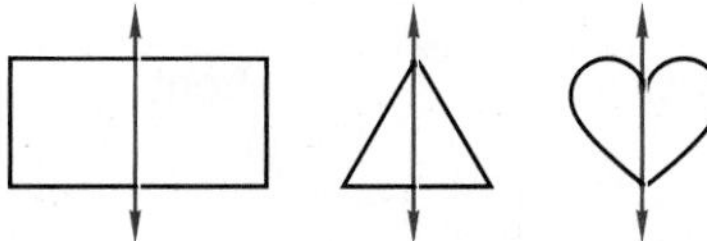

lines of symmetry not **lines of symmetry**

---

line plot *diagrama de puntos* *(Inv. 4)* A method of plotting a set of numbers by placing a mark above a number on a number line each time it occurs in the set.

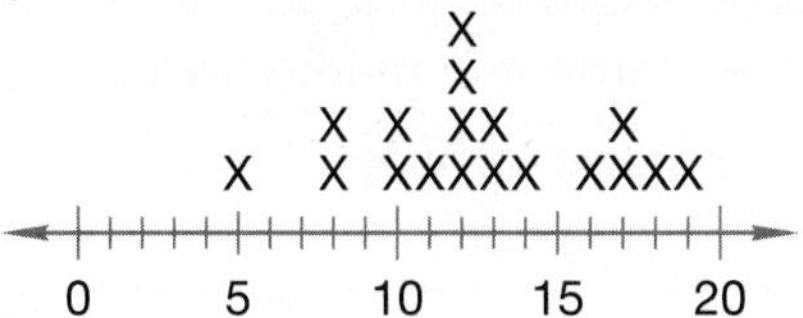

*This is a **line plot** of the numbers 5, 8, 8, 10, 10, 11, 12, 12, 12, 12, 13, 13, 14, 16, 17, 17, 18, and 19.*

M

---

mass *masa* *(102)* The amount of matter in an object.

*Grams and kilograms are units of **mass.***

---

mean *media* *(18)* *See* **average.**

---

median *mediana* *(Inv. 5)* The middle number (or the average of the two central numbers) of a list of data when the numbers are arranged in order from the least to the greatest.

*In the data at right, 7 is the **median.*** *1, 1, 2, 5, 6, 7, 9, 15, 24, 36, 44*

metric system *sistema métrico* (7) — An international system of measurement based on multiples of ten. Also called *International System.*

Centimeters and kilograms are units in the ***metric system.***

minuend *minuendo* (1) — A number from which another number is subtracted.

12 − 8 = 4 The ***minuend*** *in this problem is 12.*

mixed number(s) *número(s) mixto(s)* (17) — A whole number and a fraction together.

The ***mixed number*** $2\frac{1}{3}$ *means "two and one third."*

mode *moda* (Inv. 5) — The number or numbers that appear most often in a list of data.

In the data at right, 5 is the ***mode.*** *5, 12, 32, 5, 16, 5, 7, 12*

multiple(s) *múltiplo(s)* (25) — A product of a counting number and another number.

The ***multiples*** *of 3 include 3, 6, 9, and 12.*

N

negative numbers *números negativos* (9) — Numbers less than zero.

−15 and −2.86 are ***negative numbers.***
19 and 0.74 are not ***negative numbers.***

net *red* (Inv. 12) — A two-dimensional representation of a three-dimensional figure.

top
back
side
front
side
bottom

nonexample *contraejemplo* (Inv. 2) — A nonexample is the opposite of an example. Nonexamples can be used to prove that a fact or a statement in mathematics is incorrect.

The integer 7 is a ***nonexample*** *of an even number, and a circle is a nonexample of a polygon.*

number line *recta numérica* (9) — A line for representing and graphing numbers. Each point on the line corresponds to a number.

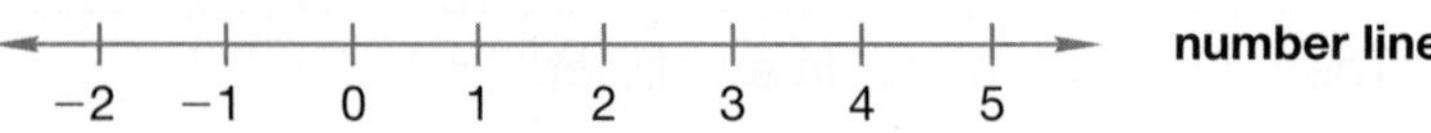

numerator *numerador* (6) — The top term of a fraction.

$\frac{9}{10}$ ← **numerator** (9) ← denominator (10)

O

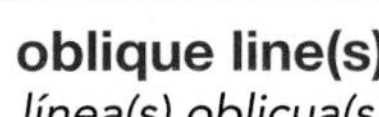

oblique line(s)
línea(s) oblicua(s)
(28)

1. A line that is neither horizontal nor vertical.

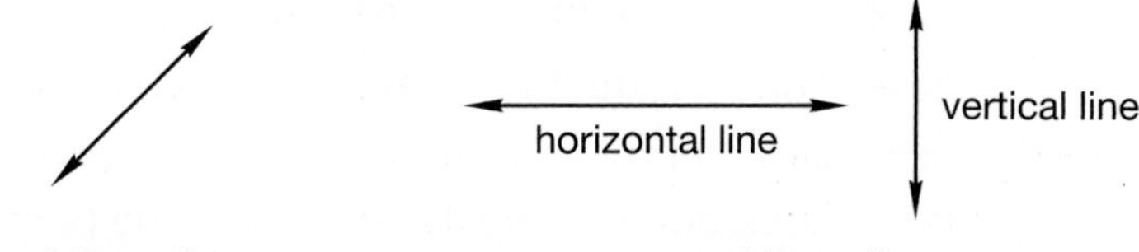

oblique line not **oblique lines**

2. Lines in the same plane that are neither parallel nor perpendicular.

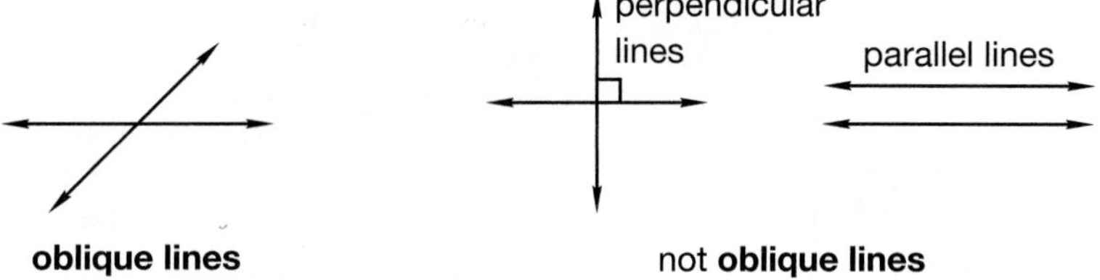

oblique lines not **oblique lines**

---

obtuse angle
ángulo obtuso
(28)

An angle whose measure is more than 90° and less than 180°.

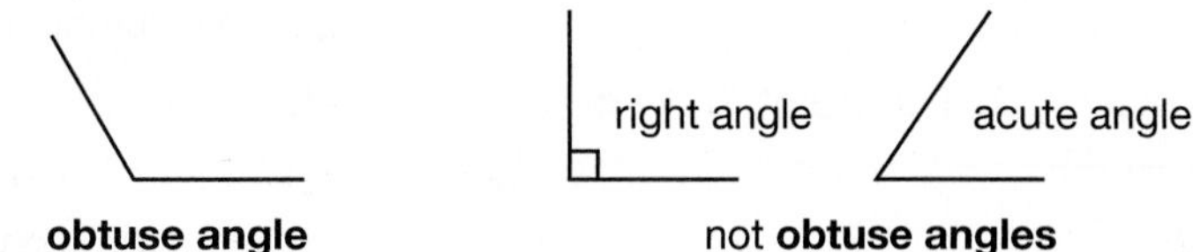

obtuse angle not **obtuse angles**

*An **obtuse angle** is larger than both a right angle and an acute angle.*

---

obtuse triangle
triángulo obtusángulo
(93)

A triangle whose largest angle measures more than 90° and less than 180°.

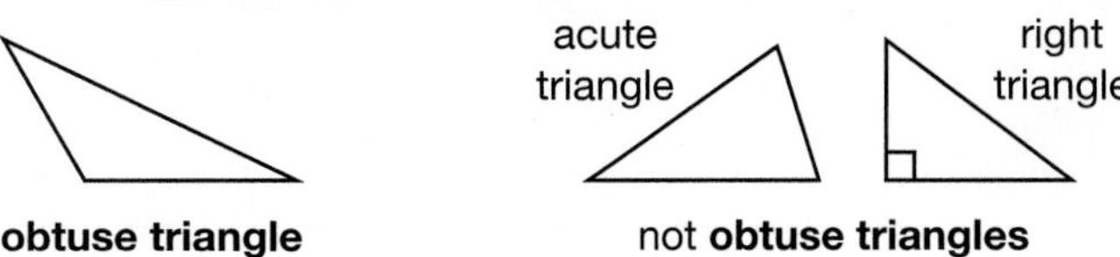

obtuse triangle not **obtuse triangles**

---

odd numbers
números impares
(10)

Numbers that have a remainder of 1 when divided by 2; the members of the set {..., −3, −1, 1, 3, ...}.

***Odd numbers** have 1, 3, 5, 7, or 9 in the ones place.*

---

open-option survey
encuesta de opinión abierta
(Inv. 1)

A survey that does not limit the possible responses.

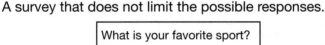

What is your favorite sport?
____

open-option survey

---

operations of arithmetic
operaciones aritméticas
(12)

The four basic mathematical operations: addition, subtraction, multiplication, and division.

$1 + 9$ $21 - 8$ 6×22 $3 \div 1$

the **operations of arithmetic**

---

opposites
opuestos
(14)

Two numbers whose sum is 0.

$$(-3) + (+3) = 0$$

*The numbers +3 and −3 are **opposites.***

order of operations
orden de las operaciones
(5)

The order in which the four fundamental operations occur.

1. Simplify powers and roots.

2. Multiply or divide in order from left to right.

3. Add and subtract in order from left to right.

With parentheses, we simplify within the parentheses, from innermost to outermost, before simplifying outside the parentheses.

---

ordered pair
par ordenado
(Inv. 7)

A pair of numbers, written in a specific order, that are used to designate the position of a point on a coordinate plane. *See also* **coordinate(s).**

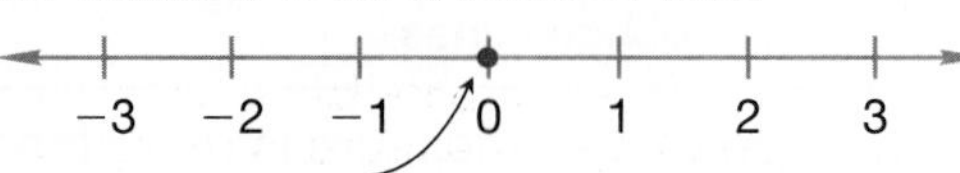

ordered pairs

---

origin
origen
(Inv. 7)

1. The location of the number 0 on a number line.

−3 −2 −1 0 1 2 3

origin on a number line

2. The point (0, 0) on a coordinate plane.

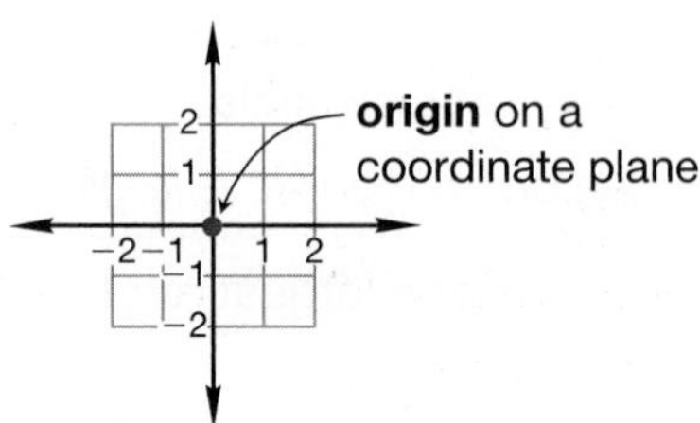

P

parallel lines
líneas paralelas
(28)

Lines in the same plane that do not intersect.

parallel lines

---

parallelogram
paralelogramo
(64)

A quadrilateral that has two pairs of parallel sides.

parallelograms

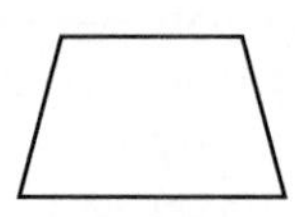

not a **parallelogram**

---

percent
por ciento
(33)

A fraction whose denominator of 100 is expressed as a percent sign (%).

$$\frac{99}{100} = 99\% = 99 \textbf{ percent}$$

---

perfect square
cuadrado perfecto
(38)

The product when a whole number is multiplied by itself.

*The number 9 is a **perfect square** because 3 × 3 = 9.*

---

perimeter
perímetro
(8)

The distance around a closed, flat shape.

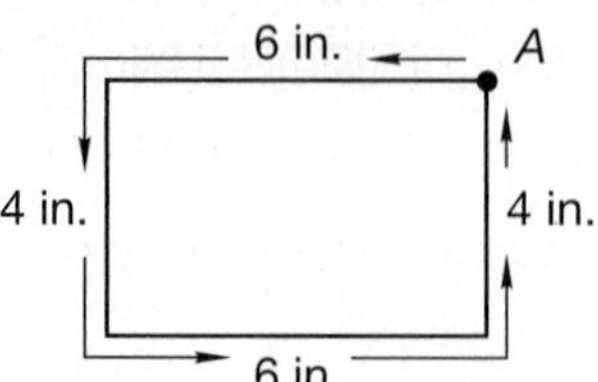

*The **perimeter** of this rectangle (from point A around to point A) is 20 inches.*

perpendicular bisector
mediatriz
(Inv. 8)

A line, ray, or segment that intersects a segment at its midpoint at a right angle, thereby dividing the segment into two congruent parts.

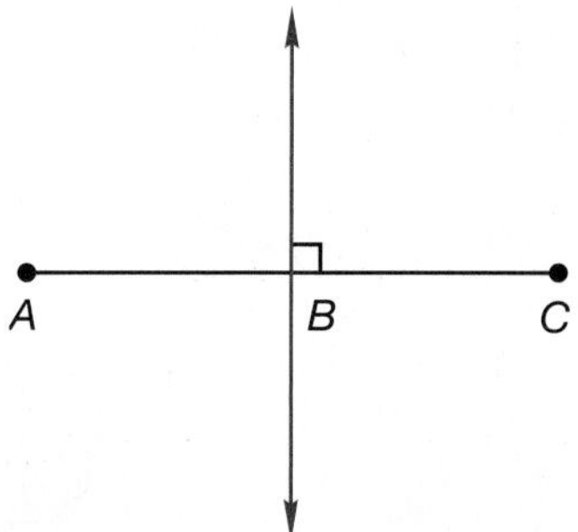

This vertical line is a ***perpendicular bisector*** *of $\overline{AC}$.*

---

perpendicular lines
líneas perpendiculares
(28)

Two lines that intersect at right angles.

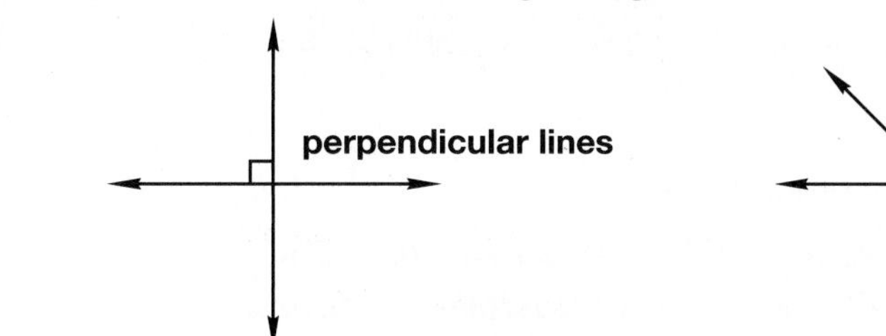

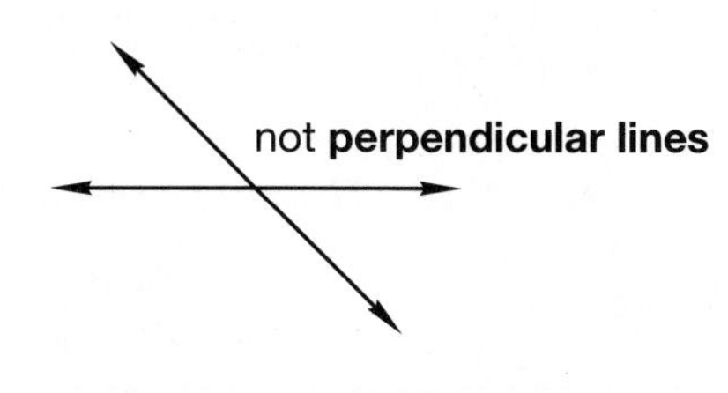

---

pi (π)
pi (π)
(47)

The number of diameters equal to the circumference of a circle.

Approximate values of ***pi*** *are 3.14 and $\frac{22}{7}$.*

---

pictograph
pictografía
(Inv. 5)

A method of displaying data that involves using pictures to represent the data being counted.

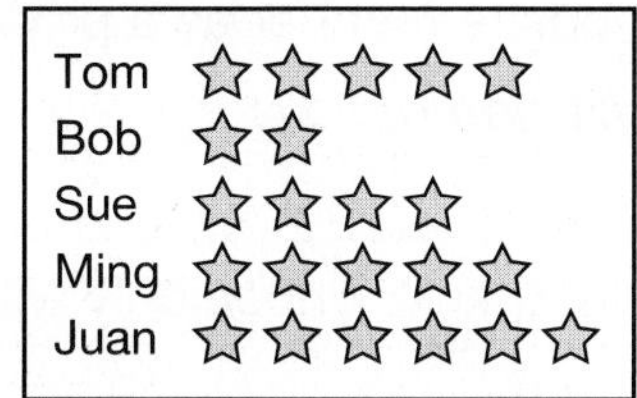

This is a ***pictograph.***
It shows how many stars each person saw.

---

pie graph
gráfica circular
(40)

See **circle graph.**

---

place value
valor posicional
(12)

The value of a digit based on its position within a number.

$$\begin{array}{r} 341 \\ 23 \\ +\quad 7 \\ \hline 371 \end{array}$$

Place value *tells us that 4 in 341 is worth four tens. In addition and subtraction problems we align digits with the same* ***place value.***

---

plane
plano
(28)

A flat surface that has no boundaries.

The flat surface of a desk is part of a ***plane.***

---

p.m.
p.m.
(32)

The period of time from noon to just before midnight.

I go to bed at 9 ***p.m.*** *I go to bed at 9 o'clock at night.*

---

point
punto
(69, Inv. 7)

An exact position on a line, on a plane, or in space.

• *A* *This dot represents* ***point*** *A.*

polygon *polígono* *(60)*

A closed, flat shape with straight sides.

polygons — not **polygons**

---

polyhedron *poliedro* *(Inv. 6)*

A geometric solid whose faces are polygons.

polyhedrons: cube, triangular prism, pyramid

not **polyhedrons**: sphere, cylinder, cone

---

population *población* *(Inv. 4)*

A certain group of people that a survey is about.

---

positive numbers *números positivos* *(14)*

Numbers greater than zero.

*0.25 and 157 are **positive numbers.***

*−40 and 0 are not **positive numbers.***

---

power(s) *potencia(s)* *(73)*

1. The value of an exponential expression.

*16 is the fourth **power** of 2 because $2^4 = 16$.*

2. An exponent.

*The expression 2^4 is read "two to the fourth **power.**"*

---

prime factorization *factorización prima* *(65)*

The expression of a composite number as a product of its prime factors.

***The prime factorization** of 60 is $2 \times 2 \times 3 \times 5$.*

---

prime number(s) *número(s) primo(s)* *(19)*

A counting number greater than 1 whose only two factors are the number 1 and itself.

*7 is a **prime number.** Its only factors are 1 and 7.*

*10 is not a **prime number.** Its factors are 1, 2, 5, and 10.*

---

principal *capital* *(116)*

The amount of money borrowed in a loan, deposited in an account that earns interest, or invested in a fund.

*If we borrow $750.00, the **principal** is $750.00.*

---

prism *prisma* *(Inv. 6)*

A polyhedron with two congruent parallel bases.

rectangular **prism** — triangular **prism**

---

probability *probabilidad* *(58)*

A way of describing the likelihood of an event; the ratio of favorable outcomes to all possible outcomes.

*The **probability** of rolling a 3 with a standard number cube is $\frac{1}{6}$.*

---

product *producto* *(2)*

The result of multiplication.

*$5 \times 4 = 20$ The **product** of 5 and 4 is 20.*

proportion *proporción* (83) A statement that shows two ratios are equal.

$$\frac{6}{10} = \frac{9}{15}$$

These two ratios are equal, so this is a ***proportion.***

protractor *transportador* (Inv. 3) A tool used to measure and draw angles.

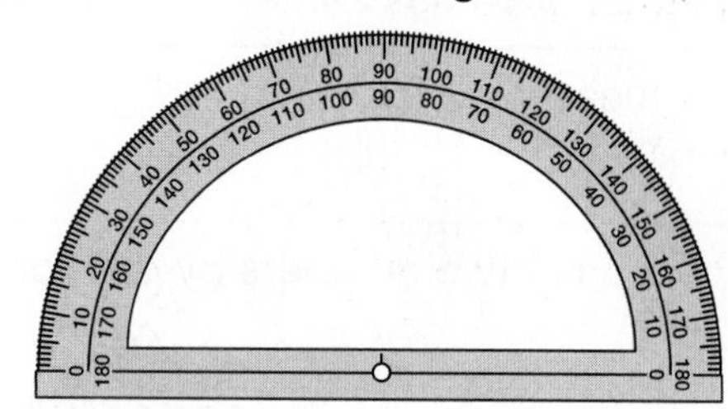

pyramid *pirámide* (Inv. 6) A three-dimensional solid with a polygon as its base and triangular faces that meet at a vertex.

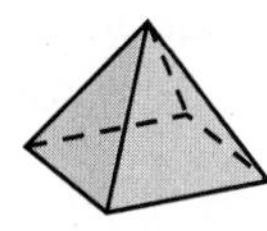

Q

quadrilateral *cuadrilátero* (60) Any four-sided polygon.

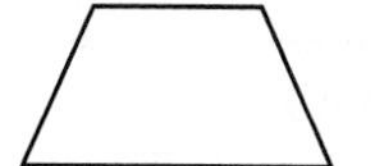 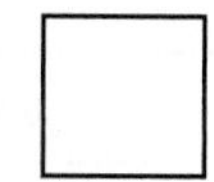 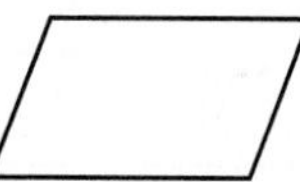 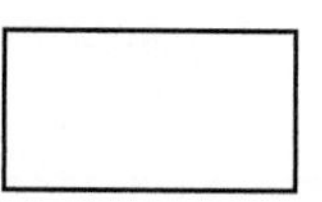 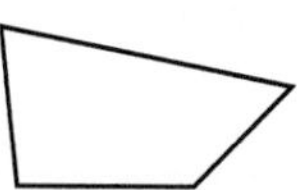

*Each of these polygons has 4 sides. They are all **quadrilaterals.***

qualitative *cualitativo* (Inv. 4) Expressed in or relating to categories rather than quantities or numbers.

***Qualitative data** are categorical: Examples include the month in which someone is born and a person's favorite flavor of ice cream.*

quantitative *cuantitativo* (Inv. 4) Expressed in or relating to quantities or numbers.

***Quantitative data** are numerical: Examples include the population of a city, the number of pairs of shoes someone owns, and the number of hours per week someone watches television.*

quotient *cociente* (2) The result of division.

$12 \div 3 = 4$ $\qquad 3\overline{)12}$ with quotient 4 $\qquad \frac{12}{3} = 4$ $\qquad$ *The **quotient** is 4 in each of these problems.*

R

radius *radio* (27) (Plural: *radii*) The distance from the center of a circle to a point on the circle.

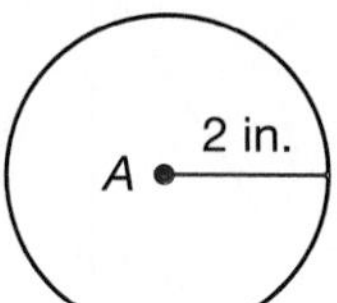

*The **radius** of circle A is 2 inches.*

range *intervalo* *(Inv. 5)* The difference between the largest number and smallest number in a list.

*To calculate the **range** of the data at right, we subtract the smallest number from the largest number. The **range** of this set of data is 29.*

5, 17, 12, 34, 29, 13

---

rate *tasa* *(23)* A ratio of measures.

---

ratio *razón* *(23)* A comparison of two numbers by division.

△△△
☆☆☆☆☆☆

*There are 3 triangles and 6 stars. The **ratio** of triangles to stars is $\frac{3}{6}$ (or $\frac{1}{2}$), which is read as "3 to 6" (or "1 to 2").*

---

rational numbers *números racionales* *(23)* Numbers that can be expressed as a ratio of two integers.

---

ray *rayo* *(7)* A part of a line that begins at a point and continues without end in one direction.

A B

ray *AB*

---

reciprocals *recíprocos* *(30)* Two numbers whose product is 1.

$\frac{3}{4} \times \frac{4}{3} = \frac{12}{12} = 1$ *Thus, the fractions $\frac{3}{4}$ and $\frac{4}{3}$ are **reciprocals.***

---

rectangle *rectángulo* *(64)* A quadrilateral that has four right angles.

rectangles not **rectangles**

---

rectangular prism *prisma rectangular* *(Inv. 6)* See **prism.**

---

reduce *reducir* *(26)* To rewrite a fraction in lowest terms.

*If we **reduce** the fraction $\frac{9}{12}$, we get $\frac{3}{4}$.*

---

reflection *reflexión* *(108)* Flipping a figure to produce a mirror image.

reflection

regular polygon *polígono regular* (60) A polygon in which all sides have equal lengths and all angles have equal measures.

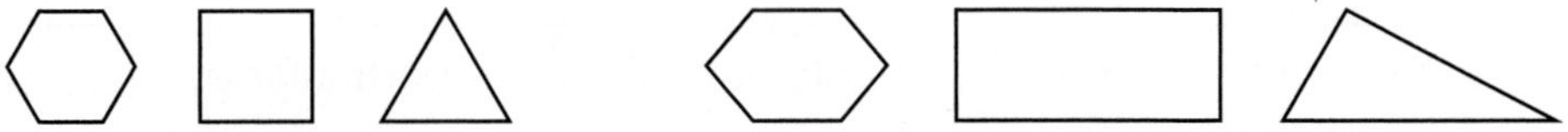

regular polygons — not **regular polygons**

rhombus *rombo* (64) A parallelogram with all four sides of equal length.

rhombuses — not **rhombuses**

right angle *ángulo recto* (28) An angle that forms a square corner and measures 90°. It is often marked with a small square.

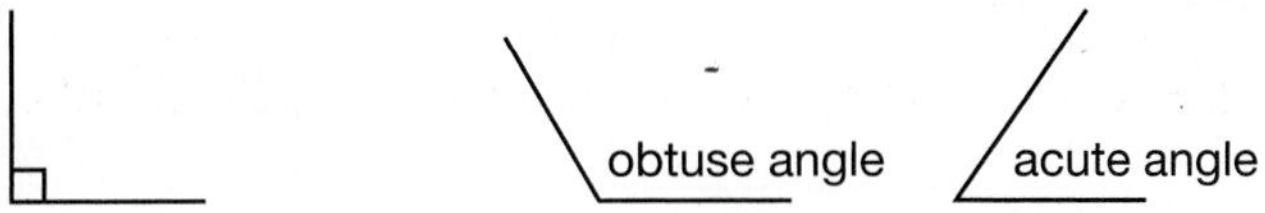

obtuse angle — acute angle

right angle — not **right angles**

right triangle *triángulo rectángulo* (93) A triangle whose largest angle measures 90°.

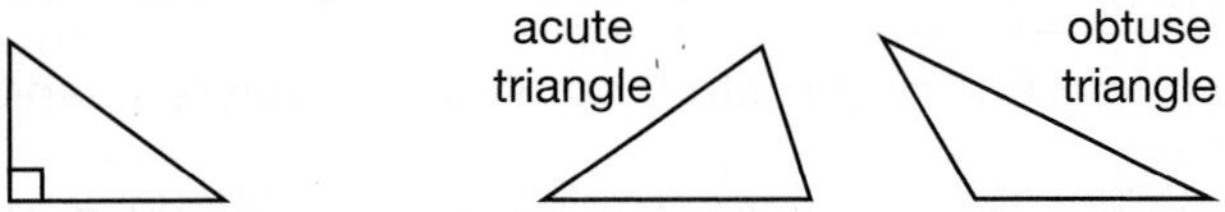

acute triangle — obtuse triangle

right triangle — not **right triangles**

rotation *rotación* (108) To rotate, or turn a figure about a specified point is called the *center of rotation.*

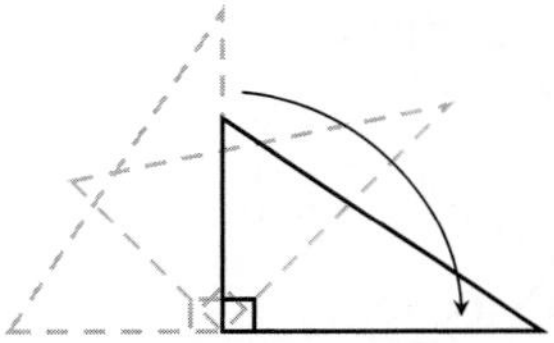

rotation

rotational symmetry *simetría rotacional* (110) A figure has rotational symmetry when it does not require a full rotation for the figure to look as if it re-appears in the same position as when it began the rotation, for example, a square or a triangle.

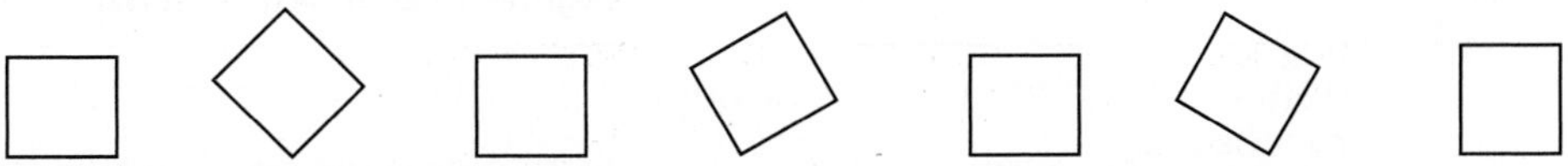

original position — 45° turn — **90° turn** — 150° turn — **180° turn** — 210° turn — **270° turn**

round *redondear* (16) A way of estimating a number by increasing or decreasing it to a certain place value. Example: 517 **rounds** to 520

S

sales tax *impuesto sobre la venta* (41) The tax charged on the sale of an item and based upon the item's purchase price.

*If the **sales-tax** rate is 7%, the **sales tax** on a $5.00 item will be $5.00 × 7% = $0.35.*

sample *muestra* *(Inv. 1)*

A smaller group of a population that a survey focuses on.

sample space *espacio muestral* *(58)*

Set of all possible outcomes of a particular event.

*The **sample space** of a 1–6 number cube is {1, 2, 3, 4, 5, 6}.*

scale *escala* *(10)*

A ratio that shows the relationship between a scale drawing or model and the actual object.

*If a drawing of the floor plan of a house has the legend 1 inch = 2 feet, the **scale** of the drawing is* $\frac{1 \text{ in.}}{2 \text{ ft}} = \frac{1}{24}$.

scale drawing *dibujo a escala* *(Inv. 11)*

A two-dimensional representation of a larger or smaller object.

*Blueprints and maps are examples of **scale drawings.***

scale factor *factor de escala* *(Inv. 11)*

The number that relates corresponding sides of similar geometric figures.

25 mm

10 mm

10 mm

4 mm

*The **scale factor** from the smaller rectangle to the larger rectangle is 2.5.*

scale model *modelo a escala* *(Inv. 11)*

A three-dimensional rendering of a larger or smaller object.

*Globes and model airplanes are examples of **scale models.***

scalene triangle *triángulo escaleno* *(93)*

A triangle with three sides of different lengths.

*All three sides of this **scalene triangle** have different lengths.*

sector *sector* *(Inv. 5)*

A region bordered by part of a circle and two radii.

*This circle is divided into 3 **sectors.***

segment *segmento* *(7)*

A part of a line with two distinct endpoints.

A B

segment *AB* **or segment** *BA*

sequence *secuencia* *(10)*

A list of numbers arranged according to a certain rule.

*The numbers 2, 4, 6, 8, ... form a **sequence.** The rule is "count up by twos."*

similar *semejante* *(109)*

Having the same shape but not necessarily the same size. Corresponding angles of similar figures are congruent. Corresponding sides of similar figures are proportional.

A D C B E F

*$\triangle ABC$ and $\triangle DEF$ are **similar.** They have the same shape but not the same size.*

simple interest *interés simple* (116)

Interest calculated as a percentage of the principal only.

Simple Interest		Compound Interest	
\$100.00	*principal*	*\$100.00*	*principal*
\$6.00	*first-year interest (6% of \$100)*	+ *\$6.00*	*first-year interest (6% of \$100)*
+ *\$6.00*	*second-year interest (6% of \$100)*	*\$106.00*	*total after one year*
\$112.00	*total after two years*	+ *\$6.36*	*second-year interest (6% of \$106)*
		\$112.36	*total after two years*

solid *sólido* (Inv. 6)

See **geometric solid.**

sphere *esfera* (Inv. 6)

A round geometric solid in which every point on the surface is at an equal distance from its center.

sphere

square *cuadrado* (64)

1. A rectangle with all four sides of equal length.

2 in.

2 in. 2 in.

2 in.

All four sides of this ***square*** *are 2 inches long.*

2. The product of a number and itself.

The ***square*** *of 4 is 16.*

square root *raíz cuadrada* (38)

One of two equal factors of a number. The symbol for the principal, or positive, square root of a number is $\sqrt{\ }$.

A ***square root*** *of 49 is 7 because* $7 \times 7 = 49$.

statistics *estadística* (Inv. 4)

The science of gathering and organizing data in such a way that conclusions can be made; the study of data.

stem-and-leaf plot *diagrama de tallo y hojas* (Inv. 5)

A method of graphing a collection of numbers by placing the "stem" digits (or initial digits) in one column and the "leaf" digits (or remaining digits) out to the right.

Stem	Leaf
2	1 3 5 6 6 8
3	0 0 2 2 4 5 6 6 8 9
4	0 0 1 1 1 2 3 3 5 7 7 8
5	0 1 1 2 3 5 8

In this ***stem-and-leaf plot,*** *3|5 represents 35.*

subtrahend *sustraendo* (1)

A number that is subtracted.

$12 - 8 = 4$ *The subtrahend in this problem is 8.*

sum *suma* (1)

The result of addition.

$7 + 6 = 13$ *The* ***sum*** *of 7 and 6 is 13.*

supplementary angles *ángulos suplementarios* (69)

Two angles whose sum is 180°.

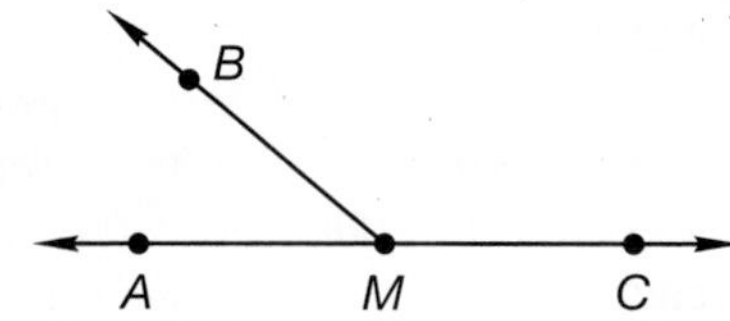

*∠AMB and ∠CMB are **supplementary.***

surface area *área superficial* (Inv. 6)

The total area of the surface of a geometric solid.

	Area of top	$= 5$ cm $\times$ 6 cm $=$	*30 cm²*
	Area of bottom	$= 5$ cm $\times$ 6 cm $=$	*30 cm²*
	Area of front	$= 3$ cm $\times$ 6 cm $=$	*18 cm²*
	Area of back	$= 3$ cm $\times$ 6 cm $=$	*18 cm²*
	Area of side	$= 3$ cm $\times$ 5 cm $=$	*15 cm²*
+	*Area of side*	$= 3$ cm $\times$ 5 cm $=$	*15 cm²*
	*Total **surface area***		$=$ *126 cm²*

survey *encuesta* (Inv. 1)

A method of collecting data about a particular population.

*Mia conducted a **survey** by asking each of her classmates the name of his or her favorite television show.*

T

term(s) *término(s)* (10, 30)

1. A number that serves as a numerator or denominator of a fraction.

$\frac{5}{6}$ → **terms**

2. A number in a sequence.

1, 3, 5, 7, 9, 11, ...

*Each number in this sequence is a **term.***

theoretical probability *probabilidad teórica* (Inv. 9)

The probability that an event will occur, as determined by analysis rather than by experimentation.

*The **theoretical probability** of rolling a three with a standard number cube is $\frac{1}{6}$.*

transformation *transformación* (108)

The changing of a figure's position through rotation, reflection, or translation.

Transformations

Movement	Name
flip	reflection
slide	translation
turn	rotation

translation *traslación* (108)

Sliding a figure from one position to another without turning or flipping the figure.

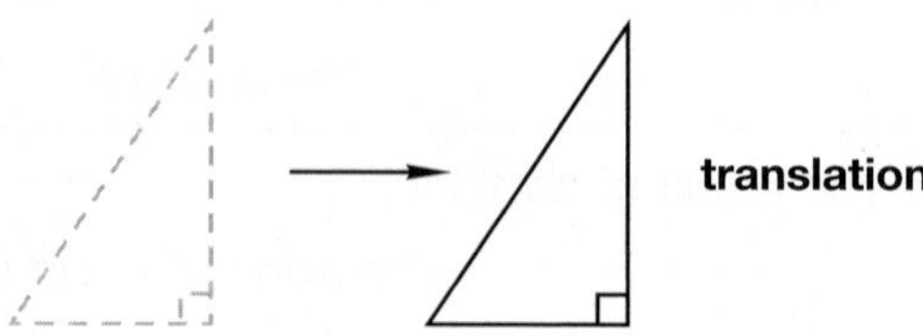

transversal *transversal* *(97)*

A line that intersects one or more other lines in a plane.

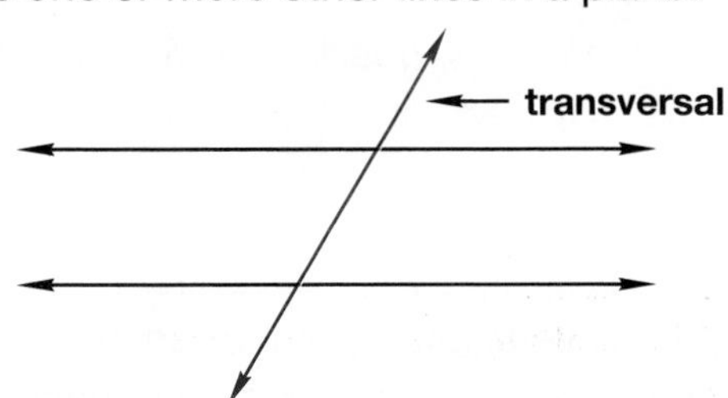

trapezium *trapezoide* *(64)*

A quadrilateral with no parallel sides.

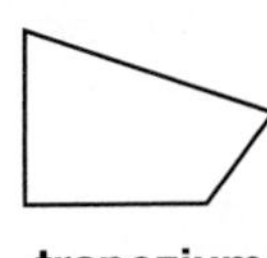

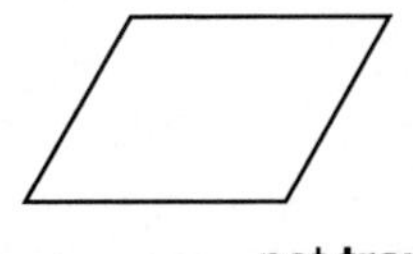

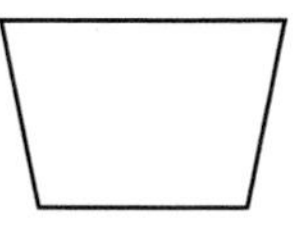

trapezium not **trapeziums**

trapezoid *trapecio* *(64)*

A quadrilateral with exactly one pair of parallel sides.

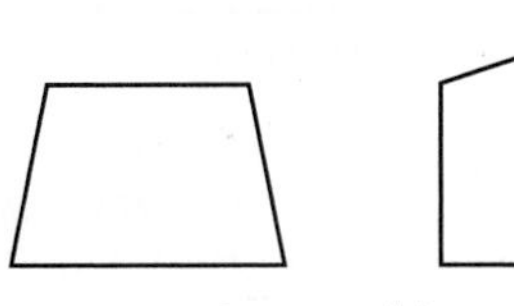

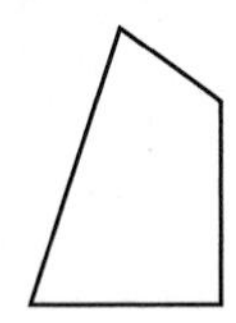

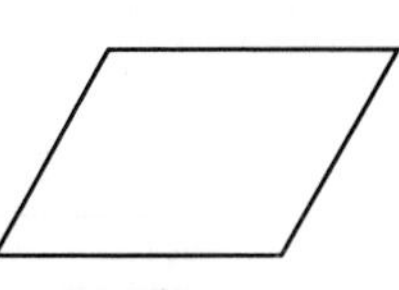

trapezoids not **trapezoids**

tree diagram *diagrama de árbol* *(Inv. 10)*

A visual representation of a compound experiment.

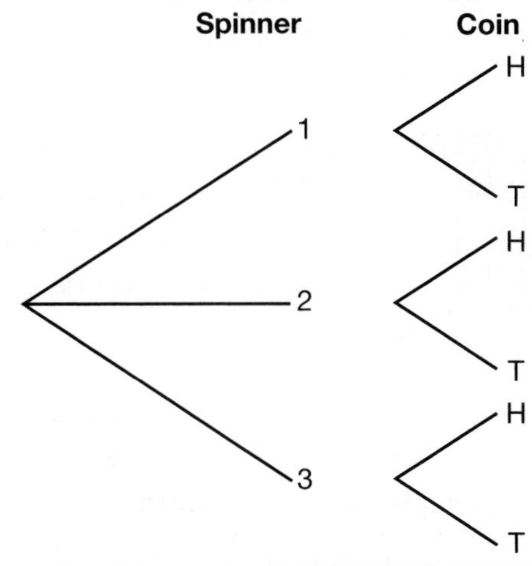

*This **tree diagram** shows all 6 possible outcomes for a spinner with 3 sectors being spun and a coin being flipped.*

triangular prism *prisma triangular* *(Inv. 6)*

See **prism.**

U

unit multiplier *factor de conversión* *(114)*

A ratio equal to 1 that is composed of two equivalent measures.

$$\frac{12 \text{ inches}}{1 \text{ foot}} = 1$$

*We can use this **unit multiplier** to convert feet to inches.*

unknown *incógnita* *(3)*

A value that is not given. A letter is frequently used to stand for an unknown number.

U.S. Customary System
Sistema usual de EE.UU.
(7)

A system of measurement used almost exclusively in the United States.

Pounds, quarts, and feet are units in the ***U.S. Customary System.***

V

vertex
vértice
(28)

(Plural: *vertices*) A point of an angle, polygon, or polyhedron where two or more lines, rays, or segments meet.

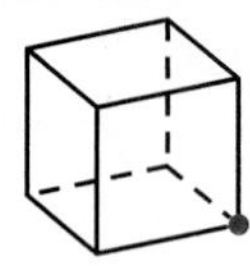

A dot is placed at one ***vertex*** *of this cube. A cube has eight* ***vertices.***

vertical
vertical
(18)

Upright; perpendicular to horizontal.

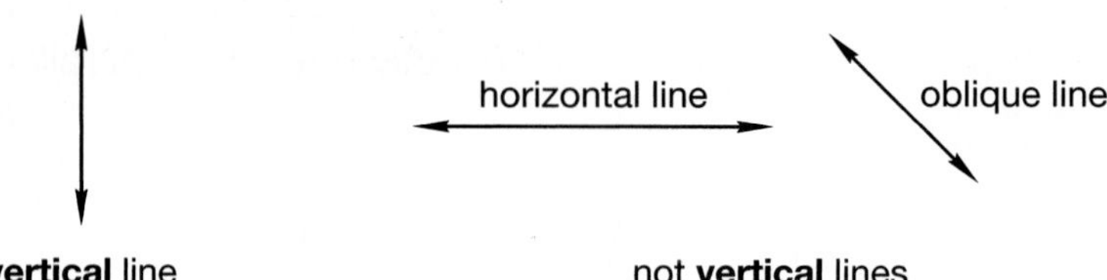

vertical line   not **vertical** lines

vertices
vértices
(Inv. 6)

See **vertex.**

volume
volumen
(Inv. 6, 82)

The amount of space a solid shape occupies. Volume is measured in cubic units.

This rectangular prism is 3 units wide, 3 units high, and 4 units deep. Its ***volume*** *is $3 \cdot 3 \cdot 4 = 36$ cubic units.*

W

weight
peso
(102)

The measure of how heavy an object is.

The ***weight*** *of the car was about 1 ton.*

whole numbers
números enteros
(9)

The members of the set {0, 1, 2, 3, 4, ...}.

0, 25, and 134 are ***whole numbers.***
-3, 0.56, and $100\frac{3}{4}$ are not ***whole numbers.***

X

x-axis
eje de las x
(Inv. 7)

The horizontal number line of a coordinate plane.

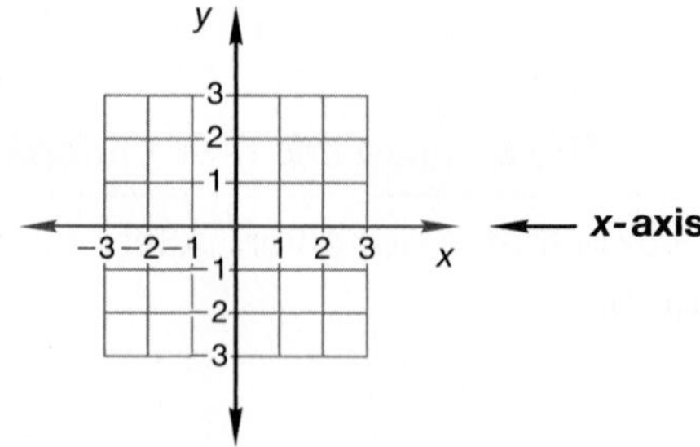

Y

y-axis *eje de las y* *(Inv. 7)* The vertical number line of a coordinate plane.

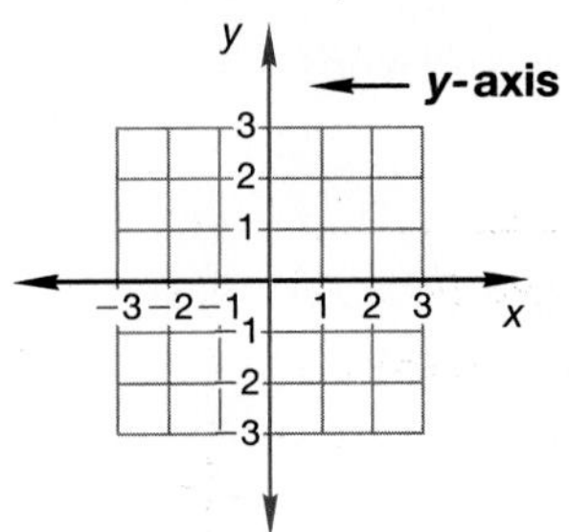

Z

Zero Property of Multiplication *propiedad del cero en la multiplicación* *(2)* Zero times any number is zero. In symbolic form, $0 \times a = 0$.

*The **Zero Property of Multiplication** tells us that 89 × 0 = 0.*

GLOSSARY

A

B

C

D

INDEX

F

G

H

I

INDEX

J

K

L

M

N

Q

R

S

INDEX

T

U

V

W

Z